IONISATION CONSTANTS OF ORGANIC ACIDS IN AQUEOUS SOLUTION

IUPAC CHEMICAL DATA SERIES

NOTICE TO READERS

Dear Reader

If your library is not already a standing order customer or subscriber to this series, may we recommend that you place a standing order or subscription order to receive immediately upon publication all new issues and volumes published in this valuable series. Should you find that these volumes no longer serve your needs your order can be cancelled at any time without notice.

The Editors and the Publisher will be glad to receive suggestions or outlines of suitable titles, reviews or symposia for consideration for rapid publication in this series.

Robert Maxwell
Publisher at Pergamon Press

INTERNATIONAL UNION OF PURE AND APPLIED CHEMISTRY
ANALYTICAL CHEMISTRY DIVISION
COMMISSION ON EQUILIBRIUM DATA

IUPAC CHEMICAL DATA SERIES - No. 23

IONISATION CONSTANTS OF ORGANIC ACIDS IN AQUEOUS SOLUTION

by

E. P. SERJEANT and BOYD DEMPSEY
University of New South Wales,
Duntroon, Australia

PERGAMON PRESS

OXFORD · NEW YORK · TORONTO · SYDNEY · PARIS · FRANKFURT

U.K.	Pergamon Press Ltd., Headington Hill Hall, Oxford OX3 0BW, England
U.S.A.	Pergamon Press Inc., Maxwell House, Fairview Park, Elmsford, New York 10523, U.S.A.
CANADA	Pergamon of Canada, Suite 104, 150 Consumers Road, Willowdale, Ontario M2 J1P9, Canada
AUSTRALIA	Pergamon Press (Aust.) Pty. Ltd., P.O. Box 544, Potts Point, N.S.W. 2011, Australia
FRANCE	Pergamon Press SARL, 24 rue des Ecoles, 75240 Paris, Cedex 05, France
FEDERAL REPUBLIC OF GERMANY	Pergamon Press GmbH, 6242 Kronberg-Taunus, Pferdstrasse 1, Federal Republic of Germany

First edition 1979

British Library Cataloguing in Publication Data

International Union of Pure and Applied Chemistry. *Commission on Equilibrium Data* Ionisation constants of organic acids in aqueous solution. - (International Union of Pure and Applied Chemistry. Chemical data series ; no. 23).
1. Ionic equilibrium - Tables 2. Acids, Organic - Tables
I. Title II. Serjeant, Eldon Percy
III. Dempsey, Boyd IV. Series
547'.1'3723   QD561   78-40988
ISBN 0-08-022339-7

In order to make this volume available as economically and as rapidly as possible the author's typescript has been reproduced in its original form. This method unfortunately has its typographical limitations but it is hoped that they in no way distract the reader.

INTERNATIONAL UNION OF PURE AND APPLIED CHEMISTRY

IUPAC Secretariat:
Bank Court Chambers, 2—3 Pound Way,
Cowley Centre, Oxford OX4 3YF, U.K.

Printed in Great Britain by A. Wheaton & Co. Ltd., Exeter

CONTENTS

F O R E W O R D

This critical compilation has been prepared by E.P. Serjeant and B. Dempsey, members of IUPAC Subcommittee V.6.2. It is a supplement to the earlier report by G. Kortum, W. Vogel and K. Andrussow, IUPAC Commission on Electrochemical Data, entitled "Dissociation Constants of Organic Acids in Aqueous Solution", which was published in 1961. The compilations of pK values of organic acids are now comparable with those for organic bases which were extensively covered in two earlier IUPAC reports published in 1965 ("Dissociation Constants of Organic Bases in Aqueous Solution", D.D. Perrin) and 1972 (Supplement 1972, D.D. Perrin).

Serjeant and Dempsey have extended the literature surveyed to the end of 1970 and have summarised data in some 4500 acids. Their index has incorporated the index of the Kortum compilation and this gives access to some 5500 acids. Users of the Tables will be indebted to the compilers who have been willing to make such massive and voluntary sacrifices of time and energy to search the literature and to screen, select and evaluate the very large amount of information they contain. The usefulness of the resulting Tables of pK values needs no comment.

D.D. Perrin

Chairman, IUPAC Subcommittee V.6.2

Commission on Equilibrium Data

Canberra

August, 1978

PREFACE

This compilation is a supplement to the report prepared by G. Kortum, W. Vogel and K. Andrussow which was published in Pure and Applied Chemistry, Vol. 1, No. 2-3, 1961, and also reprinted in monograph form as <u>Dissociation Constants of Organic Acids in Aqueous Solution</u>, Butterworths (1961). The Kortum compilation surveyed the literature to the end of 1956. The present volume extends the survey to the end of 1970. It contains data on some 4,500 acids set out to the same general format for individual acids as that used in the Kortum compilation. However, the arrangement of compounds in the Tables differs from the system used by Kortum: acids are arranged in order of molecular formula rather than by the type of acidic functional group. In some instances several values from different sources are given for a single compound. Two criteria have been used in choosing values for inclusion in the Tables: reliability of the measurement, and range of experimental conditions. Where references are given to "other values", these generally duplicate conditions for determinations quoted in the Tables. The Index lists a total of about 5,500 acids the pK values of which are contained in either or both compilations. Compounds with numbers less than 2000 are in the Kortum compilation whilst those with numbers greater than 2000 are contained in this volume. No attempt has been made, however, to cross-index any of the amphiprotic compounds included in this compilation with identical compounds contained in the companion compilations of D.D. Perrin, <u>Dissociation Constants of Organic Bases in Aqueous Solution</u>, (1965) and its Supplement (1972), published by Butterworths (London). It is inevitable that some duplication must occur when amphiprotic substances are included in compilations of acids and bases but these duplications are relatively few in number. We have attempted to avoid quoting pK values from literature sources already cited by D.D. Perrin. It is emphasized that because the compilations on bases were published before we began data collection, the compilations of D.D. Perrin should be regarded as the major source of pK values for amphiprotic compounds. Although it was intended originally that this supplement should survey the literature to the end of 1973, it soon became evident as data collection progressed that it would be advantageous to combine future compilations of pK values for organic acids and bases. If separate compilations were prepared for acids and bases, with different starting and cut-off dates, future compilers would be faced with a considerable additional task of checking whether a given reference to amphiprotic compounds had been used in companion compilations on acids or bases. In order to avoid this and also to avoid perpetuating a redundant distinction between acids and bases, we have surveyed the literature to the end of 1970, as was done in the Supplement (1972) of **Perrin**. Subsequent compilations could with advantage combine acids and bases. However, we have some entries from post-1970 literature which had been included before the decision on the new cut-off date was made.

The assessment categories used by D.D. Perrin have been adopted plus a further category, "very uncertain", where large errors are known or suspected to exist. Such values may be a useful guide in subsequent determinations.

In general, the main name given to a compound in this compilation is based on the principal group as defined in IUPAC Organic Nomenclature Rules C, 1969, and appears in the Tables and in the Index as a stem name, and where appropriate, followed by a comma of inversion and the named substituents. Some trivial and trade names and alternative systematic names have been given both in the Tables and in the Index. In the Tables these names are given in parentheses.

Use of alternative systematic names has been made, for example, in the case of polyaminopolycarboxylic acids and analogous compounds. These compounds have been named systematically both as assemblies of identical units and by replacement nomenclature. Thus, the systematic name of the polyaminopolycarboxylic acid, EDTA, by the method of assemblies of identical units is ethylenedinitrilotetraacetic acid, and by replacement nomenclature is 3,6-bis(carboxymethyl)-3,6-diazaoctanedioic acid. Our preference is for replacement nomenclature which has been used extensively throughout this compilation.

In the case of phosphorus acids, we have arbitrarily assigned their acid groups a priority for citation as suffix below that of sulphur and other Group VI acids. In the naming of phosphorus acids, we have been guided by the usage in J.H. Fletcher, O.C. Derner and R.B. Fox (ed.), <u>Nomenclature of Organic Compounds</u>, American Chemical Society (1974). The hydroxy and mercapto derivatives of nitrogen heterocycles have generally been named as the corresponding "-ones" in accordance with the higher priority given to ketones and thiones. In general, where it has not been possible to accommodate a tautomeric hydrogen on a ring nitrogen, hydroxyl and mercapto groups have been retained.

Values in mixed solvents have been quoted for mixtures containing up to 10 per cent non-aqueous component.

We gratefully acknowledge the use of the various libraries of the Australian National University, the library of the Commonwealth Scientific and Industrial Research Organisation, Canberra, the library of the Royal Society of New South Wales, Sydney, and the Bridges Library, Duntroon. The advice of the staff of the Medical Chemistry Group, John Curtin School of Medical Research, during the tenure of a Visiting Fellowship granted to one of us (EPS) in 1977 by the Australian National University is also acknowledged. In particular, we acknowledge the interest and advice of the Chairman of IUPAC Subcommittee V.6.2, Dr. D.D. Perrin, throughout all stages of this project. We also wish to thank Mrs. J. Archbald for assistance in the preparation of the Index and Mrs. E. Laker and Mrs. M. Keys who typed the camera-ready manuscript.

I.   HOW TO USE THE TABLES

ORDER OF COMPOUNDS

Compounds, other than macromolecules, are arranged in order of molecular formula using the Beilstein system. Macromolecules are arranged alphabetically at the end of the Tables following the molecular formula sequence.

The Beilstein system and its application may be summarised as follows:

(1)   Molecular formulae are written with the elements arranged in the order:

$$C, H, O, N, Cl, Br, I, F, S, P . . .$$

followed by the other elements in the alphabetical order of their symbols.

(2)   All molecular formulae with $\underline{x}$ carbons precede all those with $\underline{x+1}$ carbons.

Thus, for example, $C_6H_{10}O_2$ precedes C_7H_8O.

(3)   Compounds containing $\underline{y}$ elements precede those with $\underline{y+1}$ elements.

Thus, for example, $C_{11}H_{10}O_4$ precedes $C_{11}H_9ON_3$.

(4)   For molecular formulae containing $\underline{x}$ carbons and $\underline{y}$ elements, two situations exist:

. compounds containing the same set of $\underline{y}$ elements;

. compounds containing different sets of $\underline{y}$ elements.

(a) When the elements involved in a pair of compounds are the same, the precedence of one compound over the other is decided by considering the elements in the order given in (1) : the compound having the lower number of atoms for the element at the first point of difference has precedence. Thus, for example, $C_6H_{10}O_4S$ precedes $C_6H_{10}O_4S_2$, the first point of difference being at sulphur.

(b) When the elements involved in a pair of compounds are not the same, the precedence may still be decided by 4(a).

Thus, for example, $C_3H_2OF_6$ precedes $C_3H_3ON_3$, the decision being based on hydrogen.

However, if the decision has to be based on elements which are not common to both compounds, precedence is given to the compound which contains the non-common element coming first in the list in (1).

Thus, for example, $C_2H_7O_4P$ precedes C_2H_7NS, the decision being based on the fact that oxygen precedes nitrogen in the list in (1).

Similarly, C_3H_8S precedes C_3OF_6, since hydrogen comes before oxygen in the list in (1).

For isomeric compounds the sequence of arrangement in the Tables is alphabetical based on the systematic name given to the compound. Stereoisomers have been allocated the same compound number. For example, data for the cis-isomer, maleic acid, and its trans-isomer, fumaric acid are given under the compound, butenedioic acid. A similar system is adopted for isotopically labelled compounds. Thus the pK value for acetic acid-0-d is given under the compound: $C_2H_4O_2$, acetic acid. Where pK values have been determined in deuterium oxide, the value is followed by (D_2O).

Where acidic groups are within a cationic species and the associated anion has been specified, the anion is included in the molecular formula entry in the Tables in round brackets. For example, the molecular formula entry for triphenyl(carboxymethyl)-phosphonium chloride is $C_{20}H_{18}O_2(Cl)P$. In those cases where the anion has not been specified, the molecular formula given is that of the cation and the word, cation, is written immediately after the molecular formula.

The pK of a hydrated species is given under the molecular formula of the anhydrous form. Thus, the pK for hydrated ethanal appears under C_2H_4O.

INDIVIDUAL ENTRIES

For each compound in the Tables, the compound number, molecular formula and name(s) are given as a general heading followed by a tabular presentation of pK values, experimental conditions and other relevant information.

The left-hand column gives the negative logarithm of the dissociation constant, the pK_a value. When the compound is monoprotic, the term, pK, is not given. For polyprotic acids, the sequence: pK_1, pK_2 ... represents decreasing acid strength. Unless otherwise noted in "Remarks", thermodynamic pK_a values are given. These have been obtained by correcting for ionic strength effects, using some form of the Debye-Huckel equation or by extrapolation to zero ionic strength.

The second column gives the stated temperature in ^{O}C. Where authors have made measurements at several temperatures, the value at 25° (or nearest) has generally been quoted separately.

The Remarks column gives details of:

. concentration units (c in mole per litre of solution; m in mole per kg of solvent).

. total ionic strength, I ($I = \frac{1}{2}\Sigma c_i z_i^2$) with details of the added electrolyte.

type of constant if not thermodynamic -

"Mixed" constants, also known as "practical" or "Bjerrum" constants,

involve activities of hydrogen ion and concentrations of other species.

"Concentration" constants, also known as "stoicheiometric" constants,

involve concentrations only and normally involve the conversion of

pH values to hydrogen ion concentrations.

special experimental conditions and/or observations.

The Method column summarises in coded symbols the general experimental method and procedures for calculations and correction. The coded symbols are generally the same as those used in the original report (Kortum, 1961) and in the companion volumes on bases (Perrin, 1965 and 1972). Since variation is possible within individual methods, the original literature should be consulted if precise details are required.

The Assessment column gives the compilers' assessment of the reliability of the quoted constant as one of the grades:

"reliable", where the estimated uncertainty is $\leqslant$ 1 per cent of the value of K,

i.e., $\Delta pK \leqslant 0.005$;

"approximate", where the estimated uncertainty lies between 1 and 10 per

cent of the value of K, i.e., ΔpK lies between ± 0.04 and ± 0.005;

"uncertain", where the estimated uncertainty is greater than 10 per cent

of the value of K, i.e., $\Delta pK > \pm 0.04$;

"very uncertain", where the uncertainty cannot be estimated but is likely

to be very great.

In many cases, values have been assessed as "uncertain" because of the absence of essential details concerning temperature, ionic strength and general experimental procedure. In addition, the pK values of very strong and very weak acids are difficult to determine with high accuracy. No assessment has been made for data taken from Chemical Abstracts or from certain non-English language journals.

The final column refers to the literature references which are given following the Tables and before the Index. Reference abbreviations are as used in Chemical Abstracts Service Source Index 1907-74.

II. METHODS OF MEASUREMENT, CALCULATION AND CORRECTION

The abbreviations in the Method column of the Tables are, with only minor additions, the same as used in D.D. Perrin, Dissociation Constants of Organic Bases in Aqueous Solution, Butterworths (1965) and in the Supplement (1972), which were essentially those used by Professor G. Kortum in the original compilation, Table of Dissociation Constants of Organic Acids, (Pure Appl. Chem. 1, 190 (1961)).

CONDUCTOMETRIC METHODS

C1   Measurements in solutions of salt and acid

C2   Measurements in solutions of acid only

C3   Calculated from the decrease in conductivity in the reaction:

$$ROH + OH^- \rightarrow RO^- + H_2O$$

(Ballinger and Long, J. Am. Chem. Soc. 81, 1050 (1959))

ELECTROMETRIC METHODS

(i)   Cells without diffusion potentials

E1a   Method of Harned and Ehlers (J. Am. Chem. Soc. 54, 1350 (1932))

Cell:

$Pt;H_2|HA,NaA,NaCl|AgCl;Ag$

for which:

$-(RT/F) \ln K' = E - E_0 + (RT/F) \ln (m_{HA}m_{Cl^-}/m_{A^-})$

A plot of log K' vs. I, extrapolated to I=0, yields log K.

E1b   Method of Harned and Owen (J. Am. Chem. Soc. 52, 5079 (1930))

Cell:

$Pt;H_2|HA(molality=M),NaCl(molality=m)|AgCl;Ag$

for which:

$E = E_0 - (RT/F) \ln m_{H^+}m_{Cl^-} \gamma_\pm^2$

Extrapolation of K' to I=0 for M constant and m variable and then to M=0, yields K provided that $m_{H^+} \ll M$.

E1cg   Determination of $[H^+]$ from cells of the type:

$glass|solution, Cl^-|AgCl;Ag$

E1ch   Determination of $[H^+]$ from cells of the type:

$$Pt;H_2|solution, Cl^-|AgCl;Ag$$

E1d   Method of Bates (J. Am. Chem. Soc. <u>70</u>, 1579 (1948))

for the determination of K_1 and K_2 of dibasic acids

E1e   Method of Bates and Pinching (J. Res. Natl. Bur. Std. <u>43</u>, 519 (1949)),

a particular case of method E1cg, in which the solution is a buffer

made from a weak acid and a weak base

(ii)   <u>Approximately symmetrical cells with diffusion potentials</u>

E2a   Method of Owen (J. Am. Chem. Soc. <u>60</u>, 2229 (1938))

E2b   Method of Larsson and Adell (Z. Physik. Chem. <u>156</u>, 352, 381 (1931))

Cell:

$$Pt;H_2|HA,NaCl|sat.KCl|HCl,NaCl|H_2;Pt$$

An approximate value of K is used to adjust to equal ionic strengths

in the half-cells. Measured E gives $[H^+]$ and hence K'; extrapolation

to I=0 gives K.

E2c   Method of Everett and Landsman (Proc. Roy. Soc. London <u>A215</u>, 403

(1952))

(iii)   <u>Unsymmetrical cells with diffusion potentials</u>

E3ag   pH measurements in buffer solutions using glass electrodes

E3ah   Similar measurements using hydrogen electrodes

E3bg   Measurements of pH changes during titrations using glass electrodes

E3bh   Similar measurements using hydrogen electrodes

E3bq   Similar measurements using quinhydrone electrodes

E3c   Differential potentiometric methods

E3d   pH measurements at equal concentrations of salt and acid

OPTICAL METHODS

O1   Direct determination of the degree of dissociation by light absorbance

measurements in solutions of weak acids and salts

O2   Colorimetric determination with an indicator of known pK

O3   Colorimetric determination with an indicator calibrated with a buffer

solution of known pH

O4   Method of von Halban and Brull (Helv. Chim. Acta <u>27</u>, 1719 (1944))

(Solutions of the acid being studied, plus indicator, are compared

with similar solutions containing strong acid and indicator)

05	Light absorbance measurements combined with electrometric measurements of pH
05a	Similar to 05 but using $p(a_{H^\gamma Cl})$ buffers
06	Light absorbance measurements using solutions of mineral acids of known concentrations and (usually) Hammett's acidity function, H_o
07	Similar to 06 but using solutions of alkalis

OTHER METHODS

Constants estimated from:

CAL	Calorimetry
CAT	Catalytic studies
DIS	Phase distribution studies
FLU	Fluorescent titrations
FLU/05	Fluorescence maxima measurements coupled with 05 studies in the ground state
KIN	Kinetic measurements
NMR	Nuclear magnetic resonance measurements
RAM	Raman spectral measurements
REF	Differential refractometry
ROT	Optical rotation measurements
SOL	Solubility measurements

CALCULATIONS

(i)   Conductance measurements

R1a   Method of Davies (The Conductivity of Solutions, Chapman and Hall, London (1930))

(By successive approximations, f_Λ is calculated from the Debye-Huckel-Onsager equation in the form:

$$f_\Lambda = 1 - A(\alpha c_o)^{\frac{1}{2}}/\Lambda_0$$

which assumes that Λ_0 can be obtained from Kohlrausch's law of independent ionic mobilities)

R1b   Method of MacInnes (J. Am. Chem. Soc. 48, 2068 (1926))

(The quantity, $\Lambda_e = f_\Lambda \Lambda_0$, is determined directly, where Λ_e is the conductance of the weak electrolyte if it were completely dissociated at the ionic strength studied: it is necessary to know Λ for strong electrolytes as a function of I)

R1c   Method of Fuoss and Krauss (J. Am. Chem. Soc. <u>55</u>, 476 (1933))

(The Debye-Huckel-Onsager equation is used in the form:

$$\Lambda_c = \alpha(\Lambda_0 - A(\alpha c_0)^{\frac{1}{2}})$$

to derive an equation relating Λ_0, c and K, which is solved by successive approximations until Λ_0 is constant)

R1d   Method of Shedlovsky (J. Franklin Inst. <u>225</u>, 739 (1938))

(This is like R1c but a different equation is used)

R1e   Method of Fuoss and Shedlovsky (J. Am. Chem. Soc. <u>71</u>, 1496 (1949))

(ii)   <u>Differential potentiometric measurements</u>

R2a   Method of Kilpi (Z. Physik. Chem. <u>173</u>, 223, 427 (1935); <u>175</u>, 239 (1936))

R2b   Method of Hahn and Klockmann (Z. Physik. Chem. <u>146</u>, 373 (1930))

R2c   Method of Grunwald (J. Am. Chem. Soc. <u>73</u>, 4934 (1951))

(iii)   <u>Constants for polybasic acids</u>

R3a   Method of Kolthoff and Bosch (Rec. Trav. Chim. <u>47</u>, 861 (1928))

R3b   Method of Auerbach and Smolczyk (Z. Physik. Chem. <u>A110</u>, 83 (1924))

R3c   Method of Britton (J. Chem. Soc. <u>127</u>, 1896 (1925))

R3d   Method of Speakman (J. Chem. Soc. <u>1940</u>, 855)

R3f   Method of Bjerrum and Anderson (Kgl. Danske Videnskab. Selskab. Mat.-Fys. Medd <u>22</u>, No. 7 (1945))

R3g   Method of Schwarzenbach (Helv. Chim. Acta <u>33</u>, 947 (1950))

R3h   Method of Rossotti, Rossotti and Sillen (Acta Chem. Scand. <u>10</u>, 203 (1956)) (by curve fitting)

(iv)   <u>Correction for ionic strength effects</u>

R4   Graphical extrapolation to I = 0

CORRECTIONS

(i)   <u>For effects due to diffusion potentials</u>

D1   Salt bridge, no attempt at elimination

D2   Method of Bjerrum (Z. Physik. Chem. <u>53</u>, 428 (1905); Z. Elecktrochem. <u>17</u>, 389 (1911))

(The strength of the potassium chloride solution in the salt bridge is varied and eliminated by extrapolation)

D3   The diffusion potentials are calculated

(ii)   <u>For hydrolysis of salts in conductivity measurements</u>

K1a   Method of MacInnes and Shedlovsky (J. Am. Chem. Soc. <u>54</u>, 1429 (1932))

(Addition of slight excess of free acid)

K1b   Method of Jeffery and Vogel (J. Chem. Soc. <u>1932</u>, 2829)

(Corrections are calculated using K_{H_2O} and an approximate K_c)

K1c   Method of Jeffery and Vogel (Phil. Mag. <u>18</u>, 901 (1934))

(Corrects for hydrolysis and for solvent)

K1d   Method of Ives (J. Chem. Soc. <u>1933</u>, 313)

(Corrects for hydrolysis and for solvent)

K1e   Method of Davies (Trans. Faraday Soc. <u>28</u>, 607 (1932))

(Corrects for hydrolysis and for solvent)

K1f   Method of Banks and Davies (J. Chem. Soc. <u>1938</u>, 73)

(Average of results obtained using method K1e from which

κ_{H_2O} has been completely eliminated)

(iii)   <u>For solvent in conductivity measurements</u>

K2a   "Normal correction", where κ_{H_2O} is not eliminated for acids but
is completely eliminated for salts

K2b   Complete elimination of κ_{H_2O} for both acid and salt solutions

K2c   Method of MacInnes and Shedlovsky (J. Am. Chem. Soc. <u>54</u>, 1429 (1932))

K2d   Method of Jeffery and Vogel (J. Chem. Soc. <u>1933</u>, 1637)

K2e   Method of Dippy and Williams (J. Chem. Soc. <u>1934</u>, 161, 1888)

III TABLES

No.	Molecular formula, name and pK value(s)	$T(^{o}C)$	Remarks	Method	Assessment	Ref
2001	CH_2O Methanal (Formaldehyde)					
	13.27	25	(pK of hydrate)	C3	Approx.	B54
2002	CH_2O_2 Methanoic acid (Formic acid)					
	3.737	25	m = 0.0003-0.05	C2,R1b	Rel.	B53
	3.739	25		E1e	Rel.	P87
	3.71	25		02	Approx.	B50
	3.76(1atm), 3.61(1000atm), 3.49(2000atm), 3.33(3000atm)	25		C1,R4	Uncert.	H7
	3.56	25	I = 1.0(KCl), mixed constant	E3bg	Approx.	S18
	3.75(H_2O), 4.20(D_2O)	25		E3bg	Uncert.	G24
	3.75(H_2O), 4.21(D_2O)	25	Based on pK = 5.28 for acetic acid in D_2O	03	Approx.	B51
	Methanoic acid-d					
	3.772	25	m = 0.0003-0.05	C2,R1b	Rel.	B53
	3.76	25		02	Approx.	B50
2003	CH_2O_3 Peroxymethanoic acid (Performic acid)					
	7.1	19.5	c = 0.1	E3bg	Uncert.	E27

No.	Molecular formula, name and pK value(s)	$T(^oC)$	Remarks	Method	Assessment	Ref
2004	CH$_4$O Methanol					
	15.5	25		C3	Uncert.	B8
	15.09	25		KIN	Uncert.	M126
2005	CH$_4$O$_2$ Methyl hydroperoxide					
	11.5	20		05	Uncert.	E27
2006	CH$_4$S Methanethiol					
	10.33	25	1% ethanol, gas solubility method		Uncert.	K57
2007	CHO$_6$N$_3$ Methane, trinitro-					
	0.14	20	In aqueous HClO$_4$,H$_o$ scale	06	Uncert.	T58a
	0.06	25.5	Mixed constant	05	Uncert.	H5
	0.05	9.6				
	0.23	5	In aqueous HCl,H$_o$ scale	06	Uncert.	N39,S82
	0.17	20				
	0.11	40				
	0.02	60				

Thermodynamic quantities are derived from the results

No.	Formula	Name	pK	t (°C)	Conditions	Method	Reliability	Ref.
2008	$CH_2O_4N_2$	Methane, dinitro-	3.63	20		05	Approx.	T58a
			3.72	5		05	Approx.	N39
			3.60	20				
			3.51	40				
			3.43	60				
					Thermodynamic quantities are derived from the results			
			3.57	25	I = 0.06	05	Approx.	A7
2009	CH_3O_2N	Methane, nitro-	10.45	10	c = 0.005-0.017, mixed constant	E3bg	Approx.	T67
			10.33	18				
			10.21	25				
			10.24	25		E3bg	Approx.	W22
2010	CH_3O_2N	Methanohydroxamic acid (Formohydroxamic acid)	8.65		I = 0.2(NaCl)	E3bg	Uncert.	C72
2011	CH_3NS_2	Methanedithioic acid, amino- (Dithiocarbamic acid)	2.95	25	c = 0.002-0.01	C2	Approx.	G4
2012	CH_4O_2S	Methanesulfinic acid, hydroxy-	1.65	20	c = 0.1, mixed constant	E3bg	Approx.	R54

No.	Molecular formula, name and pK value(s)	T(°C)	Remarks	Method	Assessment	Ref.
2013	CH_4O_2Ge Ethanoic acid, 2-germa-					
	3.5	25	Compound decomposes	E3bg	Uncert.	K71
2014	CH_4O_3S Methanesulfonic acid					
	-1.86	25		RAM	Uncert.	C64
	-1.2	25		NMR	Uncert.	C77
2015	$CH_4O_7P_2$ Methylenediphosphonic acid, oxo- (Carbonyldiphosphonic acid)					
	pK_3 5.81, pK_4 8.42	25	c = 0.005, I = 0.032-0.25	E3bg	Approx.	G36
2016	CH_5O_3P Methylphosphonic acid					
	pK_1 1.87, pK_2 7.19	15	I = 0.2(NH_4ClO_4)			B106
	pK_1 2.12, pK_2 7.29	25				
	pK_1 2.29, pK_2 7.23	45				
	pK_1 2.41	25	c = 0.001, I = 0.1(KCl), concentration	E3bg	Uncert.	D68
	pK_2 7.35		constants		Approx.	
	pK_1 2.48, pK_2 7.34	not stated	c ≃ 0.05	E3bg	Uncert.	B61
	pK_1 2.35, pK_2 7.1	not stated	c = 0.02	E3bg	Uncert.	C38
2017	CH_5O_3As Methylarsonic acid					
	pK_2 8.53	25	c = 0.05, I = 1.0(KCl), mixed constant	E3bg	Uncert.	S19
	pK_1 4.58, pK_2 7.82	not stated	Mixed constants	E3bg	Uncert.	C42

2018 CH₅O₄P   Methyl dihydrogen phosphate

pK$_1$ 1.54	25	m ≈ 0.04, mixed constants	E3bg	Uncert.	K67
pK$_2$ 6.31				Approx.	
pK$_1$ 1.52	22		E3bg	Uncert.	B174
pK$_2$ 6.58	22.5			Approx.	

Variation with temperature

T	35	58	60	72	
pK$_1$	1.45	1.46	1.43	1.60	
T	1.1	11	34.2	58	74.5
pK$_2$	6.75	6.61	6.59	6.70	6.71

2019 CH₅N₃S   Thiosemicarbazide

pK$_1$ 1.88	25	c = 0.1, I ≈ 1.0(KCl), mixed constant	E3bg	Approx.	S18
pK$_2$ 12.81(H$_2$O), pK$_2$ 13.42(D$_2$O)	25	I = 1.0(KCl), mixed constants	05	Uncert.	

2020 CH₆O₆P₂   Methylenediphosphonic acid

pK$_2$ 2.49, pK$_3$ 6.87, pK$_4$ 10.54	25	I = 0.5[(CH$_3$)$_4$NCl], concentration constants	E3bg	Approx.	C17
pK$_1$ < 2, pK$_2$ 2.6	20	I = 0.1(KCl)	E3bg	Uncert.	S48
pK$_3$ 6.87, pK$_4$ 10.33					
pK$_2$ 3.05	25	c = 0.005, I = 0.03-0.25	E3bg	Approx.	G36
pK$_3$ 7.35, pK$_4$ 10.96				Uncert.	
pK$_1$ 1.7, pK$_2$ 2.75	25	I = 0.1(KCl)	E3bg,R3f	Uncert.	K4
pK$_3$ 7.33, pK$_4$ 10.42				Approx.	

(Contd)

No.	Molecular formula, name and pK value(s)	T(°C)	Remarks	Method	Assessment	Ref.
2020 (Contd)	$CH_6O_6P_2$ Methylenediphosphonic acid					
	pK_2 2.87, pK_3 7.45, pK_4 10.69	25	$I = 0.1-1.0[(CH_3)_4NBr]$	E3bg	Uncert.	I6
	pK_2 2.90, pK_3 7.62, pK_4 10.51	37				
	pK_2 3.08, pK_3 7.57, pK_4 10.47	50				
2021	$CH_6O_7P_2$ Methylenediphosphonic acid, hydroxy					
	pK_2 2.74, pK_3 7.05, pK_4 10.56	25	$c = 0.005$, $I = 0.032-0.25$	E3bg	Approx.	G36
2022	$CH_7O_{10}P_3$ Methyl tetrahydrogen triphosphate					
	6.45	20	$I = 0.1(NaClO_4)$, mixed constant	E3bg	Approx.	S24
2023	CHO_2NCl_2 Methane, dichloronitro-					
	5.99	25	$I = 0.06$	05	Approx.	A7
	5.98	40	Thermodynamic quantities are derived from the results			
2024	CHO_2NF_2 Methane, difluoronitro-					
	12.4	not stated			V.uncert.	K39
2025	CHO_4N_2Cl Methane, chlorodinitro-					
	3.80	25	$I = 0.06$	05	Approx.	A7
	3.67	5	Mixed constant	05	Approx.	I24
	3.53	20				
	3.36	40				
	3.28	60				

(contd)

No.	Formula	Name		Temp (°C)	Notes		Uncert./Approx.	Ref.
2025 (Contd)	CHO_4N_2Cl	Methane, chlorodinitro-	3.62	9.3	Mixed constant	05	Uncert.	H5
			3.80	25.5				
2026	CHO_4N_2Br	Methane, bromodinitro-	3.6	20		05	Uncert.	I24
			3.64	10.3	Mixed constant	05	Uncert.	H5
			3.47	25.3	Mixed constant			
			3.58	20		05	Uncert.	T58a
2027	CHO_4N_2I	Methane, iododinitro-	3.19	20		05	Uncert.	T58a
2028	CHO_4N_2F	Methane, fluorodinitro-	8.31	5	Thermodynamic quantities are derived from the results	05	Approx.	S80
			7.70	20				
			6.99	40				
			6.37	60				
2029	CH_2O_2NCl	Methane, chloronitro-	7.20	25	$I = 0.06$	05	Approx.	A7
			7.16	40	Thermodynamic quantities are derived from the results			

No.	Molecular formula, name and pK value(s)	T(^{o}C)	Remarks	Method	Assessment	Ref.
2030	CH$_2$O$_2$F$_3$P Methylphosphonous acid, trifluoro-					
	1.01	25	c = 0.0004-0.035	C2,R1a	Approx.	E16
2031	CH$_2$O$_3$Cl$_3$P Methylphosphonic acid, trichloro-					
	pK$_2$ 4.28	25	c = 0.05, I = 1.0(KCl), mixed constant	E3bg	Uncert.	S19
	pK$_1$ 1.63, pK$_2$ 4.71	not stated		E3	Uncert.	M51
2032	CH$_2$O$_3$F$_3$P Methylphosphonic acid, trifluoro-					
	pK$_1$ 1.17, pK$_2$ 3.93	not stated	c ≈ 0.05	E3bg	Uncert.	B61
	pK$_1$ 1.17, pK$_2$ 3.92	25	c = 0.0007-0.059	C2,R1a	Approx.	E16
2033	CH$_2$O$_3$F$_3$As Methylarsonic acid, trifluoro-					
	pK$_1$ 1.12	25	c = 0.00025-0.05 Calculated from data in E15	C2,R1a	Approx.	E16
	pK$_2$ 5.5				Uncert.	
2034	CH$_3$O$_3$Cl$_2$P Methylphosphonic acid, dichloro-					
	4.97	25	c = 0.05, I = 1.0(KCl), mixed constant	E3bg	Uncert.	S19
	pK$_1$ 1.14, pK$_2$ 5.58	not stated		E3	Uncert.	M51
2035	CH$_4$O$_3$ClP Methylphosphonic acid, chloro-					
	pK$_1$ 1.51	25	c = 0.001, I = 0.1(KCl), concentration constants	E3bg	Uncert.	D68
	pK$_2$ 6.17				Approx.	

No.	Formula	Name / pK values	Temp	Conditions			Ref
2036	$CH_4O_6Cl_2P_2$	Methylenediphosphonic acid, dichloro- pK_3 6.11, pK_4 9.78	25	c = 0.005, I = 0.032-0.25 Possible hydrolysis during titration	E3bg	Uncert.	G36
2037	$CH_5O_6BrP_2$	Methylenediphosphonic acid, bromo- pK_2 2.20, pK_3 6.55, pK_4 10.15	25	I = 0.032	E3bg	Uncert.	G36
2038	CH_6O_3NP	Methylphosphonic acid, amino- pK_1 1.85, pK_2 5.35, pK_3 10.00	not stated	c = 0.02	E3bg	Uncert.	C38
2039	CHO_2NClF	Methane, chlorofluoronitro- 10.14	25	I = 0.06	05	Approx.	A7
		9.92	40	Thermodynamic quantities are derived from the results			
2040	$C_2H_2O_3$	Acetic acid, 2-oxo- (Glyoxylic acid) 3.18	20	I = 0.1($NaClO_4$), concentration constant	E3bg	Approx.	P82
		3.46	25	Thermodynamic data are also given	E3bg	Uncert.	08
		2.98	25	I = 0.5(KCl), mixed constant	E3bg	Approx.	L28
2041	$C_2H_2O_4$	Ethanedioic acid (Oxalic acid) pK_1 1.244	0		E1b,R3d	Rel.	M3
		1.252	15				
		1.252	25				
		1.286	35				
		1.295	45				

(contd)

No.	Molecular formula, name and pK value(s)	T(^{o}C)	Remarks	Method	Assessment	Ref
2041 (Contd)	$C_2H_2O_4$ Ethanedioic acid (Oxalic acid)					
	pK$_1$ 1.1	25	c $\simeq$ 0.005, I = 1.0(KNO$_3$), concentration constants	E3bg	Uncert.	R5
	pK$_2$ 3.62				Approx.	
	pK$_1$ 1.13, pK$_2$ 3.85	25		E3bg	Approx.	N2
	pK$_1$ 1.00	25	c = 0.02, I = 0.5(LiClO$_4$), concentration constants	E3bg	Uncert.	D33
	pK$_2$ 3.50				Approx.	
	pK$_1$ 1.32, pK$_2$ 3.63	5	I = 0.5(NaClO$_4$), concentration constants	E3ag	Uncert.	B36
	pK$_1$ 1.23, pK$_2$ 3.65	15				
	pK$_1$ 1.20, pK$_2$ 3.67	25				
	pK$_1$ 1.37, pK$_2$ 3.81	25	I = 0.1(NaClO$_4$), mixed constants	E3bg	Approx.	M2
	pK$_1$ 1.08, pK$_2$ 3.55	25	c $\simeq$ 0.02, I = 1.0(NaClO$_4$)	E3bg	Approx.	M101
	pK$_1$ 1.26, pK$_2$ 3.80		I = 3.0(NaClO$_4$), concentration constants			
	pK$_1$ 1.32		I = 0.1(NaClO$_4$)			
	Other values in D44, K69, L37 and L77					
2042	C_2H_4O Ethanal (Acetaldehyde)					
	13.57	25	pK of hydrate	C3	Approx.	B54
	13.48	25	Based on a hydrate content of 55%	C3	Uncert.	B52

2043 $C_2H_4O_2$ Acetic acid   (Ethanoic acid)

Value	Temp	Conditions			Ref.
4.757	25		E3bg	Approx.	W29
4.53	25	$I = 0.1(NaClO_4)$, concentration constant	E3bg	Approx.	Y12
4.48	25	$c = 0.02$, $I = 0.5(LiClO_4)$, concentration constant	E3bg	Approx.	D33
4.73(H_2O), 5.25(D_2O)	25		E3bg	Uncert	G24
4.76	25	$m = 0.01$	C2,R1b	Approx.	E13

Variation with temperature

	50	100	150	200	225
	4.79	4.95	5.21	5.51	5.74

4.75	25		C1	Approx.	E14

Variation with pressure(atm)

	500	1000	2000
	4.65	4.56	4.41

4.76	25		C	Approx.	L67

Variation with temperature and pressure(bar)

T	1bar	1000bar	2000bar	3000bar
25	4.76	4.58	4.42	4.28
50	4.79	4.60	4.43	4.30
75	4.84	4.64	4.47	4.33
100	4.94	4.71	4.52	4.36
150	5.19	4.90	4.68	4.50

(Contd)

No.	Molecular formula, name and pK value(s)	T(°C)	Remarks	Method	Assessment	Ref.
2043 (contd)	$C_2H_4O_2$ Acetic acid (Ethanoic acid)					

Variation with temperature and pressure(bar) (contd)

	200	5.49	5.15	4.89	4.66
	225	5.68	5.32	5.04	4.79

Values also given for 400, 1400 and 2400 bar. Thermodynamic quantities are derived from the results

Acetic acid-0-d

		T(°C)	Remarks	Method	Assessment	Ref.
	5.313	25	Molality scale	Ele	Rel.	G3

Variation with temperature

5	10	15	20	25	30	35	40	45	50
5.346	5.334	5.324	5.317	5.312	5.310	5.309	5.312	5.317	5.324

Thermodynamic quantities are derived from the results

Acetic acid-2-d$_3$

		T(°C)	Remarks	Method	Assessment	Ref.
	4.772	25	m = 0.005-0.05	Ele	Rel.	P2

Variation with temperature

0	5	10	15	20	30	35	40	45	50
4.796	4.785	4.777	4.775	4.771	4.774	4.776	4.782	4.790	4.799

Thermodynamic quantities are derived from the results

	4.76	≈ 25	c = 0.008-0.115	Cl	Approx.	S120

(contd)

No.	Formula	Name / pK	t (°C)	Conditions	Ele	Rel.	Ref
2043 (contd)	C₂H₄O₂	Acetic acid d₃-0d					
		5.325	25	m = 0.005-0.05			P3

Variation with temperature

t (°C)	5	10	15	20	30	35	40	45	50
pK	5.360	5.348	5.338	5.331	5.323	5.323	5.326	5.330	5.336

Thermodynamic quantities are derived from the results

Other values for acetic acid are given in F7,H7,H17,M82,S18,S88 and V4

No.	Formula	Name / pK	t (°C)	Conditions	Ele	Rel.	Ref
2044	C₂H₄O₃	Acetic acid, hydroxy- (Glycolic acid)					
		3.70	30	I = 0.1(KCl), mixed constant	E3bg	Approx.	C23
		3.62	25	I = 1.0(NaClO₄)	E3bg	Approx.	M54
		3.83(H_2O), 4.26(D_2O)	25	I <0.07, based on pK of 5.28 for acetic acid in D_2O	O3	Approx.	B51
		4.27(D_2O)	25	c = 0.02 in D_2O, I = 0.11, pH of 0.01M DCl in 0.1M NaCl taken as 1.70. Mixed constant	E3bg	Approx.	L38
2045	C₂H₄O₃	Peroxyacetic acid (Peracetic acid)					
		8.2	20	c = 0.1	E3bg	Uncert.	E27
2046	C₂H₆O	Ethanol					
		15.93	25		KIN	Uncert.	M126
		15.9		Extrapolated value from the experimentally determined values of other primary alcohols. The estimated error is ±0.2 pK		Uncert.	B8
		16.0			C3	Uncert.	L63

No.	Molecular formula, name and pK value(s)	$T(^{o}C)$	Remarks	Method	Assessment	Ref.
2047	$C_2H_6O_2$ 1,2-Ethanediol (Ethylene glycol)					
	15.4	25	Experimental value.	C3	Uncert.	B8
	15.1		Corrected for statistical factor			
2048	$C_2H_6O_2$ Ethyl hydroperoxide					
	11.8	20		O5	Uncert.	E27
2049	C_2H_6S Ethanethiol					
	10.61	25	$c \approx 1 \times 10^{-4}$	O5	Approx.	I20
			Heat of ionization = 6.42 kcal $mole^{-1}$			
	10.50	20	$c = 0.01$	E3bg	Uncert.	D8
2050	$C_2H_6S_2$ 1,2-Ethanedithiol					
	pK_1 8.96, pK_2 10.54	30	$I = 0.1$(KCl)	E3ag	Uncert.	L25
	pK_1 9.05, pK_2 10.56	25	$I = 0-1.0$(NaCl)	E3bg	Uncert.	A71
2051	$C_2H_7N_5$ Biguanide					
	pK_1 3.2, pK_2 13.3			O5	Uncert.	H43
2052	C_2HOCl_3 Ethanal, trichloro- (Chloral)					
	9.66	30		E3bg	Uncert.	G6
	10.04	25		E3bg	Approx.	B54
	9.95	30	$c = 0.01$, $I = 0.1$(KCl), mixed constant (pK values of hydrate)	E3bg	Approx.	E18

2053 C_2HOF_3 Ethanal, trifluoro-

	Temp	Conditions			
10.21	25	I = 0.01	E3bg	Approx.	S114
10.19		I = 0.05			
10.01		I = 1.00			
		(pK of hydrate)			

Ethanal-1-d, trifluoro-

	Temp	Conditions			
10.22	25	I = 0.01	E3bg	Approx.	S114
10.03		I = 1.00			

2054 $C_2HO_2Cl_3$ Acetic acid, trichloro-

	Temp	Conditions			
0.512	25	c = 0.07-0.2	E3ag	Approx.	K69

Variation with temperature

15	20	30	35	40	45	50	55
0.532	0.518	0.513	0.513	0.516	0.527	0.528	0.538

Thermodynamic quantities are derived from the results

	Temp	Conditions			
-0.51	25	c ≈ 0.3-3.4	RAM	Uncert.	B100

2055 $C_2HO_2Br_3$ Acetic acid, tribromo-

	Temp	Conditions			
0.72	25	c = 0.16	E3bg	Uncert.	B113

2056 $C_2HO_2F_3$ Acetic acid, trifluoro-

	Temp	Conditions			
1.1	22		REF	Uncert.	G50
0.52	25	c = 0.1	E3ag	Approx.	K69

(contd)

No.	Molecular formula, name and pK value(s)	T(°C)	Remarks	Method	Assessment	Ref.
2056 (contd)	$C_2HO_2F_3$ Acetic acid, trifluoro-					

Variation with temperature

15	20	30	35	40	45	50
0.50	0.52	0.51	0.48	0.47	0.51	0.52

Thermodynamic quantities are derived from the results

No.	Molecular formula, name and pK value(s)	T(°C)	Remarks	Method	Assessment	Ref.
2057	$C_2HO_4N_3$ Ethanenitrile, 2,2-dinitro-					
	-6.2	25	In aq. H_2SO_4, H_- scale	O6	Uncert.	H5
	-6.22	20	In aq. $HClO_4$, H_- scale	O6	Uncert.	T58a
	-6.40	20	In aq. H_2SO_4, H_- scale	O6	Uncert.	I24
2058	$C_2H_2O_2Cl_2$ Acetic acid, dichloro-					
	1.35	25	c = 0.1	E3ag	Approx.	K69

Variation with temperature

15	20	30	35	40	45	50
1.31	1.33	1.38	1.40	1.42	1.44	1.45

Thermodynamic quantities are derived from the results

	1.30	25	c = 0.16	E3bg	Uncert.	B113

Values in mixed solvents also given

2059	$C_2H_2O_2Br_2$ Acetic acid, dibromo-					
	1.48	25	c = 0.08	E3bg	Uncert.	B113

Values in mixed solvents also given

2060 C$_2$H$_2$O$_2$F$_2$ Acetic acid, difluoro-

1.13	25	I = 1.0(KCl), mixed constant		E3bg	Approx.	S18
1.34	25			E3ag	Approx.	K69

Variation with temperature

15	20	30	35	40	45	50
1.29	1.31	1.37	1.39	1.41	1.43	1.46

Thermodynamic quantities are derived from the results

2061 C$_2$H$_2$O$_2$S$_2$ Ethanebis(thioic acid) (Dithiooxalic acid)

pK$_1$ 0.91, pK$_2$ 2.71	Titration of K-salt	E	P65

2062 C$_2$H$_3$OCl$_3$ Ethanol, 2,2,2-trichloro-

12.24	25	c = 0.01-0.11	C3	Approx.	B8
12.25	25		KIN	Uncert.	M126

2063 C$_2$H$_3$OF$_3$ Ethanol, 2,2,2-trifluoro-

12.37(in H$_2$O)	25	c = 0.004-0.2	C3	Approx.	B6
13.02(in 99.5% D$_2$O)	25	c = 0.009-0.07			
12.8	25	Mixed constant	E3bg	Uncert.	K37
12.5	25		E3c	Uncert.	M107
12.39	25		KIN	Uncert.	M126

Other values in D65, F14, M4 and R26.

No.	Molecular formula, name and pK value(s)	T($^{\circ}$C)	Remarks	Method	Assessment	Ref.
2064	$C_2H_3O_2Cl$ Acetic acid, chloro-					
	2.866	25		C2,Rle	Rel.	S89
			Variation with temperature			
			15 35 50 60 75 90			
			2.845 2.900 2.955 3.004 3.062 3.134			
			Equation given for this variation.			
			Thermodynamic quantities are derived from the results.			
	2.84(H_2O)	25	I < 0.07	O3	Approx.	B51
	3.33(D_2O)		Based on pK = 5.28 for acetic acid in D_2O			
	2.70	25	I = 1.0(KCl), mixed constant	E3bg	Approx.	S18
			Other values are given in C83, M82 and S88			
2065	$C_2H_3O_2Br$ Acetic acid, bromo-					
	2.92(H_2O)	25	I < 0.07	O3	Approx.	B51
	3.46(D_2O)		Based on pK = 5.28 for acetic acid in D_2O			
	2.93	40		E3bg	Approx.	C50
	2.901	25				S88

No.	Formula / Name	pK	T (°C)	Conditions	Code	Reliability	Ref.
2066	$C_2H_3O_2I$ Acetic acid, iodo-	3.124	10		E3bg	Approx.	C50
		3.211	40				
		3.19(H_2O)	25	I < 0.07	03	Approx.	B51
		3.61(D_2O)		Based on pK = 5.28 for acetic acid in D_2O			
		3.174	25				S88
2067	$C_2H_3O_2F$ Acetic acid, fluoro-	2.72(H_2O)	25	I < 0.07	03	Approx.	B51
		3.13(D_2O)		Based on pK = 5.28 for acetic acid in D_2O			
		2.584	25			S88	
2068	$C_2H_3O_4N$ Acetic acid, nitro-	pK$_1$ 1.48	23.5	c ≈ 0.01, I ≈ 0.1, concentration constant. Concentration corrected for decarboxylation	05	Uncert.	F15
		pK$_2$ 8.90		c ≈ 0.0001, I ≈ 0.1, concentration constant	05	Approx.	
2069	$C_2H_3O_5N$ Acetic acid, nitrooxy-	2.26	not stated	c ≈ 0.025	E3bg	Uncert.	M9

No.	Molecular formula, name and pK value(s)	T(°C)	Remarks	Method	Assessment	Ref.
2070	$C_2H_3O_5N_3$ Acetamide, 2,2-dinitro-					
	1.30	20	In $HClO_4$ solutions, H_o scale	06	Uncert.	T58a
2071	$C_2H_4OCl_2$ Ethanol, 2,2-dichloro-					
	12.89	25		C3	Approx.	B8
2072	C_2H_4OS Ethanethioic acid					
	3.62	25	$c \approx 0.0008$	05	Approx.	I20
			Heat of ionization = 0.56 kcal mole^{-1}			
2073	$C_2H_4OS_2$ Methanedithioic acid, methoxy- (Xanthic acid, methyl-)					
	1.55	not stated		05	Uncert.	Z2
	2.29	0	Extrapolated to zero time	E3d		G5
2073a	$C_2H_4O_2N_2$ Ethanal oxime, 1-nitroso-					
	5.8	21	$I \approx 0.3$	05	Uncert.	A74
2074	$C_2H_4O_2N_2$ Ethanedial dioxime (Glyoxime)					
	8.88	25		E3bg	Uncert.	B10
	9.1	25	Unstable in aqueous solution	05	Uncert.	U10
	9.05	25		E3bg	Uncert.	

2075 $C_2H_4O_2S$ Acetic acid, mercapto- (Thioglycolic acid)

pK$_1$ 3.42, pK$_2$ 10.48	0	I = 0.1(KCl) concentration constants	E3bg	Approx.	L30
pK$_1$ 3.55, pK$_2$ 10.35	15				
pK$_1$ 3.59, pK$_2$ 10.07	35				
pK$_1$ 3.66, pK$_2$ 9.95	40				
pK$_1$ 3.581	25	c ≈ 0.001, I = 0.1(KNO$_3$) concentration constant	E3bg	Approx.	P40
pK$_1$ 3.55	30	I = 0.1(KCl), mixed constant	E3bg	Approx.	C23
pK$_1$ 3.677, pK$_2$ 10.550	25		E3bg	Approx.	M127
pK$_1$ 3.52, pK$_2$ 10.20	20	I = 0.1(NaClO$_4$), mixed constants	E3bg	Approx.	P32
pK$_1$ 3.55, pK$_2$ 10.22	25	c = 0.01, I = 0.1(KCl), mixed constants	E3bg	Approx.	L75
pK$_1$ 3.60, pK$_2$ 9.74	60				
pK$_1$ 3.42, pK$_2$ 10.20	25	I = 0.1(KCl), concentration constants	E3bg	Approx.	L24

Other values in B59, D8, F42, J5, I20, K69, L24, L68, P79 and W46

2076 $C_2H_4O_2Se$ Acetic acid, hydroseleno- (Selenoglycolic acid)

pk$_a$ 4.7, pk$_b$ 4.1	25	Microscopic constants. pk$_a$ is assumed	05	Uncert.	K69
pk$_c$ 6.7, pk$_d$ 7.3		to be equal to pK of HSeCH$_2$COOCH$_3$			
		(See No.2218)			

No.	Molecular formula, name and pK value(s)		$T(^{o}C)$	Remarks	Method	Assessment	Ref.
2077	$C_2H_4O_3N_2$	Acetamide, 2-nitro-					
		5.18	25	I = 0.06	05	Approx.	A7
		5.15	40	Thermodynamic quantities are derived from the results.			
2078	$C_2H_4O_3N_2$	Ethanal oxime, 1-nitro-					
		7.4	21	I ≈ 0.3	05	Uncert.	A74
2079	$C_2H_4O_4N_2$	Ethane, 1,1-dinitro-					
		5.30	20		05	Approx.	T58a
		5.21	20		05		S83
		5.13	25	I = 0.05(phthalate buffers), mixed constant	05	Approx.	S77
		5.28	25	I ≈ 0.002	E3bg	Approx.	B45
				Derived from equilibrium values of pH achieved in 10-60 minutes.			
		5.38	5	Microscopic constants are also given.	05	Approx.	N39

(Contd)

No.	Formula	Name	pK	Temp	Conditions	Method	Assessment	Ref.
2079 (contd)	$C_2H_4O_4N_2$	Ethane, 1,1-dinitro-	5.21	20	Thermodynamic quantities are derived from the results.			
			5.06	40				
			4.91	50				
			5.25	9.6	Mixed constant	05	Uncert.	H5
			5.13	25.1				
2080	$C_2H_4O_4N_2$	Ethanedihydroxamic acid (Oxalodihydroxamic acid)	pK_1 6.97, pK_2 9.00	20	I = 0.005 - 0.01	E3bg	Approx.	M98
			pK_1 6.55, pK_2 8.63	30	c = 0.002, I = 0.1($NaClO_4$), concentration constants	E3bg	Uncert.	D62
			pK_1 6.90	30.5	c = 0.001, I = 0.1(KNO_3), mixed constant	E3d	Uncert.	S137
2081	$C_2H_4O_4N_2$	Ethanedihydroximic acid	pK_1 6.81, pK_2 8.66	25		E3bg	Uncert.	B10
2082	$C_2H_4O_5N_2$	Ethanol, 2,2-dinitro-	4.24	20		05	Approx.	T58a
2083	$C_2H_4O_5S$	Acetic acid, sulfo-	pK_2 4.20	25	c = 0.002, I = 0.011, mixed constant	E3bg	Approx.	B58
			pK_2 4.07	not stated		E	Uncert.	B21
2084	$C_2H_4N_2S_2$	Ethanebis(thioamide) (Rubeanic acid)	pK_1 10.9	not stated		E3bg	Uncert.	Y1

No.	Molecular formula, name and pK value(s)		T(°C)	Remarks	Method	Assessment	Ref.
2085	C_2H_5OCl	Ethanol, 2-chloro-					
		14.31(in H_2O)	25		C3	Uncert.	B8
		14.99(in 99.5% D_2O)	25				
2086	$C_2H_5O_2N$	Acetic acid, amino- (Glycine)					
	pK_1 2.33, pK_2 9.60		25	I = 0.2($NaClO_4$), concentration constants	E3bg	Approx.	C44
	pK_1 2.78, pK_2 10.10		10	I = 0.65, mixed constants	E3bg	Approx.	L31
	pK_1 2.46, pK_2 9.70		25				
	pK_1 2.23, pK_2 9.28		40				
	pK_1 2.48		25	I = 1.0(KCl), mixed constant	E3bg	Approx.	S18
	pK_2 10.25		0.35	I = 0.09(KCl), concentration constant	E3bg	Approx.	M125
	pK_2 9.44		30.0				
	pK_2 9.00		48.8	Thermodynamic quantities are derived from the results			
	pK_2 9.53		30	I = 0.1(KCl), mixed constant	E3bg	Approx.	C23
		Other values in L27, P28, S109 and T23					
2087	$C_2H_5O_2N$	Acetohydroxamic acid					
		8.70	25(?)	c = 0.02, I = 0.1(KCl), mixed constant	E3bg	Uncert.	M66
		9.4	20	c = 0.01	E3d	Uncert.	W39

No.	Formula / Name	pK	Temp.	Conditions	Method	Reliability	Ref.
2088	$C_2H_5O_2N$ Ethane, nitro-	8.60	25		E3bg	Approx.	W22
		8.49	18	c = 0.003-0.007, mixed constant	E3bg	Approx.	T67
		8.46	25	pH equilibrium requires <u>ca</u> 5 min.			
		8.44	30				
2089	$C_2H_5O_2N_3$ Biuret	13.2	25		05	Uncert.	H43
2090	$C_2H_5O_5P$ Acetyl dihydrogen phosphate	pK$_1$ 1.2, pK$_2$ 4.9	21		E3bg	Uncert.	L54
2091	C_2H_5NS Ethanethioamide (Thioacetamide)	13.4	25		07	Uncert.	E5
2092	$C_2H_5NS_2$ Methanedithioic acid, methylamino- (Dithiocarbamic acid, methyl-)	2.89	25		KIN	Uncert.	M79
2093	$C_2H_5N_3S_2$ Dithiobiuret	pK$_1$ 9.09, pK$_2$ 10.43	20	c = 0.001, I = 0.1(KNO_3), mixed constants	E3bg	Uncert.	U12
2094	$C_2H_6ON_2$ Acetohydrazide	pK$_1$ 3.24	25	mixed constant(?)	E3bg	Uncert.	T43
		pK(NH) 13.04	25		05	Uncert.	

No.	Molecular formula, name and pK value(s)	T(°C)	Remarks	Method	Assessment	Ref.
2095	$C_2H_6OH_4$ Methanamide, guanidino-					
	8.0			05	Uncert.	H43
2096	C_2H_6OS Ethanol, 2-mercapto					
	9.72	25	c ≈ 0.00012	05	Approx.	I20
			Heat of ionization = 6.21 kcal mole^{-1}			
	9.44	25	I = 0-1(NaCl)	E3bg	Approx.	A71
	9.48	20		05	Approx.	D8
	9.32	30				
	9.17	40				
	9.43	25		E3bg	Approx.	K58
2097	$C_2H_6O_2N_2$ Acetohydroxamic acid, amino-					
	7.35	25(?)	c = 0.02, I = 0.1(KCl), mixed constant	E3bg	Uncert.	M66
	7.70	not stated	I ≈ 0.2(NaCl)	E3bg	Uncert.	C72
	7.40	25	c = 0.01, I = 0.1(KCl), mixed constant	E3bg	Approx.	G38
2098	$C_2H_6O_2N_4$ Ethanediamide dioxime					
	pK$_1$ 2.95, pK$_2$ 11.37	20		E3bg/05	Uncert.	W18
	pK$_2$ 10.62	26		E3bg	Uncert.	U11

No.	Formula / Name	pK	T (°C)	Conditions	Method	Reliability	Ref.
2099	$C_2H_6O_3S_2$ Ethanesulfonic acid, 2-mercapto-						
		pK_2 9.53	20		E3bg	Uncert.	D8
		9.08	20		O5	Uncert.	
2100	$C_2H_7O_2P$ Phosphinic acid, dimethyl-						
		3.13	not stated	c = 0.005, I ⊅ 0.025, mixed constant, 7% ethanol	E3bg	Uncert.	M57
2101	$C_2H_7O_2As$ Arsinic acid, dimethyl- (Cacodylic acid)						
		6.15	25	c = 0.05, I = 1.0(KCl), mixed constant	E3bg	Uncert.	S19
2102	$C_2H_7O_3P$ Ethylphosphonic acid						
		pK_2 7.60	25	c = 0.05, I = 1.0(KCl), mixed constant	E3bg	Uncert.	S19
		pK_1 2.45, pK_2 7.85	not stated	c = 0.02	E3bg	Uncert.	C38
2103	$C_2H_7O_3As$ Ethylarsonic acid						
		pK_1 4.72, pK_2 8.00	not stated	Mixed constants	E3bg	Uncert.	C42
2104	$C_2H_7O_4P$ Dimethyl hydrogen phosphate						
		1.22	20	Concentration constant	E3bg	Uncert.	A15
		1.25	20	7% ethanol	E	Uncert.	K3
		1.29	25	m ≈ 0.04, mixed constant	E3bg	Uncert.	K67
		1.25	not stated	c = 0.005, I ⊅ 0.025, mixed constant, 7% ethanol	E3bg	Uncert.	M57

No.	Molecular formula, name and pK value(s)	$T(^{o}C)$	Remarks	Method	Assessment	Ref.
2105	C₂H₇O₄P Ethyl dihydrogen phosphate					
	pK₁ 1.60, pK₂ 6.62	25	m ≈ 0.035, mixed constants	E3bg	Uncert.	K67
2106	C₂H₇NS Ethanethiol, 2-amino					
	pK₁ 8.19, pK₂ 10.75	25	c = 0.01, mixed constants	E3bg	Approx.	B147
	pK₁ 8.19, pK₂ 10.73	25	c = 0.001, I = 0.1(KNO₃) concentration constants	E3bg	Approx.	T59
	pK₁ 8.35	23	c = 0.002	05	Approx.	B59
	pK₁ 8.23 (-SH)	25	c ≈ 0.00015, heat of ionization = 7.43 kcal mole⁻¹	05	Approx.	I20
2107	C₂H₈O₆P₂ 1,1-Ethylidenediphosphonic acid					
	pK₂ 3.14, pK₃ 7.49, pK₄ 11.97	25	c = 0.005, I = 0.032-0.25	E3bg	Uncert. / Approx.	G36
	pK₂ 2.66, pK₃ 7.18, pK₄ 11.54	25	I = 0.5[(CH₃)₄NCl] concentration constants	E3bg	Approx.	C17
2108	C₂H₈O₆P₂ 1,2-Ethylenediphosphonic acid					
	pK₁ 1.5, pK₂ 3.18, pK₃ 7.62, pK₄ 9.28		I = 0.1-1.0[(CH₃)₄NBr]. Average of values at 25, 37 and 50°C	E3bg	Uncert.	I6

2109 $C_2H_8O_7P_2$ 1,1-Ethylidenediphosphonic acid, 1-hydroxy-

pK$_1$ 1.7, pK$_2$ 2.47, pK$_3$ 7.28, pK$_4$ 10.29, pK$_5$ 11.13

    25   E3bg,R3f Uncert. K4   I = 0.1(KCl)

pK$_2$ 2.54, pK$_3$ 6.97, pK$_4$ 11.41

    25   E3bg Uncert. C17   I = 0.5[$(CH_3)_4NCl$]concentration constants

pK$_2$ 3.03, pK$_3$ 7.31

    25   E3bg Approx. G36   c = 0.005, I = 0.032-0.25

pK$_4$ 11.52

                     Uncert.

2110 $C_2H_{12}O_4B_{12}$ Dodecaborane (10)-1,12-dicarboxylic acid

pK$_1$ 9.07, pK$_2$ 10.236

    25   CAL Approx. C49

Thermodynamic quantities are also given

2111 C_2HONF_6 N,N-Bis(trifluoromethyl)hydroxylamine

8.82   E3bg Uncert. B187   No details

2112 $C_2HO_2NF_4$ Nitroethane, 1,2,2,2-tetrafluoro-

9.10   not stated E3bg Uncert. K39

2113 $C_2HO_2ClF_2$ Acetic acid, chlorodifluoro-

0.46   25   E3ag Approx. K69   From pH measurements on the acid, c=0.1

Variation with temperature

15	20	25	30	35	40	45	50
0.42	0.46	0.46	0.48	0.49	0.51	0.54	0.56

Thermodynamic quantities are derived from the results

No.	Molecular formula, name and pK value(s)	$T(^oC)$	Remarks	Method	Assessment	Ref.
2114	$C_2HO_2F_6As$ Arsinic acid, bis(trifluoromethyl)					
	1.42	25	c = 0.025-0.05	C2,R1a	Approx.	E15
2115	$C_2H_2ONF_3$ Ethanal oxime, 2,2,2-trifluoro-					
	8.9	25		E3bg	Uncert.	D65
2116	$C_2H_2O_2NF_3$ Ethane, 1,1,1-trifluoro-2-nitro-					
	7.40	not stated	pH corrected graphically to zero time	E3bg	Uncert.	K39
2117	$C_2H_2O_2N_2Cl_2$ Ethanedial dioxime, 1,2-dichloro-					
	2.95	25		E3bg	Uncert.	U10
2118	$C_2H_3O_2NCl$ Acetohydroxamic acid, chloro-					
	8.40		I = 0.2(NaCl)	E3bg	Uncert.	C72
2119	$C_2H_3O_2N_2Cl$ Ethanedial dioxime, 1-chloro-					
	8.35(amphi-form)	25		E3bg	Uncert.	U10
	3.40(anti-form)			05		
	8.13(amphi-form)					
	3.92(anti-form)					
	(Unstable in aqueous solution)					

No.	Formula / Name	pK	T (°C)	Conditions		Reliability	Ref.
2120	$C_2H_3O_3N_2Cl$ Acetamide, 2-chloro-2-nitro-	3.50	25		I = 0.06	Approx.	A7
		3.56	40		Thermodynamic quantities are derived from the results		
2121	$C_2H_3O_3N_2F$ Acetamide, 2-fluoro-2-nitro-	5.89	25		I = 0.06	Approx.	A7
		5.82	40		Thermodynamic quantities are derived from the results		
2122	$C_2H_4ONF_3$ N-Ethylhydroxylamine, 2,2,2-trifluoro-	11.8	25	E3bg	Mixed constant	Uncert.	K38
		11.3	25	E3bg		Uncert.	D65
2123	$C_2H_4ON_2S$ Acetamide, 2-amino-2-thioxo- (Monothiooxamide)	11.42	25	05,	I = 0.1(KCl), mixed constant	Approx.	C13
		11.53		05,R4	(Values for I = 0.02, 0.05, 0.25, 0.50, 1.00 are also given)	Approx.	
2124	$C_2H_7O_2NS$ Ethanesulfinic acid, 2-amino- (Hypotaurine)	pK_1 2.16, pK_2 9.56					C22
2125	$C_2H_7O_2S_2P$ O,O-Dimethyl hydrogen phosphorodithioate	1.55	20	E3bg	c = 0.005, 7% ethanol, mixed constant	Uncert.	K5
2126	$C_2H_7O_3NS$ Ethanesulfonic acid, 2-amino- (Taurine)	9.07	20	E3d	c = 0.025, mixed constant	Uncert.	H17

No.	Molecular formula, name and pK value(s)	T(°C)	Remarks	Method	Assessment	Ref.
2127	C$_2$H$_7$O$_3$SP O,O-Dimethyl O-hydrogen phosphorothioate					
	1.18	20	7% ethanol	E		K3
2128	C$_2$H$_8$O$_3$NP Ethylphosphonic acid, 2-amino-					
	pK$_1$ 2.45, pK$_2$ 7.00, pK$_3$ 10.8	not stated	c = 0.02	E3bg	Uncert.	C38
2129	C$_2$H$_8$O$_4$NP Ethyl dihydrogen phosphate, 2-amino- (O-Phosphonoethanolamine)					
	pK$_2$ 5.77, pK$_3$ 10.26	20	I = 0.1[(C$_3$H$_7$)$_4$NI]	E3bg	Uncert.	H31
	pK$_2$ 5.57, pK$_3$ 10.13	25	I = 0.15(KCl), concentration constants	E3bg	Approx.	O24
2130	C$_3$H$_2$O$_2$ Propynoic acid (Propiolic acid)					
	1.84	25	c = 0.01, I = 0.1(NaCl), mixed constant	E3bg	Approx.	M48
	1.887	25		C2	Rel.	G51
			Variation with temperature			
			10 15 20 30 35 40 45			
			1.791 1.867 1.867 1.936 1.952 1.963 1.978			
			Thermodynamic quantities are derived from the results			
2131	C$_3$H$_2$N$_2$ Propanedinitrile (Malononitrile)					
	11.19	25		E3bg	Uncert.	P18
	11.20	25		E3bg	Approx.	B118
			Thermodynamic quantities are also given			

2132	C$_3$H$_3$N$_5$	1,2,3-Triazole-5-carbonitrile, 4-amino-					
		6.15	20	c = 0.00012	05	Approx.	A41
2133	C$_3$H$_4$O	Prop-2-yn-1-ol (Propargyl alcohol)					
		13.55	25		C3	Uncert.	B8
2134	C$_3$H$_4$O$_2$	Propanal, 2-oxo-					
		11.0	30	c = 0.01, I = 0.1(KCl), mixed constant	E3bg	Uncert.	E18
				(pK of hydrate)			
2135	C$_3$H$_4$O$_2$	Propenoic acid (Acrylic acid)					
		4.25	25	c = 0.01	E3bg	Approx.	B112
		4.247	25		C2	Rel.	G51

Variation with temperature

15	20	30	35	40	45
4.267	4.246	4.250	4.267	4.301	4.311

Thermodynamic quantities are derived from the results

2136	C$_3$H$_4$O$_3$	Propanoic acid, 2-oxo- (Pyruvic acid)					
		2.39	10	I = 0.65, mixed constant	E3bg	Approx.	L31
		2.39	25				
		2.22	40				
		2.60	25		E3bg	Uncert.	08
		2.105	25	I = 1.0(NaCl)	E3bg	Approx.	T3

Other value in L27

No.	Molecular formula, name and pK value(s)	$T(^{o}C)$	Remarks	Method	Assessment	Ref.
2137	$C_3H_4O_4$ Propanedioic acid (Malonic acid)					
	pK_1 2.847, pK_2 5.696	25	I = 0.01	Elch	Rel.	I27

Variation with temperature

	5	10	15	20	30	35	40	45
pK_1	2.882	2.869	2.857	2.851	2.849	2.855	2.863	2.876
pK_2	5.663	5.666	5.673	5.683	5.710	5.729	5.751	5.774

	Molecular formula, name and pK value(s)	$T(^{o}C)$	Remarks	Method	Assessment	Ref.
	pK_1 2.847, pK_2 5.696	25		E1	Rel.	D11

Variation with temperature

	5	10	15	20	30	35	40	45
pK_1	2.881	2.867	2.856	2.849	2.849	2.855	2.862	2.875
pK_2	5.664	5.666	5.673	5.683	5.711	5.730	5.752	5.775

Thermodynamic quantities are derived from the results

	Molecular formula, name and pK value(s)	$T(^{o}C)$	Remarks	Method	Assessment	Ref.
	pK_1 2.80, pK_2 5.60	35	c ≈ 0.02, mixed constants	E3bg,R3f	Approx.	D60
	pK_1 2.61, pK_2 5.27	25	I = 0.1(KNO_3),concentration constants	E3bg,R3f	Approx.	P78

pK_2 is extrapolated value at c = 0

	Molecular formula, name and pK value(s)	$T(^{o}C)$	Remarks	Method	Assessment	Ref.
	pK_1 2.87, pK_2 5.70	25	c = 0.005, I = 0.35	O3,R3d	Approx.	J19
	pK_1 2.30, pK_2 5.7	20	c = 0.005	E3bg,R3a	Uncert.	D44

Values are also given in mixed solvents

	Molecular formula, name and pK value(s)	$T(^{o}C)$	Remarks	Method	Assessment	Ref.
	pK_1 2.87, pK_2 5.69(H_2O)	25	c = 0.02, I = 0.11(NaCl)	E3bg	Approx.	G23
	pK_1 3.37, pK_2 6.12(D_2O)					
	pK_1 3.30(D_2O)	25	pH of 0.01M DCl in 0.1 M NaCl taken as 1.70.	E3bg	Approx.	L38

(Other values in D33,L33,N3,N32,R5,W40 and Y12.)

2138 C$_3$H$_4$O$_5$ Propanedioic acid, hydroxy- (Tartronic acid)

pK$_1$ 1.89, pK$_2$ 4.28	20	c = 0.025	E3bg	Uncert.	Z17
pK$_1$ 2.37, pK$_2$ 4.74	25		E3bg	Uncert.	P19
pK$_1$ 2.38, pK$_2$ 4.76	37				

2139 C$_3$H$_4$N$_2$ Imidazole

pK$_1$ 7.15   25   I = 0.55   E3bg Approx. S108

Variation with temperature

20	30	35	40
7.27	7.08	7.00	6.90

pK$_2$ 14.44   25   I = 0.5(NaCl). Calculated from hydrolytic data   07 Uncert. G14

Variation with temperature

15.3	25.3	35.2
14.89	14.43	14.02

Thermodynamic quantities are derived from the results

pK$_2$ 14.17   25   In aqueous KOH, H$_-$ scale   07 Uncert. Y3

2140 C$_3$H$_4$N$_2$ Pyrazole

14.21   25   In aqueous KOH, H$_-$ scale   07 Uncert. Y3.

2141 C$_3$H$_6$O 2-Propen-1-ol (Allyl alcohol)

15.5   25   C3 Uncert. B8

No.	Molecular formula, name and pK value(s)	T(°C)	Remarks	Method	Assessment	Ref.
2142	C$_3$H$_6$O Propanone (Acetone) 20.0			KIN	Uncert.	B47
2143	C$_3$H$_6$O$_2$ Propanoic acid					
	4.875	25		C2	Rel.	G51
			Variation with temperature			
			15 20 30 35 40 45			
			4.873 4.870 4.880 4.891 4.901 4.912			
			Thermodynamic quantities are derived from the results			
	4.88	25	m = 0.01	C2,R1b	Approx.	E13
			Variation with temperature		*Uncert.	
			50 100 150 200 225			
			4.90 5.07 5.30* 5.64* 5.92*			
			Equation derived for variation with temperature			
	4.90	25	Variation with pressure(atm.)	C1	Uncert.	H7
			1000 2000 3000			
			4.69 4.50 4.35			
	4.66	20	I = 0.1(NaClO$_4$), concentration constant	E3bg	Approx.	P79
	4.86	25	c = 0.01, mixed constant. Value in mixed solvent also given	E3bg	Approx.	M82
	4.69	25	I = 0.1(KCl), concentration constant	E3bg	Approx.	S1

(Contd)

No.	Formula	Name	pK	T	Conditions	Method	Assessment	Ref.
2143	$C_3H_6O_2$	Propanoic acid (Contd)	4.89					
			4.70	25	I = 0.0025-1.8(KCl)	E3bg,R4		
					c = 0.05, I = 1.0(KCl), mixed constant	E3bg	Uncert.	S19
2144	$C_3H_6O_3$	Acetic acid, methoxy-	3.570	25	I = 0.02-0.096, molality scale	E1a	Rel.	K29

Variation with temperature

5	10	15	20	30	35	40	45	50
3.538	3.544	3.551	3.559	3.583	3.597	3.613	3.631	3.651

Thermodynamic quantities are derived from the results

No.	Formula	Name	pK	T	Conditions	Method	Assessment	Ref.
			3.57	25	c ≈ 0.01. Values in mixed solvents also given	E3bg	Approx.	B113
			3.31	20	I = 0.1(NaClO$_4$),concentration constant	E3bg	Approx.	P79
			3.36	20	I = 1.0(NaClO$_4$), concentration constant	E3bq	Approx.	S8
			3.40	25	I = 1.0(KCl), mixed constant	E3bg	Approx.	S18
2145	$C_3H_6O_3$	Peroxypropanoic acid	8.1	23	c = 0.1	E3bg	Uncert.	E27
2146	$C_3H_6O_3$	Propanoic acid, 2-hydroxy- (Lactic acid)	3.83	31		E3bg	Approx.	C83
2146a	$C_3H_6O_3$	Propanoic acid, 3-hydroxy-	4.52	31		E3bg	Approx.	C83
			4.49	30	I = 0.1(KCl), mixed constant	E3bg	Approx.	C23

No.	Molecular formula, name and pK value(s)	$T(^oC)$	Remarks	Method	Assessment	Ref.
2147	C_3H_6S 2-Propene-1-thiol 10.0	25	Gas solubility method		V.Uncert.	K58
2148	C_3H_8O 1-Propanol 16.1	25		KIN	Uncert.	M126
2149	C_3H_8O 2-Propanol 17.1	25		KIN	Uncert.	M126
2150	$C_3H_8O_2$ Ethanol, 2-methoxy- 14.8	25		C3	Uncert.	B8
2151	$C_3H_8O_2$ 1,2-Propanediol 14.9	25		KIN	Uncert.	M126
2152	$C_3H_8O_2$ 1,3-Propanediol 15.1	25		KIN	Uncert.	M126
2153	$C_3H_8O_2$ Isopropyl hydroperoxide 12.1	20		O5	Uncert.	E27
2154	$C_3H_8O_3$ 1,2,3-Propanetriol (Glycerol) pK_1 14.4			C3		L63

2155 C_3H_8S 2-Propanethiol

10.86   25   05   Approx.   I20

c ≈ 0.0001

Heat of ionization = 5.38 kcal mole⁻¹

2156 C_3OF_6 Propanone, 1,1,1,3,3,3-hexafluoro

6.58   Uncert.   M73

Conditions not stated (pK of hydrate)

2157 C_3HO_2Cl Propynoic acid, 3-chloro-

1.845   25   C2   Rel.   G51

Variation with temperature

10   15   20   30

1.767   1.796   1.820   1.864

Thermodynamic quantities are derived from the results

2158 C_3HO_2Br Propynoic acid, 3-bromo-

1.855   25   C2   Rel.   G51

Variation with temperature

10   15   20   30   35   40   45

1.786   1.814   1.839   1.875   1.900   1.915   1.924

Thermodynamic quantities are derived from the results

2159 $C_3HO_2F_5$ Propanoic acid, pentafluoro-

-0.41   not stated   c = 0.06-1.0   NMR(^{19}F) Uncert.   H35

No.	Molecular formula, name and pK value(s)	T(°C)	Remarks	Method	Assessment	Ref.
2160	$C_3H_2OF_4$ Propanone, 1,1,3,3-tetrafluoro-					
	8.79		Conditions not stated (pK of hydrate)		Uncert.	M73
2161	$C_3H_2OF_6$ 2-Propanol, 1,1,1,3,3,3-hexafluoro-					
	9.3		Conditions not stated		Uncert.	F14
	9.3	25	Mixed constant	E3bg	Uncert.	K38
	9.3	25		E3bg	Uncert.	D65
	9.30		Conditions not stated		Uncert.	M73
2162	$C_3H_3ON_3$ 1,2,4-Triazin-5(2H)-one					
	7.15	Not stated	c < 0.0001, I = 0.01, mixed constant	O5	Uncert.	B137
2163	$C_3H_3O_2N$ Acetic acid, cyano-					
	2.471	25	c = 0.00032-0.00477, λ_0 = 393.23	C2,R1c	Rel.	F6

Variation with temperature

T	5	10	15	20	30	35	40	45
pK	2.445	2.447	2.452	2.460	2.482	2.496	2.511	2.528
λ_0	275.89	305.32	334.77	364.06	421.97	450.21	478.10	505.48

Thermodynamic quantities are derived from the results

		T(°C)	Remarks	Method	Assessment	Ref.
	2.33	25	I = 1.0(KCl), mixed constant	E3bg	Approx.	S18
	2.50(H_2O)	25	I < 0.07			B51
	2.94(D_2O)		Based on pK = 5.28 for acetic acid in D_2O			

No.	Formula	Name / pK values	T (°C)	Conditions	Method	Assessment	Ref.
2164	$C_3H_3O_2N_3$	1,2,4-Triazine-3,5(2H,4H)-dione (6-Azauracil)					
		pK$_1$ 7.00	25	c ≈ 0.001	E3bg	Approx.	G62
		pK$_2$ 12.9				Uncert.	
		pK$_1$ 7.00	25	Titrant(CH$_3$)$_4$NOH	05	Approx.	J16
		pK$_2$ 12.9				Uncert.	
		pK$_1$ 7.00	25	c = 0.01	E3bg	Uncert.	K9
		pK$_1$ 6.9	not stated	c = 0.005	E3bg	Uncert.	C28
2165	$C_3H_3O_2N_3$	1,2,4-Triazine-5,6(1H,2H)-dione					
		pK$_1$ 7.21, pK$_2$ 12.07	not stated	c < 0.0001, I = 0.01, mixed constants	05	Uncert.	B137
2166	$C_3H_3O_2N_3$	1,3,5-Triazine-2,4(1H,3H)-dione					
		6.5	20	c = 0.02	E3bg	Uncert.	A36
		6.73	25	Titrant(CH$_3$)$_4$NOH, constant corrected for hydrolysis	E3bg	Approx.	J16
2167	$C_3H_3O_2N_3$	1,2,3-Triazole-4-carboxylic acid					
		pK$_1$ 3.22, pK$_2$ 8.73	25		E3bg	Uncert.	H13
2168	$C_3H_3O_2Cl$	Propenoic acid, 3-chloro-					
		3.79(trans)	25	c = 0.01	E3bg	Approx.	B112
		3.45(cis)					
2169	$C_3H_3O_2Br$	Propenoic acid, 3-bromo-					
		3.71(trans)	25	c = 0.01	E3bg	Approx.	B112
		3.32(cis)					

No.	Molecular formula, name and pK value(s)	T(°C)	Remarks	Method	Assessment	Ref.
2170	$C_3H_3O_2I$ Propenoic acid, 3-iodo- 3.74(trans) 3.42(cis)	25	c = 0.01	E3bg	Approx.	B112
2171	$C_3H_3O_3N_3$ 1,2,4-Triazine-3,5,6(1H,2H,4H)-trione 2.95	not stated	c = 0.005	E3bg	Uncert.	C29
2172	$C_3H_3O_3N_3$ 1,3,5-Triazine-2,4,6(1H,3H,5H)-trione pK_1 6.5, pK_2 10.6	not stated		05	Uncert.	H43
2173	$C_3H_3O_4N$ Propenoic acid, 3-nitro- 2.58(trans)	25	c = 0.01	E3bg	Approx.	B112
2174	$C_3H_3O_4N_3$ Propanenitrile, 3,3-dinitro- 2.27 2.34	25 20	I = 0.05(phthalate buffers),mixed constant	05 05	Approx. Approx.	S77 T58a
2175	$C_3H_3O_4N_3$ Pyrimidine-2,4,6(1H,3H,5H)-trione, 5-hydroxyimino- (Violuric acid) pK_1 4.41, pK_2 9.66			05	Uncert.	T13
2176	$C_3H_3O_5N_3$ Pyrimidine-2,4(1H,3H)-dione, 6-hydroxy-5-nitro- (Dilituric acid) pK_2 10.25			05	Uncert.	T13
2177	$C_3H_3N_3S_2$ 1,2,4-Triazine-3,5(2H,4H)-dithione (2,4-Dithio-6-azauracil) 5.66	25	c ≈ 0.001	E3bg	Approx.	G63

No.	Formula / Name	pK	Temp (°C)	Conditions	Method	Reliability	Ref.
2178	$C_3H_4OI_4$ 1,2,4-Triazine-5(2H)-one, 3-amino-						
	pK₁ 1.55		25		05	Approx.	P70
2179	$C_3H_4OCl_2$ Propanone, 1,1-dichloro-						
	14.9				KIN	Uncert.	B47
2180	$C_3H_4OF_4$ 1-Propanol, 2,2,3,3-tetrafluoro-						
	12.74		25		C3	Approx.	B8
2181	$C_3H_4O_2N_2$ Imidazolidine-2,4-dione (Hydantoin)						
	8.93		25	I = 0.1(KNO₃)	E3bg	Approx.	C10
	9.15		25	c ≈ 0.001	E3bg	Approx.	G62
	9.16		18±3	I = 0.01	05	Uncert.	E4
2182	$C_3H_4O_2N_4$ 1,2,4-Triazine-3,5(2H,4H)-dione, 6-amino-						
	7.35		not stated	c = 0.005	E3bg	Uncert.	C29
2183	$C_3H_4O_2N_4$ 1,3,5-Triazine-2,4(1H,3H)-dione, 6-amino- (Ammelide)						
	pK₁ 1.8, pK₂ 6.9, pK₃ 13.5		not stated		05	Uncert.	H43
2184	$C_3H_4O_3N_2$ Propanone, 1,3-bis(hydroxyimino)-						
	pK₁ 7.65		30	c = 0.01, I = 0.1(KCl), mixed constant	E3bg	Approx.	E18
	pK₁ 7.58, pK₂ 8.85		not stated		E3bg	V.Uncert.	M103
2185	$C_3H_4O_3S$ Acetic acid, 2-methylthio-2-oxo-						
	0.54		30	I = 1.0, mixed constant (Estimated thermodynamic value ≈ 1)	05	Uncert.	H35

No.	Molecular formula, name and pK value(s)	T(°C)	Remarks	Method	Assessment	Ref.
2186	$C_3H_4O_6N_2$ Methyl 2,2-dinitroacetate					T58a
	0.98	20	In aqueous $HClO_4$, H_0 scale	06	Uncert.	
2187	$C_3H_4O_8N_4$ Propane, 1,1,3,3-tetranitro-					N39
	pK_1 1.15	5	In aqueous HCl solutions, H_- scale	06	Uncert.	
	pK_1 1.11	20		06	Uncert.	
	pK_2 5.04	5		05	Approx.	
	pK_2 4.96	20			Approx.	
2188	$C_3H_4N_2S_2$ 1,2,4-Thiadiazole-5(2H)-thione, 3-methyl-					G30
	5.18	25	c = 0.01, mixed constant	E3bg	Approx.	
2189	$C_3H_4N_2S_2$ 1,3,4-Thiadiazole-2(3H)-thione, 5-methyl-					S9
	4.90	25	c ≃ 0.004, mixed constant	E3bg	Approx.	
2190	$C_3H_5ON_3$ Acetohydrazide, 2-cyano-					T43
	pK_1 2.47, pK_2 11.04	25	Mixed constant	E3bg	Uncert.	
	pK_1 2.3, pK_2 11.15			0		Z14
2191	$C_3H_5ON_5$ 1,3,5-Triazin-2(1H)-one, 4,6-diamino- (Ammeline)					H43
	pK_1 4.5, pK_2 9.4			05	Uncert.	
2192	C_3H_5OCl Propanone, 1-chloro-					B47
	16.5			KIN	Uncert.	

53

No.	Formula	Name / pK	T	Conditions	KIN	Uncert.	Ref.
2193	C$_3$H$_5$OBr	Propanone, 1-bromo-					
		16.1				Uncert.	B47
2194	C$_3$H$_5$O$_2$N	Propanone, 1-hydroxyimino-					
		8.35	30	c = 0.01, I = 0.1(KCl), mixed constant	E3bg	Approx.	E18
		8.30	25	c ≈ 0.01, I = 0.1(KCl), mixed constant	E3bg	Approx.	G39
2195	C$_3$H$_5$O$_2$N$_3$	1,2,4-Triazine-3,5(2H,4H)-dione, 1,6-dihydro- (6-Azauracil, 5,6-dihydro-)					
		10.3	25	c ≈ 0.001	E3bg	Uncert.	G62
		10.3	25	Titrant (CH$_3$)$_4$NOH	E3bg	Uncert.	J16
2196	C$_3$H$_5$O$_2$N$_3$	1,3,5-Triazine-2,4-(1H,3H)-dione, 5,6-dihydro- (5-Azauracil, 5,6-dihydro-)					
		10.7	25	Titrant N(Me)$_4$OH	E3bg	Uncert.	J16
2197	C$_3$H$_5$O$_2$Cl	Propanoic acid, 2-chloro-					
		2.89	25	c ≈ 0.01 Values in mixed solvents also given	E3bg	Approx.	B113
2198	C$_3$H$_5$O$_2$Cl	Propanoic acid, 3-chloro-					
		4.08	25	c ≈ 0.01 Values in mixed solvents also given	E3bg	Approx.	B113
		3.93	25	I = 1.0(KCl), mixed constant	E3bg	Approx.	S18
2199	C$_3$H$_5$O$_2$Br	Propanoic acid, 2-bromo-					
		3.01	25	I = 0.1(KCl), concentration constant	E3bg	Approx.	S1

I.C.O.A.A.—C

No.	Molecular formula, name and pK value(s)	$T(^{o}C)$	Remarks	Method	Assessment	Ref.
2200	$C_3H_5O_2Br$ Propanoic acid, 3-bromo-					
	4.01	25	$c \approx 0.01$	E3bg	Approx.	B113
			Value in mixed solvents also given			
	4.07	25	$I = 0.1(KCl)$, concentration constant	E3bg	Approx.	S1
2201	$C_3H_5O_3N$ Propanone, 1-nitro-					
	5.10	25		E3bg	Approx.	P18
2202	$C_3H_5O_4N$ Propanedioic acid, amino-					
	pK_1 3.32, pK_2 9.83	20	$c = 0.0027$	E3bg	Approx.	S45
2203	$C_3H_5O_4N_3$ Propanedial dioxime, 2-nitro-	not stated				
	pK_2 10.4, pK_3 11.5			05	Uncert.	D36
2204	$C_3H_5O_5N$ Propanoic acid, 2-nitrooxy-	not stated				
	2.39		$c \approx 0.025$	E3bg	Uncert.	M9
2205	$C_3H_5O_5N$ Propanoic acid, 3-nitrooxy-	not stated				
	3.97		$c \approx 0.025$	E3bg	Uncert.	M9
2206	$C_3H_5O_5N_3$ Propanamide, 3,3-dinitro-					
	3.41	20		05	Approx.	T58a

		T/°C	Conditions / Notes	Method	Assessment	Ref.
2306	$C_4H_4O_4$ Butenedioic acid					
	cis-Isomer (Maleic acid)					
	pK_1 1.910, pK_2 6.332(H_2O)	25	I = 0.02–0.1, molality scale	E1a,R4	Rel.	D2
	pK_1 2.535, pK_2 6.711(D_2O)					
	pK_1 1.92	25	c = 0.05–0.09	E3bg	Approx.	M11
	pK_2 6.34		c = 0.005–0.009			
	pK_1 1.65	25	c ≃ 0.004, I = 1.0(KNO_3),	E3bg	Uncert.	Rt
	pK_2 5.61		concentration constants		Approx.	
	pK_2 5.79	25	I = 0.1($NaClO_4$), concentration constant	E3bg	Approx.	Y12
			Other values in D44, G24 and R46			
	trans-Isomer (Fumaric acid)					
	pK_1 3.095, pK_2 4.602(H_2O)	25	I = 0.02–0.1, molality scale	E1a,R4	Rel.	D2
	pK_1 3.557, pK_2 5.025(D_2O)					
2307	$C_4H_4O_5$ Butanedioic acid, oxo- (Oxaloacetic acid)					
	pK_1 2.22, pK_2 3.89	25	I = 0.1(KCl)	E3bg,R3d	Approx.	T11
	pK_3 13.0		Microscopic and tautomeric constants are	O5	Uncert.	
			also given			
2308	$C_4H_4O_6$ Butenedioic acid, dihydroxy-					
	pK_1 1.57, pK_2 3.36(trans)	25		E3bg	Approx.	H24

No.	Formula	Name / pK	T	Conditions	Method	Reliability	Ref.
2207	$C_3H_5O_6N_3$	Propane, 1,1,3-trinitro-					
		3.24	25	$I = 0.05$(phthalate buffers), mixed constant	05	Approx.	S77
		3.37	20		05	Approx.	T58a
2208	$C_3H_5O_6P$	Propenoic acid, 2-phosphonooxy-					
		pK$_2$ 3.4, pK$_3$ 6.35		$I = 0.1$ [$(n$-$C_3H_7)_4$NI]	E3bg	Uncert.	W41
		pK$_2$ 3.5, pK$_3$ 6.4		$I = 0.4$			
2209	$C_3H_5O_7N_3$	Propane, 1,1-dinitro-3-nitrooxy-					
		3.67	20		05	Approx.	T58a
2210	C_3H_6OS	Propanone, 1-mercapto-					
		7.86	25		E3bg	Approx.	K58
2211	$C_3H_6OS_2$	Methanedithioic acid, ethoxy- (Xanthic acid, ethyl-)					
		pK$_2$ 7.68	20	$c \simeq 0.0002$	KIN	V.uncert.	I30
		pK$_1$ 2.74	0	Extrapolated to zero time	E3d		G5
		pK$_1$ 2.47					H27
		pK$_1$ 1.41			05	Uncert.	Z2
2212	$C_3H_6O_2N_2$	Propanone oxime, 1-hydroxyimino-					
		9.7	25		E3bg	Uncert.	U10
2213	$C_3H_6O_2S$	Acetic acid, methylthio-					
		3.655	25	$c \simeq 0.001$, $I = 0.10$(KNO$_3$), concentration constant	E3bg	Approx.	P40

No.	Molecular formula, name and pK value(s)	T(°C)	Remarks	Method	Assessment	Ref.
2214	C$_3$H$_6$O$_2$S Methyl 2-mercaptoacetate (Methylthioglycolate)					
	8.03	25		05	Approx.	K69
	7.68	20		E3bg	Uncert.	D8
2215	C$_3$H$_6$O$_2$S Propanoic acid, 2-mercapto- (Thiolactic acid)					
	pK$_1$ 3.63, pK$_2$ 10.24	20	I = 0.1(NaClO$_4$), c = 0.01, mixed constants	E3bg	Approx.	S17
	pK$_1$ 3.66, pK$_2$ 10.18	30				
	pK$_1$ 3.70, pK$_2$ 10.11	40				
	pK$_2$ 10.26	20	c = 0.001	Polarography		
	pK$_2$ 10.19	30				
	pK$_2$ 10.0	40				
2216	C$_3$H$_6$O$_2$S Propanoic acid, 3-mercapto-					
	pK$_2$ 10.84	25	c ≈ 0.00015 heat of ionization = 6.10 kcal mole^{-1}	05	Approx.	I20
	pK$_1$ 4.38, pK$_2$ 10.38	25	I = 0.2, mixed constants	E3bg	Approx.	F9
	pK$_1$ 4.27	30	I = 0.7(KCl), mixed constant	E3bg	Approx.	C23
	pK$_2$ 10.20	20		05	Approx.	D8
	pK$_2$ 10.03	30				
	pK$_2$ 9.85	40				
	pK$_2$ 10.05	30	c = 0.005, I = 0.3(KCl)	E3bg	Approx.	F42

No.	Formula / Name	pK	T (°C)	Notes	Code	Assessment	Ref.
2217	$C_3H_6O_2S_2$ Propanoic acid, 2,3-dimercapto- pK$_1$ (COOH) 3.45, pK$_2$ 9.54, pK$_3$ 11.55			Microconstants calculated	E3bg		T9
2218	$C_3H_6O_2Se$ Methyl 2-hydroselenoacetate	4.70	25		05	Approx.	K69
2219	$C_3H_6O_3N_2$ Propanal oxime, 3-nitro- pK$_1$ 4.5, pK$_2$ 10.5				05	Uncert.	D36
2220	$C_3H_6O_3S$ 2-Propene-1-sulfonic acid (Allylsulfonic acid)	-0.89		I = 2.0(HCl-NaCl)	06	Uncert.	M76
2221	$C_3H_6O_4N_2$ Propane, 1,1-dinitro-	5.65	25	I ≈ 0.002 Derived from equilibrium values of pH achieved in 10-60mins. Microscopic constants are also given	E3ag	Approx.	B45
		5.61	20		05	Approx.	T58a
		5.70	5		05	Approx.	N39
		5.53	20				
		5.36	40	Thermodynamic quantities are derived			
		5.22	60	from the results			
		5.52	20	Thermodynamic quantities are also given	05	Uncert.	S83

No.	Molecular formula, name and pK value(s)	T(°C)	Remarks	Method	Assessment	Ref.
2222	$C_3H_6O_4S_2$ 1,3-Dithiacyclopentane 1,1,3,3-tetraoxide					
	13.9	25	No details		Uncert.	C76
2223	$C_3H_6O_5N_2$ Ethane, 1-methoxy-2,2-dinitro-					
	3.48	25	I = 0.05(phthalate buffers), mixed constant	05	Approx.	S77
	3.56	20		05	Approx.	T58a
2224	$C_3H_6O_5N_2$ 1-Propanol, 3,3-dinitro-					
	4.73	20		05	Approx.	T58a
	4.68	5		05	Uncert.	S81
	4.52	20	Thermodynamic quantities are derived from the results			
	4.37	40				
	4.24	60				
2225	$C_3H_6O_5S$ Propanoic acid, 3-sulfo-					
	pK_2 4.74(-COOH)	25	c ≈ 0.002, I = 0.01, mixed constant	E3bg	Approx.	B58
	pK_2 4.52	not stated		E	Uncert.	B21
2226	$C_3H_6O_6N_4$ Ethylamine, N-methyl-N,2,2-trinitro-					
	2.69	20		05	Approx.	T58a
2227	$C_3H_6O_6N_4$ Propylamine, N,3,3-trinitro-					
	3.71	20		05	Approx.	T58a

No.	Formula	pK / Name	Conditions	T (°C)	Method	Reliability	Ref.
2228	C_3H_7ON	Propanone oxime					
		12.42		25	KIN	Uncert.	K28
2229	$C_3H_7O_2N$	Acetic acid, methylamino- (Sarcosine)					
		pK_1 2.30, pK_2 10.14	I = 0.5(KCl), mixed constants	25	E3bg	Approx.	L27
		pK_2 9.73	I = 0.1(K_2SO_4)	25	E3bg	Approx.	J5
		pK_2 10.22	c - 0.02, mixed constant	20	E3bg	Approx.	P28
		pK_2 10.45	c = 0.025, mixed constant	20	E3d	Uncert.	H17
2230	$C_3H_7O_2N$	Propane, 1-nitro-					
		8.98		25	E3bg	Approx.	W22
2231	$C_3H_7O_2N$	Propane, 2-nitro-					
		7.7-7.8		25	E3bg	Uncert.	W22
		7.68	c = 0.05-0.06, equilibrium requires 12-24 hours	25	E3bg	Approx.	T67
		7.63		30			
2232	$C_3H_7O_2N$	Propanohydroxamic acid					
		9.5	c = 0.01	20	E3d	Uncert.	W39

No.	Molecular formula, name and pK value(s)	T(°C)	Remarks	Method	Assessment	Ref.
2233	$C_3H_7O_2N$ Propanoic acid, 2-amino- (α-Alanine)					
	pK_1 2.35, pK_2 9.59	25	I = 0.1($NaClO_4$), concentration constants	E3bg	Approx.	M61
	pK_1 2.20	20	c = 0.002, I = 0.1(KCl)	E3bg	Uncert.	G16
	pK_2 9.84				Approx.	
	pK_1 2.31, pK_2 9.73	25	I = 0.2($NaClO_4$),concentration constants	E3bg	Approx.	C44
	pK_1 2.39	25	Change of optical rotation with pH,		Uncert.	R12
	pK_2 9.84		mixed constant			
	Other values in H17, J5, L27, M72 and P28					
2234	$C_3H_7O_2N$ Propanoic acid, 3-amino- (β-Alanine)					
	pK_1 3.51, pK_2 10.20	25	I = 0.2($NaClO_4$), concentration constants	E3bg	Approx.	C44
	pK_2 10.55	30	I = 0.1(KCl),mixed constant	E3bg	Approx.	C23
	pK_2 10.17	25	I = 0.5(KNO_3),concentration constant	E3bg	Approx.	T23
	Other values in H17, L27 and S109					
2235	$C_3H_7O_2P$ 1-Phosphacyclobutan-1-ol 1-oxide (Trimethylenephosphinic acid)					
	3.07	25	c = 0.01, mixed constant	E3bg	Uncert.	K50
2236	$C_3H_7O_3N$ Propanohydroxamic acid, 2-hydroxy- (Lactohydroxamic acid)					
	9.35(L isomer)		I = 0.2(NaCl)	E3bg	Uncert.	C72

I.C.O.A.A.—C*

No.	Formula	Name	pK values	Conditions	Temp.	Method	Assessment	Ref.
2237	$C_3H_7O_3N$	Propanoic acid, 2-amino-3-hydroxy- (Serine)	pK_1 2.18, pK_2 8.84	$I = 0.15(KNO_3)$, mixed constants	37	E3bg	Approx.	P34
			pK_1 2.30, pK_2 9.34	$I = 0.2(KNO_3)$	15	E3bg	Approx.	R7
			pK_1 2.29, pK_2 9.12		25			
			pK_1 2.27, pK_2 8.78		40			
2238	$C_3H_7O_3As$	2-Propene-1-arsonic acid	pK_1 4.48, pK_2 7.51	Mixed constants	not stated	E3bg	Uncert.	C42
2239	$C_3H_7O_4N_3$	Propylamine, 3,3-dinitro	pK_1 2.71		20	05	Approx.	T58a
2240	$C_3H_7O_7P$	Propanoic acid, 3-hydroxy-2-phosphonooxy- (Glyceric acid-2-phosphate)	pK_2 3.55, pK_3 7.0	$I = 0.1[(n-C_3H_7)_4NI]$		E3bg	Uncert.	W41
			pK_2 3.6, pK_3 7.1	$I = 0.4$				
2241	$C_7H_7NS_2$	Methanedithioic acid, dimethylamino- (Dithiocarbamic acid, N,N-dimethyl-)	3.36		25	KIN	Uncert.	M79
2242	$C_7H_7NS_2$	Methanedithioic acid, ethylamino- (Dithiocarbamic acid, N-ethyl-)	3.04		25	KIN	Uncert.	M79
2243	$C_3H_8OS_2$	1-Propanol, 2,3-dimercapto- (BAL)	pK_1 8.62, pK_2 10.57	$I = 0-1(NaCl)$	25	E3bg,R4	Approx.	A71
			pK_1 8.69, pK_2 10.72	$I = 0.1(KCl)$	30	E3bg	Approx.	L29

No.	Molecular formula, name and pK value(s)	$T(^oC)$	Remarks	Method	Assessment	Ref.
2244	$C_3H_8O_2N_2$ Acetohydrazide, 2-methoxy-					
	pK_1 2.76	25	mixed constant(?)	E3bg	Uncert.	T43
	pK_2(NH)12.14	25		05	Uncert.	
2245	$C_3H_8O_2N_2$ Propanohydroxamic acid, 2-amino-					
	7.25	25(?)	c = 0.02, I = 0.1(KCl), mixed constant	E3bg	Uncert.	M66
2246	$C_3H_8O_2N_2$ Propanohydroxamic acid, 3-amino-					
	7.90	25(?)	c = 0.02, I = 0.1(KCl), mixed constant	E3bg	Uncert.	M66
2247	$C_3H_8O_2N_2$ Propanoic acid, 2,3-diamino-					
	pK_1 1.30, pK_2 6.79, pK_3 9.51	25	I = 0.1(KCl), mixed constants	E3bg	Approx.	H25
	pK_1 1.23, pK_2 6.69, pK_3 9.50	20	c = 0.01, mixed constants	E3bg	Approx.	A16
2248	$C_3H_8O_3S$ 1,2-Propanediol, 3-mercapto-					
	9.46	25	I = 0-1(NaCl)	E3bg,R4	Approx.	A71
	9.51	25		E3bg	Approx	K58
2249	$C_3H_8O_3S_3$ 1-Propanesulfonic acid, 2,3-dimercapto- (Unithiol)					
	pK_2 8.84, pK_3 11.20			E3b		P67
	pK_2 8.93	25	c = 0.004-0.01, I = 0-1(KCl)	E3bg,R4	Approx.	A70
	pK_3 11.94	25	I = 0.1(KCl)			

No.	Formula	Name / pK values		Temp.	Conditions	Method	Assessment	Ref.
2250	$C_3H_8O_3S_3$	2-Propanesulfonic acid, 1,3-dimercapto-				E3b		P67
		pK_2 8.74, pK_3 11.67						
2251	$C_3H_8O_4S_2$	2,4-Dithiapentane-2,2,4,4-tetraoxide	[Methane, bis(methylsulfonyl-)]			E3bg	Approx.	H42
		12.54		25	c = 0.006-0.023			
2252	$C_3H_9O_3P$	Ethyl hydrogen methylphosphonate				E3bg	Uncert.	N28
		2.25		20	c = 0.005			
2253	$C_3H_9O_3P$	Isopropylphosphonic acid				E3bg	Uncert.	C38
		pK_1 2.55, pK_2 7.75		not stated	c = 0.02			
2254	$C_3H_9O_3P$	Methyl hydrogen ethylphosphonate				E3bg	Uncert.	A15
		1.88		20	Concentration constant			
2255	$C_3H_9O_3As$	Isopropylarsonic acid				E3bg	Uncert.	C42
		pK_1 4.86, pK_2 8.36		not stated	Mixed constants			
2256	$C_3H_9O_4P$	Isopropyl dihydrogen phosphate				E3bg	pK_1,Uncert. pK_2, Approx.	K64
		pK_1 1.72, pK_2 7.03		14.3	c = 0.02			
		pK_1 1.75, pK_2 7.02		22.2				
		pK_1 1.79, pK_2 7.03		31.2				
		pK_1 1.86, pK_2 7.07		41.4				
		pK_1 1.93, pK_2 7.07		50.3				

No.	Molecular formula, name and pK value(s)	$T(^{o}C)$	Remarks	Method	Assessment	Ref.
2257	$C_3H_9O_4P$ Propyl dihydrogen phosphate					
	pK_1 1.88	25	$m \simeq 0.04$, mixed constants	E3bg	Uncert.	K67
	pK_2 6.67				Approx.	
2258	$C_3H_9O_6P$ 2,3-Dihydroxypropyl dihydrogen phosphate (Glycerol-1-phosphoric acid)					
	pK_2 6.656	25		E1a	Rel.	D12

Variation with temperature

	5	10	15	20	30	35	37	40	45	50
	6.642	6.641	6.643	6.648	6.667	6.679	6.685*	6.694	6.713	6.733

* calculated value

Thermodynamic quantities are derived from these results

	pK_2 6.07	20	$I = 0.1(KCl)$	E3bg	Approx.	S37
2259	$C_3H_9O_6P$ (2-Hydroxy-1-hydroxymethyl)ethyl dihydrogen phosphate (Glycerol-2-phosphoric acid)					
	pK_2 6.00	25	$c = 0.05$, $I = 1.0(KCl)$, mixed constant	E3bg	Uncert.	S19
2260	C_3H_9NS 1-Propanethiol, 2-amino-					
	pK_1 8.14, pK_2 10.81	25	$c = 0.01$, mixed constants	E3bg	Approx.	B147
2261	C_3H_9NS 2-Propanethiol, 1-amino-					
	pK_1 8.10, pK_2 10.12	25	$c = 0.01$, mixed constants	E3bg	Approx.	B147

2262 $C_3H_{10}O_6P_2$ Isopropylidenediphosphonic acid

	Temp	Conditions	Method	Reliability	Ref.
pK_2 2.94, pK_3 7.75	25	$I = 0.5\ [(CH_3)_4NCl]$, concentration constants	E3bg	Approx.	C17
pK_4 12.4				Uncert.	
pK_2 3.16, pK_3 8.04	25	$c = 0.005$, $I = 0.032-0.25$	E3bg	Approx.	G36
pK_4 12.10				Uncert.	

2263 $C_3H_{10}O_6P_2$ 1,2-Propanediyldiphosphonic acid

	Temp	Conditions	Method	Reliability	Ref.
pK_1 <2, pK_2 2.6	20	$I = 0.1(KCl)$	E3bg	Uncert.	S48
pK_3 7.00, pK_4 9.27				Approx.	

2264 $C_3H_{10}O_6P_2$ 1,3-Propanediyldiphosphonic acid (Trimethylenediphosphonic acid)

	Temp	Conditions	Method	Reliability	Ref.
pK_1 <2, pK_2 2.6	20	$I = 0.1(KCl)$	E3bg	Uncert.	S48
pK_3 7.34, pK_4 8.35				Approx.	
pK_2 2.6	20	$I = 0.1(KCl)$, mixed constant	E3bg	Uncert.	S53
pK_3 7.34, pK_4 8.35				Approx.	
pK_1 1.6, pK_2 3.06, pK_3 7.65, pK_4 8.63		Average of values at 25, 37 and 50°, extrapolated from values at $I = 0.1-1.0$	E3bg	Uncert.	I6

2265 $C_3H_{12}O_2B_{10}$ 1,2-Dicarbadodecaborane(12)-1-carboxylic acid (Barenecarboxylic acid)

	Temp	Conditions	Method	Reliability	Ref.
2.49	25	Mixed constant	E3bg	Uncert.	Z3

2266 $C_3H_{12}O_2B_{10}$ 1,7-Dicarbadodecaborane(12)-1-carboxylic acid (Neobarenecarboxylic acid)

	Temp	Conditions	Method	Reliability	Ref.
3.20		No details given			S107

No.	Molecular formula, name and pK value(s)	$T({}^{\circ}C)$	Remarks	Method	Assessment	Ref.
2267	$C_3OCl_2F_4$ Propanone, 1,3-dichloro-1,1,3,3-tetrafluoro- 6.67		Conditions not stated (pK of hydrate)		Uncert.	M73
2268	$C_3OCl_3F_3$ Propanone, 1,1,3-trichloro-1,3,3-trifluoro- 6.48		Conditions not stated (pK of hydrate)		Uncert.	M73
2269	$C_3OCl_4F_2$ Propanone, 1,1,3,3-tetrachloro-1,3-difluoro- 6.42		Conditions not stated (pK of hydrate)		Uncert.	M73
2270	C_3HONF_6 Propanone oxime, hexafluoro- 6.0	25	Mixed constant	E3bg	Uncert.	K38
	6.0	25		E3bg	Uncert.	D65
2271	C_3HOClF_4 Propanone, 1-chloro-1,1,3,3-tetrafluoro- 7.90		Conditions not stated (pK of hydrate)		Uncert.	M73
2272	$C_3HOBr_2F_3$ Propanone, 1,1-dibromo-3,3,3-trifluoro- 7.69		Conditions not stated (pK of hydrate)		Uncert.	M73
2273	$C_3H_2O_2N_3Cl$ 1,2,4-Triazine-3,5($2\underline{H}$,$4\underline{H}$)-dione, 6-chloro- (6-Azauracil, 5-chloro-) 5.80	not stated	c = 0.005	E3bg	Uncert.	C29
2274	$C_3H_2O_2N_3Br$ 1,2,4-Triazine-3,5($2\underline{H}$,$4\underline{H}$)-dione, 6-bromo- (6-Azauracil, 5-bromo-) 6.05	not stated	c = 0.005	E3bg	Uncert.	C29
	6.00	25	Titrant $(CH_3)_4NOH$	E3bg	Approx.	J16

No.	Formula	Name	pK	Temp.	Conditions	Method	Reliability	Ref.
2275	$C_3H_2O_2N_3I$	1,2,4-Triazine-3,5($\underline{2H},\underline{4H}$)-dione, 6-iodo- (6-Azauracil, 5-iodo-) 6.27		not stated	c = 0.005	E3bg	Uncert.	C29
2276	$C_3H_3ONF_6$	Ethylhydroxylamine, 2,2,2-trifluoro-1-trifluoromethyl- 8.5		25		E3bg	Uncert.	D65
2277	C_3H_3ONS pK$_1$ -0.33	Isothiazol-3($\underline{2H}$)-one		25	In aq.H_2SO_4 solutions, H_o scale	06	Uncert.	C82
2278	$C_3H_3ONS_2$	Thiazolidin-4-one, 2-thioxo- (Rhodanine) 5.18		20	I = 0.1($NaClO_4$)	DIS	Uncert.	N17
2279	$C_3H_3ON_3S$	1,2,4-Triazin-3($\underline{2H}$)-one, 4,5-dihydro-5-thioxo- (6-Azauracil,5-mercapto-) 6.33		25	c $\approx$ 0.001	E3bg	Approx.	G63
2280	$C_3H_3ON_3S$	1,2,4-Triazin-5($\underline{2H}$)-one, 3,4-dihydro-3-thioxo- (6-Azauracil, 2-mercapto-) 5.98		25	c $\approx$ 0.001-0.005	E3bg	Approx.	G63
		5.98		25	c = 0.004	E3bg	Approx.	K9
2281	$C_3H_3O_2NS$	Thiazolidine-2,4-dione 6.44		not stated	I = 0.01 Value in mixed solvent also given	05	Uncert.	T14
2282	$C_3H_3O_2F_3S$	Acetic acid, trifluoromethylthio- 2.95		25		C2	Approx.	020

No.	Molecular formula, name and pK value(s)	T(°C)	Remarks	Method	Assessment	Ref.
2283	$C_3H_3O_3F_3S$ Acetic acid, trifluoromethylsulfinyl- 2.06	25		C2	Approx.	O20
2284	$C_3H_3O_4F_3S$ Acetic acid, trifluoromethylsulfonyl- 1.88	25	Compound unstable in aq.solution	C2	Uncert.	O20
2285	$C_3H_4ON_2S$ 1,3,4-Oxadiazole-2(3H)-thione, 5-methyl- 5.01	25	c ≈ 0.004, mixed constant	E3bg	Approx.	S9
2286	$C_3H_5O_2NF_2$ Propanoic acid, 3-difluoroamino- 3.74	25	Mixed constant	E3bg	Approx.	B37
2287	$C_3H_5O_2NS_2$ Acetic acid, (dithiocarboxy)amino- pK$_1$ 3.12, pK$_2$ 3.84	30	I = 0.1, mixed constants	E3b	V.uncert.	B169
2288	$C_3H_6ONF_3$ 2-Propanol, 1-amino-3,3,3-trifluoro- pK$_2$ 12.29			E3bg	Uncert.	R26
2289	$C_3H_7O_2NS$ Propanoic acid, 2-amino-3-mercapto- (Cysteine) pK$_2$ 8.33, pK$_3$ 10.50 (pK$_1$ ≈ 2)	20	I = 0.1(NaClO$_4$), concentration constants	E3bg	Approx.	P33
2290	$C_3H_8O_3BrP$ Propylphosphonic acid, 3-bromo- pK$_1$ 2.25, pK$_2$ 7.3	not stated	c = 0.02	E3bg	Uncert.	C38

2291 $C_3H_8O_6NP$ Propanoic acid, 2-amino-3-phosphonooxy-   (O-Phosphonoserine)

pK_3 5.80, pK_4 9.90

20   I = 0.1[$(C_3H_7)_4NI$]   E3bg   Uncert.   H31

2292 $C_3H_9O_2SP$ O-Ethyl O-hydrogen methylphosphonothioate

1.82

20   c = 0.005 in 7% ethanol   E3bg   V.uncert.   K6

81% thione tautomer

2293 $C_3H_9O_2SP$ S-Ethyl hydrogen methylphosphonothioate

2.03

20   c = 0.005, mixed constant   E3bg   Uncert.   N28

2294 $C_3H_{10}O_3NP$ Ethylphosphonic acid, 1-amino-1-methyl

pK_1 (POH) 1.65

25   c = 0.001, I = 0.1(KCl), concentration constant   E3bg   Uncert.   D68

pK_2 (POH) 5.85, pK_3 (NH_3^+) 10.31   Approx.

2295 $C_3H_{10}O_3N_3P$ 1-Guanidinophosphonic acid, 3,3-dimethyl-

pK_1 -0.31   30.5   I = 0.2(NaCl)   KIN   Uncert.   A48

pK_2 4.31   E3bg   Approx.

pK_3 11.3   E3bg   Uncert.

2296 $C_3H_{12}O_9NP_3$ 2-Azapropane-1,3-diphosphonic acid, 2-phosphonomethyl-   [Nitrilotris(methylenephosphonic acid)]

pK_1 1.9, pK_2 2.0, pK_3 4.60, pK_4 5.92, pK_5 7.35, pK_6 10.9

20   I = 0.1[$(CH_3)_4NCl$], concentration constants   E3bg   Uncert   H30

pK_1 < 2, pK_2 < 2, pK_3 4.30, pK_4 5.46, pK_5 6.66, pK_6 12.34

25   I = 1.0(KNO_3), concentration constants   E3bg   Uncert.   C18

No.	Molecular formula, name and pK value(s)	$T(^\circ C)$	Remarks	Method	Assessment	Ref.
2297	$C_3H_{12}O_{10}NP_3$ 2-Azapropane-1,3-diphosphonic acid 2-oxide, 2-phosphonomethyl- [Nitrilotris(methylenephosphonic acid)-N-oxide] $pK_1 < 2$, $pK_2 < 2$, pK_3 3.28, pK_4 5.26, pK_5 6.95, pK_6 12.05	25	$I = 1.0(KCl)$, concentration constants	E3bg	Uncert.	C19
2298	$C_3H_{10}O_5NSP$ O,O-Dimethylphosphoramidate, N-methylsulfonyl- 2.20	20	$c = 0.005$, 7% ethanol, mixed constant	E3bg	Uncert.	G19

No.	Formula		Temp	Conditions	Method	Uncert.	Ref.
2299	C_4HN_3	1,2,3-Propanetrinitrile					
		−5.1	25		06	Uncert.	B116
2300	$C_4H_2O_4$	Butynedioic acid					
		pK$_1$ 1.04, pK$_2$ 2.50	18.8	c = 0.01	E3bg	Approx.	B109
		pK$_1$ 1.23, pK$_2$ 2.53	28.2				
		pK$_1$ 1.30, pK$_2$ 2.58	36.6				
2301	$C_4H_2O_4$	Cyclobutenedione, 3,4-dihydroxy- (Squaric acid)					
		pK$_1$ 0.55	25	c = 0.12-0.17	E3bg	V.uncert.	S32
		pK$_2$ 3.480				Approx.	

Variation of pK$_2$ with temperature

10	20	30	40	50
3.351	3.444	3.511	3.575	3.624

Thermodynamic quantities are derived from the results

		Temp	Conditions	Method	Uncert.	Ref.
pK$_1$ 1.2		25	c = 0.17	E3bg	V.uncert.	M13
pK$_2$ 3.48					Approx.	
pK$_2$ 2.89		25	I = 0.5(NaCl or NaClO$_4$), mixed constant	E3bg	Approx.	T15
pK$_2$ 3.09		0	c ≈ 0.02, I = 0.1(KCl),mixed constant	E3bg	Approx	I7
pK$_2$ 3.17		25				

(Contd)

No.	Molecular formula, name and pK value(s)	$T(^oC)$	Remarks	Method	Assessment	Ref.
2301(Contd)	$C_4H_2O_4$ Cyclobutenedione,3,4-dihydroxy- (Squaric acid)					
	pK$_2$ 3.28	50				
	pK$_1$ 1.50, pK$_2$ 2.93	20		E3bg	Uncert.	B126
2302	$C_4H_4O_2$ 2,3-Butadienoic acid					
	3.69	25	c = 0.01, I = 0.1(NaCl), mixed constant	E3bg	Approx.	M48
2303	$C_4H_4O_2$ 2-Butynoic acid (Tetrolic acid)					
	2.59	25	c = 0.01, I = 0.1(NaCl), mixed constant	E3bg	Approx.	M48
	2.616	25		C2	Rel.	G51
			Variation with temperature			
			10 15 20 30 35 40 45			
			2.618 2.616 2.611 2.618 2.621 2.631 2.647			
			Thermodynamic quantities are derived from the results			
2304	$C_4H_4O_2$ 3-Butynoic acid					
	3.32	25	c = 0.01, I = 0.1(NaCl), mixed constant	E3bg	Approx.	M48
2305	$C_4H_4O_3$ Furan-2(5H)-one, 4-hydroxy- (Tetronic acid)					
	3.6		Conditions not stated		Uncert.	C45

No.	Molecular formula, name and pK value(s)	$T(^{o}C)$	Remarks	Method	Assessment	Ref.
2309	Pyrrole					
	C_4H_5N					
	17.51	25	In aq.KOH,H_ scale	07	Uncert.	Y3
2310	$C_4H_6O_2$ 2-Butenoic acid					
	cis-Isomer (Isocrotonic acid)					
	4.42	25	c = 0.01	E3bg	Approx.	B112
	4.44	23	c = 0.01	E3bg	Approx.	D45
			Values in mixed solvents are also given			
	4.70	25		E3bg	Approx.	M12
	trans-Isomer (Crotonic acid)					
	4.70	25	c = 0.01	E3bg	Approx.	B112
	4.70	23	c = 0.01	E3bg	Approx.	D45
			Values in mixed solvents are also given			
	4.74	25		E3bg	Approx.	M12
2311	$C_4H_6O_2$ 3-Butenoic acid (Vinylacetic acid)					
	4.37	23	c = 0.01	E3bg	Approx.	D45
			Values in mixed solvents are also given			
2312	$C_4H_6O_2$ Propenoic acid, 2-methyl- (Methacrylic acid)					
	4.65		Conditions not stated	E3bg	Uncert.	A43

No.	Formula	Name	pK values	T (°C)	Conditions	Method	Uncert./Approx.	Ref.
2313	$C_4H_6O_3$	Butanoic acid, 2-oxo-	2.50	25		E3bg	Uncert.	O8
		Thermodynamic data are also given						
2314	$C_4H_6O_3$	Butanoic acid, 3-oxo- (Acetoacetic acid)						
			pK_1 3.6	25		E3bg	Uncert.	C57
			pK_2 7.8			KIN	Uncert.	B47
2315	$C_4H_6O_3$	Propenoic acid, 3-methoxy-						
			4.85 (trans)	25	$c = 0.01$	E3bg	Approx.	B112
2316	$C_4H_6O_4$	Butanedioic acid (Succinic acid)						
			pK_1 4.22, pK_2 5.70	25	$c = 0.005$, $I = 0.035$	O3,R3d	Approx.	J19
			pK_1 4.206, pK_2 5.639	25		E3bg	Approx.	W29
			pK_1 4.21, pK_2 5.72	25	$c = 0.0005-0.0007$	E3bg,R3d	Approx.	D41
			pK_1 4.14, pK_2 5.67	20	$c = 0.005$	E3bg,R3a	Approx.	D43
		Values in mixed solvents are also given						
			pK_1 3.98, pK_2 5.20	25	$c \approx 0.005$, $I = 1.0(KNO_3)$, concentration constants	E3bg	Approx.	R5
			pK_1 4.753, pK_2 6.171(D_2O)	25	$m \approx 0.01$	E1d,R4	Approx.	R36
		Calculated from data given in C78						
		Other values in D33, F7, N32 and Y12						

No.	Molecular formula, name and pK value(s)	T(°C)	Remarks	Method	Assessment	Ref.
2317	$C_4H_6O_4$ Propanedioic acid, methyl-					
	pK$_1$ 2.5	20	c = 0.005	E3bg	Uncert.	D44
	pK$_2$ 5.78			R3a	Approx.	
	Values in mixed solvents are also given					
	pK$_1$ 3.05, pK$_2$ 5.76	25		E3bg	Uncert.	W40
2318	$C_4H_6O_5$ Butanedioic acid, hydroxy- (Malic acid)					
	pK$_1$ 3.459, pK$_2$ 5.097	25	m ≈ 0.01-0.1	E1d,R3d	Rel.	E2

Variation with temperature

	0	5	10	15	20	30	35	40	45	50
pK$_1$	3.538	3.520	3.494	3.482	3.473	3.452	3.447	3.444	3.446	3.445
pK$_2$	5.119	5.109	5.099	5.096	5.096	5.099	5.105	5.118	5.133	5.149

Thermodynamic quantities are derived from the results

Molecular formula, name and pK value(s)	T(°C)	Remarks	Method	Assessment	Ref.
pK$_1$ 3.22, pK$_2$ 4.78	30	I = 0.1(KCl), mixed constants	E3bg,R3f	Approx.	C23
pK$_1$ 3.28, pK$_2$ 4.72	20	I = 0.1(NaClO$_4$)	E3bg	Approx.	T41
pK$_1$ 2.77, pK$_2$ 4.44	25	c = 0.017, I = 1.0(KNO$_3$), concentration constants	E3bg	Approx.	R3
pK$_1$ 3.23, pK$_2$ 4.77	25	I = 0.1(NaClO$_4$), mixed constants	E3ag	Uncert.	R46
pK$_1$ 3.45, pK$_2$ 5.21					

2319 $C_4H_6O_5$ 2-Oxapentanedioic acid   (Oxydiacetic acid; Diglycolic acid)

pK_1 3.06, pK_2 4.22	20	c = 0.005-0.01	E3bg,R3c Approx.	K47
pK_1 2.97, pK_2 4.37	25		E3ag Approx.	S92
pK_1 2.90, pK_2 4.03	30	I = 0.1(KCl), mixed constants	E3bg Approx.	T32
pK_1 2.77, pK_2 3.92	25	c ≈ 0.003, I = 0.1($NaClO_4$), concentration constants	E3bg,R3h Approx.	S134

Other values in C8, S93 and Y12

2320 $C_4H_6O_6$ Butanedioic acid, 2,3-dihydroxy-   (Tartaric acid)

pK_1 3.17, pK_2 4.91(meso)	25	m ≈ 0.01	E1ch,R3d Rel.	P88

Thermodynamic quantities are also given

pK_1 3.03, pK_2 4.46(D)	25	c ≈ 0.001	E3bg,R3d Approx.	D41
pK_1 3.03 (D and DL)	25	c ≈ 0.004-0.008	C2,R1a Approx.	
pK_1 3.29, pK_2 4.92 (meso)	25	c ≈ 0.001	E3bg,R3d Approx	
pK_1 3.22 (meso)	25	c ≈ 0.0036-0.008	C2,R1a Approx.	
pK_1 2.89, pK_2 4.52	20		E3bg Approx.	F41
pK_1 2.60, pK_2 3.77	25	c = 0.004, I = 1.0(KNO_3), concentration constants	E3bg Approx.	R3
pK_3 14.4	25	I = 1(KCl-NaOH)	SOL Uncert.	D14
pK_3 14.3		I = 3		

Other values in B108 and T40

No.	Molecular formula, name and pK value(s)	T(°C)	Remarks	Method	Assessment	Ref.
2321	$C_4H_6O_8$ Butanedioic acid, tetrahydroxy-					
	pK₁ 1.95, pK₂ 4.00	25		E3bg	Approx.	P19
2322	C_4H_8O Propanal, 2-methyl-					
	13.77	25	c = 0.107 (pK of hydrate)	O5	Uncert.	H41
2323	$C_4H_8O_2$ Butanoic acid (Butyric acid)					
	4.83	25	c = 0.01	E3bg	Approx.	Y5
	4.82	25	m = 0.01	C2,R1b	Approx.	E13
			Variation with temperature			
			50 100 150 200 225			
			4.91 5.09* 5.34* 5.70* 6.00* *Uncert.			
			Equation given for temperature variation			
	4.80	23	c = 0.01	E3bg	Approx.	D45
	4.820	25	Values in mixed solvents are also given			S88
	4.8	25		E3bg	Uncert.	C57
2324	$C_4H_8O_2$ Propanoic acid, 2-methyl- (Isobutyric acid)					
	4.64	25	I = 0.5(NaClO₄),concentration constant	E3bg	Approx.	S105
	4.598	25	I = 0.1, concentration constant	E3bg	Approx.	C60
			Values in mixed solvents are also given			

No.	Formula	Name / Conditions	Value	Temp	Conditions	Method	Quality	Ref
2325	$C_4H_8O_3$	Acetic acid, ethoxy-						
			3.46	20	I = 1.0(NaClO$_4$),concentration constant	E3bg	Approx.	S8
			3.84	25	c = 0.01, mixed constant	E3bg	Approx.	M82
		Value in mixed solvent also given						
2326	$C_4H_8O_3$	Butanoic acid, 2-hydroxy-						
			3.81	31		E3bg	Approx.	C83
2327	$C_4H_8O_3$	Butanoic acid, 3-hydroxy-						
			4.40	31		E3bg	Approx.	C83
2328	$C_4H_8O_3$	Peroxybutanoic acid						
			8.2	21.5	c = 0.1	E3bg	Uncert.	E27
2329	$C_4H_8O_3$	Propanoic acid, 2-hydroxy-2-methyl-						
			3.776	25	I = 0.1, concentration constant	E3bg	Approx.	C60
		Values in mixed solvents are also given						
			4.00	25	I = 0.05-2.0	E3bg,R4	Approx.	D22
			3.75	25	I = 0.5(NaClO$_4$),concentration constant	E3bg	Approx.	S105
			3.61	25	I = 0.5, mixed constant	E3bg	Uncert.	L77
2330	$C_4H_8O_5$	Propanoic acid, 2,3-dihydroxy-2-hydroxymethyl-						
			3.29	25	I = 0.5(NaClO$_4$),concentration constant	E3bg	Approx.	S105

No.	Molecular formula, name and pK value(s)	$T(^{\circ}C)$	Remarks	Method	Assessment	Ref.
2331	$C_4H_{10}O$ 1-Butanol 16.1	25		KIN	Uncert.	M126
2332	$C_4H_{10}O$ 2-Butanol 17.6	25		KIN	Uncert.	M126
2333	$C_4H_{10}O$ 2-Propanol, 2-methyl- 19.2	25		KIN	Uncert.	M126
2334	$C_4H_{10}O_2$ 1,4-Butanediol 15.1	25		KIN	Uncert.	M126
2335	$C_4H_{10}O_2$ tert-Butyl hydroperoxide 12.8	20		05	Uncert.	E27
2336	$C_4H_{10}O_2$ Isobutyl hydroperoxide 12.8	20		05	Uncert.	E27
2237	$C_4H_{10}S$ 2-Propanethiol, 2-methyl- 11.22	25	$c \approx 0.0001$ heat of ionization = 5.30 kcal mole^{-1}	05	Approx.	I20
	11.0	25	1% ethanol	05	Uncert.	K57

No.	Formula	Name / pK	T (°C)	Conditions	Method	Assessment	Ref.
2338	C_4HOF_9	2-Propanol,1,1,1,3,3,3-hexafluoro-2-trifluoromethyl-					
		5.2		Conditions not stated		Uncert.	F14
		5.4	25		E3bg	Uncert.	D65
		5.4	25	Mixed constant	E3bg	Uncert.	K37
2339	$C_4H_2O_3Cl_2$	2-Butenoic acid, 2,3-dichloro-4-oxo- (Mucochloric acid)					
		4.20	25		E		V11
2340	$C_4H_2O_4N_4$	1,2,4,5-Tetrazine-3,6-dicarboxylic acid					
		2.8	not stated		E3bg	Uncert.	M56
2341	$C_4H_3O_4N_3$	1,2,3-Triazole-4,5-dicarboxylic acid					
		pK_1 1.86, pK_2 5.90, pK_3 9.30	25		E3bg	Uncert.	H13
2342	$C_4H_3O_4N_3$	Pyrimidine-2,4(1H,3H)-dione, 5-nitro- (5-Nitrouracil)					
		pK_1 5.55	20	c = 0.001	E3bg	Approx.	B129
		pK_2 11.3				Uncert.	
		pK_1 5.56	25	titrant $(CH_3)_4NOH$	E3bg	Approx.	J16
2343	$C_4H_4OF_6$	2-Propanol, 1,1,1,3,3,3-hexafluoro-2-methyl-					
		9.6		Conditions not stated		Uncert.	F14
2344	$C_4H_4O_2N_2$	4-Imidazolecarboxylic acid					
		pK_1 2.01, pK_2 6.26	20	c = 0.01, I = 0.10(NaCl)	E3bg	Uncert.	L58
		pK_2 6.02	25	c ≈ 0.0001, I = 0.10(NaCl), mixed constant	O5	Approx.	

No.	Molecular formula, name and pK value(s)	$T(^{\circ}C)$	Remarks	Method	Assessment	Ref.
2345	$C_4H_4O_2N_2$ 3-Pyrazolecarboxylic acid					
	3.74	Room temp	c = 0.01	E3bg	Uncert.	H1
2346	$C_4H_4O_2N_2$ Pyridazine-3,6(1$\underline{H}$,2$\underline{H}$)-dione (Maleic hydrazide)					
	pK$_1$ 5.67	25	c = 0.0002	E3bg	Approx.	M32
	pK$_2$ 13.3	25		0	Uncert.	
2347	$C_4H_4O_2N_2$ Pyrimidine-2,4(1$\underline{H}$,3$\underline{H}$)-dione (Uracil)					
	pK$_1$ 9.38, pK$_2$ 12	20	c = 0.01	E3bg	Uncert.	A36
	pK$_1$ 9.45		Conditions not stated	05	Uncert.	G33
	pK$_1$ 9.43	25	c $\approx$ 0.001	E3bg	Approx.	G62
	pK$_2$ $\approx$ 13.3-14.2					
	pK$_1$ 9.43	25	Titrant (CH$_3$)$_4$NOH	E3bg	Approx.	J16
	pK$_2$ >13.2			05		
	pK$_1$ 9.51	25		05	Approx.	N8
2348	$C_4H_4O_2N_2$ Pyrimidin-4(1$\underline{H}$)-one, 6-hydroxy-					
	5.4	20	c = 0.02	E3bg	Uncert.	A36
			(Unstable to alkali)			
2349	$C_4H_4O_3N_2$ Pyrimidine-2,4(1$\underline{H}$,3$\underline{H}$)-dione, 5-hydroxy- (Isobarbituric acid)					
	pK$_1$ 8.11, pK$_2$ 11.48	20	c = 0.01	E3bg	Approx.	A36

2350 C$_4$H$_4$O$_3$N$_2$ Pyrimidine-2,4(1$\underline{H}$,3$\underline{H}$)-dione, 6-hydroxy- (Barbituric acid)

	Temp	Conditions			Ref
pK$_1$ 3.990	25		05	Approx.	B124

Variation with temperature

15	20	30	35	40	45	50
3.969	3.980	3.999	4.008	4.017	4.025	4.032

Thermodynamic quantities are derived from the results

	Temp	Conditions			Ref
pK$_1$ 4.02	25	c = 0.000009, I $\approx$ 0.003-0.03	03	Approx.	A67
pK$_1$ 4.12		Conditions not stated	E3bg	Uncert.	M68
pK$_1$ 3.9, pK$_2$ 12.5	25	Conditions not stated	05	Uncert.	F32

2351 C$_4$H$_4$O$_3$N$_4$ Pyrimidin-2(1$\underline{H}$)-one, 4-amino-5-nitro-

	Temp	Conditions			Ref
pK$_1$ 7.43	20	c = 0.001	E3bg	Approx.	B129

2352 C$_4$H$_4$O$_3$N$_4$ Pyrimidin-4(1$\underline{H}$)-one, 2-amino-5-nitro-

	Temp	Conditions			Ref
pK$_1$ 6.70	20	c = 0.001	E3bg	Approx.	B129

2353 C$_4$H$_4$O$_4$N$_2$ Methyl 2-cyano-2-nitroacetate

	Temp	Conditions			Ref
pK$_1$ -5.20	20	In aq.H$_2$SO$_4$, H$_-$ scale	06	Uncert.	I24

2354 C$_4$H$_4$O$_4$Cl$_2$ Butanedioic acid, 2,3-dichloro-

	Temp	Conditions			Ref
pK$_1$ 1.68, pK$_2$ 3.18 (DL)	25	c $\approx$ 0.005	E3bg,R3d	Approx.	D41
pK$_1$ 1.60 (DL)	25	c = 0.06-0.11	C2,R1a		
pK$_1$ 1.74, pK$_2$ 3.24 (meso)	25	c $\approx$ 0.005	E3bg,R3d		
pK$_1$ 1.65 (meso)	25	c = 0.07-0.12	C2,R1a		

No.	Molecular formula, name and pK value(s)	$T(^{O}C)$	Remarks	Method	Assessment	Ref.
2355	$C_4H_4O_4Br_2$ Butanedioic acid, 2,3-dibromo-					
	pK$_1$ 1.48 (DL*)	25	c = 0.078-0.102	C2,R1a	Approx.	D41
	pK$_1$ 1.42, pK$_2$ 2.97 (meso)	25	c ≈ 0.005	E3bg,R3d		
	pK$_1$ 1.42 (meso)			C2,R1a		
	* DL-isomer reacts with water. Rate increases as pH is increased					
2356	$C_4H_4N_2Se_2$ Pyrimidine-2,4(1$\underline{H}$,3$\underline{H}$)-diselone					
	5.5		c = 0.005	E3bg	Uncert.	M67
2357	$C_4H_5ON_5$ 1H-1,2,3- Triazolo[4,5-d]pyrimidin-5(4$\underline{H}$)-one, 6,7-dihydro-					
	pK$_1$ -1.36, pK$_2$ 8.03, pK$_3$ 14.48	20	c = 0.00002	05	Uncert.	A33
2358	$C_4H_5O_2N$ 2-Butynoic acid, 4-amino-					
	pK$_1$ 1.80, pK$_2$ 8.34	20	Mixed constants	E3bg	Approx.	B39
2359	$C_4H_5O_2N$ 2,4-Pyrrolidinedione (Tetramic acid)					
	6.4		Conditions not stated	05	Uncert.	M109
2360	$C_4H_5O_2N$ 2,5-Pyrrolidinedione (Succinimide)					
	9.70			E3bg	Uncert.	C62
	9.62	25	Thermodynamic quantities are also given	E1cg	Approx.	W6
	9.38	25	I = 0.1(KNO$_3$)	E3bg	Approx.	C10

No.	Formula	Compound / alt. name	pK	T (°C)	Conditions	Method	Reliability	Ref.
2361	$C_4H_5O_2N_3$	Pyrimidin-2(1H)-one, 4-hydroxyamino-	pK_1 2.8		Conditions not stated	05	Uncert.	J7
2362	$C_4H_5O_2N_3$	1,2,4-Triazine-3,5(2H,4H)-dione, 2-methyl- (6-Azauracil, 1-methyl-)	pK_1 7.42					
			6.99	25	$c \approx 0.001$	E3bg	Approx.	G62
			6.99	25	Titrant $(CH_3)_4NOH$	E3bg	Approx.	J16
2363	$C_4H_5O_2N_3$	1,2,4-Triazine-3,5(2H,4H)-dione, 4-methyl- (6-Azauracil, 3-methyl-)	pK_1 9.52					
			9.52	25	$c \approx 0.001$	E3bg	Approx.	G62
			9.25	25	Titrant $(CH_3)_4NOH$	E3bg	Approx.	J16
2364	$C_4H_5O_2N_3$	1,2,4-Triazine-3,5(2H,4H)-dione, 6-methyl- (6-Azauracil, 5-methyl-)	pK_1 7.48, pK_2 13.4					
			8.15	25	$c = 0.01$	E3bg	Approx.	K9
				25	Titrant $(CH_3)_4NOH$	E3bg	Approx.	J16
			pK_1 7.6	not stated	$c = 0.005$	05	Uncert.	C28
2365	$C_4H_5O_2N_3$	1,3,5-Triazine-2,4(1H,3H)-dione, 1-methyl- (5-Azauracil, 1-methyl-)	pK_1 8.15	25	Titrant $(CH_3)_4NOH$	E3bg	Approx.	J16
2366	$C_4H_5O_2N_3$	1,3,5-Triazine-2,4(1H,3H)-dione, 3-methyl- (5-Azauracil, 3-methyl-)	pK_1 6.58	25	Titrant $(CH_3)_4NOH$	E3bg	Approx.	J16
2367	$C_4H_5O_2N_3$	1,3,5-Triazine-2,4(1H,3H)-dione, 6-methyl- (5-Azauracil, 6-methyl-)	pK_1 7.23	25		E3bg	Approx.	P69
			7.21	25	Titrant $(CH_3)_4NOH$	E3bg	Approx.	J16

No.	Molecular formula, name and pK value(s)	T(°C)	Remarks	Method	Assessment	Ref.
2368	$C_4H_5O_2N_3$ 1,3,5-Triazin-2(1$\underline{H}$)-one, 4-methoxy-					
	6.92	25		E3bg	Approx.	P69
2369	$C_4H_5O_2Cl$ 2-Butenoic acid, 2-chloro-					
	2.80(cis)	23	c = 0.01	E3bg	Approx.	D45
	3.22(trans)					
	Values in mixed solvents are also given					
2370	$C_4H_5O_2Cl$ 2-Butenoic acid, 3-chloro-					
	3.91(trans)	25	c = 0.01	E3bg	Approx.	L17
	4.07(cis)					
	3.95(trans)	25	c = 0.01	E3bg	Approx.	E112
	4.10(cis)					
	3.93(trans)	23	c = 0.01	E3bg	Approx.	D45
	4.09(cis)					
	Values in mixed solvents are also given					
2371	$C_4H_5O_2Cl$ 2-Butenoic acid, 4-chloro-					
	4.15(trans)	23	c = 0.01	E3bg	Approx.	D45
	Values in mixed solvents are also given					

2372	$C_4H_5O_2Cl$	3-Butenoic acid, 2-chloro-					
		2.55	23	c = 0.01	E3bg	Approx.	D45
		Values in mixed solvents are also given					
2373	$C_4H_5O_2Cl$	3-Butenoic acid, 4-chloro-					
		4.07(cis)	23	c = 0.01	E3bg	Approx.	D45
		4.12(trans)					
		Values in mixed solvents are also given					
2374	$C_4H_5O_2Cl$	Propenoic acid, 3-chloro-2-methyl-					
		4.04(trans)	not stated		E3bg	Uncert.	F11
		3.04(cis)					
2375	$C_4H_5O_2Br$	2-Butenoic acid, 3-bromo-					
		3.74(trans)	25	c = 0.01	E3bg	Approx.	L17
		3.98(cis)					
2376	$C_4H_5O_2I$	2-Butenoic acid, 3-iodo-					
		3.77(trans)	25	c = 0.01	E3bg	Approx.	L17
		3.90(cis)					
2377	$C_4H_5O_3N$	2,5-Pyrrolidinedione, 1-hydroxy- (Succinimide, N-hydroxy-)					
		6.0	Conditions not stated		E3bg	Uncert.	A50

88

No.	Molecular formula, name and pK value(s)	T(°C)	Remarks	Method	Assessment	Ref.
2378	$C_4H_5O_3N_5$ 1,3,5-Triazine-2-carbaldehyde oxime, 3,4,5,6-tetrahydro-6-hydroxyimino-4-oxo-					
	pK_1 1.66	25	$I = 0.3(NaClO_4)$	05	Uncert.	C2
	pK_2 9.15				Approx.	
2379	$C_4H_5O_4N_3$ Butanenitrile, 4,4-dinitro-					
	3.45	25	$I = 0.05$(phthalate buffers), mixed constant	05	Approx.	S77
	2.50	20		05	Approx.	T58a
2380	$C_4H_5O_8N_3$ Butanoic acid, 4,4,4-trinitro-					
	3.64	25	$c = 0.01$	E3bg	Uncert.	F5
2381	$C_4H_5O_9N_3$ Butanoic acid, 2-hydroxy-4,4,4-trinitro-					
	3.05	25	$c = 0.01$	E3bg	Approx.	F5
2382	$C_4H_5N_3S_2$ 1,2,4-Triazine-3,5(2H,4H)-dithione, 2-methyl-					
	5.76	25	$c \approx 0.001$	E3bg	Approx.	G63
2383	$C_4H_5N_3S_2$ 1,2,4-Triazine-3,5(2H,4H)-dithione, 4-methyl-					
	7.37	25	$c \approx 0.001$	E3bg	Approx.	G63
2384	$C_4H_5N_3S_2$ 1,2,4-Triazine-3,5(2H,4H)-dithione, 6-methyl-					
	6.10	25	$c \approx 0.001$	E3bg	Approx.	J17

No.	Formula / Name	pK	Temp.	Conditions	Method	Reliability	Ref.
2385	$C_4H_6ON_4$ 1,2,4-Triazin-5(2H)-one, 3-amino-6-methyl-	pK₁ 2.19	25		05	Approx.	P70
2386	$C_4H_6ON_4$ 1,2,4-Triazin-5(4H)-one, 3-amino-4-methyl-	pK₁ 3.95	25		05	Approx.	P70
2387	$C_4H_6OS_2$ Methanedithioic acid, allyloxy- (Xanthic acid, allyl-)	1.52	not stated		05	Uncert.	Z2
2388	$C_4H_6O_2N_2$ 3H-Diazurin-3-ylacetic acid, 3-methyl- (Butanoic acid, 3,3-(NN')azo-)	3.9	25		E3bg	Uncert.	C57
2389	$C_4H_6O_2N_2$ Imidazolidine-2,4-dione, 1-methyl- (Hydantoin, 1-methyl-)	9.09	25	I = 0.1(KNO₃)	E3bg	Approx.	C10
		9.20	25	c ≈ 0.001	E3bg	Approx.	G62
2390	$C_4H_6O_2N_2$ Imidazolidine-2,4-dione, 3-methyl- (Hydantoin, 3-methyl-)	> 11	25	c ≈ 0.001	E3bg		G62
2391	$C_4H_6O_2N_2$ Imidazolidine-2,4-dione, 5-methyl- (Hydantoin, 5-methyl-)	9.07	25	I = 0.1(KNO₃)	E3bg	Approx.	C10
2392	$C_4H_6O_2N_2$ Propanoic acid, 2-amino-3-cyano-	pK₁ 1.7, pK₂ 7.4	20	c = 0.05	E3bg	Uncert.	R21

No.	Molecular formula, name and pK value(s)	T(°C)	Remarks	Method	Assessment	Ref
2393	$C_4H_6O_2N_2$ 2,5-Pyrrolediol, 3-amino- (Succinimide, 3-amino-) 9.0		Conditions not stated	E3bg	Uncert.	C62
2394	$C_4H_6O_2N_2$ Pyrimidine-2,4(1H,3H)-dione, 5,6-dihydro- (5,6-Dihydrouracil) 11.74			0		P83
2395	$C_4H_6O_2N_6$ 1H-1,2,3-Triazolo[4,5-d]pyrimidin-5(4H)-one, 6,7-dihydro-7-hydroxyamino- pK$_1$ -1.43 pK$_2$ 6.45	20		06 05	Uncert. Approx.	A35
2396	$C_4H_6O_2F_4$ 1,4-Butanediol, 2,2,3,3-tetrafluoro- pK$_1$ 12.1, pK$_2$ 13.7	25		E3bg	V.uncert.	M4
2397	$C_4H_6O_2S_2$ 3-Thiapentanoic acid, 4-thioxo- 2.90 2.99	25	I = 0.1(NaClO$_4$),concentration constant	E3bg 02	Uncert. Uncert.	J9
2398	$C_4H_6O_3S$ 3-Thiapentanoic acid, 4-oxo- 3.23	25	I - 0.1(NaClO$_4$),concentration constant	02	Approx.	J9
2399	$C_4H_6O_4N_2$ 1-Butene, 4,4-dinitro- 5.15 4.99 4.83 4.69	5 20 40 60		05	Uncert.	S81

Thermodynamic quantities are derived from the results   (Contd)

No.	Formula / Name	pK values	T	Conditions	05	Approx.	Ref.
2399(Contd)	$C_4H_6O_4N_2$ 1-Butene, 4,4-dinitro	4.92	20			Approx.	T58a
2400	$C_4H_6O_4S$ Butanedioic acid, mercapto- (Thiomalic acid)						
		pK_1 3.15, pK_2 4.68	30	$I = 0.1(KCl)$, mixed constants	E3bg,R3f	Approx.	C23
		pK_1 3.165, pK_2 4.670, pK_3 10.502	25	$c = 0.01$, mixed constants	E3bg	Approx.	P73
		pK_1 3.64, pK_2 4.64, pK_3 10.37	25	$c \simeq 0.0015$, $I = 0.1(KNO_3)$	E3bg	Approx.	L18
		pK_1 2.53, pK_2 4.44, pK_3 9.73		$I = 1.0(KNO_3)$concentration constants			
		pK_1 3.01, pK_2 4.51	31	$I = 0.1(NaClO_4)$,mixed constants	E3bg	Approx.	R11
		Other values in C41 and K11					
2401	$C_4H_6O_4S$ 3-Thiapentanedioic acid (Thiodiacetic acid)						
		pK_1 3.15, pK_2 4.13	25	$c \simeq 0.003$, $I = 0.1(NaClO_4)$, concentration constants	E3bg,R3h	Approx.	S134
		pK_1 3.24, pK_2 4.56	25	$c < 0.001$	E3bg	Approx.	S93
		pK_1 3.26, pK_2 4.29	30	$I = 0.1(KCl)$, mixed constants	E3bg	Approx.	T32
		Other values in C8 and S92					
2402	$C_4H_6O_4S_2$ 3,4-Dithiahexanedioic acid (Dithiodiacetic acid)						
		pK_1 2.88, pK_2 3.81	25	$c \simeq 0.005$, $I = 0.1(NaClO_4)$, concentration constants	E3bg,R3f	Approx.	S131
		pK_1 3.15, pK_2 4.13	25	$I = 0.1(NaClO_4)$,concentration constants	E3bg	Approx.	Y12
		pK_1 3.15, pK_2 4.24	25				L68

No.	Molecular formula, name and pK value(s)	T(^{o}C)	Remarks	Method	Assessment	Ref.
2403	$C_4H_6O_4S_2$ Butanedioic acid, 2,3-dimercapto- (meso)					
	pK$_1$ 2.71, pK$_2$ 3.48, pK$_3$ 8.89, pK$_4$ 10.79	25		E3bg	Approx.	L18
	pK$_1$ 2.40, pK$_2$ 3.46, pK$_3$ 9.44, pK$_4$ 11.82	20	c $\approx$ 0.0015, I = 0.1(KNO$_3$)	E3bg	Approx.	A12
			c $\approx$ 0.0016, I = 0.1(KCl), concentration constants.			
2404	$C_4H_6O_4Se$ 3-Selenapentanedioic acid (Selenodiacetic acid)					
	pK$_1$ 3.17, pK$_3$ 4.31	25	c $\approx$ 0.003, I = 0.1(NaClO$_4$), concentration constants	E3bg,R3h	Approx.	S134
2405	$C_4H_6O_5N_2$ 3-Azapentanedioic acid, 3-nitroso- (Iminodiacetic acid, N-nitroso-)					
	pK$_1$ 2.28, pK$_2$ 3.38	30	I = 0.1(KCl), mixed constants	E3bg	Approx.	T32
2406	$C_4H_6O_6N_2$ 3-Azapentanedioic acid, 3-nitro- (Iminodiacetic acid, N-nitro-)					
	pK$_1$ 2.21, pK$_2$ 3.33	30	I = 0.1(KCl), mixed constants	E3bg	Approx.	T32
2407	$C_4H_6O_6N_2$ Butanoic acid, 4,4-dinitro-					
	pK$_2$ 5.35	20		05	Approx.	T58a
2408	$C_4H_6O_6N_2$ Methyl 3,3-dinitropropanoate					
	3.08	20		05	Approx.	T58a
	3.08	25	I = 0.05(phthalate buffers), mixed constant	05	Approx.	S77

No.	Formula	Name	pK	T(°C)	Conditions	Reliability		Ref.
2409	$C_4H_6O_7N_2$	Butanoic acid, 2-hydroxy-3,3-dinitro-						
		pK_1 2.37, pK_2 5.42		25	c = 0.01	Uncert.	E3bg	F5
		(Existence of two pK values said to be due to tendency of compound to dissociate to form 1,1-dinitroethane)						
2410	$C_4H_6O_6N_4$	Butane, 1,1,3,3-tetranitro-						
			1.37	20	In aq.$HClO_4$ solutions, H_0 scale	Uncert.	06	T58a
			1.44	5	In aq.HCl solutions, H_0 scale	Uncert.	06	N39
			1.36	20				
			1.32	40				
			1.32	60				
			1.37	80				
2411	C_4H_7ON	2-Pyrrolidinone (γ-Butyrolactam)						
			14.7	25		Uncert.	C1	H13
2412	$C_4H_7ON_3$	Pyrimidin-2(1$\underline{H}$)-one, 2-amino-5,6-dihydro- (Dihydrocytosine)						
			6.3	25	Conditions not stated	Uncert.	05	J6
2413	$C_4H_7O_2N$	2-Butanone, 1-hydroxyimino-						
			8.37	25	c ≃ 0.01, I = 0.1(KCl), mixed constant	Approx.	E3bg	G39
2414	$C_4H_7O_2N$	2-Butanone, 3-hydroxyimino- (Diacetylmonoxime)						
			9.28	25	c = 0.0025, I = 0.1(KCl), concentration constant	Approx.	E3bg	M40

(Contd)

94

No.	Molecular formula, name and pK value(s)	T(°C)	Remarks	Method	Assessment	Ref.
2414(Contd)	$C_4H_7O_2N$ 2-Butanone, 3-hydroxyimino- (Diacetylmonoxime)					
	9.51	25		E3bg,R4	Approx.	M40
	9.27	25	c = 0.0025-0.005, I = 0.1(KNO_3)	E		U9
	9.30	25	c ≈ 0.01, I = 0.1(KCl), mixed constant	E3bg	Approx.	G39
2415	$C_4H_7O_2N$ 2-Butenohydroxyamic acid (Crotonohydroxyamic acid)					
	8.90	30.5	c = 0.001, I = 0.1(KNO_3), mixed constant	E3d	Uncert.	S137
2416	$C_4H_7O_2N_3$ 1,2,4-Triazine-3,5(1H,2H)-dione, 4,6-dihydro-1-methyl- (6-Azauracil, 5,6-dihydro-6-methyl-)					
	10.6	25	Titrant $N(Me)_4OH$	E3bg	Uncert.	J16
2417	$C_4H_7O_2N_3$ 1,2,4-Triazine-3,5(1H,2H)-dione, 4,6-dihydro-2-methyl- (6-Azauracil, 5,6-dihydro-1-methyl-)					
	10.6	25	c ≈ 0.001	E3bg	Uncert.	G62
2418	$C_4H_7O_2N_3$ 1,2,4-Triazine-3,5(1H,2H)-dione, 4,6-dihydro-4-methyl- (6-Azauracil, 5,6-dihydro-3-methyl-)					
	> 11	25	Titrant $N(Me)_4OH$	E3bg		J16
2419	$C_4H_7O_2Cl$ Butanoic acid, 3-chloro-					
	4.2	25		E3bg	Uncert.	C57
2420	$C_4H_7O_3N$ 2-Butanone, 1-hydroxy-3-hydroxyimino-					
	9.07	25	c = 0.0025-0.005, I = 0.1(KNO_3) (probably a mixed constant)	E		U9

No.	Formula	Name / pK	t	Conditions			Ref.
2421	$C_4H_7O_4N$	3-Azapentanedioic acid (Iminodiacetic acid)					
		pK$_1$ 2.98, pK$_2$ 9.89	20	c = 0.002	E3bg	Approx.	S45
		pK$_1$ 2.50, pK$_2$ 9.40	25	I = 0.1, concentration constants	E3bg	Approx.	R2
		pK$_1$ 2.55, pK$_2$ 9.38		I = 1.0			
		pK$_1$ 2.65, pK$_2$ 9.45	20	I = 0.1(KNO$_3$)	E3bg	Approx.	A54
		Thermodynamic quantities are also given					
		pK$_2$ 9.44	20	I = 0.1[(CH$_3$)$_4$NCl],concentration constant	E3bg	Approx.	A55
				Values in other inert salt solutions are also given			
		pK$_1$ 3.01, pK$_2$ 9.60	25	I = 0.3(KCl)	E3bg	Approx	M85
		pK$_1$ 2.84, pK$_2$ 10.70	0.4	I = 0.1(KNO$_3$),concentration constants	E3bg	Approx	K25
2422	$C_4H_7O_4N$	Butanedioic acid, amino- (Aspartic acid)					
		pK$_2$ 3.87, pK$_3$ 9.87	25	c = 0.001, I = 0.1(KCl)	E3bg	Approx.	B35
		pK$_1$ 1.92, pK$_2$ 3.69	25	I = 0.2	E3bg	Approx.	F7
		pK$_2$ 3.79, pK$_3$ 9.63	30	I = 0.1(KCl),mixed constants	E3bg	Approx.	C23
		pK$_2$ 3.92, pK$_3$ 9.95	20	c = 0.025	E3d	Uncert.	H17
2423	$C_4H_7O_4N$	Butanoic acid, 4-hydroxyamino-4-oxo- (Succinamic acid, N-hydroxy-)					
		pK$_1$ 4.5, pK$_2$ 9.6		I = 0.06	E3bg	Uncert.	A50
2424	$C_4H_7O_4N$	Ethyl 2-nitroacetate					
		5.75	25		05	Approx.	A7
		5.81	25		E3bg	Approx.	P18

No.	Molecular formula, name and pK value(s)	$T(^oC)$	Remarks	Method	Assessment	Ref.
2425	$C_4H_7O_5N$ Butanoic acid, 2-nitrooxy-					
	2.39	not stated	$c \approx 0.025$	E3bg	Uncert.	M9
2426	$C_4H_7O_5N_3$ Acetamide, N-(2,2-dinitroethyl)-					
	4.00	20		05	Approx.	T58a
2427	$C_4H_7O_5N_3$ Butanamide, 4,4-dinitro-					
	4.49	20		05	Approx.	T58a
	4.41	25	$c \approx 0.00005$	05	Approx.	S79
			Variation with temperature			
			5 20 40 60			
	4.53	4.42 4.31 4.22				
			Thermodynamic quantities are derived from the results			
2428	$C_4H_7O_6N_3$ Butane, 1,1,3-trinitro-					
	2.94	20		05	Approx.	T58a
2429	$C_4H_7N_3S$ 1,2,4-Triazole-3($\underline{2H}$)-thione, 4,5-dimethyl-					
	8.19	25	$c \approx 0.004$, mixed constant	E3bg	Approx.	S9
2430	$C_4H_8OS_2$ Methanedithioic acid, isopropoxy- (Xanthic acid, isopropyl-)					
	1.77	not stated		05	Uncert.	Z2
	2.16	0	Extrapolated to zero time	E3d		G5

2431 $C_4H_8OS_2$ Methanedithioic acid, propoxy- (Xanthic acid, propyl-)

1.43	not stated		O5	Uncert.	Z2
3.07					H27

2432 $C_4H_8O_2N_2$ 2,3-Butanedione dioxime (Dimethylglyoxime)

pK_1 10.6, pK_2 11.9	20-25		O5	Uncert.	B18
pK_1 10.60, pK_2 11.85	25	concentration constants	E3bg	Uncert.	H4
pK_1 10.42	25	$c = 0.0025-0.005$, $I = 0.1(KNO_3)$	E3	Uncert.	U9
pK_1 10.54, pK_2 12.05			O		S15

Other values in B10 and U10

2433 $C_4H_8O_2S$ Ethyl 2-mercaptoacetate

7.95	25		E3bg	Approx.	K58

2434 $C_4H_8O_2S$ 3-Thiapentanoic acid (Acetic acid, ethylthio-)

3.712	25	$c \approx 0.001$, $I = 0.10(KNO_3)$, concentration constant	E3bg	Approx.	P40
3.61	20	$I = 1.0(NaClO_4)$, concentration constant	E3bg	Approx.	S8
4.11	25	$c = 0.01$, mixed constant	E3bg	Approx.	M82

Value in mixed solvent is also given

2435 $C_4H_8O_3N_2$ 3-Azapentanoic acid, 5-amino-4-oxo- (Glycylglycine)

pK_2 8.21	21	$c = 0.02$, mixed constant	E3bg	Approx.	P28
pK_1 3.20, pK_2 8.22	25	$I = 0.06$, mixed constants	E3bg	Approx.	V1

(Contd)

No.	Molecular formula, name and pK value(s)	T(°C)	Remarks	Method	Assessment	Ref.
2435 (Contd)	$C_4H_8O_3N_2$ 3-Azapentanoic acid, 5-amino-4-oxo- (Glycylglycine)					
			Variation with temperature			
		10 30 35 40				
	pK₁ 3.21 3.20 3.19 3.19					
	pK₂ 8.64 8.11 7.97 7.86					
			Thermodynamic quantities are derived from the results			
	pK₁ 3.08, pK₂ 8.09	25	I = 0.1(NaClO₄),concentration constants	E3bg	Approx.	B148
			Thermodynamic quantities are also given			
	pK₁ 3.15, pK₂ 8.10	25	I = 0.1(KCl), concentration constants	E3bg	Approx.	D47
	pK₁ 3.23, pK₂ 8.16	25	I = 0.16	E3bg	Approx.	B152
	pK₂ 8.71	0.35	I = 0.09(KCl), concentration constant	E3bg	Approx.	M125
	pK₂ 8.01	30	Thermodynamic quantities are derived from			
	pK₂ 7.50	48.8	the results			
2436	$C_4H_8O_3N_2$ 2,3-Butanedione dioxime, 1-hydroxy-					
	10.05	25	c = 0.0025-0.005, I = 0.1(KNO₃)	E		U9
2437	$C_4H_8O_3S$ Propanone, 1-methylsulfonyl-					
	10.03	20	c ≃ 0.004	E3bg	Uncert.	S42
2438	$C_4H_8O_4N_2$ 3-Azapentanedioic acid, 3-amino- (Iminodiacetic acid, N-amino-)					
	pK₁ 2.54, pK₂ 9.30	25	I = 0.1(KCl)	E3bg	Approx.	V9

No.	Formula	Name / pK	T	Conditions		Assessment	Ref.
2439	$C_4H_8O_4N_2$	Butane, 1,1-dinitro-					
		5.40	20		05	Uncert.	S83
				Thermodynamic quantities are given			
		5.55	5		05	Approx.	N39
		5.39	20				
		5.21	40				
		5.08	60				
				Thermodynamic quantities are derived from the results			
2440	$C_4H_8O_4N_2$	3,4-Diazahexanedioic acid (Hydrazine-N,N'-diacetic acid)					
		pK_1 3.02, pK_2 7.14	30	$I = 0.1(KCl)$, mixed constants	E3bg	Approx.	T32
2441	$C_4H_8O_4N_2$	Propane, 2-methyl-1,1-dinitro-					
		6.77	20		05	Approx.	T58a
2442	$C_4H_8O_4N_2$	Propane, 3-methyl-1,1-dinitro-					
		6.75	not stated		E3bg	Uncert.	N36
2443	$C_4H_8O_4S_2$	1,3-Dithiacyclohexane-1,1,3,3-tetraoxide					
		12.61	25	No details		Uncert.	C76
2444	$C_4H_8O_5N_2$	1-Butanol, 4,4-dinitro-					
		5.06	20		05	Approx.	T58a

No.	Molecular formula, name and pK value(s)	T(°C)	Remarks	Method	Assessment	Ref.
2445	$C_4H_8O_5N_4$ Butanamide, N-amino-4,4-dinitro-					
	pK_1 4.15	20		05	Approx.	T58a
2446	$C_4H_8O_5S$ Butanoic acid, 4-sulfo-					
	pK_2 4.91 (-COOH)	25	c ≈ 0.02, I ≈ 0.08, mixed constant	E3bg	Approx.	B58
2447	$C_4H_8O_6N_4$ Ethylamine, N-ethyl-N,2,2-trinitro-					
	2.44	20		05	Approx.	T58a
2448	$C_4H_8O_6N_4$ Propylamine, N-methyl-N,3,3-trinitro-					
	3.45	20		05	Approx.	T58a
2449	C_4H_9ON 2-Butanone oxime					
	12.45	25		KIN	Uncert.	K28
2450	$C_4H_9O_2N$ Butanohydroxamic acid					
	9.5	20	c = 0.01	E3d	Uncert.	W39
2451	$C_4H_9O_2N$ Butanoic acid, 2-amino-					
	pK_2 9.62	20	c = 0.02, mixed constant	E3bg	Approx.	P28
	pK_1 2.1, pK_2 9.0	20	c = 0.05	E3bg	Uncert.	R21
	pK_2 9.56	25	I = 0.16(KNO_3), mixed constant	E3bg	Approx.	M72
			Thermodynamic quantities are also given			

		pK	T (°C)	Conditions			Ref.
2452	$C_4H_9O_2N$ Butanoic acid, 4-amino-						
		pK$_1$ 4.14, pK$_2$ 10.55		c = 0.005	E3bg	Uncert.	S87
		pK$_1$ 4.01	25	c = 0.01	E3bg	Approx.	Y5
		pK$_2$ 10.48	25	I = 0.5(NaCl)	E3bg	Approx.	S109
		pK$_2$ 10.46	25	I = 0.5(KNO$_3$), concentration constant	E3bg	Approx.	T23
2453	$C_4H_9O_2N$ Propanoic acid, 2-amino-2-methyl-						
		pK$_1$ 2.48, pK$_2$ 10.22	25	I = 0.5(KCl), mixed constants	E3bg	Approx.	L28
		pK$_2$ 10.24	19	c = 0.02, mixed constant	E3bg	Approx.	P28
2454	$C_4H_9O_2S$ (cation) Acetic acid, dimethylsulfonio- (Dimethylacetothetin)						
		1.5			E3bg	Uncert.	K34
		Determined from infra-red measurements in D_2O and corrected to the corresponding value in water by subtracting 0.5 pK unit.					
2455	$C_4H_9O_2P$ 1-Phosphacyclopentan-1-ol-1-oxide (Tetramethylenephosphinic acid)						
		2.51	25	c = 0.01, mixed constant	E3bg	Uncert.	K50
2456	$C_4H_9O_3N$ Butanoic acid, 2-amino-3-hydroxy- (Threonine)						
		pK$_1$ 2.32, pK$_2$ 9.26	15	I = 0.2(KNO$_3$)	E3bg	Approx.	R7
		pK$_1$ 2.32, pK$_2$ 9.03	25				
		pK$_1$ 2.30, pK$_2$ 8.71	40				
2457	$C_4H_9O_3N$ Butanoic acid, 4-amino-3-hydroxy-						
		3.80	25	c = 0.01	E3bg	Approx.	Y5

No.	Molecular formula, name and pK value(s)	$T(^oC)$	Remarks	Method	Assessment	Ref.
2458	$C_4H_9O_4N_3$ Butylamine, 4,4-dinitro- pK_1 3.93	20		O5	Approx.	T58a
2459	$C_4H_9O_5P$ Butyryl dihydrogen phosphate pK_1 1.2, pK_2 5.1	24		E3bg	Uncert.	L54
2460	$C_4H_9NS_2$ Methanedithioic acid, isopropylamino- 2.91 2.79	(Dithiocarbamic acid, N-isopropyl-) 21	$c \simeq 0.01$-0.00001, concentration constant (Compound unstable in acid solution)	O5 E	Uncert. Uncert.	T66
2461	$C_4H_9NS_2$ Methanedithioic acid, propylamino- 3.10	(Dithiocarbamic acid, N-propyl-) 25		KIN	Uncert.	M79
2462	$C_4H_{10}OS$ Ethanethiol, 2-ethoxy- 9.38	25		E3bg	Approx.	K58
2463	$C_4H_{10}OS$ 1-Propanol, 2-mercapto-2-methyl- 9.85	25	1% ethanol	E3bg	Uncert.	K57
2464	$C_4H_{10}O_2N_2$ Acetohydroxamic acid, 2-dimethylamino- 7.10	25 (?)	c = 0.02, I = 0.1(KCl), mixed constant	E3bg	Uncert.	M66
2465	$C_4H_{10}O_2N_2$ Butanoic acid, 2,4-diamino- pK_1 1.85, pK_3 10.44 pK_2 8.24	20	c = 0.01, mixed constants	E3bg	Uncert. Approx.	A16

No.	Formula / Name	pK	T (°C)	Conditions	Method	Reliability	Ref.
2466	$C_4H_{10}O_2S$ Butanesulfinic acid	2.11	20	c = 0.1, mixed constant	E3bg	Approx.	R54
2467	$C_4H_{10}O_2S_2$ 2,3-Butanediol, 1,4-dimercapto- (1,4-Dithiothreitol)	pK_1 8.3, pK_2 9.5		c = 0.05	E3bg,R3d	Uncert.	Z1
2468	$C_4H_{10}O_4S_2$ 1,2,3,4-Butanetetrol, 1,4-dimercapto- (1,4-Dithioerythritol)	pK_1 9.0, pK_2 9.9		c = 0.05	E3bg,R3d	Uncert.	Z1
2469	$C_4H_{11}O_2P$ Diethyl phosphinic acid	3.10	20	c = 0.0004-0.05, concentration constant	C2	Approx.	K63
2470	$C_4H_{11}O_3P$ Propyl hydrogen methylphosphonate	2.38	20	c = 0.005	E3bg	Uncert.	N28
2471	$C_4H_{11}O_3As$ Butylarsonic acid	pK_1 4.76	not stated	Mixed constant	E3bg	Uncert.	C42
2472	$C_4H_{11}O_3As$ Isobutylarsonic acid	pK_1 4.79, pK_2 8.18	not stated	Mixed constants	E3bg	Uncert.	C42
2473	$C_4H_{11}O_4P$ Butyl dihydrogen phosphate	pK_1 1.89, pK_2 6.84	25	m = 0.036, mixed constants	E3bg	Uncert. Approx.	K67

No.	Molecular formula, name and pK value(s)	T(°C)	Remarks	Method	Assessment	Ref.
2474	$C_4H_{11}O_4P$ Diethyl hydrogen phosphate					
	1.37	not stated	c = 0.005, I < 0.025, 7% ethanol	E3bg	Uncert	M57
	1.37	20	7% ethanol	E		K3
	1.39	25	m ≃ 0.04, mixed constant	E3bg	Uncert.	K67
	0.73	25	I = 0.1–1(NaClO₄), concentration constant	DIS	Uncert.	D78
2475	$C_4H_{11}NS$ 2-Butanethiol, 1-amino-					
	pK₁ 8.19, pK₂ 10.91	25	c = 0.01, mixed constants	E3bg	Approx.	B147
2476	$C_4H_{11}NS$ 2-Propanethiol, 1-amino-2-methyl-					
	pK₁ 8.07, pK₂ 10.77	25	c = 0.01, mixed constants	E3bg	Approx.	B147
2477	$C_4H_{11}S_2P$ Diethylphosphinodithioic acid					
	0.98	20	c = 0.0004–0.05, concentration constant	C	Uncert.	K63
	1.71	20	c = 0.005 in 7% ethanol	E3bg	Uncert.	K6
2478	$C_4H_{12}O_6P_2$ 1,4-Butanediyldiphosphonic acid (Tetramethylenediphosphonic acid)					
	pK₁ < 2, pK₂ 2.7	20	I = 0.1(KCl), mixed constants	E3bg	Uncert.	S48
	pK₃ 7.54, pK₄ 8.38				Approx.	
	pK₁ 1.7, pK₂ 3.19, pK₃ 7.78, pK₄ 8.58		Average of values at 25°,37°, 50°, extrapolated from values at I = 0.1–1.0[(CH₃)₄NBr]	E3bg	Uncert.	I6

No.	Formula	Name	t (°C)	pK	Conditions	Method	Assessment	Ref.
2479	$C_4H_{14}O_2B_{10}$	1,2-Dicarbadodecaborane(12)-1-carboxylic acid, 2-methyl- (Barenecarboxylic acid, methyl-)	25	2.53	Mixed constant	E3bg	Uncert.	Z3
2480	$C_4H_{14}O_2B_{10}$	1,2-Dicarbadodecaboran-1-yl(12)acetic acid (Bareneacetic acid)	25	4.06	Mixed constant	E3bg	Uncert.	Z3
2481	C_4HOClF_8	2-Propanol, 1-chloro-1,1,3,3,3-pentafluoro-2-trifluoromethyl-		5.3	Conditions not stated		Uncert.	F14
2482	$C_4HOCl_3F_6$	2-Propanol, 1,1,1-trichloro-3,3,3-trifluoro-2-trifluoromethyl-		5.1	Conditions not stated		Uncert.	F14
2483	$C_4HO_3NF_8$	2-Propanol, 1,1,1,3,3-pentafluoro-3-nitro-2-trifluoromethyl-	25	3.9	Mixed constant	E3bg	Uncert.	K37
			25	3.9		E3bg	Uncert.	D65
2484	$C_4H_2ONF_9$	N-(2,2,2-Trifluoroethyl)hydroxylamine, 1,1-bis(trifluoromethyl)-	25	5.8-6.0				K38
2485	$C_4H_3ON_2Cl$	Pyrimidin-4(1H)-one, 6-chloro-	20	7.43	c = 0.002	E3bg	Approx.	B130
2486	$C_4H_3O_2N_2Cl$	Pyrimidine-2,4(1H,3H)-dione, 5-chloro-	20-22	7.95	I = 0.05-0.1, mixed constant	05	Uncert.	B64

No.	Molecular formula, name and pK value(s)	T (°C)	Remarks	Method	Assessment	Ref.
2487	$C_4H_3O_2N_2Br$ Pyrimidine-2,4(1H,3H)-dione, 5-bromo- (Uracil, 5-bromo-)					
	8.05	20-22	I = 0.05-0.1, mixed constant	05	Uncert.	B64
2488	$C_4H_3O_2N_2I$ Pyrimidine-2,4(1H,3H)-dione, 5-iodo- (Uracil, 5-iodo-)					
	8.25	20-22	I = 0.05.0.1, mixed constant	05	Uncert.	B64
2489	$C_4H_3O_2N_2F$ Pyrimidine-2,4(1H,3H)-dione, 5-fluoro- (Uracil, 5-fluoro-)					
	8.00	20-22	I = 0.05-0.1, mixed constant	05	Uncert.	B64
	8.04	25	Titrant $(CH_3)_4NOH$	E3bg	Approx.	J16
	8.15	25	Conditions not stated	05	Uncert.	G33
2490	$C_4H_4ON_2S$ Pyrimidin-4(1H)-one, 2,3-dihydro-2-thioxo- (2-Thiouracil)					
	7.96	25	c ≈ 0.001	E3bg	Approx.	j17
	7.74	25	I = 0.16	E3bg	Approx.	C48
	7.52	25	c = 0.0001, mixed constant	05	Uncert.	G2
			Other values (V.uncert.) quoted at 35° and 45°			G2
2491	$C_4H_4ON_2S$ Pyrimidin-4(1H)-one, 6-mercapto-					
	pK_1 4.33	20	c = 0.005	E3bg	Approx.	B130
	pK_2 10.52				Uncert.	
2492	$C_4H_4ON_2Se$ Pyrimidin-4(1H)-one, 2,3-dihydro-2-selenoxo-					
	7.18		c = 0.005	E3bg	Uncert.	M67

No.	Formula	Name	pK	T (°C)	Conditions	Method	Assessment	Ref.
2493	$C_4H_4O_2N_2S$	Pyrimidine-2,4(1H,3H)-dione, 5-mercapto- (Uracil, 5-mercapto-)	pK$_1$ 5.3, pK$_2$ 10.6, pK$_3$ >13		Conditions not stated	05	Uncert.	B23
2494	$C_4H_4O_2N_2S$	Pyrimidin-4-(1H)-one, 2,3-dihydro-6-hydroxy-2-thioxo- (2-Thiobarbituric acid)	pK$_1$ 2.25, pK$_2$ 10.72	28	2% ethanol	05	Uncert.	S14
			pK$_1$ 3.75		Conditions not stated	E3bg	Uncert.	M68
2495	$C_4H_4O_2N_2Se$	Pyrimidin-4(1H)-one, 2,3-dihydro-6-hydroxy-2-selenoxo- (2-Selenobarbituric acid)	3.74		Conditions not stated	E3bg	Uncert.	M68
2496	$C_4H_4O_5N_2S$	Pyrimidine-4-sulfonic acid, 1,2,3,6-tetrahydro-2,6-dioxo- (Uracil, 6-sulfo-)	7.26	25	c = 0.005	E3bg	Approx.	G45
2497	$C_4H_5ON_3S$	1,2,4-Triazin-3(2H)-one, 4,5-dihydro-2-methyl-5-thioxo-	6.25	25	c ≈ 0.001	E3bg	Approx.	G63
2498	$C_4H_5ON_3S$	1,2,4-Triazin-3(2H)-one, 4,5-dihydro-4-methyl-5-thioxo-	8.57	25	c ≈ 0.001	E3bg	Approx.	G63
2499	$C_4H_5ON_3S$	1,2,4-Triazin-3(2H)-one, 5-methylthio-	9.18	25	c ≈ 0.001	E3bg	Approx.	G63
2500	$C_4H_5ON_3S$	1,2,4-Triazin-5(2H)-one, 3,4-dihydro-2-methyl-3-thioxo-	6.24	25	c ≈ 0.001	E3bg	Approx.	G63
2501	$C_4H_5ON_3S$	1,2,4-Triazin-5(2H)-one, 3,4-dihydro-4-methyl-3-thioxo-	8.12	25	c ≈ 0.001	E3bg	Approx.	G63

No.	Molecular formula, name and pK value(s)	$T(^{o}C)$	Remarks	Method	Assessment	Ref.
2502	$C_4H_5ON_3S$ 1,2,4-Triazin-5(2H)-one, 3,4-dihydro-6-methyl-3-thioxo-					
	6.47	25	c ≈ 0.001	E3bg	Approx.	J17
	6.39	25	c ≈ 0.005	E3bg	Uncert.	K9
2503	$C_4H_5ON_3S$ 1,2,4-Triazin-5(2H)-one, 3-methylthio-					
	pK$_1$ -1.17	not stated		06	Uncert.	B137
	pK$_2$ 5.92	not stated	c ≈ 0.001, I = 0.01, mixed constant	05	Uncert.	
	pK$_2$ 5.94	25	c ≈ 0.001	E3bg	Approx.	G63
2504	$C_4H_5O_2SB$ 2-Thiopheneboronic acid					
	8.11	25		E3bg	Approx.	B143

Variation with temperature

	0.2	5	10	15	20	30	35	40	45	50
	8.24	8.21	8.18	8.15	8.13	8.09	8.08	8.07	8.06	8.06

Thermodynamic quantities are calculated from the results

No.	Molecular formula, name and pK value(s)	$T(^{o}C)$	Remarks	Method	Assessment	Ref.
2505	$C_4H_5O_2SB$ 3-Thiopheneboronic acid					
	8.77	25		E3bg	Approx.	B143

Variation with temperature

	0.2	5	10	15	20	30	35	40	45	50
	8.91	8.88	8.85	8.82	8.80	8.74	8.71	8.69	8.68	8.67

No.	Formula	Name	pK	T (°C)	Conditions	Method	Reliability	Ref.
2506	$C_4H_5O_3NS$	Thiazolidine-4-carboxylic acid, 2-oxo-	3.32	22	Mixed constant	E3bg	Approx.	M17
2507	$C_4H_5O_4N_3S$	Pyrimidine-4-sulfonamide, 1,2,3,6-tetrahydro-2,6-dioxo-	5.43	25	c = 0.005	E3bg	Approx.	G45
2508	$C_4H_6ON_2S$	1,2,4-Thiadiazol-5(2H)-one, 3-ethyl-	6.89	25	c = 0.01, mixed constant	E3bg	Approx.	G30
2509	$C_4H_6O_2NCl$	2-Butanone, 1-chloro-3-hydroxyimino-	8.8	25	c = 0.0025-0.005, I = 0.1(KNO_3)	E		U9
2510	$C_4H_6O_2NBr$	2-Butanone, 1-bromo-3-hydroxyimino-	8.75	25	c = 0.0025-0.005, I = 0.1(KNO_3)	E		U9
2511	$C_4H_6O_2NI$	2-Butanone, 3-hydroxyimino-1-iodo-	8.78	25	c = 0.0025-0.005, I = 0.1(KNO_3)	E		U9
2512	$C_4H_6O_2N_2S$	2-Thiazoline-4-carboxylic acid, 2-amino-	pK_1 2.03 pK_2 8.48	25	c = 0.03, mixed constant	E3bg	Uncert. Approx.	G7
			pK_1 2.93, pK_2 8.36	22	Mixed constants	E3bg	Approx.	M17
2513	$C_4H_6O_4NCl$	Ethyl 2-chloro-2-nitroacetate	4.16	25	I = 0.06	O5	Approx.	A7

No.	Molecular formula, name and pK value(s)	T(^{o}C)	Remarks	Method	Assessment	Ref.
2514	$C_4H_6O_4NF$ Ethyl 2-fluoro-2-nitroacetate					
	6.28	25	I = 0.06	05	Approx.	A7
2515	$C_4H_6O_6N_3Cl$ Butane, 2-chloro-2,4,4-trinitro-					
	2.32	20		05	Approx	T58a
2516	$C_4H_7O_2NS_2$ 5-Thia-3-azahexanoic acid, 4-thioxo-					
	3.19	25	I = 0.1($NaClO_4$),concentration constant	02	Approx.	J9
2517	C_4H_9ONS Acetamide, N-(2-mercaptoethyl)-					
	9.92	25	c ≈ 0.00015	05	Approx.	I20
			Heat of ionization = 6.26 kcal mole^{-1}			
2518	$C_4H_{11}OSP$ Diethylphosphinothioic O-acid					
	2.54	20	c = 0.0004-0.05, concentration constant	C2	Approx.	K6
	2.80	20	c = 0.005 in 7% ethanol, 99% thione tautomer	E3bg	Uncert.	K63
2519	$C_4H_{11}OS_2P$ O-Propyl S-hydrogen methylphosphonodithioate					
	1.74	20	c = 0.005, 7% ethanol, mixed constant	E3bg	V.uncert.	K5
2520	$C_4H_{11}O_2SP$ O-Ethyl O-hydrogen ethylphosphonothioate					
	1.88	20	c = 0.005 in 7% ethanol, 86% thione tautomer	E3bg	V.Uncert.	K6
2521	$C_4H_{11}O_2SP$ S-Isopropyl hydrogen methylphosphonothioate					
	2.02	20	c = 0.005, mixed constant	E3bg	Uncert.	N28

No.	Formula	Name / pK values	Temp (°C)	Conditions	Method	Reliability	Ref.
2522	$C_4H_{11}O_2SP$	O-Propyl O-hydrogen methylphosphonothioate	20	$c = 0.005$ in 7% ethanol, 85% thione tautomer	E3bg	V.uncert.	K6
		1.87					
2523	$C_4H_{11}O_2SP$	S-Propyl hydrogen methylphosphonothioate	20	$c = 0.005$, mixed constant	E3bg	Uncert.	N28
		2.03					
2524	$C_4H_{11}O_2S_2P$	O,O-Diethyl S-hydrogen phosphorodithioate	20	$c = 0.005$, 7% ethanol, mixed constant	E3bg	V.Uncert.	K5
		1.62					
2525	$C_4H_{11}O_3SP$	S-Butyl dihydrogen phosphorothioate	25	$c = 0.04$, $I = 0.9$, mixed constants	E3bg	V.uncert.	D42
		$pK_1 \approx 1$, pK_2 5.50		Compound is subject to hydrolysis			
2526	$C_4H_{11}O_3SP$	O,O-Diethyl S-hydrogen phosphorothioate	20	$c = 0.005$ in 7% ethanol, 64% thiol tautomer	E3bg	V.uncert.	K6
		1.49					
2527	$C_4H_{11}O_3PSe$	O,O-Diethyl hydrogen phosphoroselenoate	25		E3ag	Uncert.	T54
		1.77					
2528	$C_4H_{11}O_8NP_2$	Acetic acid, N,N-bis(phosphonomethyl)amino-	25	$I = 0.1(KNO_3)$, concentration constants	E3bg	Uncert. Approx.	W21
		pK_1 1.73, pK_2 2.00					
		pK_3 5.01, pK_4 6.37, pK_5 10.80					
2529	$C_4H_{12}O_3NP$	Butylphosphonic acid, 4-amino-	not stated	$c = 0.02$	E3bg	Uncert.	C38
		pK_1 2.55, pK_2 7.55, pK_3 10.9					

No.	Molecular formula, name and pK value(s)	T(°C)	Remarks	Method	Assessment	Ref.
2530	$C_4H_{13}O_6NP_2$ 2-Azapropane-1,3-diphosphonic acid, 2-ethyl- [Iminodi(methylphosphonic acid), N-ethyl-]					
	pK$_3$ <2 , pK$_4$ 4.70, pK$_5$ 5.92	25	I = 1(KNO$_3$),concentration constants	E3bg	Approx.	C18
	pK$_6$ 12.42				Uncert.	
2531	$C_4H_{14}O_6N_2P_2$ 2,5-Diazahexane-1,6-diphosphonic acid [Ethylenediiminobis(methylphosphonic acid)]					
	pK$_1$ 4.61, pK$_2$ 5.72, pK$_3$ 8.02, pK$_4$ 10.47	25	I = 0.1(KCl),concentration constants	E3bg,R3g	Approx.	D67,D70
2532	C_5H_6 Cyclopentadiene					
	15			05	Uncert.	W10a
2533	$C_5H_2O_5$ 4-Cyclopentene-1,2,3-trione, 4,5-dihydroxy- (Croconic acid)					
	pK$_1$ 0.89	20	I = 1.0, mixed constant	05	Approx.	A2
	pK$_2$ 3.06		I = 0.2(KCl),mixed constant	E3bg	Approx.	
	pK$_1$ 0.32, pK$_2$ 1.51	25	I = 2.0(HCl-NaCl), concentration constants	05	Uncert.	C14
	pK$_1$ 0.6, pK$_2$ 1.8		I = 0.1(NaCl)	E3bg	Uncert.	
2534	$C_5H_4O_2$ Cyclobutenedione, 3-hydroxy-4-methyl-					
	0.24		c = 0.0003-0.0005	06	Uncert.	P15
2535	$C_5H_4O_2$ 4-Cyclopentene-1,3-dione					
	10.5	not stated	1% methanol	E3bg	Uncert.	G46

No.	Formula	Name	pK	T	Conditions			Ref.
2536	$C_5H_4O_2$	2-Penten-4-ynoic acid						
		3.75(trans)		25	c = 0.01, I = 0.1(NaCl), mixed constant	E3bg	Approx.	M48
2537	$C_5H_4O_3$	2-Furancarboxylic acid (2-Furoic acid)						
		3.16		25		E3bg	Approx.	L73
		3.20		30				
		3.22		35				
		3.24		40				
		Thermodynamic quantities are derived from the results						
		3.224		10	c ≈ 0.005	C1	Approx.	K35
		3.272		25				
		3.321		44				
		Thermodynamic quantities are derived from the results						
		2.98		25	I = 0.1(NaClO$_4$),concentration constant	E3ag	Uncert.	R45
		3.20						
2538	$C_5H_4O_3$	4H-Pyran-4-one, 3-hydroxy- (Pyromeconic acid)						
		pK$_1$ (CO) -1.28		25	In aq.H$_2$SO$_4$ solutions,H$_0$ scale	06	Uncert.	C47
		pK$_2$ (OH) 7.99		25	I = 0.50 to 0.01(NaCl)	E3bg,R4	Approx.	
		pK$_2$ 7.91		24	I = 0.13	E3bg	Approx.	T60
2539	$C_5H_4O_4$	4H-Pyran-4-one, 3,5-dihydroxy-						
		pK$_1$ 7.68		24	I = 0.12	E3bg	Approx.	T60
		pK$_2$ 10.54		24	I = 0.13			

No.	Molecular formula, name and pK value(s)	$T(^oC)$	Remarks	Method	Assessment	Ref.
2540	$C_5H_4N_4$ Purine					
	pK_2 (-NH) 8.98	20	I = 0.1	05	Uncert.	L34
	8.72	30				
	8.50	40	Thermodynamic quantities are derived from the results			
	8.35	50				
2541	$C_5H_5N_5$ 6-Purinylamine (Adenine)					
	pK_2 (-NH) 9.87	20	I = 0.005	E3bg	Approx.	L34
	9.70	30	Thermodynamic quantities are derived from the results			
	9.45	40				
	9.21	50				
	pK_1 4.22, pK_2 9.96,	20	c ≈ 0.003, I = 0.006-0.012,	E3bg	Approx.	L35
	pK_1 4.12, pK_2 9.67	30	mixed constants			
	pK_1 4.06, pK_2 9.49	40				
			Thermodynamic quantities are derived from the results			
	pK_1 3.7	20	I = 1.0($NaNO_3$)	E3bg	Uncert.	F19
2542	$C_5H_6O_2$ 2-Cyclopentenone, 2-hydroxy-					
	9.14	20	100% enol form	E3bh	Approx.	S52

No.	Formula	Name	pK	T (°C)	Conditions	C1	Rel.	Ref.
2543	$C_5H_6O_2$	Methanol, 2-furyl- (Furfuryl alcohol)	9.55			C1		B22
2544	$C_5H_6O_2$	2-Pentynoic acid	2.61	25	c = 0.01, I = 0.1(NaCl), mixed constant	E3bg	Approx.	M48
2545	$C_5H_6O_2$	3-Pentynoic acid	3.59	25	c = 0.01, I = 0.1(NaCl), mixed constant	E3bg	Approx.	M48
2546	$C_5H_6O_2$	4-Pentynoic acid	4.20	25	c = 0.01, I = 0.1(NaCl), mixed constant	E3bg	Approx.	M48
2547	$C_5H_6O_3$	Furan-2(5H)-one, 3-hydroxy-4-methyl-	7.77	20	100% enol	E3bh	Approx.	S52
2548	$C_5H_6O_3$	Furan-2(5H)-one, 4-hydroxy-3-methyl- (Tetronic acid, 3-methyl-)	4.15	25		C	Rel.	B49
			3.6		Conditions not stated		Uncert.	C45
2549	$C_5H_6O_3$	2-Pentenoic acid, 4-oxo-	3.24	25			Uncert.	S11
2550	$C_5H_6O_4$	Butanedioic acid, 2-methylene (Itaconic acid) pK_1 3.629, pK_2 4.998		25	I = 1.0($NaClO_4$), values extrapolated to c = 0, concentration constants	E3bg*,R3d.	Rel.	S31

*Reference half cell $Ag;AgCl|NaCl(c=0.01)$, $NaClO_4(c=0.99)||$, salt bridge:- 1.00M $NaClO_4$

(Contd)

No.	Molecular formula, name and pK value(s)	T(°C)	Remarks	Method	Assessment	Ref.
2550(Contd)	$C_5H_6O_4$ Butanedioic acid, 2-methylene (Itaconic acid)					
	pK$_1$ 3.61, pK$_2$ 5.08	31	I = 0.1(NaClO$_4$), mixed constants	E3bg	Approx.	R11
	pK$_1$ 3.68, pK$_2$ 5.14	25	I = 0.1(NaClO$_4$),concentration constants	E3bg	Approx.	Y12
	pK$_1$ 4.05, pK$_2$ 5.40	10		E3bg,R3d	Approx.	K61
	pK$_1$ 3.95, pK$_2$ 5.54	18	Equation derived for variation with			
	pK$_1$ 3.92, pK$_2$ 5.73	28.4	temperature			
	pK$_1$ 3.87, pK$_2$ 5.82	36.4				
2551	$C_5H_6O_4$ Butenedioic acid, methyl- (Citraconic acid)					
	pK$_1$ 2.2 (cis)	25	I = 0.1(NaClO$_4$), concentration constant	E3bg	Uncert.	Y12
	pK$_2$ 5.60 (cis)				Approx.	
	pK$_1$ 2.8, pK$_2$ 6.1 (cis)	20		E3bg	Uncert.	E1
			pK$_1$ was corrected for the acid - anhydride equilibrium			
	pK$_1$ 2.55, pK$_2$ 6.30 (cis)	30		05	Uncert.	S60
2552	$C_5H_6O_4$ 1,2-Cyclopropanedicarboxylic acid					
	pK$_1$ 3.56, pK$_2$ 6.65 (cis)	20		E3bg	Approx.	M12
	pK$_1$ 3.80, pK$_2$ 5.08 (trans)					
2553	$C_5H_6O_4$ Oxolane-3-carboxylic acid, 2-oxo- (Tetrahydrofuran-3-carboxylic acid, 2-oxo-)					
	3.75	20	c = 0.001-0.002	E3bg	Approx.	I4

No.	Formula	Name / pK	T (°C)	Method	Assessment	Ref.
2554	$C_5H_6N_6$	1H-1,2,3-Triazolo[4,5-d]pyrimidin-7(6H)-imine, 6-methyl-				
		pK$_1$ 3.25, pK$_2$ 9.12	20	05	Approx.	A21
		c = 0.0001, I = 0.01, mixed constants				
2555	C_5H_8O	Cyclopentanone				
		enol 11.8	25	KIN	Uncert.	B55
		keto 16.7				
2556	$C_5H_8O_2$	2-Butenoic acid, 3-methyl-				
		5.12	25	E3bg	Approx.	B112
		c = 0.01				
2557	$C_5H_8O_2$	Cyclopropanecarboxylic acid, 2-methyl-				
		5.02 (cis)	25	E3bg	Approx.	M12
		5.00 (trans)				
2558	$C_5H_8O_2$	2,4-Pentanedione (Acetylacetone)				
		9.10	10	E3bg	Approx.	I31
		9.02	20	Thermodynamic quantities are derived from		
		8.94	30	the results		
		8.86	40			
		9.09	15	E3bg	Approx.	L3a
		9.03	25	Thermodynamic quantities are derived from		
		8.97	35	the results		
		8.91	45			
		9.17	5	05	Approx.	C4
		c = 0.00005, 0.5-1.0% ethanol, mixed constant				

(Contd)

No.	Molecular formula, name and pK value(s)		T(°C)	Remarks	Method	Assessment	Ref.
2558 (Contd)	$C_5H_8O_2$	2,4-Pentanedione (Acetylacetone)					
		9.01	25	Thermodynamic quantities are derived			C5
		8.89	45	from the results for both enol and keto forms			C5
	pK(enol)8.24, pK(keto)8.93		25	Calculated using absorbance data	E3bg	Uncert.	C5
		8.76	20	I = 0.03	O2	Approx.	B47
		8.78		I = 0.03	O2		
		8.84	20	c = 0.00002, I = 0.1($NaClO_4$)	O5	Approx.	K49
		8.88	25	I = 1.0(KCl), concentration constant	E3bg	Approx.	S20
				Other values in E11, I32, M50, R52 and S42			
2559	$C_5H_8O_2$	2-Pentenoic acid					
		4.70 (cis)	25		E3bg	Approx.	M12
		4.74 (trans)					
2560	$C_5H_8O_3$	Butanoic acid, 3-formyl-					
		5.02			E		P68
2561	$C_5H_8O_3$	Methyl 3-oxobutanoate					
		10.72	20.8		E3bg	Uncert.	R53
2562	$C_5H_8O_3$	Pentanoic acid, 4-oxo-					
		4.6	25		E3bg	Uncert.	C57

2563 $C_5H_8O_4$ Acetic acid, 2-ethoxycarbonyl-   (Ethyl hydrogen malonate)

3.23	7	Mixed constant	E3bg	Uncert.	L33
3.25	25				
3.31	37				
3.45(H_2O)	25		E3bg	Approx.	G23
3.90(D_2O)					

2564 $C_5H_8O_4$ Butanedioic acid, methyl-

pK_1 3.88, pK_2 5.35	25	I = 0.1($NaClO_4$)	E3ag	Uncert.	R46
pK_1 4.0, pK_2 5.79					

2565 $C_5H_8O_4$ Pentanedioic acid   (Glutaric acid)

pK_1 4.35, pK_2 5.40	25	c = 0.005, I = 0.005	O3,R3d	Approx.	J19
pK_1 4.30, pK_2 5.65	20	c = 0.005	E3bg,R3a	Approx.	D43
pK_1 4.42, pK_2 5.44	25	c = 0.00125	E3bg,R3c	Approx.	B146
pK_1 4.14, pK_2 5.01	25	c $\approx$ 0.003, I = 0.1($NaClO_4$), concentration constants	E3bg,R3h	Approx.	S134
pK_1 4.09, pK_2 4.84	25	c = 0.02, I = 0.5($LiClO_4$), concentration constants	E3bg	Approx.	D33

Other values in N32 and Y12

No.	Molecular formula, name and pK value(s)	T(°C)	Remarks	Method	Assessment	Ref.
2566	C$_5$H$_8$O$_4$ Propanedioic acid, dimethyl-					
	pK$_2$ 5.44	25	I = 1.0(KCl), mixed constant	E3bg	Approx.	S18
	pK$_1$ 3.03, pK$_2$ 5.73	25		E3bg	Uncert.	L33
	pK$_1$ 3.06, pK$_2$ 5.80	37				
2567	C$_5$H$_8$O$_4$ Propanedioic acid, ethyl-					
	pK$_1$ 2.96, pK$_2$ 5.81(H$_2$O)	25		E3bg	Approx.	G23
	pK$_1$ 3.46, pK$_2$ 6.28 (D$_2$O)					
	pK$_1$ 2.99, pK$_2$ 5.83	25	c ≈ 0.01	E3bg	Uncert.	W40
2568	C$_5$H$_8$O$_5$ 3-Oxapentanedioic acid, 2-methyl-					
	pK$_1$ 3.03, pK$_2$ 4.58	25		E3ag	Approx.	S92
2569	C$_5$H$_8$O$_7$ Pentanedioic acid, 2,3,4-trihydroxy- (Xylotrihydroxyglutaric acid)					
	pK$_1$ 3.08, pK$_2$ 4.21	20	c = 0.005, concentration constants	E3bg	Uncert.	D13
	pK$_1$ 3.29, pK$_2$ 4.52	25		C		G47
	pK$_1$ 3.07, pK$_2$ 3.94	23-25	c = 0.005, I = 0.2(KCl)	E3bg,R3d	Uncert.	D15
2570	C$_5$H$_9$N$_3$ Ethylamine, 2-(4'-imidazolyl)- (Histamine)					
	pK$_1$ 6.02, pK$_2$ 9.70	25	I = 0.1(KCl), concentration constants	E3bg	Approx.	D47
	pK$_1$ 6.12, pK$_2$ 9.88	25	I = 0.12(KCl)	E3bg,R3f	Approx.	C39
	pK$_1$ 5.86, pK$_2$ 9.52	35	Thermodynamic quantities are derived			
	pK$_1$ 5.65, pK$_2$ 9.20	45	from the results			

No.	Formula	Name / pK	T (°C)	Method	Assessment	Ref.	Remarks
2571	$C_5H_{10}O_2$	Butanoic acid, 2-methyl-					
		4.761	25	E3bg	Approx.	C54	
2572	$C_5H_{10}O_2$	Pentanoic acid					
		4.763	10	E3bg	Approx.	C54	
		4.861	40				
		4.8	25	E3bg	Uncert.	C57	
2573	$C_5H_{10}O_2$	Propanoic acid, 2,2-dimethyl- (Pivalic acid)					
		5.04	25	C1	Approx.	S120	c = 0.004-0.01, mixed constant
		5.032	25	E3bg	Approx.	W29	
		5.03 (H_2O)	25	O3	Approx.	B51	I < 0.07
		5.52 (D_2O)					Based on pK = 5.28 for acetic acid in D_2O
		Propanoic acid, 2,2-dimethyl-2,3-d_9-					
		5.05	25	C1	Approx.	S120	c = 0.004-0.01, mixed constant
2574	$C_5H_{10}O_3$	Butanoic acid, 2-hydroxy-2-methyl-					
		3.73	25	E3bg	Approx.	P76	I = 0.1(KNO_3), extrapolated to c = 0, concentration constant. Variation of K_c with c given by $K_c = (1.86-2.4c) \times 10^{-4}$ up to c = 0.03 at I = 0.1
2575	$C_5H_{10}O_3$	Pentanoic acid, 2-hydroxy-					
		3.59	25	E3bg	Approx.	G35	I = 1.0($NaClO_4$), extrapolated to c = 0, concentration constant

No.	Molecular formula, name and pK value(s)	$T(^oC)$	Remarks	Method	Assessment	Ref.
2576	$C_5H_{10}O_4$ Propanoic acid, 2,2-bis(hydroxymethyl)-					
	4.39	25	$I = 0.1(NaClO_4)$	E3ag	Uncert.	R47
	4.61					
2577	$C_5H_{10}O_4$ Ribose, 2-deoxy-					
	12.98	10	c = 0.01	CAL	Approx.	C53
	12.61	25	Thermodynamic quantities are also given	CAL	Approx.	I34
	12.67	25	Thermodynamic quantities are also given			
2578	$C_5H_{10}O_5$ Arabinose					
	12.79	10	c = 0.01	CAL	Approx.	C53
	12.34	25	Thermodynamic quantities are also given	CAL	Approx.	I34
	12.54	25	Thermodynamic quantities are also given			

No.	Formula	Name	pK	Conditions	T (°C)	Method	Reliability	Ref.
2579	$C_5H_{10}O_5$	Lyxose	12.11	Thermodynamic quantities are given	25	CAL	Approx.	T34
			12.48	c = 0.01	10	CAL	Approx.	C53
			12.11	Thermodynamic quantities are also given	25			
2580	$C_5H_{10}O_5$	Ribose	12.22	Thermodynamic quantities are also given	25	CAL	Approx.	I34
			12.54	c = 0.01	10	CAL	Approx.	C53
			12.11	Thermodynamic quantities are also given	25			
2581	$C_5H_{10}O_5$	Xylose	12.29	Thermodynamic quantities are also given	25	CAL	Approx.	I34
			12.61	c = 0.01	10	CAL	Approx.	C53
			12.15	Thermodynamic quantities are also given	25			
2582	$C_5H_{12}O_4$	1,3-Propanediol, 2,2-bis(hydroxymethyl)- (Pentaerythritol)	pK_1 14.1			C3		L63
2583	$C_5H_{12}S$	1-Propanethiol, 1,1-dimethyl-	11.3	1% ethanol	25	O5	Uncert.	K57
2584	$C_5H_3ON_5$	Pyrimidino[5,4-e]-1,2,4-triazin-5(1H)-one	6.59	c < 0.0001, I = 0.01, mixed constant	20	O5	Approx.	B141

No.	Molecular formula, name and pK value(s)	$T(^{o}C)$	Remarks	Method	Assessment	Ref
2585	$C_5H_3O_3N_5$ 1,2,4-Triazolo[4,3-c]pyrimidin-4(1H)-one, 7-nitro-					
	4.11	20	c < 0.001, I = 0.01, mixed constant	05	Approx.	B141
2586	$C_5H_3O_6N_3$ 4-Pyrimidinecarboxylic acid, 1,2,3,6-tetrahydro-5-nitro-2,6-dioxo-					
	pK_1 < 1.5	25	c = 0.002, I = 0.1[$(CH_3)_4$NBr]	E3bg	Uncert.	T63
	pK_2 4.94				Approx.	
2587	$C_5H_4ON_4$ Purin-6(1H)-one (Hypoxanthine)					
	pK_1 1.79	25		CAL	Approx.	C52
	pK_2 8.91	25	Thermodynamic quantities are also given	E3bg	Approx	
	pK_3 12.64	10		CAL		
	pK_3 12.07	25				
	pK_3 11.81	40				
	pK_1 1.9, pK_2 8.8, pK_3 12.0	25	Thermodynamic quantities are also given	E3bg	Uncert.	W45
	pK_2 8.6, $pK_3 \approx$ 12.5		Conditions not stated	05	Uncert.	B66
2588	$C_5H_4ON_4$ Purin-8(7H)-one					
	pK_1 7.9, pK_2 13.5		Conditions not stated	05	Uncert.	L42
2589	$C_5H_4ON_4$ 1H-Pyrazolo[3,4-d]pyrimidin-4(2H)-one					
	9.38	20	c = 0.0001 in 0.01 M buffer	05	Approx.	L78
2590	$C_5H_4O_2N_2$ Methyl 2,2-dicyanoacetate					
	-2.8	25	Thermodynamic quantities are also given	06	Uncert.	B118

No.	Compound	pK	Temp (°C)	Notes	Method	Reliability	Ref.
2591	$C_5H_4O_2N_4$ Purine-2,6(1H,3H)-dione (Xanthine)						
		pK$_1$ 7.53	25		E3bg	Approx.	C52
		pK$_2$ 12.36	10	Thermodynamic quantities are also given	CAL	Approx.	
		11.84	25				
		11.51	40				
		pK$_1$ 7.44	20	c = 0.001	E3bg	Approx.	A32
		pK$_1$ 7.70, pK$_2$ 11.94	20		E3bg/05	Uncert.	P55
		pK$_1$ 7.5, pK$_2$ 11.0		Conditions not stated	05	Uncert.	L39
2592	$C_5H_4O_2N_4$ Purine-2,8(1H,7H)-dione						
		7.45	20		05	Uncert.	A23
2593	$C_5H_4O_2N_4$ Purine-6,8(1H,7H)-dione						
		pK$_1$ 7.65, pK$_2$ 9.87	20	c = 0.002	E3bg	Uncert.	A23
2594	$C_5H_4O_2S$ 2-Thiophenecarboxylic acid						
		3.49	25	c = 0.01	E3bg	Approx.	O28
		3.53	30		E3bg	Approx.	L73
		3.55	35	Thermodynamic quantities are derived from the results			
		3.56	40				
		3.54	25		E3ag	Uncert.	R45
		3.53	25	c = 0.005, mixed constant.Value in mixed solvent also given	E3bg	Approx.	B188
		3.49	25			Uncert.	I3

No.	Molecular formula, name and pK value(s)	$T(^{o}C)$	Remarks	Method	Assessment	Ref.
2595	$C_5H_4O_2Se$ Selenofuran-2-carboxylic acid					S99
	pK 3.60	25				
2596	$C_5H_4O_3N_2$ 4-Pyrimidinecarboxylic acid, 1,2-dihydro-2-oxo-			E3bg	Uncert.	C58
	pK₁ 2.8, pK₂ 8.4	not stated	c = 0.01			
2597	$C_5H_4O_3N_4$ Purine-2,6,8(1H,3H,7H)-trione (Uric acid)					
	pK₁ 5.4, pK₂ 10.6	not stated		05	Uncert.	J13
	pK₁ 5.75, pK₂ 10.3	not stated		05	Uncert.	B65
2598	$C_5H_4O_4N_2$ 4,5-Imidazoledicarboxylic acid			0		S13
	pK₁ 2.93, pK₂ 8.04					
2599	$C_5H_4O_4N_2$ 4-Pyrimidinecarboxylic acid, 1,2,3,6-tetrahydro-2,6-dioxo- (Orotic acid)					
	pK₁ 2.07	25	c = 0.002, I = 0.1(KCl), mixed constant	E3bg	Uncert.	T63
	pK₂ 9.45				Approx.	
	pK₁ 1.8, pK₂ 9.55	25	Thermodynamic quantities are also given	E3bg	Uncert.	W45
	pK₁ 2.4, pK₂ 9.45	not stated	c = 0.01	E3bg	Uncert.	C58
2600	$C_5H_4O_4N_2$ 5-Pyrimidinecarboxylic acid, 1,2,3,4-tetrahydro-2,4-dioxo-					
	pK₁ 4.16, pK₂ 8.89	25	c = 0.002, I = 0.1(KCl), mixed constants	E3bg	Approx.	T63
2601	$C_5H_4O_4N_2$ 2-Pyrrolecarboxylic acid, 4-nitro-					
	pK 3.37	25	c = 0.01	E3bg	Approx.	F43

No.	Formula	Name / pK	Temp.	Conditions	Method	Assessment	Ref.
2602	$C_5H_4O_4N_2$	2-Pyrrolecarboxylic acid, 5-nitro-					
		3.22	25	c = 0.01	E3bg	Approx.	F43
2603	$C_5H_4N_4S$	Purine-6(1H)-thione					
		pK_1 7.77, pK_2 10.84	20	c = 0.0015	E3bg	Approx.	A23
		pK_1 7.5		Conditions not stated	05	Uncert.	B66
2604	$C_5H_4N_4S$	Purine-8(1H)-thione					
		pK_1 6.64	20	c = 0.0025	E3bg	Approx.	A23
		pK_2 11.16				Uncert.	
2605	$C_5H_4N_4S$	1H-Pyrazolo[3,4-d]pyrimidine-4(2H)-thione					
		pK_1 -0.66	20	c < 0.001, I = 0.01, mixed constant	06	Uncert.	B134
		pK_2 8.34	20		05	Uncert.	
		pK_3 11.65	20		05	Approx.	
2606	$C_5H_4N_4Se$	Purine-6(1H)-selone					
		7.33		c = 0.005	E3bg	Uncert.	M67
2607	C_5H_5ON	3-Pyridinol					
		5.37	25	I = 0.1(KCl), mixed constant	E3bg	Approx.	E20
2608	$C_5H_5ON_5$	1,2,4-Triazolo[4,3-b]-1,2,4-triazin-7(1H)-one, 6-methyl-					
		5.40	25	c = 0.01	E3bg	Uncert.	K9

No.	Molecular formula, name and pK value(s)	T(^{o}C)	Remarks	Method	Assessment	Ref.
2609	$C_5H_5O_2N$ 2-Furancarbaldehyde oxime					
	11.16 (trans)	25	c = 0.00005, I = 0.1	O5	Approx.	T6
2610	$C_5H_5O_2N$ Pyridin-2(1H)-one, 1-hydroxy-					
	pK_1 -0.9	25	In aq. H_2SO_4, H_0 scale	O6	Uncert.	S127
	pK_2 5.98	25	I = 0.005, concentration constant	E3bg	Approx.	S63
	pK_2 5.9		Condition not stated		Uncert.	
2611	$C_5H_5O_2N$ Pyridin-2(1H)-one, 3-hydroxy-					
	pK_2 8.62	25	I = 0.1($NaClO_4$)	E3bg	Approx.	G27
2612	$C_5H_5O_2N$ Pyridin-2(1H)-one, 5-hydroxy-					
	8.51	20	c = 0.005	E3bg	Approx.	A18
2613	$C_5H_5O_2N$ 2-Pyrrolecarboxylic acid					
	4.39	20	c = 0.01, mixed constant	E3bg	Approx.	K26
	4.43			E3bg	Approx.	
	4.50	25	c = 0.01	E3bg	Approx.	F43
	4.45	25		E3bg	Approx.	L73
	4.40			E		S55

No.	Formula	Name / pK	pK	T (°C)	Conditions	Method	Assessment	Ref.
2614	$C_5H_5O_2N$	3-Pyrrolecarboxylic acid	5.00	20	c = 0.01, mixed constant	E3bg	Approx.	K26
			5.07		Conditions not stated		Uncert.	R16
			4.95			E		S55
2615	$C_5H_5O_2N_3$	2-Pyrazinecarbaldehyde, 3-amino-4,5-dihydro-5-oxo-	6.61	20	c = 0.005	E3bg	Approx.	B136
2616	$C_5H_5O_2N_3$	2-Pyrazinecarbohydroxamic acid	8.1	not stated	c = 0.04, I = 0.1(KCl)	E3bg	Uncert.	H2
2617	$C_5H_5O_2N_3$	2-Pyrimidinecarbohydroxamic acid	7.88	25	c = 0.01, I = 0.1(KCl), mixed constant	E3bg	Approx.	G38
2618	$C_5H_5O_2F_3$	2,4-Pentanedione, 1,1,1-trifluoro-	6.3	room temp.		E3bg	Uncert.	R19
2619	$C_5H_5O_3N$	Pyridin-2(1$\underline{H}$)-one, 4,6-dihydroxy- pK_1 4.6, pK_2 9.0, pK_3 >13		20	c = 0.02	E3bg	Uncert.	A36
2620	$C_5H_5O_4N_3$	4-Pyrimidinecarboxylic acid, 5-amino-1,2,3,6-tetrahydro-2,6-dioxo- (Orotic acid, 5-amino-) pK_1 2.63, pK_2 8.72		25	I = 0.1 [(CH$_3$)$_4$NBr], mixed constant	E3bg	Approx.	T64
2621	$C_5H_5O_4N_3$	Pyrimidine-2,4(1$\underline{H}$,3$\underline{H}$)-dione, 1-methyl-5-nitro-	7.35	25	Titrant N(Me)$_4$OH	E3bg	Approx.	J16
			7.20	20		E3bg	Approx.	B133

No.	Molecular formula, name and pK value(s)	$T(^{o}C)$	Remarks	Method	Assessment	Ref.
2622	$C_5H_5O_4N_3$ Pyrimidine-2,4(1H,3H)-dione, 3-methyl-5-nitro-					
	5.70	25	Titrant N(Me)$_4$OH	E3bg	Approx.	J16
2623	$C_5H_6ON_6$ 1H-1,2,3-Triazolo[4,5-d]pyrimidin-5(4H)-one, 7-amino-1-methyl-					
	pK$_1$ 2.94	20	c = 0.000025	05	Approx.	A40
	pK$_2$ 9.04	20	c = 0.00012			
2624	$C_5H_6ON_6$ 1H-1,2,3-Triazolo[4,5-d]pyrimidin-7(4H)-one, 5-amino-1-methyl-					
	pK$_1$ 1.58	20	c = 0.000033	05	Approx.	A40
	pK$_2$ 8.25	20	c = 0.000047			
2625	$C_5H_6ON_6$ 2H-1,2,3-Triazolo[4,5-d]pyrimidin-5(4H)-one, 7-amino-2-methyl-					
	pK$_1$ 3.94	20	c = 0.000025	05	Approx.	A40
	pK$_2$ 9.98	20	c = 0.00012			
2626	$C_5H_6ON_6$ 2H-1,2,3-Triazolo[4,5-d]pyrimidin-7(6H)-one, 5-amino-2-methyl-					
	pK$_1$ 1.86	20	c = 0.000024	05	Approx.	A40
	pK$_2$ 8.64	20	c = 0.00003			
2627	$C_5H_6O_2N_2$ Acetic acid, 2-(4-imidazolyl)-					
	pK$_2$ 7.22	25	I = 0.16(KCl)	E3bg	Approx.	K41
2628	$C_5H_6O_2N_2$ Acetic acid, 2-(1-pyrazolyl)-					
	3.31	room temp.	c = 0.01	E3bg	Uncert.	H1

No.	Formula	Name	pK	T (°C)	Conditions	Method	Assessment	Ref.
2629	$C_5H_6O_2N_2$	5-Pyrazolecarboxylic acid, 1-methyl-	3.27	room temp.	c = 0.01, 5% methanol	E3bg	Uncert.	H1
2630	$C_5H_6O_2N_2$	5-Pyrazolecarboxylic acid, 3-methyl-	3.79	room temp.	C = 0.01	E3bg	Uncert.	H1
2631	$C_5H_6O_2N_2$	Pyrimidine-2,4($1\underline{H},3\underline{H}$)-dione, 1-methyl-	9.72	25	c ≈ 0.001	E3bg	Approx.	G62
			9.77	25		05	Approx.	N8
			9.43	25	Titrant $(CH_3)_4NOH$	E3bg	Approx.	J16
2632	$C_5H_6O_2N_2$	Pyrimidine-2,4($1\underline{H},3\underline{H}$)-dione, 3-methyl-	9.35	25	c ≈ 0.001	E3bg	Approx.	G62
			10.00	25		05	Approx.	N8
			9.85	25	Titrant $(CH_3)_4NOH$	E3bg	Approx.	J16
2633	$C_5H_6O_2N_2$	Pyrimidine-2,4($1\underline{H},3\underline{H}$)-dione, 5-methyl- (Thymine)	9.82		Conditions not stated	05	Uncert.	G33
			9.90	25	Titrant $(CH_3)_4NOH$	E3bg	Approx.	J16
			8.89	20	I = 0.005	E3bg	Approx.	L34
			9.10	30	Thermodynamic quantities are derived			
			9.55	40	from the results			
			9.34	50				

No.	Molecular formula, name and pK value(s)	T(°C)	Remarks	Method	Assessment	Ref.
2634	$C_5H_6O_2N_2$ Pyrimidine-2,4(1H,3H)-dione, 6-methyl- 9.68	25	Titrant $(CH_3)_4NOH$	E3bg	Approx.	J16
2635	$C_5H_6O_3N_2$ Cyclopentanone, 2,5-bis(hydroxyimino)- 7.95	30	c = 0.01, I = 0.1(KCl), mixed constant	E3bg	Approx.	E18
2636	$C_5H_6O_3N_2$ Pyrimidine-2,4(1H,3H)-dione, 5-hydroxymethyl- 9.40		Conditions not stated	05	Uncert.	F21
2637	$C_5H_6O_3N_2$ Pyrimidine-2,4(1H,3H)-dione, 6-hydroxy-1-methyl- pK_1 4.2, pK_2 12.8		Conditions not stated	05	Uncert.	F32
2638	$C_5H_6O_3N_2$ Pyrimidine-2,4(1H,3H)-dione, 6-hydroxy-5-methyl- 4.40	25	Thermodynamic quantities are also given	E3bg	Approx.	M80
2639	$C_5H_6O_3N_4$ 4-Imidazolecarboxylic acid, 5-ureido- pK_1 2.0, pK_2 4.9, pK_3 12.2		Conditions not stated	05	Uncert.	R1
2640	$C_5H_6O_4N_2$ Acetic acid, 2-amino-2-(2,3-dihydro-3-oxo-5-isoxazolyl)- (Ibotenic acid) pK_1 2, pK_2 5.1, pK_3 8.2		Conditions not stated		Uncert.	T2
2641	$C_5H_6N_2S_2$ Pyrimidine-2,4(1H,3H)-dithione, 1-methyl- (Dithiouracil, 1-methyl-) 7.47	20	c = 0.0005	05	Approx.	B130

No.	Formula	Name	pK	Temp.	Method	Assessment	Ref.
2642	$C_5H_7ON_3$	Pyrimidin-2(1H)-one, 4-methylamino- (Cytosine, N-methyl-)	pK_1 4.40, pK_2 12.7	22	05	Uncert.	S141
2643	$C_5H_7O_2N$	Isoxazol-3(2H)-one, 4,5-dimethyl-	6.1	20	E3bg	Uncert.	B110
2644	$C_5H_7O_2N$	Propanoic acid, 2-cyano-2-methyl-	2.422	25	C2	Rel.	I26

2644 Variation with temperature

	5	10	15	20	30	35	40	45
	2.342	2.360	2.379	2.400	2.446	2.471	2.498	2.525

Thermodynamic quantities are derived from the results

No.	Formula	Name	pK	Conditions	Method	Assessment	Ref.
2645	$C_5H_7O_2N$	Pyrrolidine-2,4-dione, 1-methyl-	7.05	Conditions not stated	05	Uncert.	M109
2646	$C_5H_7O_2N$	Pyrrolidine-2,4-dione, 5-methyl-	6.42	Conditions not stated	05	Uncert.	M109
2647	$C_5H_7O_2N_3$	Pyrimidin-2(1H)-one, 4-amino-5-hydroxymethyl- (Cytosine, 5-hydroxymethyl-)	pK_1 4.32, pK_2 13	Conditions not stated	05	Uncert.	F21
2648	$C_5H_7O_2N_3$	Pyrimidin-2(1H)-one, 4-hydroxyamino-1-methyl-	pK_1 2.9, pK_2 10.4	Conditions not stated	05	Uncert.	J7

134

No.	Molecular formula, name and pK value(s)	$T(^oC)$	Remarks	Method	Assessment	Ref.
2649	$C_5H_7O_2N_3$ Pyrimidin-2(1H)-one, 4-hydroxyamino-5-methyl- pK$_1$ 2.8	Conditions not stated		05	Uncert.	J7
2650	$C_5H_7O_2N_3$ 1,2,4-Triazine-3,5(2H,4H)-dione, 2,6-dimethyl- 7.49	25	Titrant $(CH_3)_4$NOH	E3bg	Approx.	J16
2651	$C_5H_7O_2N_3$ 1,2,4-Triazine-3,5(2H,4H)-dione, 4,6-dimethyl- 10.1	25	Titrant $(CH_3)_4$NOH	E3bg	Uncert.	J16
2652	$C_5H_7O_2N_3$ 1,2,4-Triazine-3,5(2H,4H)-dione, 2-ethyl- 7.07	25	Titrant $(CH_3)_4$NOH	E3bg	Approx.	J16
2653	$C_5H_7O_2N_3$ 1,2,4-Triazine-3,5(2H,4H)-dione, 4-ethyl- 9.59	25	Titrant $(CH_3)_4$NOH	E3bg	Approx.	J16
2654	$C_5H_7O_2N_3$ 1,2,4-Triazine-3,5(2H,4H)-dione, 6-ethyl- 7.47	not stated	c = 0.005	E3bg	Uncert.	C28
2655	$C_5H_7O_2N_5$ Pyrimidin-2(1H)-one, 4-semicarbazido- 3.20	Conditions not stated		05	Uncert.	J8
2656	$C_5H_7O_3N$ Oxazolidine-2,4-dione, 5,5-dimethyl- 6.2	Conditions not stated		05	Uncert.	B189

No.	Formula	Name	pK	T (°C)	Conditions	Ref	Reliability	Method
2657	$C_5H_7O_3N$	2,4-Pentanedione, 3-hydroxyimino-	7.50	30	$c = 0.01$, $I = 0.1(KCl)$, mixed constant	E3bg	Approx.	E18
			7.38	25	$c \approx 0.01$, $I = 0.1(KCl)$, mixed constant	E3bg	Approx.	G39
2658	$C_5H_7O_3N$	2,6-Piperidinedione, 1-hydroxy-	7.6		Conditions not stated	E3bg	Uncert.	A50
2659	$C_5H_7O_3N_3$	1-Imidazolecarboxamide, 2,4-dihydroxy-5-methyl-	7.0	not stated	$I \approx 0.1$	05	Uncert.	S110
2660	$C_5H_7O_3N_3$	Pyrimidin-2(1H)-one, 4-hydroxyamino-5-hydroxymethyl-	2.2		Conditions not stated	05	Uncert.	J7
2661	$C_5H_7O_4N_3$	Butanenitrile, 2-methyl-4,4-dinitro-	3.27	20		05	Approx.	T58a
2662	$C_5H_7O_4N_3$	1-Imidazolecarboxamide, 2,4-dihydroxy-5-hydroxymethyl-	6.9	not stated	$I \approx 0.1$	05	Uncert.	S110
2663	$C_5H_7O_4N_3$	Pentanenitrile, 5,5-dinitro-	4.34	20		05	Approx.	T58a
2664	$C_5H_8ON_4$	1,2,4-Triazin-5(2H)-one, 3-amino-2,6-dimethyl-	pK_1 1.90	25		05	Approx.	P70
2665	$C_5H_8ON_4$	1,2,4-Triazin-5(2H)-one, 6-methyl-3-methylamino-	pK_1 2.07	25		05	Approx.	P70
2666	$C_5H_8ON_4$	1,2,4-Triazin-5(4H)-one, 3-amino-4,6-dimethyl-	pK_1 4.52	25		05	Approx.	P70

136

No.	Molecular formula, name and pK value(s)	T(°C)	Remarks	Method	Assessment	Ref.
2667	$C_5H_8O_2N_2$ 1,2-Cyclopentanedione dioxime 10.1	20	I = 0.2(NaNO₃)	E		M46
2668	$C_5H_8O_2N_2$ Propanoic acid, 3-(3-methyl-3H-diazirin-3-yl)- 4.4	25	[Pentanoic acid, 4,4-(NN')azo-]	E3bg	Uncert.	C57
2669	$C_5H_8O_2N_2$ Pyrimidine-2,4(1H,3H)-dione, 5,6-dihydro-1-methyl- ≈ 11.9		Conditions not stated	05	Uncert.	J6
2670	$C_5H_8O_2N_2$ Pyrimidine-2,4(1H,3H)-dione, 5,6-dihydro-5-methyl- 11.87			0		P83
2671	$C_5H_8O_2N_2$ Pyrimidine-2,4(1H,3H)-dione, 5,6-dihydro-6-methyl- 11.7 11.60		Conditions not stated	05 0	Uncert.	J6 P83
2672	$C_5H_8O_2N_6$ 1H-1,2,3-Triazolo[4,5-d]pyrimidin-5(4H)-one, 6,7-dihydro-7-methoxyamino- pK₁ 1.45, pK₂ 6.48	20		05	Approx.	A35
2673	$C_5H_8O_2S$ Acetic acid, allylthio- 3.683	25	c ≈ 0.001, I = 0.10(KNO₃),concentration constant	E3bg	Approx.	P40
2674	$C_5H_8O_2S_3$ 3,5-Dithiaheptanoic acid, 4-thioxo- 2.91 2.97	25	I = 0.1(NaClO₄), concentration constant	E3bg 02	Approx. Approx.	J9
2675	$C_5H_8O_3S_2$ 5-Oxa-3-thiaheptanoic acid, 4-thioxo- 3.09 3.07	25	I = 0.1(NaClO₄), concentration constant	E3bg 02	Approx. Approx.	J9

No.	Compound	pK	T (°C)	Conditions	Method	Assessment	Ref.
2676	$C_5H_8O_4N_2$ 2-Pentene, 5,5-dinitro-						
		5.26	20		05	Approx.	T58a
2677	$C_5H_8O_4S$ 3-Thiahexanedioic acid						
		pK_1 3.64, pK_2 4.70	20	$I = 0.1(NaClO_4)$, mixed constants	E3bg	Uncert.	C8
2678	$C_5H_8O_4S$ 3-Thiapentanedioic acid, 2-methyl-						
		pK_1 3.38, pK_2 4.45	25±0.5	c < 0.001	E3bg	Approx.	S93
				Microconstants are calculated from data on the two corresponding monoesters			
2679	$C_5H_8O_4S_2$ 3,5-Dithiaheptanedioic acid						
		pK_1 3.37, pK_2 4.19	20	$I = 0.1(NaClO_4)$, mixed constants	E3bg	Uncert.	O21
		pK_1 3.27		$I = 0.3$			C11
		pK_1 3.30, pK_2 4.08		$I = 0.05$			
		pK_2 4.04		$I = 0.1$			
2680	$C_5H_8O_5N_2$ 2-Pentanone, 5,5-dinitro-						
		4.86(C-H)	5	c ≈ 0.00005	05	Approx.	S79
		4.71	20				
		4.65	25				
		4.55	40				
		4.40	60	Thermodynamic quantities are derived from the results			
		4.73	20		05	Approx.	T58a

No.	Molecular formula, name and pK value(s)	T(°C)	Remarks	Method	Assessment	Ref.
2681	$C_5H_8O_6N_2$ 3,3-Dinitropropyl acetate 4.19	20		05	Approx.	T58a
2682	$C_5H_8O_6N_2$ Methyl 4,4-dinitrobutanoate 4.43 4.34	20 25	 I = 0.05(phthalate buffers), mixed constant	05 05	Approx. Approx.	T58a S77
2683	$C_5H_8O_6N_2$ Pentanoic acid, 4,4-dinitro- 3.98	25	c = 0.01	E3bg	Uncert.	F5
2684	$C_5H_8O_6N_2$ 2-Pentanone, 3-hydroxy-5,5-dinitro- 3.98	20		05	Approx.	T58a
2685	$C_5H_8O_4N_2$ Acetic acid, 2-amino-2-(2-oxo-5-isoxolidinyl)- pK_1 2, pK_2 6.0, pK_3 8.6,		Conditions not stated		Uncert.	T1a
2686	$C_5H_8O_8N_4$ Pentane, 1,1,3,3-tetranitro- 1.47	20		05	Approx.	T58a
2687	$C_5H_8O_8N_4$ Pentane, 1,1,4,4-tetranitro- 3.72	20		05	Approx.	T58a
2688	$C_5H_8O_9N_4$ Butane, 1-methoxy-2,2,4,4-tetranitro- 1.45	20		05	Approx.	T58a

No.	Formula	Name	pK	T (°C)	Conditions	Method	Reliability	Ref.
2639	$C_5H_8N_2S_2$	Imidazolidine-2,4-dithione, 5,5-dimethyl-	7.6	18-21	Mixed constant	05	Uncert.	E3
2690	$C_5H_9O_2N$	2-Pentanone, 3-hydroxyimino-	9.38	25	$c \approx 0.01$, $I = 0.1(KCl)$, mixed constant	E3bg	Approx.	G39
2691	$C_5H_9O_2N$	2-Pyrrolidinecarboxylic acid (Proline)	pK_1 1.80	20	$C = 0.002$, $I = 0.1(KCl)$	E3bg	Uncert.	G16
			pK_2 10.63				Approx.	
			pK_2 10.46	17	$c = 0.02$	E3bg	Approx.	P28
			pK_2 10.83	20	$c = 0.025$, mixed constant	E3d	Uncert.	H17
2692	$C_5H_9O_2Cl$	Pentanoic acid, 4-chloro-	4.4	25		E3bg	Uncert.	C57
2693	$C_5H_9O_3N$	Pentanoic acid, 5-amino-4-oxo-	pK_1 4.05, pK_2 8.90	22		E3bg	Uncert.	L7
2694	$C_5H_9O_3N$	2-Pyrrolidinecarboxylic acid, 4-hydroxy- (L-Hydroxyproline)	pK_2 9.70	17	$c = 0.02$, mixed constant	E3bg	Approx.	P28
2695	$C_5H_9O_4N$	3-Azapentanedioic acid, 3-methyl- (Iminodiacetic acid, N-methyl-)	pK_1 2.146, pK_2 10.088	20		E1	Rel.	O3

(Contd)

No.	Molecular formula, name and pK value(s)	T(°C)	Remarks	Method	Assessment	Ref.
2695 (Contd)	$C_5H_9O_4N$ 3-Azapentanedioic acid, 3-methyl- (Iminodiacetic acid, N-methyl-)					
	Variation with temperature					
			0	10	30	40				
	pK_1 2.138 2.142 2.150 2.154					
	pK_2 10.474 10.287 9.920 9.763					
	Thermodynamic quantities are derived from the results					
	pK_1 2.12, pK_2 9.65	20	I = 0.1(KCl),concentration constant	E3bg	Approx.	S40
	pK_2 9.57	25	I = 0.1(KCl),concentration constant	E3bg	Approx.	N11
	Other values in V10, S36 and S45					
2696	$C_5H_9O_4N$ 3-Oxa-5-azahexanoic acid, 5-methyl-4-oxo-					
	3.01	25	I = 0.1($NaClO_4$),concentration constant	E3bg	Approx.	J9
2697	$C_5H_9O_4N$ Pentanedioic acid, 2-amino- (Glutamic acid)					
	pK_2 4.23, pK_3 9.46	30	I = 0.1(KCl), concentration constants	E3bg	Approx.	N41
	pK_1 2.06, pK_2 4.26	20	I = 0.2	E3bg	Approx.	F7
	pK_2 4.35, pK_3 9.85	20	c = 0.025, mixed constants	E3d	Uncert.	H17
2698	$C_5H_9O_4N$ Pentanoic acid, 5-hydroxyimino-5-oxo- (Glutaramic acid, N-hydroxy-)					
	pK_1 4.6, pK_2 9.6		Conditions not stated	E3bg	Uncert.	A50
2699	$C_5H_9O_5N_3$ Acetamide, N-(3,3-dinitropropyl)-					
	4.28	20		O5	Approx.	T58a

No.	Formula	Name / Value		T	Conditions		Method	Ref.
2700	$C_5H_9O_5N_3$	Butanamide, N-methyl-4,4-dinitro-						
		4.30 (C-H)		25	$c \approx 0.00005$		Approx.	S79

Variation with temperature

	5	20	40	60
	4.41	4.33	4.23	4.15

Thermodynamic quantities are derived from the results

No.	Formula	Name / Value	T		Method	Ref.
		4.41	20	05	Approx.	T58a
2701	$C_5H_9O_5N_3$	Butanamide, 2-methyl-4,4-dinitro-				
		5.24	20	05	Approx.	T58a
2702	$C_5H_9O_5N_3$	Pentanamide, 5,5-dinitro-				
		4.75	20	05	Approx.	T58a
2703	$C_5H_9O_5N_3$	2-Pentanone oxime, 5,5-dinitro-				
		4.46	20	05	Approx.	T58a
2704	$C_5H_9O_5N_3$	3-Azapentanedioic acid, 3-ureido- (Iminodiacetic acid, N-ureido-)				
		pK_1 2.96, pK_2 4.04	30	$c \approx 0.002$, $I = 0.1$(KCl), concentration constants	E3bg Approx.	G26
2705	$C_5H_9O_6N_3$	Methyl carbamate, N-(3,3-dinitropropyl)-				
		4.23	20	05	Approx.	T58a
2706	$C_5H_9O_6N_3$	Pentane, 1,1,3-trinitro-				
		2.96	20	05	Approx.	T58a

No.	Molecular formula, name and pK value(s)	T(°C)	Remarks	Method	Assessment	Ref.
2707	$C_5H_9NS_2$ Pyrrolidine-1-carbodithioic acid					
	3.29	25	Decomposition observed during titration.	E3bg	Uncert.	L2
2708	$C_5H_9NS_3$ 1-Thia-4-azacyclohexane-4-carbodithioic acid (Thiamorpholine-4-carbodithioic acid)					D25
	7.11	15	From pH and conductivity measurements on			
	6.33	25	Na-salt			
	7.74	40				
2709	$C_5H_{10}OS_2$ Methanedithioic acid, butoxy- (Xanthic acid, butyl-)					
	3.03	0	Extrapolated to zero time	E3d		G5
	3.28					H27
	1.72		Conditions not stated	05	Uncert.	Z2
2710	$C_5H_{10}OS_2$ Methanedithioic acid, sec-butoxy- (Xanthic acid, sec-butyl-)					
	1.82		Conditions not stated	05	Uncert.	Z2
2711	$C_5H_{10}OS_2$ Methanedithioic acid, isobutoxy- (Xanthic acid, isobutyl-)					
	1.66		Conditions not stated	05	Uncert.	Z2
2712	$C_5H_{10}O_2S$ Ethyl 3-mercaptopropanoate					
	9.49	25	1% ethanol	E3bg	Uncert.	K57

No.	Formula	Name / pK values	Temp (°C)	Conditions	Method	Assessment	Reference
2713	$C_5H_{10}O_2S$	Acetic acid, propylthio- 3.729	25	$c \approx 0.001$, $I = 0.10(KNO_3)$, concentration constant	E3bg	Approx.	P40
2714	$C_5H_{10}O_3N_2$	3-Azapentanoic acid, 5-amino-2-methyl-4-oxo- (Glycyl-DL-alanine) pK_1 3.21, pK_2 8.21	25	$I = 0.1(NaClO_4)$	E3bg	Approx.	S139
		pK_1 3.22, pK_2 8.24	25	$I = 0.16$	E3bg	Approx.	B152
2715	$C_5H_{10}O_3N_2$	3-Azapentanoic acid, 5-amino-5-methyl-4-oxo- (L-Alanylglycine) pK_1 3.23, pK_2 8.15	25	$I = 0.16$	E3bg	Approx.	B152
2716	$C_5H_{10}O_3N_2$	Pentanoic acid, 2,5-diamino-5-oxo- (Glutamine) pK_1 2.17, pK_2 9.13		Conditions not stated		Uncert.	T1a
2717	$C_5H_{10}O_3N_2$	Pentanoic acid, 4,5-diamino-5-oxo- pK_1 3.81, pK_2 7.88		Conditions not stated		Uncert.	T1a
2718	$C_5H_{10}O_4N_2$	Butane, 3-methyl-1,1-dinitro- 5.40	20		05	Approx.	T58a
		5.40	not stated		E3bg	Uncert.	N36
2719	$C_5H_{10}O_4N_2$	Pentane, 1,1-dinitro- 5.45	20	$c = 0.005$-0.015, mixed constant	05	Approx.	T58a
		5.54	5	Thermodynamic quantities are derived from the results	05	Approx.	N38
		5.39	20				
		5.20	40				
		5.04	60				

No.	Molecular formula, name and pK value(s)	T(°C)	Remarks	Method	Assessment	Ref.
2720	$C_5H_{10}O_4S_2$ 1,3-Dithiacycloheptane-1,1,3,3-tetroxide					
	11.75	25		Uncert.		C76
2721	$C_5H_{10}O_5S$ Pentanoic acid, 5-sulfo-					
	pK_2 5.04 (-COOH)	25	I = 0.007-0.04, mixed constant	E3bg	Approx.	B58
2722	$C_4H_{10}N_2S_2$ 1-Piperazinecarbodithioic acid					
	7.97	15	From pH and conductivity measurements			P86
	7.63	25	on Na-salt			
	7.85	40				
2723	$C_5H_{11}ON$ 3-Pentanone oxime					
	12.60	25		KIN	Uncert.	K28
2724	$C_5H_{11}O_2N$ Butanoic acid, 2-amino-3-methyl-	(Valine)				
	pK_1 2.24	20	c = 0.002, I = 0.1(KCl)	E3bg	Uncert.	G16
	pK_2 9.65				Approx.	
	pK_1 2.35, pK_2 9.62	25		ROT	Approx.	R12
	pK_2 9.59	19.5	c = 0.02, mixed constant	E3bg	Approx.	P28
	pK_2 9.44	25	I = 0.16(KNO_3), mixed constant	E3bg	Approx.	M72
			Thermodynamic quantities are also given			
2725	$C_5H_{11}O_2N$ Butanoic acid, 4-amino-2-methyl-					
	pK_1 4.17, pK_2 10.45	not stated	c = 0.005	E3bg	Uncert.	S87

No.	Formula	Name / pK	T (°C)	Method	Assessment	Remarks	Ref.
2726	$C_5H_{11}O_2N$	Pentanohydroxamic acid					
		9.37	30.5	E3d	Uncert.	c = 0.001, I = 0.1(KNO_3), mixed constant	S137
2727	$C_5H_{11}O_2N$	Pentanoic acid, 2-amino- (Norvaline)					
		pK_1 9.87	20	E3bg	Approx.	c = 0.001, mixed constant	P28
2728	$C_5H_{11}O_2N$	Pentanoic acid, 5-amino- (Valeric acid, 5-amino-)					
		pK_1 4.26	10	E3bg	Approx.		C50
		pK_1 4.25	40				
		pK_2 10.71	25	E3bg	Uncert.	I = 0.5(NaCl)	S109
		pK_2 10.66	25	E3bg	Approx.	I = 0.5(KNO_3), concentration constant	T23
2729	$C_5H_{11}O_2S$ (cation)	Propanoic acid, 3-dimethylsulfonio- (Dimethylpropiothetin)					
		3.35		E3bg	Uncert.	Conditions not stated	K34
2730	$C_5H_{11}O_2P$	1-Phosphacyclohexan-1-ol 1-oxide (Pentamethylenephosphinic acid)					
		2.73	25	E3bg	Uncert.	c = 0.01, mixed constant	K50
2731	$C_5H_{11}O_4N_3$	Trimethylammonio-N-(2-nitroethyl-2-nitronate)					
		-1.87	25	06	Uncert.	In aq.HCl solution, H_0 scale	S77
2732	$C_5H_{11}O_8P$	D-Ribos-5-yl dihydrogen phosphate (Ribose-5-phosphoric acid)					
		pK_3 13.05	25	CAL	Uncert.	Thermodynamic quantities also given	I34
2733	$C_5H_{11}NS_2$	Methanedithioic acid, butylamino- (Dithiocarbamic acid, N-butyl-)					
		2.90	21	05	Uncert.	c ≈ 0.01-0.00001, concentration constant	T66
		2.75		E	Uncert.	Compound unstable in acid solution	

No.	Molecular formula, name and pK value(s)	$T(^\circ C)$	Remarks	Method	Assessment	Ref.
2734	$C_5H_{11}NS_2$ Methanedithioic acid, diethylamino- (Dithiocarbamic acid, N,N-diethyl-)					
	3.38	25		KIN	Uncert.	A79
	pK$_1$ 7.5, pK$_2$ 8.0	30	By extrapolation from values in NaClO$_4$ solutions	E3bg	Uncert.	B74
	6.0	20	I = 0.01-0.015. Unstable; calculated from absorbances extrapolated to zero time	O5	Uncert.	K42
	3.65	25		KIN	Uncert.	M79
	pK$_1$ 7.5, pK$_2$ 8.4			C		S96
	5.4		c = 0.1	E		T65
2735	$C_5H_{12}ON_2$ 2-Butanone oxime, 3-amino-3-methyl-					
	9.09	25	I = 0.002 Thermodynamic quantities are also given	CAL	Uncert.	W8
2736	$C_5H_{12}O_2N_2$ Pentanoic acid, 2,5-diamino- (Ornithine)					
	pK$_2$ 8.98, pK$_3$ 10.73	25	I = 0.1(KNO$_3$), concentration constants	E3bg	Approx.	C63
2737	$C_5H_{12}O_2N_2$ Propanohydroxamic acid, 2-dimethylamino-					
	6.80	25(?)	c = 0.02, I = 0.1(KCl), mixed constant	E3bg	Uncert.	M66
2738	$C_5H_{12}O_2N_2$ Propanohydroxamic acid, 3-dimethylamino-					
	7.85	25(?)	c = 0.02, I = 0.1(KCl), mixed constant	E3bg	Uncert.	M66

No.	Formula	Name / Notes	T (°C)	pK values	Method	Remarks	Ref.
2739	$C_5H_{12}O_3N_2$	Pentanoic acid, 2-amino-5-hydroxyamino-	25	pK_2 5.6, pK_3 9.4	E3bg	Uncert.	R40
2740	$C_5H_{12}O_3S_4$	3-Thiahexanesulfonic acid, 5,6-dimercapto-		pK_2 8.80, pK_3 12.03	E3b		P67
				pK_2 8.80, pK_3 11.14			P66
2741	$C_5H_{12}O_4N_3$(cation)	N,N,N-Trimethyl-N-(2,2-dinitroethyl)ammonium cation	20	Aqueous $HClO_4$, H_o value −1.87	06	Uncert.	T58a
2742	$C_5H_{12}O_4S_3$	3-Oxahexanesulfonic acid, 5,6-dimercapto-		pK_2 8.82, pK_3 11.48			P66
				pK_2 8.91, pK_3 11.35	E3b		P67
2743	$C_5H_{12}O_5S_4$	3-Thiahexanesulfonic acid 3,3-dioxide, 5,6-dimercapto-		pK_2 8.83, pK_3 10.85	E3b		P67
2744	$C_5H_{13}NS$	2-Butanethiol, 1-amino-2-methyl-	25	c = 0.01, mixed constants. pK_1 8.19, pK_2 10.94	E3bg	Approx.	B147
2745	$C_5H_{13}NS$	2-Pentanethiol, 1-amino-	25	c = 0.01, mixed constants. pK_1 8.32, pK_2 11.05	E3bg	Approx.	B147
2746	$C_5H_{14}O_2B_{10}$	1,2-Dicarbadodecaborane(12)-1-carboxylic acid, 2-vinyl- (Barenecarboxylic acid, vinyl-)	25	mixed constant 2.72	E3bg	Uncert.	Z3

148

No.	Molecular formula, name and pK value(s)	$T(^{o}C)$	Remarks	Method	Assessment	Ref.
2747	$C_5H_{16}O_3B_{10}$ 1,2-Dicarbadodecaborane(12)-1-carboxylic acid, 2-methoxymethyl- (Barenecarboxylic acid, methoxymethyl-)					
	2.33	25	Mixed constant	E3bg	Uncert.	Z3
2748	$C_5H_2O_2NCl_3$ 2-Pyridinone, 3,3,5-trichloro-2,3-dihydro-6-hydroxy- (Glutaconimide, α,α,γ-trichloro-)					
	6.53	20	$c = 10^{-4}$, $I = 0.01$	05	Approx.	S100
2749	$C_5H_2O_2NBr_3$ 2-Pyridinone, 3,3,5-tribromo-2,3-dihydro-6-hydroxy- (Glutaconimide, α,α,γ-tribromo-)					
	6.78	20	$c = 0.0001$, $I = 0.01$		Approx.	S100
2750	$C_5H_3O_2N_2F_3$ Pyrimidine-2,4($1\underline{H},3\underline{H}$)-dione, 5-trifluoromethyl-					
	7.35			05	Uncert.	G33
2751	$C_5H_3O_2N_2F_3$ Pyrimidine-2,4($1\underline{H},3\underline{H}$)-dione, 6-trifluoromethyl-					
	pK_1 5.7, $pK_2 \simeq 13$			05	Uncert.	G20
2752	$C_5H_3O_2ClS$ 2-Thiophenecarboxylic acid, 5-chloro-					
	3.41	25			Uncert.	I3
	3.22	25	$c = 0.005$, mixed constant Value in mixed solvent is also given	E3bg	Approx.	B188
2753	$C_5H_3O_2ClSe$ 2-Selenofurancarboxylic acid, 5-chloro-					
	3.30	25				S99
2754	$C_5H_3O_2BrS$ 2-Thiophenecarboxylic acid, 4-bromo-					
	3.11	25	$c = 0.01$	E3bg	Approx.	O28

No.	Formula	Name	T	pK	Method	Assessment	Ref.	Notes
2755	$C_5H_3O_2BrS$	2-Thiophenecarboxylic acid, 5-bromo-	25	3.30		Uncert.	I3	
			25	3.19		Approx.	B188	c = 0.005, mixed constant. Value in mixed solvent is also given
2756	$C_5H_3O_2BrSe$	2-Selenofurancarboxylic acid, 4-bromo-	25	3.14	E3bg		S99	
2757	$C_5H_3O_2BrSe$	2-Selenofurancarboxylic acid, 5-bromo	25	3.27			S99	
2758	$C_5H_3O_4NS$	2-Thiophenecarboxylic acid, 4-nitro-	25	2.86	E3bg	Approx.	028	c = 0.01
			25	2.81		Uncert.	I3	
2759	$C_5H_3O_4NS$	2-Thiophenecarboxylic acid, 5-nitro-	25	2.68		Uncert.	I3	
2760	$C_5H_3O_4NSe$	2-Selenofurancarboxylic acid, 4-nitro-	25	2.74			S99	
2761	$C_5H_3O_4NSe$	2-Selenofurancarboxylic acid, 5-nitro-	25	2.63			S99	

No.	Molecular formula, name and pK value(s)	$T(^oC)$	Remarks	Method	Assessment	Ref.
2762	$C_5H_3O_4N_2Br$ 4-Pyrimidinecarboxylic acid, 5-bromo-1,2,3,6-tetrahydro-2,6-dioxo-					
	pK_1 2.38, pK_2 7.33	25	$I = 0.1 [(CH_3)_4NBr]$, mixed constants	E3bg	Approx.	T64
	pK_1 2.21, pK_2 7.59		Conditions not stated	E3bg	Uncert.	C81
2763	$C_5H_3O_4N_2I$ 4-Pyrimidinecarboxylic acid, 1,2,3,6-tetrahydro-5-iodo-2,6-dioxo-					
	pK_1 1.88, pK_2 7.63	25	$I = 0.1 [(CH_3)_4NBr]$, mixed constants	E3bg	Approx.	T64
2764	$C_5H_4ON_4S$ Purin-2(1H)-one, 3,6-dihydro-6-thioxo-					
	pK_1 6.2, pK_2 11.4		Conditions not stated	05	Uncert.	L40
2765	$C_5H_4O_2NCl$ 2-Pyrrolecarboxylic acid, 4-chloro-					
	4.07	25	$c = 0.01$	E3bg	Approx.	F43
2766	$C_5H_4O_2NCl$ 2-Pyrrolecarboxylic acid, 5-chloro-					
	4.32	25	$c = 0.01$	E3bg	Approx.	F43
2767	$C_5H_4O_2NBr$ 2-Pyrrolecarboxylic acid, 4-bromo-					
	4.06	25	$c = 0.01$	E3bg	Approx.	F43
2768	$C_5H_4O_2NBr$ 2-Pyrrolecarboxylic acid, 5-bromo-					
	4.17	25	$c = 0.01$	E3bg	Approx.	F43

No.	Formula / Name / pK	T (°C)	Conditions	Method	Assessment	Ref.
2769	C_5H_5ONS Pyridine-2(1H)-thione, 1-hydroxy-					
	pK_1 -1.95		$c = 0.00005$	05	Uncert.	J18
	pK_2 4.67	not stated	$c = 0.024$	E3bg	Uncert.	
	pK_2 4.60	25	$I = 0.005$, concentration constant	E3bg	Uncert.	S127
2770	C_5H_5ONS Pyridine-4(1H)-thione, 1-hydroxy-					
	pK_1 1.53	not stated	$c = 6.8 \times 10^{-5}$	05	Uncert.	J18
	pK_2 3.82		$c = 0.021$	E3bg		
2771	C_5H_5ONS 2-Thiophenecarbaldehyde oxime					
	10.76 (trans)	25	$c = 5 \times 10^{-5}$, $I = 0.1$	05	Approx.	T6
2772	$C_5H_5ON_5S$ 1H-1,2,3-Triazolo[4,5-d]pyrimidin-7(4H)-one, 5,6-dihydro-1-methyl-5-thioxo-					
	6.73	20	$c = 0.00004$	05	Approx.	A40
2773	$C_5H_5O_2NS$ 2-Thiophenecarbohydroxamic acid					
	7.77	not stated	Mixed constant	E3d	Uncert.	M84
2774	$C_5H_5O_2N_2Br$ Pyrimidine-2,4(1H,3H)-dione, 5-bromo-1-methyl-					
	8.30	20-22	$I = 0.05-0.1$, mixed constant	05	Uncert.	B64
2775	$C_5H_5O_2N_2Br$ Pyrimidine-2,4(1H,3H)-dione, 5-bromo-3-methyl-					
	8.30	20-22	$I = 0.05-0.1$, mixed constant	05	Uncert.	B64

No.	Molecular formula, name and pK value(s)	T(^{o}C)	Remarks	Method	Assessment	Ref.
2776	C$_5$H$_6$ON$_2$S Pyrimidin-4(1H)-one, 2,3-dihydro-5-methyl-2-thioxo-					
	8.29	25	c ≈ 0.001	E3bg	Approx.	J17
	7.80	25	c = 0.0001, mixed constant	05	Uncert.	G2
			Values(v.uncert.) are given at other temperatures.			
2777	C$_5$H$_6$ON$_2$S Pyrimidin-4(1H)-one, 2,3-dihydro-6-methyl-2-thioxo-					
	7.94	25	c = 0.0001, mixed constant	05	Uncert.	G2
			Values (v.uncert.)are given at other temperatures			
2778	C$_5$H$_6$ON$_2$S Pyrimidin-4(1H)-one, 6-methylthio-					
	8.52	20	c = 0.005	E3bg	Approx.	B130
2779	C$_5$H$_6$O$_2$N$_2$S Pyrimidine-2,4(1H,3H)-dione, 5-methylthio-					
	pK$_1$ 8.5, pK$_2$ > 13	not stated		05	Uncert.	B23
2780	C$_5$H$_6$O$_4$NP Pyridin-3-yl dihydrogen phosphate					
	pK$_1$ 3.86, pK$_2$ 5.64	25	I = 0.1(KNO$_3$)	E3bg	Approx.	M118
	pK$_1$ 3.85, pK$_2$ 5.62	50				M113
2781	C$_5$H$_6$O$_4$N$_2$S Pyrimidine-2,4(1H,3H)-dione, 6-mesyl-					
	4.68	25	Titrant (CH$_3$)$_4$NOH	E3bg	Approx.	J16
	4.73	25	c = 0.005	E3bg	Approx.	G45

No.	Formula / Name / pK	T (°C)	Notes			Ref.
2782	$C_5H_7O_4N_2Cl_3$ Pentane, 1,1,1-trichloro-5,5-dinitro-					
	4.79	20		05	Approx.	T58a
2783	$C_5H_8ON_2S$ Imidazolidin-4-one, 5,5-dimethyl-2-thioxo-					
	8.71	18±3	I = 0.01	05	Uncert.	E4
2784	$C_5H_8O_2N_2F_4$ Pentanoic acid, 4,4-bis(difluoramino)-					
	4.01	25	mixed constant	E3bg	Approx.	B37
2785	$C_5H_8O_4N_4F_4$ Pentane, 2,2-bis(difluoramino)-5,5-dinitro-					
	3.87	25	mixed constant	05	Approx.	B37
2786	$C_5H_9ONS_2$ 1-Oxa-4-azacyclohexane-4-carbodithioic acid (Morpholine-4-carbodithioic acid)					
	5.65	15	From pH measurements on			
	5.19	25	different concentrations of			
	4.73	40	Na-salt			
	5.45	15	From pH and conductivity			
	5.16	25	measurements on Na-salt			
	4.65	40	solutions			D24
2787	$C_5H_9O_2NS_2$ 3-Thia-5-azahexanoic acid, 5-methyl-4-thioxo-					
	3.32	25	I = 0.1(NaClO$_4$), concentration constant	E3bg	Approx.	J9
	3.38			02	Approx.	

No.	Molecular formula, name and pK value(s)	$T(^oC)$	Remarks	Method	Assessment	Ref.
2788	$C_5H_9O_2NS_2$ 5-Thia-3-azahexanoic acid, 3-methyl-4-thioxo-					J9
	2.93	25	$I = 0.1(NaClO_4)$, concentration constant	E3bg	Uncert.	
	3.04			O2	Uncert.	
2789	$C_5H_9O_3NS$ 3-Oxa-5-azahexanoic acid, 5-methyl-4-thioxo-					J9
	2.87	25	$I = 0.1(NaClO_4)$, concentration constant	E3bg	Approx.	
2790	$C_5H_9O_3NS$ Propanoic acid, 2-acetylamino-3-mercapto- (Cysteine, N-acetyl-)					F42
	9.52(-SH)	30	$c = 0.005$, $I = 0.3(KCl)$	E3bg	Approx.	
2791	$C_5H_9O_3NS$ 3-Thia-5-azahexanoic acid, 5-methyl-4-oxo-					J9
	3.49	25	$I = 0.1(NaClO_4)$,concentration constant	E3bg	Approx.	
	3.55			O2	Approx.	
2791a	$C_5H_8O_4N_3S$ 3-Azapentanedioic acid, 3-thioureido- (Iminodiacetic acid, N-thioureido-)					G26
	pK_1 2.94, pK_2 4.07	30	$c \approx 0.002$, $I = 0.1(KCl)$, concentration constants	E3bg	Approx.	
2791b	$C_5H_{10}ONF_3$ 2-Propanol, 1-dimethylamino-3,3,3-trifluoro-					R26
	12.56		Conditions not stated	E3bg	Uncert.	
2791c	$C_5H_{10}O_2N_2S$ 3,5-Diazahexanoic acid, 5-methyl-4-thioxo-					J9
	3.71	25	$I = 0.1(NaClO_4)$, concentration constant	O2	Approx.	

2792   C$_5$H$_{10}$O$_7$NP   3-Azapentanedioic acid, 3-phosphonomethyl-   (Iminodiacetic acid, N-phosphonomethyl-)
pK$_3$ 5.80, pK$_4$ 10.64   30   I = 0.1(KCl), concentration constants   E3bg   Approx.   O4

pK$_1$ 2.0, pK$_2$ 2.25, pK$_3$ 5.57, pK$_4$ 10.76   20   I = 0.1(KCl), concentration constants   E3bg   Approx.   S36

2793   C$_5$H$_{11}$ONS   O-Isopropyl methylaminomethanethioate   (O-Isopropyl thiocarbamate, N-methyl-)
pK(NH) 11.56   20   c = 0.01   E   L61

2794   C$_5$H$_{11}$O$_2$NS   Butanoic acid, 2-amino-3-mercapto-3-methyl-   (Penicillamine)
pK$_2$ 8.03, pK$_3$ 10.68   20   I = 0.1(NaClO$_4$), concentration constants   E3bg   Approx.   P33
(pK$_1 \simeq$ 2)

2795   C$_5$H$_{11}$O$_2$NS   Butanoic acid, 2-amino-4-methylthio-   (L-Methionine)
pK$_1$ 2.17, pK$_2$ 9.20   20   I = 0.15(NaClO$_4$), concentration constants   E3bg   Approx.   H23
pK$_1$ 2.2   Uncert.   K34
pK$_2$ 9.2   E3bg/05

pK$_1$ was determined from infra-red measurements in D$_2$O and corrected to the corresponding
value in water by subtracting 0.5 pK unit

2796   C$_5$H$_{11}$O$_7$N$_2$P   3-Azahexanoic acid, 5-amino-4-oxo-6-phosphonooxy-   (O-Phosphono-D,L-serylglycine)
pK$_2$ 3.13, pK$_3$ 5.41, pK$_4$ 8.02   25   I = 0.15(KCl), concentration constants   E3bg   Approx.   O24

2797   C$_5$H$_{11}$O$_7$N$_2$P   3-Azapentanoic acid, 5-amino-4-oxo-2-phosphonooxymethyl-   (Glycyl-O-phosphono-D,L-serine)
pK$_2$ 2.91, pK$_3$ 6.03, pK$_4$ 8.42   25   I = 0.15(KCl), concentration constants   E3bg   Approx.   O24

No.	Molecular formula, name and pK value(s)	$T(^{o}C)$	Remarks	Method	Assessment	Ref.
2798	$C_5H_{12}O_2N_2S$ Butanohydroxamic acid, 2-amino-4-methylthio- 6.60	25(?)	c = 0.02, I = 0.1(KCl), mixed constant	E3bg	Uncert.	M66
2799	$C_5H_{12}O_2N_2S_2$ 4,5-Dithiaheptanoic acid, 2,7-diamino- pK_2 8.28, pK_3 9.30	20	I = 0.15($NaClO_4$), concentration constants	E3bg,R3d	Uncert.	H23
2800	$C_5H_{13}OS_2P$ 0-Butyl hydrogen methylphosphonodithioate 1.73	20	c = 0.005 in 7% ethanol	E3bg	V.uncert.	K6
2801	$C_5H_{13}O_2N_2(Br)$ Acetohydroxamic acid, 2-trimethylammonio-, bromide 6.70	25(?)	c = 0.02, I = 0.1(KCl), mixed constant	E3bg	Uncert.	M66
2802	$C_5H_{13}O_2SP$ 0-Butyl 0-hydrogen methylphosphonothioate 1.95	20	c = 0.005 in 7% ethanol, 88% thione tautomer.	E3bg	V.uncert.	K6
2803	$C_5H_{13}O_2SP$ 0-Ethyl 0-hydrogen propylphosphonothioate 2.00	20	c = 0.005 in ethanol, 88% thione tautomer.	E3bg	V.uncert.	K6
2804	$C_5H_{13}O_2SP$ 0-Isobutyl 0-hydrogen methylphosphonothioate 1.81	20	c = 0.005 in 7% ethanol,84% thione tautomer.	E3bg	V.uncert.	K6
2805	$C_5H_{13}O_2SP$ S-Propyl hydrogen ethylphosphonothioate 2.13	20	c = 0.005, mixed constant	E3bg	Uncert.	N28

No.	Formula / Name / pK values	T (°C)	Conditions	Method	Assessment	Ref.
2806	$C_5H_{14}O_3NP$ Methylphosphonic acid, diethylamino- pK_5 5.79, pK_6 12.32	25	I = 1(KNO$_3$), concentration constants	E3bg	Uncert.	C18
2807	$C_5H_{14}O_3NP$ Propylphosphonic acid, 5-amino- pK_1 2.6, pK_2 7.65, pK_3 11.00	not stated	c = 0.02	E3bg	Uncert.	C38
2808	$C_5H_{14}O_5NP$ 2-Azabutylphosphonic acid, 4-hydroxy-2-(2-hydroxyethyl)- (Methylphosphonic acid, N,N-bis(2-hydroxyethyl)amino-) pK_1 4.98, pK_2 9.46	20	I = 0.1(NaClO$_4$)	E	Uncert.	K12
2809	$C_5H_{14}O_5NSP$ O,O-Diethylphosphoramidate, N-mesyl- 2.36	20	c = 0.005, 7% ethanol, mixed constant	E3bg	Uncert.	G19
2810	$C_6H_2O_6$ 5-Cyclohexene-1,2,3,4-tetrone, 5,6-dihydroxy- (Rhodizonic acid) 5.06	25	I = 0.1(NaClO$_4$), mixed constant	E3bg	Uncert.	B14
	pK_1 4.1, pK_2 4.5	30	Obtained from the points of discontinuity in the linear relations observed when the concentration - dependent reduction potential is plotted against the measured pH.	E	Uncert.	P84
	pK_1 4.25, pK_2 4.72	not stated	c = 0.00005, I = 0.1(KCl) Measured constants include hydration equilibria. Values for anhydrous acid cannot be measured. Anhydrous values possibly pK_1 < 2, pK_2 < 4.7	05	Uncert.	P14
2811	$C_6H_4O_2$ 2,4-Hexadiynoic acid 1.94	25	c = 0.01, I = 0.1(NaCl), mixed constant	E3bg	Approx.	M48

No.	Molecular formula, name and pK value(s)	T(°C)	Remarks	Method	Assessment	Ref.
2812	$C_6H_4O_4$ 1,4-Benzoquinone, 2,5-dihydroxy-					
	pK$_1$ 2.81, pK$_2$ 5.22	25	I = 0.5(HCl-NaCl), concentration constants	05	Approx.	B42
	pK$_1$ 2.73, pK$_2$ 5.18	25		05	Approx.	W2
2813	$C_6H_4O_4$ 2H-Pyran-5-carboxylic acid, 2-oxo-					
	pK$_1$ 3.26	25		C2	Approx.	S10
2814	$C_6H_4O_4$ 4H-Pyran-2-carboxylic acid, 4-oxo-					
	pK$_1$ 1.50	25		C2	Approx.	S10
2815	$C_6H_4O_5$ 2,3-Furandicarboxylic acid					
	pK$_1$ 2.45, pK$_2$ 7.25		Conditions not stated	E3bg	Uncert.	01
2816	$C_6H_4O_5$ 2,4-Furandicarboxylic acid					
	pK$_1$ 2.63, pK$_2$ 3.77		Conditions not stated	E3bg	Uncert.	01
2817	$C_6H_4O_5$ 2,5-Furandicarboxylic acid					
	pK$_1$ 2.60, pK$_2$ 3.55		Conditions not stated	E3bg	Uncert.	01
2818	$C_6H_4O_5$ 3,4-Furandicarboxylic acid					
	pK$_1$ 2.51, pK$_2$ 7.43		Conditions not stated	E3bg	Uncert.	01
	pK$_1$ 1.44, pK$_2$ 7.84	25	c = 0.005-0.09	E3bg	Approx.	M11

No.	Formula	Name / pK	Temp	Conditions	Method	Reliability	Ref.
2819	$C_6H_4O_5$	4H-Pyran-2-carboxylic acid, 3-hydroxy-4-oxo-					
		7.67	25	I = 0.14	E3bg	Approx.	T60
2820	$C_6H_4O_5$	4H-Pyran-2-carboxylic acid, 5-hydroxy-4-oxo- (Comenic acid)					
		pK_2 7.29	25	I = 0.5(NaCl), mixed constant	E3bg and O5	Approx.	C46
		pK_1 (CO) -2.34		In aq.H_2SO_4 solutions, H_o scale	06	Uncert.	C47
		pK_2 7.75		I = 0.01-0.5(NaCl)	E3bg	Approx	
		pK_2 7.68	room temp	Mixed constant	E3bg	Uncert.	E9
2821	$C_6H_4O_6$	1,4-Benzoquinone, 2,3,5,6-tetrahydroxy-					
		pK_1 4.8, pK_2 6.8	30	Method as for No.2810, Ref.P84	E	Uncert.	P84
2822	$C_6H_4N_6$	1H-1,2,3-Triazolo[4,5-b]pyridine-6-carbonitrile, 5-amino-					
		pK_1 -0.06	20	c = 0.000055	06	Uncert.	A34
		pK_2 6.53	20		05	Approx.	
2823	C_6H_5N	2,4-Cyclopentadiene-1-carbonitrile					
		9.78		Conditions not stated	05	Uncert.	W10a
2824	$C_6H_5N_3$	1H-Benzotriazole					
		pK_2 8.37	20		E & 05		L62
		pK_2 8.11	20	I = 0.4(KCl)			

No.	Molecular formula, name and pK value(s)	$T(^{o}C)$	Remarks	Method	Assessment	Ref.
2825	C_6H_6O Phenol					
	9.994	25	m = 0.0004-0.0008, I = 0.01-05	O5a	Rel.	F12
	10.020	25	c = 0.00004, I = 0.1	O1	Approx.	C40
			Variation with temperature			
		5 10 15 20 30 38				
		10.334 10.248 10.167 10.091 9.952 9.853				
			Thermodynamic quantities are derived from the results			
	9.97	25	m = 0.04-0.23, I = 0.04-0.21	E1a	Approx.	B78
	9.90	30				
	9.78	40				
	9.66	50				
	9.55	60				
	9.86	25	I = 0.1, mixed constant	E3d	Approx.	Z4
			Variation with temperature			
		20 30 35 40 45 50 55 60				
		9.92 9.80 9.74 9.69 9.64 9.59 9.54 9.49				
			Thermodynamic quantities are derived from the results			
	9.98	25	c = 0.004-0.04	CAL	Approx.	F8
	9.78	25	I = 0.1(KCl)	E3bg	Approx.	E20
			Thermodynamic quantities are also given			

(Contd)

2825 (Contd) C_6H_6O Phenol

pK	T (°C)	Code	Quality	Ref.	Remarks
9.80	25	05	Approx.	D34	$I = 0.1(NaClO_4)$, concentration constant. Thermodynamic quantities are given
9.98	25	05	Approx.	M19	$c \approx 0.0002$
9.78					$I = 0.1$, mixed constant
10.04	20	E3bg	Approx.	D31	$c = 0.01$
9.79	25	E3bg	Approx.	M5	$I = 0.5(KNO_3)$, mixed constant
9.85	25	05	Uncert.	W12	$I = 0.027(NaClO_4)$, mixed constant
10.00 (H_2O)					
10.62 (D_2O)					
3.62	25	FLU/05	V.uncert.	B25	pK of electronically excited singlet state calculated assuming pK = 9.99 in ground state

Other values in A11, B96, H32, H33, H51, K40, M128 and W32

2826 $C_6H_6O_2$ 1,2-Benzenediol   (Catechol)

pK	T (°C)	Code	Quality	Ref.	Remarks
pK_1 9.34,	30	E3bg	Approx.	A80	$c = 0.004$, $I = 0.1(NaClO_4)$, mixed constants
$pK_2 \approx 12.6$					
pK_1 9.37	20	E3bg	Approx.	B31	$I = 0.1(KNO_3)$, mixed constants
pK_2 13.7			Uncert.		
pK_1 9.43	20	E3bg	Approx.	P30	$I = 0.1-0.15(NaClO_4)$, mixed constants
$pK_2 \approx 13$		05	Uncert.		Dianion oxidizes readily in air

(Contd).

No.	Molecular formula, name and pK value(s)	T(°C)	Remarks	Method	Assessment	Ref.
2826 (Contd)	$C_6H_6O_2$ 1,2-Benzenediol (Catechol)					
	pK$_1$ 9.45	25		E3bg,R3d	Approx.	T38
	pK$_2$ 12.8				Uncert.	
	pK$_1$ 9.229	25	c = 0.02, I = 1.0(KNO$_3$), concentration constants	E3bg	Approx.	T69
	pK$_2$ 13.05				Uncert.	
	pK$_1$ 9.30	25	I = 0.1(KCl), mixed constant	E3bg	Approx.	E20
	pK$_1$ 9.23	31.5	I = 0.2(KCl), mixed constant	E3bg	Approx.	S33
	Other values in A1, B29, C44, H15, L37, M59, M70, M112, P63 and S128					
2827	$C_6H_6O_2$ 1,3-Benzenediol (Resorcinol)					
	pK$_1$ 9.20, pK$_2$ 11.27	20	I = 0.1, mixed constants (?)	E3bg	Uncert.	S128
	pK$_1$ 9.30, pK$_2$ 11.06	20-25	I = 0.1(KNO$_3$)	E3bg	Uncert.	B29
	pK$_1$ 9.32	25	I = 0.1(KCl), mixed constant	E3bg	Approx.	E20
2828	$C_6H_6O_2$ 1,4-Benzenediol (Hydroquinone)					
	pK$_1$ 10.01, pK$_2$ 11.66	13	I = 0.1(KCl), mixed constants	05	Approx.	S111
	9.91, 11.56	22	Thermodynamic quantities are derived from the results			
	pK$_1$ 9.91, pK$_2$ 11.56	20	I = 0.1, mixed constants	E3bg	Uncert.	S128
	pK$_1$ 9.97, pK$_2$ 11.49	13.2	I = 0.65(NaClO$_4$), concentration constants	05	Approx.	B38
	9.91 -	18.7				
	- 11.44	19.6				
	9.85 -	25.0	Thermodynamic quantities are derived from the results			

(Contd)

No.	Formula	Name / pK values	t(°C)	Conditions	Method	Assessment	Ref.
2828 (Contd)	$C_6H_6O_2$	1,4-Benzenediol (Hydroquinone)	25.9				
		pK_2 11.39					
		pK_1 9.82	25	I = 0.1(KCl), mixed constant	E3bg	Approx.	E20
2829	$C_6H_6O_2$	2-Hexen-4-ynoic acid	25	c = 0.005, I = 0.1(NaCl), mixed constant	E3bg	Approx.	M48
		4.20(trans)					
2830	$C_6H_6O_2$	4-Hexen-2-ynoic acid	25	c = 0.005, I = 0.1(NaCl), mixed constant	E3bg	Approx.	M48
		2.67					
2831	$C_6H_6O_2$	5-Hexen-3-ynoic acid	25	c = 0.01, I = 0.1(NaCl), mixed constant	E3bg	Approx.	M48
		3.39					
2832	$C_6H_6O_3$	1,2,3-Benzenetriol (Pyrogallol)					
		pK_1 9.28, pK_2 11.34	20	I = 0.1, mixed constants	E3bg	Uncert.	S128
		pK_1 9.05, pK_2 11.19, $pK_3 \simeq 14$	20	I = 0.1(KNO_3), mixed constants	E3bg	Approx.	B31
		pK_1 9.12	25	I = 0.1(KNO_3), mixed constant	E3bg	Approx.	E19
		pK_1 9.12	31.5	I = 0.2(KCl), mixed constant	E3bg	Approx.	S33
2833	$C_6H_6O_3$	1,2,4-Benzenetriol					
		pK_1 9.08, pK_2 11.82	20	I = 0.1, mixed constants	E3bg	Uncert.	S128
2834	$C_6H_6O_3$	1,3,5-Benzenetriol (Phloroglucinol)					
		pK_1 7.97, pK_2 9.23	20	I = 0.1, mixed constants (?)	E3bg	Uncert.	S128

No.	Molecular formula, name and pK value(s)	T(°C)	Remarks	Method	Assessment	Ref.
2835	$C_6H_6O_3$ 4H-Pyran-4-one, 3-hydroxy-2-methyl-					
	8.36	25	I = 0.5(NaCl)	E3bg & O5	Approx.	C46
	pK$_1$ (CO) -0.71	25	In aq. H_2SO_4, H_0 scale	O6	Uncert.	B41
	pK$_2$ 8.36		I = 0.5(NaCl), concentration constant	E3bg & O5	Approx.	C47
	pK$_1$(CO) -0.71	25	in aq. H_2SO_4, H_0 scale	O6	Uncert.	
	pK$_2$ 8.66		I = 0.01-0.5(NaCl)	E3bg,R4	Approx.	C47
	pK$_2$ 7.88	25		E3bg	Approx	M7
			Variation with I			
			0.03 0.23 0.53 1.03			
			7.83 7.77 7.73 7.81			
2836	$C_6H_6O_3$ 4H-Pyran-4-one, 3-hydroxy-6-methyl- (Allomaltol)					
	7.98	25	I = 0.5(NaCl)	E3bg & O5	Approx.	C46
	pK$_1$ (CO) -0.72	25	In aq. H_2SO_4 solutions, H_0 scale	O6	Uncert.	C47
	pK$_2$ 8.28	25	I = 0.01-0.5	E3bg,R4	Approx.	
	pK$_2$ 8.28	25	I = 0.12	E3bg	Approx.	T60
2837	$C_6H_6O_4$ 1-Cyclobutene-1,2-dicarboxylic acid					
	pK$_1$ 1.12, pK$_2$ 7.63	25	c = 0.05-0.09	E3bg	Approx.	M11

No.	Compound	pK	Temp (°C)	Method	Assessment	Ref.	Remarks
2838	$C_6H_6O_4$ 4H-Pyran-4-one, 2-hydroxy-5-hydroxymethyl- (Kojic acid)						
		pK_2 7.66	25	E3bg & 05	Approx.	C46	I = 0.5(NaCl)
		pK_1 -1.38	25	06	Uncert.	C47	In aq H_2SO_4 solutions, H_0 scale
		pK_2 7.96	25	E3bg,R4	Approx.	M110	I = 0.01-0.5(NaCl)
		pK_2 7.68	25	E3bg	Approx.	T60	I = 0.1(KNO_3)mixed constant
		pK_2 7.90	25	E3bg	Approx.	E9	I = 0.12
		pK_2 8.09	room temp	E3bg	Uncert.		
2839	$C_6H_6O_4$ 4H-pyran-4-one, 3-hydroxy-5-hydroxymethyl-						
		pK_2 7.75	20-25	E3bg	Uncert.	B29	I = 0.1(KNO_3)
2840	$C_6H_6O_6$ Benzenehexol						
		9.0	30	E	Uncert.	P84	Obtained from the points of discontinuity in the linear relation observed when the concentration dependent reduction potential is plotted against pH
2841	$C_6H_6O_6$ Oxolane-2,3-dicarboxylic acid, 5-oxo- (Tetrahydrofuran-2,3-dicarboxylic acid, 5-oxo-)						
		pK_1 2.26, pK_2 4.50 (cis)	28	E3bg	Approx.	G8	c = 0.034
		pK_1 2.13, pK_2 3.95 (trans)					c = 0.037, mixed constants
2842	C_6H_6S Benzenethiol (Thiophenol)						
		6.615	25	05a	Rel.	D28	m = 0.0001
		6.62	23	05	Approx.	D9	
		6.43	35	05	Approx.	M81	I = 1.0 . (Other values in K57 and K58)

No.	Molecular formula, name and pK value(s)	T(°C)	Remarks	Method	Assessment	Ref.
2843	$C_6H_8O_2$ Bicyclo[1.1.1]pentane-1-carboxylic acid					
	4.09	25	c = 0.01	E3dg	Approx.	W28
2844	$C_6H_8O_2$ 2-Cyclobutenone, 3-hydroxy-2,4-dimethyl-					
	2.8	not stated	c = 0.1	E3bg	Uncert.	W44
2845	$C_6H_8O_2$ 1,2-Cyclohexanedione					
	10.30	20	40% enol form	E3bh	Approx.	S52
2846	$C_6H_8O_2$ 1,3-Cyclohexanedione					
	5.25	25		E3bh	Approx.	S47
2847	$C_6H_8O_2$ Cyclopentanecarbaldehyde, 2-oxo-					
	5.83	20	c = 0.01	E3bg	Uncert.	S42
2848	$C_6H_8O_2$ 2-Cyclopentenone, 2-hydroxy-5-methyl-					
	9.60	20	100% enol	E3bh	Approx.	S52
2849	$C_6H_8O_2$ 1,2-Cyclopropanedicarboxylic acid, 1-methyl-					
	pK_1 3.67, pK_2 6.47 (cis)	20		E3bg	Approx.	M12
	pK_1 3.93, pK_2 5.30 (trans)					
2850	$C_6H_8O_2$ 2,4-Hexadienoic acid					
	4.51 (trans-trans)	25	c = 0.005, I = 0.1(NaCl), mixed constant	E3bg	Approx.	M48
	4.49(cis-cis)					

No.	Formula / Name / pK	T (°C)	Conditions / Remarks	Method	Assessment	Ref.
2851	$C_6H_8O_2$ 5-Hexynoic acid					
	4.59	25	c = 0.01, I = 0.1(NaCl), mixed constant	E3bg	Approx.	M48
2852	$C_6H_8O_3$ Cyclopentanecarboxylic acid, 3-oxo-					
	4.25	20	c = 0.001-0.002	E3bg	Approx.	I4
2853	$C_6H_8O_3$ Furan-2(5<u>H</u>)-one, 3-ethyl-4-hydroxy- (Tetronic acid, 3-ethyl-)					
	4.15	25		C	Rel.	B49
	4.0		Conditions not stated		Uncert.	C45
2854	$C_6H_8O_3$ Furan-2(5<u>H</u>)-one, 4-hydroxy-3,5-dimethyl-					
	4.0		Conditions not stated		Uncert.	C45
2855	$C_6H_8O_4$ Butenedioic acid, dimethyl- (Maleic acid, dimethyl-)					
	pK_1 3.1, pK_2 6.1	20	pK_1 was corrected for the acid - anhydride equilibrium.	E3bg	Uncert.	E1
2856	$C_6H_8O_4$ 1,2-Cyclobutanedicarboxylic acid					
	pK_1 4.16, pK_2 6.23 (cis)	25	c = 0.005	E3bg	Approx.	B85
	pK_1 3.94, pK_2 5.55 (trans)					
2857	$C_6H_8O_4$ 1,3-Cyclobutanedicarboxylic acid					
	pK_1 4.08, pK_2 5.12 (cis)	25	c = 0.005	E3bg	Approx.	B85
	pK_1 4.11, pK_2 5.15 (trans)					

No.	Molecular formula, name and pK value(s)	T(°C)	Remarks	Method	Assessment	Ref.
2858	$C_6H_8O_4$ Propenoic acid, 3-ethoxycarbonyl-					
	3.396 (H_2O) (trans)	25	molality scale, I = 0.02-0.1	E1a,R4,	Rel.	D2
	3.847 (D_2O)					
	3.077 (H_2O) (cis)					
	3.532 (D_2O)					
2859	$C_6H_8O_5$ Butanedioic acid, 2-acetyl-					
	pK_1 2.86, pK_2 4.57	25	I = 0.1, concentration constants	E3ag	Approx.	P20
2860	$C_6H_8O_5$ Butanedioic acid, 2,2-dimethyl-3-oxo-					
	pK_1 1.77, pK_2 4.62	25		E3bg,R3d	Approx.	G12
2861	$C_6H_8O_6$ l-Ascorbic acid					
	pK_1 4.49, pK_2 12.72	0.4	I = 0.1(KNO_3),	E3bg	pK_1 Approx. pK_2 Uncert.	K25
	pK_1 4.04, pK_2 11.34	25	concentration constants			
	pK_1 4.04, pK_2 11.34	25	I = 0.1(KNO_3),	E3bg	Approx.	K26
	pK_1 3.57, pK_2 11.98	40	concentration constants			
	pK_1 4.56, pK_2 12.22 (D_2O)	25	I = 0.1(KNO_3), concentration constants			
2862	$C_6H_8O_6$ Pentanedioic acid, 2-hydroxy-2-methyl-4-oxo-					
	pK_1 1.73, pK_2 3.72	25	I = 1.0(NaCl), mixed constants Determined by titration of mixtures of monomeric pyruvate and dimeric pyruvate of known composition.	E3bg	Approx.	T3

No.	Formula / Name	pK values	T(°C)	Conditions	Methods	Assessment	Ref.
2863	$C_6H_8O_6$ 1,2,3-Propanetricarboxylic acid (Tricarballylic acid)						
		pK_1 3.47, pK_2 4.54, pK_3 5.89	20	$I = 0.1(NaClO_4)$, concentration constants	E3bg	Approx.	C9
		pK_1 4.03, pK_2 5.28, pK_3 6.55 (D_2O)	25	$c = 0.02$, $I = 0.11$. pH of 0.01M DCl in 0.1M NaCl taken as 1.70	E3bg,R3f	Uncert.	L38
2864	$C_6H_8O_7$ 1,2,3-Propanetricarboxylic acid, 1-hydroxy- (Isocitric acid)						
		pK_1 3.287, pK_2 4.714, pK_3 6.396	25		E3ah,R4	Approx.	H45
		pK_1 3.244, pK_2 4.693, pK_3 6.373	38				
2865	$C_6H_8O_7$ 1,2,3-Propanetricarboxylic acid, 2-hydroxy- (Citric acid)						
		pK_1 2.96, pK_2 4.38, pK_3 5.68	20	$I = 0.1(NaClO_4)$	E3bg	Approx.	T41
		pK_1 2.79, pK_2 4.30, pK_3 5.65	25	$c \approx 0.004$, $I = 0.1(KNO_3)$	E3bg	Approx.	R4
		pK_1 2.63, pK_2 4.11, pK_3 5.34		$I = 1.0(KNO_3)$, concentration constants			
		pK_1 2.88, pK_2 4.36, pK_3 5.84	25	$I = 0.1[(CH_3)_4NCl]$, titrant $(CH_3)_4NOH$, mixed constants	E3bg,R3d	Approx.	T12
		pK_1 2.87, pK_2 4.35, pK_3 5.68	20	$I = 0.1(NaClO_4)$, concentration constants	E3bg,R3f	Approx.	C9
		pK_1 3.13, pK_2 4.78, pK_3 6.43	25	$c = 0.01$	E3bg,R3d	Approx.	L59
		pK_1 3.44, pK_2 5.02, pK_3 6.55 (D_2O)	25	$c = 0.02$, $I = 0.11$, concentration constants pH of 0.01M DCl in 0.1M NaCl taken as 1.7	E3bg	Uncert.	L38

No.	Molecular formula, name and pK value(s)	$T(^{o}C)$	Remarks	Method	Assessment	Ref.
2866	$C_6H_8O_7$ D-Saccharic acid 1,4-monolactone			E3		M26
	2.72					
2867	$C_6H_8O_7$ D-Saccharic acid 3,6-monolactone			E3		M26
	2.76					
2868	$C_6H_8O_7$ D-Talomucic acid 1,4-monolactone	23 ± 2	I = 0.01-0.03, mixed constant	E3	Uncert.	M21
	2.81					
2869	$C_6H_8O_7$ D-Talomucic acid 3,6-monolactone	23 ± 2	I = 0.01-0.03, mixed constant	E3	Uncert.	M21
	2.84					
2870	$C_6H_{10}O$ Cyclohexanone	25		KIN	Uncert.	B55
	11.3(enol), 16.7(keto)					
2871	$C_6H_{10}O_2$ 2,4-Hexanedione	5	c = 0.0005, 0.5-1% ethanol. Apparent pK for keto-enol mixture	O5	Approx.	C4
	9.63					
	9.38	25				C5
	9.28	45	Thermodynamic quantities for the keto and enol forms are derived from the results			C4
	8.49(enol), 9.32(keto)	25	Calculated using absorbance data		Uncert.	C5
	9.36	20	c = 0.01	E3bg & O5	Approx.	R52

No.	Formula	Name / value	Temp (°C)	Conditions	Method	Reliability	Ref.
2872	$C_6H_{10}O_2$	2,5-Hexanedione					
		18.7			KIN	Uncert.	B47
2873	$C_6H_{10}O_2$	2-Hexenoic acid					
		4.63(cis)	25		E3bg	Approx.	M10
		4.75(trans)					
2874	$C_6H_{10}O_2$	2,4-Pentanedione, 3-methyl-					
		10.95	15	c = 0.01	E3bg	Approx.	L3a
		10.87	25	Thermodynamic quantities are derived			
		10.80	35	from the results			
		10.74	45				
		10.70	20	c = 0.01	E3bg & O5	Approx.	R52
		11.06	20	c = 0.008-0.023	E3bg	Uncert.	S42
		Enol form					
		9.16 (H_2O)	25	I = 0.2	KIN	Uncert.	D1
		9.82 (D_2O)					
2875	$C_6H_{10}O_3$	Cyclopentanecarboxylic acid, 1-hydroxy-					
		4.155	25		E3ag	Approx.	P81

Variation of the concentration constant with I:

I	2.00	1.00	0.50	0.20	0.10	0.05	0.02	0.01
pK_c	4.041	3.924	3.900	3.928	3.967	4.005	4.050	4.076

		3.97	25	I = 0.1($NaClO_4$), concentration constant	E3bg	Approx.	P77

No.	Molecular formula, name and pK value(s)	T(°C)	Remarks	Method	Assessment	Ref.
2876	$C_6H_{10}O_3$ Ethyl 3-oxobutanoate (Ethyl acetoacetate)					
	10.64	25.7		O5	Uncert.	R53
	10.68	25	c = 0.003	E3bg	Approx.	E11
	10.49	20	c ≃ 0.02	E3bg	Uncert.	S42
	10.7			KIN	Uncert.	B47
2877	$C_6H_{10}O_3$ Hexanoic acid, 5-oxo-					
	4.8	25		E3bg	Uncert.	C57
2878	$C_6H_{10}O_3$ Tetrahydropyran-2-carboxylic acid					
	3.80	20		E3bg	Approx.	D55
2879	$C_6H_{10}O_4$ Butanedioic acid, 2,3-dimethyl-					
	pK_1 3.92, pK_2 6.00 (DL)	25	c ≃ 0.0005-0.0013	E3bg,R3d	Approx.	D41
	pK_1 3.77, pK_2 5.36 (meso)		c ≃ 0.0006-0.0015			
	pK_1 3.82, pK_2 5.93 (DL)	25	m 0.01	Elch,R3d	Rel.	P88
	pK_1 3.67, pK_2 5.30 (meso)		Equation relating pK to temperature is given together with thermodynamic quantities			
2880	$C_6H_{10}O_4$ Hexanedioic acid (Adipic acid)					
	pK_1 4.42, pK_2 5.41	20	c = 0.005 Values in mixed solvents are also given	E3bg/R3a	Approx.	D43
	pK_1 4.44, pK_2 5.44	25	c = 0.005, I = 0.035	O3,R3d	Approx.	J19

(Contd)

2880(Contd) $C_6H_{10}O_4$ Hexanedioic acid (Adipic acid)

	Temp	Conditions	Method		Ref
pK$_1$ 4.28, pK$_2$ 5.00	25	I = 0.1(NaClO$_4$), concentration constant	E3bg,R3c	Uncert.	Y12
pK$_1$ 4.42, pK$_2$ 5.41	25-30	c ≈ 0.0005	E3bg,R3d	Uncert.	H52
pK$_1$ 4.42, pK$_2$ 5.43	20	Recalculated from data of J.C.Speakman, J.Chem.Soc., 1940, 855		Approx.	L59
pK$_1$ 4.44, pK$_2$ 5.45	25	I = 0.1(KNO$_3$)	E3bg	Approx.	N32

2881 $C_6H_{10}O_4$ Pentanedioic acid, 2-methyl-

pK$_1$ 4.36, pK$_2$ 5.37	25	c = 0.005, I = 0.035	O3,R3d	Approx.	J19

2882 $C_6H_{10}O_4$ Pentanedioic acid, 3-methyl-

pK$_1$ 4.35, pK$_2$ 5.44	25	c = 0.00125, mixed constants	E3bg,R3c	Approx.	B146
pK$_1$ 4.27, pK$_2$ 5.37	25	c = 0.005, I = 0.035	O3,R3d	Approx.	J19

2883 $C_6H_{10}O_4$ Propanedioic acid, 2-ethyl-2-methyl-

pK$_1$ 2.2	20	c = 0.005	E3bg,R3d	Uncert.	D44
pK$_2$ 6.55		Values are also given in mixed solvents		Approx.	

2884 $C_6H_{10}O_4$ Propanedioic acid, isopropyl-

pK$_1$ 2.92, pK$_2$ 5.88 (H$_2$O)	25		E3bg	Approx.	G23
pK$_1$ 3.43, pK$_2$ 6.36 (D$_2$O)					
pK$_1$ 2.94, pK$_2$ 5.88	25	c ≈ 0.01	E3bg	Uncert.	W40

2885 $C_6H_{10}O_4$ Propanedioic acid, propyl-

pK$_1$ 3.00, pK$_2$ 5.84	25	c ≈ 0.01	E3bg	Uncert.	W40

No.	Molecular formula, name and pK value(s)	$T(^oC)$	Remarks	Method	Assessment	Ref.
2886	$C_6H_{10}O_5$ 3-Oxapentanedioic acid, 2,2-dimethyl- pK$_1$ 3.44, pK$_2$ 4.87	25±0.5	c < 0.001	E3bg	Approx.	S93
2887	$C_6H_{10}O_5$ 3-Oxapentanedioic acid, 2,4-dimethyl- (2,2'-Oxydipropionic acid) pK$_1$ 3.15, pK$_2$ 5.24$_5$ (meso) pK$_1$ 3.09, pK$_2$ 4.60 (DL)	25±0.5	c < 0.001	E3bg	Approx.	S93
2888	$C_6H_{10}O_5$ Propanedioic acid, 2-hydroxy-2-propyl- pK$_1$ 2.54, pK$_2$ 4.64					G37
2889	$C_6H_{10}O_6$ 3,6-Dioxaoctanedioic acid (Ethylenedioxydiacetic acid) pK$_1$ 3.15, pK$_2$ 3.97	20	I = 0.1(NaClO$_4$), mixed constants	E3bg	Uncert.	021
2890	$C_6H_{10}O_8$ Galactosaccharic acid (Mucic acid) pK$_1$ 3.08, pK$_2$ 3.63 pK$_1$ 3.29, pK$_2$ 4.41	25	I = 1.0(NaNO$_3$), concentration constants	E3bg E3	Uncert.	B107 M26
2891	$C_6H_{10}O_8$ D-Idosaccharic acid pK$_1$ 3.26, pK$_2$ 4.13	23±2	I = 0.01-0.03, mixed constants	E3	Uncert.	M21
2892	$C_6H_{10}O_8$ D-Mannosaccharic acid pK$_1$ 2.94, pK$_2$ 3.88			E3		M26
2893	$C_6H_{10}O_8$ D-Saccharic acid pK$_1$ 3.01, pK$_2$ 3.94			E3		M26

No.	Compound	pK	Temp. (°C)	Conditions	Method	Reliability	Ref.
2894	$C_6H_{10}O_8$ D-Talomucic acid	pK_1 3.45, pK_2 4.25	23±2	I = 0.01-0.03, mixed constants	E3	Uncert.	M21
2895	$C_6H_{10}O_8$ DL-Talomucic acid	pK_1 3.47, pK_2 4.29	23±2	I = 0.01-0.03, mixed constants	E3	Uncert.	M21
2896	$C_6H_{12}O_2$ Butanoic acid, 2,2-dimethyl-	4.93	25		CAL	Uncert.	C54
2897	$C_6H_{12}O_2$ Hexanoic acid	4.8	25		E3bg	Uncert.	C57
		4.35	not stated	I = 0.01(KCl)	E3bg	Uncert.	C21
2898	$C_6H_{12}O_2$ Pentanoic acid, 2-methyl-	4.742	10		E3bg	Approx.	C54
		4.782	25				
		4.877	40				
2899	$C_6H_{12}O_2$ Pentanoic acid, 3-methyl-	4.752	10		E3bg	Approx.	C54
		4.766	25				
		4.821	40				
2900	$C_6H_{12}O_5$ Glucose, 2-deoxy-	12.89	10	c = 0.01	CAL	Approx.	C53
		12.52	25	Thermodynamic quantities are also given			

(Contd)

No.	Molecular formula, name and pK value(s)	$T(^oC)$	Remarks	Method	Assessment	Ref.
2900 (Contd)	$C_6H_{12}O_5$ Glucose, 2-deoxy-					
	12.28	40				
	12.51	25	Thermodynamic quantities are also given	CAL	Approx.	I34
2901	$C_6H_{12}O_6$ Fructose					
	12.53	10	c = 0.01	CAL	Approx.	C53
	12.03	25				
	12.27	25	Thermodynamic quantities are also given	CAL	Approx.	I34
2902	$C_6H_{12}O_6$ Galactose					
	12.82	10	c = 0.01	CAL	Approx.	C53
	12.35	25				
	12.48	25	Thermodynamic quantities are also given	CAL	Approx.	I34
2903	$C_6H_{12}O_6$ Glucose					
	12.46	25	Thermodynamic quantities are also given	CAL	Approx.	I34
	12.72	10	c = 0.01	CAL	Approx.	C53
	12.28	25	Thermodynamic quantities are also given	CAL	Approx.	
	12.38	25	c = 0.2-0.8	C3	Approx.	B174
	12.34	25		E3c	Uncert.	
	11.80	45				

No.	Formula / Name	pK	T	Remarks	Method	Reliability	Ref.
2904	$C_6H_{12}O_6$ Mannose	12.45	10	c = 0.01	CAL	Approx.	C53
		12.08	25	Thermodynamic quantities are also given			
		11.81	40				
		12.08	25	Thermodynamic quantities are also given	CAL	Approx.	I34
2905	$C_6H_{12}O_7$ D-gluconic acid	3.76	17		E		L60
2906	$C_6H_{14}O_6$ D-Mannitol	13.29	25		KIN	Uncert.	M126
2907	$C_6H_{14}O_6$ Sorbitol	13.00	60		KIN	Uncert.	R58
2908	$C_6H_{14}S$ Hexanethiol	10.3	25	Mixed constant	SOL	Uncert.	D4
2909	C_6HOCl_5 Phenol, pentachloro-	4.5	not stated		05	Uncert.	P9
2910	C_6HOF_5 Phenol, pentafluoro-	5.53	25	Mixed constant	E3d	Uncert.	B79
		5.2	not stated		05	Uncert.	P9

No.	Molecular formula, name and pK value(s)	$T(^{\circ}C)$	Remarks	Method	Assessment	Ref.
2911	$C_6HO_2F_3$ 1,4-Benzoquinone, 2,3,6-trifluoro-					
	1.73	25	Mixed constant	05	Approx.	W2
2912	$C_6HO_3Cl_3$ 1,4-Benzoquinone, 3,5,6-trichloro-2-hydroxy-					
	pK_1 (CO) -8.8	25±0.2	In aq.H_2SO_4; H_0 scale	06	Uncert.	B41
	pK_2 (OH) 1.05	25±0.2	I = 0.5(NaCl), concentration constant	E3bg		
	pK_2 1.09	25	Mixed constant	05	Approx.	W2
2913	$C_6HO_3Br_3$ 1,4-Benzoquinone, 3,5,6-tribromo-2-hydroxy-					
	1.10	25	Mixed constant	05	Approx.	W2
2914	$C_6H_2O_2Cl_4$ 1,2-Benzenediol, tetrachloro-					
	pK_1 5.80, pK_2 10.10	25	c = 0.00015, I = 0.1(KCl)	05,E3bg	Approx.	M70
2915	$C_6H_2O_4Cl_2$ 1,4-Benzoquinone, 2,5-dichloro-3,6-dihydroxy- (Chloranilic acid)					
	pK_1 0.66, pK_2 2.37	25	I = 0.5(NaCl-HCl), concentration constants	05	Approx.	B40
	pK_1 0.76, pK_2 2.58	25	I = 0.5(NaClO$_4$), concentration constants	05	Approx.	K43
	pK_1 0.97, pK_2 2.55		I = 2.0(NaClO$_4$)			
	pK_1 1.0	25	c = 0.00003, I = 1.0(NaClO$_4$), mixed constants	05	Uncert.	B86
	pK_2 2.64				Approx.	
	pK_1 0.81, pK_2 2.72	25	I = 0.15, concentration constants	06	Uncert.	C1
	pK_1 0.56	20	c = 0.00002, I = 0.1, concentration constants	05	Uncert.	N21
	pK_2 2.97				Approx.	
	pK_1 0.73, pK_2 3.08	25		05	Uncert.	W2
	pK_2 2.75	25	I = 0.37(NaClO$_4$)	05	Uncert.	L13

No.	Formula	Name / pK values	Temp	Conditions		Ref. code	Reliability	Ref.
2916	$C_6H_2O_4Br_2$	1,4-Benzoquinone, 2,5-dibromo-3,6-dihydroxy-						
		pK_1 0.80, pK_2 3.10	25	Mixed constants		05	Uncert.	W2
2917	$C_6H_2O_4F_2$	1,4-Benzoquinone, 2,5-difluoro-3,6-dihydroxy-						
		pK_1 1.40, pK_2 3.30	25	Mixed constants		05	Approx.	W2
2918	$C_6H_2O_8N_2$	1,4-Benzoquinone, 2,5-dihydroxy-3,6-dinitro-						
		pK_2 -0.5	25	Conditions not stated			Uncert.	B42
		pK_1 -3.0, pK_2 -0.5	25			06	Uncert.	W2
2919	$C_6H_3OCl_3$	Phenol, 2,4,6-trichloro-						
		6.23	25	c = 0.0016		E3bg	Approx.	F17
2920	$C_6H_3OCl_3$	Phenol, 3,4,5-trichloro-						
		7.839	25	c = 0.0002, I = 0.02-0.06		05a	Rel.	B90

Variation with temperature

	5	10	15	20	30	35	40	45	50
	8.102	8.032	7.967	7.902	7.786	7.728	7.672	7.624	7.580

No.	Formula	Name / pK values	Temp	Conditions	Ref. code	Reliability	Ref.
2921	$C_6H_3O_7N_3$	Phenol, 2,4,6-trinitro- (Picric acid)					
		0.33	25	c = 0.00005	05	Approx.	D20
		0.27	20	c ≈ 0.0001, aq.$HClO_4$, H_0 values	06	Uncert.	T58
		0.373	25	I = 0.1(NH_4ClO_4)	06	Uncert.	W30
				Other values in M100 and P16			

No.	Molecular formula, name and pK value(s)	$T(^{\circ}C)$	Remarks	Method	Assessment	Ref.
2922	$C_6H_3O_8N_3$ 1,3-Benzenediol, 2,4,6-trinitro-	(Styphnic acid)				
	pK$_1$ 0.06	25	$c = 0.00005$, $I = 0.1(NH_4ClO_4)$	06	Uncert.	W30
	pK$_2$ 4.227		$c = 0.00008$, $I = 0.05$-0.005	05	Approx.	
	pK$_1$ 1.74, pK$_2$ 4.86	not stated	$c = 0.003$, $I = 0.5(NaClO_4)$	05	Uncert.	P16
2923	$C_6H_3O_9N_3$ 1,3,5-Benzenetriol, 2,4,6-trinitro-	(Phloroglucinol, trinitro-)				
	pK$_1$ 1.26, pK$_2$ 4.16, pK$_3$ 7.66	not stated	$c = 0.003$, $I = 0.5(NaClO_4)$	05	Uncert.	P16
2924	$C_6H_4ON_4$ Pteridin-2(1H)-one					
	pK$_3$ 13.03	20	Equilibrium between hydrated mononanion and a dianion	05	Uncert.	B170
	pK$_3$ 14.11	20	Equilibrium between the anhydrous/hydrated monoanion mixture and a dianion		Uncert.	
2925	$C_6H_4ON_4$ Pteridin-6(5H)-one					
	pK$_3$ 12.70	20	Equilibrium between hydrated monoanion and a dianion	05	Uncert.	B170
	pK$_3$ 14.2	20	Equilibrium between the anhydrous/ hydrated monoanion mixture and a dianion		Uncert.	
2926	$C_6H_4OCl_2$ Phenol, 2,3-dichloro-					
	7.696	25	$c \simeq 0.0001$	05a	Approx.	R28

2927 C$_6$H$_4$OCl$_2$ Phenol, 2,4-dichloro-

7.892	25	c ≈ 0.00006	05a	Approx.	R28
7.70	25	c ≈ 0.0001-0.002, I ≈ 0.1, mixed constant	05	Approx.	S76

2928 C$_6$H$_4$OCl$_2$ Phenol, 2,5-dichloro-

7.508	25	c ≈ 0.00005	05a	Approx.	R28

2929 C$_6$H$_4$OCl$_2$ Phenol, 2,6-dichloro-

6.786	25	c ≈ 0.00005	05a	Rel.	P1
6.791	25	c ≈ 0.00005	05a	Rel.	R28
6.81	25	c ≈ 0.005	E3bg	Approx.	F17

2930 C$_6$H$_4$OCl$_2$ Phenol, 3,4-dichloro-

8.585	25	c ≈ 0.00006	05a	Approx.	R28
8.630	25	c = 0.00014, I = 0.02-0.06	05a	Rel.	B90

Variation with temperature

5	10	15	20	30	35	40	45	50
8.930	8.848	8.774	8.701	8.566	8.500	8.437	8.380	8.329

Thermodynamic quantities are derived from the results

2931 C$_6$H$_4$OCl$_2$ Phenol, 3,5-dichloro-

8.185	25	c ≈ 0.0002	05a	Rel.	R28
8.179	25	m = 0.00013	05a	Rel.	B92

(Contd)

No.	Molecular formula, name and pK value(s)	T(°C)	Remarks	Method	Assessment	Ref.
2931 (Contd)	$C_6H_4OCl_2$ Phenol, 3,5-dichloro-		Variation with temperature			
		5	8.447			
		10	8.376			
		15	8.311			
		20	8.243			
		30	8.120			
		35	8.063			
		40	8.013			
		45	7.965			
		50	7.922			
			Thermodynamic quantities are derived from the results			
2932	$C_6H_4OBr_2$ Phenol, 2,4-dibromo- 7.790	25	c ≈ 0.0002	05a	Rel.	R29
2933	$C_6H_4OBr_2$ Phenol, 2,6-dibromo- 6.674	25	c ≈ 0.00003	05a	Approx.	R29
2934	$C_6H_4OBr_2$ Phenol, 3,5-dibromo- 8.056	25	m = 0.000012	05a	Rel.	B92
			Variation with temperature			
		5	8.375			
		10	8.284			
		15	8.200			
		20	8.125			
		30	7.995			
		35	7.941			
		40	7.893			
		45	7.850			
		50	7.813			
			Thermodynamic quantities are derived from the results			

2935 $C_6H_4OI_2$ Phenol, 3,5-diiodo-

    8.103   25   m = 0.000016

Variation with temperature

5	10	15	20	30	35	40	45	50	05a	Rel.	B92
8.432	8.338	8.252	8.175	8.040	7.983	7.933	7.889	7.851			

Thermodynamic quantities are derived from the results

2936 $C_6H_4O_2N_2$ 2,1,3-Benzoxadiazol-4(1H)-one

    6.83   25   c ≈ 0.0001, mixed constant   05   Approx.   D5

2937 $C_6H_4O_2N_2$ 2,1,3-Benzoxadiazol-5(3H)-one

    7.28   25   c ≈ 0.0001, mixed constant   05   Approx.   D5

2938 $C_6H_4O_2N_2$ 2,1,3-Benzoxadiazol-6(1H)-one

    7.28   25   0   D6

2939 $C_6H_4O_2N_2$ 2,1,3-Benzoxadiazol-7(3H)-one

    6.83   25   0   D6

2940 $C_6H_4O_2N_4$ Pteridine-2,4(1H,3H)-dione

    7.95   20   c = 0.001   E3bg   Approx.   P41

    7.91   20   c = 0.01   E3bg   Approx.   A24

No.	Molecular formula, name and pK value(s)	$T(^{\circ}C)$	Remarks	Method	Assessment	Ref.
2941	$C_6H_4O_2N_4$ Pteridine-2,6($1\underline{H},5\underline{H}$)-dione pK_1 5.58, pK_2 8.64	20	Anhydrous. c = 0.001	05	Approx.	A28
	pK_1 8.55, pK_2 9.69		Equilibrium pK, partial covalent hydration			
	pK_1 8.84, pK_2 10.29	20	Covalently hydrated species		Approx.	I5
2942	$C_6H_4O_2N_4$ Pteridine-2,7($1\underline{H},3\underline{H}$)-dione pK_1 5.83, pK_2 10.07	20	c = 0.005	E3bg	Approx.	A29
2943	$C_6H_4O_2N_4$ Pteridine-4,6($1\underline{H},5\underline{H}$)-dione pK_1 6.05, pK_2 9.51	20	"Anhydrous" species	E3bg	Approx.	I5
	pK_1 6.40, pK_2 9.51		Equilibrium pK, partial covalent hydration			
	pK_1 8.34, pK_2 10.08		Covalently hydrated species			
2944	$C_6H_4O_2N_4$ Pteridine-4,7($1\underline{H},8\underline{H}$)-dione pK_1 6.08, pK_2 9.62	20	c = 0.0013	E3bg	Approx.	A22
2945	$C_6H_4O_2Cl_2$ 1,4-Benzenediol, 2,6-dichloro- pK_1 7.38	13.6	I = 0.65($NaClO_4$), concentration constants	05	Approx.	B38
	pK_1 7.34	20.2				
	pK_1 7.30	26.1				
	pK_2 10.16	12.7				
	pK_2 10.06	19.4				
	pK_2 9.99	25.1	Thermodynamic quantities are derived from these results			

No.	Formula / Name / pK	Temp	Conditions	Method	Assessment	Ref.
2946	C₆H₄O₃N₄ Pteridine-2,4,6(1H,3H,5H)-trione					
	pK₁ 5.85, pK₂ 9.43	20		E3bg	Approx.	P43
2947	C₆H₄O₃N₄ Pteridine-2,4,7(1H,3H,8H)-trione					
	pK₁ 3.61	20	c = 0.0005	E3bg	Approx.	A29
	pK₂ 9.80	20	c = 0.001	E3bg	Approx.	P42
2948	C₆H₄O₃N₄ 1H-Pyrazolo[4,3-d]pyrimidine-3-carboxylic acid, 4,7-dihydro-7-oxo-					
	3.6		Conditions not stated	E3bg	Uncert.	R27
2949	C₆H₄O₃S 1-Thiacyclohexa-2,5-diene-3-carboxylic acid, 4-oxo-					
	4.10	25		C2	Uncert.	S11
2950	C₆H₄O₄N₂ 2,3-Pyrazinedicarboxylic acid					
	pK₁ 0.9, pK₂ 3.57		Conditions not stated	O5 & E3bg	Uncert.	M56
	pK₁ 2.20			0		S12
2951	C₆H₄O₄N₂ 2,5-Pyrazinedicarboxylic acid					
	pK₁ 2.29			0		S12
2952	C₆H₄O₄N₂ 4,5-Pyridazinedicarboxylic acid					
	3.30		Conditions not stated	E3bg	Uncert.	M56
2953	C₆H₄O₄S 3,4-Thiophenedicarboxylic acid					
	pK₁ 2.81, pK₂ 6.93		Conditions not stated	E3bg	Uncert.	O1

No.	Molecular formula, name and pK value(s)	T(^{o}C)	Remarks	Method	Assessment	Ref.
2954	$C_6H_4O_5N_2$ Phenol, 2,3-dinitro-					
	4.96	25	c $\approx$ 0.0002	05	Approx.	R34
	4.98			E3bg		
2955	$C_6H_4O_5N_2$ Phenol, 2,4-dinitro-					
	4.073	25		C2,Rle	Rel.	S90
			Variation with temperature			
			35 50 60 75 90			
			4.017 3.947 3.910 3.873 3.854			
			Equation given for temperature variation.			
			Thermodynamic quantities are derived from the results			
	4.10	25	c $\approx$ 0.0002	05	Approx.	M19
	3.93		I = 0.1			
	4.07	25	c = 0.002, I $\approx$ 0.5, mixed constant	05	Approx.	R15
	4.12 (H_2O)	25		E3bg	Approx.	G21
	4.68 (D_2O)					
	4.07 (H_2O)	25	I < 0.07.	03	Approx.	B51
	4.59 (D_2O)					
	4.06 (H_2O)	25	Based on pK = 5.28 for acetic acid in D_2O	05	Uncert.	W12
	4.55 (D_2O)					

2956 C$_6$H$_4$O$_5$N$_2$ Phenol, 2,5-dinitro-

pK	T	Conditions		Reliability		Ref.
5.21	25	c ≈ 0.0001		05	Approx.	R34
5.04	25	c ≈ 0.00001-0.0001, I ≈ 0.1, mixed constant		05	Approx.	H11
5.212	25					S88
5.20 (H$_2$O)	25	I < 0.07.		03	Approx.	B51
5.73 (D$_2$O)		Based on pK = 5.28 for acetic acid in D$_2$O				
5.19 (H$_2$O)	25			05	Uncert.	W12
5.70 (D$_2$O)						

2957 C$_6$H$_4$O$_5$N$_2$ Phenol, 2,6-dinitro-

pK	T	Conditions		Reliability		Ref.
3.695	25			C2,R1e	Rel.	S90

Variation with temperature

35	50	60	75	90
3.652	3.611	3.593	3.582	3.578

Equation given for temperature variation.

Thermodynamic quantities are derived from the results.

pK	T	Conditions		Reliability		Ref.
3.73 (H$_2$O)	25	I < 0.07		03	Approx.	B51
4.22 (D$_2$O)		Based on pK = 5.28 for acetic acid in D$_2$O				

2958 C$_6$H$_4$O$_5$N$_2$ Phenol, 3,4-dinitro-

pK	T	Conditions		Reliability		Ref.
5.42	25	c ≈ 0.00007		05	Approx.	R34

No.	Molecular formula, name and pK value(s)	T(°C)	Remarks	Method	Assessment	Ref.
2959	C$_6$H$_4$O$_5$N$_2$ Phenol, 3,5-dinitro-					
	6.684 (H$_2$O)	25	m ≈ 0.0004	05a	Rel.	R31
	7.305 (D$_2$O)	25				
	6.69	25	c ≈ 0.0005	05	Approx.	R34
				E3bg		
	6.66					
	6.732	25	m = 0.00022	05a	Rel.	B92

Variation with temperature

T(°C)	5	10	15	20	30	35	40	45	50	55	60
	6.937	6.884	6.831	6.780	6.686	6.644	6.604	6.567	6.531	6.495	6.466

Thermodynamic quantities are derived from the results

No.	Molecular formula, name and pK value(s)	T(°C)	Remarks	Method	Assessment	Ref.
2960	C$_6$H$_4$O$_5$N$_2$ 2-Pyridinecarboxylic acid N-oxide, 4-nitro-					
	1.78		c = 0.04 or 0.001 (?)	E		K51
2961	C$_6$H$_4$O$_5$N$_2$ 3-Pyridinecarboxylic acid N-oxide, 4-nitro-					
	pK (N$^+$-OH) 0.50		Conditions not stated	E	Uncert.	D52
2962	C$_6$H$_4$O$_6$N$_2$ 1,2-Benzenediol, 3,4-dinitro-					
	pK$_1$ 4.02, pK$_2$ 8.24	30	I = 0.1(NaClO$_4$), mixed constant	E	Uncert.	M59
2963	C$_6$H$_4$O$_6$N$_2$ 1,2-Benzenediol, 3,5-dinitro-					
	pK$_1$ 3.32, pK$_2$ 9.86	30	I = 0.1(NaClO$_4$), mixed constants	E	Uncert.	M59
	pK$_1$ 3.39, pK$_2$ 10.03	25	I = 0.1(NaClO$_4$) in 4% ethanol	05	Uncert.	N24

No.		Temp.		Method		
2964	$C_6H_4O_{10}S_2$ 1,4-Cyclohexadiene-1,4-disulfonic acid, 2,5-dihydroxy-3,6-dioxo- (Euthiochronic acid)					
	pK$_1$ 0.75, pK$_2$ 3.81	25	I = 0.5(HCl-NaCl), concentration constants		Uncert.	B42
2965	$C_6H_4N_4S$ Pteridine-2(1H)-thione					
	6.52	20	"Anhydrous" species	E3bg	Approx.	I5
	9.00	20	Equilibrium pK, partial covalent hydration		Approx.	
	9.72	20	Covalently hydrated species		Approx.	
2966	$C_6H_4N_4S$ Pteridine-4(1H)-thione					
	6.81	20	I = 0.0016	E3bg	Approx.	A24
2967	$C_6H_4N_4S$ Pteridine-7(1H)-thione					
	5.50	20	c = 0.002	E3bg	Approx.	A26
2968	C_6H_5ON 2-Pyridinecarbaldehyde					
	pK$_1$ 3.84	25	c = 0.001	05	Approx.	G41

Variation of pK$_1$ with temperature

5	15	30	40	50	60
4.13	4.00	3.76	3.57	3.42	3.25

Thermodynamic quantities are derived from the results

	pK$_2$ 12.68	25	c = 0.0002 (pK of hydrated aldehyde)	05	Uncert.	G43
2969	C_6H_5ON 3-Pyridinecarbaldehyde					
	13.0	30	I = 0.1(KCl), mixed constant	E3bg	V.uncert.	E18
	(pK of hydrated species)					

No.	Molecular formula, name and pK value(s)	T(°C)	Remarks	Method	Assessment	Ref.
2970	C_6H_5ON 4-Pyridinecarbaldehyde 12.05	30	I = 0.1(KCl), mixed constant (pK of hydrated species)	E3bg	Uncert.	E18
2971	$C_6H_5ON_5$ Pteridin-4(1H)-one, 2-amino- pK_1 2.20, pK_2 7.86	20		E3bg & 05	Uncert.	P52
2972	$C_6H_5ON_5$ Pyrimidino[5,4-e]-1,2,4-triazin-5(1H)-one, 3-methyl- 7.00	20	c < 0.0001, I = 0.01, mixed constant	05	Approx.	B141
2973	C_6H_5OCl Phenol, 2-chloro- 8.555	25	m = 0.0002	05a	Rel.	B94

Variation with temperature

5	10	15	20	30	35	40	45	50
8.827	8.747	8.675	8.611	8.507	8.465	8.430	8.400	8.377

Thermodynamic quantities are derived from the results

	8.527	25	c = 0.00004, I = 0.01	01	Approx.	C40

Variation with temperature

5	10	15	20	30	35
8.770	8.698	8.634	8.572	8.473	8.404

Thermodynamic quantities are derived from the results

(Contd)

2973 (Contd) C$_6$H$_5$OCl Phenol, 2-chloro-

pK	Temp	Conditions	Method	Reliability	Ref.
8.48	25		CAL	Approx.	O6
		Thermodynamic quantities are also given			
8.25	25	I = 0.1(NaClO$_4$), concentration constant	05	Approx.	D34
		Thermodynamic quantities are also given			
8.53	25	c = 0.0002	05	Approx.	B77
8.34	25	c ≈ 0.0001-0.0002, I ≈ 0.1, mixed constant	05	Approx.	S76
3.3		pK of electronically excited singlet state, calculated assuming pK = 8.5 in ground state	FLU/05	V.uncert.	B25

Other values in F8 and J1

2974 C$_6$H$_5$OCl Phenol, 3-chloro-

pK	Temp	Conditions	Method	Reliability	Ref.
9.119	25	m = 0.00018	05a	Rel.	B95

Variation with temperature

10	15	20	30	35	40	45	50	55	60
9.332	9.258	9.186	9.055	8.995	8.938	8.883	8.833	8.784	8.739

Thermodynamic quantities are derived from the results

pK	Temp	Conditions	Method	Reliability	Ref.
9.08	25	Thermodynamic quantities are also given	CAL	Approx.	O6
9.13	25	c = 0.0002	05	Approx.	B77
8.76	25	I = 0.1(NaClO$_4$), concentration constant	E3bg	Approx.	J1
8.80	25	I = 0.1(NaClO$_4$), concentration constant	05	Approx.	D34

Thermodynamic quantities are also given

(Contd)

No.	Molecular formula, name and pK value(s)	$T(^oC)$	Remarks	Method	Assessment	Ref.
2974(Contd)	C_6H_5OCl Phenol, 3-chloro-					
	9.13	25	I = 0.1(KCl), mixed constant	E3bg	Approx.	H51
	4.0		pK of electronically excited singlet state,	FLU/O5	V.Uncert.	B25
			calculated assuming pK = 0.1 in ground state			
2975	C_6H_5OCl Phenol, 4-chloro-					
	9.406	25.9	m = 0.0002	O5a	Rel.	B96
			Variation with temperature			
		15.9 37.6 47.2 57.2 59.3				
		9.564 9.261 9.164 9.041 9.045				
			Thermodynamic quantities are derived from the results			
	9.38	25	c = 0.004-0.04	CAL	Approx.	F8, O6
			Thermodynamic quantities are also given			
	9.42	25	c = 0.0002	O5	Approx.	B77
	9.16	25	I = 0.1($NaClO_4$), concentration constant	O5	Approx.	D34
			Thermodynamic quantities are also given			
	9.10	25	I = 0.1($NaClO_4$), concentration constant	E3bg	Approx	J1
	9.35	25	I = 0.1(KCl), mixed constant	E3bg	Approx.	H51
	3.5		pK of electronically excited singlet state,	FLU/O5	V.uncert.	B25
			calculated assuming pK = 9.4 in ground state			

2976 C_6H_5OBr Phenol, 2-bromo-

| 8.452 | 25 | m = 0.0002 | 05a | Rel. | B94 |

Variation with temperature

5	10	15	20	25	m = 0.0002	30	35	40	45	50
8.705	8.634	8.569	8.508			8.400	8.352	8.308	8.268	8.231

Thermodynamic quantities are derived from the results

8.44	25	c = 0.00029	05	Approx.	B77
8.22	25	I = 0.1($NaClO_4$), concentration constant	E3bg	Approx.	J1
8.33	25	I = 0.1(KCl), mixed constant	E3bg	Approx.	H51

2977 C_6H_5OBr Phenol, 3-bromo

| 9.031 | 25 | m = 0.00017 | 05a | Rel. | B95 |

Variation with temperature

10	15	20	25	m = 0.00017	30	35	40	45	50	55	60
9.249	9.172	9.099			8.967	8.907	8.853	8.801	8.753	8.708	8.666

Thermodynamic quantities are derived from the results

9.03	25	c = 0.0003	05	Approx.	B77
9.00	25	c = 0.00023	03	Approx.	A67
8.75	25	I = 0.1($NaClO_4$), concentration constant	E3bg	Approx.	J1
9.06	25	I = 0.1(KCl), mixed constant	E3bg	Approx.	H51

No.	Molecular formula, name and pK value(s)	T(°C)	Remarks	Method	Assessment	Ref.
2978	C_6H_5OBr Phenol, 4-bromo-					
	9.366	23.6	m = 0.00014	05a	Rel.	B96

Variation with temperature

T(°C)	5.7	9.4	15.1	31.9	39.7	45.3	50.8	56.0	58.6
pK	9.641	9.581	9.481	9.244	9.152	9.073	9.019	8.971	8.947

Thermodynamic quantities are derived from the results

pK	T(°C)	Remarks	Method	Assessment	Ref.
9.36	25	c = 0.0003	05	Approx.	B77
9.06	25	I = 0.1($NaClO_4$), concentration constant	E3bg	Approx.	J1
9.17	25	I = 0.5(KNO_3), mixed constant	E3bg	Approx.	M5
9.35 (H_2O)	25		05	Uncert.	W12
9.94 (D_2O)					
2.9		pK of electronically excited singlet state, calculated assuming pK = 9.3 in ground state	FLU/05	V.uncert.	B25

Other values in E20 and H51

No.	Molecular formula, name and pK value(s)	T(°C)	Remarks	Method	Assessment	Ref.
2979	C_6H_5OI Phenol, 2-iodo-					
	8.513	25	m = 0.00021	05a	Rel.	B94

Variation with temperature

T(°C)	5	10	15	20	30	35	40	45	50
pK	8.765	8.694	8.628	8.568	8.463	8.418	8.377	8.340	8.306

Thermodynamic quantities are derived from the results

pK	T(°C)	Remarks	Method	Assessment	Ref.
8.51	25	c = 0.00012	05	Approx.	B77
8.44	25	I = 0.1(KCl), mixed constant	E3bg	Approx.	H51

2980 C_6H_5OI Phenol, 3-iodo-

9.033	25	m = 0.00012	05a	Rel.	B95

Variation with temperature

10	15	20	30	35	40	45	50	55	60
9.261	9.180	9.104	8.967	8.905	8.848	8.794	8.744	8.697	8.653

Thermodynamic quantities are derived from the results

9.06	25	c = 0.00023	05	Approx.	B77
8.74	25	I = 0.1($NaClO_4$), concentration constant	E3bg	Approx.	J1

2981 C_6H_5OI Phenol, 4-iodo-

9.327	24.5	m = 0.000022	05a	Rel.	B96

Variation with temperature

10.0	11.7	16.5	33.8	41.9	50.0	55.7
9.538	9.519	9.444	9.206	9.126	9.053	8.994

Thermodynamic quantities are derived from the results

9.30	25	c = 0.00023	05	Approx.	B77
9.21	25	I = 0.1(KCl), mixed constant	E3bg	Approx.	H51

2982 C_6H_5OF Phenol, 2-fluoro-

8.70	25	c = 0.00045	05	Approx.	B77
8.49	25	I = 0.1($NaClO_4$), concentration constant	E3bg	Approx.	J1
8.73	25	Thermodynamic quantities are also given	05	Approx.	C80

No.	Molecular formula, name and pK value(s)	T(°C)	Remarks	Method	Assessment	Ref.
2983	C_6H_5OF Phenol, 3-fluoro-					
	9.21	25	c = 0.00045	05	Approx.	B77
	9.29	25	Thermodynamic quantities are also given	05	Approx.	C80
	8.81	25	I = 0.1($NaClO_4$), concentration constant	E3bg	Approx.	J1
2984	C_6H_5OF Phenol, 4-fluoro-					
	9.91	25	c = 0.0002	05	Approx.	B77
	9.89	25	Thermodynamic quantities are also given	05	Approx.	C80
	9.75	25	I = 0.1(KCl), mixed constant	E3bg	Approx.	E20
	9.46	25	I = 0.1($NaClO_4$), concentration constant	E3bg	Approx.	J1
	3.5		pK of electronically excited singlet state, calculated assuming pK = 9.8 in ground state	FLU/05	V.uncert.	B25
2985	$C_6H_5O_2N$ 1,4-Benzoquinone 1-oxime					
	6.35	not stated	Mixed constant	05	Uncert.	D35
	6.25	not stated	I = 0.2(NaCl)	E3bg	Uncert.	C72
2986	$C_6H_5O_2N$ Phenol, 4-nitroso-					
	6.361	25	m = 0.00007, I = 0.01-0.04	05a	Rel.	F12

No.	Compound	pK	T (°C)	Method	Reliability	Ref.	Conditions
2987	$C_6H_5O_2N$ 2-Pyridinecarboxylic acid (Picolinic acid)						
	pK$_1$ 1.03	25	E3bg	Uncert.	D61	$c \approx 0.01$, I = 0.1(NaClO$_4$), mixed constants
	pK$_2$ 5.30			Approx.		
	pK$_1$ 1.07	25	E3bg	Uncert.	M105	$c = 0.05$, I = 0.5(NaClO$_4$), concentration constants
	pK$_2$ 5.25			Approx.		
	pK$_1$ 1.36	25	E3bg	Uncert.	P6	I = 3.0(KBr)
	pK$_2$ 5.80			Approx.		
	pK$_1$ 1.5, pK$_2$ 5.43	18±2	05	Uncert.	H50	I = 0.5(NaCl), concentration constants
	pK$_2$ 5.20	25	E3bg	Approx.	T28	$c \approx 0.01$ or 0.001, I = 0.1(KNO$_3$), concentration constant
						Other values in K51 and S135
2988	$C_6H_5O_2N$ 3-Pyridinecarboxylic acid (Nicotinic acid)						
	pK$_2$ 4.78	25	05	Uncert.	S19	$c = 0.00015$, I = 1.0, mixed constant
	pK$_2$ 4.89		E		K51	
2989	$C_6H_5O_2N$ 4-Pyridinecarboxylic acid (Isonicotinic acid)						
	4.90		E		K51	
2990	$C_6H_5O_2N_5$ Pteridine-4,6(1H,5H)-dione, 2-amino- (Xanthopterin)						
	pK$_1$ 6.25	20	E3bg	Uncert.	A24	$c = 0.001$
	pK$_2$ 9.23			Approx.		

No.	Molecular formula, name and pK value(s)	$T(^{o}C)$	Remarks	Method	Assessment	Ref.
2991	$C_6H_5O_2N_5$ Pteridine-4,7(1H,8H)-dione, 2-amino- (Isoxanthopterin)					
	pK_1 7.34	20		05	Approx.	P57
	pK_2 10.06				Uncert.	
2992	$C_6H_5O_2Cl$ 1,2-Benzenediol, 4-chloro-					
	pK_1 8.62, pK_2 12.55	25	I = 0.1(KCl), concentration constants	05	Approx.	P63
	pK_1 8.43, pK_2 11.54	30	I = 0.1(KNO$_3$), mixed constants	E3bg	Approx.	M123
	pK_1 8.7, pK_2 12.4	25	I = 0.1(KCl), concentration constants	E3bg & 05	Uncert.	A1
			Other value in M70			
2993	$C_6H_5O_2Cl$ 1,4-Benzenediol, 2-chloro-					
	pK_1 8.91, pK_2 10.86	13	I = 0.1(KCl), mixed constants	05	Approx.	S111
	pK_1 8.81, pK_2 10.78	23				
			Thermodynamic quantities are derived from the results			
2994	$C_6H_5O_2Br$ 1,2-Benzenediol, 4-bromo-					
	pK_1 8.7	25	I = 0.1(KCl), concentration constants	E3bg	Uncert.	A1
	pK_2 12.4			05		
2995	$C_6H_5O_2Br$ 1,4-Benzenediol, 2-bromo-					
	pK_1 8.78, pK_2 10.77	13	I = 0.1(KCl), mixed constants	05	Approx.	S111
	pK_1 8.67, pK_2 10.68	22				
			Thermodynamic quantities are derived from the results			

2996 $C_6H_5O_3N$ Phenol, 2-nitro-

7.230   25   c = 0.0002   05a   Rel.   R37

Variation with temperature

5	10	15	20	30	35	40	45	50	55	60
7.499	7.424	7.353	7.293	7.180	7.135	7.085	7.043	6.993	6.966	6.931

Thermodynamic quantities are derived from the results

7.22   25   c = 0.004-0.04   CAL   F8

Thermodynamic quantities are also given

7.21   25   05   Approx.   B77

7.25   25   c ≈ 0.0002   05   Approx.   M19

7.06   I = 0.1, mixed constant

7.08   25   I = 0.1(KCl), mixed constant   E3bg   Approx.   E20

7.04   25   I = 0.1($NaClO_4$), concentration constant   E3bg   Approx.   J1

7.08   25   c = 0.0002, I ≈ 0.5, mixed constant   05   Approx.   R15

7.22 (H_2O)   25   E3bg   Approx.   G21

7.82 (D_2O)

Other values in H11, H17, H51 and S88

2997 $C_6H_5O_3N$ Phenol, 3-nitro-

8.355   25   c = 0.0006   05a   Rel.   R38

Variation with temperature

5	10	15	20	30	35	40	45	50
8.663	8.547	8.475	8.423	8.284	8.232	8.167	8.111	8.078

Thermodynamic quantities are derived from the results   (Contd)

No.	Molecular formula, name and pK value(s)		T(°C)	Remarks	Method	Assessment	Ref.
2997 (Contd)	$C_7H_5O_3N$	Phenol, 3-nitro					
	8.360		25	c = 0.00038	O5a	Rel.	B97

Variation with temperature

T	5	10	15	20	30	35	40	45	50
	8.624	8.553	8.485	8.421	8.303	8.249	8.198	8.149	8.103

Thermodynamic quantities are derived from the results

8.35		25	c = 0.004-0.04		CAL	Approx.	F8

Thermodynamic quantities are also given

8.12		25	I = 0.1($NaClO_4$), concentration constant		O5	Approx.	D34

Thermodynamic quantities are also given

8.04		25	I = 0.1($NaClO_4$), concentration constant	E3bg	Approx.	J1
8.23		25	I = 0.1(KCl), mixed constant	E3bg	Approx.	E20
8.18		25	I = 0.5(KNO_3), mixed constant	E3bg	Approx.	M5

Other values in B77, H17 and H51

No.	Molecular formula, name and pK value(s)		T(°C)	Remarks	Method	Assessment	Ref.
2998	$C_6H_5O_3N$	Phenol, 4-nitro-					
	7.156		25		O5	Rel.	A47

Variation with temperature

T	0	5	10	15	20	30	35	40	45	50	55	60
	7.501	7.425	7.350	7.281	7.216	7.101	7.046	7.001	6.958	6.914	6.875	6.839

Thermodynamic quantities are derived from the results

(Contd)

2998(Contd) $C_6H_5O_3N$ Phenol, 4-nitro-

			CAL		
7.14	25	c = 0.004-0.04		Approx.	F8
		Thermodynamic quantities are given			
7.151	25	m = 0.00007	O5a	Rel.	F12
7.16	25	c ≈ 0.0002	O5	Approx.	M19
6.98		I = 0.1, mixed constant			
7.02	25	I = 0.1($NaClO_4$), concentration constant	E3bg	Approx.	J1
6.89	25	I = 0.1($NaClO_4$), concentration constant	O5	Approx.	D34
		Thermodynamic quantities are given			
6.96	25	I = 0.5(KNO_3), mixed constant	E3bg	Approx.	M5
7.14 (H_2O)	25		E3bg	Approx.	G21
7.72 (D_2O)					

Other values in B27, B77, E20, H17, H51, S88, W12 and W32

2999 $C_6H_5O_3N$ 2-Pyridinecarboxylic acid, 3-hydroxy-

pK_2 5.17, pK_3 10.76	25	c ≈ 0.01, I = 0.1($NaClO_4$), mixed constants	E3bg	Approx.	D61

3000 $C_6H_5O_3N$ 3-Pyridinecarboxylic acid, 6-hydroxy-

3.82	20	c = 0.005	E3bg	Approx.	A18

3001 $C_6H_5O_3N$ 4-Pyridinecarboxylic acid, 3-hydroxy-

pK_1 0.10, pK_3 11.30	25	I = 0.1(KCl), mixed constants	O5	Approx.	F35a
pK_2 4.83			E3bg		

No.	Molecular formula, name and pK value(s)	T(°C)	Remarks	Method	Assessment	Ref.
3002	$C_6H_5O_3N$ 2-Pyridinecarboxylic acid N-oxide (Picolinic acid N-oxide) 3.64			E		K51
3003	$C_6H_5O_3N$ 3-Pyridinecarboxylic acid N-oxide (Nicotinic acid N-oxide) pK(N$^+$-OH) 2.74		Conditions not stated	E	Uncert.	D52
	2.63			E		K51
3004	$C_6H_5O_3N$ 4-Pyridinecarboxylic acid N-oxide (Isonicotinic acid N-oxide) 2.90			E		K51
3005	$C_6H_5O_3N_5$ Pteridine-4,6,7(1H,5H,3H)-trione, 2-amino- (Leucopterin) pK$_1$ 7.6, pK$_3$ 13.6 pK$_2$ 9.78	20		05/06	Uncert. Approx.	P56
3006	$C_6H_5O_3N_5$ 1,2,4-Triazolo[4,3-c]pyrimidin-4(1H)-one, 3-methyl-7-nitro- 4.21	20	c < 0.0001, I = 0.01, mixed constant	05	Approx.	B141
3007	$C_6H_5O_3N_5$ 1,2,4-Triazolo[4,3-c]pyrimidin-4(1H)-one, 6-methyl-7-nitro- 4.86	20	c < 0.0001, I = 0.01, mixed constant	05	Approx.	B141
3008	$C_6H_5O_3Cl$ 4H-Pyran-4-one, 6-chloromethyl-3-hydroxy- (Chlorokojic acid) pK$_2$ 7.40	25	I = 0.5(NaCl), mixed constant	E3bg & 05	Approx.	C46
	pK$_1$ (CO) -1.71	25	In aq.H_2SO_4 solutions, H_0 scale	06	Uncert.	C47
	pK$_2$ (OH) 7.70		I = 0.01-0.5(NaCl)	E3bg	Approx.	

No.	Formula, name, pK	T (°C)	Conditions	Method	Reliability	Ref.
3009	$C_6H_5O_3I$ 4H-Pyran-4-one, 3-hydroxy-6-iodomethyl- (Iodokojic acid)					
	pK_2 7.50	25	I = 0.5(NaCl), mixed constant	E3bg & 05	Approx.	C46
	$pK_1(CO)$ -1.60	25	In aq.H_2SO_4 solution, H_0 scale	06	Uncert.	C47
	$pK_2(OH)$ 7.80		I = 0.01-0.5(NaCl)	E3bg	Approx.	
3010	$C_6H_5O_4N$ 1,2-Benzenediol, 3-nitro-					
	pK_1 6.50, pK_2 11.48	30	I = 0.1($NaClO_4$), mixed constants	E	Uncert.	M59
	pK_1 6.49, pK_2 11.83	25	I = 0.1($NaClO_4$)	05	Uncert.	S128
	pK_1 6.76, pK_2 11.78	20	I = 0.1, mixed constants	E3bg	Uncert.	N24
3011	$C_6H_5O_4N$ 1,2-Benzenediol, 4-nitro-					
	pK_1 6.78, pK_2 10.90	25	I = 0.1(KCl), concentration constants	05	Approx.	P63
	pK_1 6.59, pK_2 10.75	30	I = 0.1(KNO_3), mixed constants	E3bg	Approx.	M123
	pK_1 7.19, pK_2 11.29	20	I = 0.1	E3bg	Uncert.	S128
	pK_1 6.63, pK_2 10.59	30	I = 0.1($NaClO_4$), mixed constants	E	Uncert.	M59
			Other values in A1, B27 and B28			
3012	$C_6H_5O_4N$ 1,3-Benzenediol, 2-nitro-					
	pK_1 6.37, pK_2 9.46	20	I = 0.1, mixed constants (?)	E3bg	Uncert.	S128
3013	$C_6H_5O_4N$ 1,3-Benzenediol, 4-nitro-					
	pK_1 6.15, pK_2 9.37	20	I = 0.1, mixed constants (?)	E3bg	Uncert.	S128
3014	$C_6H_5O_4N$ 1,3-Benzenediol, 5-nitro-					
	pK_1 6.59, pK_2 9.24	20	I = 0.1, mixed constants (?)	E3bg	Uncert.	S128

No.	Molecular formula, name and pK value(s)	T(°C)	Remarks	Method	Assessment	Ref.
3015	C$_6$H$_5$O$_4$N 1,4-Benzenediol, 2-nitro-					
	pK$_1$ 7.57	10	I = 0.1(KCl)	05	Approx.	S111
	pK$_1$ 7.47, pK$_2$ 10.11	20				
	pK$_1$ 7.37	30				
			Thermodynamic quantities are derived from the results			
	pK$_1$ 7.47, pK$_2$ 10.11	20	I = 0.1, mixed constants (?)	E3bg	Uncert.	S128
3016	C$_6$H$_5$O$_4$N 3-Pyridinecarboxylic acid N-oxide, 4-hydroxy-					
	pK(N$^+$-OH) 3.08		Conditions not stated	E	Uncert.	D52
3017	C$_6$H$_5$O$_4$N$_3$ 2-Pyrazinecarboxylic acid, 5-amino-6-formyl-3,4-dihydro-3-oxo-					
	pK$_1$ 3.71, pK$_2$ 7.63	20	c = 0.0001 in 0.01M buffer	05	Approx.	A39
3018	C$_6$H$_5$O$_4$Cl Phenol, 3-perchloryl-					
	8.00	30	Mixed constant	E3d	Uncert.	G1
3019	C$_6$H$_5$O$_4$Cl Phenol, 4-perchloryl-					
	7.62	30	Mixed constant	E3d	Uncert	G1
3020	C$_6$H$_5$O$_5$N$_3$ Phenol, 2-amino-4,6-dinitro					
	1.00	not stated	c = 0.003, I = 0.5(NaClO$_4$)	05	Uncert.	P16
3021	C$_6$H$_5$ClS Benzenethiol, 3-chloro-					
	5.780	25	m = 0.0001	05a	Rel.	D28

205

3022 C_6H_5ClS Benzenethiol, 4-chloro-

6.135	25		05a	Rel.	D28
5.70	35	m = 0.001	05	Approx.	M81
		I = 1.0			
5.9		Conditions not stated	05	Uncert.	D9

3023 C_6H_5BrS Benzenethiol, 4-bromo-

6.020	25	m = 0.0001	05a	Rel.	D28

3024 $C_6H_6ON_2$ 2-Pyridinecarbaldehyde oxime

pK_1 3.56, pK_2 10.17	25	c = 0.0001	05	Approx.	G41

Variation with temperature

	5	15	30	40	50	60
pK_1	3.88	3.70	3.51	3.42	3.39	3.38
pK_2	10.25	10.21	10.13	10.08	10.00	9.91

Thermodynamic quantities are derived from the results

pK_1 3.42, pK_2 10.22	25		05	Approx.	H8
pK_1 3.34, pK_2 10.06	34	Thermodynamic quantities are derived from			
pK_1 3.51	15	the results			
pK_2 10.29	18.5				
pK_1 3.29	41.5				
pK_1 3.69, pK_2 10.02	24	I = 0.1(KNO_3), mixed constants	E3bg	Approx.	B98

Other values in B177, G39, L65 and W36

No.	Molecular formula, name and pK value(s)	$T(^oC)$	Remarks	Method	Assessment	Ref.
3025	$C_6H_6ON_2$ 3-Pyridinecarbaldehyde oxime					
	pK_1(NH) 4.1, pK_2(NOH) 10.1	25	$c \approx 0.005-0.01$, $I = 0.5$(KNO$_3$),mixed constants	E3bg	Uncert.	L65
	pK_2 10.2	not stated	$c \approx 0.04$	E3bg	Uncert.	W36
3026	$C_6H_6ON_2$ 4-Pyridinecarbaldehyde oxime					
	pK_1(NH) 4.9 , pK_2(NOH) 9.6	25	$c \approx 0.005-0.01$, $I = 0.5$(KNO$_3$),mixed constants	E3bg	Uncert.	L65
	pK_2 10.2	not stated	$c \approx 0.04$	E3bg	Uncert.	W36
3027	$C_6H_6ON_2$ (Isonicotinamide) 4-Pyridinecarboxamide					
	11.47	25	$c = 0.0002$, $I = 0.07$(KCl), mixed constant	05	Approx.	T42
	11.58	25				
3028	$C_6H_6ON_4$ Purin-6(1$\underline{H}$)-one, 1-methyl-					
	8.8	not stated	Mixed constant(?)	05	Uncert.	B66
3029	$C_6H_6ON_4$ Purin-6(1$\underline{H}$)-one, 7-methyl-					
	9.4	not stated	Mixed constant(?)	05	Uncert.	B66
3030	$C_6H_6ON_4$ Purin-6(1$\underline{H}$)-one, 9-methyl-					
	10.3	not stated	Mixed constant(?)	05	Uncert.	B66
3031	$C_6H_6ON_4$ Purin-6(3$\underline{H}$)-one, 3-methyl-					
	8.4	not stated	Mixed constant(?)	05	Uncert.	B66

No.	Formula	Name / pK	T (°C)	Conditions	Method	Reliability	Ref.
3032	$C_6H_6ON_4$	Purin-8(1H)-one, 1-methyl-					
		9.8		Conditions not stated	05	Uncert.	L42
3033	$C_6H_6ON_4$	Purin-8(3H)-one, 3-methyl-					
		10.5		Conditions not stated	05	Uncert.	L42
3034	$C_6H_6ON_4$	Purin-8(7H)-one, 7-methyl-					
		8.0		Conditions not stated	05	Uncert.	L42
3035		Purin-8(7H)-one, 9-methyl-					
		9.2		Conditions not stated	05	Uncert.	L42
3036	$C_6H_6ON_4$	1H-Pyrazolo[3,4-d]pyrimidin-4(2H)-one, 1-methyl-					
		9.26	20	c = 0.0001 in 0.01M buffer	05	Approx.	L78
3037	$C_6H_6ON_4$	1,2,4-Triazolo[1,5-a]pyrimidin-7(1H)-one, 5-methyl-					
		6.22	20	I = 0.1(KNO$_3$)	E3bg	Approx.	O22
		6.19	25				F3
		6.09	35				
		6.01	45				
3038	$C_6H_6ON_6$	Pyrimidino[5,4-e]-1,2,4-triazin-5(1H)-one, 7-amino-3-methyl-					
		pK$_1$ 1.14, pK$_2$ 6.95	20	I = 0.01, mixed constants	05	Uncert.	B139
3039	$C_6H_6ON_6$	1H-1,2,3-Triazolo[4,5-b]pyridine-6-carboxamide, 5-amino-					
		5.35	20	c = 0.00025	05	Approx.	A34

No.	Molecular formula, name and pK value(s)	T(°C)	Remarks	Method	Assessment	Ref.
3040	$C_6H_6O_2N_2$ 1,2-Benzoquinone dioxime 6.93					B178
3041	$C_6H_6O_2N_2$ 2-Pyrazinecarboxylic acid, 5-methyl- 2.85			0		S12
3042	$C_6H_6O_2N_2$ 2-Pyridinecarbaldehyde oxime, 3-hydroxy- 8.1(-OH group)	25	c ≈ 0.04	E3bg	Uncert.	W36
3043	$C_6H_6O_2N_2$ 2-Pyridinecarbohydroxamic acid (Picolinoylhydroxamic acid)					
	8.50	25	c = 0.01, I = 0.1(KCl), mixed constant	E3bg	Approx.	G38
	8.7	not stated	c = 0.04, I = 0.1(KCl)	E3bg	Uncert.	H2
3044	$C_6H_6O_2N_2$ 3-Pyridinecarbohydroxamic acid (Nicotinohydroxamic acid)					
	pK_2 8.09	25	I = 0.1($NaClO_4$)	05	Approx.	R48
	pK_1 3.75, pK_2 7.60	30	I = 0.1($NaClO_4$)	E3bg	Approx.	D63
	pK_2 8.30	25	c = 0.01, I = 0.1(KCl), mixed constant	E3bg	Approx.	G38
	pK_2 8.3	not stated	c = 0.04, I = 0.1(KCl)	E3bg	Uncert.	H2
3045	$C_6H_6O_2N_2$ 4-Pyridinecarbohydroxamic acid (Isonicotinoylhydroxamic acid)					
	7.85	25	c = 0.01, I = 0.1(KCl), mixed constant	E3bg	Approx.	G38
	7.8	not stated	c = 0.04, I = 0.1(KCl)	E3bg	Uncert.	H2

No.	Formula / pK	Name	T (°C)	Conditions	Method	Assessment	Ref.
3046	$C_6H_6O_2N_4$ 5.84	Pteridine-2,7(1$\underline{H}$,3$\underline{H}$)-dione, 5,6-dihydro-	20	$I = 0.01$	05	Approx.	A31
3047	$C_6H_6O_2N_4$ 9.14	Pteridine-4,6(1$\underline{H}$,5$\underline{H}$)-dione, 7,8-dihydro-	20	$I = 0.01$	05	Approx.	A31
3048	$C_6H_6O_2N_4$ 8.45	Pteridine-4,7(1$\underline{H}$,8$\underline{H}$)-dione, 5,6-dihydro-	20	$c = 0.002$	E3bg	Approx.	A31
3049	$C_6H_6O_2N_4$ 11.05	Pteridin-2(1$\underline{H}$)-one, 3,4-dihydro-4-hydroxy-	20	$c = 0.0001$	05	Approx.	A28
3050	$C_6H_6O_2N_4$ 9.90	Pteridin-6(5$\underline{H}$)-one, 7,8-dihydro-7-hydroxy-	20	$c = 0.0001$	05	Approx.	A28
3051	$C_6H_6O_2N_4$ pK_1 7.9, pK_2 12.2 pK_1 7.9, pK_2 11.8	Purine-2,6(1$\underline{H}$,3$\underline{H}$)-dione, 1-methyl-	20	Conditions not stated	E3bg/05 / 05	Uncert. / Uncert.	P55 / L39
3052	$C_6H_6O_2N_4$ pK_1 8.45, pK_2 11.92 pK_1 8.5, pK_2 11.5	Purine-2,6(1$\underline{H}$,3$\underline{H}$)-dione, 3-methyl-	20	Conditions not stated	E3bg / 05	Approx. / Uncert.	P55 / L39

No.	Molecular formula, name and pK value(s)	$T(^{\circ}C)$	Remarks	Method	Assessment	Ref.
3053	$C_6H_6O_2N_4$ Purine-2,6(1H,3H)-dione, 7-methyl-					
	pK_1 8.42	20		E3bg	Approx.	P55
	$pK_2 > 13$			05		
	pK_1 8.4, $pK_2 \approx 10.5$		Conditions not stated	05	Uncert.	L39
3054	$C_6H_6O_2N_4$ Purine-2,6(1H,3H)-dione, 9-methyl-					
	pK_1 6.12	20		E3bg	Approx.	P55
	$pK_2 > 13$			05		
	pK_1 5.9, $pK_2 \approx 10.5$		Conditions not stated	05	Uncert.	L39
3055	$C_6H_6O_2N_4$ Purine-2,8(1H,7H)-dione, 1-methyl-					
	7.94	20	c = 0.0025	E3bg	Approx.	B128
3056	$C_6H_6O_2N_4$ Purine-2,8(3H,7H)-dione, 3-methyl-					
	8.39	20	c = 0.0025	E3bg	Approx.	B128
3057	$C_6H_6O_2N_4$ Purine-2,8(3H,7H)-dione, 9-methyl-					
	pK_1 8.74	20	c = 0.0020	E3bg	Approx.	B128
	pK_2 11.5				Uncert.	
3058	$C_6H_6O_2F_8$ 1,6-Hexanediol, 2,2,3,3,4,4,5,5-octafluoro-					
	pK_1 12.1, pK_2 12.8	25		E3bg	V.uncert.	M4

No.	Formula	Name / pK	T °C	Remarks	Reliability	Ref.
3059	$C_6H_6O_2S$	Acetic acid, 2-(2-thienyl)-				
		3.89	25		Uncert.	I3
3060	$C_6H_6O_2S$	Benzenesulfinic acid				
		1.29	20	c = 0.02, mixed constant	Approx.	R54
		2.76	25	I ≈ 0.005	Uncert.	D23
		pK_1 1.84*	25		Uncert.	B180
		pK_1 2.16				
				* Some oxidation said to occur on drying		
3061	$C_6H_6O_2S$	2-Thiophenecarboxylic acid, 4-methyl-				
		3.59	25	c = 0.005, mixed constant	Approx.	B188
				Value in mixed solvent also given		
		3.56	25	c = 0.01	Approx.	O28
		3.76	25			I3
3062	$C_6H_6O_2S$	2-Thiophenecarboxylic acid, 5-methyl-				
		3.70	25	c = 0.005, mixed constant	Approx.	B188
				Value in mixed solvent is also given		
		3.76	25			I3
3063	$C_6H_6O_2Se$	Benzeneseleninic acid				
		4.70	25	I ≈ 0.005, in 4% methanolic solution	Approx.	D23
3064	$C_6H_6O_2Se$	2-Selenofurancarboxylic acid, 5-methyl-				
		3.82	25			S99

No.	Molecular formula, name and pK value(s)	$T(^{o}C)$	Remarks	Method	Assessment	Ref.
3065	$C_6H_6O_3N_2$ 3,6-Diazabicyclo[3.3.0]octane-2,4,7-trione 8.15(cis)			E3bg	Uncert.	C62
3066	$C_6H_6O_3N_2$ Methyl (1,2-dihydro-2-oxopyrimidin-5-yl)carboxylate pK$_1$ 0.65, pK$_2$ 7.29		Conditions not stated		Uncert.	S136
3067	$C_6H_6O_3N_2$ Phenol, 4-amino-2-nitro- 7.81	25	c = 0.0002, I ≈ 0.5, mixed constant	05	Approx.	R15
3068	$C_6H_6O_3N_2$ 3-Pyridinecarboxylic acid N-oxide, 4-amino- pK(N$^+$-OH) 3.98		Conditions not stated	E	Uncert.	D52
3069	$C_6H_6O_3N_4$ Pteridine-2,6(1H,5H)-dione, 7,8-dihydro-7-hydroxy- pK$_1$ 9.44, pK$_2$ 11.50	20	c = 0.0001	05	Approx.	A28
3070	$C_6H_6O_3N_4$ Pteridine-2,4,6(1H,3H,5H)-trione, 7,8-dihydro- 7.07	20	I = 0.01	05	Approx.	A31
3071	$C_6H_6O_3N_4$ Purine-2,6,8(1H,3H,7H)-trione, 1-methyl- pK$_1$ 5.75, pK$_2$ 10.6	not stated	Mixed constants	05	Uncert.	B65
3072	$C_6H_6O_3N_4$ Purine-2,6,8(1H,3H,7H)-trione, 3-methyl- pK$_1$ 6.2		Conditions not stated	05	Uncert.	J13
	pK$_1$ 5.75, pK$_2$ > 12	not stated	Mixed constants	05	Uncert.	B65

No.	Compound / pK values	Temp.	Conditions		Reliability	Ref.
3073	C₆H₆O₃N₄ Purine-2,6,8(1H,3H,7H)-trione, 7-methyl-					
	pK₁ 5.5, pK₂ 10.6		Conditions not stated	05	Uncert.	J13
	pK₁ 5.6, pK₂ 10.3	not stated	Mixed constants	05	Uncert.	D38
3074	C₆H₆O₃S 2-Thiophenecarboxylic acid, 5-methoxy-					
	3.80	25	c = 0.005, mixed constant	E3bg	Approx.	B188
			Value in mixed solvent also given			
3075	C₆H₆O₄N₂ Acetic acid, 2-(1,2,3,4-tetrahydro-2,4-dioxo-5-pyrimidinyl)-					
	pK₁ 4.31, pK₂ 9.97, pK₃ ca 14.2	not stated		05	Uncert.	F22
3076	C₆H₆O₄N₂ 4,5-Imidazoledicarboxylic acid, 2-methyl-					
	pK₁ 4.25, pK₂ 8.28		Conditions not stated	0		S13
3077	C₆H₆O₄N₂ Methyl (1,2,3,6-tetrahydro-2,6-dioxopyrimidin-4-yl)carboxylate (Methyl orotate)					
	7.93	25	Titrant (CH₃)₄NOH	E3bg	Approx.	J16
3078	C₆H₆O₄N₂ 4-Pyrimidinecarboxylic acid, 1,2,3,6-tetrahydro-1-methyl-2,6-dioxo-					
	pK₁ < 1, pK₂ 10.52			05	Uncert.	F34
3079	C₆H₆O₄N₂ 4-Pyrimidinecarboxylic acid, 1,2,3,6-tetrahydro-3-methyl-2,6-dioxo-					
	pK₁ 0.7, pK₂ 9.82		Conditions not stated	05	Uncert.	F34
3080	C₆H₆O₄S Benzenesulfonic acid, 2-hydroxy-					
	pK₂ 9.35	25	I = 0.1(KCl), concentration constant	E3bg	Uncert.	A1

No.	Molecular formula, name and pK value(s)	$T(^\circ C)$	Remarks	Method	Assessment	Ref.
3081	$C_6H_6O_4S$ Benzenesulfonic acid, 4-hydroxy-					
	pK$_2$ 8.95	25	I = 0.1(KCl), concentration constant	E3bg	Uncert.	A1
	pK$_2$ 8.56	25	I = 0.5(KNO$_3$), mixed constant	E3bg	Approx.	M5
	pK$_2$ 8.97(H$_2$O)	25		O5	Uncert.	W12
	pK$_2$ 9.52(D$_2$O)					
3082	$C_6H_6O_5S$ Benzenesulfonic acid, 3,4-dihydroxy-					
	pK$_2$ 8.26	20	I = 0.1(KNO$_3$)	E3bg	Approx.	B30
	pK$_3$ 12.8				Uncert.	
	pK$_2$ 8.50, pK$_3$ 12.8	20-25	I = 0.1(KNO$_3$)	E3bg	Uncert.	B29
	pK$_2$ 8.26, pK$_3$ 12.16	30	I = 0.1(KCl), mixed constants	E3bg	Approx.	M112
			Data from potassium salt			
	pK$_2$ 8.1, pK$_3$ 12.2	25	I = 0.1(KCl), concentration constants	E3bg & O5	Uncert.	A1
3083	$C_6H_6O_6S$ Benzenesulfonic acid, 3,4,5-trihydroxy-					
	pK$_2$ 8.39	20	I = 0.1(KNO$_3$), mixed constants	E3bg	Approx.	B27
	pK$_3$ 11.4				Uncert.	
3084	No entry					

3085 $C_6H_6O_8S_2$ 1,3-Benzenedisulfonic acid, 4,5-dihydroxy- (Tiron)

pK	T	Conditions	Method	Remarks	Ref.
pK$_3$ 7.81, pK$_4$ 11.96	25	I = 0.1(KCl)	E3bg	pK$_3$ Approx. pK$_4$ Uncert.	O9
pK$_3$ 7.52, pK$_4$ 12.2		I = 0.35(KCl)			
pK$_3$ 8.33	25	c = 0.003	E3bg	Approx.	N15
pK$_4$ 12.45	25	c = 0.0001, I = 1.0(KCl), concentration constant	O5	Uncert.	
pK$_3$ 7.60, pK$_4$ 12.60	25	I = 0.1(KNO$_3$). Values extrapolated to	E3bg	pK$_3$ Approx. pK$_4$ Uncert.	M99
pK$_3$ 7.49, pK$_4$ 12.08	34.9	c = 0, concentration constants			
pK$_3$ 7.66	20	Mixed constants	E1d	Approx.	S51
pK$_4$ 12.6				Uncert.	
pK$_3$ 7.66, pK$_4$ 12.6	20-22	I = 0.1(KCl)	E3bg	Uncert.	W31

Other values in B29, D80 and L37

3086 $C_6H_6N_2S$ 2-Pyridinecarbothioamide (Picolinethioamide)

pK	T	Conditions	Method	Remarks	Ref.
12.55	25	c = 0.0001	O5	Approx.	T42
12.44		c = 0.0001, I = 0.07(KCl), mixed constant			

3087 $C_6H_6N_2S$ 3-Pyridinecarbothioamide (Nicotinethioamide)

pK	T	Conditions	Method	Remarks	Ref.
11.84	25	c = 0.0002	O5	Approx.	T42
11.73		c = 0.0002, I = 0.07(KCl), mixed constant			

3088 $C_6H_6N_4S$ Pteridine-2(1H)-thione, 3,4-dihydro-

pK	T	Conditions	Method	Remarks	Ref.
10.95	20	c = 0.0001	O5	Approx.	A30

No.	Molecular formula, name and pK value(s)	T(°C)	Remarks	Method	Assessment	Ref.
3089	$C_6H_6N_4S$ Purine-6(1H)-thione, 1-methyl					
	8.8	not stated	Mixed constant(?)	05	Uncert.	B66
3090	$C_6H_6N_4S$ Purine-6(1H)-thione, 2-methyl-					
	8.2	not stated	Mixed constant(?)	05	Uncert.	B66
3091	$C_6H_6N_4S$ Purine-6(1H)-thione, 7-methyl-					
	7.9	not stated	Mixed constant(?)	05	Uncert.	B66
3092	$C_6H_6N_4S$ Purine-6(1H)-thione, 8-methyl-					
	7.8	not stated	Mixed constant(?)	05	Uncert.	B66
3093	$C_6H_6N_4S$ Purine-6(1H)-thione, 9-methyl-					
	7.8	not stated	Mixed constant(?)	05	Uncert.	B66
3094	$C_6H_6N_4S$ Purine-6(3H)-thione, 3-methyl-					
	7.8	not stated	Mixed constant(?)	05	Uncert.	B66
3095	C_6H_7ON Phenol, 2-amino-					
	9.75	25	I = 0.1(KCl), mixed constant	E3bg	Approx.	E20
	10.05	20	c = 0.025, mixed constant	E3d	Uncert.	H17
3096	C_6H_7ON Phenol, 3-amino-					
	9.86	25	I = 0.1(KCl), mixed constant	E3bg	Approx.	E20
	9.96	20	c = 0.025, mixed constant	E3d	Uncert.	H17

No.	Formula	Name / pK	Temp.	Conditions	Method		Ref.
3097	C_6H_7ON	Phenol, 4-amino-					
		10.46	25	I = 0.1(KCl), mixed constant	E3bg	Approx.	E20
		10.44	25	I = 0.1(KCl), mixed constant	E3bg	Approx.	H51
3098	$C_6H_7ON_3$	3-Pyridinecarboxamide, 2'-amino- (Nicotinic acid hydrazide)					
		pK(NH) 11.49			0		T44
3099	$C_6H_7ON_3$	4-Pyridinecarboxamide, 2'-amino- (Isonicotinic acid hydrazide)					
		pK$_1$(NHNH$_2$) 1.75, pK$_2$ (=N-) 3.57, pK$_3$ (-NH) 10.75			0		Z14
3100	$C_6H_7ON_5$	Ethanenitrile, 2-(4-amino-1,6-dihydro-6-oxo-5-pyrimidinyl)amino- (Pyrimidine, 4-amino-5-cyanomethylamino-6-hydroxy-)					
		9.79	20	c = 0.01	E3bg	Approx.	B136
3101	$C_6H_7ON_5$	1,2,4-Triazolo[4,3-b]-1,2,4-triazin-7(1H)-one, 3,6-dimethyl-					
		5.85	25	c = 0.01	E3bg	Approx.	K9
3102	$C_6H_7O_2N$	3-Azabicyclo[3.2.0]heptane-2,4-dione					
		9.63(cis)		Conditions not stated	E3bg	Uncert.	C62
3103	$C_6H_7O_2N$	1,2-Benzenediol, 3-amino-					
		pK$_1$ 7.70, pK$_2$ 10.50, pK$_3$ 11.90	20	I = 0.1, mixed constants(?)	E3bg	Uncert.	S128
3104	$C_6H_7O_2N$	1,2-Benzenediol, 4-amino-					
		pK$_1$ 8.12, pK$_2$ 10.65, pK$_3$ 12.11	20	I = 0.1, mixed constants(?)	E3bg	Uncert.	S128
3105	$C_6H_7O_2N$	1,3-Benzenediol, 2-amino-					
		pK$_1$ 8.09, pK$_2$ 9.35, pK$_3$ 11.58	20	I = 0.1, mixed constants(?)	E3bg	Uncert.	S128

No.	Molecular formula, name and pK value(s)	$T(^{o}C)$	Remarks	Method	Assessment	Ref.
3106	$C_6H_7O_2N$ 1,3-Benzenediol, 4-amino-					
	pK_1 7.91, pK_2 9.16, pK_3 11.24	20	I = 0.1, mixed constants(?)	E3bg	Uncert.	S128
3107	$C_6H_7O_2N$ 1,3-Benzenediol, 5-amino-					
	pK_1 7.88, pK_2 9.23, pK_3 11.16	20	I = 0.1, mixed constants(?)	E3bg	Uncert.	S128
3108	$C_6H_7O_2N$ 1,4-Benzenediol, 2-amino-					
	pK_2 10.18, pK_3 11.30	14	I = 0.1(KCl), mixed constants	05	Approx.	S111
	pK_2 10.05, pK_3 11.25	22	Thermodynamic quantities are derived from the results			
	pK_1 8.01, pK_2 10.05, pK_3 11.25	20	I = 0.1, mixed constants(?)	E3bg	Uncert.	S128
3109	$C_6H_7O_2N$ 2-Pyrrolecarboxylic acid, 3-methyl-					
	5.10			E		S55
3110	$C_6H_7O_2N$ 2-Pyrrolecarboxylic acid, 4-methyl-					
	4.60			E		S55
3111	$C_6H_7O_2N$ 2-Pyrrolecarboxylic acid, 5-methyl-					
	4.88	25	c = 0.01	E3bg	Approx.	F43
	5.0			E		S55
3112	$C_6H_7O_2N$ 3-Pyrrolecarboxylic acid, 2-methyl-					
	5.80			E		S55
	5.75		Conditions not stated		Uncert.	K22

No.	Formula	Name	pK	T	Conditions	Method	Assessment	Ref.
3113	$C_6H_7O_2N$	3-Pyrrolecarboxylic acid, 4-methyl-	5.65			E		S55
3114	$C_6H_7O_2N$	3-Pyrrolecarboxylic acid, 5-methyl-	5.35			E		S55
3115	$C_6H_7O_2N_3$	2-Pyrazinecarbaldehyde, 4,5-dihydro-3-methylamino-5-oxo-	6.44	20	c = 0.0025	E3bg	Approx.	B136
3116	$C_6H_7O_2N_5$	1H-1,2,3-Triazolo[4,5-d]pyrimidin-7(4H)-one, 5-methoxy-1-methyl-	7.17	20	c = 0.000039	05	Approx.	A40
3117	$C_6H_7O_2N_5$	2H-1,2,3-Triazolo[4,5-d]pyrimidin-7(6H)-one, 5-methoxy-2-methyl-	7.69	20	c = 0.000039	05	Approx.	A40
3118	$C_6H_7O_2P$	Phenylphosphinic acid	1.75	25	c ≈ 0.05, mixed constant	E3bg	Uncert.	Q1
3119	$C_6H_7O_2B$	Benzeneboronic acid	8.83	25		E3bg	Approx.	B143

Variation with temperature

0.2	5	10	15	20	30	35	40	45	50
9.00	8.97	8.93	8.90	8.86	8.81	8.79	8.78	8.77	8.77

Thermodynamic quantities are derived from the results

(Contd)

No.	Molecular formula, name and pK value(s)	T(°C)	Remarks	Method	Assessment	Ref.
3119(Contd)	$C_6H_7O_2B$ Benzeneboronic acid					
	8.835	25.5		E3bg	Approx.	E6
			Variation with temperature			
		19.5 20.0 25.9 33.0 33.5 36.5 38.0 38.5				
		8.855 8.850 8.823 8.805 8.795 8.788 8.788 8.767				
			Thermodynamic quantities are derived from the results			
3120	$C_6H_7O_3N_3$ Acetic acid, 2-(1,2-dihydro-2-oxo-4-pyrimidinyl)amino- (Pyrimidine, 4-carboxymethylamino-2-hydroxy-)					
	pK_2 4.40, $pK_3 \simeq$ 12.6	25	I = 0.009-1.8	05	Uncert.	J8
3121	$C_6H_7O_3N_3$ 4-Pyrimidinecarboxamide, 1,2,3,6-tetrahydro-3-methyl-2,6-dioxo-					
	9.10		Conditions not stated	05	Uncert.	F34
3122	$C_6H_7O_3P$ Phenylphosphonic acid					
	pK_2 7.43	25	I = 0.009-1.8	E3bg,R4	Approx.	M34
3123	$C_6H_7O_3As$ Phenylarsonic acid					
	pK_1 3.65, pK_2 8.77	25	c ≈ 0.01	E3bg	Approx.	E10
3124	$C_6H_7O_4N_3$ Acetic acid, 2-(5-amino-4-carboxyimidazol-1-yl)-					
	pK_1 3.25, pK_2 6.83	20	c ≈ 0.0001 I = 0.10(NaCl), mixed constants	05	Uncert.	L58
3125	$C_6H_7O_4N_3$ Ethyl 2,3,4,5-tetrahydro-3,5-dioxo-1,2,4-triazine-6-carboxylate					
	6.34	25	Titrant $(CH_3)_4NOH$	E3bg	Approx.	J16

No.	Formula / Name	pK	T (°C)	Conditions	Method	Assessment	Ref.
3126	$C_6H_7O_4N_3$ Pyrimidine-2,4,6(1$\underline{H}$,3$\underline{H}$,5$\underline{H}$)-trione, 5-hydroxyimino-1,3-dimethyl-	pK_1 4.72		Conditions not stated	O5	Uncert.	T13
3127	$C_6H_7O_4N_3$ Pyrimidin-2(1$\underline{H}$)-one, 4-ethoxy-5-nitro-	6.6	20	c = 0.0025	E3bg	Uncert.	B129
3128	$C_6H_7O_4N_3$ Pyrimidin-4(1$\underline{H}$)-one, 2-ethoxy-5-nitro-	4.88	20	c = 0.0025	E3bg	Approx.	B129
3129	$C_6H_7O_4Cl$ r-1,c-2-Cyclopropanedicarboxylic acid, 1-chloro-t-2-methyl-	pK_1 2.66, pK_2 5.41	20		E3bg	Approx.	M12
3130	$C_6H_7O_4Cl$ r-1,t-2-Cyclopropanedicarboxylic acid, 1-chloro-c-3-methyl-	pK_1 2.61, pK_2 4.46	20		E3bg	Approx.	M12
3131	$C_6H_7O_4Cl$ 1,2-Cyclopropanedicarboxylic acid, 2-chloro-1-methyl-	pK_1 2.95, pK_2 5.25 (cis) pK_1 2.72, pK_2 4.27 (trans)	20		E3bg	Approx.	M12
3132	$C_6H_7O_4P$ Phenyl dihydrogen phosphate (Phenylphosphate)	pK_2 6.28	25	I = 0.01-2.0	E3bg,R4	Approx.	M34
		pK_1 1.46, pK_2 6.29	25	c = 0.005, I < 0.025, mixed constants	E3bg	Uncert.	M57
		pK_1 1.00	26	c = 0.1, I = 0.1, mixed constants	E3bg	Uncert.	C31
		pK_2 5.88		c = 0.01, I = 0.1		Approx.	

pK_1 was determined in 10% dioxan-water.

Value corrected to 100% water

No.	Molecular formula, name and pK value(s)	T(°C)	Remarks	Method	Assessment	Ref.
3133	$C_6H_7O_4As$ Phenylarsonic acid, 2-hydroxy- pK_1 4.04, pK_2 7.92 pK_3 13.27	25	c ≈ 0.01	E3bg 05	Approx. Uncert.	E10
3134	$C_6H_7O_4As$ Phenylarsonic acid, 4-hydroxy- pK_1 3.85, pK_2 8.63 pK_3 10.13	25	c ≈ 0.01	E3bg 05	Approx.	E10
3135	$C_6H_7O_5As$ Phenylarsonic acid, 2,4-dihydroxy- pK_1 4.22, pK_2 7.98, pK_3 9.82 pK_4 14.27	25	c ≈ 0.01	E3bg 05	Approx. Uncert.	E10
3136	$C_6H_7O_6N_3$ 2,5-Pyrrolidinedione, 1-(2,2-dinitroethyl)- 2.92	20		05	Approx.	T58a
3137	$C_6H_7O_8N_5$ Hexanenitrile, 4,4,6,6-tetranitro- 0.90	20	In aq.$HClO_4$,H_o scale	06	Uncert.	T58a
3138	$C_6H_7O_{10}N_3$ Butanedioic acid, 2-(2,2,2-trinitroethyl)- pK_1 3.03, pK_2 4.42	25	c = 0.01	E3bg	Uncert.	F5
3139	$C_6H_7O_{10}N_3$ Butanoic acid, 2-acetoxy-4,4,4-trinitro- 2.02	25	c = 0.01	E3bg	Approx.	F5

No.	Formula	Name	pKa	Temp.	Conditions	Method	Reliability	Ref.
3140	C_6H_7NS	Benzenethiol, 2-amino-						
			pK$_1$ 3.00, pK$_2$ 6.59	20		E3bg	Uncert.	D8
			pK$_2$ 6.5	25		E	Uncert.	V13
3141	C_6H_7NS	Methanethiol, 2-pyridyl-						
			8.82	25	1% ethanol	E3bg	Uncert.	K57
3142	$C_6H_8O_2N_2$	3-Pyrazolecarboxylic acid, 1,5-dimethyl-						
			4.24	room temp.	c = 0.01	E3bg	Uncert.	H1
3143	$C_6H_8O_2N_2$	5-Pyrazolecarboxylic acid, 1,3-dimethyl-						
			3.31	room temp.	c = 0.01	E3bg	Uncert.	H1
3144	$C_6H_8O_2N_2$	Pyrimidine-2,4(1H,3H)-dione, 5-ethyl-						
			9.90	20		05	Uncert.	S138
3145	$C_6H_8O_2N_4$	2-Pyrazinecarboxamide, 4,5-dihydro-3-methylamino-5-oxo-						
			7.48	20	c = 0.0025	E3bg	Approx.	B136
3146	$C_6H_8O_3N_2$	Imidazolidine-2,4-dione, 1-acetyl-5-methyl-						
			7.2	not stated	I ≅ 0.1	05	Uncert.	S110
3147	$C_6H_8O_3N_2$	Pyrimidine-2,4(1H,3H)-dione, 5-(2-hydroxyethyl)-						
			9.68	Conditions not stated		05	Uncert.	F21
3148	$C_6H_8O_3N_2$	2,5-Pyrrolidinedione, 3-acetylamino-						
			8.85	Conditions not stated		E3bg	Uncert.	C62

No.	Molecular formula, name and pK value(s)	T(°C)	Remarks	Method	Assessment	Ref.
3149	$C_6H_8O_3N_4$ Pyrimidine-2,4(1H,3H)-dione, 5-hydroxyimino-6-imino-1,3-dimethyl- 8	not stated	Titration of Na-salt	E3bg	V.uncert.	B176
3150	$C_6H_8O_4N_2$ 3-Azapentanedioic acid, 3-cyanomethyl- pK$_1$ 3.06, pK$_2$ 4.34	20	(Iminodiacetic acid, N-cyanomethyl-) I = 0.1(KCl), concentration constants	E3bg	Approx.	S40
3151	$C_6H_8O_4Se$ 2,5-Selenacyclopentanedicarboxylic acid pK$_1$ 3.24, pK$_2$ 4.47(cis)	25	(2,5-Tetrahydroselenoledicarboxylic acid) c ≈ 0.003, I = 0.1(NaClO$_4$), concentration constants	E3bg,R3h	Approx.	S132
3152	$C_6H_8O_8N_2$ Butanoic acid, 2-acetoxy-3,3-dinitro- 1.85	25	c = 0.01	E3bg	Uncert.	F5
3153	$C_6H_8O_9N_4$ Butanoic acid, 2-acetylamino-4,4,4-trinitro- pK$_1$ 2.35, pK$_2$ 3.43	25	c = 0.01 Decomposition reported, see No.2409	E3bg	V.Uncert.	F5
3154	$C_6H_9ON_3$ Pyrimidin-2(1H)-one, 4-dimethylamino- pK$_1$ 4.15, pK$_2$ 12.8	22		05	Uncert.	S141
3155	$C_6H_9ON_3$ Pyrimidin-2(1H)-one, 1-methyl-4-methylamino- 4.40	22		05	Uncert.	S141

3156 C₆H₉O₂N Butanoic acid, 2-cyano-3-methyl-

$C_6H_9O_2N$ Butanoic acid, 2-cyano-3-methyl-

2.401   25   C2   Rel.   I26

Variation with temperature

5	10	15	20	30	35	40	45
2.299	2.320	2.343	2.365	2.427	2.452	2.481	2.511

Thermodynamic quantities are derived from the results

3157 $C_6H_9O_2N$ 3-Pentenal oxime, 4-methyl-2-oxo-

8.90   30   c = 0.01, I = 0.1(KCl), mixed constant   E3bg   Approx.   E18

3158 $C_6H_9O_2N_3$ Propanoic acid, 2-amino-3-(4-imidazolyl)-

pK_1 2.000, pK_2 6.371, pK_3 9.670   15   I = 0.2(KNO₃), mixed constants   E3bg   Approx.   R7

pK_1 2.025, pK_2 6.219, pK_3 9.416   25   Thermodynamic quantities are derived from

pK_1 2.061, pK_2 5.992, pK_3 9.082   40   the results

pK_1 2.24, pK_2 5.96, pK_3 8.92   37   I = 0.15(KNO₃), mixed constants   E3bg   Approx.   P35

pK_2 6.05, pK_3 9.16   25   I = 0.12(KCl)   E3bg,R3f   Approx.   C39

pK_2 5.87, pK_3 8.85   35   Thermodynamic quantities are derived from

pK_2 5.60, pK_3 8.60   45   the results

pK_4 14.37   25   In aq. KOH, H_ scale   07   Uncert.   Y3

Other values in D47, H17 and J5

3159 $C_6H_9O_2N_3$ Propanoic acid, 2-amino-3-(1-pyrazolyl)-

2.2   Conditions not stated   Uncert.   N33

No.	Molecular formula, name and pK value(s)	T(°C)	Method	Assessment	Ref.	Remarks
3160	$C_6H_9O_2N_3$ Pyrimidin-2(1H)-one, 4-amino-5-(2-hydroxyethyl)- pK$_1$ 4.52, pK$_2$ 13		05	Uncert.	F21	Conditions not stated
3161	$C_6H_9O_2N_3$ Pyrimidin-2(1H)-one, 4-hydroxyamino-1,5-dimethyl- pK$_1$ 2.9, pK$_2$ 11.1		05	Uncert.	J7	Conditions not stated
3162	$C_6H_9O_2N_3$ Pyrimidin-2(1H)-one, 4-hydroxyamino-1,6-dimethyl- pK$_1$ 3.65, pK$_2$ 10.7		05	Uncert.	J7	Conditions not stated
3163	$C_6H_9O_2N_3$ 1,2,4-Triazine-3,5(2H,4H)-dione, 6-isopropyl- 7.45	not stated	E3bg	Uncert.	C28	c = 0.005
3164	$C_6H_9O_2N_3$ 1,2,4-Triazine-3,5(2H,4H)-dione, 6-propyl- 7.5	not stated	E3bg	Uncert.	C28	c = 0.005
3165	$C_6H_9O_2N_5$ Pyrimidin-2(1H)-one, 1-methyl-4-semicarbazido- 3.10		05	Uncert.	J8	Conditions not stated
3166	$C_6H_9O_2N_5$ Pyrimidin-4(1H)-one, 2,6-diamino-5-(N-methylformamido)- pK$_1$ 3.8		E3bg	Uncert.	H3	Conditions not stated
3167	$C_6H_9O_3N$ Oxazolidine-2,4-dione, 5-ethyl-5-methyl- 6.1		05	Uncert.	B189	Conditions not stated
3168	$C_6H_9O_3N$ 2-Pyrroline-5-carboxylic acid, 1-hydroxy-5-methyl- 2.95		E3bg	Uncert.	B101	Conditions not stated

No.	Formula	Name	pK	T (°C)	Conditions	Method	Reliability	Ref.
3169	$C_6H_9O_3N_3$	1,2,3-Cyclohexanetrione trioxime	8.0	25	c = 0.01, I = 0.1(KCl)	E3bg	Uncert.	J11
3170	$C_6H_9O_3N_3$	Pyrimidine-2,6(1H,3H)-dione, 4-hydroxyamino-1,3-dimethyl-	9.58	20	c = 0.001	E3bg	Approx.	P51
3171	$C_6H_9O_3Br$	Ethyl 2-bromo-3-oxopropanoate	8.5			KIN	Uncert.	B47
3172	$C_6H_9O_4N$	Ethyl 2-hydroxyimino-3-oxobutanoate	7.20	30	c = 0.01, I = 0.1(KCl), mixed constant	E3bg	Approx.	E18
3173	$C_6H_9O_4N$	3-Hydroxyimino-2-oxopropyl acetate	8.72	25	c = 0.0025-0.005, I = 0.1(KNO_3), probably a mixed constant	E		U9
3174	$C_6H_9O_4N_3$	Hexanenitrile, 6,6-dinitro	5.00	20		05	Approx.	T58a
3175	$C_6H_9O_6N$	3-Azapentanedioic acid, 3-carboxymethyl- (Nitrilotriacetic acid)	pK_1 0.8, pK_2 1.71 pK_3 2.47, pK_4 9.71	20	c ≈ 0.008, I = 0.1(KCl), concentration constants	E3bg	Uncert. Approx.	I15
			pK_2 1.65 pK_3 2.940, pK_4 10.334	20	Molality scale	E1b	Approx. Rel.	H57

(Contd)

No.	Molecular formula, name and pK value(s)	T(°C)	Remarks	Method	Assessment	Ref.
3175(Contd)	$C_6H_9O_6N$ 3-Azapentanedioic acid, 3-carboxymethyl- (Nitrilotriacetic acid)					
			Variation with temperature			
				0	10	30	40			
		pK$_2$	1.69	1.65	1.66	1.69			
		pK$_3$	2.953	2.948	2.956	2.978			
		pK$_4$	10.594	10.454	10.230			
			Thermodynamic quantities are derived from the results			
	pK$_2$ 1.9, pK$_3$ 2.5	20	$I = 0.1(KNO_3)$	E3bg	Uncert.	A55
	pK$_4$ 9.73			E3bh	Approx.	
	pK$_2$ 1.7, pK$_3$ 2.4		$I = 1.0 [(CH_3)_4NCl]$	E3bh	Uncert.	
	pK$_4$ 9.67		concentration constants		Approx.	
	pK$_2$ 2.30, pK$_3$ 3.00, pK$_4$ 10.76	0.4	$I = 0.1(KNO_3)$, concentration constants	E3bg	Approx.	K25
	pK$_4$ 9.75	25	$c \approx 0.01$, $I = 0.1(KNO_3)$, concentration constant	E3bg	Approx.	M95
			Variation with temperature			
				15	20	30	35	40			
			9.86	9.80	9.70	9.62	9.58			
			Thermodynamic quantities are derived from the results			
			Other values in S36, S40 and S43			

229

No.	Formula	Name / pK	T	Conditions	Method	Reliability	Ref.
3176	$C_6H_9O_7N$	3-Azapentanedioic acid 3-oxide, 3-carboxymethyl- (Nitrilotriacetic acid N-oxide)					
		pK_2 2.57, pK_3 7.89 ($pK_1 < 2$)	25	$I = 1.0(KCl)$, concentration constants	E3bg	Approx.	C19
3177	$C_6H_9O_9N_5$	Hexanamide, 4,4,6,6-tetranitro-					
		1.16	20		05	Approx.	T58a
3178	$C_6H_{10}ON_4$	1,2,4-Triazin-5(2H)-one, 3-dimethylamino-6-methyl-					
		1.6	25		05	Uncert.	P70
3179	$C_6H_{10}O_2N_2$	Butanoic acid, 3-(3-methyl-3H-diazirin-3-yl)- [Hexanoic acid, 5,5-(NN')azo-]					
		4.5	25		E3bg	Uncert.	C57
3180	$C_6H_{10}O_2N_2$	1,2-Cyclohexanedione dioxime (Nioxime)					
		pK_1 10.6, pK_2 12.4	20-25		05	Uncert.	B18
		pK_1 10.68, pK_2 11.92	25	Concentration constants	E3d	Uncert.	H4
		pK_1 10.70, pK_2 12.16	0				S15
		pK_1 10.1	25		E3bg	Uncert.	B10
3181	$C_6H_{10}O_2N_2$	Pyrimidine-2,4(1H,3H)-dione, 5,6-dihydro-5,5-dimethyl-					
		11.92	0				P83
3182	$C_6H_{10}O_2N_2$	Pyrimidine-2,4(1H,3H)-dione, 5,6-dihydro-5,6-dimethyl-					
		11.73(cis)	0				P83
		11.67(trans)					

No.	Molecular formula, name and pK value(s)	T(°C)	Remarks	Method	Assessment	Ref.
3183	$C_6H_{10}O_2N_2$ Pyrimidine-2,4(1H,3H)-dione, 5,6-dihydro-6,6-dimethyl- 11.44			0		P83
3184	$C_6H_{10}O_2N_4$ 1-Pyrroline-5-carboxylic acid, 2-guanidino- (Viomycidine) pK$_1$ 1.3, pK$_2$ 5.50, pK$_3$ 12.6		Conditions not stated		Uncert.	D75
3185	$C_6H_{10}O_2S$ Acetic acid, 3-butenylthio- 3.717	25	c ≈ 0.001, I = 0.10(KNO$_3$), concentration constant	E3bg	Approx.	P40
3186	$C_6H_{10}O_4N_2$ 2,3-Bis(hydroxyimino)propyl acetate 9.78	25	c = 0.0024-0.005, I = 0.1(KNO$_3$), probably a mixed constant	E		U9
3187	$C_6H_{10}O_4S$ 4-Thiaheptanedioic acid (3,3'-Thiodipropionic acid) pK$_1$ 3.84, pK$_2$ 4.66	25	c ≈ 0.005, I = 0.1(NaClO$_4$),concentration constants	E3bg,R3f	Approx.	S131
	pK$_1$ 4.09, pK$_2$ 4.91	20	I = 0.1(NaClO$_4$), mixed constants	E3bg	Uncert.	C8
3188	$C_6H_{10}O_4S$ 3-Thiapentanedioic acid, 2,2-dimethyl- pK$_1$ 3.74, pK$_2$ 4.30	25	c < 0.001 Microscopic constants are calculated from data on the two corresponding monoesters	E3bg	Approx.	S93

No.	Formula, Name, pK values	Temp	Conditions	Method	Reliability	Ref.
3189	$C_6H_{10}O_4S$ 3-Thiapentanedioic acid, 2,4-dimethyl- pK$_1$ 3.40, pK$_2$ 4.84	25±0.5	c < 0.001, mixture of stereoisomers	E3bg	Uncert.	S93
3190	$C_6H_{10}O_4S$ 3-Thiapentanoic acid, 5-ethoxy-5-oxo- 3.66	25	c = 0.00075	E3bg	Approx.	S93
3191	$C_6H_{10}O_4S_2$ 3,6-Dithiaoctanedioic acid [Ethylenebis(thioacetic acid)] pK$_1$ 3.16, pK$_2$ 4.10	25	I = 0.1(NaClO$_4$), concentration constants	E3bg,R3f	Approx.	S133
	pK$_1$ 3.19, pK$_2$ 3.81	25	I = 0.1(K$_2$SO$_4$)	E3bg	Approx.	J5
	pK$_1$ 3.39, pK$_2$ 4.21	20	I = 0.1(NaClO$_4$), mixed constants	E3bg	Approx.	S2
3192	$C_6H_{10}O_4S_2$ 4,5-Dithiaoctanedioic acid pK$_1$ 3.88, pK$_2$ 4.47	20	I = 0.15(NaClO$_4$), concentration constants	E3bg,R3d	Uncert.	H23
	pK$_1$ 4.02, pK$_2$ 4.77	20	I = 0.1(NaClO$_4$), mixed constants	E3bg	Uncert.	O21
3193	$C_6H_{10}O_5N_2$ 3-Azapentanedioic acid, 3-carbamoylmethyl- (Iminodiacetic acid, N-carbamoylmethyl) pK$_1$ 2.3 pK$_2$ 6.60	20	I = 0.1(KCl), concentration constants	E3bg	Uncert. / Approx.	S40
3193a	$C_6H_{10}O_6N_2$ 4-Azaheptanedioic acid, 4-nitro- pK$_1$ 3.78, pK$_2$ 5.02	21	c = 0.01, mixed constants	E3bg	Uncert.	N37
3194	$C_6H_{11}ON_7$ 1,2,3-Triazole-5-carbaldehyde dimethylhydrazone, 4-ureido- pK$_1$ 2.49, pK$_2$ 7.75	20		05	Approx.	A35

No.	Molecular formula, name and pK value(s)	T(°C)	Remarks	Method	Assessment	Ref.
3195	$C_6H_{11}O_2N$ 2-Pentanone, 3-hydroxyimino-4-methyl- 9.50	25	$c \approx 0.01$, $I = 0.1(KCl)$, mixed constant	E3bg	Approx.	G39
3196	$C_6H_{11}O_2N$ 2-Piperidinecarboxylic acid (Pipecolic acid) pK_1 2.29, pK_2 10.77	20	$c = 0.004$	E3bg	Approx.	C3
3197	$C_6H_{11}O_2N$ 2-Pyrrolidinone, 1-hydroxy-5,5-dimethyl- 8.7		Conditions not stated	E3bg	Uncert.	B101
3198	$C_6H_{11}O_2N_3$ Imidazolidine-2,4-dione, 5-(3-aminopropyl)- pK_1 8.8, pK_2 10.5	25		E3bg	Uncert.	R40
3199	$C_6H_{11}O_2Cl$ Hexanoic acid, 5-chloro- 4.6	25		E3bg	Uncert.	C57
3200	$C_6H_{11}O_3N$ Ethyl 2-amino-3-oxobutanoate pK_1 5.20, pK_2 11.0	22		E3bg	Uncert.	L7
3201	$C_6H_{11}O_3N_3$ Imidazolidine-2,4-dione, 5-(3-hydroxyaminopropyl)- pK_1 5.6, pK_2 9.0	25		E3bg	Uncert.	R40
3202	$C_6H_{11}O_4N$ 3-Azapentanedioic acid, 3-ethyl (Iminodiacetic acid, N-ethyl-) pK_1 2.55, pK_2 8.56	21	$I = 0.1$, concentration constants	E3bg	Approx.	V10

233

3203 $C_6H_{11}O_4N_3$ 3,6-Diazaoctanoic acid, 8-amino-4,7-dioxo-   (Glycylglycylglycine)

pK values	Temp	Conditions	Method	Reliability	Ref
pK$_1$ 3.27, pK$_2$ 7.90	24.9	I = 0.1(KNO$_3$), concentration constants	E3bg	Approx.	K27
pK$_1$ 3.18, pK$_2$ 7.87	25	I = 0.1(NaClO$_4$), concentration constants.	E3bg	Approx.	B148
		Thermodynamic quantities are given			
pK$_1$ 3.30, pK$_2$ 7.96	25	I = 0.16(KCl), mixed constants	E3bg	Approx.	B151
pK$_1$ 3.71, pK$_2$ 8.55	25	I = 3.0(NaClO$_4$),concentration constants	E3bg	Approx.	026
pK$_2$ 8.57	0.35	I = 0.09(KCl), concentration constant	E3bg	Approx.	M125
pK$_2$ 7.74	30				
pK$_2$ 7.51	48.8				

Thermodynamic quantities are derived from the results

3204 $C_6H_{11}O_5N$ 3-Azapentanedioic acid, 3-(2-hydroxyethyl)-   (Iminodiacetic acid, N-(2-hydroxyethyl)-)   (HIMDA)

pK values	Temp	Conditions	Method	Reliability	Ref
pK$_1$ 2.16, pK$_2$ 8.72	25	c = 0.001, I = 0.1, mixed constants	E3bg	Approx.	Z5
pK$_3$(OH) 13.7		c = 0.04, I = 0.7		Uncert.	
pK$_1$ 2.44, pK$_2$ 9.92	0.4	I = 0.1(KNO$_3$), concentration constants	E3bg	Approx.	K25
pK$_1$ 2.20, pK$_2$ 8.72	25	I = 0.1(KNO$_3$)	E3bg	Approx.	R2
pK$_1$ 2.22, pK$_2$ 8.67		I = 1.0(KNO$_3$)			
pK$_1$ 2.2	20	I = 0.1(KCl), concentration constants	E3bg	Uncert.	S40
pK$_2$ 8.73				Approx.	
pK$_1$ 1.91, pK$_2$ 8.72	25	I = 0.1(KNO$_3$), concentration constants	E3bg	Approx.	T31

Other values in J15 and V10

No.	Molecular formula, name and pK value(s)	T(°C)	Remarks	Method	Assessment	Ref.
3205	$C_6H_{11}O_5N_3$ Butanamide, NN-dimethyl-4,4-dinitro-					
	4.49	20		05	Approx.	T58a
3206	$C_6H_{11}NS_2$ Methanedithioic acid, cyclopentylamino- (Dithiocarbamic acid, N-cyclopentyl-)					
	3.00	21	$c \simeq 0.01\text{-}0.00001$, concentration constant	05	Uncert.	T66
	2.84		(Compound unstable in acid solution)	E	Uncert.	
3207	$C_6H_{11}NS_2$ 1-Piperidinecarbodithioic acid					
	6.32	15	From pH measurements on the Na-salt			P86
	5.64	25				
	7.70	40				
3208	$C_6H_{12}OS_2$ Methanedithioic acid, 3-methylbutoxy- (Xanthic acid, isoamyl-)					
	2.85	0	Extrapolated to zero time	E3d		G5
	1.82	not stated		05	Uncert.	Z2
3209	$C_6H_{12}OS_2$ Methanedithioic acid, neopentyloxy- (Xanthic acid, neopentyl)-					
	1.62	not stated		05	Uncert.	Z2
3210	$C_6H_{12}OS_2$ Methanedithioic acid, pentyloxy- (Xanthic acid, pentyl-)					
	1.80	not stated		05	Uncert.	Z2
	2.96	0	Extrapolated to zero time	E3d		G5
	3.44					H27

3211 2-Butanone, 4-dimethylamino-3-hydroxyimino-

$C_6H_{12}O_2N_2$
pK$_1$ 7.05 — 30 — c = 0.01, I = 0.1(KCl), mixed constant — E3bg — Approx. — E18
pK$_1$ 6.92, pK$_2$ 10.35 — 25 — c = 0.0025–0.005, I = 0.1(KNO$_3$), probably mixed constants — E — — U9

3212 Acetic acid, butylthio-

$C_6H_{12}O_2S$
3.739 — 25 — c ≈ 0.001, I = 0.10(KNO$_3$), concentration constant — E3bg — Approx. — P40

3213 Acetic acid, sec-butylthio-

$C_6H_{12}O_2S$
3.768 — 25 — c ≈ 0.001, I = 0.10(KNO$_3$), concentration constant — E3bg — Approx. — P40

3214 3-Azahexanoic acid, 5-amino-2-hydroxymethyl-4-oxo-   (Dl-α-Alanyl-DL-serine)

$C_6H_{12}O_4N_2$
pK$_1$ 2.98, pK$_2$ 8.25 — 25 — I = 0.1(NaClO$_4$) — E3bg — Approx. — S139

3215 3-Azapentanedioic acid, 3-(2-aminoethyl)-   [Iminodiacetic acid, N-(2-aminoethyl)-]

$C_6H_{12}O_4N_2$
pK$_1$ 5.58, pK$_2$ 11.05 — 20 — I = 0.1(KCl), concentration constants — E3bg — Approx. — S40

3216 Butane, 3,3-dimethyl-1,1-dinitro-

$C_6H_{12}O_4N_2$
5.05 — 20 — — 05 — Approx. — T58a

3217 3,6-Diazaoctanedioic acid   (Ethylenediiminodiacetic acid)   (EDDA)

$C_6H_{12}O_4N_2$
pK$_1$ 6.48, pK$_2$ 9.57 — 25 — I = 0.1(KNO$_3$), concentration constants — E3bg — Approx. — T26
pK$_1$ 6.42, pK$_2$ 9.46 — 25 — I = 0.1(KCl) — E3bg,R3g — Approx. — D67
pK$_1$ 6.55, pK$_2$ 9.62 — 25 — c ≈ 0.005–0.01, I = 0.1(KNO$_3$), concentration constants. Thermodynamic quantities are also given. — E3bg — Approx. — D27

No.	Molecular formula, name and pK value(s)	T(°C)	Remarks	Method	Assessment	Ref.
3218	Hexane, 1,1-dinitro- $C_6H_{12}O_4N_2$					
	5.60	5	c = 0.005-0.015, mixed constant	05	Approx.	N39,N38
	5.41	20				
	5.23	40				
	5.07	60				
			Thermodynamic quantities are derived from the results			
3219	1,3-Dithiacyclooctane 1,1,3,3-tetroxide $C_6H_{12}O_4S_2$					
	10.99	25	No details		Uncert.	C76
3220	3-Azapentanoic acid, 5-hydroxyamino-3(2-hydroxyethyl)-5-oxo- $C_6H_{12}O_5N_2$					
	pK_1 5.69, pK_2 9.18	20	c = 0.1($NaClO_4$), mixed constant	E3bg	Uncert.	M24
3221	1-Piperazinecarbodithioic acid, 4-methyl- $C_6H_{12}N_2S_2$					
	6.03	15	From pH measurements on Na-salt		Uncert.	P86
	6.62	25				
	7.42	40				
3222	Hexanohydroxamic acid $C_6H_{13}O_2N$					
	9.75	not stated	I = 0.2(NaCl)	E3bg	Uncert.	C72
	9.48	30.5	c = 0.001, I = 0.1(KNO_3), mixed constant	E3d	Uncert.	S137

No.	Formula		T(°C)	Conditions	Method		Ref.
3223	$C_6H_{13}O_2N$	Hexanoic acid, 2-amino- (Norleucine)					
	pK$_2$ 9.92		20	c = 0.001, mixed constant	E3bg	Approx.	P28
	pK$_2$ 9.86			I = 0.1	E3bg	Approx.	M102
3224	$C_6H_{13}O_2N$	Hexanoic acid, 6-amino-					
	pK$_2$ 10.77		25	I = 0.5(KNO$_3$), concentration constant	E3bg	Approx.	T23
	pK$_2$ 10.85		25	I = 0.5(NaCl), mixed constant	E3bg	Approx.	S109
3225	$C_6H_{13}O_2N$	Pentanoic acid, 2-amino-3-methyl- (Isoleucine)					
	pK$_1$ 2.66, pK$_2$ 9.70		25	I = 0.5(KCl), mixed constants	E3bg	Approx.	L27
	pK$_2$ 9.86		20	c = 0.02, mixed constant	E3bg	Approx.	P28
	pK$_2$ 9.76(DL)		25	I = 0.1	E3bg	Approx.	M102
	pK$_2$ 9.73[D(−)]						
	pK$_2$ 9.69[L(+)]						
3226	$C_6H_{13}O_2N$	Pentanoic acid, 2-amino-4-methyl- (Leucine)					
	pK$_2$ 9.71(DL)		25	I = 0.1	E3bg	Approx.	M102
	9.69[D(−)]						
	9.76[D(+)]						
	pK$_2$ 9.54		25	I = 0.16(KNO$_3$), mixed constant. Thermodynamic quantities are also given	E3bg	Approx.	M72
	pK$_2$ 9.92		20	c = 0.02, mixed constant	E3bg	Approx.	P28

No.	Molecular formula, name and pK value(s)	$T(^\circ C)$	Remarks	Method	Assessment	Ref.
3227	$C_6H_{13}O_2N_3$ 2,3-Butanedione dioxime, 1-dimethylamino- pK_1 8.4, pK_2 10.45	25	$c = 0.0025-0.005$, $I = 0.1(KNO_3)$ (probably mixed constants)	E		U9
3228	$C_6H_{13}O_3N$ Butanoic acid, 4-dimethylamino-3-hydroxy- (Norcarnitine) 3.81	25	$c = 0.01$	E3bg	Approx.	Y5
3229	$C_6H_{13}O_3N_3$ Pentanoic acid, 2-amino-5-ureido- (Citrulline) pK_2 9.71	25	$I = 0.1(KNO_3)$, concentration constant	E3bg	Approx.	C63
3230	$C_6H_{13}O_3P$ Cyclohexylphosphonic acid pK_1 2.2, pK_2 8.3	19	$c = 0.05$	E3bg	Uncert.	L21
3231	$C_6H_{13}O_4N$ 3-Azapentanoic acid, 5-hydroxy-3-(2-hydroxyethyl)- (Acetic acid, N,N-bis(2-hydroxyethyl)amino-) pK_1 1.68, pK_2 8.14 pK_1 2.50, pK_2 8.11 pK_1 (COOH) 1.99, pK_2 (NH$^+$) 8.41	20 25	$I = 0.1(NaClO_4)$ $I = 0.5$, concentration constants	05 E3bg E	Approx. Approx.	S102 T52 J15
3232	$C_6H_{13}O_5P$ Acetic acid, 2-(diethoxyphosphinyl)- 3.30	25	$c = 0.01$, mixed constant	E3bg	Uncert.	M52
3233	$C_6H_{13}O_6N$ Gluconic acid, 2-amino-2-deoxy- 9.24	20		E		G15

No.	Formula / Name	pK	T (°C)	Conditions	Method	Reliability	Ref.
3234	$C_6H_{13}O_7N$ Gluconohydroxamic acid						
		8.94	30.5(?)			Uncert.	E17
3235	$C_6H_{13}O_9P$ Fructose 6-(dihydrogen phosphate)						
		5.84	20	$I = 0.1(KCl)$	E3bg	Approx.	S37
3236	$C_6H_{13}O_9P$ Glucose 6-(dihydrogen phosphate)						
		pK_1 1.49, pK_2 6.22	10.3	$c = 0.01$	E3bg	pK_1 Uncert. pK_2 Approx.	D26
		pK_1 1.50, pK_2 6.22	20.0				
		pK_1 1.54, pK_2 6.24	30.5				
		pK_1 1.61, pK_2 6.28	40.1				
		pK_1 1.67, pK_2 6.31	50.0				
		pK_1 1.65, pK_2 6.31	25	$c = 0.02$	E3bg	pK_1 Uncert. pK_2 Approx.	B171
		pK_1 1.6, pK_2 6.2	70			Uncert.	
		pK_3 11.71	25		CAL	Approx.	I34
							Thermodynamic quantities are also given
3237	$C_6H_{13}NS_2$ Methanedithioic acid, isopentylamino- (Dithiocarbamic acid, N-isopentyl-)						
		2.91	21	$c \approx 0.01-0.00001$, concentration constant	05	Uncert.	T66
		2.73		(Compound unstable in acid solution)	E	Uncert.	
3238	$C_6H_{14}ON_2$ 2-Butanone oxime, 3-methyl-3-methylamino-						
		9.24	25	$I = 0.002$	CAL	Uncert.	W8
							Thermodynamic quantities are also given

No.	Molecular formula, name and pK value(s)	T(°C)	Method	Assessment	Ref.	Remarks
3239	$C_6H_{14}O_2N_2$ Acetohydroxamic acid, 2-diethylamino-					
	7.20	25(?)	E3bg	Uncert.	M66	c = 0.02, I = 0.1(KCl), mixed constant
3240	$C_6H_{14}O_2N_2$ Butanohydroxamic acid, 4-dimethylamino-					
	8.40	25(?)	E3bg	Uncert.	M66	c = 0.02, I = 0.1, mixed constant
3241	$C_6H_{14}O_2N_4$ Pentanoic acid, 2-amino-5-guanido- (Arginine)					
	pK_2 9.36	25	E3bg	Approx.	C63	I = 0.1(KNO$_3$), concentration constants
	pK_3 11.5			Uncert.		
3242	$C_6H_{14}O_3N_2$ Hexanoic acid, 2-amino-6-hydroxyamino-					
	pK_2 5.7, pK_3 9.5	25	E3bg	Uncert.	R40	
3243	$C_6H_{14}O_5S$ D-Sorbitol, 1-thiol-					
	9.50	20	O5	Approx.	D8	
	9.35	30				
	9.20	40				
3244	$C_6H_{14}O_{12}P_2$ Fructose 1,6-bis(dihydrogen phosphate) (Fructose 1,6-diphosphoric acid)					
	pK_3 6.43, pK_4 7.28	25	E3bg	Approx.	M15	
	pK_3 6.14, pK_4 6.93	25				I = 0.05
	pK_3 6.08, pK_4 6.91	30				I = 0.05
	pK_3 6.07, pK_4 6.91	35				I = 0.05
	pK_3 5.94, pK_4 6.75	40				I = 0.096

No.	Formula / Name / pK	Temp (°C)	Conditions	Method	Reliability	Ref
3245	$C_6H_{15}O_2N_3$ Hexanohydroxamic acid, 2,6-diamino- (Lysine hydroxamic acid)					
	7.93		I = 0.2(NaCl), mixture of D and L isomers	E3bg	Uncert.	C72
3246	$C_6H_{15}O_2P$ Dipropylphosphinic acid					
	3.26	20	c = 0.0004-0.05, concentration constant	C2	Approx.	K63
3247	$C_6H_{15}O_3P$ Hexylphosphonic acid					
	pK_1 2.4, pK_2 8.2	17	c = 0.02	E3bg	Uncert.	L21
	pK_1 2.6, pK_2 7.9	not stated	c = 0.02	E3bg	Uncert.	C38
3248	$C_6H_{15}O_4P$ Dipropyl hydrogen phosphate					
	1.59	25	m ≈ 0.04	E3bg	Uncert.	K67
	1.52	not stated	c = 0.005, I < 0.025, 7% ethanol	E3bg	Uncert.	M57
	1.52	20	7% ethanol	E		K3
3249	$C_6H_{15}S_2P$ Di-isopropylphosphinodithioic acid					
	1.64	20	c = 0.005, 7% ethanol, mixed constant	E3bg	Uncert.	K5,K6
3250	$C_6H_{15}S_2P$ Dipropylphosphinodithioic acid					
	1.84	20	c = 0.005, 7% ethanol, mixed constant	E3bg	Uncert.	K5,K6
3251	$C_6H_{16}O_6P_2$ 1,6-Hexanediyldiphosphonic acid (Hexamethylene diphosphonic acid) (HEDPA)					
	pK_1 1.8	25	I = 0.1(KCl)	E3bg,R3f	Uncert.	K4
	pK_2 3.07, pK_3 7.65, pK_4 8.37				Approx.	
	pK_1 1.8, pK_2 3.07, pK_3 7.65, pK_4 8.37	25	c = 0.001, I = 0.1(KCl), concentration constants	E3bg	Uncert.	D69

(Contd)

No.	Molecular formula, name and pK value(s)	T(^{o}C)	Remarks	Method	Assessment	Ref.
3251 (Contd)	$C_6H_{16}O_6P_2$ 1,6-Hexanediyldiphosphonic acid (Hexamethylene diphosphonic acid) (HEDPA)					
	pK_1 1.8, pK_2 3.12, pK_3 7.73, pK_4 8.56	Average of values at 25, 37 and 50^{0}. Extrapolated	E3bg	Uncert.	I6
			from values at I = 0.1-1.0[$(CH_3)_4$!Br]			
3252	$C_6H_{18}O_2B_{10}$ 1,2-Dicarbadodecaborane(12)-1-carboxylic acid, 2-isopropyl- (Barenecarboxylic acid, isopropyl-)					
	2.56	25	Mixed constant	E3bg	Uncert.	Z3
3253	$C_6HO_7N_3Cl_2$ Phenol, 3,5-dichloro-2,4,6-trinitro-					
	-0.7	not stated	In aq.$HClO_4$ solutions, H_o scale	06	Uncert.	P16
3254	$C_6H_2O_7N_3Cl$ Phenol, 3-chloro-2,4,6-trinitro-					
	-0.2	not stated	In aq.$HClO_4$ solutions, H_o scale	06	Uncert.	P16
3255	$C_6H_2O_7N_3Br$ Phenol, 3-bromo-2,4,6-trinitro-					
	-0.05	not stated	In aq.$HClO_4$ solutions, H_o scale	06	Uncert.	P16
3256	$C_6H_2O_7N_3I$ Phenol, 3-iodo-2,4,6-trinitro-					
	0.15	not stated	I = 0.5($NaClO_4$), c = 0.003	06	Uncert.	P16
3257	$C_6H_2O_8N_3Cl$ 1,3-Benzenediol, 5-chloro-2,4,6-trinitro-					
	pK_1 1.13	not stated	c = 0.003, I = 0.5($NaClO_4$)	05	Uncert.	P16
3258	$C_6H_3ON_4F_3$ Purin-2(1H)-one, 8-trifluoromethyl-					
	pK_1 5.35, pK_2 10.92	20	c = 0.0001, I $\approx$ 0.01	05	Approx.	A19

No.	Formula, Name, pK	T	Conditions	Method	Assessment	Ref.
3259	$C_6H_3ON_4F_3$ Purin-6(1$\underline{H}$)-one, 8-trifluoromethyl- pK_1 ca 5, pK_2 10.9			05	Uncert.	G20
3260	$C_6H_3OCl_2Br$ Phenol, 4-bromo-2,6-dichloro- 6.21	25	c = 0.001	E3bg	Approx.	F17
3261	$C_6H_3O_3NCl_2$ Phenol, 2,6-dichloro-4-nitro- 3.549	25	c ≈ 0.00002	05a	Approx.	R31
	3.56	25	c = 0.0032	E3bg	Approx.	F17
3262	$C_6H_3O_3NBr_2$ Phenol, 2,4-dibromo-6-nitro- 4.708	25	c ≈ 0.00005	05a	Rel.	R30
3263	$C_6H_3O_3NBr_2$ Phenol, 2,6-dibromo-4-nitro- 3.392	25	c ≈ 0.00002	05a	Approx.	R30
3264	$C_6H_3O_3NI_2$ Phenol, 2,6-diiodo-4-nitro- 3.324	25	c ≈ 0.000025	05a	Approx.	R30
3265	$C_6H_3O_5N_2Cl$ Phenol, 2-chloro-4,6-dinitro- 2.100	25	c ≈ 0.00001	05a	Approx.	R30
	2.01	not stated	c = 0.003, I = 0.5($NaClO_4$)	05	Uncert.	P16
3266	$C_6H_3O_5N_2Cl$ Phenol, 4-chloro-2,6-dinitro- 2.96(H_2O)	25	I < 0.07	03	Approx.	B51
	3.45(D_2O)				in D_2O	
	3.482(D_2O)	25	c ≈ 0.00008	05	Approx.	R36

244

No.	Molecular formula, name and pK value(s)	T(°C)	Remarks	Method	Assessment	Ref.
3267	$C_6H_3O_5N_2Br$ Phenol, 2-bromo-4,6-dinitro-					
	2.11	25	$c \approx 0.00005$, $I = 0.1$	05	Approx.	D50
	2.35	not stated	$c \approx 0.003$, $I = 0.5(NaClO_4)$	05	Uncert.	P16
3268	$C_6H_4ON_2S$ 2,1,3-Benzothiadiazol-4(1H)-one					
	7.86	25	$c \approx 0.0001$, mixed constant	05	Approx.	D5
3269	$C_6H_4ON_2S$ 2,1,3-Benzothiadiazol-5(3H)-one					
	8.16	25	$c \approx 0.0001$, mixed constant	05	Approx.	D5
3270	$C_6H_4ON_2Se$ 2,1,3-Benzoselenadiazol-4(1H)-one					
	8.16	25	$c \approx 0.0001$, mixed constant	05	Approx.	D5
3271	$C_6H_4ON_2Se$ 2,1,3-Benzoselenadiazol-5(3H)-one					
	8.06	25	$c \approx 0.0001$, mixed constant	05	Approx.	D5
3272	C_6H_4OClBr Phenol, 4-bromo-2-chloro-					
	7.64	25	$c = 0.0001-0.0002$, $I \approx 0.1$, mixed constant	05	Approx.	S76
3273	$C_6H_4O_2NCl$ 1,4-Benzoquinone 1-oxime, 2-chloro-					
	5.70	not stated	Mixed constant	05	Uncert.	D35
3274	$C_6H_4O_2NCl$ 1,4-Benzoquinone 1-oxime, 3-chloro-					
	5.60	not stated	Mixed constant	05	Uncert.	D35

No.	Formula	Name	pK	T	Conditions	Method	Reliability	Ref.
3275	$C_6H_4O_2NCl$	2-Pyridinecarboxylic acid, 4-chloro-	4.09		c = 0.04 or 0.001(?)	E		K51
3276	$C_6H_4O_2NCl$	3-Pyridinecarboxylic acid, 2-chloro-	2.54		c = 0.04 or 0.001(?)	E		K51
3277	$C_6H_4O_2NCl$	3-Pyridinecarboxylic acid, 4-chloro-	4.40		c = 0.04 or 0.001(?)	E		K51
3278	$C_6H_4O_2NCl$	4-Pyridinecarboxylic acid, 2-chloro-	2.70		c = 0.04 or 0.001(?)	E		K51
3279	$C_6H_4O_2NBr$	1,4-Benzoquinone 1-oxime, 2-bromo-	5.70	not stated	Mixed constant	05	Uncert.	D35
3280	$C_6H_4O_2NBr$	1,4-Benzoquinone 1-oxime, 3-bromo-	5.55	not stated	Mixed constant	05	Uncert.	D35
3281	$C_6H_4O_3NCl$	Phenol, 2-chloro-4-nitro-	5.45	25	c = 0.000058, I = 0.04-0.2	05a	Approx.	B115
			5.31	25	$c \approx$ 0.0001-0.0002, $I \approx$ 0.1, mixed constant	05	Approx.	S76
3282	$C_6H_4O_3NCl$	Phenol, 2-chloro-6-nitro-	5.483	25	$c \approx$ 0.00004	05a	Approx.	R30
3283	$C_6H_4O_3NCl$	Phenol, 3-chloro-6-nitro-	6.05	25	$c \approx$ 0.0001, $I \approx$ 0.1, mixed constant	05	Approx.	H11

No.	Molecular formula, name and pK value(s)	$T(^oC)$	Remarks	Method	Assessment	Ref.
3284	$C_6H_4O_3NCl$ Phenol, 4-chloro-2-nitro-					
	6.458	25	c ≈ 0.00005	05a	Approx.	P1
	6.46	25	c = 0.000233, I = 0.002-0.2	05a	Approx.	B115
	6.36	25	c = 0.0002, I ≈ 0.5, mixed constant	05	Approx.	R15
3285	$C_6H_4O_3NCl$ 3-Pyridinecarboxylic acid N-oxide, 4-chloro-					
	$pK(N^+-OH)$ 1.80		Conditions not stated	E	Uncert.	D52
3286	$C_6H_4O_3NF$ Phenol, 5-fluoro-2-nitro-					
	6.07	25	c ≈ 0.0001, I ≈ 0.1, mixed constant	05	Approx.	H11
3287	$C_6H_4O_4Cl_3P$ Phenyl dihydrogen phosphate, 2,4,5-trichloro-					
	pK_1 0.53	20		05	Approx.	P60
	pK_2 5.47		I = 0.02	E3bg		
	pK_2 5.48		I = 0.1-0.2	05		
3288	$C_6H_4O_8Br_2S_2$ 1,4-Benzenedisulfonic acid, 2,3-dibromo-5,6-dihydroxy-					
	pK_3 9.15	25	I = 0.1(KCl),	E3bg	Uncert.	A1
	pK_4 12.6		concentration constants	05		
3289	$C_6H_5O_2NS$ Benzenethiol, 3-nitro-					
	5.241	25	m = 0.0001, I < 0.15	05a	Rel.	D28

No.	Formula / Name		pK	T (°C)	Conditions	Method	Assessment	Ref.
3290	$C_6H_5O_2NS$	Benzenethiol, 4-nitro-						
			4.715	25	m = 0.0001, I < 0.15	05a	Rel.	D28
			4.42	35	I = 1.0	05	Approx.	M81
3291	$C_6H_5O_2N_4Br$	Purine-2,6(1H,3H)-dione, 8-bromo-9-methyl-						
			5.45	20	c = 0.001	E3bg	Approx.	P49
3292	$C_6H_5O_2ClS$	Acetic acid, 2-(5-chloro-2-thienyl)-						
			3.89	25			Uncert.	I3
3293	$C_6H_5O_2ClS$	Benzenesulfinic acid, 3-chloro-						
			2.68	25	I ≈ 0.005	E3bg	Uncert.	D23
3294	$C_6H_5O_2ClS$	Benzenesulfinic acid, 4-chloro-						
			2.76	25	I ≈ 0.005	E3bg	Uncert.	D23
			1.81	25		E3bg	Uncert.	B180
3295	$C_6H_5O_2ClSe$	Benzeneseleninic acid, 4-chloro-						
			4.30	25	I ≈ 0.005, in 4% methanolic solution	E3bg	Approx.	D23
3296	$C_6H_5O_2BrS$	Benzenesulfinic acid, 4-bromo-						
			3.08	25	I ≈ 0.005	E3bg	Uncert.	D23
			1.89	25		E3bg	Uncert.	B180
3297	$C_6H_5O_2BrSe$	Benzeneseleninic acid, 4-bromo-						
			4.28	25	I ≈ 0.005, in 4% methanolic solution	E3bg	Approx.	D23

No.	Molecular formula, name and pK value(s)		T(°C)	Remarks	Method	Assessment	Ref.
3298	$C_6H_5O_4NS$	Benzenesulfinic acid, 3-nitro-					
	2.81		25	I ≈ 0.005	E3bg	Uncert.	D23
	1.88		25		E3bg	Uncert.	B180
3299	$C_6H_5O_4NS$	Benzenesulfinic acid, 4-nitro-					
	1.86		25		E3bg	Uncert.	B180
	2.77		25	I ≈ 0.005	E3bg	Uncert.	D23
3300	$C_6H_5O_4Cl_2P$	Phenyl dihydrogen phosphate, 2,4-dichloro-					
	pK_1 0.65		20		05	Approx.	P60
	pK_2 5.76			I = 0.02	E3bg		
	pK_2 5.68			I = 0.1-0.2	05		
3301	$C_6H_5O_8N_2P$	Phenyl dihydrogen phosphate, 2,4-dinitro-					
	pK_2 ≈ 4.5		1	Very unstable	E3bg	Uncert.	B172
3302	$C_6H_5O_9NS_2$	1,3-Benzenedisulfonic acid, 4,5-dihydroxy-6-nitroso-					
	pK_3 4.45		not stated	c = 0.0001, I = 0.1(NaCl)	05	Uncert.	B63
	pK_3 4.41			c = 0.001, I = 0.1(NaCl)	E3bg	Approx.	
	pK_4 10.6			c = 0.0001, I = 0.1(NaCl)	05	V.Uncert.	
				Concentration constants			
3303	C_6H_6ONCl	Phenol, 4-amino-2-chloro-					
	pK_2 9.16		25	c ≈ 0.0001-0.0002, I ≈ 0.1, mixed constant	05	Approx.	S76

No.	Formula	Name	pK values	Conditions	T	c			Ref.
3304	C_6H_6ONCl	Pyridin-2(1H)-one, 6-chloro-4-methyl-	pK_1 0.2, pK_2 8.06		20	c = 0.00005	05	Approx.	S74
3305	$C_6H_6ON_4S$	Pteridine-2(1H)-thione, 3,4-dihydro-4-hydroxy-	9.72		20	c = 0.0001	05	Approx.	A30
3306	$C_6H_6ON_4S$	Purin-2(1H)-one, 3,6-dihydro-1-methyl-6-thioxo-	pK_1 6.7, pK_2 11.0	Conditions not stated			05	Uncert.	L40
3307	$C_6H_6ON_4S$	Purin-2(1H)-one, 3,6-dihydro-3-methyl-6-thioxo-	pK_1 7.9, pK_2 11.2	Conditions not stated			05	Uncert.	L40
3308	$C_6H_6ON_4S$	Purin-2(1H)-one, 3,6-dihydro-7-methyl-6-thioxo-	pK_1 6.8, pK_2 12.1	Conditions not stated			05	Uncert.	L40
3309	$C_6H_6ON_4S$	Purin-2(1H)-one, 3,6-dihydro-9-methyl-6-thioxo-	pK_1 4.9, pK_2 12.6	Conditions not stated			05	Uncert.	L40
3310	$C_6H_6ON_4S$	Purin-2(1H)-one, 8-mercapto-1-methyl-	6.61		20	c = 0.001	E3bg	Approx.	B128
3311	$C_6H_6ON_4S$	Purin-2(1H)-one, 8-mercapto-9-methyl-	pK_1 7.00, pK_2 11.21		20	c = 0.000025	05	Uncert.	B128
3312	$C_6H_6ON_4S$	Purin-2(1H)-one, 6-methylthio-	pK_1 1.4, pK_2 7.7, pK_3 11.7	Conditions not stated			05	Uncert.	L41

No.	Molecular formula, name and pK value(s)	T(°C)	Remarks	Method	Assessment	Ref.
3313	$C_6H_6ON_4S$ Purin-6(1H)-one, 2-methylthio- pK$_1$ 2.4, pK$_2$ 7.6, pK$_3$ 12		Conditions not stated	05	Uncert.	R18
3314	$C_6H_6O_2ClP$ Phenylphosphonous acid, 3-chloro- 1.35	25	c ≈ 0.05, mixed constant	E3bg	Uncert.	Q1
3315	$C_6H_6O_2ClP$ Phenylphosphonous acid, 4-chloro- 1.57	25	c ≈ 0.05, mixed constant	E3bg	Uncert.	Q1
3316	$C_6H_6O_2BrP$ Phenylphosphonous acid, 3-bromo- 1.39	25	c ≈ 0.05, mixed constant	E3bg	Uncert.	Q1
3317	$C_6H_6O_3N_2S$ Acetic acid, 2-(1,2,3,4-tetrahydro-4-oxo-2-thioxo-5-pyrimidinyl)- (Pyrimidine, 5-carboxymethyl-4-hydroxy-2-mercapto-) pK$_1$ 4.15, pK$_2$ 8.44, pK$_3$ ≈ 13.5	not stated		05	Uncert.	F22
3318	$C_6H_6O_4N_2S$ Benzenesulfonamide, 3-nitro- 9.20	20	c = 0.0025, I = 0.1(KCl), mixed constant	E3bg	Approx.	W33
3319	$C_5H_6O_4N_2S$ Benzenesulfonamide, 4-nitro- 9.14	20	c = 0.0025, I = 0.1(KCl), mixed constant	E3bg	Approx.	W33
3320	$C_6H_6O_4ClP$ Phenyl dihydrogen phosphate, 4-chloro- pK$_1$ 0.70 pK$_2$ 5.89 pK$_2$ 5.83	20	I = 0.02 I = 0.1-0.2	05 E3bg 05	Approx.	P60

No.	Formula	Name	pK	T (°C)	Conditions	Method	Assessment	Ref.
3321	$C_6H_6O_5N_2S$	Benzenesulfonic acid, 2-amino-5-nitro-	pK_3 17.47	20	c = 0.001, in aq.NaOH, H_- scale	07	Uncert.	H6
3322	$C_6H_6O_5N_2S$	Benzenesulfonic acid, 4-amino-3-nitro-	pK_3 17.65	20	c = 0.001, in aq.NaOH, H_- scale	07	Uncert.	H6
3322a	$C_6H_6O_6NP$	Phenyl dihydrogen phosphate, 2-nitro-	pK_2 5.6	32	c = 0.01	E3bg	Uncert.	B173
3323	$C_6H_7ON_5S$	2H-1,2,3-Triazolo[4,5-d]pyrimidin-7(6H)-one, 2-methyl-5-methylthio-	7.28	20	c = 0.00004	05	Approx.	A40
3324	$C_6H_7O_2NS$	Benzenesulfonamide	10.00	20	c = 0.0025, I = 0.1(KCl), mixed constant	E3bg	Approx.	W33
3325	$C_6H_7O_3NS$	Benzenesulfonic acid, 2-amino- (Orthanilic acid)	2.49	25		05	Approx.	S21
3326	$C_6H_7O_3NS$	Benzenesulfonic acid, 3-amino- (Metanilic acid)	3.75	25		05	Approx.	S21
			3.738	25				S88
3327	$C_6H_7O_3NS$	Benzenesulfonic acid, 4-amino- (Sulfanilic acid)	3.34	15		05	Approx.	S23
			3.25	25				
			2.93	20	c = 0.025, mixed constant	E3d	Uncert.	H17
			3.228	25				S88

No.	Molecular formula, name and pK value(s)	T(°C)	Remarks	Method	Assessment	Ref.
3328	$C_6H_7O_3N_5S$ 1H-1,2,3-Triazolo[4,5-d]pyrimidin-7(4H)-one, 1-methyl-5-methylsulfonyl-					A40
	2.96	20	c = 0.00004	05	Approx	
3329	$C_6H_7O_3N_5S$ 2H-1,2,3-Triazolo[4,5-d]pyrimidin-7(6H)-one, 2-methyl-5-methylsulfonyl-					A40
	3.54	20	c = 0.000019	05	Approx.	
3330	$C_6H_7O_3SP$ 5-Phenyl dihydrogen phosphorothioate					M81
	pK$_2$ 5.26	15	I = 1.0	E3bg	Uncert.	
	pK$_2$ 5.42	25				
	pK$_2$ 5.21	15	Titrated as the dicyclohexylammonium salt			
3331	$C_6H_7O_4NS$ Methanesulfonic acid, 1-hydroxy-1-(2-pyridyl)-					B15
	pK$_1$ 4.05, pK$_2$ 9.70	25	I = 0.1(NaClO$_4$), mixed constants	E3bg	Approx.	
3332	$C_6H_7O_4NS$ Methanesulfonic acid, 1-hydroxy-1-(3-pyridyl)-					B15
	pK$_1$ 4.60, pK$_2$ 9.53	25	I = 0.1(NaClO$_4$), mixed constants	E3bg	Approx.	
3333	$C_6H_7O_4NS$ Methanesulfonic acid, 1-hydroxy-1-(4-pyridyl)-					B15
	pK$_1$ 4.95, pK$_2$ 10.05	25	I = 0.1(NaClO$_4$), mixed constants	E3bg	Approx.	
3334	$C_6H_8ON_2S$ Pyrimidin-4(1H)-one, 2,3-dihydro-5,6-dimethyl-2-thioxo-					G2
	8.08	25	Mixed constant	E3d	Uncert.	
	8.06	35				
	7.76	45				

No.	Compound / pK	Temp (°C)	Conditions	Method	Uncert.	Ref. (G2)
3335	$C_6H_8ON_2S$ Pyrimidin-4(1H)-one, 2-ethylthio-					
	7.01	25	Mixed constant	E3d		
3336	$C_6H_8O_2N_2S$ Benzenesulfonamide, 4-amino-					
	10.58	20	$c = 0.0025$, $I = 0.1$(KCl), mixed constant	E3d	Approx.	W33
3337	$C_6H_8O_4NP$ 2-Pyridylmethyl dihydrogen phosphate					
	pK_2 4.42, pK_3 6.29	25	$c = 0.001-0.002$, $I = 0.1$(KNO$_3$)	E3bg	Approx.	M122
	pK_2 4.15, pK_3 6.54	80	$c = 0.002$, $I = 0.1$(NaClO$_4$)	E3bg	Approx.	M121
	pK_1 1.8	25	$I = 0.1$(KNO$_3$)	E3bg	Uncert.	M120
	pK_2 4.42, pK_3 6.30				Approx.	
3338	$C_6H_8O_4NP$ 3-Pyridylmethyl dihydrogen phosphate					
	pK_1 4.86, pK_2 6.23	25	$c = 0.001-0.002$, $I = 0.1$(KNO$_3$)	E3bg	Approx.	M122
	pK_1 4.43, pK_2 6.48	80	$c = 0.002$, $I = 0.1$(NaClO$_4$)	E3bg	Approx.	M121
3339	$C_6H_8O_4NP$ 4-Pyridylmethyl dihydrogen phosphate					
	pK_1 5.14, pK_2 6.25	25	$c = 0.001-0.002$, $I = 0.1$(KNO$_3$)	E3bg	Approx.	M122
	pK_1 4.73, pK_2 6.42	80	$c = 0.002$, $I = 0.1$(NaClO$_4$)	E3bg	Approx.	M121
3340	$C_6H_8O_5NP$ 2-Pyridylmethyl dihydrogen phosphate, 3-hydroxy-					
	pK_1 4.34, pK_2 5.75, pK_3 9.67	25	$I = 0.1$	E3bg	Approx.	M114
3341	$C_6H_9O_4N_2P$ 3-Pyridylmethyl dihydrogen phosphate, 2-amino-					
	pK_1 5.42, pK_2 7.03	25	$c = 0.002$, $I = 0.1$	E3bg	Approx.	M117
	pK_1 5.11, pK_2 7.40	70				

No.	Molecular formula, name and pK value(s)	$T(^oC)$	Remarks	Method	Assessment	Ref.
3342	$C_6H_{10}ON_2S$ Imidazolidin-4-one, 5-isopropyl-2-thioxo- 8.70	18±3	I = 0.01	05	Uncert.	E4
3343	$C_6H_{10}ON_2S$ Imidazolidin-4-one, 3,5,5-trimethyl-2-thioxo- 10.30	18±3	I = 0.01	05	Uncert.	E4
3344	$C_6H_{10}O_2N_2F_4$ Hexanoic acid, 5,5-bis(difluoramino)- 4.62	25	Mixed constant	E3bg	Approx.	B37
3345	$C_6H_{11}O_2NF_2$ Pentanoic acid, 4-difluoramino-4-methyl- 4.35	25	Mixed constant	E3bg	Approx.	B37
3346	$C_6H_{11}O_2NS$ Thiazolidine-4-carboxylic acid, 5,5-dimethyl- 1.6		I = 0.1(KCl)	E3bg	Uncert.	W13
3347	$C_6H_{11}O_4NS$ 3-Azapentanedioic acid, 3-(2-mercaptoethyl)- (Iminodiacetic acid, N-(2-mercaptoethyl)-) pK$_1$ 2.14 pK$_2$ 8.17, pK$_3$ 10.79	20	I = 0.1(KCl), concentration constants	E3bg	Uncert. Approx.	S40
3348	$C_6H_{11}O_7NS$ 3-Azapentanedioic acid, 3-(2-sulfoethyl)- (Iminodiacetic acid, N-(2-sulfoethyl)-) pK$_1$ 1.9 pK$_2$ 2.28, pK$_3$ 8.16	20	I = 0.1(KCl), concentration constants	E3bg	Uncert. Approx.	S36

No.	Formula	Name			T(°C)	Conditions	Ref. code	Remark	Ref.
3349	$C_6H_{12}O_2N_2S_2$	3,6-Diazaoctane-4,5-dithione, 1,8-dihydroxy-	[Ethanedithioamide, N,N'-bis(2-hydroxyethyl)-]						P75
			[Dithiooxamide, N,N'-bis(2-hydroxyethyl)-]						
		pK$_1$ 10.71			25	c = 0.0001, I = 0.5(NaClO$_4$), mixed constant	05	Approx.	
		pK$_1$ 11.04			25	I = 0.01-0.5	05,R4	Approx.	
		pK$_2$ 13.92			25	c = 0.0001, I = 0.5(NaClO$_4$), mixed constant	07	Uncert.	
3350	$C_6H_{12}O_4N_2S_2$	4,5-Dithiaoctanedioic acid, 2,7-diamino-	(L-Cystine)						H23
		pK$_3$ 8.03, pK$_4$ 8.80			20	I = 0.15(NaClO$_4$), concentration constants	E3bg,R3d	Uncert.	
3351	$C_6H_{12}O_6N_2S_4$	3,6-Diazaoctane-1,8-disulfonic acid, 4,5-dithioxo-	[Dithiooxamide, N,N'-bis(2-sulfoethyl)-]						G28
		pK$_1$ 11.73			25	By extrapolation from values at I = 0.01-1.0(KCl)	05	Approx.	
		pK$_2$ 13.33			25	I = 1.0(KCl), concentration constant		Uncert.	
3352	$C_6H_{12}O_7NP$	3-Azahexanedioic acid, 3-phosphonomethyl-	(NAPMP)						W21
		pK$_2$ 2.72, pK$_3$ 3.48, pK$_4$ 5.59, pK$_5$ 10.41			25	I = 0.1(KNO$_3$), concentration constants	E3bg	Approx.	
3353	$C_6H_{12}O_7NP$	3-Azapentanedioic acid, 3-(2-phosphonoethyl)-	[Iminodiacetic acid, N-(2-phosphonoethyl)-]						S36
		pK$_1$ 1.9			20	I = 0.1(KCl), concentration constants	E3bg	Uncert.	
		pK$_2$ 2.45, pK$_3$ 6.54, pK$_4$ 10.46						Approx.	
3354	$C_6H_{12}O_7N_2S_3$	3,6-Diazaoctane-1,8-disulfonic acid, 4-oxo-5-thioxo-	[Monothiooxamide, N,N'-bis(2-sulfoethyl)-]						G29
		11.97			25	c = 0.0001, I = 0.01-0.10(KCl)	05	Approx.	
3355	$C_6H_{13}ONS$	3-Oxa-5-azaheptane-4-thione, 2-methyl-	(O-Isopropyl thiocarbamate, N-ethyl-)						L61
		pK(NH) 11.52			20	c =0.01	E		

No.	Molecular formula, name and pK value(s)	T(°C)	Remarks	Method	Assessment	Ref.
3356	$C_6H_{14}O_2NS$(cation) Butanoic acid, 2-amino-4-dimethylsulfonio- (L-Methionine, S-methyl-)					
	pK₁ 1.9		Determined from infra-red measurements in D_2O and corrected to the corresponding value in water by subtracting 0.5 pK unit		Uncert.	K34
	pK₂ 7.9			E3bg & 05		
3357	$C_6H_{14}O_4NP$ Diethyl phosphoramidate, N-acetyl-					
	pK(NH) 9.46		Conditions not stated	E3bg	Uncert.	K2
3358	$C_6H_{15}OSP$ Di-isopropylphosphinothioic 0-acid					
	3.03	20	c = 0.005 in 7% ethanol, 99.5% thione tautomer	E3bg	Uncert.	K6
3359	$C_6H_{15}OSP$ Dipropylphosphinothioic 0-acid					
	2.58	20	c = 0.0004-0.05, concentration constant	C2	Approx.	K63
	2.83	20	c = 0.005 in 7% ethanol, 99% thione tautomer	E3bg	Uncert.	K6
3360	$C_6H_{15}O_2N_2$(I) Propanohydroxamic acid, 2-trimethylammonio-iodide ([(1-Hydroxycarbamoyl)ethyl]trimethylammonium iodide)					
	6.65	25(?)	c = 0.02, I = 0.1(KCl), mixed constant	E3bg	Uncert.	M66
3361	$C_6H_{15}O_2N_2$(I) Propanohydroxamic acid, 3-trimethylammonio-,iodide ([2-(Hydroxycarbamoyl)ethyl]trimethylammonium iodide)					
	8.0	25(?)	c = 0.02, I = 0.1(KCl), mixed constant	E3bg	Uncert.	M66

No.	Formula	Name	pK	Temp.	Method	Uncertainty	Ref.	Notes
3362	$C_6H_{15}O_2SP$	0-Butyl 0-hydrogen ethylphosphonothioate	2.21	20	E3bg	V.uncert.	K6	c = 0.005 in 7% ethanol, 91% thione tautomer
3363	$C_6H_{15}O_2SP$	0-Ethyl 0-hydrogen butylphosphonothioate	2.11	20	E3bg	V.uncert.	K6	c = 0.005 in 7% ethanol, 90% thione tautomer
3364	$C_6H_{15}O_2S_2P$	0,0-Di-isopropyl hydrogen phosphorodithioate	1.82	20	E3bg	V.uncert.	K6,K5	c = 0.005 in 7% ethanol
3365	$C_6H_{15}O_2S_2P$	0,0-Dipropyl hydrogen phosphorodithioate	1.75	20	E3bg	V.uncert.	K5,K6	c = 0.005, 7% ethanol, mixed constant
3366	$C_6H_{15}O_3SP$	0,0-Di-isopropyl 0-hydrogen phosphorothioate	1.59	20	E3bg	V.uncert.	K6	c = 0.005 in 7% ethanol, 54% thiol tautomer
3367	$C_6H_{15}O_3SP$	0,0-Dipropyl hydrogen phosphorothioate	1.55	20	E3bg	V.uncert.	K6	c = 0.005 in 7% ethanol, 50% thiol tautomer
3368	$C_6H_{20}O_{12}N_2P_4$	2,5-Diazahexane-1,6-diphosphonic acid, 2,5-bis(phosphonomethyl)- [Ethylenedinitrilotetrakis(methylenephosphonic acid)]	pK_1 1.46, pK_2 2.72, pK_3 5.05, pK_4 6.18, pK_5 6.63, pK_6 7.43, pK_7 9.22, pK_8 10.60	25	E3bg	Uncert. Approx.	W21	I = 0.1(KNO_3), concentration constants

(Contd)

No.	Molecular formula, name and pK value(s)	$T(^oC)$	Remarks	Method	Assessment	Ref.
3368 (Contd)	$C_6H_{20}O_{12}N_2P_4$ 2,5-Diazahexane-1,6-diphosphonic acid, 2,5-bis(phosphonomethyl)- pK$_3$ 3.0, pK$_4$ 5.23, pK$_5$ 6.54, pK$_6$ 8.08, pK$_7$ 10.18, pK$_8$ 12.10	25	c = 0.0018, I = 0.1(KCl)	E3bg	Uncert.	K1
3369	$C_5H_5ON_2(F_4B)$ Benzenediazonium fluoborate, 4-hydroxy- 3.47	0	Mixed constant	E3bg	Uncert.	L36
	3.40	25		05		
3370	$C_6H_6O_2NClS$ Benzenesulfonamide, 4-chloro- 9.77	20	c = 0.0025, I = 0.1(KCl), mixed constant	E3bg	Approx.	W33
3371	$C_6H_6O_2NBrS$ Benzenesulfonamide, N-bromo- 4.95	25	c = 0.00125	E3bg	Approx.	H16
3372	$C_6H_6O_3ClSP$ 5-(4-Chlorophenyl) dihydrogen phosphorothioate pK$_2$ 5.26	15	I = 1.0; titrated as the dicyclohexylammonium salt	E3bg	Uncert.	M81
3373	$C_6H_7O_4NClP$ 2-Chloropyridin-3-ylmethyl dihydrogen phosphate pK$_1$ 2.17, pK$_2$ 5.97	25	I = 0.1(KNO$_3$)	E3bg	Approx.	M116
	pK$_1$ 1.8, pK$_2$ 6.3	80				
3374	$C_6H_{11}O_2NClF$ Acetic acid, N-(2-chloroethyl)-N-(2-fluoroethyl)amino- pK$_1$ 2.65, pK$_2$ 6.20		Conditions not stated		Uncert.	B84

259

3375　$C_6H_{11}O_4NCl_3P$　Diethyl phosphoramidate, N-(trichloroacetyl)-

pK(NH)4.77　Conditions not stated　E3bg　Uncert.　K2

3376　$C_6H_{11}O_4NF_3P$　Diethyl phosphoramidate, N-(trifluoroacetyl)-

pK(NH) 4.16　Conditions not stated　E3bg　Uncert.　K2

3377　$C_6H_{12}O_4NCl_2P$　Diethyl phosphoramidate, N-(dichloroacetyl)-

pK(NH) 6.05　Conditions not stated　E3bg　Uncert.　K2

3378　$C_6H_{13}O_4NClP$　Diethyl phosphoramidate, N-(chloroacetyl)-

pK(NH) 7.66　Conditions not stated　E3bg　Uncert.　K2

No.	Molecular formula, name and pK value(s)	T(°C)	Remarks	Method	Assessment	Ref.
3379	$C_7H_2O_2$ 2,4,6-Heptatriynoic acid					
	1.67	25	c = 0.005, I = 0.1(NaCl), mixed constant	E3bg	Approx.	M48
3380	$C_7H_4O_6$ 4H-Pyran-2,6-dicarboxylic acid, 4-oxo- (Chelidonic acid)					
	pK$_2$ 2.36	25	c = 0.001, mixed constant	E3bg	Approx.	M86
3381	$C_7H_4O_7$ 4H-Pyran-2,6-dicarboxylic acid, 3-hydroxy-4-oxo- (Meconic acid)					
	pK$_1$ 1.83, pK$_2$ 2.11, pK$_3$ 11.25	20-25	I = 0.1(KNO$_3$)	E3bg	Uncert.	B29
	pK$_2$ 2.13, pK$_3$ 10.10	25	c = 0.001, mixed constants	E3bg	Approx.	M86
	pK$_3$ 10.08		c = 0.005, mixed constant	05	Approx.	
	pK$_3$ 10.30	25	I = 0.01-0.5(NaCl)	E3bg,R4	Approx.	C47
	pK$_3$ 9.35	25	I = 0.5(NaCl)	E3bg,05	Approx.	C46
	pK$_3$ 10.81	24	I = 0.19	E3bg	Approx.	T60
3382	$C_7H_4N_2$ 1,3-Cyclopentadiene-1,2-dicarbonitrile					
	1.11		Conditions not stated	05	Uncert.	W10a
3383	$C_7H_4N_2$ 1,3-Cyclopentadiene-1,3-dicarbonitrile					
	2.52		Conditions not stated	05	Uncert.	W10a
3384	C_7H_6O Benzaldehyde					
	14.90	25	c = 0.0001, in 1% ethanol. pK of hydrate	07,R4	Uncert.	B111

3385 $C_7H_6O_2$ Benzaldehyde, 2-hydroxy- (Salicylaldehyde)

8.374	25	05	Approx.	R35
8.37	25	05	Approx.	G40
8.34	20	E3bg	Approx.	P30
8.22	25	E3bg	Approx.	L26

Variation with I: $pK = 8.37-0.78 \ (I)^{\frac{1}{2}}$

$I = 0.1-0.15(NaClO_4)$, mixed constant

$I = 0.5(KCl)$, mixed constant

Other values in A11, D50, M19 and W24.

3386 $C_7H_6O_2$ Benzaldehyde, 3-hydroxy-

pK_1 8.983	25	05a	Rel.	B97

$m = 0.00006$

Variation with temperature

5	10	15	20	30	35	40	45	50	55	60
9.264	9.186	9.114	9.046	8.924	8.868	8.817	8.768	8.724	8.682	8.643

Thermodynamic quantities are derived from the results

pK_1 8.994	25	E3d	Approx.	L52

$I = 0.0017$

Equation relating pK to temperature for 25 values from 14.90 to 55.96. Thermodynamic quantities are given

pK_1 9.016	25	05	Approx.	R35
pK_1 8.88	25	E3bg	Approx.	E20
pK_2 15.81	25	07	Uncert.	B111

$I = 0.1(KCl)$, mixed constant

$c = 0.0001$ in 1% ethanol solutions

No.	Molecular formula, name and pK value(s)	$T(^oC)$	Remarks	Method	Assessment	Ref.
3387	$C_7H_6O_2$ Benzaldehyde, 4-hydroxy-					
	7.76	10		05,R4	Approx.	K17
	7.61	25	Thermodynamic quantities are derived			
	7.47	40	from the results			
	7.35	55				
	7.61	25	$I = 0.0017$. Equation for variation of pK	E3d	Approx.	L52
			with temperature for 12 values in the			
			temperature range 14.9 to 55.96.			
			Thermodynamic quantities are also given			
	7.615	25		05	Approx.	R35
	7.60	25		05	Approx.	C70
	7.62	25	$c \approx 0.0002$	05	Approx.	M19
	7.45		$I = 0.1$			
	7.38	25	$I = 0.1(NaClO_4)$, concentration constant	05	Approx.	D34
			Thermodynamic quantities are given			
			Other value in V2			

3388 $C_7H_6O_2$ Benzoic acid

pK	T (°C)	Conditions			Ref.
4.205	25		C2,R1b	Rel.	F18

Variation with pressure (bar)

	500	1000	1500	2000	2500	3000
	4.107	4.021	3.943	3.865	3.795	3.726

Other values at high pressures given in C61

pK	T (°C)	Conditions			Ref.
4.204*	25		05a	Rel.	B91

*Value calculated from thermodynamic data given in reference

pK	T (°C)	Conditions			Ref.
4.205	25	m = 0.005-0.02	E1cg	Rel.	K30
4.21	25	m = 0.01	C2,R1b	Approx.	E13

Variation with temperature

	50	100	150	200	225
	4.24	4.35	4.55	4.85	5.05

Equation derived for temperature variation

pK	T (°C)	Conditions			Ref.
4.202	25		05	Approx.	W38

Variation with temperature

	15	20	30	35	40
	4.218	4.208	4.205	4.210	4.219

Thermodynamic quantities are derived from the results

pK	T (°C)	Conditions			Ref.
4.01	25	I = 0.1($NaClO_4$), concentration constant	E3bg	Approx.	Y12
4.01	31	I = 0.1($NaClO_4$), mixed constant	E3bg	Approx.	R11
3.99	20	I = 0.1(KCl), mixed constant	E3bg	Approx.	W32

(contd)

No.	Molecular formula, name and pK value(s)	$T(^oC)$	Remarks	Method	Assessment	Ref.
3388 (Contd)	$C_7H_6O_2$ Benzoic acid					
	4.16(H_2O)	25	Probably mixed constant	O5	Uncert.	W12
	4.65(D_2O)					
	Other values in H17, H33, L32, P4, P24, P25, R33, S88, S115, S120, T61 and W29					
	Benzoic acid-2,6-d_2					
	4.20	25	c = 0.004-0.011, mixed constant	C1	Approx.	S120
	Benzoic acid-d_5					
	4.21	24.6	c = 0.002-0.008, mixed constant	C1	Approx.	S120
3389	$C_7H_6O_2$ 2,4,6-Cycloheptatrien-1-one, 2-hydroxy- (Tropolone)					
	pK_1(CO) -0.53	25	In aq.H_2SO_4, H_o scale	O6	Uncert.	B41
	pK_2(OH) 6.67	25	I = 0.5(NaCl), concentration constant		Approx.	
			Mean of E3bg and O5 determination			
	pK_2 7.00	20		E3bg	Approx.	C75
	pK_2 6.694	25	c = 0.002, I = 0.1(KNO$_3$), concentration constant	E3bg	Approx.	C7
	pK_2 6.42	25	I = 2.0	O5	Approx.	O13
	Other values in O10 and Y15					
3390	$C_7H_6O_2$ 2,4-Heptadiynoic acid					
	1.90	25	c = 0.01, I = 0.1(NaCl), mixed constant	E3bg	Approx.	M48

No.	Formula	Name / pK values	T (°C)	Method	Quality	Conditions	Ref.
3391	$C_7H_6O_2$	3,5-Heptadiynoic acid					
		3.23	25	E3bg	Approx.	$c = 0.01$, $I = 0.1(NaCl)$, mixed constant	M48
3392	$C_7H_6O_3$	Benzaldehyde, 2,3-dihydroxy-					
		pK_1 7.73, pK_2 10.91	20	E3bg	Uncert.	$I = 0.15(NaClO_4)$, 4% ethanol	K8
3393	$C_7H_6O_3$	Benzaldehyde, 2,4-dihydroxy-					
		pK_1 7.10, pK_2 9.42	20	E3bg	Uncert.	$I = 0.15(NaClO_4)$, 4% ethanol	K8
		pK_1 7.0	20	05	Uncert.	1-4% ethanol, mixed constant	W24
3394	$C_7H_6O_3$	Benzaldehyde, 2,5-dihydroxy-					
		pK_1 8.28, pK_2 10.29	20	E3bg	Uncert.	$I = 0.15(NaClO_4)$, 4% ethanol	K8
3395	$C_7H_6O_3$	Benzaldehyde, 2,6-dihydroxy-					
		pK_1 7.0	20	05	Uncert.	1-4% ethanol, mixed constant	W24
3396	$C_7H_6O_3$	Benzaldehyde, 3,4-dihydroxy-					
		pK_1 7.190	25	E3bg	Approx.	$c \approx 0.01$, $I = 0.1(KCl)$, concentration constant	A69
		pK_1 7.27, pK_2 11.4		E and 0			H15
3397	$C_7H_6O_3$	Benzoic acid, 2-hydroxy- (Salicylic acid)					
		pK_1 3.08, pK_2 13.55	1	05	pK_1 Approx. pK_2 Uncert.	$c = 0.002$, $I = 0.1(KCl)$, mixed constants	H33
		pK_1 2.99, pK_2 12.95	25			Thermodynamic quantities are derived from	
		pK_1 2.91, pK_2 12.48	45			the results	

(contd)

No.	Molecular formula, name and pK value(s)	T(°C)	Remarks	Method	Assessment	Ref.
3397 (Contd)	$C_7H_6O_3$ Benzoic acid, 2-hydroxy- (Salicylic acid)					
	pK$_1$ 3.015	20		05	Approx.	G11
	pK$_1$ 2.991	30				
	pK$_1$ 2.978	40	Thermodynamic quantities are derived from the			
	pK$_1$ 2.975	50	results			
	pK$_1$ 2.93, pK$_2$ 13.65	25	I = 0.6-3.0($NaCl,KCl,NaNO_3,KNO_3,NaClO_4$)	0		V5
	pK$_1$ 3.093	5		C1		
	pK$_1$ 2.959	40	Thermodynamic quantities are derived from the			
	pK$_1$ 3.047	70	results			
	pK$_1$ 3.00	25	c = 0.0005-0.011	C2,R1b	Approx.	C61
		Variation with pressure (atm)				
		1010 2010 2780				
		2.87 2.76 2.66				
	pK$_1$ 2.70, pK$_2$ 13.90	25	c = 0 by extrapolation, I = 0.1(KNO_3),	E3bg pK$_1$	Approx.	M99
	pK$_1$ 2.45, pK$_2$ 13.40	34.9	concentration constants	pK$_2$	Uncert.	
	pK$_1$ 2.97	25	c = 0.001	05	Rel.	E24
	pK$_2$ 13.6			07	Uncert.	
	pK$_1$ 2.98	20	I = 0.1-0.15($NaClO_4$), mixed constants	E3bg	Approx.	P30
	pK$_2$ 13.61			07	Uncert.	
	pK$_1$ 3.15	25	I = 0.1(KCl), concentration constants	E3bg	Uncert.	A1
	pK$_2$ 13.2			05		

(contd)

3397 (Contd) $C_7H_6O_3$ Benzoic acid, 2-hydroxy- (Salicylic acid)

pK_1 2.81	37	I = 0.15(KNO_3), mixed constant	E3bg	Approx.	P34
pK_1 2.82	25	I = 0.1($NaClO_4$)	E3bg	Approx.	P7
pK_1 2.82	25	I = 0.5($NaClO_4$), concentration constant	E3bg	Approx.	A18
pK_1 3.01(H_2O)	25		E3bg	Approx.	G21
pK_1 3.57(D_2O)	25				

Other values in A11, B31, B123, C23, D56, L37, P24, P25, S88, W32 and Y3

3398 $C_7H_6O_3$ Benzoic acid, 3-hydroxy-

pK_1 3.96, pK_2 9.61	20	I = 0.1(KCl), mixed constants	E3bg	Approx.	W32
pK_1 4.3, pK_2 10.25	25	I = 0.1(KCl), concentration constants	E3bg	Uncert.	A1
pK_1 4.08	25	c = 0.0003-0.002	C1	Approx.	B123
pK_2 9.98	25		O5	Approx.	C71
pK_2 9.73	25	I = 0.1(KCl), mixed constant	E3bg	Approx.	E20

3399 $C_7H_6O_3$ Benzoic acid, 4-hydroxy-

pK_1 4.67, pK_2 9.37	25	Thermodynamic quantities are given	E1 and O5	Approx.	P4
pK_1 4.60, pK_2 9.39	1	c = 0.002, I = 0.1(KCl), mixed constants	O5	Approx.	H33
pK_1 4.50, pK_2 9.11	25	Thermodynamic quantities are derived			
pK_1 4.42, pK_2 9.02	45	from the results			
pK_1 4.36, pK_2 8.99	20	I = 0.1(KCl), mixed constants	E3bg	Approx.	W32
pK_1 4.53	25	c = 0.0003-0.004	C1	Approx.	B123
pK_2 9.32	25	I = 0.1(KCl), mixed constant	E3bg	Approx.	H51

(contd)

No.	Molecular formula, name and pK value(s)	T(°C)	Remarks	Method	Assessment	Ref.
3399 (Contd)	$C_7H_6O_3$ Benzoic acid, 4-hydroxy- pK_2 9.39	25		05	Approx.	C70
	Other values in A1 and E20					
3400	$C_7H_6O_3$ 1,4-Benzoquinone, 2-hydroxy-6-methyl- 4.04	20	c = 0.0001, mixed constant	05	Uncert.	M128
3401	$C_7H_6O_3$ 2,4,6-Cycloheptatrien-1-one, 2,3-dihydroxy- pK_1 6.72, pK_2 11.52	25		E3bg	Uncert.	Y16
3402	$C_7H_6O_3$ 2,4,6-Cycloheptatrien-1-one, 2,4-dihydroxy- pK_1 5.76, pK_2 9.52	25		E3bg	Uncert.	Y16
3403	$C_7H_6O_3$ 2,4,6-Cycloheptatrien-1-one, 2,5-dihydroxy- pK_1 6.47, pK_2 10.10	25		E3bg	Uncert.	Y16
3404	$C_7H_6O_4$ Benzoic acid, 2,3-dihydroxy- 2.914	25	c = 0.0002-0.0012	C1,R1a, K1a,K2e	Rel.	D40
3405	$C_7H_6O_4$ Benzoic acid, 2,4-dihydroxy- pK_1 3.11, pK_2 8.55	25	c ≈ 0.005, I = 0.25($NaClO_4$), mixed constants	E3bg and 05	Approx.	M74
	pK_3 14.0		(contd)	07	Uncert.	

	Temp (°C)	Conditions	Method	Assessment	Ref.
3405 (Contd) $C_7H_6O_4$ Benzoic acid, 2,4-dihydroxy-					
pK_1 3.30, pK_2 9.12	25		05	Approx.	M65
pK_3 15.6				Uncert.	
pK_1 3.33, pK_2 8.91	30	I = 0.01. Thermodynamic quantities are also given	0		G59
pK_1 3.325	25	c = 0.00029-0.0051	C1,R1a, K1a,K2e	Rel.	D40
pK_1 3.10	25	I = 0.1($NaClO_4$)	E3bg	Approx.	P7
Other values in D56 and R14					
3406 $C_7H_6O_4$ Benzoic acid, 2,5-dihydroxy-					
2.951	25	c = 0.00028-0.0027	C1,R1a, K1a,K2e	Rel.	D40
2.98	25	c = 0.00022, I = 0.01	05	Approx.	D56
3407 $C_7H_6O_4$ Benzoic acid, 2,6-dihydroxy-					
1.051	25	c = 0.00028-0.0021	C1,R1a, K1a,K2e	Approx.	D40
1.07	25	I = 0.1($NaClO_4$)	E3bg	Uncert.	P7
3408 $C_7H_6O_4$ Benzoic acid, 3,4-dihydroxy-					
pK_1 4.26, pK_2 8.64	25	c ≈ 0.005, I = 0.25($NaClO_4$), mixed constants	E3bg and 05	Approx.	M74
pK_3 13.1			07	Uncert.	

(contd)

No.	Molecular formula, name and pK value(s)	T(°C)	Remarks	Method	Assessment	Ref.
3408 (Contd)	C$_7$H$_6$O$_4$ Benzoic acid, 3,4-dihydroxy-					
	pK$_1$ 4.38, pK$_2$ 8.67, pK$_3$ 11.94	30	I = 0.1(KCl), mixed constants	E3bg	Approx. Uncert.	H12
	pK$_1$ 4.40, pK$_2$ 8.83, pK$_3$ 12.6	30	c = 0.004, I = 0.1(NaClO$_4$), mixed constants	E3bg	Approx. Uncert.	A80
	pK$_1$ 4.491	25	c = 0.00023-0.0016	C1,R1a, K1a,K2e	Rel.	D40
	pK$_2$ 9.15	25	I = 0.1(KCl), concentration constant	E3bg	Uncert.	A1
	Other values in H15 and J10					
3409	C$_7$H$_6$O$_4$ Benzoic acid, 3,5-dihydroxy-					
	4.039	25	c = 0.0004-0.0016	C1,R1a, K1a,K2e	Rel.	D40
3410	C$_7$H$_6$O$_5$ Benzoic acid, 3,4,5-trihydroxy- (Gallic acid)					
	pK$_1$ 3.13, pK$_2$ 8.84, pK$_3$ 12.4	20	I = 0.1(KNO$_3$), mixed constants	E3bg	Approx. Uncert.	B27
	pK$_1$ 4.27, pK$_2$ 8.68	25	I = 0.1(KCl)	E		A72

(contd)

3410 (Contd) C$_7$H$_6$O$_5$ Benzoic acid, 3,4,5-trihydroxy-   (Gallic acid)

	Temp	Conditions			
pK$_1$ 4.46			E3bg	Approx.	R9

Variation with temperature

	40	50	60	70	80
	4.49	4.53	4.56	4.64	4.77
	30				

Thermodynamic quantities are derived from the results

pK$_2$ 8.78	25	I = 0.1(KNO$_3$), mixed constant	E3bg	Approx.	E19

3411 C$_7$H$_6$O$_5$ 3-Furancarboxylic acid, 2-methoxycarbonyl-

3.09		Conditions not stated	E3bg	Uncert.	O1

3412 C$_7$H$_6$O$_5$ 3-Furancarboxylic acid, 4-methoxycarbonyl-

3.75	25	c = 0.005-0.009	E3bg	Approx.	M11
3.75		Conditions not stated	E3bg	Uncert.	O1

3413 C$_7$H$_6$O$_5$ 4H-Pyran-2-carboxylic acid, 5-methoxy-4-oxo-

1.53	25		C1	Uncert.	S11

3414 C$_7$H$_6$N$_2$ Benzimidazole

pK$_2$ 13.2	20	c = 0.0001 in 0.01M buffer	O5	Uncert.	B127
pK$_2$ 12.86	25	c $\approx$ 0.0001	O5	Uncert.	T21
pK$_2$ 12.86	25	In aq.KOH, H$_-$ scale	O7	Uncert.	Y2

No.	Molecular formula, name and pK value(s)	T(°C)	Remarks	Method	Assessment	Ref.
3415	C$_7$H$_6$N$_2$ 1H-Indazole					
	13.80	25	c ≈ 0.0001	07	Uncert.	T21
	13.80	25	In aq.KOH, H_ scale	07	Uncert.	Y2
3416	C$_7$H$_8$O Methanol, phenyl- (Benzyl alcohol)					
	15.4	25		KIN	Uncert.	M126
3417	C$_7$H$_8$O Phenol, 2-methyl- (o-Cresol)					
	10.22	25	Variation with temperature	E1a,R4	Approx.	L66
			30 35 45 55 65 70			
			10.15 10.095 9.99 9.90 9.82 9.78			
	10.333	25	c = 0.00004	01	Approx.	C40
			Variation with temperature			
			5 10 15 20 30 38			
			10.655 10.566 10.488 10.409 10.268 10.169			
			Thermodynamic quantities are derived from the results			
	10.32	25	c = 0.0002	05	Approx.	H32
	10.29	25		05	Approx.	B77
	10.10	25	I = 0.1(KCl), mixed constant	E3bg	Approx.	E20
	5.3		pK of electronically excited singlet state calculated assuming pK = 10.28 in ground state	FLU/05	V.Uncert.	B25

3418 $C_7H_8O_2$ Phenol, 3-methyl- (m-Cresol)

10.098 25 c = 0.00004 01 Approx. C40

Variation with temperature

5	10	15	20	30	38
10.406	10.320	10.240	10.165	10.032	9.936

Thermodynamic quantities are derived from the results

10.09	25	05	Approx.	B77
10.09	25 c = 0.0002	05	Approx.	H32
9.82	25 I = 0.1(KCl), mixed constant	E3bg	Approx.	E20
4.2	pK of electronically excited singlet state calculated	FLU/05	V.Uncert.	B25

assuming pK = 10.09 in ground state

Other values in H51 and M128

3419 C_7H_8O Phenol, 4-methyl- (p-Cresol)

10.276 25 c = 0.00005 01 Approx. C40

Variation with temperature

5	10	15	20	30	38
10.597	10.501	10.423	10.349	10.214	10.118

Thermodynamic quantities are derived from the results

10.26	25	05	Approx.	B77
10.27	25 c = 0.00017	05	Approx.	H32
10.10	25 I = 0.1(KCl), mixed constant	E3bg	Approx.	E20

(contd)

No.	Molecular formula, name and pK value(s)	T(^{o}C)	Remarks	Method	Assessment	Ref.
3419 (Contd)	C_7H_8O Phenol, 4-methyl- (p-Cresol)					
	10.02	25	I = 0.5(KNO_3), mixed constant	E3bg	Approx.	M5
	10.30	20	c = 0.005	E3bg	Approx.	D31
	4.1		pK of electronically excited singlet state, calculated assuming pK = 10.26 in ground state	FLU/05	V.Uncert.	B25
3420	$C_7H_8O_2$ 1,2-Benzenediol, 3-methyl- pK_1 9.51, pK_2 10.90	20	I = 0.1, mixed constants(?)	E3bg	Uncert.	S128
3421	$C_7H_8O_2$ 1,2-Benzenediol, 4-methyl- pK_1 9.67 pK_2 12.77	20	I = 0.1, mixed constants	E3bg	Uncert.	S128
	pK_1 9.7 pK_2 12.95	25	I = 0.1(KCl), concentration constants	E3bg 05	Uncert.	A1
	pK_1 9.44, pK_2 11.9			E and 0		H15
3422	$C_7H_8O_2$ 1,3-Benzenediol, 2-methyl- pK_1 9.61, pK_2 11.98	20	I = 0.1, mixed constants(?)	E3bg	Uncert.	S128
3423	$C_7H_8O_2$ 1,3-Benzenediol, 4-methyl- pK_1 9.46, pK_2 11.42	20	I = 0.1, mixed constants(?)	E3bg	Uncert.	S128

3424 C7H8O2 1,3-Benzenediol, 5-methyl-

	Temp	Method	Assessment	Ref	Remarks
pK1 9.36					
pK2 11.66	20	E3bg	Uncert.	S128	I = 0.1, mixed constants(?)
pK1 9.48			Approx.		
pK2 11.20	20	E3bg	Uncert.	M128	c = 0.02, mixed constants

3425 C7H8O2 1,4-Benzenediol, 2-methyl-

	Temp	Method	Assessment	Ref	Remarks
pK1 10.20	13.1	05	Approx.	B38	I = 0.65(NaClO4), concentration constants
pK1 10.10	20.1				
pK1 10.05	25.1				Thermodynamic quantities are derived
pK2 11.72	13.2				from the results
pK2 11.67	19.6				
pK2 11.62	25.9				
pK1 10.25, pK2 11.85	14	05	Approx.	S111	I = 0.1(KCl)
pK1 10.15, pK2 11.75	21				Thermodynamic quantities are derived
					from the results
pK1 10.15, pK2 11.75	20	E3bg	Uncert.	S128	I = 0.1, mixed constants(?)

3426 C7H8O2 Phenol, 2-hydroxymethyl-

	Temp	Method	Assessment	Ref	Remarks
pK 9.84	25	E3d	Approx.	Z4	c = 0.1

Variation with temperature

20	30	35	40	45	50	55	60
9.91	9.77	9.71	9.65	9.59	9.53	9.48	9.42

Thermodynamic quantities are derived from the results

(contd)

No.	Molecular formula, name and pK value(s)	$T(^{o}C)$	Remarks	Method	Assessment	Ref.
3426 (Contd)	$C_7H_8O_2$ Phenol, 2-hydroxymethyl-					
	2.9		pK of electronically excited singlet state, calculated assuming pK = 9.92 in ground state	FLU/05	V.Uncert.	B25
3427	$C_7H_8O_2$ Phenol, 3-hydroxymethyl-					
	3.0		pK of electronically excited singlet state, calculated assuming pK = 9.83 in ground state	FLU/05	V.Uncert.	B25
3428	$C_7H_8O_2$ Phenol, 4-hydroxymethyl-					
	9.73	25	c = 0.1	E3d	Approx.	Z4
			Variation with temperature			
			20 30 35 40 45 50 55 60			
			9.79 9.68 9.63 9.58 9.53 9.48 9.43 9.39			
			Thermodynamic quantities are derived from the results			
	3.0		pK of the electronically excited singlet state, calculated assuming pK = 9.82 in the ground state	FLU/05	V.Uncert.	B25
3429	$C_7H_8O_2$ Phenol, 2-methoxy-					
	9.98	25		05	Approx.	B77
	9.90	25	I = 0.1(KCl), mixed constant	E3bg	Approx.	H51
	9.85	25	I = 0.1(KCl), mixed constant	E3bg	Approx.	E20
	5.2		pK of electronically excited singlet state, calculated assuming pK = 9.98 in ground state	FLU/05	V.Uncert.	B25

3430 $C_7H_8O_2$ Phenol, 3-methoxy-

pK	T	Notes	Rel.		
9.652	25	m = 0.00026		05a	B97

Variation with temperature

	5	10	15	20	25	30	35	40	45	50	55	60
	9.936	9.858	9.784	9.716		9.593	9.537	9.486	9.438	9.393	9.352	9.313

Thermodynamic quantities are derived from the results

pK	T	Notes	Rel.		
9.65	25		Approx.	05	B77
9.50	25	I = 0.1(KCl), mixed constant	Approx.	E3bg	E20
9.62(H_2O)	25		Uncert.	05	W12
10.20(D_2O)	25				
2.7		pK of electronically excited singlet state, calculated assuming pK = 9.65 in ground state	V.Uncert.	FLU/05	B25

3431 $C_7H_8O_2$ Phenol, 4-methoxy-

pK	T	Notes	Rel.		
10.21	25		Approx.	05	B77
10.13	10	I = 0.1(KCl)	Approx.	05	S111
10.00	20	Thermodynamic quantities are derived from the results			
9.89	30				
10.12	25	I = 0.1(KCl), mixed constant	Approx.	E3bg	H51
10.05	25	I = 0.1(KCl), mixed constant	Approx.	E3bg	E20
10.24(H_2O)	25		Uncert.	05	W12
10.85(D_2O)	25				

(contd)

No.	Molecular formula, name and pK value(s)	T(°C)	Remarks	Method	Assessment	Ref.
3431 (Contd)	$C_7H_8O_2$ Phenol, 4-methoxy- 4.7		pK of electronically excited singlet state, calculated assuming pK = 10.20 in ground state	FLU/05	V.Uncert.	B25
3432	$C_7H_8O_4$ 1-Cyclobutene-1-carboxylic acid, 2-methoxycarbonyl- 3.19	25	c = 0.005-0.009	E3bg	Approx.	M11
3433	$C_7H_8O_4$ 1-Cyclopentene-1,2-dicarboxylic acid pK_1 1.64, pK_2 7.27 pK_1 3.36 pK_2 7.18	25 20	c = 0.005-0.09 c = 0.001-0.002	E3bg E3bg	Approx. Uncert. Approx.	M11 I4
3434	$C_7H_8O_4$ 1-Cyclopentene-1,4-dicarboxylic acid pK_1 4.27, pK_2 5.10	20	c = 0.001-0.002	E3bg	Approx.	I4
3435	$C_7H_8O_4$ 1-Cyclopentene-1,5-dicarboxylic acid pK_1 4.03, pK_2 5.92	20	c = 0.001-0.002	E3bg	Approx.	I4
3436	$C_7H_8O_5$ 1,2-Cyclopentanedicarboxylic acid, 4-oxo- pK_1 3.90, pK_2 5.60(cis) pK_1 3.59, pK_2 5.34(trans)	20	c = 0.001-0.002	E3bg	Approx.	I4

No.	Formula	Name	pK	T	Conditions			
3437	C_7H_8S	Benzenethiol, 2-methyl-	6.64	26-7		05	Uncert.	D9
3438	C_7H_8S	Benzenethiol, 3-methyl-	6.660	25	m = 0.0001, I < 0.15	05a	Rel.	D28
			6.58	25		05	Approx.	D9
3439	C_7H_8S	Benzenethiol, 4-methyl-	6.820	25	m = 0.0001, I < 0.15	05a	Rel.	D28
			6.50	35	I = 1.0	05	Approx.	F13
			6.52	23		05	Approx.	D9
3440	C_7H_8S	Methanethiol, phenyl-	9.43	25		05	Approx.	K58
3441	$C_7H_{10}O_2$	Bicyclo[2.1.1]hexane-1-carboxylic acid	4.46	25	c = 0.01	E3d	Approx.	W27,W28
3442	$C_7H_{10}O_2$	Cyclohexanecarbaldehyde, 2-oxo-	6.35	20	c ≃ 0.008	E3bg	Approx.	S42
3443	$C_7H_{10}O_2$	2-Cyclohexen-1-one, 2-hydroxy-6-methyl-	11.23	20	60% enol form	E3bh	Approx.	S52
3444	$C_7H_{10}O_2$	Cyclopentanone, 2-acetyl-	7.82	20	c = 0.007-0.009	E3bg	Approx.	S42

No.	Molecular formula, name and pK value(s)	T(°C)	Remarks	Method	Assessment	Ref.
3445	$C_7H_{10}O_2$ Cyclopropene-1-carboxylic acid, 2,3,3-trimethyl- 3.7		Conditions not stated		Uncert.	C65
3446	$C_7H_{10}O_2$ 2-Heptynoic acid 2.60	25	c = 0.01, I = 0.1(NaCl), mixed constant	E3bg	Approx.	M48
3447	$C_7H_{10}O_2$ 6-Heptynoic acid 4.57	25	c = 0.01, I = 0.1(NaCl), mixed constant	E3bg	Approx.	M48
3448	$C_7H_{10}O_2$ 2-Pentynoic acid, 4,4-dimethyl- 2.66	25	c = 0.01, I = 0.1(NaCl), mixed constant	E3bg	Approx.	M48
3449	$C_7H_{10}O_3$ Acetic acid, 2-(2-oxocyclopentyl)- 4.58	20	c = 0.001-0.002	E3bg	Approx.	I4
3450	$C_7H_{10}O_3$ Acetic acid, 2-(3-oxocyclopentyl)- 4.64	20	c = 0.001-0.002	E3bg	Approx.	I4
3451	$C_7H_{10}O_3$ Cyclohexanecarboxylic acid, 4-oxo- 4.38	25	c = 0.01 Values in mixed solvents are also given	E3bg	Approx.	S69

No.	Compound / pK values	Temp	Conditions			Ref.
3452	$C_7H_{10}O_4$ Butenedioic acid, 2-ethyl-3-methyl-					
	pK_1 3.2, pK_2 6.1	20		E3bg	Uncert.	E1
	pK_1 was corrected for the acid-anhydride equilibrium					
3453	$C_7H_{10}O_4$ 1,2-Cyclopentanedicarboxylic acid					
	pK_1 4.42, pK_2 6.57(cis)	20	c = 0.001-0.002	E3bg	Approx.	I4
	pK_1 4.14, pK_2 5.99(trans)					
3454	$C_7H_{10}O_4$ 1,2-Cyclopropanedicarboxylic acid, 1,2-dimethyl-					
	pK_1 4.06, pK_2 6.56(cis)	20		E3bg	Approx.	M12
	pK_1 3.70, pK_2 5.24(trans)					
3455	$C_7H_{10}O_4$ cis-1,2-Cyclopropanedicarboxylic acid, 3,3-dimethyl- (cis-Caronic acid)					
	pK_1 2.61, pK_2 8.17	10.3	I = 0.111, mixed constants	E3bg	Approx.	H20
	pK_1 2.30, pK_2 8.01	25	I = 0.115			
	pK_1 2.38, pK_2 8.25(H_2O)	25		E3bg	Approx.	H21
	pK_1 3.04, pK_2 8.57(D_2O)					
3456	$C_7H_{10}O_4$ r-1,c-2-Cyclopropanedicarboxylic acid, 1,c-2-dimethyl-					
	pK_1 3.09, pK_2 7.91	20		E3bg	Approx.	M12
3457	$C_7H_{10}O_4$ r-1,c-2-Cyclopropanedicarboxylic acid, 1,t-3-dimethyl-					
	pK_1 3.69, pK_2 6.82	20		E3bg	Approx.	M12

No.	Molecular formula, name and pK value(s)	T(°C)	Remarks	Method	Assessment	Ref.
3458	$C_7H_{10}O_4$ r-1,t-2-Cyclopropanedicarboxylic acid, 1,c-3-dimethyl-					
	pK_1 3.93, pK_2 5.51	20		E3bg	Approx.	M12
3459	$C_7H_{10}O_4$ r-1,t-2-Cyclopropanedicarboxylic acid, 1,t-3-dimethyl-					
	pK_1 4.23, pK_2 5.65	20		E3bg	Approx.	M12
3460	$C_7H_{10}O_4$ Propanoic acid, 3-(2-oxo-3-oxolanyl)-					
	4.55	20	c = 0.001-0.002	E3bg	Approx.	I4
3461	$C_7H_{12}O_2$ Cyclohexanecarboxylic acid					
	4.89	25	c = 0.01	E3d	Approx.	W27
	4.90	25	c = 0.01	E3bg	Approx.	S69
	Values in mixed solvents are also given					
3462	$C_7H_{12}O_2$ 2,4-Heptanedione					
	9.43	5	c = 0.0005, 0.5-1.0% ethanol. Apparent pK for keto-enol mixture	O5	Approx.	C4
	9.23	25				C5
	9.14	45	Thermodynamic quantities for the keto and enol forms are derived from the results			C4
	pK(enol) 8.43, pK(keto) 9.15	25	Calculated using absorbance data		Uncert.	C5

No.	Formula	Name / pK	T	Conditions	Method	Reliability	Ref.
3463	$C_7H_{12}O_2$	3,5-Heptanedione					
		9.55	20	c = 0.00002, I = 0.1($NaClO_4$), mixed constant	05	Approx.	K49
		9.83	20	c = 0.01	E3bg and 05	Approx.	R52
3464	$C_7H_{12}O_2$	2-Heptenoic acid					
		4.12(cis)	25		E3bg	Approx.	M10
		4.88(trans)					
3465	$C_7H_{12}O_2$	2,4-Hexanedione, 5-methyl-					
		9.42	20	c = 0.01	E3bg and 05	Approx.	R52
		9.46	22	c = 0.01	E3bg	Approx.	L3a
		9.43	25	Thermodynamic quantities are derived			
		9.40	35	from the results			
		9.33	45				
		9.57	5	c = 0.005, 0.5-1.0% ethanol. Apparent pK	05	Approx.	C4
		9.40	25	for keto-enol mixture. Thermodynamic			C5
		9.33	45	quantities for the keto and enol forms are			C4
				derived from the results			
		pK(enol) 8.66, pK(keto) 9.31	25	Calculated from absorbance data		Uncert.	C5

No.	Molecular formula, name and pK value(s)	T(°C)	Remarks	Method	Assessment	Ref.
3466	$C_7H_{12}O_2$ Hexanoic acid, 2-methylene-					
	4.80		Conditions not stated	E3bg	Uncert.	C21
3467	$C_7H_{12}O_2$ 2-Hexenoic acid, 2-methyl-					
	5.13(trans)		Conditions not stated	E3bg	Uncert.	C21
	4.44(cis)					
3468	$C_7H_{12}O_2$ 2,4-Pentanedione, 3-ethyl-					
	11.41	15	c = 0.01	E3bg	Approx.	L3a
	11.34	25	Thermodynamic quantities are derived			
	11.28	35	from these results			
	11.25	45				
	11.33	20	c = 0.01	E3bg and 05	Approx.	R52
3469	$C_7H_{12}O_3$ Cyclohexanecarboxylic acid, 1-hydroxy-					
	4.159	25	Variation of the concentration constant with ionic strength	E3ag	Approx.	P77
3470	$C_7H_{12}O_3$ Cyclohexanecarboxylic acid, 2-hydroxy-					
	4.00	25	I = 1.0(NaClO$_4$), concentration constant	E3bg	Approx.	S30

Variation of the concentration constant with ionic strength (entry 3469):

I	2.00	1.00	0.50	0.20	0.10	0.05	0.02	0.01
pK_c	4.051	3.932	3.910	3.939	3.979	4.013	4.059	4.081

No.	Formula	Name / pK	T (°C)	Conditions	Method	Reliability	Ref.
3471	$C_7H_{12}O_3$	Cyclohexanecarboxylic acid, 4-hydroxy-					
		4.69(trans)	25	c = 0.01	E3bg	Approx.	S69
		Values in mixed solvents are also given					
3472	$C_7H_{12}O_3$	Ethyl 2-methyl-3-oxobutanoate					
		12.42	25.7		05	Uncert.	R53
3473	$C_7H_{12}O_3$	Ethyl 3-oxopentanoate					
		11.07	25.7		05	Uncert.	R53
3474	$C_7H_{12}O_4$	Butanoic acid, 2-ethoxycarbonyl-					
		3.43(H_2O)	25		E3bg	Approx.	G23
		3.90(D_2O)					
3475	$C_7H_{12}O_4$	Diethyl propanedioate (Diethyl malonate)					
		12.9			KIN	Uncert.	B47
3476	$C_7H_{12}O_4$	Heptanedioic acid (Pimelic acid)					
		pK_1 4.46, pK_2 5.58	25	c = 0.01, mixed constants	E3bg	Uncert.	N37
		pK_1 4.51, pK_2 5.51	25	I = 0.1(KNO_3)	E3bg		N32
3477	$C_7H_{12}O_4$	Pentanedioic acid, 3,3-dimethyl-					
		pK_1 3.85, pK_2 6.45	25	c = 0.00125, mixed constants	E3bg	Approx.	B146
		pK_1 3.73, pK_2 6.71	25	c = 0.005, I = 0.035	03,R3d	Approx.	J19

No.	Molecular formula, name and pK value(s)	T(°C)	Remarks	Method	Assessment	Ref.
3478	$C_7H_{12}O_4$ Propanedioic acid, tert-butyl-					
	pK_1 2.92, pK_2 7.04	25	c ≈ 0.01	E3bg	Uncert.	W40
3479	$C_7H_{12}O_4$ Propanedioic acid, diethyl-					
	pK_1 2.151, pK_2 7.417	25	I = 0.01	Elch*	Rel.	I28

(*Hg_2Cl_2/Hg replaces Ag/AgCl of Elch)

Variation with temperature

	5	10	15	20	30	35	40	45
pK_1	2.123	2.129	2.136	2.144	2.160	2.172	2.187	2.207
pK_2	7.401	7.400	7.401	7.408	7.428	7.441	7.458	7.472

Thermodynamic quantities are derived from the results

	T(°C)	Remarks	Method	Assessment	Ref.
pK_1 1.96	25	I = 0.1(KNO_3), concentration constants	E3bg	Uncert.	P78
pK_2 6.98		pK_2 is extrapolated value at c = 0		Approx.	
pK_1 2.15, pK_2 7.05	25	I = 0.1(KCl), mixed constants	E3bg	Approx.	M77
pK_1 2.21, pK_2 7.33(H_2O)	25		E3bg	Approx.	G23
pK_1 2.70, pK_2 7.74(D_2O)					

Other values in D44, L33 and O23

3480	$C_7H_{12}O_4$ Propanoic acid, 2-ethoxycarbonyl-2-methyl-					
	3.49	25	mixed constant	E3bg	Uncert.	L33
	3.55	37				

No.	Formula	Name	pK	T(°C)	Conditions		Assessment	Ref.
3481	$C_7H_{12}O_5$	Cyclohexanecarboxylic acid, 3,4,5-trihydroxy- (Dihydroshikimic acid)	4.39 4.25	25	$c = 0.0025$ $I = 0.05(KNO_3)$	E3bg	Approx.	T39
3482	$C_7H_{12}O_5$	3-Oxapentanedioic acid, 2,2,4-trimethyl-	pK$_1$ 3.32	25±0.5	$c < 0.001$	E3bg	Uncert.	S93
3483	$C_7H_{12}O_5$	Propanedioic acid, 2-butyl-2-hydroxy-	pK$_1$ 2.59, pK$_2$ 4.75					G37
3484	$C_7H_{12}O_6$	Cyclohexanecarboxylic acid, 1,3,4,5-tetrahydroxy- (Quinic acid)	3.58 3.50	25	$c = 0.0025$ $I = 0.05(KNO_3)$	E3bg	Approx.	T39
3485	$C_7H_{14}O_2$	Heptanoic acid	4.794 4.88 4.8	10 40 25		E3bg E3bg	Approx. Uncert.	C54 C57
3486	$C_7H_{14}O_2$	Hexanoic acid, 2-methyl-	5.04	not stated	$I = 0.01(KCl)$	E3bg	Uncert.	C21

No.	Molecular formula, name and pK value(s)	$T({}^{o}C)$	Remarks	Method	Assessment	Ref.
3487	$C_7H_{14}O_2$ Pentanoic acid, 2,2-dimethyl-					C54
	5.021	10		E3bg	Approx.	
	4.969	25				
	5.088	40				
3488	$C_7HO_2F_5$ Benzoic acid, pentafluoro-					R56
	1.75	25	$c = 0.02$	E3bg	Approx.	
3489	$C_7H_2O_2F_4$ Benzoic acid, 2,3,5,6-tetrafluoro-					R56
	1.87	25	Low solubility. Estimated from Hammett σ constants			
3490	$C_7H_2O_3F_4$ Benzoic acid, 2,3,5,6-tetrafluoro-4-hydroxy-					R56
	pK_1 2.27, pK_2 5.17	25	$c = 0.02$	E3bg	Approx.	
3491	$C_7H_3O_2Br_3$ 2,4,6-Cycloheptatrien-1-one, 3,5,7-tribromo-2-hydroxy-					Y15
	4.27	25		E3bg	Uncert.	
3492	$C_7H_3O_2F_4$ Benzoic acid, 4-amino-2,3,5,6-tetrafluoro-					R56
	pK_1 2.17(COOH)		Low solubility. Estimated from OH stretching frequencies		Uncert.	

No.	Formula	Name / pK	T (°C)	c	Method	Assessment	Ref.
3493	$C_7H_3O_{10}N_5$	Methane, (2,4,6-trinitrophenyl)dinitro-					
		-3.50	20	$c \approx 0.0001$	06	Uncert.	T58
		(H_0 in $HClO_4$ solutions)					
3494	$C_7H_4O_2Cl_2$	Benzaldehyde, 3,5-dichloro-4-hydroxy-					
		4.25	25	$c = 0.0011$	E3bg	Approx.	F18
3495	$C_7H_4O_2Cl_2$	Benzoic acid, 2,3-dichloro-					
		2.67	22	$c = 0.001$	E3bg	Approx.	M60
3496	$C_7H_4O_2Cl_2$	Benzoic acid, 2,4-dichloro-					
		2.680	25	$c = 0.00016-0.00052$	C1,R1a, K1a,K2e	Rel.	D40
		2.75	25	$c = 0.00005$	05	Approx.	D16
		2.76	25	$c = 0.01$	E3bg	Approx.	
3497	$C_7H_4O_2Cl_2$	Benzoic acid, 2,5-dichloro-					
		2.466	25	$c = 0.00015-0.00018$	C1,R1a, K1a,K2e	Rel.	D40
3498	$C_7H_4O_2Cl_2$	Benzoic acid, 2,6-dichloro-					
		1.594	25	$c = 0.00039-0.0022$	C1,R1a, K1a,K2e	Rel.	D40
		1.82	25	$c = 0.002$	E3bg	Approx.	D16

No.	Molecular formula, name and pK value(s)	$T(^{o}C)$	Remarks	Method	Assessment	Ref.
3499	$C_7H_4O_2Cl_2$ Benzoic acid, 3,4-dichloro- 3.64	25	c = 0.00004, I = 0.002-0.0005	05	Approx.	D16
3500	$C_7H_4O_2Cl_2$ Benzoic acid, 3,5-dichloro- 3.54	22	c = 0.0005	E3bg	Approx.	M60
3501	$C_7H_4O_2Br_2$ 2,4,6-Cycloheptatrien-1-one, 3,5-dibromo-2-hydroxy- 5.18	25		E3bg	Uncert.	Y15
3502	$C_7H_4O_2Br_2$ 2,4,6-Cycloheptatrien-1-one, 3,7-dibromo-2-hydroxy- 4.68	25		E3bg	Uncert.	Y15
3503	$C_7H_4O_3N_2$ Benzenecarbonitrile, 2-hydroxy-5-nitro- (Phenol, 2-cyano-4-nitro-) 4.1	30	I = 0.1(KCl)	05	Uncert.	C20
3504	$C_7H_4O_3N_2$ Benzenecarbonitrile, 2-hydroxy-6-nitro- (Phenol, 2-cyano-3-nitro-) 5.2	30	I = 0.1(KCl)	05	Uncert.	C20
3505	$C_7H_4O_3N_4$ 6-Pteridinecarbaldehyde, 1,4,7,8-tetrahydro-4,7-dioxo- pK_1 5.93, pK_2 9.31	20	c = 0.01	E3bg	Approx.	A22
3506	$C_7H_4O_3N_4$ 6-Pteridinecarboxylic acid, 1,7-dihydro-7-oxo- pK_1 1.50, pK_2 6.73	20	c = 0.0001 in 0.01M buffer	05	Approx.	A39

3507 $C_7H_4O_3Cl_2$ Benzoic acid, 3,5-dichloro-2-hydroxy-

pK$_1$	2.204	20	O5,R4	Approx.	G11
	2.166	30			
	2.138	40			
	2.120	50			

Thermodynamic quantities are derived from the results.

3508 $C_7H_4O_3Cl_2$ Benzoic acid, 3,5-dichloro-4-hydroxy-

5.97(OH)	25	E3bg,R3d	Approx.	F17

c = 0.0012, mixed constant

3509 $C_7H_4O_3Br_2$ Benzoic acid, 3,5-dibromo-2-hydroxy-

pK$_1$	2.162	25	O5,R4	Approx.	G11

Variation with temperature

20	30	35	40	45	50
2.178	2.149	2.138	2.130	2.124	2.121

Thermodynamic quantities are derived from the results

3510 $C_7H_4O_3I_2$ Benzoic acid, 2-hydroxy-3,5-diiodo-

pK$_1$	2.299	25	O5,R4	Approx.	G11
	2.288	30			
	2.279	35			
	2.273	40			

Thermodynamic quantities are derived from the results

No.	Molecular formula, name and pK value(s)	T(°C)	Remarks	Method	Assessment	Ref.
3511	$C_7H_4O_4N_4$ 6-Pteridinecarboxylic acid, 1,4,7,8-tetrahydro-4,7-dioxo-					
	pK_1 ca.3	20	c = 0.0013	E3bg	Uncert.	A22
	pK_2 6.69, pK_3 10.05	20	c = 0.0013	E3bg	Approx.	
3512	$C_7H_4O_4N_4$ 7-Pteridinecarboxylic acid, 1,4,5,6-tetrahydro-4,6-dioxo-					
	pK_1 2.3, pK_2 6.6, pK_3 9.85	Conditions not stated		E3bg	Uncert.	P48
3513	$C_7H_4O_5N_4$ 1H-Indazol-7-ol, 4,6-dinitro-					
	2.50	25	c = 0.01	E3bg	Approx.	S129
3514	$C_7H_4O_5N_4$ 6-Pteridinecarboxylic acid, 1,2,3,4,7,8-hexahydro-2,4,7-trioxo-					
	pK_1 2.0	20		E3bg	Uncert.	P44
	pK_2 5.98, pK_3 9.90			E3bg	Approx.	
3515	$C_7H_4O_5N_4$ 7-Pteridinecarboxylic acid, 1,2,3,4,5,6-hexahydro-2,4,6-trioxo-					
	pK_1 1.7	20		E3bg	Uncert.	P45
	pK_2 7.20, pK_3 9.63				Approx.	
3516	$C_7H_4O_6N_2$ Benzaldehyde, 2-hydroxy-3,5-dinitro-					
	2.27	25	c = 0.0002	O5	Approx.	M19
	2.09		I = 0.1, mixed constant			

No.	Formula	Name / pK	pK value	T	c	Remarks	Qualifier	Ref.
3517	$C_7H_4O_6N_2$	Benzoic acid, 3,5-dinitro-	2.60	10	05		Approx.	G56
			2.73	20				
			2.85	30				
			2.96	40				
			3.07	50				
						Thermodynamic quantities are derived from these results		
			2.94	30.3	05	$c = 0.001$	Uncert.	G55
3518	$C_7H_4O_7N_2$	Benzoic acid, 2-hydroxy-3,5-dinitro-						
		pK_2 7.40		25	E3bg	$I = 0.1(KNO_3)$, mixed constant	Approx.	B26
		pK_2 7.34		25	05	Value calculated from equation relating pK and T. Data given for 10-60°.		G57
						Thermodynamic quantities are derived from the results		
		pK_1 0.45, pK_2 7.6				Conditions not stated	Uncert.	A9
		pK_1 0.70		25	C1	$c = 0.0016$–0.0053	Uncert.	B123
3519	$C_7H_4O_8N_4$	...ilethane, (2,4-dinitrophenyl)dinitro-						
		0.03		20	06	$c \approx 0.0001$ (H_0 in $HClO_4$ solutions)	Uncert.	T58
3520	$C_7H_4O_8N_4$	Methane, (2,6-dinitrophenyl)dinitro-						
		-2.70		20	06	$c \approx 0.0001$ (H_0 in $HClO_4$ solutions)	Uncert.	T58

No.	Molecular formula, name and pK value(s)	$T(^oC)$	Remarks	Method	Assessment	Ref.
3521	C_7H_5ON Benzenecarbonitrile, 2-hydroxy- (Phenol, 2-cyano-)					
	6.86	25	$I = 0.1(NaClO_4)$, concentration constant	05	Approx.	D34
	6.9	30	$I = 0.1(KCl)$	05	Uncert.	C20
			Thermodynamic quantities are given			
3522	C_7H_5ON Benzenecarbonitrile, 3-hydroxy- (Phenol, 3-cyano-)					
	8.608	25	$m = 0.00007$, $I = 0.02-0.1$	05a	Rel.	F12
	8.57	25		05	Approx.	K40
	8.34	25	$I = 0.1(NaClO_4)$, concentration constant	05	Approx.	D34
			Thermodynamic quantities are given			
3523	C_7H_5ON Benzenecarbonitrile, 4-hydroxy- (Phenol, 4-cyano-)					
	7.967	25	$m = 0.00007$, $I = 0.02-0.07$	05a	Rel.	F12
	7.97	25		05	Approx.	K40
	7.85	25	$I = 0.1(KCl)$, mixed constant	E3bg	Approx.	H51
	7.71	25	$I = 0.1(NaClO_4)$, concentration constant	05	Approx.	D34
			Thermodynamic quantities are also given			
3524	C_7H_5OCl Benzaldehyde, 3-chloro-					
	13.92	25	$c = 0.0001$, 1% ethanol	07,R4	Approx.	B111
			pK of hydrate			

No.	Formula	Name	pK	T (°C)	Conditions	Notes	Method	Reliability	Ref.
3525	C_7H_5OCl	Benzaldehyde, 4-chloro-	14.44	25	$c = 0.0001$, in 1% ethanol	pK of hydrate	07,R4	Uncert.	B111
3526	$C_7H_5OF_3$	Phenol, 3-trifluoromethyl-	8.950	25	$I = 0.0017$	Half-neutralised p-nitrophenol used as calibration standard	E3d	Uncert.	L53
3527	$C_7H_5OF_3$	Phenol, 4-trifluoromethyl-	8.675	25	$I = 0.0017$	Half-neutralised p-nitrophenol used as calibration standard. Compound unstable; results extrapolated to zero time	E3d	Uncert.	L53
3528	$C_7H_5O_2N$	Benzoxazol-2(3H)-one	8.9	23	$I = 0.5$		05	Uncert.	H59
3529	$C_7H_5O_2N_3$	Benzimidazole, 5-nitro-	10.84	25	$c \approx 0.0001$		05	Approx.	T21
3530	$C_7H_5O_2N_3$	1H-Indazole, 4-nitro-	11.57	25	$c = 0.0001$		05	Approx.	T21
3531	$C_7H_5O_2N_3$	1H-Indazole, 5-nitro-	11.69	25	$c \approx 0.0001$		05	Approx.	T21

No.	Molecular formula, name and pK value(s)	$T(^{o}C)$	Remarks	Method	Assessment	Ref.
3532	$C_7H_5O_2N_3$ 1H-Indazole, 6-nitro- 11.67	25	$c \approx 0.0001$	05	Approx.	T21
3533	$C_7H_5O_2N_3$ 1H-Indazole, 7-nitro- 12.48	25	$c \approx 0.0001$	05	Approx.	T21
3534	$C_7H_5O_2Cl$ Benzaldehyde, 2-chloro-4-hydroxy- 6.60	25	$c = 0.0001$, $I = 0.1(NaClO_4)$, concentration constant	05	Approx.	P74
3535	$C_7H_5O_2Cl$ Benzaldehyde, 2-chloro-6-hydroxy- 8.26	25	$c = 0.0001$, $I = 0.1(NaClO_4)$, concentration constant	05	Approx.	P74
3536	$C_7H_5O_2Cl$ Benzaldehyde, 3-chloro-2-hydroxy- 6.61	25	$c = 0.0001$, $I = 0.1(NaClO_4)$, concentration constant	05	Approx.	P74
3537	$C_7H_5O_2Cl$ Benzaldehyde, 3-chloro-6-hydroxy- 7.41	25	$c = 0.0001$, $I = 0.1(NaClO_4)$, concentration constant	05	Approx.	P74

No.	Formula	Name	pK	T(°C)	Conditions	Method	Reliability	Ref.
3538	$C_7H_5O_2Cl$	Benzaldehyde, 4-chloro-2-hydroxy-	7.18	25	c = 0.0001, I = 0.1($NaClO_4$), concentration constant	05	Approx.	P74
3539	$C_7H_5O_2Cl$	Benzoic acid, 2-chloro-	2.900	25	c = 0.0006-0.0029	C2,R1c	Rel.	M58
			3.03	20	c = 0.001, 1% ethanol	E3bg	Approx.	P24,P25
			2.91(H_2O)	25		05	Uncert.	W12
			3.38(D_2O)					
3540	$C_7H_5O_2Cl$	Benzoic acid, 3-chloro-	3.839	25		05a	Rel.	B91
					Value calculated from thermodynamic data given in the reference			
			3.82	25	c = 0.0006, mixed constant	E3bg	Approx.	S115
3541	$C_7H_5O_2Cl$	Benzoic acid, 4-chloro-	3.988	25		05a	Rel.	B91
					Value calculated from thermodynamic data given in the reference			
3542	$C_7H_5O_2Cl$	2,4,6-Cycloheptatrien-1-one, 5-chloro-2-hydroxy-	5.62	25	I = 2.0	05	Approx.	O13
3543	$C_7H_5O_2Br$	Benzoic acid, 2-bromo-	2.88	20	c = 0.001, 1% ethanol	E3bg	Approx.	P24,P25

No.	Molecular formula, name and pK value(s)	T(°C)	Remarks	Method	Assessment	Ref.
3544	C₇H₅O₂Br Benzoic acid, 3-bromo-					
	3.813	25		05a	Rel.	B91
	Value calculated from thermodynamic data given in the reference					
3545	C₇H₅O₂Br Benzoic acid, 4-bromo-					
	3.963	25		05a	Rel.	B91
	Value calculated from thermodynamic data given in the reference					
3546	C₇H₅O₂Br 2,4,6-Cycloheptatrien-1-one, 3-bromo-2-hydroxy-					
	5.14	25	I = 2.0	05		015
	5.96	25		E3bg	Approx.	Y15
3547	C₇H₅O₂Br 2,4,6-Cycloheptatrien-1-one, 4-bromo-2-hydroxy-					
	5.72	25	I = 2.0	05	Approx.	016
3548	C₇H₅O₂Br 2,4,6-Cycloheptatrien-1-one, 5-bromo-2-hydroxy-					
	5.54	25	I = 2.0	05	Approx.	013
	6.32	25		E3bg	Uncert.	Y15
3549	C₇H₅O₂I Benzoic acid, 2-iodo-					
	2.93	20	c = 0.001, 1% ethanol	E3bg	Approx.	P24,P25

3550 $C_7H_5O_2I$ Benzoic acid, 4-iodo-

3.997   25   05a   Rel.   B91

Value calculated from thermodynamic data given in the reference

3.93   25 $c = 0.0000176$ 01   Approx.   R32

3551 $C_7H_5O_2F$ Benzoic acid, 3-fluoro-

3.862   25   C2,R1a/b   Approx.   F18

Variation with pressure(bar)

500	1000	1500	2000	2500	3000
3.776	3.694	3.620	3.546	3.477	3.412

3552 $C_7H_5O_2F$ Benzoic acid, 4-fluoro-

4.153   25   C2,R1a/b   Rel.   F18

Variation with pressure(bar)

500	1000	1500	2000	2500	3000
4.061	3.976	3.900	3.822	3.753	3.687

3553 $C_7H_5O_3N$ Benzaldehyde, 3-nitro-

13.04   25 $c = 0.0001$, 1% ethanol   07,R4   Approx.   B111

(pK of hydrate)

3554 $C_7H_5O_3N$ Benzaldehyde, 4-nitro-

12.79   25 $c = 0.0001$, 1% ethanol   07,R4   Approx.   B111

(pK of hydrate)

No.	Molecular formula, name and pK value(s)	T(°C)	Remarks	Method	Assessment	Ref.
3555	$C_7H_5O_3Cl$ Benzoic acid, 2-chloro-6-hydroxy-					
	pK$_1$ 2.63	25	c = 0.0003-0.0010	C1	Approx.	B123
3556	$C_7H_5O_3Cl$ Benzoic acid, 3-chloro-4-hydroxy-					
	7.52(-OH)	25	c ≈ 0.0001-0.0002, I ≈ 0.1, mixed constant	O5	Approx.	S76
3557	$C_7H_5O_3Cl$ Benzoic acid, 5-chloro-2-hydroxy-					
	pK$_1$ 2.65	25	c = 0.01	O5	Approx.	E24
	pK$_2$ 13.0				Uncert.	
	pK$_1$ 2.8	25	I = 0.1(KCl), concentration constants	E3bg	Uncert.	A1
	pK$_2$ 13.05	18		O5		
	pK$_1$ 2.676	30		O5,R4	Approx.	G11
	pK$_1$ 2.646	40	Thermodynamic quantities are derived			
	pK$_1$ 2.632	50	from the results			
	pK$_1$ 2.627					
	pK$_1$ 2.63	25	c = 0.0003-0.0014	C1	Approx.	B123
	pK$_1$ 2.70	25	c = 0.0003, I = 0.01	O5	Approx.	D56
3558	$C_7H_5O_3Cl$ Peroxybenzoic acid, 2-chloro-					
	6.54	25				K14

3559	$C_7H_5O_3Cl$	Peroxybenzoic acid, 3-chloro-				
		7.57	25			K14

3560	$C_7H_5O_3Cl$	Peroxybenzoic acid, 4-chloro-				
		7.64	25			K14

3561	$C_7H_5O_3Br$	Benzoic acid, 3-bromo-2-hydroxy-				
	pK_1 2.516		20	O5,R4	Approx.	G11
	2.487		30			
	2.468		40			
	2.459		50			

Thermodynamic quantities are derived from the results

3562	$C_7H_5O_3Br$	Benzoic acid, 5-bromo-2-hydroxy-				
	pK_1 2.66	c = 0.01	25	O5	Approx.	E24
	pK_2 12.8				Uncert.	
	pK_1 2.75	I = 0.1(KCl), concentration constants	25	E3bg	Uncert.	A1
	pK_2 12.85			O5		
	pK_1 2.627		25	O5,R4	Approx.	G11

Variation with temperature

20	30	40	45	50
2.638	2.615	2.602	2.600	2.598

Thermodynamic quantities are derived from the results

(contd)

No.	Molecular formula, name and pK value(s)	$T(^\circ C)$	Method	Assessment	Ref.	Remarks
3562 (Contd)	C$_7$H$_5$O$_3$Br Benzoic acid, 5-bromo-2-hydroxy-					
	pK$_1$ 2.61	25	05	Approx.	D56	c = 0.0003, I = 0.01
	pK$_1$ 2.61	25	C1	Approx.	B123	c = 0.0002-0.0013
3563	C$_7$H$_5$O$_3$Br Benzoic acid, 4-bromo-2-hydroxy-					
	pK$_1$ 2.669	25	05,R4	Approx.	G11	

Variation with temperature

20	30	35	40	45
2.678	2.662	2.659	2.656	2.656

Thermodynamic quantities are derived from the results

3564	C$_7$H$_5$O$_3$I 1,2-Benziodoxol-3(1H)-one, 1-hydroxy-					
	pK$_3$ 7.35	23±3	05	V.Uncert.	W42	Mixed constant
	pK$_2$ -0.58		06	Uncert.		In aq.H$_2$SO$_4$, H$_0$ scale
	pK$_1$ -5.75		06	Uncert.		In aq.H$_2$SO$_4$, H$_0$ scale
3565	C$_7$H$_5$O$_3$I Benzoic acid, 2-hydroxy-3-iodo-					
	pK$_1$ 2.520	25	05,R4	Approx.	G11	

Variation with temperature

20	30	35	40	45
2.536	2.507	2.496	2.488	2.482

Thermodynamic quantities are derived from the results

3566 $C_7H_5O_3I$ Benzoic acid, 2-hydroxy-4-iodo-

pK$_1$ 2.750   25   05,R4   Approx.   G11

Variation with temperature

20	30	35	40	45
2.761	2.742	2.736	2.733	2.732

Thermodynamic quantities are derived from the results

3567 $C_7H_5O_3I$ Benzoic acid, 2-hydroxy-5-iodo-

pK$_1$ 2.619   25   05,R4   Approx.   G11

Variation with temperature

20	30	35	40	45
2.632	2.608	2.600	2.594	2.591

Thermodynamic quantities are derived from the results

pK$_1$ 2.70   25 $c = 0.0003$, $I = 0.01$ 05   Approx.   D56

3568 $C_7H_5O_3F$ Peroxybenzoic acid, 2-fluoro-

6.77   E   K15

3569 $C_7H_5O_3F$ Peroxybenzoic acid, 3-fluoro-

7.57   E   K15

3570 $C_7H_5O_3F$ Peroxybenzoic acid, 4-fluoro-

7.76   E   K15

No.	Molecular formula, name and pK value(s)	T(°C)	Remarks	Method	Assessment	Ref.
3571	C$_7$H$_5$O$_4$N Benzaldehyde, 2-hydroxy-3-nitro- 5.41 5.21	25	c ≈ 0.0002, I = 0.1 I = 0.1, mixed constant	05	Approx.	M19
3572	C$_7$H$_5$O$_4$N Benzaldehyde, 2-hydroxy-5-nitro- 5.51 5.42	25	c ≈ 0.0002, I = 0.1 I = 0.1, mixed constant	05	Approx.	M19
3573	C$_7$H$_5$O$_4$N Benzaldehyde, 3-hydroxy-4-nitro- 6.00	25	c ≈ 0.00001-0.0001, I ≈ 0.1, mixed constant	05	Approx.	H11
3574	C$_7$H$_5$O$_4$N Benzoic acid, 2-nitro- 2.47 2.21	20 25	c = 0.001 (1% ethanol) c = 0.001-0.01	E3bg C2,R1a/b	Approx. Approx.	P24,P25 C61
	Variation with pressure(atm) 1010 2010 2780 2.02 1.87 1.78					
	2.17(H$_2$O) 2.60(D$_2$O)	25		05	Uncert.	W12

305

3575 $C_7H_5O_4N$ Benzoic acid, 3-nitro-

3.462	25	05a	Rel.	B91

Value calculated from thermodynamic data given in the reference

3.473	25	C2,R1a/b	Approx.	F18

Variation with pressure(bar)

500	1000	1500	2000	2500	3000
3.397	3.328	3.258	3.192	3.130	3.070

3.51	25	C2,R1a/b	Approx.	C61

$c = 0.0005-0.0098$

Variation with pressure(atm)

1010	2010	2780
3.35	3.22	3.13

3576 $C_7H_5O_4N$ Benzoic acid, 4-nitro-

3.426	25	05a	Rel.	B91

Value calculated from thermodynamic data given in the reference

3.422	25	C2,R1a/b	Approx.	F18

Variation with pressure(bar)

500	1000	1500	2000	2500	3000
3.347	3.275	3.209	3.142	3.081	3.022

3.48	25	C2,R1a/b	Approx.	C61

$c = 0.0005-0.01$

Variation with pressure(atm)

1010	2010	2780
3.32	3.20	3.09

No.	Molecular formula, name and pK value(s)	T($^{\circ}$C)	Method	Assessment	Ref.	Remarks
3577	C$_7$H$_5$O$_4$N 2,4,6-Cycloheptatrien-1-one, 2-hydroxy-5-nitro-					
	3.21	25	05		O10	I = 0.1
	2.64	25	05	Approx.	O13	I = 2.0
3578	C$_7$H$_5$O$_4$N 2,3-Pyridinedicarboxylic acid (Quinolinic acid)					
	pK$_1$ 2.43, pK$_2$ 4.78	25	E3bg	Approx.	D61	c ≈ 0.01, I = 0.1(NaClO$_4$), mixed constants
	pK$_1$ 2.36, pK$_2$ 4.72	25	E3bg	Approx.	Y11	I = 0.1(KNO$_3$), mixed constants
3579	C$_7$H$_5$O$_4$N 2,4-Pyridinedicarboxylic acid (Lutidinic acid)					
	pK$_1$ 2.23, pK$_2$ 4.79	25	E3bg	Approx.	Y11	I = 0.1(KNO$_3$)
3580	C$_7$H$_5$O$_4$N 2,5-Pyridinedicarboxylic acid (Isocinchomeronic acid)					
	pK$_1$ 2.17, pK$_2$ 4.58	25	E3bg*	Approx.	N14	I = 0.5(NaClO$_4$), concentration constants
	*Reference electrode is Ag; AgCl/0.01M Ag$^+$, 0.49M Na$^+$, 0.5M NaClO$_4$					
3581	C$_7$H$_5$O$_4$N 2,6-Pyridinedicarboxylic acid (Dipicolinic acid)					
	pK$_1$ 2.16, pK$_2$ 4.76	30	E3bg	Approx.	T32	I = 0.1(KCl), mixed constants
	pK$_1$ 2.24, pK$_2$ 4.67	25	E3bg	Approx.	B62	I = 0.1
	pK$_1$ 2.09, pK$_2$ 4.53	25	E3bg	Approx.	N13	I = 0.5(NaClO$_4$), concentration constants
3582	C$_7$H$_5$O$_4$N 3,4-Pyridinedicarboxylic acid					
	pK$_1$ 1.50	25	05	Approx.	F35a	I = 0.1(KCl), mixed constants
	pK$_2$ 2.95, pK$_3$ 5.07		E3bg			

3583 $C_7H_5O_5N$ Benzoic acid, 2-hydroxy-3-nitro-

pK_1 2.3, pK_2 10.35	25	I = 0.1(KCl), concentration constants	E3bg	Uncert.	A1
pK_2 10.3			05		
pK_1 1.87	25	c = 0.0005-0.0043	C2	Approx.	B123
pK_2 10.33	25	c = 0.0002	05	Approx.	E25

3584 $C_7H_5O_5N$ Benzoic acid, 2-hydroxy-4-nitro-

pK_1 2.23	25	c = 0.0007-0.0024	C1	Approx.	B123
pK_1 2.31	25	c = 0.00008, I = 0.01	05	Approx.	D56
pK_2 2.63		Conditions not stated	E3bg	Uncert	V3

3585 $C_7H_5O_5N$ Benzoic acid, 2-hydroxy-5-nitro-

pK_1 2.12	25	c = 0.0008-0.0036	C1	Approx.	B123
pK_1 2.32	25	c = 0.00009, I = 0.01	05	Approx.	D56
pK_2 10.34	25	c = 0.00005	05	Approx.	E24
pK_1 2.1, pK_2 10.6	25	I = 0.1(KCl), concentration constants	E3bg	Uncert.	A1

3586 $C_7H_5O_5N$ Benzoic acid, 2-hydroxy-6-nitro-

pK_1 2.24	25	c = 0.00035-0.0032	C1	Approx.	B123

3587 $C_7H_5O_5N$ Benzoic acid, 4-hydroxy-3-nitro-

pK_2 6.41	25	c = 0.0002, I ≈ 0.5, mixed constant	05	Approx.	R15

No.	Molecular formula, name and pK value(s)	$T(^{\circ}C)$	Remarks	Method	Assessment	Ref.
3588	$C_7H_5O_5N$ 2,6-Pyridinedicarboxylic acid, 1,4-dihydro-4-oxo- (Chelidamic acid)					
	pK_1 1.4	20	$c = 0.0017$, $I = 0.1(NaNO_3)$,	E3bg	Uncert.	A52
	pK_2 3.11, pK_3 10.88		mixed constants	E3bg	Approx.	
	pK_2 3.2, pK_3 10.8	25	$I = 0.1(NaClO_4)$, mixed constants	05	Uncert.	B1
	pK_2 3.47, pK_3 11.4		$I \approx 0.005-0.01$	E3bg		
	$(pK_1 <2)$					
3589	$C_7H_5O_6N_3$ Methane, (2-nitrophenyl)dinitro-					
	0.98	20	$c \approx 0.0001$	06	Uncert.	T58
			(H_0 in $HClO_4$ solutions)			
3590	$C_7H_5O_7N_3$ Phenol, 3-methyl-2,4,6-trinitro-					
	0.81	25	$c \approx 0.00005$	05	Approx.	D19
	1.00	not stated	$c = 0.003$, $I = 0.5(NaClO_4)$	05	Uncert.	P16
3591	$C_7H_5O_8N_3$ 1,3-Benzenediol, 5-methyl-2,4,6-trinitro-					
	pK_1 1.17, pK_2 5.04	not stated	$c = 0.003$, $I = 0.5(NaClO_4)$	05	Uncert.	P16
3592	$C_7H_5O_8N_3$ Phenol, 3-methoxy-2,4,6-trinitro-					
	0.37	not stated	$c = 0.003$, $I = 0.5(NaClO_4)$	06	Uncert.	P16
3593	$C_7H_5NS_2$ Benzthiazol-2(3H)-one					
	6.93	20±1		DIS	Uncert.	N18

No.	Formula	Name	pK	T (°C)	Notes	Method	Reliability	Ref.
3594	$C_7H_6ON_2$	1H-Indazol-5-ol	10.05	25	c = 0.01	E3bg	Approx.	S129
3595	$C_7H_6ON_2$	1H-Indazol-7-ol	8.60	25	c = 0.01	E3bg	Approx.	S129
3596	$C_7H_6ON_2$	1H-Indazol-4(2H)-one	8.65	25	c = 0.01	E3bg	Approx.	S129
3597	$C_7H_6ON_2$	1H-Indazol-6(2H)-one	9.35	25	c = 0.01	E3bg	Approx.	S129
3598	$C_7H_6ON_4$	Pteridin-2(1H)-one, 4-methyl-	10.85(hydrated)	20	c = 0.001	E3bg	Uncert.	A27
			8.2(anhydrous)		rapid titration			
			9.00		Equilibrium value of two forms		Approx.	I5
					The anhydrous form is the stable form			
3599	$C_7H_6ON_4$	Pteridin-2(1H)-one, 7-methyl-	8.07	20	"Anhydrous" species	E3bg	Approx.	I5
			9.60	20	Equilibrium pK, partial covalent hydration			
			10.85	20	Covalently hydrated species			
3600	$C_7H_6ON_4$	Pteridin-4(1H)-one, 6-methyl-	8.19	20	c = 0.01	E3bg	Approx.	A25

No.	Molecular formula, name and pK value(s)	T(°C)	Remarks	Method	Assessment	Ref.
3601	$C_7H_6ON_4$ Pteridin-4(1$\underline{H}$)-one, 7-methyl- 8.09	20	c = 0.01	E3bg	Approx.	A25
3602	$C_7H_6ON_4$ Pteridin-7(1$\underline{H}$)-one, 6-methyl- 6.97	20	c = 0.02	E3bg	Approx.	A38
3603	$C_7H_6OCl_2$ Phenol, 2,4-dichloro-6-methyl- 8.14	20	c = 0.0001-0.00001, I = 0.1(NaClO$_4$), 1% ethanol	05	Approx.	S124
3604	$C_7H_6OCl_2$ Phenol, 2,6-dichloro-4-methyl- 7.19	25	c = 0.0012	E3bg	Approx.	F17
3605	C_7H_6OS 2,4,6-Cycloheptatrien-1-one, 2-mercapto- 5.90	25		E3bg	Uncert.	Y16
3606	$C_7H_6O_2N_2$ Benzaldehyde oxime, 1-nitroso- 4.9	21	I ≈ 0.3, unstable	05	Uncert.	A74
3607	$C_7H_6O_2N_2$ Ethanal oxime, 2-oxo-2-(3-pyridyl)- (Pyridine, 3-hydroxyiminoacetyl-) 7.8	not stated	c ≈ 0.04	E3bg	Uncert.	W36
3608	$C_7H_6O_2N_2$ Ethanal oxime, 2-oxo-2-(4-pyridyl)- (Pyridine, 4-hydroxyiminoacetyl-) 7.8	not stated	c ≈ 0.04	E3bg	Uncert.	W36

3609 $C_7H_6O_2N_4$ Pteridine-2,4($1\underline{H},3\underline{H}$)-dione, 1-methyl-

8.57	20	c = 0.005	E3bg	Approx.	B135
8.45	20	c = 0.0001	E3bg	Approx.	P41

3610 $C_7H_6O_2N_4$ Pteridine-2,4($1\underline{H},3\underline{H}$)-dione, 3-methyl-

8.00	20	c = 0.001	E3bg	Approx.	P41
8.0	20		E3bg	Uncert.	B135

3611 $C_7H_6O_2N_4$ Pteridine-2,7($1\underline{H},3\underline{H}$)-dione, 4-methyl-

6.31	20	c = 0.00004 in 0.01M buffer	05	Approx.	J4

3612 $C_7H_6O_2N_4$ Pteridine-4,6($3\underline{H},5\underline{H}$)-dione, 3-methyl-

6.26	20	c = 0.001	E3bg	Approx.	P48

Probably an equilibrium pK involving hydrated and anhydrous species

3613 $C_7H_6O_2N_4$ Pteridine-4,7($3\underline{H},8\underline{H}$)-dione, 3-methyl-

6.19	20	c = 0.001	E3bg	Approx.	P48

3614 $C_7H_6O_2N_4$ Pteridine-4,7($3\underline{H},8\underline{H}$)-dione, 6-methyl-

pK_1 6.82	20	c = 0.0013	E3bg	Approx.	A22
pK_2 10.02				Uncert.	

3615 $C_7H_6O_2N_4$ Pteridine-6,7($1\underline{H},5\underline{H}$)-dione, 4-methyl-

pK_1 7.12, pK_2 10.17		c = 0.0025	E3bg	Uncert.	A42

No.	Molecular formula, name and pK value(s)	T(°C)	Remarks	Method	Assessment	Ref.
3616	$C_7H_6O_2N_4$ Pteridin-4(8H)-one, 6-hydroxy-8-methyl-					
	8.67	20	c = 0.005	E3bg	Approx.	J4
3617	$C_7H_6O_2N_4$ Pteridin-6(5H)-one, 7-hydroxy-5-methyl-					
	7.02	20	c = 0.01	E3bg	Approx.	B140
3618	$C_7H_6O_2S$ Benzoic acid, 2-mercapto-					
	pK_1 3.54, pK_2 8.60		I = 0.3	0		T10
	pK_2 8.88	25	c ≈ 0.00007	05	Approx.	I20
	Thermodynamic quantity is given					
	pK_2 8.20	27		05	Approx.	D9
	pK_2 7.79	35		05	Approx.	F13
	pK_2 8.40		Conditions not stated	E3bg	Uncert.	S25
3619	$C_7H_6O_2S$ Benzoic acid, 3-mercapto-					
	pK_2 6.32	20		E3bg	Approx.	D9
	6.20	30		E3bg		
	6.15	28		05		
3620	$C_7H_6O_2S$ Benzoic acid, 4-mercapto-					
	pK_2 5.90	25		05	Approx.	D9
	pK_2 5.80	25	c ≈ 0.00007	05	Approx.	I20
	pK_2 5.86	35		05	Approx.	F13

313

No.	Formula	Name	pK	T (°C)	Notes	Method	Assessment	Ref.
3621	$C_7H_6O_3N_4$	Pteridine-2,4,6(1$\underline{H}$,3$\underline{H}$,5$\underline{H}$)-trione, 1-methyl-	pK_1 5.62, pK_2 9.85	20		E3bg	Approx.	P43
3622	$C_7H_6O_3N_4$	Pteridine-2,4,6(1$\underline{H}$,3$\underline{H}$,5$\underline{H}$)-trione, 3-methyl-	pK_1 5.96, pK_2 9.72	20		E3bg	Approx.	P43
3623	$C_7H_6O_3N_4$	Pteridine-2,4,6(1$\underline{H}$,3$\underline{H}$,5$\underline{H}$)-trione, 7-methyl-	pK_1 6.57, pK_2 9.45	20		E3bg	Approx.	P43
3624	$C_7H_6O_3N_4$	Pteridine-2,4,7(1$\underline{H}$,3$\underline{H}$,8$\underline{H}$)-trione, 1-methyl-	pK_1 3.31, pK_2 10.51	20	c = 0.001	E3bg	Approx.	P42
3624a	$C_7H_6O_3N_4$	Pteridine-2,4,7(1$\underline{H}$,3$\underline{H}$,8$\underline{H}$)-trione, 3-methyl-	pK_1 3.60, pK_2 10.26	20	c = 0.001	E3bg	Approx.	P42
3625	$C_7H_6O_3N_4$	Pteridine-2,4,7(1$\underline{H}$,3$\underline{H}$,8$\underline{H}$)-trione, 6-methyl-	pK_1 4.13, pK_2 10.09	20		E3bg	Approx.	P42
3626	$C_7H_6O_3N_4$	Pteridine-2,4,7(1$\underline{H}$,3$\underline{H}$,8$\underline{H}$)-trione, 8-methyl-	11.70	20		E3bg	Uncert.	P42
3627	$C_7H_6O_4N_2$	Benzohydroxamic acid, 2-nitro-	pK_1 1.45, pK_2 7.05; pK_2 8.2	20; 30	c = 0.0025, I = 1.0(KNO_3), mixed constants	E3bg	Uncert.	D64, S117

No.	Molecular formula, name and pK value(s)	$T(^{o}C)$	Remarks	Method	Assessment	Ref.
3628	$C_7H_6O_4N_2$ Benzohydroxamic acid, 3-nitro-					
	pK_1 1.85, pK_2 8.40	20±1	c = 0.0025, I = 1.0(KNO_3), mixed constants	E3bg	Uncert.	D64
3629	$C_7H_6O_4N_2$ Benzohydroxamic acid, 4-nitro-					
	pK_1 1.80, pK_2 8.35	20	c = 0.0025, I = 1.0(KNO_3), mixed constants	E3bg	Uncert.	D64
	pK_2 8.12	25	c = 0.01(?) (Possible error in detail given)	E3bg	Uncert.	A10
	pK_2 8.01	30.5	c = 0.0006, I = 0.1(KNO_3), mixed constant	E3d	Uncert.	S137
	pK_2 8.01	not stated	c = 0.04, I = 0.1(KCl)	E3bg	Uncert.	H2
3630	$C_7H_6O_4N_2$ Benzoic acid, 2-amino-4-nitro-					
	pK_1 0.65, pK_2 3.70	25	c = 0.0004, I = 0.1(KCl), mixed constants	05	Uncert.	L15
3631	$C_7H_6O_4N_2$ Benzoic acid, 3-amino-5-nitro-					
	pK_1 1.55	25	c = 0.01	05	Approx.	S58
	pK_2 3.55	25	c = 0.00008	E3bg	Approx.	
3632	$C_7H_6O_4N_4$ 4-Pteridinecarboxylic acid, 1,2,5,6,7,8-hexahydro-2,6-dioxo-					
	12.04(OH)	20	c = 0.009, mixed constant	E3bg	Uncert.	C59
3633	$C_7H_6O_4N_4$ Pteridine-2,4,6,7($1\underline{H}$,$3\underline{H}$,$5\underline{H}$,$8\underline{H}$)-tetrone, 1-methyl-					
	pK_1 3.49, pK_2 9.60	20	c = 0.001	E3bg	Approx.	P46
3634	$C_7H_6O_4N_4$ Pteridine-2,4,6,7($1\underline{H}$,$3\underline{H}$,$5\underline{H}$,$8\underline{H}$)-tetrone, 3-methyl-					
	pK_1 4.02, pK_2 9.57	20	c = 0.001	E3bg	Approx.	P46

No.	Formula	Name	pK	T (°C)	Conditions	Method	Reliability	Ref.
3635	$C_7H_6O_4N_4$	Pteridine-2,4,6,7(1H,3H,5H,8H)-tetrone, 8-methyl-	pK_1 4.36, pK_2 9.65	20	c = 0.001		Uncert.	P46
3636	$C_7H_6O_4S$	Benzoic acid, 3-sulfino-	4.16(COOH)	25	I = 0.002, concentration constant	E3bg	Approx.	L47
			4.09		I = 0.010, concentration constant			
			4.05		I = 0.025, concentration constant			
			4.01		I = 0.05, concentration constant			
			3.95		I = 0.10, concentration constant			
			4.24		I = 0 (by extrapolation)			
3637	$C_7H_6O_4S$	Benzoic acid, 4-sulfino-	4.20(COOH)	25	I = 0.002, concentration constant	E3bg	Approx.	L47
			4.14		I = 0.004, concentration constant			
			4.11		I = 0.010, concentration constant			
			4.06		I = 0.025, concentration constant			
			4.00		I = 0.05, concentration constant			
			3.86		I = 0.10, concentration constant			
			4.27		I = 0			
3638	$C_7H_6O_5N_2$	Phenol, 2-methyl-4,6-dinitro-	4.70	25	c ≃ 0.00008	O5a	Approx.	R30

No.	Molecular formula, name and pK value(s)	T(^{o}C)	Remarks	Method	Assessment	Ref.
3639	$C_7H_6O_5N_2$ Phenol, 4-methyl-2,6-dinitro-					R30
	4.231	25	c ≈ 0.00002	O5a	Rel.	
3640	$C_7H_6O_5N_4$ Benzohydrazide, 3,5-dinitro-					T43
	10.28(NH)	25	Mixed constant(?)	E3bg	Uncert.	
3641	$C_7H_6O_5S$ Benzoic acid, 4-sulfo-					W12
	pK$_2$ 3.65(H$_2$O)	25		O5	Uncert.	
	pK$_2$ 4.12(D$_2$O)					
3642	$C_7H_6O_5S$ 1,3,6-Cycloheptatrienesulfonic acid, 4-hydroxy-5-oxo- (Tropolone-5-sulfonic acid)					
	pK$_2$ 4.68	25	I = 2.0	O5	Approx.	O12
	pK$_2$ 4.92	25	I = 0.1	O5		O10
3643	$C_7H_6O_6S$ Benzoic acid, 2-hydroxy-5-sulfo- (5-Sulfosalicylic acid)					
	pK$_2$ 2.48, pK$_3$ 11.97	20	I = 0.1(KNO$_3$), mixed constants	E3bg	Approx.	B31
	pK$_2$ 2.62	20	I = 0.1-0.15(NaClO$_4$), mixed constants	E3bg	Approx.	P30
	pK$_3$ 11.95			O5		
	pK$_2$ 2.50	25	I = 0.1(KNO$_3$), concentration constants	E3bg	Approx.	L37
	pK$_3$ 11.7				Uncert.	
	pK$_2$ 2.67, pK$_3$ 11.67	25	I = 0.1(KNO$_3$), concentration constants,	E3bg	Approx.	M99
	pK$_2$ 2.36, pK$_3$ 11.59	34.9	extrapolated to c = 0			

(contd)

No.	Formula / Name	pK	T (°C)	Conditions	Method	Assess.	Ref.
3643 (Contd)	$C_7H_6O_6S$ Benzoic acid, 2-hydroxy-5-sulfo- (5-Sulfosalicylic acid)						
		pK$_2$ 2.30, pK$_3$ 11.41	25	$I = 0.5(NaClO_4)$, concentration constants	E3bg	Approx.	M8
		pK$_2$ 2.30, pK$_3$ 11.47	25	$I = 1.0(NaClO_4)$, mixed constants	E3bg	Approx.	M55
		pK$_2$ 2.23-2.31	25	$c = 0.01$, $I = 1.0(KCl,KNO_3,NaClO_4)$, concentration constant	E3bg	Uncert.	M6
		pK$_3$ 12.00	25	$c = 0.0005$, $I = 0.1$ ⎫	05	Approx.	N3
		pK$_3$ 11.96	25	$I = 0.1$ ⎬ in various buffers			
		pK$_3$ 12.23	25	$I = 0.16-0.19$ ⎭			
	Other values in A11, B20, R2 and T22						
3644	$C_7H_6N_2S$ Benzimidazole-2(3H)-thione						
		pK$_1$ 9.18, pK$_2$ 10.98	20	$I = 0.4$	E and 05	Uncert.	L62
		10.24	20	$c = 0.000012$ in 0.01M buffers	05	Approx.	B127
		10.33	25	$I = 0.2$	05		L10
3645	$C_7H_6N_4S$ Pteridine-4(1H)-thione, 7-methyl-						
		7.02	20	$c = 0.0006$	E3bg	Approx.	A26
3646	$C_7H_6N_4S$ Tetrazole-5(2H)-thione, 1-phenyl- (1-Phenyltetrazole-5-thiol)						
		2.8	25	$I = 0.1$	E and 0		S113
3647	$C_7H_7ON_5$ Pteridin-4(1H)-one, 2-amino-1-methyl-						
		2.86	20		05	Uncert.	P52

No.	Molecular formula, name and pK value(s)	T(°C)	Remarks	Method	Assessment	Ref.
3648	$C_7H_7ON_5$ Pteridin-4(3H)-one, 2-amino-3-methyl- pK$_1$ 2.18	20		05	Approx.	P52
3649	$C_7H_7ON_5$ Pteridin-4(1H)-one, 2-amino-6-methyl- pK$_1$ 2.8 pK$_2$ 8.3	22	I ≈ 0.1 c = 0.01-0.02	05 E3bg	Uncert.	W26
3650	$C_7H_7ON_5$ Pteridin-4(1H)-one, 2-methylamino- pK$_1$ 1.95, pK$_2$ 7.95	20		E3bg/05	Uncert.	P52
3651	No entry					
3652	C_7H_7OCl Phenol, 2-chloro-4-methyl- 8.74	20	c = 0.0001-0.00001, I = 0.1(NaClO$_4$), 1% ethanol	05	Approx.	S124
3653	C_7H_7OCl Phenol, 2-chloro-6-methyl- 8.69	20	c = 0.0001-0.00001, I = 0.1(NaClO$_4$), 1% ethanol	05	Approx.	S124
3654	C_7H_7OCl Phenol, 4-chloro-2-methyl- 9.706	25	c ≈ 0.0003	05a	Approx.	R29
	9.55	20	c = 0.00001-0.0001, I = 0.1(NaClO$_4$), 1% ethanol	05	Approx.	S124

3655   C$_7$H$_7$OCl   Phenol, 4-chloro-3-methyl-

9.549   25   m = 0.00017   Rel.   05a   B90

Variation with temperature

5	10	15	20	30	35	40	45	50	55	60
9.860	9.776	9.696	9.621	9.489	9.429	9.372	9.318	9.270	9.218	9.173

Thermodynamic quantities are derived from the results

9.43   20   c = 0.0001-0.00001, I = 0.1(NaClO$_4$),   Approx.   05   S124
                      1% ethanol

3656   C$_7$H$_7$O$_2$N   Benzaldehyde oxime, 2-hydroxy-

pK$_2$ 9.30, pK$_3$ 12.10   25   E3bg,R3h, Approx.   M45
                                                                  R4

Variation with I(KCl)

	0.01	0.025	0.05	0.075	0.100
pK$_2$	9.20	9.05	8.98	8.90	8.85
pK$_3$	11.90	11.85	11.62	11.50	11.07

pK$_1$ 1.4   I ≈ 0.26   25   E3bg   Uncert.   L72

pK$_2$ 9.18, pK$_3$ 12.11   Approx.

pK$_2$ 9.17   c ≈ 0.01, I = 0.1(KCl), mixed constant   25   E3bg   Approx.   G39

3657   C$_7$H$_7$O$_2$N   Benzaldehyde oxime, 4-hydroxy-

8.93   c ≈ 0.01, I = 0.1(KCl), mixed constant   25   E3bg   Approx.   G39

No.	Molecular formula, name and pK value(s)	$T(^{\circ}C)$	Remarks	Method	Assessment	Ref.
3658	$C_7H_7O_2N$ Benzenecarboxamide, 2-hydroxy-					
	8.7	20	1-4% ethanol, mixed constant	05	Uncert.	W24
	8.89	25	$I = 1.0(NaClO_4)$, mixed constant	05	Uncert.	A11
3659	$C_7H_7O_2N$ Benzenecarboxamide, 3-hydroxy-					
	9.30	25		05	Approx.	C71
3660	$C_7H_7O_2N$ Benzenecarboxamide, 4-hydroxy-					
	8.56	25		05	Approx.	C70
3661	$C_7H_7O_2N$ Benzohydroxamic acid					
	pK_1 2.25, pK_2 8.43	30	$I = 0.1(NaClO_4)$	E3bg	Approx.	D63
	pK_1 1.85, pK_2 8.805	20	$c = 0.0025$, $I = 1.0(KNO_3)$, mixed constants	E3bg	Uncert.	D64
	pK_1 2.25	30	$I = 0.1(NaClO_4)$, concentration constant	E3bg	Approx.	S59
	pK_2 8.91	25	$c \approx 0.01$, $I \approx 0.01$	E3bg	Approx.	S67
	pK_2 8.43	30	$c = 0.01$, $I = 0.1(NaClO_4)$, concentration constant	E3bg	Approx.	D62
	Other values in A10, A44, B24, S117, S137 and W39					
3662	$C_7H_7O_2N$ Benzoic acid, 2-amino- (Anthranilic acid)					
	pK_1 2.296, pK_2 4.955	10		E3bg	Approx.	S22
	pK_1 1.89, pK_2 4.866	25				
	pK_1 2.11, pK_2 4.95	25		E3g	Approx.	L71
	(contd)					

3662 (Contd) $C_7H_7O_2N$ Benzoic acid, 2-amino- (Anthranilic acid)

	T (°C)	Conditions	Method	Assessment	Ref.
pK_1 2.17	25	$c = 0.00066$, $I = 0.1(KCl)$, mixed constants	05	Approx.	L15
pK_2 4.85				Uncert.	
pK_2 4.86	30	$I = 0.1(KCl)$	E3bg	Approx.	C23
pK_2 4.77	20	$c \approx 0.001$, 1% ethanol	E3bg	Approx.	P24,P25

3663 $C_7H_7O_2N$ Benzoic acid, 3-amino-

	T (°C)	Conditions	Method	Assessment	Ref.
pK_1 3.12, pK_2 4.74	25		E3g	Approx.	L71
pK_2 4.80	20	$c = 0.025$, mixed constant	E3d	Uncert.	H17

3664 $C_7H_7O_2N$ Benzoic acid, 4-amino-

	T (°C)	Conditions	Method	Assessment	Ref.
pK_1 2.62, pK_2 4.730	10		E3bg	Approx.	S22
pK_1 2.204, pK_2 4.680	25				
pK_1 2.41, pK_2 4.85	25		E3g	Approx.	L71
pK_2 4.60	20	$c = 0.025$, mixed constant	E3d	Uncert.	H17

3665 $C_7H_7O_2N$ 1,4-Benzoquinone 1-oxime, 2-methyl-

	T (°C)	Conditions	Method	Assessment	Ref.
6.95	not stated	Mixed constant	05	Uncert.	D35

3666 $C_7H_7O_2N$ 1,4-Benzoquinone 1-oxime, 3-methyl-

	T (°C)	Conditions	Method	Assessment	Ref.
6.90	not stated	Mixed constant	05	Uncert.	D35

3667 $C_7H_7O_2N$ 2,4,6-Cycloheptatrien-1-one, 4-amino-2-hydroxy-

	T (°C)	Conditions	Method	Assessment	Ref.
7.47	25	$I = 2.0$	05	Approx.	O16

No.	Molecular formula, name and pK value(s)	T(°C)	Remarks	Method	Assessment	Ref.
3668	C$_7$H$_7$O$_2$N 2-Pyridinecarboxylic acid, 6-methyl-					
	pK$_1$ 0.9, pK$_2$ 5.77	18±2	I = 0.5(NaCl), concentration constants	05	Uncert.	H50
3669	C$_7$H$_7$O$_2$N$_3$ 2,6-Pyridinedicarboxaldehyde dioxime					
	pK$_2$ 10.08, pK$_3$ 10.88	25	c = 0.000022	05	Approx.	H10
	(pK$_1$ ≈ 2)					
	pK$_2$ 10.04, pK$_3$ 10.68	20.3	I = 0.05, mixed constants			
	pK$_2$ 10.00, pK$_3$ 10.63	25				
	pK$_2$ 9.92, pK$_3$ 10.51	30	Thermodynamic values are derived from			
	pK$_2$ 9.85, pK$_3$ 10.45	35	the results			
	pK$_1$ 2.34	25	I = 0.005	E3bg	Uncert.	B2
	pK$_2$ 9.7, pK$_3$ 10.7		I = 0.01			
3670	C$_7$H$_7$O$_2$N$_5$ Pteridine-4,7(1H,8H)-dione, 2-amino-6-methyl-					
	pK$_1$ 7.98, pK$_2$ 10.15	20		05	Uncert.	P57
3671	C$_7$H$_7$O$_2$N$_5$ Pteridine-4,7(1H,8H)-dione, 2-amino-8-methyl-					
	8.10	20		05	Approx.	P57
3672	C$_7$H$_7$O$_2$N$_5$ Pteridine-4,7(3H,8H)-dione, 2-amino-3-methyl-					
	7.45	20		05	Uncert.	P57

323

No.	Formula	Name	pK	T	Conditions		Assessment	Ref.
3673	$C_7H_7O_2N_5$	Pteridine-4,7(3H,8H)-dione, 2-amino-8-methyl-	8.10	20		05	Uncert.	P57
3674	$C_7H_7O_2Br$	Phenol, 2-bromo-4-methoxy-	8.81	10	$I = 0.1(KCl)$	05	Approx.	S111
			8.69	20				
			8.57	30				
		Thermodynamic quantities are derived from the results						
3675	$C_7H_7O_2Br$	Phenol, 3-bromo-4-methoxy-	9.30	10	$I = 0.1(KCl)$	05	Approx.	S111
			9.16	20				
			9.02	30				
		Thermodynamic quantities are derived from the results						
3676	$C_7H_7O_3N$	Benzenecarboxamide, 2,4-dihydroxy-	pK_1 7.5	20	1-4% ethanol, mixed constant	05	Uncert.	W24
3677	$C_7H_7O_3N$	Benzenecarboxamide, 2,6-dihydroxy-	pK_1 7.1	20	1-4% ethanol, mixed constant	05	Uncert.	W24
3678	$C_7H_7O_3N$	Benzohydroxamic acid, 2-hydroxy- (Salicylohydroxamic acid)	pK_1 2.15, pK_2 7.46	30	$I = 0.1(NaClO_4)$	E3bg	Approx.	D63
			pK_2 7.46, pK_3 9.72	30	$c = 0.01$, $I = 0.1(NaClO_4)$, concentration constants	E3bg	Approx.	D62

(contd)

324

No.	Molecular formula, name and pK value(s)	T(°C)	Remarks	Method	Assessment	Ref.
3678 (Contd)	$C_7H_7O_3N$ Benzohydroxamic acid, 2-hydroxy- (Salicylohydroxamic acid)					
	pK$_2$ 7.56, pK$_3$ 9.54	33	c = 0.0025, I = 1.0(KNO$_3$), mixed constants	E3bg	Uncert.	B73
	pK$_1$ 2.15	30	I = 0.1(NaClO$_4$), concentration constant	E3bg	Approx.	S59
	pK$_2$ 7.43	25	c ≈ 0.01, I = 0.1(KCl), mixed constant	E3bg	Approx.	G39
	Other values in S117 and W39					
3679	$C_7H_7O_3N$ Benzohydroxamic acid, 4-hydroxy-					
	8.93	25	I = 0.2(NaCl)	E3bg	Uncert.	C72
3680	$C_7H_7O_3N$ Benzoic acid, 2-amino-5-hydroxy-					
	pK$_1$ 2.72, pK$_2$ 5.37	25	c = 0.00085, I = 0.1(KCl), mixed constants	05	Approx.	L15
3681	$C_7H_7O_3N$ Benzoic acid, 4-amino-2-hydroxy- (PAS)					
	pK$_1$ 2.05, pK$_2$ 3.66	25	I = 0.5(NaClO$_4$), concentration constants	E3bg	Uncert.	M8
	pK$_2$ 3.88		Conditions not stated	E3bg	Uncert.	V3
	pK$_2$ 4.36	25	I = 1.0(KNO$_3$)	E3bg	Uncert.	G49
	pK$_3$ 13.74	25	I = 1.0(NaClO$_4$), mixed constant	07	Uncert.	A11
3682	$C_7H_7O_3N$ Methyl 1,6-dihydro-6-oxo-3-pyridylcarboxylate (3-Pyridinecarboxylic acid methyl ester, 1,6-dihydro-6-oxo-)					
	9.92	20	c = 0.01	E3bg	Approx.	A18

No.	Formula	Name	pK	T (°C)	Conditions		Assessment	Ref.
3683	$C_7H_7O_3N$	Phenol, 4-methoxy-2-nitroso-	6.592	25	I = 0.0016-0.55	05,R4	Approx.	M33
3684	$C_7H_7O_3N$	Phenol, 2-methyl-5-nitro-	8.592	25	c ≈ 0.0004	05a	Approx.	R29
3685	$C_7H_7O_3N$	Phenol, 5-methyl-2-nitro-	7.409	25	c = 0.000055	05a	Rel.	R37

Variation with temperature

5	10	15	20	30	35	40	45	50	55	60
7.671	7.596	7.525	7.466	7.354	7.303	7.257	7.215	7.176	7.133	7.099

Thermodynamic quantities are derived from the results

No.	Formula	Name	pK	T (°C)	Conditions		Assessment	Ref.
			7.25	25	c ≈ 0.0001-0.00001, I ≈ 0.1, mixed constant	05	Approx.	H11
3686	$C_7H_7O_3N$	Phenol, 4-methyl-2-nitro-	7.597	25	c ≈ 0.0005	05a	Approx.	R29
			7.40	25	c = 0.0002, I ≈ 0.5, mixed constant	05	Approx.	R15
3687	$C_7H_7O_3N$	Phenol, 4-methyl-3-nitro-	8.622	25	c ≈ 0.0005	05a	Approx.	R29
3688	$C_7H_7O_3N$	2-Pyridinecarboxylic acid, 6-hydroxymethyl- pK_1 1.2, pK_2 4.61		25	Insufficient details	E3bg	Uncert.	G42

No.	Molecular formula, name and pK value(s)	T(°C)	Remarks	Method	Assessment	Ref.
3689	$C_7H_7O_3N$ 3-Pyridinecarboxylic acid, 1,6-dihydro-1-methyl-6-oxo-					A18
	3.84	20	c = 0.005	E3bg	Approx.	
3690	$C_7H_7O_3N_3$ Benzohydrazide, 2-nitro-					T43
	pK$_1$ 2.48	25	Mixed constant(?)	E3bg	Uncert.	
	11.34(NH)	25		05	Uncert.	
3691	$C_7H_7O_3N_3$ Benzohydrazide, 3-nitro-					T43
	pK$_1$ 2.73	25	Mixed constant(?)	E3bg	Uncert.	
	11.36(NH)	25		05	Uncert.	
3692	$C_7H_7O_3N_3$ Benzohydrazide, 4-nitro-					T43
	pK$_1$ 2.69	25	Mixed constant(?)	E3bg	Uncert.	
	11.26(NH)	25		05	Uncert.	
3693	$C_7H_7O_3N_5$ Pteridine-4,6,7(1H,5H,8H)-trione, 2-amino-8-methyl-					P56
	pK$_1$ 7.95, pK$_2$ 10.20	20	c ≈ 0.00005	05	Approx.	
3694	$C_7H_7O_3N_5$ Pteridine-4,6,7(3H,5H,8H)-trione, 2-amino-3-methyl-					P56
	pK$_1$ 7.60, pK$_2$ 10.95	20	c ≈ 0.00005	05	Approx.	
3695	$C_7H_7O_3N_5$ 1,2,4-Triazolo[4,3-c]pyrimidin-4(1H)-one, 3,6-dimethyl-7-nitro-					B141
	5.04	20	c < 0.001, I = 0.01, mixed constant	05	Approx.	

No.	Formula	Name	pK	T (°C)	Conditions	Method	Reliability	Ref.
3696	$C_7H_7O_4N$	Phenol, 2-methoxy-3-nitro-	7.83	20	I = 0.1, mixed constant(?)	E3bg	Uncert.	S128
3697	$C_7H_7O_4N$	Phenol, 2-methoxy-4-nitro-	6.63	20	I = 0.1, mixed constant(?)	E3bg	Uncert.	S128
3698	$C_7H_7O_4N$	Phenol, 2-methoxy-5-nitro-	8.00	20	I = 0.1, mixed constant(?)	E3bg	Uncert.	S128
3699	$C_7H_7O_4N$	Phenol, 2-methoxy-6-nitro-	6.84	20	I = 0.1, mixed constant(?)	E3bg	Uncert.	S128
3700	$C_7H_7O_4N$	Phenol, 3-methoxy-4-nitro-	6.10	20	I = 0.1, mixed constant(?)	E3bg	Uncert.	S128
3701	$C_7H_7O_4N$	Phenol, 5-methoxy-2-nitro-	7.09	25	c $\approx$ 0.0001-0.00001, I $\approx$ 0.1, mixed constant	05	Approx.	H11
			6.54	20	I = 0.1, mixed constant(?)	E3bg	Uncert.	S128
3702	$C_7H_7O_4N$	Phenol, 4-methoxy-2-nitro-	7.31	25	c = 0.0002, I $\approx$ 0.5, mixed constant	05	Approx.	R15
			7.54	10	I = 0.1(KCl)	05	Approx.	S111
			7.47	20	Thermodynamic quantities are derived from the results			
			7.40	30				
			7.47	20	I = 0.1, mixed constant(?)	E3bg	Uncert.	S128

No.	Molecular formula, name and pK value(s)	T(°C)	Remarks	Method	Assessment	Ref.
3703	C₇H₇O₄N Phenol, 4-methoxy-3-nitro-					
	8.80	10	I = 0.1(KCl)	05	Approx.	S111
	8.68	20				
	Thermodynamic quantities are derived from the results					
	8.68	20	I = 0.1, mixed constant(?)	E3bg	Uncert.	S128
3704	C₇H₇O₄N 2-Pyridinecarboxylic acid N-oxide, 4-methoxy-					
	4.40		c = 0.04 or 0.001(?)	E		K51
3705	C₇H₇O₄N 3-Pyridinecarboxylic acid N-oxide, 4-methoxy-					
	2.88(N⁺-OH)		Conditions not stated	E	Uncert.	D52
3706	C₇H₇O₄P Benzoic acid, 2-dihydroxyphosphino-					
	pK₁ 1.32, pK₂ 4.06	25	c ≃ 0.05, mixed constants	E3bg	Uncert.	Q1
3707	C₇H₇O₆P Benzoic acid, 2-phosphonooxy-					
	pK₁ 1.35, pK₂ 3.87, pK₃ 6.74	27		E3d	Uncert.	C32
	pK₁ 1.30	26	I = 0.1	E3bg	Uncert.	C33
	pK₂ 3.73, pK₃ 6.51				Approx.	
	pK₂ 3.69, pK₃ 6.61	25	I = 0.1(KNO₃)	E3bg	Approx.	M99
	pK₂ 3.60, pK₃ 6.45	34.9				

No.	Formula	Name / pK values	T (°C)	Conditions	Method	Assessment	Ref.
3708	C$_7$H$_7$O$_6$P	Benzoic acid, 3-phosphonooxy-					
		pK$_1$ 0.85	26	c = 0.1, I = 0.1	E3bg	Uncert.	C31
		pK$_2$ 3.91, pK$_3$ 5.90		c = 0.01, I = 0.1		Approx.	
		pK$_1$ was determined in 10% dioxan-water. Value was corrected to 100% water					
3709	C$_7$H$_7$O$_6$P	Benzoic acid, 4-phosphonooxy-					
		pK$_1$ 0.83	26	c = 0.1, I = 0.1	E3bg	Uncert.	C31
		pK$_2$ 4.20, pK$_3$ 5.90		c = 0.01, I = 0.1		Approx.	
		pK$_1$ was determined in 10% dioxan-water. Value was corrected to 100% water					
3710	C$_7$H$_7$NS	Benzenecarbothioamide (Thiobenzamide)					
		12.8	not stated	c = 0.00001	05	Uncert.	W4
3711	C$_7$H$_8$ON$_2$	Benzenecarboxamide oxime					
		pK$_1$ 5.1, pK$_2$ 11.83		I = 0.2	E		M47
3712	C$_7$H$_8$ON$_2$	Benzohydrazide					
		pK$_1$ 3.05	25	Mixed constant(?)	E3bg	Uncert.	T43
		12.52(NH)	25		05	Uncert.	
3713	C$_7$H$_8$ON$_6$	Pyrimidino[5,4-e]-1,2,4-triazin-5(1H)-one, 7-amino-3-ethyl-					
		pK$_1$ 1.15, pK$_2$ 7.00	20	I = 0.01, mixed constants	05	Approx.	B139
3714	C$_7$H$_8$OS	Benzenethiol, 3-methoxy-					
		6.385	25	m = 0.0001, I < 0.15	05a	Rel.	D28

No.	Molecular formula, name and pK value(s)	T(°C)	Remarks	Method	Assessment	Ref.
3715	C_7H_8OS Benzenethiol, 4-methoxy-	25	m = 0.001, I < 0.15	O5a	Rel.	D28
	6.775					
3716	C_7H_8OS Phenol, 2-methylthio-		Conditions not stated	O5	Uncert.	B179
	ca 9.3					
3717	C_7H_8OS Phenol, 3-methylthio-	25		E3ag	Approx.	B104
	9.42					
3718	C_7H_8OS Phenol, 4-methylthio-	25		E3ag	Approx.	B104
	9.47					
	9.53	25	I = 0.1(KCl), mixed constant	E3bg	Approx.	H51
3719	$C_7H_8O_2N_2$ Benzenecarboxamide, 3-amino-6-hydroxy-	25	I = 1.0(NaClO$_4$), mixed constant	O5	Uncert.	A11
	9.11					
3720	$C_7H_8O_2N_2$ Benzohydroxamic acid, 2-amino-	30		E3bg	Uncert.	S117
	9.0					
3721	$C_7H_8O_2N_2$ Benzohydroxamic acid, 4-amino-	25	c = 0.01(?)	E3bg	Approx.	A10
	9.42					
	9.32	not stated	c = 0.04, I = 0.1(KCl)	E3bg	Uncert.	H2
	9.32	not stated	I = 0.2(NaCl)	E3bg	Uncert.	C72

No.	Formula	Name / pK	Temp.	Conditions	Method	Assessment	Ref.
3722	$C_7H_8O_2N_2$	1,4-Benzoquinone dioxime 1-0-methyl ether					
		9.20	not stated	Mixed constant	05	Uncert.	D35
3723	$C_7H_8O_2N_4$	Pteridine-4,7(1H,8H)-dione, 5,6-dihydro-6-methyl-					
		pK$_1$ 8.43	20	c = 0.002	E3bg	Approx.	A22
		(pK$_2$ > 11)					
3724	$C_7H_8O_2N_4$	Purine-2,6(1H,3H)-dione, 1,3-dimethyl-					
		8.68	20		E3bg	Approx.	P55
		8.5		Conditions not stated	05	Uncert.	L39
3725	$C_7H_8O_2N_4$	Purine-2,6(1H,3H)-dione, 1,7-dimethyl-					
		8.6		Conditions not stated	05	Uncert.	L39
3726	$C_7H_8O_2N_4$	Purine-2,6(1H,3H)-dione, 1,9-dimethyl-					
		5.99	20		E3bg	Approx.	P55
		6.3		Conditions not stated	05	Uncert.	L39
3727	$C_7H_8O_2N_4$	Purine-2,6(1H,3H)-dione, 3,7-dimethyl-					
		10.00	20		E3bg	Approx.	P55
		11.0		Conditions not stated	05	Uncert.	L39
3728	$C_7H_8O_2N_4$	Purine-2,6(1H,3H)-dione, 3,9-dimethyl-					
		10.14	20		E3bg	Approx.	P55
		10.5		Conditions not stated	05	Uncert.	L39

No.	Molecular formula, name and pK value(s)	T(°C)	Remarks	Method	Assessment	Ref.
3729	$C_7H_8O_2N_4$ Purine-2,6(1H,3H)-dione, 8,9-dimethyl-					
	6.20	20	c = 0.001	E3bg	Approx.	P49
3730	$C_7H_8O_2N_4$ Purine-2,6(1H,3H)-dione, 9-ethyl-					
	6.14	20	c = 0.001	E3bg	Approx.	P53
3731	$C_7H_8O_2N_4$ 1H-Pyrazolo[3,4-d]pyrimidine-4,6(5H,7H)-dione, 5,7-dimethyl-					
	9.26	20	c = 0.001	E3bg	Approx.	P59
3732	$C_7H_8O_2S$ Benzenesulfinic acid, 4-methyl-					
	2.80	25	I ≃ 0.005	E3bg	Uncert.	D23
	1.99	25		E3bg	Uncert.	B179
3733	$C_7H_8O_2S$ Methanesulfinic acid, phenyl-					
	1.45	20	c = 0.1, mixed constant	E3bg	Approx.	R54
3734	$C_7H_8O_2S$ Phenol, 2-methylsulfinyl-					
	7.60	25		E3bg	Approx.	F25
	ca 7.8		Conditions not stated	O5	Uncert.	B179
3735	$C_7H_8O_2S$ Phenol, 3-methylsulphinyl-					
	8.75	25	I ≃ 0.01	E3bg	Approx.	B105
	8.79	25		E3bg	Approx.	F25

No.	Formula	Name / pK	T(°C)	Method	Assessment	Ref.	Notes
3736	$C_7H_8O_2S$	Phenol, 4-methylsulphinyl-					
		8.28	25	E3bg	Approx.	B105	Mixed constant
		8.43	25	E3bg	Approx.	F25	
3737	$C_7H_8O_2S$	2-Thiophenecarboxylic acid, 5-ethyl-					
		3.70	25	E3bg	Approx.	B188	c = 0.005; mixed constant
							Value in mixed solvent is also given
3738	$C_7H_8O_2Se$	Benzeneseleninic acid, 4-methyl-					
		4.55	25	E3bg	Approx.	D23	I ≈ 0.005, 4% methanol
3739	$C_7H_8O_3N_2$	Pyrimidine-2,4(1H,3H)-dione, 5-allyl-6-hydroxy-					
		4.78	25	E3bg	Approx.	M80	
							Thermodynamic quantities are also given
3740	$C_7H_8O_3N_4$	2-Pyrazinecarboxamide, 5-amino-6-formyl-3,4-dihydro-N-methyl-3-oxo-					
		6.01	20	05	Approx.	A39	c = 0.0001, I ≈ 0.01, mixed constant
3741	$C_7H_8O_3N_4$	Purine-2,6(1H,3H)-dione, 9-(2-hydroxyethyl)-					
		5.93	20	E3bg	Approx.	P53	c = 0.001
3742	$C_7H_8O_3N_4$	Purine-2,6,8(1H,3H,7H)-trione, 1,3-dimethyl-					
		5.75	not stated	05	Uncert.	B65	Mixed constant

No.	Molecular formula, name and pK value(s)	T(°C)	Remarks	Method	Assessment	Ref.
3743	$C_7H_8O_3N_4$ Purine-2,6,8(1H,3H,7H)-trione, 1,7-dimethyl-					
	pK$_1$ 5.5, pK$_2$ 10.6	not stated	Mixed constants	05	Uncert.	D38
	pK$_1$ 5.7, pK$_2$ 10.9		Conditions not stated	05	Uncert.	J13
3744	$C_7H_8O_3N_4$ Purine-2,6,8(1H,3H,7H)-trione, 3,7-dimethyl-					
	pK$_1$ 5.5, pK$_2$ > 12	not stated	Mixed constant	05	Uncert.	B65
3745	$C_7H_8O_3S$ Benzenesulfinic acid, 4-methoxy-					
	2.72	25	I ≃ 0.005	E3bg	Uncert.	D23
3746	$C_7H_8O_3S$ Benzenesulfonic acid, 4-methyl-					
	-1.34	25	Saturated aq. solution	NMR	V.Uncert.	D39
3746a	$C_7H_8O_3S$ Phenol, 4-mesyl-					
	7.72	25	No details		Uncert.	C76
3747	$C_7H_8O_4N_2$ 4-Pyrimidinecarboxylic acid, 1,2,3,6-tetrahydro-1,3-dimethyl-2,6-dioxo-					
	0.8		Conditions not stated	05	Uncert.	F34
3748	$C_7H_8N_2S$ 2-Pyridinecarbothioamide, 6-methyl-					
	12.62	25	c = 0.0001, I = 0.07(KCl), mixed constant	05	Approx.	T42
	12.75					

No.	Formula, Name	pK	T	Conditions			Ref.
3749	$C_7H_8N_4S$. Purine, 8-methyl-2-methylthio-	pK$_1$ -1.03	20	c < 0.0001, I = 0.01	06	Uncert.	B138
		pK$_2$ 2.83, pK$_3$ 9.58			05	Approx.	
3750	C_7H_9ON Phenol, 2-aminomethyl-	8.89(OH)	25	I = 0.1(KCl), mixed constant	05	Approx.	E20
3751	C_7H_9ON Phenol, 3-aminomethyl-	9.06(OH)	25	I = 0.1(KCl), mixed constant	05	Approx.	E20
3752	C_7H_9ON Phenol, 4-aminomethyl-	9.14(OH)	25	I = 0.1(KCl), mixed constant	05	Approx.	E20
3753	$C_7H_9ON_3$ Acetohydrazide, 2-(4-pyridyl)-	11.07(NH)	25	Mixed constant(?)	E3bg	Uncert.	T43
3754	$C_7H_9ON_3$ Triazen-3-ol, 3-methyl-1-phenyl-	12.11	25	c = 0.00004, I = 0.1(KCl), mixed constant	05	Uncert.	D54
3755	$C_7H_9ON_5$ Pteridin-4(1$\underline{H}$)-one, 2-amino-7,8-dihydro-6-methyl-	pK$_1$ 3.2	22	I ≈ 0.1, mixed constants	05	Uncert.	W26
		pK$_2$ 10.6		c = 0.01-0.02	E3bg		
3756	$C_7H_9O_2N$ 3-Azabicyclo[3.3.0]octane-2,4-dione	9.53(cis)		Conditions not stated	E3bg	Uncert.	C62

No.	Molecular formula, name and pK value(s)	T(°C)	Remarks	Method	Assessment	Ref.
3757	$C_7H_9O_2N$ 1,2-Benzenediol, 3-aminomethyl- pK_1 8.30	25	I = 0.1(KNO_3), mixed constant	E3bg	Approx.	E19
3758	$C_7H_9O_2N$ 1,2-Benzenediol, 4-aminomethyl- pK_1 8.60	25	I = 0.1(KNO_3), mixed constant	E3bg	Approx.	E19
3759	$C_7H_9O_2N$ Phenol, 2-amino-4-methoxy- 10.20 10.07 9.94 10.07	10 20 30 20	I = 0.1(KCl), mixed constant Thermodynamic quantities are derived from the results I = 0.1, mixed constant(?)	05 E3bg	Approx. Uncert.	S111 S128
3760	$C_7H_9O_2P$ Methylphenylphosphinic acid 2.96		Conditions not stated	E3d	Uncert.	E28
3761	$C_7H_9O_2P$ Phenylphosphonous acid, 4-methyl- 1.83	25	c ≈ 0.05, mixed constant	E3bg	Uncert.	Q1
3762	$C_7H_9O_3N$ Pyridin-2(1H)-one, 1,6-dihydroxy-4,5-dimethyl- pK_1 5.05, pK_2 10.35		Conditions not stated	E3bg	Uncert.	A50
3763	$C_7H_9O_3N_7$ Propanediamide, 2-(4,5,6,7-tetrahydro-5-oxo-1H-1,2,3-triazolo[4,5-d]pyrimidin-7-yl)- 7.84	20	c = 0.00002	05	Approx.	A33

3764 $C_7H_9O_3P$ Benzylphosphonic acid

pK$_1$ 1.8, pK$_2$ 7.4   17   c = 0.02   E3bg   Uncert.   L21
pK$_1$ 2.3, pK$_2$ 7.55   not stated   c = 0.05   E3bg   Uncert.   C38

Value in mixed solvent is also given

3764a $C_7H_9O_3P$ Phenyl hydrogen methylphosphonate

1.39   not stated   c = 0.1   E3bg   Uncert.   B44
1.47   c = 0.001   05   Uncert.

3765 $C_7H_9O_3As$ Benzylarsonic acid

pK$_1$ 4.43, pK$_2$ 7.51   Not stated   Mixed constants   E3bg   Uncert.   C42

3766 $C_7H_9O_4N$ Ethyl 4-hydroxy-2-oxo-3-pyrroline-3-carboxylate

2.34   Conditions not stated   05   Uncert.   M109

3767 $C_7H_9O_4Cl$ r-1,c-2-Cyclopropanedicarboxylic acid, 1-chloro-t-2,t-3-dimethyl-

pK$_1$ 2.91, pK$_2$ 5.41   20   E3bg   Approx.   M12

3768 $C_7H_9O_4P$ Benzyl dihydrogen phosphate

pK$_1$ 1.59, pK$_2$ 6.12   c = 0.0008-0.0011   02   Approx.   K66

3769 $C_7H_9O_4P$ Phenoxymethylphosphonic acid

pK$_1$ 1.37   20   05   Approx.   P60
pK$_2$ 6.84   I = 0.02   E3bg
pK$_2$ 6.82   I = 0.01-0.02   05
pK$_1$ 1.38, pK$_2$ 6.75   not stated   Mixed constants(?)   M51

No.	Molecular formula, name and pK value(s)	T(°C)	Remarks	Method	Assessment	Ref.
3770	C7H9O4P p-Tolyl dihydrogen phosphate pK1 1.64, pK2 6.45	not stated	c = 0.005, I ≠ 0.025, mixed constants, 7% ethanol	E3bg	Uncert.	M57
3771	C7H10O2N2 Acetic acid, 2-(3,5-dimethyl-1-pyrazolyl)- 3.90	room temp.	c = 0.01	E3bg	Uncert.	H1
3772	C7H10O2N2 Pyrimidine-2,4(1H,3H)-dione, 1,5,6-trimethyl- 10.4					D48
3773	C7H10O2N2 Pyrimidine-2,4(1H,3H)-dione, 3,5,6-trimethyl- 10.8					D48
3774	C7H10O2N4 Propanoic acid, 2-amino-3-(2-amino-4-pyrimidinyl)- (Lathytine) pK1 2.4, pK2 4.1, pK3 9		Conditions not stated	E3bg	Uncert.	B46
3775	C7H10O3N2 1,2,3-Cyclohexanetrione 1,3-dioxime, 5-methyl- 8.5	25	c = 0.01, I = 0.1(KCl)	E3bg	Uncert.	J11
3776	C7H10O3N2 2-Pyrrolidinecarboxylic acid, 5-cyano-1-hydroxy-2-methyl- 3.6		Conditions not stated	E3bg	Uncert.	B101
3777	C7H10O4N2 Cyclohexanecarboxylic acid, 3,4-bis(hydroxyimino)- pK1 4.85 pK2 10.45, pK3 12.37	not stated		E3bg 05	Uncert.	B19

No.	Formula	Name / pK	T	Method	Assessment	Notes	Ref.
3778	$C_7H_{10}O_4N_2$	Methyl 3-(2,4-dioxo-4-imidazolidinyl)propanoate 8.7	25	E3bg	Uncert.		R40
3779	$C_7H_{10}O_8N_2$	Heptanedioic acid, 4,4-dinitro- pK_1 3.52, pK_2 4.96	21	E3bg	Uncert.	c = 0.01, mixed constants	N37
3780	$C_7H_{10}O_{10}N_4$	Methyl 4,4,6,6-tetranitrohexanoate 1.13	20	06	Uncert.	Aq.$HClO_4$ solution, H_o scale	T58a
3781	$C_7H_{11}ON_3$	Pyrimidin-2($1\underline{H}$)-one, 1,5-dimethyl-4-methylamino- pK_1 4.57	20	05	Uncert.		K65
3782	$C_7H_{11}ON_3$	Pyrimidin-2($1\underline{H}$)-one, 4-dimethylamino-1-methyl- 4.20	22	05	Uncert.		S141
3783	$C_7H_{11}ON_5$	Pteridin-4($1\underline{H}$)-one, 2-amino-5,6,7,8-tetrahydro-6-methyl- pK_1 5.4, pK_2 10.5	22	E3bg	Uncert.	c = 0.01-0.02	W26
3784	$C_7H_{11}O_2N_3$	Methyl 2-amino-3-(4-imidazolyl)propanoate pK_1 5.34, pK_2 7.35 pK_1 5.22, pK_2 7.14 pK_1 5.12, pK_2 6.96	25 35 45	E3bg,R3f	Approx.	I = 0.12(KCl) Thermodynamic quantities are derived from the results	C39
3785	$C_7H_{11}O_2N_5$	Pyrimidin-2($1\underline{H}$)-one, 1,5-dimethyl-4-semicarbazido- 3.60		05	Uncert.	Conditions not stated	J8

No.	Molecular formula, name and pK value(s)	T(°C)	Remarks	Method	Assessment	Ref.
3786	$C_7H_{11}O_2Cl$ Cyclohexanecarboxylic acid, 4-chloro-					
	4.58(trans)	25	c = 0.01	E3bg	Approx.	S69
	Values in mixed solvents are also given					
3787	$C_7H_{11}O_3N_3$ 1,2,3-Cyclohexanetrione trioxime, 5-methyl-					
	pK$_1$ 8.16, pK$_2$ 11.30	25	Concentration constants	E3d	Uncert.	H4
	pK$_1$ 8.0	25	c = 0.01, I = 0.1(KCl)	E3bg	Uncert.	J11
3788	$C_7H_{11}O_3N_3$ Pyrimidine-2,4(1H,3H)-dione, 6-methoxyamino-1,3-dimethyl-					
	10.58	20	c = 0.001	E3bg	Approx.	P51
3789	$C_7H_{11}O_4N$ 3-Azapentanedioic acid, 3-allyl- (Iminodiacetic acid, N-allyl-)					
	pK$_1$ 2.43, pK$_2$ 9.28	25	c = 0.01, I = 0.1(KCl), concentration constants	E3bg	Approx.	S97
3790	$C_7H_{11}O_4N$ 2,6-Piperidinedicarboxylic acid					
	pK$_1$ 2.87, pK$_2$ 9.92	30	I = 0.1(KCl), mixed constants	E3bg	Approx.	T32
3791	$C_7H_{11}O_5N$ 3-Azapentanedioic acid, 3-acetonyl- (Iminodiacetic acid, N-acetonyl-)					
	pK$_1$ 2.62, pK$_2$ 7.71	25	I = 0.1(KNO$_3$)	E3bg	Approx.	A63,A64
3792	$C_7H_{11}O_6N$ 3-Azahexanedioic acid, 3-carboxymethyl- (Iminodiacetic acid, N-(2-carboxyethyl)-)					
	pK$_1$ 1.90, pK$_2$ 3.71, pK$_3$ 9.61	25	I = 0.1(KNO$_3$), concentration constants	E3bg	Approx.	U8
	pK$_1$ 2.24, pK$_2$ 3.60, pK$_3$ 9.35	30	I = 0.1(KCl), concentration constants (contd)	E3bg	Uncert	C24

No.	Formula	Name	Temp	Conditions	Code	Uncert.	Ref.
3792 (Contd)	$C_7H_{11}O_6N$	3-Azahexanedioic acid, 3-carboxymethyl- (Iminodiacetic acid, N-(2-carboxyethyl)-)					
		pK$_1$ 2.1	20	I = 0.1(KCl), concentration constants	E3bg	Uncert. Approx.	S36
		pK$_2$ 3.69, pK$_3$ 9.66					
3793	$C_7H_{11}O_6N$	3-Azapentanedioic acid, 3-carboxymethyl-2-methyl- (Iminodiacetic acid, N-(1-carboxyethyl)-)					
		pK$_1$ 1.57	20	I = 0.1(KCl), concentration constants	E3bg	Uncert. Approx.	I14
		pK$_2$ 2.46, pK$_3$ 10.47					
3794	$C_7H_{11}O_6N$	Heptanedioic acid, 4-nitro-					
		pK$_1$ 4.00, pK$_2$ 5.24	21	c = 0.01, mixed constant	E3bg	Uncert.	N37
3795	$C_7H_{11}O_7N$	3-Azapentanedioic acid, 3-carboxymethyl-2-hydroxymethyl- (DL-Serine-N,N-diacetic acid)					
		pK$_1$ 2.2, pK$_2$ 2.62, pK$_3$ 9.2	20		E		E22
3796	$C_7H_{12}OS_2$	Methanedithioic acid, cyclohexyloxy- (Xanthic acid, cyclohexyl-)					
		2.05		Conditions not stated	05	Uncert.	Z2
3797	$C_7H_{12}O_2N_2$	1,2-Cycloheptanedione dioxime					
		pK$_1$ 10.7, pK$_2$ 12.3	20-25		05	Uncert.	B18
		pK$_1$ 10.65, pK$_2$ 12.21			0		S15
3798	$C_7H_{12}O_2N_2$	Pentanoic acid, 5-(3-methyl-3H-diazirin-3-yl)- (Heptanoic acid, 6,6-(N,N')azo-)					
		4.7	25		E3bg	Uncert.	C57

No.	Molecular formula, name and pK value(s)	$T(^{o}C)$	Remarks	Method	Assessment	Ref.
3799	C$_7$H$_{12}$O$_2$S Acetic acid, 4-pentenylthio- 3.739	25	c ≈ 0.001, I = 0.10(KNO$_3$), concentration constant	E3bg	Approx.	P40
3800	C$_7$H$_{12}$O$_4$S 3-Thiapentanedioic acid, 2,2,4-trimethyl- pK$_1$ 3.60, pK$_2$ 4.31	25±0.5	c < 0.001	E3bg	Approx.	S93
3801	C$_7$H$_{12}$O$_4$S 3-Thiapentanoic acid, 5-ethoxy-2-methyl-5-oxo- 3.71	25		E3ag	Approx.	S92,S93
3802	C$_7$H$_{12}$O$_4$S 3-Thiapentanoic acid, 5-ethoxy-4-methyl-5-oxo- 3.77	25±0.5	c = 0.00075	E3bg	Approx.	S93,S92
3803	C$_7$H$_{12}$O$_4$S 1-Trimethylsulfonio-1,1-bis(methoxycarbonyl)methanide 1.81		Conditions not stated Values in mixed solvents are also given	05	Uncert.	N31
3804	C$_7$H$_{12}$O$_4$S$_2$ 3,7-Dithianonanedioic acid pK$_1$ 3.44, pK$_2$ 4.25		I = 0.1(NaClO$_4$), mixed constants	E3bg	Uncert.	O21
3805	C$_7$H$_{12}$O$_4$S$_2$ 4,6-Dithianonanedioic acid pK$_1$ 4.08, pK$_2$ 4.86	20	I = 0.1(NaClO$_4$), mixed constants	E3bg	Uncert.	O21

(contd)

3805 C$_7$H$_{12}$O$_4$S$_2$ 4,6-Dithiononanedioic acid
(Contd)
pK$_1$ 3.94 — I = 0.3 — — — C11
pK$_1$ 3.97, pK$_2$ 4.75 — I = 0.05
pK$_2$ 4.69 — I = 0.1

3806 C$_7$H$_{13}$O$_2$N Cyclohexanecarbohydroxamic acid
9.75 — not stated — I = 0.2(NaCl) — E3bg — Uncert. — C72

3807 C$_7$H$_{13}$O$_2$N 2-Piperidinecarboxylic acid, 4-methyl-
pK$_1$ 2.27, pK$_2$ 10.74(cis) — 20 — c = 0.004 — E3bg — Approx. — C3
pK$_1$ 2.28, pK$_2$ 10.74(trans)

3808 C$_7$H$_{13}$O$_2$N 2-Pyrrolidinone, 1-hydroxy-4,5,5-trimethyl-
8.85 — Conditions not stated — E3bg — Uncert. — B101

3809 C$_7$H$_{13}$O$_2$N$_3$ Imidazolidine-2,4-dione, 5-(4-aminobutyl)-
pK$_1$ 8.9, pK$_2$ 10.6 — 25 — E3bg — Uncert. — R40

3810 C$_7$H$_{13}$O$_3$N$_3$ Imidazolidine-2,4-dione, 5-(4-hydroxyaminobutyl)-
pK$_1$ 5.7, pK$_2$ 9.1 — 25 — E3bg — Uncert. — R40

3811 C$_7$H$_{13}$O$_4$N 3-Azapentanedioic acid, 3-propyl- (Iminodiacetic acid, N-propyl-)
pK$_1$ 2.44, pK$_2$ 10.09 — 25 — c = 0.01, I = 0.1(KCl), concentration constants — E3bg — Approx. — S97

No.	Molecular formula, name and pK value(s)	T(°C)	Remarks	Method	Assessment	Ref.
3812	$C_7H_{13}O_4N_3$ 3,6-Diazanonanoic acid, 8-amino-4,7-dioxo- (L-Alanylglycylglycine) pK$_1$ 3.36, pK$_2$ 8.05	25	I = 0.16(KCl)	E3bg	Approx.	B151
3813	$C_7H_{13}O_4N_3$ 3,6-Diazaoctanoic acid, 8-amino-2-methyl-4,7-dioxo- (Glycylglycyl-L-alanine) pK$_1$ 3.30, pK$_2$ 8.02	25	I = 0.16(KCl)	E3bg	Approx.	B151
3814	$C_7H_{13}O_4N_3$ 3,6-Diazaoctanoic acid, 8-amino-5-methyl-4,7-dioxo- (Glycyl-L-alanylglycine) pK$_1$ 3.46, pK$_2$ 8.08	25	I = 0.16(KCl)	E3bg	Approx.	B151
3815	$C_7H_{13}O_5N$ 3-Azapentanedioic acid, 3-(3-hydroxypropyl)- (Iminodiacetic acid, N-(3-hydroxypropyl)-) pK$_1$ 1.96, pK$_2$ 8.78	30	I = 0.1(KCl), concentration constants	E3bg	Approx.	C25
3816	$C_7H_{13}O_5N$ 3-Azapentanedioic acid, 3-(2-methoxyethyl)- (Iminodiacetic acid, N-(2-methoxyethyl)-) pK$_1$ 2.2 pK$_2$ 8.96	20	I = 0.1(KCl), concentration constants	E3bg	Uncert. Approx.	S40
3817	$C_7H_{13}O_6N$ 3-Azapentanedioic acid, 3-(2,3-dihydroxypropyl)- (Iminodiacetic acid, N-(2,3-dihydroxypropyl)-) pK$_1$ 3.12, pK$_2$ 8.97	20		E		E22
3818	$C_7H_{13}NS_2$ Methanedithioic acid, cyclohexylamino- (Dithiocarbamic acid, N-cyclohexyl-) 2.68 2.54	21	c ≈ 0.01-0.00001, concentration constant (Compound unstable in acid solution)	05 E	Uncert.	T66

No.	Formula	Name	pK	T	Conditions	Method	Reliability	Ref.
3819	$C_7H_{14}OS_2$	Methanedithioic acid, hexyloxy- (Xanthic acid, hexyl-)	3.59					H27
3820	$C_7H_{14}O_2S$	Acetic acid, pentylthio-	3.753	25	$c \approx 0.001$, $I = 0.10(KNO_3)$, concentration constant	E3bg	Approx.	P40
3821	$C_7H_{14}O_3N_2$	3-Azapentanoic acid, 5-amino-2-isopropyl-4-oxo- (Glycyl-L-valine)	pK_1 3.26, pK_2 8.20	25	$I = 0.16$	E3bg	Approx.	B152
3822	$C_7H_{14}O_4N_2$	Heptane, 1,1-dinitro-	5.50		Conditions not stated	E3bg	Uncert.	N36
3823	$C_7H_{15}ON$	1-Pyrrolidinol, 2,2,3-trimethyl-	5.9		Conditions not stated	E3bg	Uncert.	B101
3824	$C_7H_{15}O_2N$	Heptanoic acid, 7-amino-	4.502	25		E3bg	Approx.	C50
3825	$C_7H_{15}NS$	Cyclohexanethiol, 1-aminomethyl-	pK_1 8.21, pK_2 11.06	25	$c = 0.01$, mixed constants	E3bg	Approx.	B147
3826	$C_7H_{15}NS_2$	Methanedithioic acid, di-isopropylamino- (Dithiocarbamic acid, N,N-di-isopropyl-)	5.62		$c = 0.1$	E		T65

No.	Molecular formula, name and pK value(s)	T(°C)	Remarks	Method	Assessment	Ref.
3827	$C_7H_{15}NS_2$ Methanedithioic acid, dipropylamino- (Dithiocarbamic acid, N,N-dipropyl-)					
	3.72	25		KIN	Uncert.	M79
3828	$C_7H_{15}NS_2$ Methanedithioic acid, hexylamino- (Dithiocarbamic acid, N-hexyl-)					
	3.34	21	$c \approx 0.01$-0.00001, concentration constant	O5	Uncert.	T65,T66
	3.58		(Compound unstable in acid solution)	E	Uncert.	
3829	$C_7H_{16}ON_2$ 2-Butanone oxime, 3-ethylamino-3-methyl-					
	9.24	25	$I = 0.002$	CAL	Uncert.	W8
	Thermodynamic quantities are also given					
3830	$C_7H_{16}O_2N_2$ Propanohydroxamic acid, 3-diethylamino-					
	8.15	25(?)	$c = 0.02$, $I = 0.1$(KCl), mixed constant	E3bg	Uncert.	M66
3831	$C_7HO_2ClF_4$ Benzoic acid, 4-chloro-2,3,5,6-tetrafluoro-					
	1.69	25	$c = 0.02$	E3bg	Approx.	R56
3832	$C_7HO_2BrF_4$ Benzoic acid, 4-bromo-2,3,5,6-tetrafluoro-					
	1.66	25	$c = 0.02$	E3bg	Approx.	R56
3833	$C_7H_2O_2F_4S$ Benzoic acid, 2,3,5,6-tetrafluoro-4-mercapto-					
	pK_1 1.67, pK_2 3.36	25	$c = 0.02$	E3bg	Approx.	R56

No.	Formula / Name	pK	T (°C)	Conditions	Method	Assessment	Ref.
3834	$C_7H_2O_8N_4Br_2$ Methane, (3,5-dibromo-2,4-dinitrophenyl)dinitro-						
		-0.86	20	$c \approx 0.0001$ (H_0 in $HClO_4$ solutions)	06	Uncert.	T58
3835	$C_7H_3ONCl_2$ Benzenecarbonitrile, 3,5-dichloro-4-hydroxy- (Phenol, 2,6-dichloro-4-cyano-)						
		4.38	25	$c = 0.0024$	E3bg	Approx.	F17
		4.23		Conditions not stated	05	Uncert.	P9
3836	$C_7H_3ONBr_2$ Benzenecarbonitrile, 3,5-dibromo-4-hydroxy- (Phenol, 2,6-dibromo-4-cyano-)						
		4.05		Conditions not stated	05	Uncert.	P9
3837	$C_7H_3ONI_2$ Benzenecarbonitrile, 4-hydroxy-3,5-diiodo- (Phenol, 4-cyano-2,6-diiodo-)						
		4.15		Conditions not stated	05	Uncert.	P9
3838	$C_7H_3O_6N_3Cl_2$ Methane, (3,4-dichloro-6-nitrophenyl)dinitro-						
		0.36	20	$c \approx 0.0001$ (H_0 in $HClO_4$ solutions)	06	Uncert.	T58
3839	$C_7H_3O_8N_4Cl$ Methane, (2-chloro-4,6-dinitrophenyl)dinitro-						
		-2.32	20	$c \approx 0.0001$ (H_0 in $HClO_4$ solutions)	06	Uncert.	T58
3840	$C_7H_3O_8N_4Cl$ Methane, (4-chloro-2,6-dinitrophenyl)dinitro-						
		-2.82	20	$c \approx 0.0001$ (H_0 in $HClO_4$ solutions)	06	Uncert.	T58

No.	Molecular formula, name and pK value(s)	T(°C)	Remarks	Method	Assessment	Ref.
3841	$C_7H_3O_8N_4Br$ Methane, (3-bromo-4,6-dinitrophenyl)dinitro- −0.39	20	$c \simeq 0.0001$ (H_o in $HClO_4$ solutions)	06	Uncert.	T58
3842	$C_7H_3O_8N_4F$ Methane, (2-fluoro-4,6-dinitrophenyl)dinitro- −1.42	20	$c \simeq 0.0001$ (H_o in $HClO_4$ solutions)	06	Uncert.	T58
3843	C_7H_4ONCl Benzenecarbonitrile, 2-chloro-6-hydroxy- 6.1	30	$I = 0.1(KCl)$	05	Uncert.	C20
3844	C_7H_4ONCl Benzenecarbonitrile, 3-chloro-6-hydroxy- 6.4	30	$I = 0.1(KCl)$	05	Uncert.	C20
3845	$C_7H_4O_4NCl$ Benzaldehyde, 3-chloro-6-hydroxy-5-nitro- 4.51	25	$c \simeq 0.00005$, $I = 0.1$	05	Approx.	D50
3846	$C_7H_4O_4NCl$ Benzoic acid, 2-chloro-4-nitro- 2.14		Conditions not stated	E3bg	Uncert.	V3
3847	$C_7H_4O_4NBr$ Benzaldehyde, 3-bromo-2-hydroxy-5-nitro- 3.48	25	$c \simeq 0.00005$, $I = 0.1$	05	Approx.	D50

No.	Formula	Name	Temp.	Value	Conditions			
3848	$C_7H_4O_4NBr$	Benzaldehyde, 3-bromo-6-hydroxy-5-nitro-	25	4.46	$c \approx 0.00005$, $I = 0.1$	05	Approx.	D50
3849	$C_7H_4O_6N_3Cl$	Methane, (2-chloro-4-nitrophenyl)dinitro-	20	0.69	$c \approx 0.0001$ (H_o in $HClO_4$ solutions)	06	Uncert.	T58
3850	$C_7H_4O_6N_3Cl$	Methane, (2-chloro-5-nitrophenyl)dinitro-	20	0.78	$c \approx 0.0001$ (H_o in $HClO_4$ solutions)	06	Uncert.	T58
3851	$C_7H_4O_6N_3Cl$	Methane, (3-chloro-6-nitrophenyl)dinitro-	20	0.55	$c \approx 0.0001$ (H_o in $HClO_4$ solutions)	06	Uncert.	T58
3852	$C_7H_4O_6N_3Cl$	Methane, (4-chloro-2-nitrophenyl)dinitro-	20	0.62	$c \approx 0.0001$ (H_o in $HClO_4$ solutions)	06	Uncert.	T58
3853	$C_7H_4O_6N_3Br$	Methane, (3-bromo-6-nitrophenyl)dinitro-	20	0.51	$c \approx 0.0001$ (H_o in $HClO_4$ solutions)	06	Uncert.	T58

No.	Molecular formula, name and pK value(s)	$T(^oC)$	Remarks	Method	Assessment	Ref.
3854	C_7H_5ONS Benzoxazol-2(3H)-thione					F2
	6.33	25				
	6.18	35				
	6.08	45				
3855	C_7H_5ONS Phenylthiocyanate, 4-hydroxy-					B104
	8.57	25		E3ag	Approx.	
3856	$C_7H_5ON_4Cl$ Pteridin-7(8H)-one, 4-chloro-8-methyl-					J4
	12	20	c = 0.00004 in 0.01M buffer	05	Uncert.	
3857	$C_7H_5O_3N_4Cl$ 4-Pteridinecarboxylic acid, 2-chloro-5,6,7,8-tetrahydro-6-oxo-					C59
	-0.19	20	c = 0.005, mixed constants; pK assignment discussed	06	Uncert.	
	2.94			E3bg	Approx.	
	12.31			07	Uncert.	
3858	$C_7H_5O_3NS$ 1,2-Benzisothiazol-3(2H)-one 1,1-dioxide (Saccharin)					D21
	1.31	25	I = 0.2(KCl-HCl)	05	Approx.	
3859	$C_7H_5O_4N_2Br$ Methane, (2-bromophenyl)dinitro-					T58
	1.45	20	c ≈ 0.0001 (H$_0$ in HClO$_4$ solutions)	06	Uncert.	

No.	Formula	Name / pK	Temp.	Concentration	Method		Ref.
3860	$C_7H_5O_5N_2Cl$	Phenol, 4-chloro-3-methyl-2,6-dinitro-					
		3.240	25	$c \approx 0.00003$	05a	Approx.	R30
3861	$C_7H_5O_5N_3S$	1H-Indazole-4-sulfonic acid, 6-hydroxy-7-nitroso-					
		5.85	25	$c = 0.01$	E3bg	Approx.	S129
3862	$C_7H_5O_5N_3S$	1H-Indazole-4-sulfonic acid, 7-hydroxy-6-nitroso-					
		6.35	25	$c = 0.01$	E3bg	Approx.	S129
3863	$C_7H_5O_5N_3S$	1H-Indazole-5-sulfonic acid, 7-hydroxy-4-nitroso-					
		8.60	25	$c = 0.01$	E3bg	Approx.	S129
3864	$C_7H_5O_5N_3S$	1H-Indazole-6-sulfonic acid, 4-hydroxy-7-nitroso-					
		7.15	25	$c = 0.01$	E3bg	Approx.	S129
3865	$C_7H_5O_5N_3S$	1H-Indazole-7-sulfonic acid, 5-hydroxy-4-nitroso-					
		7.75	25	$c = 0.01$	E3bg	Approx.	S129
3866	$C_7H_6O_2NCl$	Benzohydroxamic acid, 2-chloro-					
		pK_1 1.55, pK_2 7.85	20±1	$c = 0.0025$, $I = 1.0(KNO_3)$, mixed constants	E3bg	Uncert.	D64
3867	$C_7H_6O_2NCl$	Benzohydroxamic acid, 4-chloro-					
		pK_1 1.90, pK_2 8.62	20	$c = 0.0025$, $I = 1.0(KNO_3)$, mixed constants	E3bg	Uncert.	D65
		pK_2 8.70	25	$c = 0.01(?)$ (Possible error in detail given)	E3bg	Uncert.	A10
		pK_2 8.6	20	$c = 0.005$	E3d	Uncert.	W39

No.	Molecular formula, name and pK value(s)	$T(^{o}C)$	Remarks	Method	Assessment	Ref.
3868	$C_7H_6O_2NCl$ Benzoic acid, 2-amino-4-chloro- pK$_1$ 1.21 pK$_2$ 4.44	25	c = 0.00029, I = 0.1(KCl), mixed constants	05	Uncert. Approx.	L15
3869	$C_7H_6O_2NCl$ Benzoic acid, 2-amino-5-chloro- pK$_1$ 1.69 pK$_2$ 4.35	25	c = 0.00038, I = 0.1(KCl), mixed constants	05	Uncert. Approx.	L15
3870	$C_7H_6O_2NBr$ Benzohydroxamic acid, 4-bromo- 8.68	25	By interpolation using Hammett equation		Uncert.	A10
3871	$C_7H_6O_2NBr$ Benzoic acid, 2-amino-4-bromo- pK$_1$ 1.24, pK$_2$ 4.37	25	c = 0.00026, I = 0.1(KCl), mixed constants	05	Approx.	L15
3872	$C_7H_6O_2NBr$ Benzoic acid, 2-amino-5-bromo- pK$_1$ 1.60, pK$_2$ 4.41	25	c = 0.00061, I = 0.1(KCl), mixed constants	05	Approx.	L15
3873	$C_7H_6O_2NI$ Benzohydroxamic acid, 4-iodo- 8.63	25	By interpolation using Hammett equation		Uncert.	A10
3874	$C_7H_6O_2NF$ Benzohydroxamic acid, 2-fluoro- pK$_1$ 1.75, pK$_2$ 8.00	20±1	c = 0.0025, I = 1.0(KNO$_3$), mixed constants	E3bg	Uncert.	D64

No.	Formula	Name / pK	Value	Temp	Conditions	Method	Reliability	Ref
3875	$C_7H_6O_2NF$	Benzohydroxamic acid, 4-fluoro-	8.81	25	c = 0.01(?) (Possible error in detail given)	E3bg	Uncert.	A10
			8.70	30.5	c = 0.00006, I = 0.1(KNO$_3$), mixed constant	E3d	Uncert.	S137
			8.70	not stated	c = 0.04, I = 0.1(KCl)	E3bg	Uncert.	H2
3876	$C_7H_6O_2NF$	Benzoic acid, 2-amino-4-fluoro- pK$_1$ 1.42 pK$_2$ 4.60		25	c = 0.00054, I = 0.1(KCl), mixed constants	05	Uncert. Approx.	L15
3877	$C_7H_6O_2NF$	Benzoic acid, 2-amino-5-fluoro- pK$_1$ 2.03, pK$_2$ 4.82		25	c = 0.0009, I = 0.1(KCl), mixed constants	05	Approx.	L15
3878	$C_7H_6O_2NP$	Phenylphosphonous acid, 3-cyano-	1.19	25	c ≈ 0.05, mixed constant	E3bg	Uncert.	Q1
3879	$C_7H_6O_3Cl_3P$	Phenoxymethylphosphinic acid, 2,4,5-trichloro-	0.94	20		05	Approx.	P60
3880	$C_7H_6O_4N_2S$	1H-Indazole-4-sulfonic acid, 5-hydroxy-	10.55	25	c = 0.01	E3bg	Approx.	S129
3881	$C_7H_6O_4N_2S$	1H-Indazole-4-sulfonic acid, 6-hydroxy-	8.85	25	c = 0.01	E3bg	Approx.	S129

No.	Molecular formula, name and pK value(s)	$T({}^{\circ}C)$	Remarks	Method	Assessment	Ref.
3882	$C_7H_6O_4N_2S$ 1H-Indazole-4-sulfonic acid, 7-hydroxy- 7.60	25	c = 0.01	E3bg	Approx.	S129
3883	$C_7H_6O_4N_2S$ 1H-Indazole-5-sulfonic acid, 7-hydroxy- 8.00	25	c = 0.01	E3bg	Approx.	S129
3884	$C_7H_6O_4N_2S$ 1H-Indazole-6-sulfonic acid, 4-hydroxy- 8.10	25	c = 0.01	E3bg	Approx.	S129
3885	$C_7H_6O_4N_2S$ 1H-Indazole-7-sulfonic acid, 4-hydroxy- 7.60	25	c = 0.01	E3bg	Approx.	S129
3886	$C_7H_6O_4N_2S$ 1H-Indazole-7-sulfonic acid, 5-hydroxy- 9.65	25	c = 0.01	E3bg	Approx.	S129
3887	$C_7H_6O_4N_2S$ 1H-Indazole-7-sulfonic acid, 6-hydroxy- 9.45	25	c = 0.01	E3bg	Approx.	S129
3888	$C_7H_6O_4Cl_3P$ Phenoxymethylphosphonic acid, 2,4,5-trichloro- pK₁ 1.26	20		05	Approx.	P60
	pK₂ 6.59		I = 0.02	E3bg		
	pK₂ 6.56		I = 0.1-0.2	05		

No.	Formula	Name	pK	Temp	Conditions	Method	Assessment	Ref.
3889	$C_7H_6O_4Cl_3P$	Phenoxymethylphosphonic acid, 2,4,6-trichloro-	pK_2 7.07	20	$I = 0.02$	E3bg	Approx.	P60
3890	$C_7H_6O_7N_2S_2$	1H-Indazole-5,7-disulfonic acid, 6-hydroxy-	9.40	25	$c = 0.01$	E3bg	Approx.	S129
3891	$C_7H_6N(Cl)S$	Thieno[2,3-c]pyridinium chloride	5.25	20		E3bg	Uncert.	D51
3892	$C_7H_7ON_2Cl$	Benzohydrazide, 3-chloro-	pK_1 2.89	25	Mixed constants	E3bg	Uncert.	T43,T44
			11.95(NH)	25		05	Uncert.	
3893	$C_7H_7ON_2Cl$	Benzohydrazide, 4-chloro-	pK_1 3.02	25	Mixed constants(?)	E3bg	Uncert.	T43,T44
			12.09(NH)	25		05	Uncert.	
3894	$C_7H_7O_2N_2Cl$	1,4-Benzoquinone dioxime 1-(O-methyl ether), 2-chloro-	8.70	not stated	Mixed constant	05	Uncert.	D35
3895	$C_7H_7O_2N_2Cl$	1,4-Benzoquinone dioxime 1-(O-methyl ether), 3-chloro-	8.60	not stated	Mixed constant	05	Uncert.	D35
3896	$C_7H_7O_2N_2Br$	1,4-Benzoquinone dioxime 1-(O-methyl ether), 2-bromo-	8.70	not stated	Mixed constant	05	Uncert.	D35

No.	Molecular formula, name and pK value(s)	T(°C)	Remarks	Method	Assessment	Ref.
3897	$C_7H_7O_2N_2Br$ 1,4-Benzoquinone dioxime 1-(O-methyl ether), 3-bromo-					
	8.60	not stated	Mixed constant	05	Uncert.	D35
3898	$C_7H_7O_3Cl_2P$ Phenoxymethylphosphinic acid, 2,4-dichloro-					
	0.98	20		05	Approx.	P60
3899	$C_7H_7O_4NS$ Benzoic acid, 3-sulfamoyl-					
	3.54	25	$c = 0.004$, $I = 0.1$(KCl), mixed constant	E3bg	Approx.	Z11
3900	$C_7H_7O_4NS$ Benzoic acid, 4-sulfamoyl-					
	3.47	25	$c = 0.04$, $I = 0.1$(KCl), mixed constant	E3bg	Approx.	Z11
3901	$C_7H_7O_4N_4Cl$ Acetic acid, 2-(5-amino-6-carboxy-2-chloro-4-pyrimidinyl)amino- (Pyrimidine, 5-amino-6-carboxy-4-carboxymethylamino-2-chloro-)					
	3.32	20	$c = 0.0005$, mixed constants, pK assignment discussed	E3bg,R3d	Uncert.	C59
	4.44					
3902	$C_7H_7O_4Cl_2P$ Phenoxymethylphosphonic acid, 2,4-dichloro-					
	pK_1 1.30	20		05	Approx.	P60
	pK_2 6.72		$I = 0.02$	E3bg		
	pK_2 6.75		$I = 0.01$–0.02	05		
3903	$C_7H_7O_4Cl_2P$ Phenoxymethylphosphonic acid, 2,6-dichloro-					
	pK_2 7.11	20	$I = 0.02$	E3bg	Approx.	P60

No.	Formula	Compound / pK	T (°C)	Method	Reliability	Ref.
3904	$C_7H_7O_5NS$	Benzoic acid, 2-amino-5-sulfo-			Uncert.	B75
		pK_2 1.67, pK_3 4.61	25	E3d		
		pK_3 4.70	33	E3d		
		pK_3 4.69	20	05		
		pK_3 4.65	30			
		pK_3 4.61	40			
3905	$C_7H_7O_5SP$	Benzoic acid, 2-phosphonothio-			Uncert.	F13
		pK_2 3.21, pK_3 5.69(H_2O)	25	E3bg		
		pK_2 3.48, pK_3 5.90(D_2O)	25	05		
3906	$C_7H_6O_6N_2P$	Phosphoramidic acid, N-(4-nitrobenzoyl)-			Approx.	Z9
		pK_1 1.97, pK_2 6.24	30	E3bg		
		I = 0.1(KCl), mixed constant				
		Solution of monocyclohexylammonium salt used				
3907	$C_7H_8ON(I)$	1-Methylpyridinium iodide, 2-formyl-			Approx.	E18
		9.94	30	E3bg		
		c = 0.01, I = 0.1(KCl), mixed constant				
		(hydrated species)				
3908	$C_7H_8ON(I)$	1-Methylpyridinium iodide, 4-formyl-			Approx.	E18
		10.72	30	E3bg		
		c = 0.01, I = 0.1(KCl), mixed constant				
		(hydrated species)				
3909	$C_7H_8ON_3Cl$	Triazen-3-ol, 1-(2-chlorophenyl)-3-methyl-			Uncert.	D54
		11.31	25	05		
		c = 0.00004, I = 0.1(KCl), mixed constant				

No.	Molecular formula, name and pK value(s)	$T(^oC)$	Remarks	Method	Assessment	Ref.
3910	$C_7H_8ON_3Cl$ Triazen-3-ol, 1-(4-chlorophenyl)-3-methyl- 11.52	25	c = 0.00004, I = 0.1(KCl), mixed constant	05	Uncert.	D54
3911	$C_7H_8ON_4S$ Purin-2(1H)-one, 3,6-dihydro-1,3-dimethyl-6-thioxo- 8.2		Conditions not stated	05	Uncert.	L40
3912	$C_7H_8ON_4S$ Purin-2(1H)-one, 3,6-dihydro-1,7-dimethyl-6-thioxo- 7.5		Conditions not stated	05	Uncert.	L40
3913	$C_7H_8ON_4S$ Purin-2(1H)-one, 3,6-dihydro-1,9-dimethyl-6-thioxo- 5.3		Conditions not stated	05	Uncert.	L40
3914	$C_7H_8ON_4S$ Purin-2(1H)-one, 3,6-dihydro-3,7-dimethyl-6-thioxo- 8.8		Conditions not stated	05	Uncert.	L40
3915	$C_7H_8ON_4S$ Purin-2(1H)-one, 3,6-dihydro-3,9-dimethyl-6-thioxo- 8.8		Conditions not stated	05	Uncert.	L40
3916	$C_7H_8ON_4S$ Purin-2(1H)-one, 1-methyl-6-methylthio- pK_1 2.5, pK_2 8.8		Conditions not stated	05	Uncert.	L41
3917	$C_7H_8ON_4S$ Purin-2(1H)-one, 7-methyl-6-methylthio- pK_1 1.3, pK_2 8.7		Conditions not stated	05	Uncert.	L41

No.	Formula	Name / pK	Temp	Conditions	Method	Assessment	Ref.
3918	$C_7H_8ON_4S$	Purin-2(1H)-one, 9-methyl-6-methylthio- pK$_1$ 1.7, pK$_2$ 6.3	05	Conditions not stated		Uncert.	L41
3919	$C_7H_8ON_4S$	Purin-2(3H)-one, 3-methyl-6-methylthio- pK$_1$ 1.4, pK$_2$ 7.7	05	Conditions not stated		Uncert.	L41
3920	$C_7H_8ON_4S$	Purin-6(1H)-one, 1-methyl-2-methylthio- 9.5	05	Conditions not stated		Uncert.	R18
3921	$C_7H_8ON_4S$	Purin-6(1H)-one, 7-methyl-2-methylthio- 8.0	05	Conditions not stated		Uncert.	R18
3922	$C_7H_8ON_4S$	Purin-6(1H)-one, 9-methyl-2-methylthio- 9.5	05	Conditions not stated		Uncert.	R18
3923	$C_7H_8ON_4S$	Purin-6(3H)-one, 3-methyl-2-methylthio- pK$_1$ 0.0 pK$_2$ 8.7	0 05	Conditions not stated		Uncert. Uncert.	R18
3924	$C_7H_8ON_4S_2$	Purin-6(1H)-one, 2,8-bis(methylthio)- pK$_1$ 1.4, pK$_2$ 7.5, pK$_3$ 10.75	05	Conditions not stated		Uncert.	R18
3925	$C_7H_8O_2ClP$	4-Chlorophenylmethylphosphinic acid 2.39	20	c = 0.005	E3bg	Uncert.	N28

No.	Molecular formula, name and pK value(s)	T(°C)	Remarks	Method	Assessment	Ref.
3926	$C_7H_8O_3N_2S$ Ethyl 1,2,3,4-tetrahydro-4-oxo-2-thioxo-5-pyrimidinecarboxylate (5-Pyrimidinecarboxylic acid ethyl ester, 1,2,3,4-tetrahydro-4-oxo-2-thioxo-)					
		25	Mixed constant	E3d	Uncert.	G2
	6.43	35			V.Uncert.	
	6.40	45			V.Uncert.	
	6.27					
3927	$C_7H_8O_3N_2S$ 5-Pyrimidinecarboxylic acid, 2-ethylthio-1,4-dihydro-4-oxo-					
	pK₁ 6.01, pK₂ 10.52	25	c = 0.002, I = 0.1((CH_3)₄NBr)	E3bg	Approx.	T63
3928	$C_7H_8O_3ClP$ Phenoxymethylphosphinic acid, 4-chloro-					
	1.00	20		05	Approx.	P60
3929	$C_7H_8O_4NP$ Phosphoramidic acid, N-benzoyl-					
	pK₁ 1.99, pK₂ 6.42	30	I = 0.1(KCl), mixed constants	E3bg	Approx.	Z9
			Solutions of monocyclohexyl ammonium salt used			
	pK₃ 14				V.Uncert.	
3930	$C_7H_8O_4N_2S$ 3-Azapentanedioic acid, 3-(2-thiazolyl)- (Iminodiacetic acid, N-(2-thiazolyl)-)					
	pK₃ 9.28, pK₄ 9.81			E	V.Uncert.	D66

No.	Compound / pK	Temp	Conditions	Method	Assessment	Ref
3931	$C_7H_8O_4ClP$ Phenoxymethylphosphonic acid, 2-chloro-					
	pK$_1$ 1.43	20		05	Approx.	P60
	pK$_2$ 6.83		I = 0.02	E3bg		
	pK$_2$ 6.80		I = 0.01-0.02	05		
3932	$C_7H_8O_4ClP$ Phenoxymethylphosphonic acid, 4-chloro-					
	pK$_2$ 6.82	20	I = 0.02	E3bg	Approx.	P60
3933	$C_7H_9ON_2(I)$ 1-Methylpyridinium iodide, 2-hydroxyiminomethyl- (PAM-2)					
	8.0(anti)		c ≈ 0.04	E3bg	Uncert.	W36
	9.9(syn)				V.Uncert.	
	8.0	25	c ≈ 0.005-0.01, mixed constant	E3bg	Uncert.	L65
	7.82	25	c = 0.01, I = 0.1(KCl), mixed constant	E3bg	Approx.	G39
3934	$C_7H_9ON_2(I)$ 1-Methylpyridinium iodide, 3-hydroxyiminomethyl- (PAM-3)					
	9.2		c ≈ 0.04	E3bg	Uncert.	W36
	9.1	25	c ≈ 0.005-0.01, I = 0.5(KNO_3), mixed constant	E3bg	Uncert.	L65
	9.10	25	c = 0.01, I = 0.1(KCl), mixed constant	E3bg	Approx.	G39
3935	$C_7H_9ON_2(I)$ 1-Methylpyridinium iodide, 4-hydroxyiminomethyl- (PAM-4)					
	8.6		c ≈ 0.04	E3bg	Uncert.	W36
	8.5	25	c ≈ 0.005-0.01, I = 0.5(KNO_3), mixed constant	E3bg	Uncert.	L65

(contd)

No.	Molecular formula, name and pK value(s)	T(°C)	Remarks	Method	Assessment	Ref.
3935 (Contd)	$C_7H_9ON_2(I)$ 1-Methylpyridinium iodide, 4-hydroxyiminomethyl- (PAM;-4)					
	8.23	25	c $\approx$ 0.01, I = 0.1(KCl), mixed constant	E3bg	Approx.	G39
3936	$C_7H_9O_2NS$ Benzenesulfonamide, 4-methyl-					
	10.17	20	c = 0.0025, I = 0.1(KCl), mixed constant	E3bg	Approx.	W33
3937	$C_7H_9O_2N_2(I)$ 1-Methylpyridinium iodide, 3-hydroxy-2-hydroxyiminomethyl-					
	4.8(-OH)		c $\approx$ 0.04	E3bg	Uncert.	W36
3938	$C_7H_9O_2N_2(I)$ 1-Methylpyridinium iodide, 2-(N-hydroxycarbamoyl)- (2-Pyridinecarbohydroxamic acid 1-methiodide)					
	5.25	30	c = 0.01, I = 0.1(KCl), mixed constant	E3bg	Approx.	E18
	5.5	not stated	c = 0.04	E3bg	Uncert.	H2
3939	$C_7H_9O_2N_2(I)$ 1-Methylpyridinium iodide, 3-(N-hydroxycarbamoyl)- (3-Pyridinecarbohydroxamic acid 1-methiodide)					
	6.41	30	c = 0.01, I = 0.1(KCl), mixed constant	E3bg	Approx.	E18
	6.5	not stated	c = 0.04	E3bg	Uncert.	H2
3940	$C_7H_9O_2N_2(I)$ 1-Methylpyridinium iodide, 4-(N-hydroxycarbamoyl)- (4-Pyridinecarbohydroxamic acid 1-methiodide)					
	5.95	30	c = 0.01, I = 0.1(KCl), mixed constant	E3bg	Approx.	E18
	6.3		c = 0.04	E3bg	Uncert.	H2
3941	$C_7H_9O_3NS$ Benzenesulfonamide, 4-methoxy-					
	10.22	20	c = 0.0025, I = 0.1(KCl), mixed constant	E3bg	Approx.	W33

No.	Formula	Name / pK values	Temp.	Conditions	Method	Assessment	Ref.
3942	$C_7H_9O_3NS$	Ethanesulfonic acid, 2-(2-pyridyl)-					
		5.14	25	$I = 0.1(NaClO_4)$, mixed constant	E3bg	Approx.	B15
3943	$C_7H_9O_3NS$	Ethyl 2-(4-oxothiazolidin-2-ylidene)acetate					
		8.02(trans)	not stated	$I = 0.01$	05	Uncert.	T14
3944	$C_7H_9O_4NS$	Methanesulfonic acid, α-hydroxy-α-(6-methyl-2-pyridyl)-					
		pK$_1$ 4.83, pK$_2$ 9.88	25	$I = 0.1(NaClO_4)$, mixed constants	E3bg	Approx.	B15
3945	$C_7H_{10}ON_2S$	Pyrimidin-4(1H)-one, 2,3-dihydro-6-propyl-2-thioxo-					
		7.76	25	Mixed constant	E3d	Uncert.	G2
		7.48	35			V.Uncert.	
		7.17	45			V.Uncert.	
		7.80	25	$c = 0.0001$, mixed constant	05	Uncert.	
3946	$C_7H_{10}O_3NP$	Methyl hydrogen 4-aminophenylphosphonate					
		pK$_1$ 3.8	not stated	Mixed constant	E3d	V.Uncert.	B175
3947	$C_7H_{10}O_4NP$	(6-Methylpyridin-2-yl)methyl dihydrogen phosphate					
		pK$_1$ 4.74, pK$_2$ 6.64	25	$I = 0.1(KNO_3)$	E3bg	Approx.	M118
		pK$_1$ 4.50, pK$_2$ 6.36	80				M115
3948	$C_7H_{10}O_4NP$	2-(2-Pyridinyl)ethyl dihydrogen phosphate					
		pK$_1$ 5.31, pK$_2$ 6.83	25	$I = 0.1(KNO_3)$	E3bg	Approx.	M118
		pK$_1$ 4.84, pK$_2$ 6.79	80				M115

No.	Molecular formula, name and pK value(s)	T(°C)	Remarks	Method	Assessment	Ref.
3949	$C_7H_{11}O_6NP_2$ Methylenediphosphonic acid, 1-amino-1-phenyl- pK_1 1.6 pK_2 5.29, pK_3 8.17 pK_4 10.29(NH^+)	25	I = 0.1(KCl), mixed constant	E3bg	Uncert. Approx. Uncert.	D71
3950	$C_7H_{13}O_4NS$ 3-Azapentanedioic acid, 3-(2-methylthioethyl)- (Iminodiacetic acid, N-(2-methylthioethyl)-) pK_1 2.1 pK_2 8.91	20	I = 0.1(KCl), concentration constants	E3bg	Uncert. Approx.	S40
3951	$C_7H_{14}O_3N_2S$ 5-Thiahexanoic acid, 2-(aminoacetylamino)- (Glycyl-DL-methionine) pK_1 3.16, pK_2 8.51	25	I = 0.1($NaClO_4$)	E3bg	Approx.	S139
3952	$C_7H_{14}O_8N_3P$ 3,6-Diazaoctanoic acid, 8-amino-4,7-dioxo-5-phosphonooxymethyl- (Glycyl-O-phosphono-DL-serylglycine) pK_2 3.29, pK_3 5.78, pK_4 8.22	25	I = 0.15(KCl), concentration constants	E3bg	Approx.	024
3953	$C_7H_{16}O_2N(Cl)$ Butanoic acid, 4-trimethylammonio-, chloride 4.02	25	c = 0.01	E3bg	Approx.	Y5
3954	$C_7H_{15}O_3N(Cl)$ Butanoic acid, 3-hydroxy-4-trimethylammonio-, chloride (Carnitine hydrochloride) 3.80	25	c = 0.01	E3bg	Approx.	Y5
3955	$C_7H_{17}O_2N_2(Br)$ Butanohydroxamic acid, 4-trimethylammonio-, bromide 8.60	25(?)	c = 0.02, I = 0.1(KCl), mixed constant	E3bg	Uncert.	M66

No.	Formula	Name	pK	Temp.	Conditions	Assessment	Reliability	Ref.
3956	$C_7H_5O_2N_2(F_4B)$	Benzenediazonium fluoborate, 2-carboxy-	1.47	0	Mixed constant	E3bg	Uncert.	L36
3957	$C_7H_5O_2N_2(F_4B)$	Benzenediazonium fluoborate, 3-carboxy-	2.55	0	Mixed constant	E3bg	Uncert.	L36
3958	$C_7H_5O_2N_2(F_4B)$	Benzenediazonium fluoborate, 4-carboxy-	2.41	0	Mixed constant	E3bg	Uncert.	L36
3959	$C_8H_4O_9$	Furantetracarboxylic acid	pK_1 1.37, pK_2 2.1, pK_3 3.73, pK_4 7.7	not stated	Mixed constants	E3bg	Uncert.	C69
3960	$C_8H_6O_2$	1,2-Benzenedicarbaldehyde	13.00	25	c = 0.0001 pK of hydrated species	07,R4	Approx.	B111
3961	$C_8H_6O_2$	Benzo[b]furan-4-ol	9.06	20	2% ethanol	05	Approx.	D32
3962	$C_8H_6O_3$	Benzoic acid, 3-formyl-	3.84	25		E3bg	Approx.	H58

No.	Molecular formula, name and pK value(s)	T(°C)	Remarks	Method	Assessment	Ref.
3963	$C_8H_6O_3$ Benzoic acid, 4-formyl-					
	3.77	25		E3bg	Approx.	H58
	3.75	25		C2	Approx.	
3964	$C_8H_6O_3$ 1,3,5-Cycloheptatriene-1-carboxylic acid, 7-oxo-					
	3.73	25	c = 0.001	E3bg	Approx.	T5
3965	$C_8H_6O_4$ 1,2-Benzenedicarboxylic acid (Phthalic acid)					
	pK_1 2.97, pK_2 5.43	35	I ≈ 0.03	E3bg,R3d	Approx.	N7
	. pK_1 2.76, pK_2 4.92	25	I = 0.1(KNO_3), concentration constants	E3bg	Approx.	Y10
	pK_1 2.63, pK_2 4.73	25	I = 1.0(KNO_3), concentration constants	E3bg	Approx.	R5
	pK_1 3.05, pK_2 4.89	25	I = 3.0($NaClO_4$)	E3bg	Uncert.	P5
	Values for pK_1 in mixed solvents given in D44					
3966	$C_8H_6O_4$ 1,3-Benzenedicarboxylic acid					
	pK_1 3.70, pK_2 4.60	25	'c = 0.00016, I = 0.014	O5	Approx.	A66
	pK_1 3.0	20	c = 0.005	E3bg	Uncert.	D44
	Values in mixed solvents are also given					
3967	$C_8H_6O_4$ 1,4-Benzenedicarboxylic acid					
	pK_2 4.34	20	I = 0.1(KCl), mixed constant	E3bg	Approx.	W32
	pK_1 3.5	20	c = 0.005	E3bg	Uncert.	D44
	Values in mixed solvents are also given					

No.	Formula	Name / pK	T (°C)	Conditions	Method	Assessment	Ref.
3968	$C_8H_6O_4$	Benzoic acid, 3,4-methylenedioxy- 4.50	25	c = 0.003, 5%(v/v) ethanol	E3bg	Uncert.	B190
3969	$C_8H_6O_4$	1,3,5-Cycloheptatriene-1-carboxylic acid, 6-hydroxy-7-oxo- pK_1 3.20, pK_2 7.98	25	c = 0.001	E3bg	Approx.	T5
3970	$C_8H_6O_4$	1,3,6-Cycloheptatriene-1-carboxylic acid, 4-hydroxy-5-oxo- pK_1 4.05, pK_2 6.41	25	c = 0.001	E3bg	Approx.	T5
3971	$C_8H_6O_4$	1,4,6-Cycloheptatriene-1-carboxylic acid, 4-hydroxy-3-oxo- pK_1 3.42, pK_2 7.03	25	c = 0.001	E3bg	Approx.	T5
3972	C_8H_7N	Indole 16.97(NH)	25	In aq.KOH, H_- scale	07	Uncert.	Y3
3973	C_8H_8O	Acetophenone 19.2			KIN	Uncert.	B47
3974	C_8H_8O	Benzaldehyde, 3-methyl- 15.00	25	c = 0.0001, pK of hydrate	07,R4	Uncert.	B111
3975	C_8H_8O	Benzaldehyde, 4-methyl- 15.39	25	c = 0.0001, pK of hydrate	07,R4	Uncert.	B111

No.	Molecular formula, name and pK value(s)	$T(^{\circ}C)$	Remarks	Method	Assessment	Ref.
3976	$C_8H_8O_2$ Acetic acid, phenyl-					
	4.307	25	m = 0.005-0.02	E1cg,R4	Rel.	K30
	4.305	25		C2,R1e	Rel.	S89
			Variation with temperature			
		15 35 50 60 75 90				
		4.287 4.329 4.374 4.407 4.462 4.529				
			Thermodynamic quantities are derived from the results			
			Equation given for variation of pK with temperature			
	4.311	25		C2,R1a/b	Approx.	F18
			Variation with pressure(bar)			
		1000 2000 3000				
		4.095 3.917 3.762				
	4.20	31	I = 0.1($NaClO_4$), mixed constant	E3bg	Approx.	R11
	4.33(H_2O)	25	I < 0.07	O3	Approx.	B51
	4.79(D_2O)		Based on pK = 5.28 for acetic acid in D_2O			
			Other values in H49 and S88			
3977	$C_8H_8O_2$ Acetophenone, 2-hydroxy-					
	10.22	25		E3bg	Approx.	F25
	10.26	25	c ≈ 0.0002	O5	Approx.	M19
	10.07		I = 0.1, mixed constant			
			(contd)			

No.	Formula	Name / pK	T (°C)	Conditions		Assessment	Ref.
3977 (Contd)	$C_8H_8O_2$	Acetophenone, 2-hydroxy-					
		10.06	20	$I = 0.1-0.15(NaClO_4)$, mixed constant	05	Approx.	P30
3978	$C_8H_8O_2$	Acetophenone, 3-hydroxy-					
		9.25	25	$c \approx 0.0003$, $I \approx 0.01$	05	Approx.	E23
3979	$C_8H_8O_2$	Acetophenone, 4-hydroxy-					
		8.05	25	$c \approx 0.0003$, $I \approx 0.01$	05	Approx.	E23
		8.01	25	$I = 0.1(KCl)$, mixed constant	E3bg	Approx.	H51
		7.87	25	$c \approx 0.0002$, $I = 0.1$, mixed constant	05	Approx.	M19
3980	$C_8H_8O_2$	Benzaldehyde, 3-methoxy-					
		14.61	25	$c = 0.001$, 1% ethanol pK of hydrate	07,R4	Uncert.	B111
3981	$C_8H_8O_2$	Benzaldehyde, 4-methoxy-					
		15.96	25	$c \approx 0.0001$, 1% ethanol pK of hydrate	07,R4	Uncert.	B111
3982	$C_8H_8O_2$	Benzoic acid, 2-methyl-					
		3.98	20	$c \approx 0.001$, 1% ethanol	E3bg	Approx.	P22
		3.93	25	$I = 0.008$	E3bg	Approx.	L76

(contd)

No.	Molecular formula, name and pK value(s)	$T(^{o}C)$	Remarks	Method	Assessment	Ref.
3982 (Contd)	$C_8H_8O_2$ Benzoic acid, 2-methyl-					
	3.898	25		05	Approx.	W38
			Variation with temperature			
		15	20 30 35 40			
		3.868	3.876 3.907 3.912 3.931			
			Thermodynamic quantities are derived from the results			
3983	$C_8H_8O_2$ Benzoic acid, 3-methyl-					
	4.254	25	Value calculated from thermodynamic data given in the reference	05a	Rel.	B91
	4.274	25		05	Approx.	W38
			Variation with temperature			
		15	20 30 35 40			
		4.303	4.286 4.255 4.240 4.238			
			Thermodynamic values are derived from the results			
	4.29	25	I = 0.008	E3bg	Approx.	L76
	4.29	20	c ≈ 0.001 in 1% ethanol	E3bg	Approx.	P26
			Other values in P24 and P27			

No.	Formula	Name / pK	Value	Temp (°C)	Conditions	Method	Rel.	Ref.
3984	$C_8H_8O_2$	Benzoic acid, 4-methyl-	4.373	25		05a		B91
		Value calculated from thermodynamic data given in the reference						
			4.367	25		05	Approx.	W38
		Variation with temperature						
				15	20	30	35	40
				4.389	4.377	4.347	4.333	4.325
		Thermodynamic quantities are derived from the results						
			4.41	25	I = 0.01	E3bg	Approx.	L76
			4.40	20	c ≈ 0.001, 1% ethanol	E3bg	Approx.	P25
			4.37		Conditions not stated	05	Uncert.	S62
3985	$C_8H_8O_2$	2,4,6-Cycloheptatrien-1-one, 2-hydroxy-3-methyl-	7.92	25		E3bg	Uncert.	Y15
3986	$C_8H_8O_2$	2,4,6-Cycloheptatrien-1-one, 2-hydroxy-4-methyl-	7.26	25		E3bg	Uncert.	Y15
3987	$C_8H_8O_2$	2,4,6-Cycloheptatrien-1-one, 2-hydroxy-5-methyl-	7.32	25		E3bg	Uncert.	Y15
3988	$C_8H_8O_3$	Acetic acid, 2-hydroxy-2-phenyl-	pK_1 3.19	25	I = 0.1(KNO_3)	E3bg	Approx.	P80
			pK_1 3.14		I = 1.0(KNO_3), concentration constant			

(contd)

No.	Molecular formula, name and pK value(s)	$T(^oC)$	Remarks	Method	Assessment	Ref.
3988 (Contd)	$C_8H_8O_3$ Acetic acid, 2-hydroxy-2-phenyl-					
	pK$_1$ 2.91, pK$_2$ 16.39			ROT	Uncert.	C84
	pK$_2$ 15.1	20	Conditions not stated	O5	Uncert.	B43
3989	$C_8H_8O_3$ Acetic acid, phenoxy-					
	3.11	20	I = 0.005	E3bg	Approx.	P60
	3.12		I = 0.001	O5		
	3.16	25		E3ag	Approx.	H48
	3.182	25	c ≈ 0.01	E3bg	Approx.	P12
	2.93	25	I = 0.1(NaClO$_4$), concentration constant	E3bg	Approx.	S133
	2.96	31	I = 0.1(NaClO$_4$), mixed constant	E3bg	Approx.	R11
			Values in mixed solvents are also given in H48, P12 and P60			
3990	$C_8H_8O_3$ Acetophenone, 2,3-dihydroxy-					
	pK$_1$ 9.00	25	By extrapolation from values in ethanol/water mixtures for I = 0.002-0.20	E3bg	Uncert.	A77
	pK$_2$ 13.4				V.Uncert.	
	pK$_1$ 8.96	25	Activity corrections based on potentiometric studies in ethanol/water	O5	Uncert.	
	pK$_2$ 13.4			O7	V.Uncert.	
	-6.96(CO)	25	In aq.H$_2$SO$_4$ solutions, H$_0$ scale	O6	Uncert.	

No.	Compound / pK	°C	Remarks		Assessment	Ref.
3991	C$_8$H$_8$O$_3$ Acetophenone, 2,4-dihydroxy-					
	pK$_1$ 7.42	25	By extrapolation from values in ethanol/water for I = 0.002-0.20	E3bg	Uncert.	A77
	pK$_2$ 12.0				V.Uncert.	
	pK$_1$ 7.43	25	Activity corrections based on potentiometric studies in ethanol/water	05 07	Uncert. V.Uncert.	
	pK$_2$ 12.4					
	-5.11(CO)	25	In aq.H$_2$SO$_4$ solutions, H$_0$ scale	06	Uncert.	
	pK$_1$ 7.1	20	1-4% ethanol, mixed constant	05	Uncert.	W24
3992	C$_8$H$_8$O$_3$ Acetophenone, 2,5-dihydroxy-					
	pK$_1$ 9.48	25	By extrapolation from values in ethanol/water for I = 0.002-0.20	E3bg	Uncert.	A77
	pK$_2$ 12.9				V.Uncert.	
	pK$_1$ 9.45	25	Activity corrections based on potentiometric studies in ethanol/water	05 07	Uncert. V.Uncert.	
	pK$_2$ 13.2					
	-6.6(CO)	25	In aq.H$_2$SO$_4$ solutions, H$_0$ scale	06	Uncert.	
3993	C$_8$H$_8$O$_3$ Acetophenone, 2,6-dihydroxy-					
	pK$_1$ 7.94	25	By extrapolation from values in ethanol/water mixtures for I = 0.002-0.20	E3bg	Uncert.	A77
	pK$_2$ 12.35				V.Uncert.	
	pK$_1$ 8.00	25	Activity corrections based on potentiometric studies in ethanol/water	05 07	Uncert. V.Uncert.	
	pK$_2$ 12.70					
	-5.35(CO)	25	In aq.H$_2$SO$_4$ solutions, H$_0$ scale	06	Uncert.	
	pK$_1$ 8.0	20	1-4% ethanol, mixed constant	05	Uncert.	W24

No.	Molecular formula, name and pK value(s)	$T(^{O}C)$	Method	Assessment	Ref.	Remarks
3994	$C_8H_8O_3$ Benzaldehyde, 4-hydroxy-3-methoxy- (Vanillin)					
	7.54	10	O5,R4	Approx.	K17	
	7.40	25				
	7.27	40				
	7.18	55				
						Thermodynamic quantities are derived from the results
3995	$C_8H_8O_3$ Benzoic acid, 2-hydroxy-3-methyl-					
	pK_1 2.95	25	O5	Approx.	E25	c = 0.001
	pK_2 14.6			Uncert.		
	pK_1 3.1	25	E3bg	Uncert.	A1	I = 0.1(KCl), concentration constants
	pK_2 13.85		O5			
	pK_1 3.014	25	O5,R4	Approx.	G11	
						Variation with temperature
						20 30 35 40 50
						3.026 3.005 2.998 2.994 2.993
						Thermodynamic quantities are derived from the results
	pK_1 2.82	25	E3bg	Approx.	P7	I = 0.1($NaClO_4$), mixed constant
						Other values in D56, P24 and P27

3996 $C_8H_8O_3$ Benzoic acid, 2-hydroxy-4-methyl-

Quantity	Value	Temp	Conditions	Method	Uncert.	Ref.
pK_1	3.4	25	$I = 0.1(KCl)$, concentration constants	E3bg	Uncert.	A1
pK_2	13.45			05		
pK_1	3.172	25		05,R4	Approx.	G11

Variation with temperature

	30	35	40	45
	3.161	3.153	3.147	3.144

Thermodynamic quantities are derived from the results

Quantity	Value	Temp	Conditions	Method	Uncert.	Ref.
pK_1	3.29	20	$c = 0.001$, 1% ethanol	E3bg	Approx.	P24,P25
pK_1	3.11	25	$c = 0.00018$, $I = 0.01$	05	Approx.	D56
pK_1	2.97	25	$I = 0.1(NaClO_4)$, mixed constant	E3bg	Approx.	P7

3997 $C_8H_8O_3$ Benzoic acid, 2-hydroxy-5-methyl-

Quantity	Value	Temp	Conditions	Method	Uncert.	Ref.
pK_1	3.15	25	$I = 0.1(KCl)$, concentration constants	E3bg	Uncert.	A1
pK_2	13.35			05		
pK_1	3.03	25	$c = 0.00025$, $I = 0.01$	05	Approx.	D56
pK_1	2.90	25	$I = 0.1(NaClO_4)$, mixed constant	E3bg	Approx.	P7
pK_1	3.14	20	$c = 0.001$, 1% ethanol	E3bg	Approx.	P24,P27

3998 $C_8H_8O_3$ Benzoic acid, 2-hydroxy-6-methyl-

Quantity	Value	Temp	Conditions	Method	Uncert.	Ref.
pK_1	3.32	25	$c = 0.0001-0.001$	C1	Approx.	B123
pK_1	3.53	20	$c \approx 0.001$, 1% ethanol	E3bg	Approx.	P23
pK_1	3.16	25	$I = 0.1(NaClO_4)$, mixed constant	E3bg	Approx.	P7

No.	Molecular formula, name and pK value(s)	T(°C)	Remarks	Method	Assessment	Ref.
3999	$C_8H_8O_3$ Benzoic acid, 3-hydroxy-2-methyl- pK_1 3.83	20	c ≈ 0.001, 1% ethanol	E3bg	Approx.	P22
4000	$C_8H_8O_3$ Benzoic acid, 3-hydroxy-4-methyl- 4.32	20	c = 0.001, 1% ethanol	E3bg	Approx.	P24
4001	$C_8H_8O_3$ Benzoic acid, 5-hydroxy-2-methyl- pK_1 3.92	20	c ≈ 0.001, 1% ethanol	E3bg	Approx.	P22
4002	$C_8H_8O_3$ Benzoic acid, 4-hydroxy-2-methyl- pK_1 4.71	20	c ≈ 0.001, 1% ethanol	E3bg	Approx.	P23
4003	$C_8H_8O_3$ Benzoic acid, 4-hydroxy-3-methyl- pK_1 4.68	20	c ≈ 0.001, 1% ethanol	E3bg	Approx.	P24,P26
4004	$C_8H_8O_3$ Benzoic acid, 2-methoxy- 3.90	25	I = 0.1(NaClO_4), concentration constant	E3bg	Approx.	Y12
	4.13	20	c ≈ 0.001, 1% ethanol	E3bg	Approx.	P24,P25
4005	$C_8H_8O_3$ Benzoic acid, 3-methoxy- 4.095	25	Value calculated from thermodynamic data given in the reference	O5a	Rel.	B91
	4.086	25		C2,R1a/b	Approx.	F18

(contd)

4005 (Contd) $C_3H_8O_3$ Benzoic acid, 3-methoxy-

Variation with pressure(bar)

500	1000	1500	2000	2500	3000
3.995	3.911	3.834	3.759	3.691	3.625

4006 $C_8H_8O_3$ Benzoic acid, 4-methoxy-

4.496	25	05a	Rel.	B91

Value calculated from thermodynamic data given in the reference

4.511	25	C2,R1a/b	F18

Variation with pressure(bar)

500	1000	1500	2000	2500	3000
4.411	4.320	4.237	4.156	4.083	4.013

4007 $C_8H_8O_3$ 2,4,6-Cycloheptatrien-1-one, 2-hydroxy-3-methoxy-

8.00	25	E3bg	Uncert.	Y16

4008 $C_8H_8O_3$ 2,4,6-Cycloheptatrien-1-one, 2-hydroxy-4-methoxy-

7.24	25	E3bg	Uncert.	Y16

4009 $C_8H_8O_3$ 2,4,6-Cycloheptatrien-1-one, 2-hydroxy-5-methoxy-

7.75	25	E3bg	Uncert.	Y16

4010 $C_8H_8O_3$ Methyl 2-hydroxybenzoate (Methyl salicylate)

9.87	20	I = 0.1-0.15($NaClO_4$), mixed constant	05	Approx.	P30
10.19	25	I = 1.0($NaClO_4$), mixed constant	05	Uncert.	A11

No.	Molecular formula, name and pK value(s)	$T(^oC)$	Remarks	Method	Assessment	Ref.
4011	$C_8H_8O_4$ Acetic acid, 2-(3,4-dihydroxyphenyl)- pK_1 4.25, pK_2 9.44 pK_3 12.0	30	$c = 0.004$, $I = 0.1(NaClO_4)$, mixed constants	E3bg	Approx. Uncert.	A80
4012	$C_8H_8O_4$ Acetic acid, 2-(2-hydroxyphenoxy)- 3.02	25	$c \approx 0.01$	E3bg	Approx.	B114
4013	$C_8H_8O_4$ Benzoic acid, 2-hydroxy-4-methoxy- 3.31	25	$c = 0.00014$, $I = 0.01$	05	Approx.	D56
4014	$C_8H_8O_4$ Benzoic acid, 2-hydroxy-5-methoxy- 2.94	25	$c = 0.0002$, $I = 0.01$	05	Approx.	D56
4015	$C_8H_8O_4$ Benzoic acid, 3-hydroxy-4-methoxy- pK_1 4.47, pK_2 10.10	(Isovanillic acid) 25		05	Approx.	N35
4016	$C_8H_8O_4$ Benzoic acid, 4-hydroxy-3-methoxy- pK_1 4.51, pK_2 9.39 pK_1 4.47 pK_1 4.43 pK_1 4.39 pK_1 4.44 pK_1 4.48	(Vanillic acid) 25 25 30 40 50 60	Thermodynamic quantities are derived from the results	05 E3bg	Approx. Approx.	N35 C35

No.	Formula	Name / pK values	T (°C)	Conditions			Ref.
4017	$C_8H_8O_4$	1,4-Cyclohexadiene-1,2-dicarboxylic acid pK_1 3.64, pK_2 4.90	20	$c = 0.001$, mixed constants	E3bg	Uncert.	M22
4018	$C_8H_8O_4$	2,4-Cyclohexadiene-1,2-dicarboxylic acid pK_1 4.08, pK_2 5.02	20	$c = 0.001$, mixed constants	E3bg	Uncert.	M22
4019	$C_8H_8O_4$	2,5-Cyclohexadiene-1,2-dicarboxylic acid pK_1 4.06, pK_2 5.03	20	$c = 0.001$, mixed constants	E3bg	Uncert.	M22
4020	$C_8H_8O_4$	Methyl 2,3-dihydroxybenzoate pK_1 9.10, pK_2 12.94	20	$I = 0.1$, mixed constants(?)	E3bg	Uncert.	S128
4021	$C_8H_8O_4$	Methyl 2,4-dihydroxybenzoate pK_1 8.02, pK_2 11.73	20	$I = 0.1$, mixed constants(?)	E3bg	Uncert.	S128
		pK_1 7.9	20	1-4% ethanol, mixed constant	05	Uncert.	W24
4022	$C_8H_8O_4$	Methyl 2,5-dihydroxybenzoate pK_1 9.63, pK_2 12.02	20	$I = 0.1$, mixed constants(?)	E3bg	Uncert.	S128
4023	$C_8H_8O_4$	Methyl 2,6-dihydroxybenzoate pK_1 8.96, pK_2 11.56	20	$I = 0.1$, mixed constants(?)	E3bg	Uncert.	S128
		pK_1 8.7	20	1-4% ethanol, mixed constant	05	Uncert.	W24
4024	$C_8H_8O_4$	Methyl 3,4-dihydroxybenzoate pK_1 8.18, pK_2 13.04	20	$I = 0.1$, mixed constants(?)	E3bg	Uncert.	S128

No.	Molecular formula, name and pK value(s)	T(°C)	Remarks	Method	Assessment	Ref.
4025	$C_8H_8O_4$ Methyl 3,5-dihydroxybenzoate					
	pK$_1$ 8.71, pK$_2$ 10.66	20	I = 0.1, mixed constants(?)	E3bg	Uncert.	S128
4026	$C_8H_8O_4$ 4H-Pyran-3-carboxylic acid, 2,6-dimethyl-4-oxo-					
	3.36	25		C1	Approx.	S10
4027	$C_8H_8O_4$ 2H-Pyran-2-one, 3-acetyl-4-hydroxy-6-methyl-		(Dehydroacetic acid)			
	5.53	20	c = 0.001-0.0025	E3bg	Approx.	A37
	5.12	25	I = 0.15, mixed constant			A1a
4028	$C_8H_8O_5$ Ethyl 5-hydroxy-4-oxo-4H-pyran-2-carboxylate		(4H-Pyran-2-carboxylic acid ethyl ester, 5-hydroxy-4-oxo-)			
	6.8	room temp.	Mixed constant	E3bg	Uncert.	E9
4029	$C_8H_8O_5$ Methyl 3,4,5-trihydroxybenzoate					
	pK$_1$ 7.88	25	I = 0.1(KNO$_3$), mixed constant	E3bg	Approx.	E19
4030	$C_8H_8N_2$ Benzimidazole, 2-methyl-					
	13.18(NH)	25	In aq. KOH, H_ scale	07	Uncert.	Y2
	13.18	25	c ≈ 0.001	07	Uncert.	T21
4031	$C_8H_{10}O$ Phenol, 2-ethyl-					
	10.20	25	I = 0.1(KCl), mixed constant	E3bg	Approx.	E20
	10.47	20	2% ethanol	05	Approx.	D29
	10.27	25	I = 0.1(KCl), mixed constant	E3bg	Approx.	H51
			(contd)			

381

I.C.O.A.A.—N*

4031 C₈H₁₀O Phenol, 2-ethyl-
(Contd)

pK	t	Notes	Method	Assessment	Ref.
4.5		pK of electronically excited singlet state, calculated assuming pK = 10.2 in ground state	FLU/05	V.Uncert.	B25

4032 C₈H₁₀O Phenol, 3-ethyl-

pK	t	Notes	Method	Assessment	Ref.
10.069	25	m = 0.00018	05a	Rel.	B97

Variation with temperature

5	10	15	20	30	35	40	45	50
10.374	10.289	10.210	10.137	10.006	9.949	9.895	9.846	9.802

Thermodynamic quantities are derived from the results

pK	t	Notes	Method	Assessment	Ref.
10.17	20	2% ethanol	05	Approx.	D29
10.07	25		05	Approx.	W11
4.5		pK of electronically excited singlet state, calculated assuming pK = 9.9 in ground state	FLU/05	V.Uncert.	B25

4033 C₈H₁₀O Phenol, 4-ethyl-

pK	t	Notes	Method	Assessment	Ref.
10.21	25		05	Approx.	W11
10.38	20	2% ethanol	05	Approx.	D29
10.18	25	I = 0.1(KCl), mixed constant	E3bg	Approx.	H51
4.3		pK of electronically excited singlet state, calculated assuming pK = 10.0 in ground state	FLU/05	Uncert.	B25

No.	Molecular formula, name and pK value(s)	T(°C)	Remarks	Method	Assessment	Ref.
4034	C$_8$H$_{10}$O Phenol, 2,3-dimethyl-					
	10.544	25	c = 0.00004	01	Approx.	C40
			Variation with temperature			
			5 10 15 20 30 38			
			10.863 10.772 10.693 10.613 10.477 10.376			
			Thermodynamic quantities are derived from the results			
	10.54	25	c = 0.00017	05	Approx.	H32
	10.57	20		05		R22
4035	C$_8$H$_{10}$O Phenol, 2,4-dimethyl-					
	10.595	25	c = 0.00004	01	Approx.	C40
			Variation with temperature			
			5 10 15 20 30 38			
			10.916 10.824 10.748 10.668 10.529 10.424			
			Thermodynamic quantities are derived from the results			
	10.60	25	c = 0.00018	05	Approx.	H32
	10.63	20		05		R22
4036	C$_8$H$_{10}$O Phenol, 2,5-dimethyl-					
	10.41	25	c = 0.00016	05	Approx.	H32
	10.404	25	c = 0.00004	01	Approx.	C40

(contd)

4036 $C_8H_{10}O$ Phenol, 2,5-dimethyl-
(Contd)

Variation with temperature

5	10	15	20	30	38
10.712	10.623	10.549	10.472	10.335	10.235

Thermodynamic quantities are derived from the results

20	10.46	05	R22

4037 $C_8H_{10}O$ Phenol, 2,6-dimethyl-

20	10.65	E3bg	Approx.	D31	$c = 0.005\text{-}0.01$
25	10.615	01	Approx.	C40	$c = 0.00004$

Variation with temperature

5	10	15	20	30	38
10.920	10.835	10.757	10.684	10.549	10.456

Thermodynamic quantities are derived from the results

25	10.63	05	Approx.	H32	$c = 0.00016$
25	10.59	E3bg	Approx.	F16	$c = 0.0064$
20	10.66	05	R22		

4038 $C_8H_{10}O$ Phenol, 3,4-dimethyl-

25	10.36	05	Approx.	H32	$c = 0.00015$
25	10.356	01	Approx.	C40	$c = 0.00004$

Variation with temperature

5	10	15	20	30	38
10.656	10.569	10.498	10.422	10.291	10.197

Thermodynamic quantities are derived from the results

(contd)

No.	Molecular formula, name and pK value(s)	T(°C)	Remarks	Method	Assessment	Ref.
4038 (Contd)	$C_8H_{10}O$ Phenol, 3,4-dimethyl-					
	10.43	20		05		R22
4039	$C_8H_{10}O$ Phenol, 3,5-dimethyl-					
	10.19	25	c = 0.00017	05	Approx.	H32
	10.203	25	c = 0.00004	01	Approx.	C40
			Variation with temperature			
			5　　10　　15　　20　　30　　38			
			10.500　10.416　10.343　10.270　10.140　10.044			
			Thermodynamic quantities are derived from the results			
	10.23	20		05		R22
4040	$C_8H_{10}O_2$ Ethanol, 2-phenoxy-					
	15.1	25		KIN	Uncert.	M126
4041	$C_8H_{10}O_2$ Phenol, 2-ethoxy-					
	10.109	25	m = 0.00017	05a	Rel.	B94
			Variation with temperature			
			5　　10　　15　　20　　30　　35　　40　　45　　50			
			10.447　10.355　10.269　10.186　10.034　9.964　9.898　9.834　9.775			
			Thermodynamic quantities are derived from the results			

No.	Formula	Compound / pK	Temp (°C)	Conditions	Method	Assessment	Ref.
4042	$C_8H_{10}O_2$	Phenol, 3-ethoxy-					
		9.655	25	m = 0.00025	05a	Rel.	B92

Variation with temperature

5	10	15	20	25	30	35	40	45	50	55	60
9.953	9.868	9.791	9.720		9.595	9.540	9.491	9.446	9.406	9.369	9.337

Thermodynamic quantities are derived from the results

No.	Formula	Compound / pK	Temp (°C)	Conditions	Method	Assessment	Ref.
		9.54	25		05	Approx.	W11
4043	$C_8H_{10}O_2$	Phenol, 4-ethoxy-					
		10.13	25		05	Approx.	W11
4044	$C_8H_{10}O_2$	Phenol, 2-methoxy-3-methyl-					
		9.74	20	I = 0.1, mixed constant(?)	E3bg	Uncert.	S128
4045	$C_8H_{10}O_2$	Phenol, 2-methoxy-4-methyl-					
		10.28	25		05a,R4	Approx.	L49
		10.21	20	I = 0.1, mixed constant(?)	E3bg	Uncert.	S128
4046	$C_8H_{10}O_2$	Phenol, 2-methoxy-5-methyl-					
		9.93	20	I = 0.1, mixed constant(?)	E3bg	Uncert.	S128
4047	$C_8H_{10}O_2$	Phenol, 2-methoxy-6-methyl-					
		10.36	25	I = 0.015	E3bg	Approx.	L48
		10.21	20	I = 0.1, mixed constant(?)	E3bg	Uncert.	S128

No.	Molecular formula, name and pK value(s)	T(°C)	Remarks	Method	Assessment	Ref.
4048	$C_8H_{10}O_2$ Phenol, 5-methoxy-2-methyl-					
	9.38	20	I = 0.1, mixed constant(?)	E3bg	Uncert.	S128
4049	$C_8H_{10}O_2$ Phenol, 4-methoxy-2-methyl-					
	10.31	10	I = 0.1(KCl)	O5	Approx.	S111
	10.21	20	Thermodynamic quantities are derived from			
	10.12	30	the results			
	10.21	20	I = 0.1, mixed constant(?)	E3bg	Uncert.	S128
4050	$C_8H_{10}O_2$ Phenol, 4-methoxy-3-methyl-					
	10.43	10	I = 0.1(KCl)	O5	Approx.	S111
	10.31	20	Thermodynamic quantities are derived from			
	10.18	30	the results			
	10.31	20	I = 0.1, mixed constant(?)	E3bg	Uncert.	S128
4051	$C_8H_{10}O_2$ Tricyclo[3.1.1.0(6,7)]heptane-6-carboxylic acid					
	4.6		Conditions not stated	O5	Uncert.	C66
4052	$C_8H_{10}O_3$ 1,3,5-Cyclohexanetrione, 2,2-dimethyl-					
	pK$_1$ 4.20, pK$_2$ 9.73	25		E3bh	Approx.	S47

4053 $C_8H_{10}O_3$ Phenol, 2,4-bis(hydroxymethyl)-

9.69	25	c = 0.1	Approx.	E3d	Z4

Variation with temperature

20	30	35	40	45	50	55	60
9.75	9.63	9.57	9.52	9.46	9.41	9.36	9.31

Thermodynamic quantities are derived from these results

4054 $C_8H_{10}O_3$ Phenol, 3,5-dimethoxy-

9.345	25	m = 0.00032, I = 0.02-0.06	Rel.	05a	B93

Variation with temperature

10	15	20	30	35	40	45	50	55	60
9.517	9.458	9.402	9.292	9.237	9.185	9.136	9.086	9.038	8.988

Thermodynamic quantities are derived from the results

4055 $C_8H_{10}O_3$ Phenol, 2-hydroxymethyl-6-methoxy-

9.97	25	I = 0.01	Approx.	E3bg	L48

4056 $C_8H_{10}O_4$ 1-Cyclohexene-1,2-dicarboxylic acid

pK_1 3.75, pK_2 5.08	20	c = 0.001, mixed constants	Uncert.	E3bg	M22
pK_1 3.3, pK_2 6.2	20		Uncert.	E3bg	E1

4057 $C_8H_{10}O_4$ 1-Cyclohexene-1,6-dicarboxylic acid

pK_1 4.34, pK_2 5.31	20	c = 0.001, mixed constants	Uncert.	E3bg	M22

No.	Molecular formula, name and pK value(s)	$T(^oC)$	Remarks	Method	Assessment	Ref.
4058	C$_8$H$_{10}$O$_4$ 4-Cyclohexene-1,2-dicarboxylic acid					
	pK$_1$ 3.75, pK$_2$ 6.77(cis)	25	c $\approx$ 0.006	E3bg,R3d	Approx.	C55
	pK$_1$ 3.89, pK$_2$ 5.82(trans)					
	pK$_1$ 4.17, pK$_2$ 5.80	20	c = 0.001, mixed constants	E3bg	Uncert.	M22
4059	C$_8$H$_{10}$O$_4$ 1-Cyclopentene-1-carboxylic acid, 2-methoxycarbonyl-					
	2.94	25	c = 0.005-0.009	E3bg	Approx.	M11
4060	C$_8$H$_{10}$O$_4$ 4-Oxabicyclo[3.3.0]octane-1-carboxylic acid, 3-oxo-					
	3.99(cis)	20	c = 0.001-0.002	E3bg	Approx.	I4
4061	C$_8$H$_{10}$O$_5$ 7-Oxabicyclo[2.2.1]heptane-2,3-dicarboxylic acid					
	pK$_1$ 4.07, pK$_2$ 6.32(exo cis)	25	c $\approx$ 0.006	E3bg,R3d	Approx.	C55
	pK$_1$ 3.46, pK$_2$ 5.13(trans)					
4062	C$_8$H$_{10}$N$_4$ Purine, 2,6,8-trimethyl-					
	pK$_1$ 4.49, pK$_2$ 9.90	20	c < 0.0001, I = 0.01	O5	Approx.	B138
4063	C$_8$H$_{12}$O$_2$ Bicyclo[2.2.1]heptane-1-carboxylic acid					
	4.89	25	c = 0.01	E3d	Approx.	W27
	4.876	25		E3bg	Approx.	W29

Values in mixed solvents are also given

No.	Formula	Name	pK	Conditions	Temp.	Method	Assessment	Ref.
4064	$C_8H_{12}O_2$	1,3-Cyclohexanedione, 5,5-dimethyl- (Dimedone)	5.27		25	C	Approx.	B49
			5.25		25	E3bh	Approx.	S47
4065	$C_8H_{12}O_2$	Cyclohexanone, 2-acetyl-	10.09	$c = 0.004$	20	E3bg	Uncert.	S42
4066	$C_8H_{12}O_2$	2-Hexynoic acid, 5,5-dimethyl-	2.53	$c = 0.01$, $I = 0.1$(NaCl), mixed constant	25	E3bg	Approx.	M48
4067	$C_8H_{12}O_3$	Propanoic acid, 3-(2-oxocyclopentyl)-	4.71	$c = 0.001\text{-}0.002$	20	E3bg	Approx.	I4
4068	$C_8H_{12}O_4$	Butanoic acid, 4-(2-oxo-3-oxolanyl)-	4.76	$c = 0.001\text{-}0.002$	20	E3bg	Approx.	I4
4069	$C_8H_{12}O_4$	Butenedioic acid, diethyl-	pK_1 3.3, pK_2 6.2		20	E3bg	Uncert.	E1
		pK_1 was corrected for the acid-anhydride equilibrium						
4070	$C_8H_{12}O_4$	1,3-Cyclobutanedicarboxylic acid, 2,2-dimethyl- (Norpinic acid)	pK_1 4.34, pK_2 5.51(cis)	$c \approx 0.0005$	25-30	E3bg,R3d	Uncert.	H52
			pK_1 4.44, pK_2 5.56(trans)					

No.	Molecular formula, name and pK value(s)	T(°C)	Remarks	Method	Assessment	Ref.
4071	$C_8H_{12}O_4$ 1,2-Cyclohexanedicarboxylic acid					
	pK$_1$ 4.44, pK$_2$ 6.89(cis)	25	c ≈ 0.0005, mixed constants	E3bg,R3d	Approx.	S68
	pK$_1$ 4.30, pK$_2$ 6.06(trans)	25	c ≈ 0.0005, mixed constants			
	pK$_1$ 4.25, pK$_2$ 6.74(cis)	25	c ≈ 0.006	E3bg,R3d	Approx.	C55
	pK$_1$ 4.10, pK$_2$ 5.96(trans)					
4072	$C_8H_{12}O_4$ 1-r,2-c-Cyclopropanedicarboxylic acid, 1,2-t,3-t-trimethyl-					
	pK$_1$ 4.23, pK$_2$ 6.74	20		E3bg	Approx.	M12
4073	$C_8H_{12}O_4$ 1-r,2-t-Cyclopropanedicarboxylic acid, 1,2-c,3-c-trimethyl-					
	pK$_1$ 3.77, pK$_2$ 5.42	20		E3bg	Approx.	M12
4074	$C_8H_{12}O_5$ Diethyl 2-oxobutanedioate					
	7.49			E3d		R51
4075	$C_8H_{12}O_6$ Pentane-1,3,5-tricarboxylic acid					
	pK$_1$ 3.99, pK$_2$ 4.88, pK$_3$ 5.63	25		E1e	Approx.	A78
4076	$C_8H_{14}O_2$ 2,4-Heptanedione, 5-methyl-					
	9.30	5	c = 0.0005, 0.5-1.0% ethanol. pK for keto-enol mixture	O5	Apparent Approx.	C4
	9.20	25				C5
	9.11	45				C4

Thermodynamic quantities for the keto and enol forms are derived from the results (contd)

No.	Formula	Name / pK	t(°C)	Conditions	Method	Uncert.	Ref.
4076 (Contd)	$C_8H_{14}O_2$	2,4-Heptanedione, 5-methyl- 8.54(enol), 9.10(keto)		Calculated from absorbance data		Uncert.	C5
4077	$C_8H_{14}O_2$	2,4-Heptanedione, 6-methyl- 8.95	20	c = 0.01	E3bg/05	Approx.	R52
4078	$C_8H_{14}O_2$	2,4-Hexanedione, 3,5-dimethyl- 11.80	20	c = 0.01	E3bg/05	Approx.	R52
4079	$C_8H_{14}O_2$	2,4-Hexanedione, 5,5-dimethyl-					
		10.00	20	c = 0.00002, I = 0.1($NaClO_4$)	05	Approx.	K49
		9.94	20	c = 0.01	E3bg/05	Approx.	R52
		10.11	15	c = 0.01	E3bg	Approx.	L3a
		10.01	25	Thermodynamic quantities are derived			
		9.89	45	from the results			
4080	$C_8H_{14}O_2$	2,4-Pentanedione, 3-isopropyl- 12.85	20	c = 0.01	E3bg/05	Uncert.	R52
4080a	$C_8H_{14}O_3$	Cycloheptanecarboxylic acid, 1-hydroxy- 4.210	25		E3ag	Approx.	P77

Variation of the concentration constant with ionic strength

I	2.00	1.00	0.50	0.20	0.10	0.05	0.02	0.01
pK_c	4.102	3.983	3.959	3.987	4.024	4.062	4.107	4.133

No.	Molecular formula, name and pK value(s)	T(°C)	Remarks	Method	Assessment	Ref.
4081	C$_8$H$_{14}$O$_3$ Cyclohexanecarboxylic acid, 4-methoxy-					
	4.66(trans)	25	c = 0.01	E3bg	Approx.	S69
	Values in mixed solvents are also given					
4082	C$_8$H$_{14}$O$_3$ Ethyl 2-ethyl-3-oxobutanoate					
	13.0	23		E3bg	Uncert.	R53
	12.87	25.7		05		
4083	C$_8$H$_{14}$O$_3$ Ethyl 2-methyl-3-oxopentanoate					
	13.06	25.7		05	Uncert.	R53
4084	C$_8$H$_{14}$O$_4$ Butanedioic acid, tetramethyl-					
	pK$_1$ 3.56, pK$_2$ 7.41(H$_2$O)	25		E3bg	Approx.	G22
	pK$_1$ 4.20, pK$_2$ 7.75(D$_2$O)					
4085	C$_8$H$_{14}$O$_4$ Octanedioic acid					
	pK$_1$ 4.52	20	c = 0.005	E3bg,R3a	Approx.	D43
	pK$_2$ 5.5				Uncert.	
	Values in mixed solvents are also given					
	pK$_1$ 4.53, pK$_2$ 5.52	25	I = 0.1(KNO$_3$)	E3bg		N32
4086	C$_8$H$_{14}$O$_4$ Pentanedioic acid, 3-isopropyl-					
	pK$_1$ 4.28, pK$_2$ 5.51	25	c = 0.00125, mixed constant	E3bg,R3c	Approx.	B146

No.	Compound		T	Method		Ref.
4087	$C_8H_{14}O_4$	Pentanedioic acid, 3-propyl-				
	pK_1 4.32, pK_2 5.46		25	E3bg,R3c	Approx.	B146
		$c = 0.00125$, mixed constants				
4088	$C_8H_{14}O_4$	Propanedioic acid, 2-ethyl-2-isopropyl-				
	pK_1 2.03, pK_2 8.10		25	E3bg	Approx.	M77
		$I = 0.1(KCl)$, mixed constants				
	pK_1 2.18, pK_2 7.93		37	E3bg	Uncert.	L33
		Mixed constants				
	pK_1 2.02, pK_2 8.29(H_2O)		25	E3bg	Approx.	G23
	pK_1 2.52, pK_2 8.73(D_2O)					
4089	$C_8H_{14}O_5$	3-Oxapentanedioic acid, 2,2,4,4-tetramethyl-				
	pK_1 3.45, pK_2 5.32		20	E3bg,R3c	Approx.	K47
		$c = 0.005-0.01$				
4090	$C_8H_{14}O_5$	Propanedioic acid, 2-hydroxy-2-isopentyl-				
	pK_1 2.72, pK_2 4.78					G37
4091	$C_8H_{14}O_8$	Ethyl hydrogen mucate				
	3.64			E3		M26
4092	$C_8HO_2F_7$	Benzoic acid, 2,3,5,6-tetrafluoro-4-trifluoromethyl-				
	1.44		25	E3bg	Approx.	R56
		$c = 0.02$				
4093	$C_8H_2O_4Cl_4$	1,2-Benzenedicarboxylic acid, tetrachloro-				
	pK_2 2.94		20	E3bg	Approx.	W16
		$c = 0.002$				
4094	$C_8H_4O_2F_4$	Benzoic acid, 2,3,5,6-tetrafluoro-4-methyl-				
	2.00		25	E3bg	Approx.	R56
		$c = 0.02$				

No.	Molecular formula, name and pK value(s)	$T(°C)$	Remarks	Method	Assessment	Ref.
4095	C_8H_5ON Benzenecarbonitrile, 3-formyl- 13.26	25	c = 0.0001, 1% ethanol pK of hydrate	07,R4	Approx.	B111
4096	C_8H_5ON Benzenecarbonitrile, 4-formyl- 13.06	25	c = 0.0001, 1% ethanol pK of hydrate	07,R4	Approx.	B111
4097	$C_8H_5OF_3$ Acetophenone, α,α,α-trifluoro- 10.00	25	I ≃ 0.1 pK of hydrate	05	Approx.	S116
4098	$C_8H_5O_2N$ 1,3,6-Cycloheptatriene-1-carbonitrile, 4-hydroxy-5-oxo- 3.72	25	I = 2.0	0		O14
4099	$C_8H_5O_2N$ 1,4,6-Cycloheptatriene-1-carbonitrile, 2-hydroxy-3-oxo- 3.41	25	I = 2.0	05	Approx.	O15
4100	$C_8H_5O_2Br_3$ 2,4,6-Cycloheptatrien-1-one, 3,5,7-tribromo-2-hydroxy-6-methyl- 5.17	25		E3bg	Uncert.	Y15
4101	$C_8H_5O_3N$ 1,2-Benzenedicarboximide, N-hydroxy- (Phthalimide, N-hydroxy-) 7.0	30		E3bg	Uncert.	S117

No.	Formula	Name	pK	T(°C)	Conditions	Method	Assessment	Ref.
4102	$C_8H_5O_3N$	Benzoic acid, 3-cyano-6-hydroxy-	2.38	25	$c = 0.00025$, $I = 0.01$	05	Approx.	D56
4103	$C_8H_5O_3N$	Benzoic acid, 4-cyano-2-hydroxy-	2.35	25	$c = 0.0002$, $I = 0.01$	05	Approx.	D56
4104	$C_8H_5O_3Cl_3$	Phenoxyacetic acid, 2,3,4-trichloro-	2.60	20	$I = 0.001$	05	Approx.	P60
		Value in mixed solvent is also given						
4105	$C_8H_5O_3Cl_3$	Phenoxyacetic acid, 2,4,5-trichloro-	2.833	25	$c = 0.00026\text{-}0.00099$	C2,R1c	Rel.	M58
			2.85	20	$I = 0.001$	05	Approx.	P60
		Value in mixed solvent is also given						
4106	$C_8H_5O_4Cl$	1,2-Benzenedicarboxylic acid, 3-chloro- pK_1 <2, pK_2 4.49		25	$I = 0.1(KNO_3)$, concentration constants	E3bg	Approx.	Y10
4107	$C_8H_5O_4Br$	1,2-Benzenedicarboxylic acid, 3-bromo- pK_1 2.27, pK_2 4.35		25	$I = 0.1(KNO_3)$, concentration constants	E3bg	Approx.	Y10
4108	$C_8H_5O_4Br$	1,2-Benzenedicarboxylic acid, 4-bromo- pK_1 2.50, pK_2 4.60		25	$I = 0.1(KNO_3)$, concentration constants	E3bg	Approx.	Y10

No.	Molecular formula, name and pK value(s)	T(°C)	Remarks	Method	Assessment	Ref.
4109	$C_8H_5O_6N$ 1,2-Benzenedicarboxylic acid, 3-nitro-					
	pK$_1$ 2.11, pK$_2$ 4.48	35	I ≈ 0.03	E3bg,R3d	Approx.	N7
	pK$_1$ <2, pK$_2$ 3.93	25	I = 0.1(KNO$_3$), concentration constants	E3bg	Approx.	Y10
4110	$C_8H_5O_6N$ 1,2-Benzenedicarboxylic acid, 4-nitro-					
	pK$_1$ <2, pK$_2$ 4.12	25	I = 0.1(KNO$_3$), concentration constants	E3bg	Approx.	Y10
4111	$C_8H_5O_6N_5$ Pyrimidine-2,4,6(1H,3H,5H)-trione, 5-(1,2,3,4-tetrahydro-6-hydroxy-2,4-dioxo-5-pyrimidinyl)imino- (Purpuric acid)					
	(Murexide = ammonium salt)					
	pK$_2$ 9.2, pK$_3$ 10.9		I = 0.005-0.1	O5	Uncert.	S44
4112	C_8H_6OS Benzo[b]thiophene-4-ol					
	9.18	20	2% ethanol	O5	Approx.	D32
4113	$C_8H_6O_2N_2$ Benzohydroxamic acid, 4-cyano-					
	8.26	25	c = 0.01(?) (Possible error in detail given)	E3bg	Approx.	A10
	8.16	30.5	c = 0.0006, I = 0.1(KNO$_3$), mixed constant	E3d	Uncert.	S137
	8.16	not stated	c = 0.04, I = 0.1(KCl)	E3bg	Uncert.	H2
4114	$C_8H_6O_2N_2$ Ethanenitrile, 2-nitro-2-phenyl-					
	0.29	20	c ≈ 0.0001, H$_o$ values in HClO$_4$ solutions	O6	Uncert.	T58
4115	$C_8H_6O_2N_2$ Indole, 3-nitro-					
	10.12	25	c ≈ 0.0001	O5	Approx.	T21

No.	Formula	Name / log K	T	Conditions	Method	Reliability	Ref.
4116	$C_8H_6O_2N_2$	Indole, 5-nitro-					
		14.75(NH)	25	In aq. KOH, H_- scale	07	Uncert.	Y2
		14.75	25	$c \approx 0.0001$	07	Uncert.	T21
4117	$C_8H_6O_2N_2$	Quinazoline-2,4(1H,3H)-dione					
		9.82	25	$c \approx 0.001$	E3bg	Approx.	G62
4118	$C_8H_6O_2N_2$	Quinoxaline-2,3(1H,4H)-dione					
		9.52	20	$c = 0.001$	E3bg	Approx.	A36
4119	$C_8H_6O_2Cl_2$	Acetophenone, 3,5-dichloro-4-hydroxy-					
		4.60	25	$c = 0.0006$	E3bg	Approx.	F17
4120	$C_8H_6O_2Br_2$	2,4,6-Cycloheptatrien-1-one, 2,6-dibromo-7-hydroxy-3-methyl-					
		5.59	25		E3bg	Uncert.	Y15
4121	$C_8H_6O_2Br_2$	2,4,6-Cycloheptatrien-1-one, 2,6-dibromo-7-hydroxy-4-methyl-					
		5.35	25		E3bg	Uncert.	Y15
4122	$C_8H_6O_3Cl_2$	Phenoxyacetic acid, 2,4-dichloro-					
		2.896	25	$c = 0.0003-0.0015$	C2,R1c	Rel.	M58
		2.86	20	$I = 0.001$	05	Approx.	P60
		Value in mixed solvent is also given					
		2.90	20	$I = 0.0005$	E3bg	Approx.	P31
		2.78		$I = 1.0(KNO_3)$			
		2.93	25		C		W19

No.	Molecular formula, name and pK value(s)	T(°C)	Remarks	Method	Assessment	Ref.
4123	$C_8H_6O_3Cl_2$ Phenoxyacetic acid, 2,5-dichloro-			05	Approx.	P60
	2.94	20	I = 0.001			
			Value in mixed solvent is also given			
4124	$C_8H_6O_3Cl_2$ Phenoxyacetic acid, 2,6-dichloro-			05	Approx.	P60
	3.30	20	I = 0.001			
			Value in mixed solvent is also given			
4125	$C_8H_6O_3Cl_2$ Phenoxyacetic acid, 3,4-dichloro-			05	Approx.	P60
	2.92	20	I = 0.001			
			Value in mixed solvent is also given			
4126	$C_8H_6O_3S$ Benzo[b]thiophen-3(2H)-one 1,1-dioxide			E3bg	Uncert.	S42
	2.88	20				
4127	$C_8H_6O_4N_4$ 7-Pteridinecarboxylic acid, 3,4,5,6-tetrahydro-3-methyl-4,6-dioxo-			E3bg	Uncert.	P48
	pK_1 2.4				Approx.	
	pK_2 6.75					
4128	$C_8H_6O_5N_4$ 6-Pteridinecarboxylic acid, 1,2,3,4,7,8-hexahydro-1-methyl-2,4,7-trioxo-			E3bg	Uncert.	P44
	pK_1 2.0	20			Approx.	
	pK_2 6.15, pK_3 10.75					

No.	Formula	Name	pK values	T (°C)	Conditions	Method	Assessment	Ref.
4129	$C_8H_6O_5N_4$	6-Pteridinecarboxylic acid, 1,2,3,4,7,8-hexahydro-3-methyl-2,4,7-trioxo-	pK₁ 1.8, pK₂ 6.00, pK₃ 10.60	20		E3bg	Uncert. Approx.	P44
4130	$C_8H_6O_5N_4$	6-Pteridinecarboxylic acid, 1,2,3,4,7,8-hexahydro-8-methyl-2,4,7-trioxo-	pK₁ 2.20, pK₃ 11.10, pK₂ 4.72	20		E3bg	Uncert. Approx.	P44
4131	$C_8H_6O_5N_4$	7-Pteridinecarboxylic acid, 1,2,3,4,5,6-hexahydro-1-methyl-2,4,6-trioxo-	pK₁ 1.8, pK₂ 7.11, pK₃ 10.32	20		E3bg	Uncert. Approx.	P45
4132	$C_8H_6O_5N_4$	7-Pteridinecarboxylic acid, 1,2,3,4,5,6-hexahydro-3-methyl-2,4,6-trioxo-	pK₁ 1.9, pK₂ 7.52, pK₃ 10.20	20		E3bg	Uncert. Approx.	P45
4133	$C_8H_6O_6N_2$	Acetophenone, 2-hydroxy-3,5-dinitro-	3.26, 3.11	25	c ≈ 0.0002, I = 0.1	05	Approx.	M19
4134	$C_8H_6O_8N_4$	Ethane, 1,1-dinitro-2-(2,4-dinitrophenyl)-	3.35	20		05	Approx.	T58a
4135	$C_8H_6O_9N_4$	Methane, (2-methoxy-3,5-dinitrophenyl)dinitro-	1.00	20	c ≈ 0.0001, H_0 values in $HClO_4$ solutions	06	Uncert.	T58

No.	Molecular formula, name and pK value(s)	$T(^oC)$	Remarks	Method	Assessment	Ref.
4136	$C_8H_6O_9N_4$ Methane, (3-methoxy-4,6-dinitrophenyl)dinitro-					
	-0.21	20	$c \approx 0.0001$, H_o values in $HClO_4$ solutions	06	Uncert.	T58
4137	C_8H_6NBr Indole, 5-bromo-					
	16.13(NH)	25	In aq.KOH, H_- scale	07	Uncert.	Y2
	16.13	25	$c \approx 0.0001$	C7	Uncert.	T21
4138	C_8H_6NF Indole, 4-fluoro-					
	16.30(NH)	25	In aq.KOH, H_- scale	07	Uncert.	Y2
4139	C_8H_6NF Indole, 5-fluoro-					
	16.30(NH)	25	In aq.KOH, H_- scale	07	Uncert.	Y2
4140	$C_8H_6N_2S_2$ Quinoxaline-2,3(1H,4H)-dithione					
	pK_1 6.9, pK_2 9.9	not stated	Calculated from solubility at various pH values		Uncert.	C43
4141	$C_8H_6N_2S_2$ 1,3,4-Thiadiazoline-2(3H)-thione, 5-phenyl-					
	4.98	25	$c \approx 0.004$, mixed constant	E3bg	Approx.	S9
4142	$C_8H_7OF_3$ Ethanol, 2,2,2-trifluoro-1-phenyl-					
	11.90	25	$I \approx 0.1$, mixed constant	05	Approx.	S116

No.	Formula	Name	pK	T(°C)	Conditions	Method	Assessment	Ref.
4143	$C_8H_7O_2N$	Benzenecarbonitrile, 2-hydroxy-5-methoxy-	7.4	30	I = 0.1(KCl)	05	Uncert.	C20
4144	$C_8H_7O_2N$	Benzenecarbonitrile, 2-hydroxy-6-methoxy-	6.6	30	I = 0.1(KCl)	05	Uncert.	C20
4145	$C_8H_7O_2N$	Ethanal oxime, 2-oxo-2-phenyl-	8.40	30	c = 0.01, I = 0.1(KCl), mixed constant	E3bg	Approx.	E18
			8.25	25	c ≈ 0.01, I = 0.1(KCl), mixed constant	E3bg	Approx.	G39
4146	$C_8H_7O_2N_5$	Acetamide, N-(1,4-dihydro-4-oxo-2-pteridinyl)- (Pteridine, 2-acetylamino-4-hydroxy-)	7.37	20		05	Uncert.	P52
4147	$C_8H_7O_2N_5$	6-Pteridinecarboxamide, 1,7-dihydro-N-methyl-7-oxo- (Pteridine, 7-hydroxy-6-methylcarbamoyl-)	5.70	20	c = 0.005	E3bg	Approx.	A39
4148	$C_8H_7O_2Cl$	Acetic acid, 2-(3-chlorophenyl)-	4.110	25	Variation with pressure(bar)	C2,R1a/b	Approx.	F18
4149	$C_8H_7O_2Cl$	Acetic acid, 2-(4-chlorophenyl)-	4.177	25		C2,R1a/b	Approx.	F18

Variation with pressure(bar)

	1000	2000	3000
	3.911	3.738	3.585

(contd)

No.	Molecular formula, name and pK value(s)	T(°C)	Remarks	Method	Assessment	Ref.
4149 (Contd)	$C_8H_7O_2Cl$ Acetic acid, 2-(4-chlorophenyl)- Variation with pressure(bar) 1000 2000 3000 3.971 3.799 3.649					
4150	$C_8H_7O_2Cl$ Benzoic acid, 2-chloro-3-methyl- 3.00	20	$c = 0.001$, 1% ethanol	E3bg	Approx.	P24,P27
4151	$C_8H_7O_2Cl$ Benzoic acid, 2-chloro-4-methyl- 3.27	20	$c = 0.001$, 1% ethanol	E3bg	Approx.	P24,P25
4152	$C_8H_7O_2Cl$ Benzoic acid, 2-chloro-5-methyl- 3.12	20	$c = 0.001$, 1% ethanol	E3bg	Approx.	P24,P27
4153	$C_8H_7O_2Cl$ Benzoic acid, 2-chloro-6-methyl- 2.75	20	$c \approx 0.001$, 1% ethanol	E3bg	Uncert.	P23
4154	$C_8H_7O_2Cl$ Benzoic acid, 3-chloro-2-methyl- 3.43	20	$c \approx 0.001$, 1% ethanol	E3bg	Approx.	P22
4155	$C_8H_7O_2Cl$ Benzoic acid, 3-chloro-4-methyl- 4.06	20	$c = 0.001$, 1% ethanol	E3bg	Approx.	P24

No.	Formula	Name	Value	T	Conc.	Solvent	Method	Assessment	Ref.
4156	$C_8H_7O_2Cl$	Benzoic acid, 5-chloro-2-methyl-	3.63	20	$c \simeq 0.001$	1% ethanol	E3bg	Approx.	P22
4157	$C_8H_7O_2Cl$	Benzoic acid, 4-chloro-2-methyl-	3.75	20	$c \simeq 0.001$	1% ethanol	E3bg	Approx.	P23
4158	$C_8H_7O_2Cl$	Benzoic acid, 4-chloro-3-methyl-	4.07	20	$c = 0.001$	1% ethanol	E3bg	Approx.	P24,P26
4159	$C_8H_7O_2Br$	Benzoic acid, 2-bromo-3-methyl-	2.90	20	$c = 0.001$	1% ethanol	E3bg	Approx.	P24,P27
4160	$C_8H_7O_2Br$	Benzoic acid, 2-bromo-4-methyl-	3.09	20	$c = 0.001$	1% ethanol	E3bg	Approx.	P24,P25
4161	$C_8H_7O_2Br$	Benzoic acid, 2-bromo-5-methyl-	3.00	20	$c = 0.001$	1% ethanol	E3bg	Approx.	P24,P27
4162	$C_8H_7O_2Br$	Benzoic acid, 2-bromo-6-methyl-	2.71	20	$c \simeq 0.001$	1% ethanol	E3bg	Uncert.	P23
4163	$C_8H_7O_2Br$	Benzoic acid, 3-bromo-2-methyl-	3.36	20	$c \simeq 0.001$	1% ethanol	E3bg	Approx.	P22
4164	$C_8H_7O_2Br$	Benzoic acid, 3-bromo-4-methyl-	3.96	20	$c = 0.001$	1% ethanol	E3bg	Approx.	P24

No.	Molecular formula, name and pK value(s)	T(°C)	Remarks	Method	Assessment	Ref.
4165	$C_8H_7O_2Br$ Benzoic acid, 3-bromo-6-methyl- 3.58	20	c ≈ 0.001, 1% ethanol	E3bg	Approx.	P22
4166	$C_8H_7O_2Br$ Benzoic acid, 4-bromo-2-methyl- 3.77	20	c ≈ 0.001, 1% ethanol	E3bg	Approx.	P23
4167	$C_8H_7O_2Br$ Benzoic acid, 4-bromo-3-methyl- 4.03	20	c ≈ 0.001, 1% ethanol	E3bg	Approx.	P24,P26
4168	$C_8H_7O_2Br$ 2,4,6-Cycloheptatrien-1-one, 3-bromo-2-hydroxy-5-methyl- 6.41	25		E3bg	Uncert.	Y15
4169	$C_8H_7O_2Br$ 2,4,6-Cycloheptatrien-1-one, 3-bromo-2-hydroxy-6-methyl- 6.30	25		E3bg	Uncert.	Y15
4170	$C_8H_7O_2Br$ 2,4,6-Cycloheptatrien-1-one, 4-bromo-7-hydroxy-3-methyl- 6.60	25		E3bg	Uncert.	Y15
4171	$C_8H_7O_2I$ Benzoic acid, 2-iodo-3-methyl- 2.92	20	c ≈ 0.001, 1% ethanol	E3bg	Approx.	P24,P27
4172	$C_8H_7O_2I$ Benzoic acid, 2-iodo-4-methyl- 3.20	20	c = 0.001, 1% ethanol	E3bg	Approx.	P24,P25

No.	Formula	Name	Value	T (°C)	Conditions	Method	Status	Ref.
4173	$C_8H_7O_2I$	Benzoic acid, 2-iodo-5-methyl-	2.95	20	c = 0.001, 1% ethanol	E3bg	Approx.	P24,P27
4174	$C_8H_7O_2I$	Benzoic acid, 2-iodo-6-methyl-	2.70	20	c ≈ 0.001, 1% ethanol	E3bg	Uncert.	P23
4175	$C_8H_7O_2I$	Benzoic acid, 3-iodo-2-methyl-	3.26	20	c ≈ 0.001, 1% ethanol	E3bg	Approx.	P22
4176	$C_8H_7O_2I$	Benzoic acid, 5-iodo-2-methyl-	3.62	20	c ≈ 0.001, 1% ethanol	E3bg	Approx.	P22
4177	$C_8H_7O_2I$	Benzoic acid, 4-iodo-2-methyl-	3.79	20	c ≈ 0.001, 1% ethanol	E3bg	Approx.	P23
4178	$C_8H_7O_2F$	Acetic acid, 2-(3-fluorophenyl)-	4.130	25	Variation with pressure(bar) 1000 2000 3000 / 3.927 3.754 3.603	C2,R1a/b	Approx.	F18
4179	$C_8H_7O_2F$	Acetic acid, 2-(4-fluorophenyl)-	4.213	25	Variation with pressure(bar) 1000 2000 3000 / 4.007 3.829 3.676	C2,R1a/b	Rel.	F18

No.	Molecular formula, name and pK value(s)	$T(^{o}C)$	Remarks	Method	Assessment	Ref.
4180	$C_8H_7O_3N$ Phenol, 3-(2-nitrovinyl)-					S115
	9.10	25	c = 0.0006-0.0012, mixed constant	E3bg and O5	Approx.	
4181	$C_8H_7O_3N$ Phenol, 4-(2-nitrovinyl)-					S115
	7.98	25	c = 0.0006-0.0012, mixed constant	E3bg and O5	Approx.	
4182	$C_8H_7O_3Cl$ Acetic acid, 2-(2-chlorophenoxy)-					P60
	2.96	20	I = 0.005	E3bg	Approx.	
	2.96		I = 0.001	O5		
			Value in mixed solvent is also given			
4183	$C_8H_7O_3Cl$ Acetic acid, 2-(3-chlorophenoxy)-					P60
	2.97	20	I = 0.005	E3bg	Approx.	
	2.97		I = 0.001	O5		
			Value in mixed solvent is also given			
	3.04	25	.Value in mixed solvent is also given	E3ag	Approx.	H48
4184	$C_8H_7O_3Cl$ Acetic acid, 2-(4-chlorophenoxy)-					M58
	3.009	25	c = 0.0006-0.0028	C2,R1c	Rel.	

(contd)

407

No.	Formula	Name / pK	T (°C)	Conditions	Method	Reliability	Ref.
4184 (Contd)	$C_8H_7O_3Cl$	Acetic acid, 2-(4-chlorophenoxy)-					
		2.99	20	$I = 0.005$	E3bg	Approx.	P60
		3.00		$I = 0.001$	05		P31
		Value in mixed solvent is also given					
		3.00	20	$I = 0.0005$	E3bg	Approx.	
		2.91	20	$I = 1.0(KNO_3)$			
4185	$C_8H_7O_3Cl$	Acetic acid, 2-(2-chlorophenyl)-2-hydroxy-					
		3.31	25	$c = 0.01$, mixed constant	E3bg	Uncert.	K36
4186	$C_8H_7O_3Cl$	Acetic acid, 2-(4-chlorophenyl)-2-hydroxy-					
		3.15	25	$c = 0.01$, mixed constant	E3bg	Uncert.	K36
4187	$C_8H_7O_3Cl$	Acetophenone, α-chloro-3,4-dihydroxy-					
		pK_1 7.45, pK_2 11.54	25	$c = 0.00015$, $I = 0.1(KCl)$	05,E3bg	Approx.	M70
4188	$C_8H_7O_3Cl$	Methyl 3-chloro-4-hydroxybenzoate					
		6.73	25	$c = 0.0001\text{-}0.0002$, $I \simeq 0.1$, mixed constant	05	Approx.	S76
4189	$C_8H_7O_3Br$	Acetic acid, 2-(3-bromophenoxy)-					
		3.04	25		E3ag	Approx.	H48
		Value in mixed solvent is also given					
4190	$C_8H_7O_3Br$	Acetic acid, 2-(2-bromophenyl)-2-hydroxy-					
		3.32	25	$c = 0.01$, mixed constant	E3bg	Uncert.	K36

No.	Molecular formula, name and pK value(s)	T(°C)	Remarks	Method	Assessment	Ref.
4191	C$_8$H$_7$O$_3$Br Acetic acid, 2-(4-bromophenyl)-2-hydroxy-					
	3.06	18	c = 0.01, I = 0.1	E3bg	Approx.	A45
	2.87		I = 1.0, mixed constant			
	3.15	25	c = 0.01, mixed constant	E3bg	Approx.	K36
4192	C$_8$H$_7$O$_3$I Acetic acid, 2-hydroxy-2-(4-iodophenyl)-					
	3.14	25	c = 0.01, mixed constant	E3bg	Approx.	K36
4193	C$_8$H$_7$O$_3$I 1H-1,2-Benziodoxin-3(4H)-one, 1-hydroxy-					
	7.54	23±3	Mixed constant	05	V.Uncert.	W42
4194	C$_8$H$_7$O$_3$F Acetic acid, 2-(2-fluorophenyl)-2-hydroxy-					
	3.30	25	c = 0.01, mixed constant	E3bg	Approx.	K36
4195	C$_8$H$_7$O$_3$F Acetic acid, 2-(4-fluorophenyl)-2-hydroxy-					
	3.19	25	c = 0.01, mixed constant	E3bg	Uncert.	K36
4196	C$_8$H$_7$O$_3$P 2-Phenylethynylphosphonic acid					
	pK$_1$ 1.14, pK$_2$ 6.30	not stated	Probably mixed constants	E3(?)		M51
4197	C$_8$H$_7$O$_4$N Acetic acid, 2-(3-nitrophenyl)-					
	3.953	25		C2,R1a/b	Approx.	F18

Variation with pressure(bar)

1000	2000	3000
3.766	3.605	3.466

(contd)

No.	Formula	Name / pK value	T(°C)	Conditions	Method	Assessment	Ref.
4197 (Contd)	$C_8H_7O_4N$	Acetic acid, 2-(3-nitrophenyl)-					
		3.97	25		E3ag	Approx.	H49
		Values in mixed solvents are also given					
4198	$C_8H_7O_4N$	Acetic acid, 2-(4-nitrophenyl)-					
		3.924	25		C2,R1a/b	Approx.	F18
		Variation with pressure(bar)					
		1000 2000 3000					
		3.740 3.580 3.437	25				
		3.89	25	2-10% ethanol, extrapolated	E3ag	Uncert.	H49
		Values in mixed solvents are also given					
4199	$C_8H_7O_4N$	Acetophenone, 2-hydroxy-3-nitro-					
		6.78	25	$c \approx 0.0002$	05	Approx.	M19
		6.59		$I = 0.1$, mixed constant			
4200	$C_8H_7O_4N$	Acetophenone, 2-hydroxy-5-nitro-					
		7.05	25	$c \approx 0.0002$	05	Approx.	M19
		6.88		$I = 0.1$, mixed constant			
4201	$C_8H_7O_4N$	Acetophenone, 4-hydroxy-3-nitro-					
		5.09	25	$c = 0.0002$, $I \approx 0.5$, mixed constant	05a	Approx.	R15
4202	$C_8H_7O_4N$	Benzoic acid, 2-methyl-3-nitro-					
		2.98	20	$c \approx 0.001$, 1% ethanol	E3bg	Approx.	P22

No.	Molecular formula, name and pK value(s)	$T(^{o}C)$	Remarks	Method	Assessment	Ref.
4203	$C_8H_7O_4N$ Benzoic acid, 2-methyl-4-nitro- 2.95	20	c ≃ 0.001, 1% ethanol	E3bg	Uncert.	P23
4024	$C_8H_7O_4N$ Benzoic acid, 2-methyl-5-nitro- 3.23	20	c ≃ 0.001, 1% ethanol	E3bg	Approx.	P22
4205	$C_8H_7O_4N$ Benzoic acid, 2-methyl-6-nitro- 2.40	20	c ≃ 0.001, 1% ethanol	E3bg	Uncert.	P23
4206	$C_8H_7O_4N$ Benzoic acid, 3-methyl-2-nitro- 2.91	20	c ≃ 0.001, 1% ethanol	E3bg	Approx.	P24,P27
4207	$C_8H_7O_4N$ Benzoic acid, 3-methyl-4-nitro- 3.65	20	c ≃ 0.001, 1% ethanol	E3bg	Approx.	P24,P26
4208	$C_8H_7O_4N$ Benzoic acid, 5-methyl-2-nitro- 2.55	20	c ≃ 0.001, 1% ethanol	E3bg	Uncert.	P24,P27
4209	$C_8H_7O_4N$ Benzoic acid, 4-methyl-2-nitro- 2.68	20	c = 0.001, 1% ethanol	E3bg	Uncert.	P24,P25
4210	$C_8H_7O_4N$ Benzoic acid, 4-methyl-3-nitro- 3.62	20	c = 0.001, 1% ethanol	E3bg	Approx.	P24

No.	Formula	Name	T (°C)	pK	Method	Assessment	Ref.
4211	$C_8H_7O_4N$	2-Pyridinecarboxylic acid, 4-methoxycarbonyl-	25	3.71	E3bg	Uncert.	O19
4212	$C_8H_7O_4N$	2-Pyridinecarboxylic acid, 5-methoxycarbonyl-	25	3.62	E3bg	Uncert.	O19
4213	$C_8H_7O_4N$	2-Pyridinecarboxylic acid, 6-methoxycarbonyl-	25	2.65	E3bg	Uncert.	O19
4214	$C_8H_7O_5N$	Acetic acid, 2-(3-nitrophenoxy)-	25	2.91	E3ag	Approx.	H48
		Value in mixed solvent is also given					
4215	$C_8H_7O_5N$	Acetic acid, 2-(4-nitrophenoxy)-	25	2.81	E3ag	Approx.	H48
		Value in mixed solvent is also given					
4216	$C_8H_7O_5N$	Benzaldehyde, 2-hydroxy-3-methoxy-5-nitro-	25	4.77	05	Approx.	L50
4217	$C_8H_7O_5N$	Benzaldehyde, 4-hydroxy-3-methoxy-5-nitro-	25	4.20	05	Approx.	L50
4218	$C_8H_7O_5N$	Benzoic acid, 2-methoxy-4-nitro-	25	3.11	E3bg	Uncert.	V3
		Conditions not stated					

No.	Molecular formula, name and pK value(s)	T(°C)	Remarks	Method	Assessment	Ref.
4219	$C_8H_7O_5N$ Methyl 3-hydroxy-4-nitrobenzoate 6.15	25	c = 0.00001-0.0001, I ≈ 0.1, mixed constant	05	Approx.	H11
4220	$C_8H_7O_5N$ Methyl 4-hydroxy-3-nitrobenzoate 5.40	25	c = 0.0002, I ≈ 0.5, mixed constant	05	Approx.	R15
4221	$C_8H_7O_6N$ Benzoic acid, 2-hydroxy-3-methoxy-5-nitro- pK_1 2.58, pK_2 10.30	25	Extrapolated from data obtained in ethanol-water	E3bg and 05	Uncert.	L50
4222	$C_8H_7O_6N$ Benzoic acid, 4-hydroxy-3-methoxy-5-nitro- pK_1 4.12 pK_2 6.0	25	c ≈ 0.0004-0.01	C2 E3bg/05	Approx. Uncert.	L50
4223	$C_8H_7O_6N_3$ Ethane, 1,1-dinitro-2-(2-nitrophenyl)- 3.89	20		05	Approx.	T58a
4224	$C_8H_7O_6N_3$ Ethane, 1,1-dinitro-2-(3-nitrophenyl)- 3.77	20		05	Approx.	T58a
4225	$C_8H_7O_7N$ Ethyl 5-hydroxy-6-nitro-4-oxo-4$\underline{H}$-pyran-2-carboxylate (4$\underline{H}$-Pyran-2-carboxylic acid ethyl ester, 5-hydroxy-6-nitro-4-oxo-) 2.36	room temp.	Mixed constant	E3bg	Uncert.	E9

No.	Formula	Name	pK	Temp. (°C)	Conditions	Method	Assessment	Ref.
4226	$C_8H_7O_7N$	(5-Hydroxy-6-nitro-4-oxo-4$\underline{H}$-pyran-2-yl)methyl acetate	2.60	room temp.	Mixed constant	E3bg	Uncert.	E9
4227	$C_8H_7O_7N_3$	Phenol, 3,5-dimethyl-2,4,6-trinitro-	1.38	25	$c \approx 0.00012$	05	Approx.	D18
			1.57	not stated	$c = 0.003$, $I = 0.5(NaClO_4)$, mixed constant(?)	05	Uncert.	P16
				Other value in M100				
4228	$C_8H_7O_7N_3$	Phenol, 3-ethyl-2,4,6-trinitro-	1.00	not stated	$c = 0.003$, $I = 0.5(NaClO_4)$	05	Uncert.	P16
				Other value in M100				
4229	$C_8H_8ON_4$	Pteridin-2(1$\underline{H}$)-one, 6,7-dimethyl-	7.95	20	"Anhydrous" species	E3bg	Approx.	I5
			9.80	20	Equilibrium pK partial covalent hydration		Approx.	
			11.15	20	Covalently hydrated species		Uncert.	
4230	$C_8H_8ON_4$	Pteridin-4(1$\underline{H}$)-one, 6,7-dimethyl-	8.39	20	$c = 0.005$	E3bg	Approx.	A24
4231	$C_8H_8OCl_2$	Phenol, 2,4-dichloro-3,5-dimethyl-	8.28	20	$c = 0.0001$-0.00001, $I = 0.1(NaClO_4)$, 1% ethanol	05	Approx.	S124

No.	Molecular formula, name and pK value(s)	$T(^oC)$	Remarks	Method	Assessment	Ref.
4232	$C_8H_8OCl_2$ Phenol, 2,4-dichloro-3,6-dimethyl- 8.37	20	c = 0.0001-0.00001, I = 0.1(NaClO$_4$), 1% ethanol	05	Approx.	S124
4233	$C_8H_8OCl_2$ Phenol, 2,4-dichloro-5,6-dimethyl- 8.45	20	c = 0.0001-0.00001, I = 0.1(NaClO$_4$), 1% ethanol	05	Approx.	S124
4234	$C_8H_8OCl_2$ Phenol, 2,6-dichloro-3,4-dimethyl- 7.59	20	c = 0.0001-0.00001, I = 0.1(NaClO$_4$), 1% ethanol	05	Approx.	S124
4235	$C_8H_8O_2N_4$ Pteridine-2,7(1H,8H)-dione, 6,8-dimethyl- 7.41	20	c = 0.0004 in 0.01M buffer	05	Approx.	J4
4236	$C_8H_8O_2N_4$ Pteridine-4,7(1H,8H)-dione, 6,8-dimethyl- 7.81	20	c = 0.00004 in 0.01M buffer	05	Approx.	J4
4237	$C_8H_8O_2S$ Acetic acid, 2-(2-mercaptophenyl)- pK$_1$ 4.28, pK$_2$ 7.67		Conditions not stated	E3bg	Uncert.	P11
4238	$C_8H_8O_2S$ Acetic acid, 2-phenylthio- 3.568	25	c ≈ 0.01	E3bg	Approx.	P12

Values in mixed solvents are also given

(contd)

No.	Formula	Name / pK	Temp.	Conditions	Method	Reliability	Ref.
4238 (Contd)	$C_8H_8O_2S$	Acetic acid, 2-phenylthio-					
		3.37	25	$I = 0.1(NaClO_4)$, concentration constant	E3bg	Approx.	S133
		3.38	20	$c \approx 0.001$, $I = 0.1(KCl)$, mixed constant	E3bg	Approx.	P39
4239	$C_8H_8O_2S$	S-4-Hydroxyphenyl ethanethioate					
		8.88	25		E3ag	Approx.	B104
4240	$C_8H_8O_2S$	Methyl 4-mercaptobenzoate					
		5.330	25	$m = 0.0001$, $I < 0.15$	O5a	Rel.	D28
4241	$C_8H_8O_2Se$	Acetic acid, 2-phenylseleno-					
		3.75	20	$c \approx 0.001$, $I = 0.1(KNO_3)$, mixed constant	E3bg	Approx.	P39
4242	$C_8H_8O_2Se$	1,3-Butanedione, 1-(2-selenofuryl)-					
		8.55	25	$I = 0.1(NaClO_4)$, mixed constant	O5	Uncert.	Z15
		8.38	room temp.	$I = 0.1(NaClO_4)$	E		E7
4243	$C_8H_8O_3N_2$	Acetophenone oxime, α-nitro-					
		pK_1 6.6, pK_2 10.6		Conditions not stated	O5	Uncert.	D36
4244	$C_8H_8O_3N_2$	3,4'-Bi-3-pyrroline-2,2'-dione, 4-hydroxy-					
		3.19		Conditions not stated	O5	Uncert.	M109
4245	$C_8H_8O_3N_4$	Acetic acid, 2-(1,2,3,4-tetrahydro-2-oxo-4-pteridinyl)- (Pteridine, 4-carboxymethyl-3,4-dihydro-2-hydroxy-)					
		4.14	20	$c = 0.005$	E3bg	Approx.	A27

No.	Molecular formula, name and pK value(s)	T(°C)	Remarks	Method	Assessment	Ref.
4246	$C_8H_8O_3N_4$ Pteridine-2,4,6(1H,3H,5H)-trione, 1,3-dimethyl- 5.83	20		E3bg	Approx.	P43
4247	$C_8H_8O_3N_4$ Pteridine-2,4,6(1H,3H,5H)-trione, 1,7-dimethyl- pK$_1$ 6.52, pK$_2$ 10.25	20		E3bg	Approx.	P43
4248	$C_8H_8O_3N_4$ Pteridine-2,4,6(1H,3H,5H)-trione, 3,7-dimethyl- pK$_1$ 6.85, pK$_2$ 9.90	20		E3bg	Approx.	P43
4249	$C_8H_8O_3N_4$ Pteridine-2,4,7(1H,3H,8H)-trione, 1,6-dimethyl- pK$_1$ 3.65, pK$_2$ 10.63	20		E3bg	Approx.	P42
4250	$C_8H_8N_4O_3$ Pteridine-2,4,7(1H,3H,8H)-trione, 3,6-dimethyl- pK$_1$ 4.17, pK$_2$ 10.42	20		E3bg	Approx.	P42
4251	$C_8H_8O_3N_4$ Pteridine-2,4,7(1H,3H,8H)-trione, 6,8-dimethyl- 4.26 13.2	20		E3bg 06	Approx. Uncert.	P54
4252	$C_8H_8O_3N_4$ Pyrimidino[4,5-c]pyridazine-4,5,7(1H,6H,8H)-trione, 6,8-dimethyl- 8.51	20	c = 0.001	E3bg	Approx.	P50
4253	$C_8H_8O_3S$ Acetic acid, 2-(5-acetyl-2-thienyl)- 3.76	25				I3

No.	Formula	Name / pK	Temp.	Conditions	Method	Assessment	Ref.
4254	$C_8H_8O_3S$	Acetic acid, 2-phenylsulfinyl-					
		2.732	25	$c \approx 0.01$	E3bg	Approx.	P12
		Values in mixed solvents are also given					
4255	$C_8H_8O_3S$	Benzo[b]thiophene-5-ol 1,1-dioxide, 2,3-dihydro-					
		7.81	25	No details		Uncert.	C76
4256	$C_8H_8O_4N_2$	Acetamide, N-(3-hydroxy-4-nitrophenyl)-					
		6.71	25	$c \approx 0.00001-0.0001$, $I \approx 0.1$, mixed constant	05	Approx.	H11
4257	$C_8H_8O_4N_2$	Acetamide, N-(4-hydroxy-3-nitrophenyl)-					
		6.78	25	$c = 0.0002$, $I \approx 0.5$, mixed constant	05	Approx.	R15
4258	$C_8H_8O_4N_2$	Ethane, 1,1-dinitro-2-phenyl-					
		4.54	20		05	Approx.	T58a
4259	$C_8H_3O_4N_2$	Methyl 4-nitroanilinomethanoate (Methyl carbamate, N-(4-nitrophenyl)-)					
		13.0	25	$I = 1.0(KCl)$	05	Uncert.	H26
4260	$C_8H_8O_4N_4$	Acetic acid, 2-(1,4,5,6,7,8-hexahydro-4,7-dioxo-6-pteridinyl)-					
		pK_1 4.49, pK_2 8.59, $pK_3 \approx 11$	20	$c = 0.002$	E3bg	Uncert.	A22
4261	$C_8H_8O_4N_4$	Pteridine-2,4,6,7(1H,3H,5H,8H)-tetrone, 1,3-dimethyl-					
		pK_1 3.63, pK_2 10.73	20	$c = 0.001$	E3bg	Uncert.	P46

No.	Molecular formula, name and pK value(s)	$T(^{\circ}C)$	Remarks	Method	Assessment	Ref.
4262	$C_8H_8O_4N_4$ Pteridine-2,4,6,7(1H,3H,5H,8H)-tetrone, 3,8-dimethyl-					
	3.83	20	c = 0.001	E3bg	Approx.	P47
4263	$C_8H_8O_4S$ Acetic acid, 2-phenylsulfonyl-					
	2.513	25	c ≈ 0.01	E3bg	Approx.	P12
	Values in mixed solvents are also given					
4264	$C_8H_8O_4S_2$ 1,3-Dithiaindan 1,1,3,3-tetroxide, 6-methyl-					
	12.63	25	No details		Uncert.	C76
4265	$C_8H_8O_5N_2$ 2-Pyridinecarboxylic acid N-oxide, 5-ethyl-4-nitro-					
	2.37		c = 0.04 or 0.001(?)	E?		K51
4266	$C_8H_8O_6S$ Benzoic acid, 2-hydroxy-3-methyl-5-sulfo-					
	pK_2 2.78, pK_3 13.08	25	c = 0.0001, I = 0.10($NaClO_4$), mixed	O5	Approx.	C34
	pK_2 2.55, pK_3 12.74		constants			
4267	C_8H_9ON Acetophenone oxime					
	11.35			O5		C6
4268	$C_8H_9ON_5$ Pteridin-4(1H)-one, 2-dimethylamino-					
	pK_1 2.26	20		E3bg/O5	Uncert.	P52
	pK_2 7.81				Approx.	

No.	Formula	Name	Conditions		pKa		Uncert.	Ref.
4269	$C_8H_9ON_5$	Pteridin-7(1H)-one, 2-ethylamino-	Conditions not stated		7.50			P58
4270	C_8H_9OCl	Phenol, 2-chloro-4,5-dimethyl-	20	$c = 0.0001-0.00001$, $I = 0.1(NaClO_4)$, 1% ethanol	8.89	05	Approx.	S124
4271	C_8H_9OCl	Phenol, 4-chloro-2,3-dimethyl-	20	$c = 0.0001-0.00001$, $I = 0.1(NaClO_4)$, 1% ethanol	9.91	05	Approx.	S124
4272	C_8H_9OCl	Phenol, 4-chloro-2,5-dimethyl-	20	$c = 0.0001-0.00001$, $I = 0.1(NaClO_4)$, 1% ethanol	9.89	05	Approx.	S124
4273	C_8H_9OCl	Phenol, 4-chloro-3,5-dimethyl-	25	$m = 0.00017$	9.702	05a	Rel.	B90

Variation with temperature

5	10	15	20	25	30	35	40	45	50	55	60
10.006	9.929	9.851	9.778		9.634	9.569	9.511	9.459	9.410	9.361	9.321

Thermodynamic quantities are derived from the results

| | | | 20 | $c = 0.0001-0.00001$, $I = 0.1(NaClO_4)$, 1% ethanol | 9.71 | 05 | Approx. | S124 |

No.	Molecular formula, name and pK value(s)	$T(^oC)$	Remarks	Method	Assessment	Ref.
4274	C_8H_9OBr Phenol, 4-bromo-2,6-dimethyl-					
	9.81	25	c = 0.0013	E3bg	Approx.	F16
4275	C_8H_9OF Phenol, 4-fluoro-2,6-dimethyl-					
	10.46	25	c = 0.0026	E3bg	Approx.	F16
4276	$C_8H_9O_2N$ Acetamide, N-(3-hydroxyphenyl)-					
	9.59	25	c ≈ 0.001	E3bg	Approx.	B155
	9.49	20	c = 0.005	E3bg	Approx.	A18
4277	$C_8H_9O_2N$ Acetic acid, anilino- (Glycine, N-phenyl-)					
	4.59(COOH)	20	c = 0.025, mixed constant	E3d	Uncert.	H17
4278	$C_8H_9O_2N$ Acetohydroxamic acid, N-phenyl-					
	8.10	30	c = 0.001, I = 0.1(NaClO$_4$), concentration constant	E3bg	Approx.	D62
4279	$C_8H_9O_2N$ Acetohydroxamic acid, 2-phenyl-					
	9.2	20	c = 0.01	E3d	Uncert.	W39
4280	$C_8H_9O_2N$ Benzohydroxamic acid, 2-methyl-					
	pK$_1$ 1.87, pK$_2$ 8.55	20±1	c = 0.0025, I = 1.0(KNO$_3$), mixed constants	E3bg	Uncert.	D64

No.	Formula / Name	pK	Temp. (°C)	Conditions	Method	Assessment	Ref.
4281	$C_8H_9O_2N$ Benzohydroxamic acid, 4-methyl-	9.05	25	c = 0.01(?) (Possible error in detail given)	E3bg	Approx.	A10
		8.93	30.5	I = 0.1(KNO_3), mixed constant	E3d	Uncert.	S137
		8.93	not stated	c = 0.04, I = 0.1(KCl)	E3bg	Uncert.	H2
4282	$C_8H_9O_2N$ Benzoic acid, 2-amino-3-methyl-	4.92	20	c ≈ 0.001, 1% ethanol	E3bg	Approx.	P27
4283	$C_8H_9O_2N$ Benzoic acid, 2-amino-4-methyl-	pK_1 2.35					
		pK_2 5.07	25	c = 0.00074, I = 0.1(KCl), mixed constants	05	Uncert. Approx.	L15
		pK_2 5.04	20	c = 0.001, 1% ethanol	E3bg	Approx.	P24,P25
4284	$C_8H_9O_2N$ Benzoic acid, 2-amino-5-methyl-	pK_1 2.48, pK_2 5.13	25	c = 0.000041, I = 0.1(KCl), mixed constants	05	Approx.	L15
		pK_2 4.89	20	c = 0.001, 1% ethanol	E3bg	Approx.	P24,P27
4285	$C_8H_9O_2N$ Benzoic acid, 2-amino-6-methyl-	3.73	20	c ≈ 0.001, 1% ethanol	E3bg	Approx.	P23
4286	$C_8H_9O_2N$ Benzoic acid, 3-amino-2-methyl-	3.94	20	c ≈ 0.001, 1% ethanol	E3bg	Approx.	P22

No.	Molecular formula, name and pK value(s)	T(°C)	Remarks	Method	Assessment	Ref.
4287	$C_8H_9O_2N$ Benzoic acid, 3-amino-4-methyl- 4.49	20	c = 0.001, 1% ethanol	E3bg	Approx.	P24
4288	$C_8H_9O_2N$ Benzoic acid, 3-amino-6-methyl- 4.16	20	c ≈ 0.001, 1% ethanol	E3bg	Approx.	P22
4289	$C_8H_9O_2N$ Benzoic acid, 4-amino-2-methyl- 5.17	20	c ≈ 0.001, 1% ethanol	E3bg	Approx.	P23
4290	$C_8H_9O_2N$ Benzoic acid, 4-amino-3-methyl- 4.81	20	c ≈ 0.001, 1% ethanol	E3bg	Approx.	P24, P26
4291	$C_8H_9O_2N$ Phenol, 2,6-dimethyl-4-nitroso- 7.56	25	c = 0.0023	E3bg	Approx.	F16
4292	$C_8H_9O_2N$ Phenol, 3,5-dimethyl-4-nitroso- 8.04	25	c ≈ 0.04, mixed constant	E3bg	Uncert.	V8
4293	$C_8H_9O_2N$ 2-Pyridinecarboxylic acid, 5-ethyl- 5.94		c = 0.04 or 0.001(?)	E		K51
4294	$C_8H_9O_2N_5$ Pteridine-4,7(1H,8H)-dione, 2-amino-6,8-dimethyl- 8.40	20		O5	Approx.	P57

No.	Formula	Name / pK	Temp	Conditions	Method	Reliability	Ref
4295	$C_8H_9O_2N_5$	Pteridine-4,7(3H,8H)-dione, 2-amino-3,6-dimethyl- 8.00	20		O5	Approx.	P57
4296	$C_8H_9O_3N$	Benzohydroxamic acid, 2-methoxy- 8.9	30		E3bg	Uncert.	S117
4297	$C_8H_9O_3N$	Benzohydroxamic acid, 4-methoxy-					
		9.15	25	c = 0.01(?) (Possible error in detail given)	E3bg	Approx.	A10
		8.91	30	c = 0.01, I = 0.1(NaClO$_4$), concentration constant	E3bg	Approx.	D62
		9.03	30.5	c = 0.0006, I = 0.1(KNO$_3$), mixed constant	E3d	Uncert.	S137
		9.00	20	c = 0.0025, I = 1.0(KNO$_3$), mixed constant	E3bg	Uncert.	D64
		Other values in H2 and W39					
4298	$C_8H_9O_3N$	Benzoic acid, 2-amino-4-methoxy- (Anthranilic acid, 4-methoxy-)					
		pK$_1$ 2.06 pK$_2$ 4.88	25	c = 0.00034, I = 0.1(KCl), mixed constants	O5	Uncert. Approx.	L15
		pK$_1$ 1.94 pK$_2$ 4.71	60	I = 0.1, mixed constants	O5	Uncert. Approx.	D57
4299	$C_8H_9O_3N$	Benzoic acid, 2-amino-5-methoxy-					
		pK$_1$ 2.37 pK$_2$ 5.57	25	c = 0.00035, I = 0.1(KCl), mixed constants	O5	Approx. Uncert.	L15

No.	Molecular formula, name and pK value(s)	T(°C)	Method	Assessment	Ref.	Remarks
4300	$C_8H_9O_3N$ Benzoic acid, 4-amino-2-methoxy-					
	pK$_1$ 2.24, pK$_2$ 4.42		05		L56	
	pK$_2$ 5.25		E3bg	Uncert.	V3	Conditions not stated
4301	$C_8H_9O_3N$ Phenol, 2,6-dimethyl-4-nitro-					
	7.07	25	E3bg	Approx.	F16	c = 0.0005
4302	$C_8H_9O_3N$ 2-Pyridinecarboxylic acid, 4-ethoxy-					
	6.62		E		K51	c = 0.04 or 0.001
4303	$C_8H_9O_3N_5$ Pteridine-4,6,7(3H,5H,8H)-trione, 2-amino-3,8-dimethyl-					
	8.75	20	05	Approx.	P56	c = 0.00005
4304	$C_8H_9O_3P$ Ethenylphosphonic acid, 2-phenyl-					
	pK$_1$ 2.00, pK$_2$ 7.1	not stated	E3bg	Uncert.	C38	c = 0.05
4305	$C_8H_9O_4N_3$ Ethyl 5-amino-6-formyl-3,4-dihydro-3-oxopyrazine-2-carboxylate					
	6.09	20	05	Approx.	A39	c = 0.0001 in 0.01M buffer
4306	$C_8H_9O_7N_3$ 3-Azapentanedioic acid, 3-(1,2,3,4-tetrahydro-6-hydroxy-2,4-dioxo-5-pyrimidinyl)- (Iminodiacetic acid, N-(1,2,3,4-tetrahydro-6-hydroxy-2,4-dioxo-5-pyrimidinyl)-) (Uramildiacetic acid)					
	pK$_1$ 1.7	20	E3bg	Uncert.	I8	I = 0.1(KNO$_3$)
	pK$_2$ 2.67			Approx.		I = 0.1(KNO$_3$)
	pK$_3$ 9.63			Approx.		I = 0.1((CH$_3$)$_4$NNO$_3$), concentration constants
	(contd)					

4306 $C_8H_9O_7N_3$ 3-Azapentanedioic acid, 3-(1,2,3,4-tetrahydro-6-hydroxy-2,4-dioxo-5-pyrimidinyl)-   (Iminodiacetic acid, N-(1,2,3,4-tetrahydro-6-hydroxy-2,4-dioxo-5-pyrimidinyl)-)   (Uramildiacetic acid) (Contd)

Variation with temperature

Thermodynamic quantities (I = 0.1) are derived from the results

	27	34	39
pK_1	1.82	1.90	2.21
pK_2	2.83	2.88	2.90
pK_3	9.47	9.38	9.31

pK_1 2.86, pK_2 3.76, pK_3 10.44	20	c = 0.0017	E3bg	Approx.	S46
pK_1 1.7, pK_2 2.67, pK_3 9.63	20	I = 0.1, titrant $(CH_3)_4NOH$	E		F38

4307 $C_8H_9NS_2$ Methanedithioic acid, benzylamino-   (Dithiocarbamic acid, N-benzyl-)

3.13	21	c ≈ 0.01-0.00001, concentration constant	05	Uncert.	T66,T65
2.99		(Compound unstable in acid solution)	E	Uncert.	F38

4308 $C_8H_{10}ON_2$ Acetamidine, 2-hydroxy-2-phenyl-

pK_1 10.82, pK_2 12.52	25	I = 0.1(KCl)	E3ag	Uncert.	G34

Decomposition occurs; results obtained from extrapolation to zero time

4309 $C_8H_{10}ON_2$ Acetohydrazide, 2-phenyl-

pK_1 3.09	25	Mixed constant(?)	E3bg	Uncert.	T43,T44
13.17(NH)	25		05	Uncert.	

4310 $C_8H_{10}ON_2$ Benzohydrazide, 4-methyl-

pK_1 3.15	25	Mixed constant	E3bg	Uncert.	T44,T43
pK_2 12.75(NH)			05		

No.	Molecular formula, name and pK value(s)	T(°C)	Remarks	Method	Assessment	Ref.
4311	$C_8H_{10}ON_2$ Quinazolin-4(1H)-one, 5,6,7,8-tetrahydro-					
	pK$_1$ 2.69, pK$_2$ 9.64	20	c < 0.001, I = 0.01, mixed constants	05	Approx.	B134
4312	$C_8H_{10}ON_4$ 1,2,4-Triazolo[1,5-a]pyrimidin-7(1H)-one, 6-ethyl-5-methyl-					
	6.84	20	I = 0.1(KNO$_3$)	E3bg	Approx.	022
4313	$C_8H_{10}ON_4$ 1,2,4-Triazolo[1,5-a]pyrimidin-7(1H)-one, 5-propyl-					
	6.31	20	I = 0.1(KNO$_3$)	E3bg	Approx.	022
4314	$C_8H_{10}O_2N_2$ Acetohydrazide, 2-phenoxy-					
	pK$_1$ 2.65	25	Mixed constants(?)	E3bg	Uncert.	T43,T44
	12.23(NH)	25		05	Uncert.	
4315	$C_8H_{10}O_2N_2$ 3-Azabutanoic acid, 4-(2-pyridyl)- (N-(2-Pyridylmethyl)glycine)					
	pK$_1$ 1.62, pK$_2$ 1.81	25	I = 0.1(KNO$_3$), concentration constants	E3bg	Uncert.	L1
	pK$_3$ 5.45, pK$_4$ 8.23				Approx.	
4316	$C_8H_{10}O_2N_2$ Benzohydrazide, 4-methoxy-					
	pK$_1$ 3.25	25	Mixed constants(?)	E3bg	Uncert.	T43,T44
	12.83(NH)	25		05	Uncert.	
4317	$C_8H_{10}O_2N_2$ 1,4-Benzoquinone dioxime 1-(O-methyl ether), 2-methyl-					
	9.60	not stated	Mixed constant	05	Uncert.	D35

4318	$C_8H_{10}O_2N_2$	1,4-Benzoquinone dioxime 1-(O-methyl ether), 3-methyl-						
		9.50		not stated	Mixed constant	05	Uncert.	D35
4319	$C_8H_{10}O_2N_4$	1H-Pyrazolo[3,4-d]pyrimidine-4,6(5H,7H)-dione, 3,5,7-trimethyl-						
		9.83		20	c = 0.001	E3bg	Approx.	P59
4320	$C_8H_{10}O_2S$	Ethanesulfinic acid, 2-phenyl-						
		1.89		20	c = 0.02, mixed constant	E3bg	Approx.	R54
4321	$C_8H_{10}O_2S$	Phenol, 2-methyl-6-methylsulfinyl-						
		8.78		25		E3bg	Approx.	F25
4322	$C_8H_{10}O_2S$	2-Thiophenecarboxylic acid, 4-isopropyl-						
		3.66		25	c = 0.01	E3bg	Approx.	028
4323	$C_8H_{10}O_3N_4$	Purine-2,6,8(1H,3H,7H)-trione, 1,3,7-trimethyl-						
		6.0		not stated	Mixed constant	05	Uncert.	B65
4324	$C_8H_{10}O_3N_4$	Purine-2,6,8(1H,3H,7H)-trione, 3,7,9-trimethyl-						
		8.35		not stated	Mixed constant	05	Uncert.	B65
4325	$C_8H_{10}O_4N_2$	Ethyl 1,2,3,6-tetrahydro-1-methyl-2,6-dioxopyrimidine-4-carboxylate						
		8.18		25	Titrant $N(Me)_4OH$	E3bg	Approx.	J16
4326	$C_8H_{10}O_4N_2$	Propanoic acid, 2-amino-3-(1,4-dihydro-3-hydroxy-4-oxo-1-pyridyl)-			(DL-Mimosine)				
		pK(COOH) 2.1, pK($-NH_3^+$) 7.2, pK(OH) 9.2			Conditions not stated	E3d	Uncert.	S98

No.	Molecular formula, name and pK value(s)	T(°C)	Remarks	Method	Assessment	Ref.
4327	$C_8H_{10}O_5N_2$ Acetic acid, 2-(1,2,3,4-tetrahydro-5-methoxycarbonyl-2-oxo-4-pyrimidinyl)- (Pyrimidine, 4-carboxymethyl-1,2,3,4-tetrahydro-5-methoxycarbonyl-2-oxo-)					S136
	4.30		No details		Uncert.	
4328	$C_8H_{10}N_2S$ Quinazoline-4(1H)-thione, 5,6,7,8-tetrahydro-					B134
	pK₁ 1.95	20	c = 0.001, I = 0.01, mixed constants	05	Uncert.	
	pK₂ 8.13				Approx.	
4329	$C_8H_{10}N_4S$ Purine, 6,8-dimethyl-2-methylthio-					B138
	pK₁ -0.60	20	c = 0.0001, I = 0.01	06	Uncert.	
	pK₂ 3.04	20		05	Approx.	
	pK₃ 9.70	20			Uncert.	
4330	$C_8H_{10}N_4S_2$ Purine, 2,8-bis(methylthio)-6-methyl-					B138
	pK₁ -1.36	20	c < 0.0001, I = 0.01	06	Uncert.	
	pK₂ 2.74, pK₃ 7.94			05	Approx.	
4331	$C_8H_{11}ON$ Phenol, 2-(2-aminoethyl)-					E20
	9.19(-OH)	25	I = 0.1(KCl), mixed constant	05	Approx.	
4332	$C_8H_{11}ON$ Phenol, 3-(2-aminoethyl)-					E20
	9.27(-OH)	25	I = 0.1(KCl), mixed constant	05	Approx.	

No.	Formula	Name	pK	T	Conditions	Method	Reliability	Ref.
4333	$C_8H_{11}ON$	Phenol, 4-(2-aminoethyl)-	9.39(-OH)	25	I = 0.1(KCl), mixed constant	05	Approx.	E20
4334	$C_8H_{11}ON$	Phenol, 2-dimethylamino-	10.62	25	I = 0.1(KCl), mixed constant	E3bg	Approx.	H51
4335	$C_8H_{11}ON$	Phenol, 3-dimethylamino-	9.85	25		E3ag	Approx.	B104
4336	$C_8H_{11}ON$	Phenol, 4-dimethylamino-	10.22	25		E3ag	Approx.	B104
			10.10	25	I = 0.1(KCl), mixed constant	E3bg	Approx.	E20
			10.08	25	I = 0.1(KCl), mixed constant	E3bg	Approx.	H51
4337	$C_8H_{11}ON_3$	Triazen-3-ol, 3-ethyl-1-phenyl-	12.28	25	c = 0.00004, I = 0.1(KCl), mixed constant	05	Uncert.	D54
4338	$C_8H_{11}ON_3$	Triazen-3-ol, 3-methyl-1-(o-tolyl)-	12.30	25	c = 0.00004, I = 0.1(KCl), mixed constant	05	Uncert	D54
4339	$C_8H_{11}ON_3$	Triazen-3-ol, 3-methyl-1-(p-tolyl)-	12.52	25	c = 0.00004, I = 0.1(KCl), mixed constant	05	Uncert.	D54
4340	$C_8H_{11}OS$(cation)	Phenol, 3-dimethylsulfonio- (as the p-bromobenzenesulfonate)	7.67	25		E3ag	Approx.	B103

No.	Molecular formula, name and pK value(s)	$T(^oC)$	Remarks	Method	Assessment	Ref.
4341	$C_8H_{11}OS$(cation) Phenol, 4-dimethylsulfonio- 7.30	25	(as the p-toluenesulfonate)	E3ag	Approx.	B103
	The same result obtained with the p-bromobenzenesulfonate					
4342	$C_8H_{11}OP$ Phenol, 3-dimethylphosphino- pK$_1$ 5.89, pK$_2$ 9.66	25	Mixed constants	E3bg	Approx.	T62
4343	$C_8H_{11}OP$ Phenol, 4-dimethylphosphino- pK$_1$ 6.75, pK$_2$ 9.41	25	Mixed constants	E3bg	Approx.	T62
4344	$C_8H_{11}O_2N$ 7-Azabicyclo[4.3.0]nonane-6,8-dione 9.72(cis)		Conditions not stated	E3bg	Uncert.	C62
4345	$C_8H_{11}O_2N$ 1,2-Benzenediol, 3-(2-aminoethyl)- pK$_1$ 8.60	25	I = 0.1(KNO$_3$), mixed constant	E3bg	Approx.	E19
4346	$C_8H_{11}O_2N$ 2-Butynoic acid, 4-pyrrolidin-1-yl- pK$_1$ 1.76, pK$_2$ 8.25	20	Mixed constants	E3bg	Approx.	B39
4347	$C_8H_{11}O_2N$ Cyclohexanecarboxylic acid, 4-cyano- 4.48	25	c = 0.01	E3bg	Approx.	S69
	Values in mixed solvents are also given					

4348 $C_8H_{11}O_2N_3$ Triazen-3-ol, 1-(4-methoxyphenyl)-3-methyl-

12.69 25 $c = 0.00004$, $I = 0.1(KCl)$, mixed constant 05 Uncert. D54

4349 $C_8H_{11}O_2Br$ Bicyclo[2.2.1]heptane-1-carboxylic acid, 4-bromo-

4.356 25 Values in mixed solvents are also given E3bg Approx. W29

4350 $C_8H_{11}O_2F_3$ 2,4-Heptanedione, 1,1,1-trifluoro-6-methyl-

6.5 room temp. E3bg Uncert. R19

4351 $C_8H_{11}O_3N$ 2-Butynoic acid, 4-morpholino-

pK_1 1.79 20 Mixed constants E3bg Approx. B39
pK_2 5.57   Uncert.

4352 $C_8H_{11}O_3N$ Ethanol, 2-amino-1-(3,4-dihydroxyphenyl)- (Noradrenaline)

pK_1 3.30, pK_2 8.82, pK_3 9.98((-)isomer) 25 $I = 0.06(KCl)$, mixed constants E3bg Approx. A65
pK_2 9.34, pK_3 10.69 0

4353 $C_8H_{11}O_3N$ Hexahydrophthalimide, N-hydroxy-

6.0(cis) 30 E3bg Uncert. S117

4354 $C_8H_{11}O_3N$ Pyridin-2(1$\underline{H}$)-one, 6-hydroxy-3-(2-hydroxyethyl)-4-methyl-

pK_1 0.34 20 $c = 0.0001$, $I = 0.01$ 06 Uncert. S100
pK_2 5.48   05 Approx.

432

No.	Molecular formula, name and pK value(s)	T(°C)	Remarks	Method	Assessment	Ref.
4355	$C_8H_{11}O_3P$ Ethylphosphonic acid, 2-phenyl- pK$_1$ 2.55, pK$_2$ 7.83	not stated	Probably mixed constants	E3(?)		M51
4356	$C_8H_{11}O_4N$ Ethyl 4-hydroxy-1-methyl-2-oxo-3-pyrroline-3-carboxylate pK$_1$ 2.55		Conditions not stated	O5	Uncert.	M109
4357	$C_8H_{11}O_4N_3$ Pyrimidine-2,4(1H,3H)-dione, 6-acetoxyamino-1,3-dimethyl- 7.01	20	c = 0.001	E3bg	Approx.	P51
4358	$C_8H_{11}O_6N_3$ 1,2,4-Triazine-3,5(2H,4H)-dione, 2-(β-D-ribofuranosyl)- (6-Azauridine) 6.70	25	c ≈ 0.001	E3bg	Approx.	G62
	6.70	25	Titrant (CH$_3$)$_4$NOH	E3bg	Approx.	J16
4359	$C_8H_{11}NS$ Ethanethiol, 2-amino-2-phenyl- pK$_1$ 7.18, pK$_2$ 9.98	25	c = 0.01, mixed constants	E3bg	Approx.	B147
4360	$C_8H_{12}O_2N_2$ 2-Butynoic acid, 4-piperazino- pK$_1$ 2.78, pK$_2$ 9.28	20	Mixed constants	E3bg	Approx.	B39
4361	$C_8H_{12}O_2N_2$ 1,3-Diazabicyclo[4.4.0]decane-2,4-dione (Cyclohexano[d]pyrimidine-2,4(1H,3H)-dione) 11.84(cis) 12.09(trans)			O		P83

4362 $C_8H_{12}O_2N_2$ 2,4-Diazanona-4,6,8-trien-3-one, 9-hydroxy-2-methyl-

pK_1 -1.8, pK_2 12.0   25   KIN   Uncert.   J14

4363 $C_8H_{12}O_3N_2$ Pyrimidine-2,4,6(1H,3H,5H)-trione, 5,5-diethyl-   (Barbital)

pK_1 8.019   25   05   Approx.   B124

Variation with temperature

15	20	30	35	40	45	50
8.169	8.093	7.950	7.878	7.806	7.738	7.675

Thermodynamic quantities are derived from the results

pK_1 7.89   25   C2,R1d   Uncert.   S121

Variation with temperature

0	5	10	15	20	30	35	40	45	50	55	60
8.23	8.18	8.11	8.03	7.96	7.82	7.75	7.68	7.63	7.59	7.54	7.50

pK_1 7.91   25   E3ag   Approx.   K53

pK_1 7.85, pK_2 12.7   not stated   05   Uncert.   F32

4364 $C_8H_{12}O_3N_4$ 3-Azahexanoic acid, 5-amino-6-(4-imidazolyl)-4-oxo-   (L-Histidylglycine)

pK_1 2.96, pK_2 5.58, pK_3 7.50   25   I = 0.16(KCl)   E3bg   Approx.   B153

4365 $C_8H_{12}O_3N_4$ 3-Azapentanoic acid, 5-amino-2-(4-imidazolyl)methyl-4-oxo-   (Glycyl-L-histidine)

pK_1 2.66, pK_2 6.77, pK_3 8.24   25   I = 0.16(KCl)   E3bg   Approx.   B153

pK_2 6.83, pK_3 8.18   25   I = 0.16(KCl)   E3bg   Approx.   K41

No.	Molecular formula, name and pK value(s)	T(^{o}C)	Remarks	Method	Assessment	Ref.
4366	$C_8H_{12}O_4P_2$ 1,4-Phenylenebis(methylphosphinic acid) 2.67		Conditions not stated	E3d	Uncert.	E28
4367	$C_8H_{12}O_5N_2$ 7-Oxabicyclo[2.2.1]heptane-2,3-dicarbohydroxamic acid 9.3(cis)	30		E3bg	Uncert.	S117
4368	$C_8H_{12}N_2S_2$ 1,3-Diazaspiro[4.5]decane-2,4-dithione 10.05	18-21	Mixed constant	05	Uncert.	E3
4369	$C_8H_{13}ON_3$ Pyrimidin-2(1$\underline{H}$)-one, 4-ethylamino-1,5-dimethyl- 4.58	20		05	Uncert.	K65
4370	$C_8H_{13}O_2N$ 2,5-Pyrrolidinedione, 3,3,4,4-tetramethyl- 9.7		Conditions not stated	E3bg	Uncert.	C62
4371	$C_8H_{13}O_2N_3$ 1,2,4-Triazine-3,5(2$\underline{H}$,4$\underline{H}$)-dione, 6-pentyl- 7.42	not stated	c = 0.005	E3bg	Uncert.	C28
4372	$C_8H_{13}O_3N$ 2-Pyrroline-2-carboxylic acid, 1-hydroxy-4,5,5-trimethyl- 2.85		Conditions not stated	E3bg	Uncert.	B101
4373	$C_8H_{13}O_6N$ 4-Azaheptanedioic acid, 4-carboxymethyl- (3,3'-Iminodipropanoic acid, N-carboxymethyl-) pK$_1$ 3.25, pK$_2$ 3.98, pK$_3$ 9.37	30	I = 0.1(KCl), concentration constants	E3bg	Uncert.	C24

4374 $C_8H_{13}O_6N$ 3-Azapentanedioic acid, 3-(1-carboxy-1-methyl)ethyl-   (Iminodiacetic acid, N-(1-carboxy-1-methyl)ethyl-)
pK$_1$ 1.51
pK$_2$ 2.52, pK$_3$ 11.86
20   I = 0.1(KCl), concentration constants
E3bg   Uncert. / Approx.   I14

4375 $C_8H_{13}O_6N$ Heptanedioic acid, 4-methyl-4-nitro-
pK$_1$ 3.90, pK$_2$ 5.46
21   c = 0.01, mixed constants
E3bg   Uncert.   N37

4376 $C_8H_{13}O_6N$ 5-Oxononanedioic acid, 2-amino-6-oxo-   (O-Succinylhomoserine)
pK$_2$ 4.4, pK$_3$ 9.5
Conditions not stated
On titration the compound is said to undergo base catalyzed 1,3 O→N acyl transfer to
2-(2-hydroxyethyl)-4-oxo-3-azaheptanedioic acid
E3bg   V.Uncert.   F23

4377 $C_8H_{14}O_4N_2$ 1,2-Cyclohexanedicarbohydroxamic acid
9.75(cis)
30
E3bg   Uncert.   S117

4378 $C_8H_{14}O_4S$ 3-Thiapentanoic acid, 5-ethoxy-2,2-dimethyl-5-oxo-
4.59
25±0.5   c = 0.00075
E3bg   Approx.   S93

4379 $C_8H_{14}O_4S$ 3-Thiapentanoic acid, 5-ethoxy-2,4-dimethyl-5-oxo-
3.76
25±0.5   c = 0.00075
E3bg   Approx.   S93

4380 $C_8H_{14}O_4S$ 3-Thiapentanoic acid, 5-ethoxy-4,4-dimethyl-5-oxo-
3.84
25±0.5   c = 0.00075
E3bg   Approx.   S93

4381 $C_8H_{14}O_4S$ 3-Thiapentanedioic acid, 2,2,4,4-tetramethyl-
pK$_1$ 3.83, pK$_2$ 5.19
25±0.5   c < 0.001
E3bg   Approx.   S93,S92

No.	Molecular formula, name and pK value(s)	$T(^{o}C)$	Remarks	Method	Assessment	Ref.
4382	$C_8H_{14}O_5N_4$ 3,6,9-Triazaundecanoic acid, 11-amino-4,7,10-trioxo- (Tetraglycine)					
	pK_1 3.17, pK_2 7.88	25	I = 0.1	E3bg	Approx.	N12
			Thermodynamic quantities are also given			
	pK_1 3.24, pK_2 7.89	24.9	I = 0.1(KNO_3), concentration constants	E3bg	Approx.	K27
4383	$C_8H_{14}O_6N_4$ Ethylamine, N-cyclohexyl-N,2,2-trinitro-					
	3.55	20		05	Approx.	T58a
4384	$C_8H_{15}O_2N$ 2-Piperidinecarboxylic acid, 1-ethyl-					
	2.04	20	c = 0.004	E3bg	Approx.	C3
4385	$C_8H_{15}O_5N$ 4-Azaheptanedioic acid, 4-(2-hydroxyethyl)- (3,3'-Iminodipropanoic acid, N-(2-hydroxyethyl)-)					
	pK_1 3.93, pK_2 8.91	30	I = 0.1(KCl), concentration constants	E3bg	Approx.	C25
4386	$C_8H_{15}O_5N_3$ Butanamide, N-tert-butyl-4,4-dinitro-					
	4.43	20		05	Approx.	T58a
4387	$C_8H_{16}OS_2$ Methanedithioic acid, heptyloxy- (Xanthic acid, heptyl-)					
	3.62					H27
4388	$C_8H_{16}O_2N_2$ 2-Butanone, 4-diethylamino-3-hydroxyimino-					
	pK_1 6.9, pK_2 11.2	25	c = 0.0025-0.005, I = 0.1(KNO_3) (probably a mixed constant)	E		U9

4389 $C_8H_{16}O_2S$ Acetic acid, hexylthio-

     3.758   25 $c \approx 0.001$, $I = 0.10(KNO_3)$, concentration constant   E3bg   Approx.   P40

4390 $C_8H_{16}O_4N_2$ 3,6-Diazaoctanedioic acid, 2,7-dimethyl-   (2,2'-(Ethylenediimino)dipropanoic acid))

     pK_1 6.69, pK_2 9.58   20 $I = 0.1(KCl)$ E3bg   Approx.   I18

     pK_1 6.59, pK_2 9.48   20 $I = 0.1(KNO_3)$ E   M25

4391 $C_8H_{16}O_4N_2$ 3,6-Diazaoctanedioic acid, 3,6-dimethyl-   (Ethylenedinitrilo-N,N'-diacetic acid, N,N'-dimethyl-)

     pK_1 6.047, pK_2 10.068   20   E1   Rel.   O3

     Variation with temperature

	0	10	30	40
pK_1	6.294	6.169	5.926	5.803
pK_2	10.446	10.268	9.882	9.684

     Thermodynamic quantities are derived from the results

4392 $C_8H_{16}O_4N_2$ Octane, 1,1-dinitro-

     5.50   not stated   Extrapolated from values in ethanol/water   E3bg   Uncert.   N36

4393 $C_8H_{16}O_4N_2$ Octanedioic acid, 2,7-diamino-

     pK_1 1.84, pK_2 2.62, pK_3 9.23, pK_4 9.89   20 $I = 0.15(NaClO_4)$, concentration constants   E3bg,R3d   Approx.   H22

No.	Molecular formula, name and pK value(s)	T(°C)	Remarks	Method	Assessment	Ref.
4394	$C_8H_{17}O_2N_3$ 2,3-Butanedione dioxime, 1-diethylamino-					U9
	pK$_1$ 8.55, pK$_2$ 10.75	25	c = 0.0025-0.005, I = 0.1(KNO$_3$) (probably mixed constants)	E		
4395	$C_8H_{17}NS$ Ethanethiol, 2-amino-1-cyclohexyl-					B147
	pK$_1$ 8.50	25	c = 0.01, mixed constant	E3bg	Approx.	
4396	$C_8H_{17}NS_2$ Methanedithioic acid, heptylamino- (Dithiocarbamic acid, N-heptyl-)					T66
	3.29	21	c ≈ 0.01-0.00001, concentration constant	O5	Uncert.	
	3.26		(Compound unstable in acid solution)	E	Uncert.	
4397	$C_8H_{18}ON_2$ 2-Butanone oxime, 3-isopropylamino-3-methyl-					W8
	9.16	25	I = 0.002	CAL	Uncert.	
			Thermodynamic quantities are also given			
4398	$C_8H_{18}ON_2$ 2-Butanone oxime, 3-methyl-3-propylamino-					W8
	9.09	25	I = 0.002	CAL	Uncert.	
			Thermodynamic quantities are also given			
4399	$C_8H_{18}N_2S_2$ Methanedithioic acid, 2-aminoheptylamino- (Dithiocarbamic acid, N-2-aminoheptyl-)					T66
	2.95	21	c ≈ 0.01-0.00001, concentration constant	O5	Uncert.	
	2.80		(Compound unstable in acid solution)	E	Uncert.	

No.	Formula	Name / pK	Value	T (°C)	Conditions	Method	Reliability	Ref.
4400	$C_8H_{19}O_2P$	Dibutylphosphinic acid						
			3.39	20	c = 0.0004-0.05, concentration constant	C2	Approx.	K63
			3.29	20±2	$c \approx$ 0.001 or 0.0001, $I \approx$ 0.1 or 1.0(NaClO$_4$), concentration constant	DIS	Uncert.	K55
			3.50	not stated	c = 0.005, $I \not> 0.025$, mixed constant, 7% ethanol	E3bg	Uncert.	M57
4401	$C_8H_{19}O_2P$	Di-isobutylphosphinic acid						
			3.70	not stated	c = 0.005, $I \not> 0.025$, mixed constant, 7% ethanol	E3bg	Uncert.	M57
4402	$C_8H_{19}O_4P$	Dibutyl hydrogen phosphate						
			1.00	25	I = 0.1 and 1.0(NaClO$_4$), concentration constant	DIS	Uncert.	D77
			1.72	25	$m \approx$ 0.04, mixed constant	E3bg	Uncert.	K67
4403	$C_8H_{19}S_2P$	Dibutylphosphinodithioic acid						
			1.79	20	c = 0.005, 7% ethanol, mixed constant	E3bg	Uncert.	K5, K6
4404	$C_8H_{20}O_6P_2$	1,1-Octanediyldiphosphonic acid						
		pK_2 7.45, pK_3 11.9		25	I = 0.5((CH$_3$)$_4$NCl)	E3bg	Uncert.	C17
4405	$C_8H_4OBrF_3$	Acetophenone, 3-bromo-α,α,α-trifluoro-						
			9.51	25	$I \approx$ 0.1, pK of hydrate	O5	Approx.	S116

No.	Molecular formula, name and pK value(s)	T(°C)	Remarks	Method	Assessment	Ref.
4406	$C_8H_4O_3NF_3$ Acetophenone, α,α,α-trifluoro-3-nitro-					
	9.18	25	I ≈ 0.1 pK of hydrate	05	Approx.	S116
4407	$C_8H_5O_2NS_2$ Thiazolidin-4-one, 5-furfurylidene-2-thioxo-					
	6.37	20	I = 0.1(NaClO$_4$)	05	Approx.	R17
4408	$C_8H_5O_2F_3S$ 1,3-Butanedione, 4,4,4-trifluoro-1-(2-thienyl)-					
	6.53	25	I = 1.0	E3bg	Uncert.	P36
	6.2	room temp.		E3bg	Uncert.	R19
4409	$C_8H_5O_2F_3Se$ 1,3-Butanedione, 4,4,4-trifluoro-1-(2-selenofuryl)-					
	6.32	25	I = 1.0	E3bg	Uncert.	P36
4410	$C_8H_5ON_2S$ 1,3,4-Oxadiazole-2(3H)-thione, 5-phenyl-					
	4.27	25	c ≈ 0.004, mixed constant	E3bg	Approx.	S9
4411	$C_8H_6OBrF_3$ Ethanol, 1-(3-bromophenyl)-2,2,2-trifluoro-					
	11.50	25	I ≈ 0.1, mixed constant	05	Approx.	S116
4412	$C_8H_6O_2N_4S$ Pyrimidin-4(1H)-one, 6-(1,4-dihydro-4-oxopyrimidin-6-yl)thio- (Bis(4-hydroxypyrimidin-6-yl)sulfide)					
	pK$_1$ 7.20, pK$_2$ 8.48		c = 0.0025	E3bg	Approx.	B130

4413 $C_8H_6O_3NF_3$ Ethanol, 2,2,2-trifluoro-1-(3-nitrophenyl)-

11.23	25	$I \simeq 0.1$, mixed constant	O5	Approx.	S116

4414 $C_8H_6O_4NBr$ Acetophenone, α-bromo-4-hydroxy-3-nitro-

4.7	25	$I = 0.05$, 2.5% acetonitrile, mixed constant	O5	Uncert.	F48

4415 $C_8H_6O_4N_4S_2$ Pyrimidine-2,4(1H,3H)-dione, 5-(1,2,3,4-tetrahydro-2,4-dioxopyrimidin-5-yl)disulfanyl-   (Bis(1,2,3,4-tetrahydro-2,4-dioxopyrimidin-5-yl)disulfane)

pK$_1$ 8.0	Conditions not stated	O5	Uncert.	B23
pK$_2$ >13				

4416 $C_8H_7ON_4Cl$ Pteridin-7(8H)-one, 4-chloro-6,8-dimethyl-

12.5	20	$c = 0.00004$, 0.1M buffer	O5	Uncert.	J4

4417 $C_8H_7O_2ClS$ Acetic acid, 2-(2-chlorophenyl)thio-

3.23	20	$c \simeq 0.001$, $I = 0.1(KNO_3)$, mixed constant	E3bg	Approx.	P39

4418 $C_8H_7O_2ClS$ Acetic acid, 2-(3-chlorophenyl)thio-

3.30	20	$c \simeq 0.001$, $I = 0.1(KNO_3)$, mixed constant	E3bg	Approx.	P39
3.60	25	$c \simeq 0.01$	E3bg	Approx.	P13

Value in mixed solvent is also given

4419 $C_8H_7O_2ClS$ Acetic acid, 2-(4-chlorophenyl)thio-

3.33	20	$c \simeq 0.001$, $I = 0.1(KNO_3)$, mixed constant	E3bg	Approx.	P39
3.53	25	$c \simeq 0.01$	E3bg	Approx.	P13

Value in mixed solvent is also given

No.	Molecular formula, name and pK value(s)	T(°C)	Remarks	Method	Assessment	Ref
4420	$C_8H_7O_2ClSe$ Acetic acid, 2-(2-chlorophenyl)seleno-					
	3.57	20	$c \approx 0.001$, $I = 0.1(KNO_3)$, mixed constant	E3bg	Approx.	P39
4421	$C_8H_7O_2ClSe$ Acetic acid, 2-(3-chlorophenyl)seleno-					
	3.64	20	$c \approx 0.001$, $I = 0.1(KNO_3)$, mixed constant	E3bg	Approx.	P39
4422	$C_8H_7O_2ClSe$ Acetic acid, 2-(4-chlorophenyl)seleno-					
	3.68	20	$c \approx 0.001$, $I = 0.1(KNO_3)$, mixed constant	E3bg	Approx.	P39
4423	$C_8H_7O_2BrS$ Acetic acid, 2-(3-bromophenyl)thio-					
	3.53	25	$c \approx 0.01$	E3bg	Approx.	P13
			Value in mixed solvent is also given			
4424	$C_8H_7O_2BrS$ Acetic acid, 2-(4-bromophenyl)thio-					
	3.33	20	$c \approx 0.001$, $I = 0.1(KNO_3)$, mixed constant	E3bg	Approx.	P39
	3.54	25	$c \approx 0.01$	E3bg	Approx.	P13
			Value in mixed solvent is also given			
4425	$C_8H_7O_2BrSe$ Acetic acid, 2-(2-bromophenyl)seleno-					
	3.58	20	$c = 0.001$, $I = 0.1(KNO_3)$, mixed constant	E3bg	Approx.	P39
4426	$C_8H_7O_2BrSe$ Acetic acid, 2-(4-bromophenyl)seleno-					
	3.70	20	$c = 0.001$, $I = 0.1(KNO_3)$, mixed constant	E3bg	Approx.	P39

4427	$C_8H_7O_2FS$	Acetic acid, 2-(3-fluorophenyl)thio-					
		3.513	25	$c \approx 0.01$	E3bg	Approx.	P13
		Value in mixed solvent is also given					
4428	$C_8H_7O_2FS$	Acetic acid, 2-(4-fluorophenyl)thio-					
		3.584	25	$c \approx 0.01$	E3bg	Approx.	P13
		Value in mixed solvent is also given					
4429	$C_8H_7O_3ClS$	Acetic acid, 2-(3-chlorophenyl)sulfinyl-					
		2.663	25	$c \approx 0.01$	E3bg	Approx.	P13
		Value in mixed solvent is also given					
4430	$C_8H_7O_3ClS$	Acetic acid, 2-(4-chlorophenyl)sulfinyl-					
		2.664	25	$c \approx 0.01$	E3bg	Approx.	P13
		Value in mixed solvent is also given					
4431	$C_8H_7O_3BrS$	Acetic acid, 2-(3-bromophenyl)sulfinyl-					
		2.674	25	$c \approx 0.01$	E3bg	Approx.	P13
		Value in mixed solvent is also given					
4432	$C_8H_7O_3BrS$	Acetic acid, 2-(4-bromophenyl)sulfinyl-					
		2.657	25	$c \approx 0.01$	E3bg	Approx.	P13
		Value in mixed solvent is also given					

No.	Molecular formula, name and pK value(s)	$T(^{o}C)$	Remarks	Method	Assessment	Ref
4433	$C_8H_7O_3FS$ Acetic acid, 2-(3-fluorophenyl)sulfinyl- 2.675	25	$c \approx 0.01$ Value in mixed solvent is also given	E3bg	Approx.	P13
4434	$C_8H_7O_3FS$ Acetic acid, 2-(4-fluorophenyl)sulfinyl- 2.653	25	$c \approx 0.01$ Value in mixed solvent is also given	E3bg	Approx.	P13
4435	$C_8H_7O_4NS$ Acetic acid, 2-(2-nitrophenyl)thio- 3.10	20	$c \approx 0.001$, $I = 0.1(KNO_3)$, mixed constant	E3bg	Approx.	P39
4436	$C_8H_7O_4NS$ Acetic acid, 2-(3-nitrophenyl)thio- 3.400	25	$c \approx 0.01$ Value in mixed solvent is also given	E3bg	Approx.	P13
4437	$C_8H_7O_4NS$ Acetic acid, 2-(4-nitrophenyl)thio- 3.09	20	$c \approx 0.001$, $I = 0.1(KNO_3)$, mixed constant	E3bg	Approx.	P39
	3.375	25	$c \approx 0.01$ Value in mixed solvent is also given	E3bg	Approx.	P13
4438	$C_8H_7O_4NS$ 3-Indolyl hydrogen sulfate (3-Indoxyl sulfate) 15.23(NH)	25	In aq. KOH, H_ scale	07	Uncert.	Y3

4439	$C_8H_7O_4NSe$	Acetic acid, 2-(2-nitrophenyl)seleno-					
		3.42	20	$c = 0.001$, $I = 0.1(KNO_3)$, mixed constant	E3bg	Approx.	P39
4440	$C_8H_7O_4NSe$	Acetic acid, 2-(3-nitrophenyl)seleno-					
		3.55	20	$c = 0.001$, $I = 0.1(KNO_3)$, mixed constant	E3bg	Approx.	P39
4441	$C_8H_7O_4NSe$	Acetic acid, 2-(4-nitrophenyl)seleno-					
		3.43	20	$c = 0.001$, $I = 0.1(KNO_3)$, mixed constant	E3bg	Approx.	P39
4442	$C_8H_7O_4ClS$	Acetic acid, 2-(3-chlorophenyl)sulfonyl-					
		2.454	25	$c \approx 0.01$	E3bg	Approx.	P13
		Value in mixed solvent is also given					
4443	$C_8H_7O_4ClS$	Acetic acid, 2-(4-chlorophenyl)sulfonyl-					
		2.424	25	$c \approx 0.01$	E3bg	Approx.	P13
		Value in mixed solvent is also given					
4444	$C_8H_7O_4BrS$	Acetic acid, 2-(3-bromophenyl)sulfonyl-					
		2.523	25	$c \approx 0.01$	E3bg	Approx.	P13
		Value in mixed solvent is also given					
4445	$C_8H_7O_4BrS$	Acetic acid, 2-(4-bromophenyl)sulfonyl-					
		2.519	25	$c \approx 0.01$	E3bg	Approx.	P13
		Value in mixed solvent is also given					

No.	Molecular formula, name and pK value(s)	$T(^oC)$		Remarks	Method	Assessment	Ref
4446	$C_8H_7O_4FS$ Acetic acid, 2-(3-fluorophenyl)sulfonyl- 2.432	25	$c \approx 0.01$	Value in mixed solvent is also given	E3bg	Approx.	P13
4447	$C_8H_7O_4FS$ Acetic acid, 2-(4-fluorophenyl)sulfonyl- 2.479	25	$c \approx 0.01$	Value in mixed solvent is also given	E3bg	Approx.	P13
4448	$C_8H_7O_5NS$ Acetic acid, 2-(3-nitrophenyl)sulfino- 2.624	25	$c \approx 0.01$	Value in mixed solvent is also given	E3bg	Approx.	P13
4449	$C_8H_7O_5NS$ Acetic acid, 2-(4-nitrophenyl)sulfino- 2.586	25	$c \approx 0.01$	Value in mixed solvent is also given	E3bg	Approx.	P13
4450	$C_8H_7O_6NS$ Acetic acid, 2-(3-nitrophenyl)sulfonyl- 2.362	25	$c \approx 0.01$	Value in mixed solvent is also given	E3bg	Approx.	P13
4451	$C_8H_7O_6NS$ Acetic acid, 2-(4-nitrophenyl)sulfonyl- 2.330	25	$c \approx 0.01$	Value in mixed solvent is also given	E3bg	Approx.	P13

No.	Formula	Name	pK	T (°C)	Conditions	Method	Assessment	Ref.
4452	$C_8H_7N(Cl)S$	Thieno[3,2-c]pyridinium chloride, 4-methyl-	6.17	20		E3bg	Uncert.	D51
4453	$C_8H_7N(Cl)S$	Thieno[2,3-c]pyridinium chloride, 7-methyl-	5.81	20		E3bg	Uncert.	D51
4454	C_8H_8ONCl	3-Azabicyclo[4.3.0]nona-1(6),4-dien-2-one, 4-chloro- (5H-1-Pyrindin-1(2H)-one, 3-chloro-6,7-dihydro-)	pK_1 0.13, pK_2 8.53	20	c = 0.00005	05	Uncert.	S74
4455	$C_8H_9ON_4S$	Pteridin-7(8H)-one, 1,2-dihydro-6,8-dimethyl-2-thioxo-	6.15	20	c = 0.0025	E3bg	Approx.	J4
4456	$C_8H_8O_2NCl$	Acetamide, N-(3-chloro-4-hydroxyphenyl)-	8.20	25	c ≈ 0.0001-0.0002, I ≈ 0.1, mixed constant	05	Approx.	S76
4457	$C_8H_8O_2N_4S$	Pteridine-4,7(3H,8H)-dione, 3-methyl-2-thiomethyl-	6.47	20	c = 0.001	E3bg	Approx.	P47
4458	$C_8H_8O_3NCl$	Acetohydroxamic acid, 2-(4-chlorophenoxy)-	8.75	not stated	I = 0.2(NaCl)	E3bg	Uncert.	C72
4459	$C_8H_8N_3ClS$	4-Chlorobenzaldehyde thiosemicarbazone	11.20(H_2O) 11.80(D_2O)	25	I = 1.0(KCl), mixed constants	05	Uncert.	S18
4460	C_8H_9ONS	Benzenethiol, 2-((N-formyl-N-methyl)amino-	5.65	25±0.5	I = 0.01	05	Uncert.	V13

No.	Molecular formula, name and pK value(s)	T(°C)	Remarks	Method	Assessment	Ref.
4461	$C_8H_9O_2NS$ Acetic acid, 2-(4-aminophenyl)thio-					
	pK$_1$ 3.10, pK$_2$ 4.59	20	c ≈ 0.001, I = 0.1(KNO$_3$), mixed constants	E3bg	Approx.	P39
4462	$C_8H_9O_2N_2Cl$ Acetohydrazide, 2-(4-chlorophenoxy)-					
	pK$_1$ 2.61	25	mixed constants(?)	E3bg	Uncert.	T43,T44
	12.19(NH)	25		05	Uncert.	W36
4463	$C_8H_9O_2N_2(I)$ 1-Methylpyridinium iodide, 3-hydroxyiminoacetyl-					
	7.20	not stated	I = 0.2(NaCl)	E3bg	Uncert.	C72
	7.2	not stated	c ≈ 0.04	E3bg	Uncert.	W36
4464	$C_8H_9O_2N_2(I)$ 1-Methylpyridinium iodide, 4-hydroxyiminoacetyl-					
	7.1	not stated	c ≈ 0.04	E3bg	Uncert.	W36
	7.10	not stated	I = 0.2(NaCl)	E3bg	Uncert.	C72
4465	$C_8H_9O_3N_2Cl$ Pyrimidine-2,4(1H,3H)-dione, 5-chloro-1-(2-oxolanyl)-					
	7.87		Conditions not stated	05	Uncert.	G17
4466	$C_8H_9O_3N_2Br$ Pyrimidine-2,4(1H,3H)-dione, 5-bromo-1-(2-oxolanyl)-					
	8.00		Conditions not stated	05	Uncert.	G17
4467	$C_8H_9O_3N_2I$ Pyrimidine-2,4(1H,3H)-dione, 5-iodo-1-(2-oxolanyl)-					
	8.17		Conditions not stated	05	Uncert.	G17

No.	Formula	Name	pK	t(°C)	Conditions	Method	Reliability	Ref.
4468	$C_8H_9O_3N_2F$	Pyrimidine-2,4(1H,3H)-dione, 5-fluoro-1-(2-oxolanyl)-	7.80		Conditions not stated	05	Uncert.	G17
4469	$C_8H_{10}ON_3Cl$	Triazen-3-ol, 1-(2-chlorophenyl)-3-ethyl-	11.44	25	c = 0.00004, I = 0.1(KCl), mixed constant	05	Uncert.	D54
4470	$C_8H_{10}ON_3Cl$	Triazen-3-ol, 1-(4-chlorophenyl)-3-ethyl-	11.65	25	c = 0.00004, I = 0.1(KCl), mixed constant	05	Uncert.	D54
4471	$C_8H_{10}ON_4S$	Purin-8(7H)-one, 6,9-dimethyl-2-methylthio-	pK_1 2.62, pK_2 9.41	20	c < 0.0001, I = 0.01	05	Approx.	B138
4472	$C_8H_{10}ON_4S_2$	Purin-6(1H)-one, 2,8-bis(methylthio)-1-methyl-	pK_1 1.7, pK_2 8.2		Conditions not stated	05	Uncert.	R18
4473	$C_8H_{10}ON_4S_2$	Purin-6(1H)-one, 2,8-bis(methylthio)-9-methyl-	pK_1 2.5, pK_2 7.5		Conditions not stated	05	Uncert.	R18
4474	$C_8H_{10}ON_4S_2$	Purin-6(3H)-one, 2,8-bis(methylthio)-3-methyl-	pK_1 0.5, pK_2 7.9		Conditions not stated	05	Uncert.	R18
4475	$C_8H_{10}O_3N_2S$	Acetamide, N-(4-aminosulfonylphenyl)-	10.02	20	c = 0.0025, I = 0.1(KCl), mixed constant	E3bg	Approx.	W33
4476	$C_8H_{10}O_4NP$	Phosphoramidic acid, N-(4-methylbenzoyl)-	pK_1 2.04, pK_2 6.20	30	I = 0.1(KCl), mixed constants. Solutions of monocyclohexyl ammonium salt used	E3bg	Approx.	Z9

No.	Molecular formula, name and pK value(s)	T(°C)	Remarks	Method	Assessment	Ref.
4477	$C_8H_{10}O_5NP$ Ethyl hydrogen 4-nitrophenylphosphonate					
	2.5	not stated	mixed constant	E3d	V.Uncert.	B175
4478	$C_8H_{10}O_5NAs$ Benzenearsonic acid, 3-acetylamino-4-hydroxy-					
	pK_1 3.73			SOL	Uncert.	H44
	pK_2 7.9, pK_3 9.3			E3bg	Uncert.	
4479	$C_8H_{10}O_5ClP$ 2-(2-Chlorophenoxy)ethyl dihydrogen phosphate					
	pK_2 6.58	20	I = 0.02	E3bg	Approx.	P60
4480	$C_8H_{10}O_5ClP$ 2-(4-Chlorophenoxy)ethyl dihydrogen phosphate					
	6.58	20	I = 0.02	E3bg	Approx.	P60
4481	$C_8H_{11}ON_2(I)$ 1-Ethylpyridinium iodide, 2-hydroxyiminomethyl-					
	8.1	not stated	c ≈ 0.04	E3bg	Uncert.	W36
4482	$C_8H_{11}ON_2(I)$ 1-Methylpyridinium iodide, 2-(1-hydroxyimino)ethyl-					
	9.0	not stated	c ≈ 0.04	E3bg	Uncert.	W36
4483	$C_8H_{11}ON_2(I)$ 1-Methylpyridinium iodide, 2-hydroxyiminomethyl-3-methyl-					
	8.4	not stated	c ≈ 0.04	E3bg	Uncert.	W36
4484	$C_8H_{11}ON_2(I)$ 1-Methylpyridinium iodide, 2-hydroxyiminomethyl-6-methyl-					
	8.1	not stated	c ≈ 0.04	E3bg	Uncert.	W36

4485 $C_8H_{11}O(I)S$ Phenylsulfonium iodide, 3-hydroxy-

7.67	not stated	I = 0.01	E3ag	Uncert.	02

4486 $C_8H_{11}O(I)S$ Phenylsulfonium iodide, 4-hydroxy-

7.33	25	I = 0.1(KCl), mixed constant	E3bg	Approx.	E20
7.30	not stated	I = 0.01	E3ag	Uncert.	02

4487 $C_8H_{11}O_2N_2(I)$ 1-Methylpyridinium iodide, 2-hydroxyiminomethyl-3-methoxy-

8.6	not stated	c ≈ 0.04	E3bg	Uncert.	W36

4488 $C_8H_{12}ON_2S$ 1,3-Diazaspiro[4.5]decan-4-one, 2-thioxo-

8.79	18±3	I = 0.01	05	Uncert.	E4

4489 $C_8H_{12}O_3NP$ Ethyl hydrogen 4-aminophenylphosphonate

pK_1 3.9	not stated	mixed constant	E3d	V.Uncert.	B175

4490 $C_8H_{12}O_4NCl$ Ethyl 2-(3-chloroacetamido)-3-oxobutanoate

8.30	22	E3bg	Uncert.	L7
8.10		05		

4491 $C_8H_{12}O_4NP$ 3-(2-Pyridyl)propyl dihydrogen phosphate

pK_1 5.60, pK_2 6.90	25	I = 0.1(KNO$_3$)	E3bg	Approx.	M115
pK_1 5.09, pK_2 6.75	80				

No.	Molecular formula, name and pK value(s)	$T(^oC)$	Remarks	Method	Assessment	Ref.
4492	$C_8H_{12}O_4N_2S_2$ 4,7-Diazadecanedioic acid, 5,6-dithioxo- (Dithiooxamide, N,N'-bis(2-carboxyethyl)-)					C13
	pK_1 3.76, pK_2 4.74	25	c = 0.07, I = 1.0(KCl), mixed constants	E3bg	Approx.	
	pK_3 11.41	25	c = 0.0001, I = 1.0(KCl), mixed constant	05	Approx.	
	pK_3 12.32	25	By extrapolation from values at	05	Approx.	
			I = 0.01-0.05			
	pK_4 14.19	25	c = 0.0001, I = 1.0(KCl), mixed constant	07	Uncert.	
4493	$C_8H_{13}O_6NS$ 3-Thia-6-azaoctanedioic acid, 6-carboxymethyl- (Iminodiacetic acid, N-[(2-carboxymethylthio)ethyl]-)					Y7
	pK_1 2.02, pK_2 3.23, pK_3 9.00	25	c = 0.0005, I = 0.1(KCl), mixed constants	E3bg	Uncert.	
	pK_1 2.09, pK_2 3.21, pK_3 8.40			E		D66
4494	$C_8H_{15}O_9N_2P$ 5-Azaoctanoic acid, 7-amino-4-carboxy-6-oxo-8-phosphonooxy- (O-Phosphono-L-seryl-L-glutamic acid)					
	pK_1 3.05, pK_2 4.40, pK_3 5.69, pK_4 8.19	25	I = 0.15(KNO$_3$), concentration constants	E3bg	Approx.	O25
	pK_1 3.02, pK_2 4.39, pK_3 5.68, pK_4 8.25	25	I = 0.15(KCl), concentration constants	E3bg	Approx.	O24
4495	$C_8H_{16}O_2N_2S_2$ 4,7-Diazadecane-5,6-dithione, 1,10-dihydroxy- (Dithiooxamide, N,N'-bis(3-hydroxypropyl)-)					J3
	pK_1 11.27	25	c = 0.0001, I = 1.0(NaClO$_4$), mixed constant	05	Approx.	
	pK_2 14.29			07	Uncert.	
	pK_1 11.37			05,R4	Approx.	
4496	$C_8H_{16}O_2N_2S_2$ 4,7-Diazadecane-5,6-dithione, 2,9-dihydroxy- (Dithiooxamide, N,N'-bis(2-hydroxypropyl)-)					J3
	pK_1 10.99	25	c = 0.0001, I = 1.0(NaClO$_4$), mixed constants	05	Approx.	
	pK_2 13.75			07	Uncert.	

(contd)

4496 $C_8H_{16}O_2N_2S_2$ 4,7-Diazadecane-5,6-dithione, 2,9-dihydroxy- (Dithiooxamide, N,N'-bis(2-hydroxypropyl)-)
(Contd)
pK$_1$ 11.11   05,R4   Approx.

4497 $C_8H_{18}ON(I)$ N,N,N-Trimethyl 2-formyl-2-methylpropylammonium iodide (Propanal, 2,2-dimethyl-3-(N,N,N-trimethylammonio-), iodide)
          13.6   30   I = 0.1(KCl), pK of hydrate   E3bg   V.Uncert.   E18

4498 $C_8H_{16}O_{10}N_2P_2$ 3,6-Diazaoctanedioic acid, 3,6-bis(phosphonomethyl)- (Ethylenedinitrilo-N,N'-bis(methylphosphonic acid)-N,N'-
diacetic acid) (EDPA)

pK$_1$ 1.5, pK$_2$ 2.30, pK$_3$ 4.65, pK$_4$ 6.13, pK$_5$ 8.36, pK$_6$ 10.34
                      25   c = 0.001, I = 0.1(KCl), concentration   E3bg,R3g Approx.   D67
                                 constants   (pK$_1$ Uncert.)

pK$_3$ 4.65(POH), pK$_4$ 6.13(POH), pK$_5$ 8.36(NH$^+$), pK$_6$ 10.43(NH$^+$)
                      25   c = 0.001, I = 0.1(KCl), concentration   E3bg,R3g Approx.   D70
                                 constants

pK$_1$ 2.83, pK$_2$ 3.52, pK$_3$ 5.46, pK$_4$ 7.72
pK$_5$ 10.6, pK$_6$ 11.2
                      25   I = 1.0(KNO$_3$), concentration constants   E3bg   Approx.   R6
                                                                              Uncert.

4499 $C_8H_{19}OSP$ Bis(sec-butyl)phosphinothioic O-acid
          3.10   20   c = 0.005, 7% ethanol   E3bg   Uncert.   K6
                                   99.6% thione tautomer

4500 $C_8H_{19}OSP$ Bis(tert-butyl)phosphinothioic O-acid
          3.91   20   c = 0.005, 7% ethanol   E3bg   Uncert.   K6
                                   100% thione tautomer

No.	Molecular formula, name and pK value(s)	$T(^{o}C)$	Remarks	Method	Assessment	Ref.
4501	$C_8H_{19}OSP$ Dibutylphosphinothioic O-acid					
	2.62	20	c = 0.004-0.05, concentration constant	C2	Approx.	K63
	2.91	20	c = 0.005, 7% ethanol	E3bg	Uncert.	K6
			99.2% thione tautomer			
4502	$C_8H_{19}OSP$ Di-isobutylphosphinothioic O-acid					
	3.17	20	c = 0.005, 7% ethanol	E3bg	Uncert.	K6
			99.5% thione tautomer			
4503	$C_8H_{19}O_2N_2(Br)$ N,N,N-Triethyl 2-hydroxyamino-2-oxo-ethylammonium bromide (Acetohydroxamic acid, 2-triethylammonio-, bromide)					
	6.60	25(?)	c = 0.02, I = 0.1(KCl), mixed constant	E3bg	Uncert.	M66
4504	$C_8H_{19}O_2SP$ O-Butyl O-hydrogen isobutylphosphonothioate					
	2.20	20	c = 0.005, 7% ethanol	E3bg	Uncert.	K6
			95% thione tautomer			
4505	$C_8H_{19}O_2S_2P$ O,O-Dibutyl hydrogen phosphorodithioate					
	1.83	20	c = 0.005, 7% ethanol, mixed constant	E3bg	Uncert.	K5,K6
4506	$C_8H_{19}O_2S_2P$ O,O-Di-isobutyl hydrogen phosphorodithioate					
	1.79	20	c = 0.005, 7% ethanol	E3bg	Uncert.	K5,K6

No.	Formula	Name / pK	Temp (°C)	Conditions	Method	Reliability	Ref.
4507	$C_8H_{19}O_3SP$	O,O-Dibutyl hydrogen phosphorothioate					
		1.65	20	c = 0.005, 7% ethanol	E3bg	V.Uncert.	K6
				62% thione tautomer			
4508	$C_8H_{19}O_3SP$	O,O-Di-isobutyl O-hydrogen phosphorothioate					
		1.55	20	c = 0.005, 7% ethanol	E3bg	Uncert.	K6
				52% thiol tautomer			
4509	$C_8H_{20}O_3N_2S_2$	4,7-Diazadecane-1,10-disulfonic acid (N,N-Bis(3-sulfopropyl)ethylenediamine) (EDPS)					
		pK$_1$ 6.65, pK$_2$ 9.8	18	c < 0.02, mixed constants	E3bg	V.Uncert.	J12
4510	$C_8H_{21}O_2NP_2$	Diethylphosphinamidate, N-diethylphosphinyl-					
		6.36	20	c = 0.005, 7% ethanol, mixed constant	E3bg	Uncert.	G19
4511	$C_8H_{21}O_6NP_2$	O,O-Diethyl phosphoramidate, N'-diethoxyphosphinyl-					
		4.45	20	c = 0.005, 7% ethanol, mixed constant	E3bg	Uncert.	G19
4512	$C_8H_{22}O_6N_2P_2$	3,6-Diazaoctane-2,7-diphosphonic acid, 2,7-dimethyl- (1,1'-(Ethylenediimino)bis(1-methylethylphosphonic acid))					
		pK$_1$ 4.95, pK$_2$ 6.00, pK$_3$ 8.55, pK$_4$ 11.68	25	I = 0.1(KCl), concentration constants	E3bg,R3g	Approx.	D67,D70
		pK$_1$ 4.87, pK$_2$ 6.06, pK$_3$ 8.55, pK$_4$ 11.68	not stated		E3bg	Uncert.	K7
4513	$C_8H_7O_2N_2(F_4B)$	Benzenediazonium fluoborate, 4-carboxymethyl-					
		3.24	25	Mixed constant	E3bg	Uncert.	L36
4514	$C_8H_7O_5NF_3P$	2,2,2-Trifluoroethyl hydrogen 4-nitrophenylphosphonate					
		2.5	not stated	Mixed constant	E3d	V.Uncert.	B175

No.	Molecular formula, name and pK value(s)	T(°C)	Remarks	Method	Assessment	Ref.
4515	$C_9H_6O_2$ Cyclopropenone, 2-hydroxy-3-phenyl- 2.01	c = 0.00004		O5	Uncert.	P15
4516	$C_9H_6O_2$ 1,3-Indandione 7.2 7.41	18±2 not stated	1% ethanol, mixed constant 1% methanol	O5 or E3d E3bg	Uncert. Uncert.	W25 G46
4517	$C_9H_6O_2$ Propynoic acid, 3-phenyl- 2.23	25	c = 0.005, I = 0.1(NaCl), mixed constant	E3bg	Approx.	M48
4518	$C_9H_6O_3$ 4H-Benzopyran-4-one, 5-hydroxy- 10.75			O		M124
4519	$C_9H_6O_4$ 1,3-Indandione, 2,2-dihydroxy (Ninhydrin) 8.82	30	c = 0.01, I = 0.1(KCl), mixed constant	E3bg	Approx.	E18
4520	$C_9H_6O_6$ 1,2,3-Benzenetricarboxylic acid (Hemimellitic acid) pK$_1$ 2.84, pK$_2$ 4.26, pK$_3$ 6.17 pK$_1$ 2.62, pK$_2$ 3.82, pK$_3$ 5.51	25 25	I = 0.1(KCl)	E3bg	Approx.	Y9
4521	$C_9H_6O_6$ 1,2,4-Benzenetricarboxylic acid (Trimellitic acid) pK$_1$ 2.64 pK$_2$ 4.15, pK$_3$ 5.67	25		E3bg	Uncert. Approx.	Y9

(contd)

No.	Formula	Name / pK	t°C	Conditions	Method	Assessment	Ref.
4521 (Contd)	$C_9H_6O_6$	1,2,4-Benzenetricarboxylic acid (Trimellitic acid)					
		pK$_1$ 2.42	25	I = 0.1(KCl)		Uncert.	
		pK$_2$ 3.71, pK$_3$ 5.01				Approx.	
4522	$C_9H_6O_6$	1,3,5-Cycloheptatriene-1,2-dicarboxylic acid, 6-hydroxy-7-oxo-					
		pK$_1$ 3.24, pK$_2$ 6.26, pK$_3$ 8.57	25	c = 0.001	E3bg	Approx.	T5
4523	$C_9H_6O_7$	1,2,4-Benzenetricarboxylic acid, 5-hydroxy-					
		pK$_1$ 1.86, pK$_2$ 3.39, pK$_3$ 5.30	25	c = 0.002, I = 0.1, concentration constants	E3bg	Uncert.	P85
4524	$C_9H_6O_7$	1,3,5-Benzenetricarboxylic acid, 2-hydroxy- (Hydroxytrimesic acid)					
		pK$_1$ 2.77, pK$_2$ 3.93, pK$_3$ 4.88	25	c = 0.002, I = 0.1, concentration constants	E3bg	Uncert.	P85
4525	$C_9H_6N_2$	5-Indolecarbonitrile					
		15.24	25	c ≈ 0.0001	07	Uncert.	T21
		15.24 (NH)	25	In aq. KOH, H$_-$ scale	07	Uncert.	Y2
4526	$C_9H_8O_2$	Benzo[b]furan-4-ol, 2-methyl-					
		9.31	20	2% ethanol	05	Approx.	D32
4527	$C_9H_8O_2$	Benzo[b]furan-4-ol, 3-methyl-					
		9.65	20	2% ethanol	05	Approx.	D32
4528	$C_9H_8O_2$	1-Indanone, 5-hydroxy-					
		7.53	25	I = 0.1(NaClO$_4$), concentration constant	05	Approx.	M18

No.	Molecular formula, name and pK value(s)	T(°C)	Remarks	Method	Assessment	Ref.
4529	$C_9H_8O_2$ 1-Indanone, 7-hydroxy-					
	8.43	25	$I = 0.1(NaClO_4)$, concentration constant	05	Approx.	M18
4530	$C_9H_8O_2$ Propenoic acid, 3-phenyl- (Cinnamic acid)					
	3.93(cis)	25		E3bg	Approx.	M10
	4.50(trans)					
	4.42(trans)	25		05,R4	Approx.	N34
4531	$C_9H_8O_3$ Benzoic acid, 2-acetyl-					
	4.126	25	$c \approx 0.00022$-0.00214	C2,R1c	Rel.	B122
	4.12	25	$I = 0.02$	E3bg	Approx.	H18
4532	$C_9H_8O_3$ Benzoic acid, 3-acetyl-					
	3.827	25	$c = 0.00023$-0.00146	C2,R1c	Rel.	B122
4533	$C_9H_8O_3$ Benzoic acid, 4-acetyl-					
	3.700	25	$c = 0.00017$-0.00174	C2,R1c	Rel.	B122
4534	$C_9H_8O_3$ 2,4,6-Cycloheptatrien-1-one, 4-acetyl-2-hydroxy-					
	5.41	25	$I = 2.0$	05	Approx.	O16
4535	$C_9H_8O_3$ Propenoic acid, 3-(2-hydroxyphenyl)- (Coumarinic acid)					
	pK_2 10.66(cis)	25	$c = 0.00012$ coumarin $\underset{H_2O}{\rightleftarrows}$ coumarinic acid anion $\underset{K_2}{\rightleftarrows}$ dianion + H^+ (contd)	05	Uncert.	M62
	anion formed from hydrolysis of coumarin;					

No.	Formula	Name	pK	T(°C)	Conditions	Method	Assessment	Ref.
4535 (Contd)	$C_9H_8O_3$	Propenoic acid, 3-(2-hydroxyphenyl)- (Coumarinic acid) pK$_1$ 3.70		30	I = 1.0(LiCl), mixed constant	O5	Approx.	H36
4536	$C_9H_8O_3$	Propenoic acid, 3-(4-hydroxyphenyl)- (p-Coumaric acid) pK$_1$ 4.64, pK$_2$ 9.45		25		O5,R4	Approx.	N34
4537	$C_9H_8O_4$	Acetic acid, benzoyloxy- (Benzoylglycolic acid) 3.04				E		C74
4538	$C_9H_8O_4$	Acetic acid, 2-(6-hydroxy-7-oxo-1,3,5-cycloheptatrien-1-yl)- pK$_1$ 4.52, pK$_2$ 8.77		25	c = 0.001	E3bg	Approx.	T5
4539	$C_9H_8O_4$	1,2-Benzenedicarboxylic acid, 4-methyl- pK$_1$ 3.04, pK$_2$ 5.26		30	I = 0.1(KCl), concentration constants	E3bg,R3d	Approx.	N6
4540	$C_9H_8O_4$	Benzoic acid, 2-acetoxy- (Aspirin) 3.38		25	I = 1.0(KCl), mixed constant	O5	Approx.	S4
4541	$C_9H_8O_4$	Benzoic acid, 4-acetoxy- 3.92		25	I = 1.0(KCl), mixed constant	O5	Approx.	S4
4542	$C_9H_8O_4$	Benzoic acid, 3,4-ethylenedioxy- 4.35		25	c = 0.003, 5% ethanol	E3bg	Uncert.	B190
4543	$C_9H_8O_4$	Bicyclo[2.2.1]hepta-2,5-diene-2,3-dicarboxylic acid pK$_1$ 1.32 pK$_2$ 7.77		25	c = 0.05-0.09 c = 0.005-0.009	E3bg	Approx.	M11

No.	Molecular formula, name and pK value(s)	T(°C)	Remarks	Method	Assessment	Ref.
4544	$C_9H_8O_4$ 1,4,6-Cycloheptatriene-1-carboxylic acid, 2-hydroxy-7-methyl-3-oxo- pK_1 2.48, pK_2 7.45	25	$I = 0.2(NaClO_4)$, concentration constants	E3bg	Approx.	G53,G54
4545	$C_9H_8O_4$ 1,3,5-Cycloheptatriene-1-carboxylic acid, 6-hydroxy-2-methyl-7-oxo- pK_1 3.15, pK_2 8.29	25	c = 0.001	E3bg	Approx.	T5
4546	$C_9H_8O_4$ Propanedioic acid, phenyl- pK_1 2.52, pK_2 5.59(H_2O) pK_1 3.02, pK_2 5.96(D_2O)	25		E3bg	Approx.	G23
4547	$C_9H_8O_4$ Propenoic acid, 3-(2,4-dihydroxyphenyl)- 4.10(cis)	30	$I = 1.0(LiCl)$, mixed constant	05	Approx.	H36
4548	$C_9H_8O_4$ Propenoic acid, 3-(2,5-dihydroxyphenyl)- 3.69(cis)	30	$I = 1.0(LiCl)$, mixed constant	05	Approx.	H36
4549	$C_9H_8O_4$ Propenoic acid, 3-(3,4-dihydroxyphenyl)- pK_1 4.62, pK_2 9.07 pK_1 4.49, pK_2 8.76	25	$I = 0.05(KNO_3)$	E3bg	Approx.	T39
4550	$C_9H_8O_5$ Acetic acid, 2-(2-carboxyphenoxy)- pK_1 2.51, pK_2 4.36	25	$I = 0.1(NaClO_4)$, concentration constants	E3bg,R3f	Approx.	S133

No.	Formula	Name / pK	Temp.	Conditions	Reliability	Method	Ref.
4551	$C_9H_8O_5$	Acetic acid, 4-hydroxybenzoyloxy-				E	C74
		3.14					
4552	$C_9H_8O_5$	1,2-Benzenedicarboxylic acid, 4-methoxy-			Approx.	E3bg	Y10
		pK_1 2.67, pK_2 5.16	25	$I = 0.1(KNO_3)$, concentration constants			
4553	$C_9H_8O_6$	4H-Pyran-2-carboxylic acid, 6-ethoxycarbonyl-4-oxo-			Approx.	C1	S10
		1.28	25				
4554	C_9H_9N	Indole, 3-methyl- (Skatole)			V.Uncert.	07	Y3
		16.60	25	In aq. KOH, H_- scale			
4555	$C_9H_{10}O$	Phenol, 2-allyl-			Approx.	05	07
		10.29	25				
4556	$C_9H_{10}O$	Phenol, 4-allyl-			Approx.	05	B120
		10.23	25	$c = 0.0003$			
4557	$C_9H_{10}O$	Phenol, 4-(1-propenyl)-			Approx.	05,R4	L49
		9.82(trans)	25				
		9.8	25	$c = 0.0003$	Uncert.	05	B120
				Decomposition reported			
4558	$C_9H_{10}O_2$	Acetic acid, 2-(p-tolyl)-			Rel.	C2,R1a/b	F18
		4.361	25				

(contd)

No.	Molecular formula, name and pK value(s)	$T(^oC)$	Remarks	Method	Assessment	Ref.
4558 (Contd)	$C_9H_{10}O_2$ Acetic acid, 2-p-tolyl)-		Variation with pressure (bar)			
			1000 2000 3000			
			4.147 3.965 3.808			
4559	$C_9H_{10}O_2$ Acetophenone, 2-hydroxy-3-methyl- 9.14	25		E3bg	Approx.	F25
4560	$C_9H_{10}O_2$ Benzaldehyde, 4-hydroxy-3,5-dimethyl- 7.74	25	c = 0.0017	E3bg	Approx.	F16
4561	$C_9H_{10}O_2$ Benzoic acid, 2,3-dimethyl- 3.716	25		05	Approx.	W38
			Variation with temperature			
			15 20 30 35 40			
			3.667 3.678 3.743 3.755 3.790			
			Thermodynamic quantities are derived from the results			
	3.80	25	I = 0.004-0.007	E3bg	Approx.	L76
4562	$C_9H_{10}O_2$ Benzoic acid, 2,4-dimethyl- 4.219	25		05	Approx.	W38

(contd)

4562
(Contd) $C_9H_{10}O_2$ Benzoic acid, 2,4-dimethyl-

| | 4.30 | Approx. | L76 |

Variation with temperature

15	20	30	35	40
4.159	4.178	4.249	4.270	4.288

Thermodynamic quantities are derived from the results

| 25 | I = 0.008 | E3bg |

4563 $C_9H_{10}O_2$ Benzoic acid, 2,5-dimethyl-

| 4.001 | | 05 | W38 |

Variation with temperature

15	20	30	35	40
3.910	3.952	4.016	4.039	4.070

| 25 |

Thermodynamic quantities are derived from the results

| 25 | I = 0.006 | E3bg |

| 4.04 | Approx. | L76 |

4564 $C_9H_{10}O_2$ Benzoic acid, 2,6-dimethyl-

| 3.354 | | 05 | W38 |

Variation with temperature

15	20	30	35	40
3.242	3.293	3.423	3.451	3.466

| 25 |

Thermodynamic quantities are derived from the results

| 25 | I = 0.05 | E3bg |

| 3.26 | Approx. | L76 |

No.	Molecular formula, name and pK value(s)	T(°C)	Remarks	Method	Assessment	Ref.
4565	$C_9H_{10}O_2$ Benzoic acid, 3,4-dimethyl-					
	4.50	25	I = 0.005	E3bg	Approx.	L76
4566	$C_9H_{10}O_2$ Benzoic acid, 3,5-dimethyl-					
	4.298	25		05	Approx.	M38
			Variation with temperature			
			15 20 30 35 40			
			4.295 4.300 4.310 4.304 4.307			
			Thermodynamic quantities are derived from the results			
	4.39	25	I = 0.06	E3bg	Approx.	L76
4567	$C_9H_{10}O_2$ 2,4-Nonadiynoic acid					
	1.89	25	c = 0.01, I = 0.1(NaCl), mixed constant	E3bg	Approx.	M48
4568	$C_9H_{10}O_2$ Propanoic acid, 2-phenyl-					
	4.3	25		E3bg	Uncert.	C57
4569	$C_9H_{10}O_2$ Propanoic acid, 3-phenyl-					
	4.664	25	m = 0.005-0.02	E1cg,R4	Rel.	K30
	4.71	25	c $\simeq$ 0.0008	E3bg	Approx.	T55
	4.67	25	Values in mixed solvents are also given	E3ag	Approx.	H48
	4.708	25	c $\simeq$ 0.01	E3bg	Approx.	P12
			Values in mixed solvents are also given			

No.	Formula	Name / pK	Conditions	T(°C)	Method	Reliability	Ref.
4570	$C_9H_{10}O_3$	Acetic acid, 2-(4-methoxyphenyl)- 4.356	Variation with pressure(bar) 1000 2000 3000 / 4.146 3.967 3.811	25		C2,R1a/b Rel.	F18
4571	$C_9H_{10}O_3$	Acetophenone, 2-hydroxy-3-methoxy- 10.00	I = 0.01	25	E3bg	Approx.	L48
4572	$C_9H_{10}O_3$	Benzaldehyde, 2-hydroxy-3-methoxy-5-methyl- 8.04	I = 0.01	25	E3bg	Approx.	L43
4573	$C_9H_{10}O_3$	Benzoic acid, 2-hydroxy-3,6-dimethyl- pK_1 3.23 pK_2 12.32	I = 0.1(KCl), mixed constants	25	E3bg 05	Approx.	F35a
4574	$C_9H_{10}O_3$	Benzoic acid, 2-methoxy-3-methyl- 3.84	c = 0.001, 1% ethanol	20	E3bg	Approx.	P24,P27
4575	$C_9H_{10}O_3$	Benzoic acid, 2-methoxy-4-methyl- 4.38	c = 0.001, 1% ethanol	20	E3bg	Approx.	P24,P25
4576	$C_9H_{10}O_3$	Benzoic acid, 2-methoxy-5-methyl- 4.20	Calculated value	20		Uncert.	P24

No.	Molecular formula, name and pK value(s)	$T(^{\circ}C)$	Remarks	Method	Assessment	Ref.
4577	$C_9H_{10}O_3$ Benzoic acid, 2-methoxy-6-methyl- 3.46	20	c = 0.001, 1% ethanol	E3bg	Approx.	P23
4578	$C_9H_{10}O_3$ Benzoic acid, 3-methoxy-2-methyl- 3.72	20	c ≈ 0.001, 1% ethanol	E3bg	Approx.	P22
4579	$C_9H_{10}O_3$ Benzoic acid, 3-methoxy-4-methyl- 4.13	20	c = 0.001, 1% ethanol	E3bg	Approx.	P24
4580	$C_9H_{10}O_3$ Benzoic acid, 5-methoxy-2-methyl- 3.84	20	c ≈ 0.001, 1% ethanol	E3bg	Approx.	P22
4581	$C_9H_{10}O_3$ Benzoic acid, 4-methoxy-2-methyl- 4.54	20	c ≈ 0.001, 1% ethanol	E3bg	Approx.	P23
4582	$C_9H_{10}O_3$ Benzoic acid, 4-methoxy-3-methyl- 4.35	20	c = 0.001, 1% ethanol	E3bg	Approx.	P24,P26
4583	$C_9H_{10}O_3$ Ethyl 2-hydroxybenzoate 10.34 9.92 9.65	1 25 45	c = 0.002, I = 0.1(KCl), mixed constant	05	Approx.	H33

Thermodynamic quantities are derived from the results

No.	Formula	Name / pK	Temperature (°C)	Conditions	Method	Reliability	Ref.
4584	$C_9H_{10}O_3$	Ethyl 3-hydroxybenzoate					
		9.10	25		05	Approx.	C71
4585	$C_9H_{10}O_3$	Ethyl 4-hydroxybenzoate					
		8.50	25		05	Approx.	C70
		8.44	1	c = 0.002, I = 0.1(KCl), mixed constant	05	Approx.	H33
		8.34	25		05		
		Thermodynamic quantities are derived from the results					
4586	$C_9H_{10}O_3$	Propanoic acid, 3-(2-hydroxyphenyl)-					
		4.75	25	c ≈ 0.01	E3bg	Approx.	B114
4587	$C_9H_{10}O_3$	Propanoic acid, 3-phenoxy-					
		4.32	25	c ≈ 0.01	E3bg	Approx.	B113
		Values in mixed solvents are also given					
4588	$C_9H_{10}O_4$	Acetic acid, 2-hydroxy-2-(2-methoxyphenyl)-					
		3.64	25	c = 0.01, mixed constant	E3bg	Approx.	K36
4589	$C_9H_{10}O_4$	Acetic acid, 2-hydroxy-2-(4-methoxyphenyl)-					
		3.42	25	c = 0.01, mixed constant	E3bg	Uncert.	K36
4590	$C_9H_{10}O_4$	Acetic acid, 2-(4-hydroxy-3-methoxyphenyl)- (Homovanillic acid)					
		pK_1 4.41	25		C1	Approx.	N34
		pK_2 10.53	25		05,R4	Approx.	

No.	Molecular formula, name and pK value(s)	T(^{o}C)	Remarks	Method	Assessment	Ref.
4591	$C_9H_{10}O_4$ Benzoic acid, 4-ethoxy-2-hydroxy- 3.20	25	c = 0.00016, I = 0.01	05	Approx.	D56
4592	$C_9H_{10}O_4$ Benzoic acid, 2,3-dimethoxy- 3.98	25 Variation with temperature 30 40 50 60 3.94 3.90 3.86 3.86	E3bg	Approx.	S103
4593	$C_9H_{10}O_4$ Benzoic acid, 2,4-dimethoxy- 4.36	25 Variation with temperature 30 40 50 60 4.32 4.28 4.26 4.24	E3bg	Approx.	S103
4594	$C_9H_{10}O_4$ Benzoic acid, 2,6-dimethoxy- 3.44	25	c = 0.01	E3bg	Approx.	D17
4595	$C_9H_{10}O_4$ Benzoic acid, 3,4-dimethoxy- 4.36	25 Variation with temperature 30 40 50 60 4.34 4.30 4.28 4.25	E3bg	Uncert.	S104

(contd)

No.	Formula	Name / pK	Value	t(°C)	Conditions	Method	Assessment	Ref.
4595 (Contd)	$C_9H_{10}O_4$	Benzoic acid, 3,4-dimethoxy-	4.43	25	c = 0.003, 5% ethanol	E3bg	Uncert.	B190
4596	$C_9H_{10}O_4$	Benzoic acid, 3,5-dimethoxy-	3.97	25		E3bg	Uncert.	S104

Variation with temperature

	30	40	50	60
	3.95	3.92	3.90	3.87

No.	Formula	Name / pK	t(°C)	Conditions	Method	Assessment	Ref.
4597	$C_9H_{10}O_4$	Bicyclo[2.2.1]hept-2-ene-2,3-dicarboxylic acid					
		pK_1 1.32	25	c = 0.05-0.09	E3bg	Approx.	M11
		pK_2 8.00	25	c = 0.005-0.009			
4598	$C_9H_{10}O_4$	Bicyclo[2.2.1]hept-5-ene-2,3-dicarboxylic acid					
		pK_1 4.20, pK_2 7.10(endo cis)	25	c ≈ 0.006	E3bg,R3d	Approx.	C55
		pK_1 3.97, pK_2 5.65(trans)					
4599	$C_9H_{10}O_4$	Ethyl 3,4-dihydroxybenzoate					
		pK_1 8.08, pK_2 11.65			E and 0		H15
4600	$C_9H_{10}O_4$	Methyl 2-hydroxy-3-methoxybenzoate					
		10.30	20	I = 0.1, mixed constant(?)	E3bg	Uncert.	S128
4601	$C_9H_{10}O_4$	Methyl 2-hydroxy-4-methoxybenzoate					
		9.85	20	I = 0.1, mixed constant(?)	E3bg	Uncert.	S128

No.	Molecular formula, name and pK value(s)	T(°C)	Remarks	Method	Assessment	Ref.
4602	Methyl 3-hydroxy-4-methoxybenzoate					
	9.21	20	I = 0.1, mixed constant(?)	E3bg	Uncert.	S128
4603	Methyl 4-hydroxy-3-methoxybenzoate					
	8.30	20	I = 0.1, mixed constant(?)	E3bg	Uncert.	S128
4604	$C_9H_{10}O_4$ Propanoic acid, 3-(3,4-dihydroxyphenyl)-					
	pK_1 4.56, pK_2 9.36	30	c = 0.004, I = 0.1($NaClO_4$), mixed constants	E3bg	Approx.	A80
	pK_3 11.6				Uncert.	
4605	$C_9H_{10}O_5$ Benzoic acid, 4-hydroxy-3,5-dimethoxy- (Syringic acid)					
	pK_1 4.34, pK_2 9.49	25		05	Approx.	N35
	pK_1 4.34	25		E3bg	Uncert.	S104
			Variation with temperature			
		35 45 55 65				
		4.28 4.25 4.32 4.38				
4606	$C_9H_{10}O_6$ cis-2-Oxabicyclo[3.3.0]octane-c-6,c-7-dicarboxylic acid, 3-oxo-					
	pK_1 3.91, pK_2 5.97	20	c = 0.001-0.002	E3bg	Approx.	I4
4607	$C_9H_{10}O_6$ cis-2-Oxabicyclo[3.3.0]octane-c-6,t-7-dicarboxylic acid, 3-oxo-					
	pK_1 3.34, pK_2 5.51	20	c = 0.001-0.002	E3bg	Approx.	I4

No.	Formula	Name	pKa	T(°C)	Notes			Ref.
4608	$C_9H_{12}O$	Phenol, 2-isopropyl-	10.47	20		E3bg	Approx.	D31
			10.49			05		
			10.31	25	I = 0.1(KCl), mixed constant	E3bg	Approx.	H51
			10.34	25	I = 0.1(KCl), mixed constant	E3bg	Approx.	E20
4609	$C_9H_{12}O$	Phenol, 3-isopropyl-	10.16	20	Extrapolated value		Uncert.	D30
4610	$C_9H_{12}O$	Phenol, 4-isopropyl-	10.24	25	c = 0.005	E3bg	Approx.	D31
			10.26			05		
			10.04	25	I = 0.1(KCl), mixed constant	E3bg	Approx.	H51
4611	$C_9H_{12}O$	Phenol, 2-propyl-	10.47	25		05	Approx.	07
			10.58	20	Extrapolated value		Uncert.	D30
4612	$C_9H_{12}O$	Phenol, 4-propyl-	10.34	20	Extrapolated value		Uncert.	D30
4613	$C_9H_{12}O$	Phenol, 2,3,5-trimethyl-	10.67	25		05	Approx.	K40
			10.71	20		05	Approx.	D30
			10.69	20		05,R4		R22

No.	Molecular formula, name and pK value(s)	$T(^oC)$	Remarks	Method	Assessment	Ref.
4614	$C_9H_{12}O$ Phenol, 2,4,5-trimethyl-					
	10.57	25		05	Approx.	K40
4615	$C_9H_{12}O$ Phenol, 2,4,6-trimethyl-					
	10.86	25	c = 0.0038	E3bg	Approx.	F16
	10.89	25		05	Approx.	K40
	10.99	20		05,R4		R22
4616	$C_9H_{12}O$ Phenol, 3,4,5-trimethyl-					
	10.25	25		05	Approx.	K40
	10.51	20		05	Approx.	D30
4617	$C_9H_{12}O_2$ Phenol, 4-methoxy-2,6-dimethyl-					
	10.84	25	c = 0.0043	E3bg	Approx.	F16
4618	$C_9H_{12}O_4$ Acetic acid, 2-(3-oxo-cis-2-oxabicyclo[3.3.0]octan-c-4-yl)-					
	4.27	20	c = 0.001-0.002	E3bg	Approx.	I4
4619	$C_9H_{12}O_4$ Acetic acid, 2-(4-oxo-cis-3-oxabicyclo[3.3.0]octan-t-6-yl)-					
	4.88	20	c = 0.001-0.002	E3bg	Approx.	I4
4620	$C_9H_{12}O_4$ Bicyclo[2.2.1]heptane-1,4-dicarboxylic acid					
	pK_1 4.197, pK_2 5.284	25		E3bg,R3d	Approx.	W29

Values in mixed solvents are also given

No.	Formula	Name / pK values	T (°C)	Conditions	Method		Ref.
4621	$C_9H_{12}O_4$	Bicyclo[2.2.1]heptane-2,3-dicarboxylic acid pK_1 3.98, pK_2 7.40(endo cis) pK_1 4.02, pK_2 5.75(trans)	25	$c \simeq 0.006$	E3bg,R3d	Approx.	C55
4622	$C_9H_{12}O_4$	cis-3-Oxabicyclo[4.3.0]nonane-t-9-carboxylic acid, 2-oxo- 4.76	20	$c = 0.001-0.002$	E3bg	Approx.	I4
4623	$C_9H_{12}O_4$	Phenol, 2,4,6-tris(hydroxymethyl)- 9.45	25	$c = 0.1$	E3d	Approx.	Z4
4624	$C_9H_{12}O_6$	cis-2,9-Dioxabicyclo[4.3.0]nonane-t-4,t-6-dicarboxylic acid pK_1 4.28, pK_2 6.74	20	$c = 0.001-0.002$	E3bg	Approx.	I4
4625	$C_9H_{12}S$	Benzenethiol, 2-isopropyl- 7.02	35		05	Approx.	F13
4626	$C_9H_{14}O_2$	Bicyclo[2.2.2]octane-1-carboxylic acid 5.084	25		E3bg	Approx.	W29
4627	$C_9H_{14}O_3$	Butanoic acid, 4-(2-oxocyclopentyl)- 4.86	20	$c = 0.001-0.002$	E3bg	Approx.	I4

Variation with temperature

	20	30	35	40	45	50	55	60
	9.51	9.39	9.34	9.29	9.24	9.19	9.15	9.10

Values in mixed solvents are also given

No.	Molecular formula, name and pK value(s)	$T(^oC)$	Remarks	Method	Assessment	Ref.
4628	$C_9H_{14}O_3$ Cyclobutanecarboxylic acid, 3-acetyl-2,2-dimethyl- (Pinononic acid)					
	4.63	25-30	c ≈ 0.001	E3bg	Uncert.	H52
4629	$C_9H_{14}O_3$ Ethyl 2-oxo-cyclohexanecarboxylate					
	10.94	25	I = 0.2(KCl)	E3bg	Uncert.	B57
4630	$C_9H_{14}O_4$ Acetic acid, 2-(3-carboxy-2,2-dimethylcyclobutyl)- (Pinic acid)					
	pK_1 4.46, pK_2 5.47(dl)	25-30	c ≈ 0.0005	E3bg,R3d	Uncert.	H52
	pK_1 4.48, pK_2 5.48(d(-))					
4631	$C_9H_{14}O_4$ 1,2-Cycloheptanedicarboxylic acid					
	pK_1 3.87, pK_2 7.60(cis)	25	c ≈ 0.0005, mixed constants	E3bg	Approx.	S68
	pK_1 4.30, pK_2 6.18(trans)	25	c ≈ 0.0005, mixed constants	E3bg,R3d	Approx.	
4632	$C_9H_{14}O_4$ Cyclohexanecarboxylic acid, 4-methoxycarbonyl-					
	4.66(trans)	25	c = 0.01	E3bg	Approx.	S69
			Values in mixed solvents are also given			
4633	$C_9H_{14}O_4$ 1,2-Cyclopropanedicarboxylic acid, 1,2-diethyl-					
	pK_1 4.16, pK_2 6.58(cis)	20		E3bg	Approx.	M12
	pK_1 3.53, pK_2 5.03(trans)					
4634	$C_9H_{14}O_5$ Diethyl 2-methyl-3-oxobutanedioate					
	7.03			E3d		R51

No.	Formula / Name	pK	T (°C)	Conditions / Notes	Method	Reliability	Ref.
4635	$C_9H_{14}O_5$ Propanedioic acid, 2-cyclohexyl-2-hydroxy-	pK_1 2.62, pK_2 4.78					G37
4636	$C_9H_{16}O_2$ 2,4-Heptanedione, 5,5-dimethyl-	9.26	25	c = 0.00005, 0.5-1.0% ethanol	05	Approx.	C5
				Mixed pK for keto-enol mixture			
	pK(enol) 8.85, pK(keto) 9.05		25	Calculated from absorbance data		Uncert.	
4637	$C_9H_{16}O_2$ 3,5-Heptanedione, 2,6-dimethyl-	9.82	20	c = 0.00002, I = 0.1(NaClO$_4$)	05	Approx.	K49
		9.86	20	c = 0.01	E3bg/u5	Approx.	R52
4638	$C_9H_{16}O_2$ 2,4-Nonanedione	9.20	20	c = 0.01	E3bg/05	Approx.	R52
4639	$C_9H_{16}O_4$ Butanoic acid, 2-ethoxycarbonyl-2-ethyl-	3.44	7	Mixed constant	E3bg	Uncert.	L33
		3.54	25				
		3.64	37				
		3.70(H_2O)	25		E3bg	Approx.	G23
		4.13(D_2O)					
4640	$C_9H_{16}O_4$ Butanoic acid, 3-methoxycarbonyl-2,2,3-trimethyl-	4.98(H_2O)	25		E3bg	Approx.	G22
		5.48(D_2O)					

No.	Molecular formula, name and pK value(s)	T(°C)	Remarks	Method	Assessment	Ref.
4641	Nonanedioic acid $C_9H_{16}O_4$					
	pK$_1$ 4.55	20	c = 0.005	E3bg,R3a	Approx.	D43
	pK$_2$ 5.5				Uncert.	
	Values are also given in mixed solvents					
	pK$_1$ 4.56, pK$_2$ 5.53	25	I = 0.1(KIO$_3$)	E3bg		N32
4642	$C_9H_{16}O_4$ Pentanedioic acid, 3,3-diethyl-					
	pK$_1$ 3.67, pK$_2$ 7.42	25	c = 0.00125, mixed constants	E3bg	Approx.	B146
4643	$C_9H_{16}O_4$ Pentanedioic acid, 3-isopropyl-3-methyl-					
	pK$_1$ 3.78, pK$_2$ 6.92	25	c = 0.00125, mixed constants	E3bg	Approx.	B146
4644	$C_9H_{16}O_4$ Propanedioic acid, 2-butyl-2-ethyl-					
	pK$_1$ 2.15, pK$_2$ 7.25	25	I = 0.1(KCl), mixed constants	E3bg	Approx.	M77
4645	$C_9H_{16}O_4$ Propanedioic acid, diisopropyl-					
	pK$_1$ 2.124, pK$_2$ 8.848	25	I = 0.01	Elch*	Rel.	I28

*|Hg;Hg$_2$Cl$_2$ replaces |AgCl;Ag of Elch

Variation with temperature

	5	10	15	20	30	35	40	45
pK$_1$	2.150	2.141	2.134	2.128	2.127	2.131	2.136	2.142
pK$_2$	8.829	8.827	8.830	8.838	8.861			

Thermodynamic quantities are derived from the results

(contd)

No.	Name / Formula	pK	T(°C)	Conditions	Method	Reliability	Ref.
4645 (Contd)	$C_6H_{16}O_4$ Propanedioic acid, diisopropyl-						
	pK₁ 2.18, pK₂ 8.60		25	I = 0.1(KCl), mixed constants	E3bg	Approx.	M77
4646	$C_9H_{16}O_4$ Propanedioic acid, dipropyl-						
	pK₁ 1.82		25	I = 0.1(KNO₃), concentration constants	E3bg	Uncert.	P78
	pK₂ 7.15			I = 0.1(KNO₃)		Approx.	
		pK₂ is extrapolated value at c = 0					
	pK₁ 1.86, pK₂ 7.18		25	c = 0.001, I = 0.1(NaClO₄), mixed constants	E3bg	Approx.	O23
	pK₁ 2.15, pK₂ 7.34		25	I = 0.1(KCl), mixed constants	E3bg	Approx.	M77
4647	$C_9H_{16}O_5$ Propanedioic acid, 2-hexyl-2-hydroxy-						
	pK₁ 2.66, pK₂ 4.82						G37
4648	$C_9H_5O_4N$ 1,3-Indandione, 2-nitro-						
	1.7		18±2	Mixed constant	05 or E3d	Uncert.	M25
4649	$C_9H_6O_2N_2$ 8-Quinolinol, 5-nitroso-						
	pK₁ 2.26, pK₂ 7.60		25		E & O		A83
4650	$C_9H_5O_2N_2$ 2-Quinoxalinecarboxylic acid						
	2.875		25	c ≈ 0.005, mixed constant	E3bg	Approx.	G52
	2.80		25	c ≈ 0.01, I = 0.1(NaClO₄), mixed constant	E3bg	Approx.	D61
	2.76		25		SOL	Uncert.	

No.	Molecular formula, name and pK value(s)	T(°C)	Remarks	Method	Assessment	Ref.
4651	$C_9H_6O_2N_2$ 5-Quinoxalinecarboxylic acid					
	4.026	25	$c \approx 0.005$, mixed constant	E3bg	Approx.	G52
			Value in mixed solvent is also given			
4652	$C_9H_6O_2N_2$ 6-Quinoxalinecarboxylic acid					
	3.64	25	$c \approx 0.005$, mixed constant	E3bg	Approx.	G52
4653	$C_9H_6O_3N_2$ 2-Quinoxalinecarboxylic acid, 3-hydroxy-					
	pK$_1$ 2.56, pK$_2$ 9.43	25	$c \approx 0.01$, $I = 0.1(NaClO_4)$, mixed constants	E3bg	Approx.	D61
4654	$C_9H_6O_4N_2$ 2-Indolecarboxylic acid, 5-nitro-					
	14.91(NH)	25	In aq. KOH, H_ scale	07	Uncert.	Y2
4655	$C_9H_6O_6N_2$ 1-Indanone, 7-hydroxy-4,6-dinitro-					
	1.94	25	$I = 0.1(NaClO_4)$	05	Approx.	M18
4656	C_9H_7ON 3-Indolecarbaldehyde					
	12.36(NH)	25	$c \approx 0.0001$	07	Uncert.	T21
	12.36	25	In aq. KOH, H_ scale	07	Uncert.	Y2
			(Base value for scale)			
4657	C_9H_7ON 8-Quinolinol (Oxine)					
	pK$_1$ 4.91, pK$_2$ 9.81	25	$c = 0.002$, $I = 0.5$	E3bg	Approx.	T57
			Thermodynamic quantities are also given			

(contd)

No.	Formula	Name / pK	Temp.	Conditions	Method	Assessment	Ref.
4657 (Contd)	C_9H_7ON	8-Quinolinol (Oxine)					
		pK_1 5.00, pK_2 9.70	18±2	I = 0.5(NaCl), concentration constants	05	Uncert.	H50
		pK_1 5.1, pK_2 9.9	not stated	Mixed constants	05	Uncert.	F4
4658	$C_9H_7OF_3$	Acetophenone, α,α,α-trifluoro-4-methyl-					
		10.15	25	I ≈ 0.1, pK of hydrate	05	Approx.	S116
4659	$C_9H_7O_2N$	Benzoic acid, 3-cyano-2-methyl-					
		3.16	20	c ≈ 0.001, 1% ethanol	E3bg	Approx.	P22
4660	$C_9H_7O_2N$	Benzoic acid, 3-cyano-6-methyl-					
		3.31	20	c ≈ 0.001, 1% ethanol	E3bg	Approx.	P22
4661	$C_9H_7O_2N$	Benzoic acid, 4-cyano-2-methyl-					
		3.14	20	c ≈ 0.001, 1% ethanol	E3bg	Approx.	P23
4662	$C_9H_7O_2N$	2-Indolecarboxylic acid					
		17.13(NH)	25	In aq. KOH, H_ scale	07	Uncert.	Y2
4663	$C_9H_7O_2N$	3-Indolecarboxylic acid					
		pK_1 3.87	25	c = 0.005, I = 0.04-2.0	E3bg	Approx.	L74
		pK_2 15.59(NH)	25	In aq. KOH, H_ scale	07	Uncert.	Y2
4664	$C_9H_7O_2N$	5-Indolecarboxylic acid					
		16.92(NH)	25	In aq. KOH, H_ scale	07	Uncert.	Y2

No.	Molecular formula, name and pK value(s)	T(°C)	Remarks	Method	Assessment	Ref.
4665	$C_9H_7O_2N$ Isoxazol-5(2$\underline{H}$)-one, 3-phenyl- 4.01	20		05	Approx.	B110
4666	$C_9H_7O_2N_3$ 1,2,4-Triazine-3,5-(2$\underline{H}$,4$\underline{H}$)-dione, 6-phenyl- 7.31	25	Titrant N(Me)$_4$OH	E3bg	Approx.	J16
4667	$C_9H_7O_2N_3$ 1,2,3-Triazole-4-carboxylic acid, 1-phenyl- 2.88	25		E3bg	Uncert.	H13
4668	$C_9H_7O_2Cl$ Propenoic acid, 3-chloro-3-phenyl- 3.55(trans) 3.57(cis)	25	c = 0.01	E3bg	Approx.	L17
4669	$C_9H_7O_2Br$ Propenoic acid, 3-(2-bromophenyl)- 4.23(trans)	25	c ≈ 0.01	E3bg	Approx.	B114
4670	$C_9H_7O_2F$ Propenoic acid, 3-(2-fluorophenyl)- 4.28(trans)	25	c ≈ 0.01	E3bg	Approx.	B114
4671	$C_9H_7O_2F_3$ Acetophenone, α,α,α-trifluoro-4-methoxy- 10.18	25	I ≈ 0.1 pK of hydrate	05	Approx.	S116

No.	Formula	Name	pK	Conditions	T (°C)	Method		Ref.
4672	$C_9H_7O_3N$	Acetic acid, 2-(4-cyanophenoxy)-	2.87		25	E3ag	Approx.	H48
		Value in mixed solvent is also given						
4673	$C_9H_7O_3Br$	Propenoic acid, 3-(5-bromo-2-hydroxyphenyl)-	3.62(cis)	$I = 1.0$(LiCl), mixed constant	30	05	Approx.	H36
4674	$C_9H_7O_4N$	Benzoic acid, 3-(2-nitrovinyl)-	3.87	$c = 0.0006$-0.0020, mixed constant	25	E3bg	Approx.	S115
4675	$C_9H_7O_4N$	Benzoic acid, 4-(2-nitrovinyl)-	3.94	$c = 0.0006$, mixed constant	25	E3bg	Approx.	S115
4676	$C_9H_7O_4N$	1-Indanone, 7-hydroxy-4-nitro-	5.27	$I = 0.1$(NaClO$_4$), concentration constant	25	05	Approx.	M18
4677	$C_9H_7O_4N$	1-Indanone, 7-hydroxy-6-nitro-	5.31	$I = 0.1$(NaClO$_4$), concentration constant	25	05	Approx.	M18
4678	$C_9H_7O_4N_3$	Benzenecarbonitrile, 3-(2,2-dinitroethyl)-	3.88		20	05	Approx.	T58a
4679	$C_9H_7O_4Cl$	Acetic acid, 2-(4-chlorobenzoyloxy)-	3.00		25	E		C74

No.	Molecular formula, name and pK value(s)	$T(^{\circ}C)$	Remarks	Method	Assessment	Ref.
4680	$C_9H_7O_5N$ Propenoic acid, 3-(2-hydroxy-5-nitrophenyl)- 3.54(cis)	30	I = 1.0(LiCl), mixed constant	O5	Approx.	H36
4681	$C_9H_7O_6N$ Acetic acid, 2-(4-nitrobenzoyloxy)- 2.93			E		C74
4682	$C_9H_8OI_2$ 2-Pyrazolin-3-one, 1-phenyl- 7.57	20	Extrapolated from values in ethanol-water	E3bg	Uncert.	T1
4683	$C_9H_8OI_2$ 2-Pyrazolin-4-one, 1-phenyl- 9.05	20	Extrapolated from values in ethanol-water	E3bg	Uncert.	T1
4684	$C_9H_8ON_2$ 2-Pyrazolin-5-one, 1-phenyl- 6.56	20	c = 0.0025	E3bg	Approx.	T1
4685	C_9H_8OS Benzo[b]thiophene-5-ol, 3-methyl- 9.94	20	2% ethanol	O5	Approx.	D32
4686	C_9H_8OS Benzo[b]thiophene-6-ol, 3-methyl- 9.70	20	2% ethanol	O5	Approx.	D32

No.	Formula	Name	pK	T(°C)	Conditions	Reliability	Assessment	Ref.
4687	C_9H_8OS	Benzo[b]thiophene-7-ol, 3-methyl-	8.72	20	2% ethanol	05	Approx.	D32
4688	$C_9H_8O_2N_2$	Quinazoline-2,4(1H,3H)-dione, 1-methyl-	9.85	25	$c \simeq 0.001$	E3bg	Approx.	G62
4689	$C_9H_8O_2N_2$	Quinazoline-2,4(1H,3H)-dione, 3-methyl-	10.6	25	$c \simeq 0.001$	E3bg	Uncert.	G2
4690	$C_9H_8O_3N_4$	Ethyl 1,7-dihydro-7-oxo-6-pteridinecarboxylate (6-Pteridinecarboxylic acid ethyl ester, 1,7-dihydro-7-oxo-)	5.53	20	$c = 0.0001$ in 0.01M buffer	05	Approx.	A39
4691	$C_9H_8O_3Cl_2$	Propanoic acid, 2-(2,4-dichlorophenoxy)-	2.86	20	$I = 0.001$	05	Approx.	P60
		Value in mixed solvent is also given						
4692	$C_9H_8O_3Cl_2$	Propanoic acid, 3-(2,4-dichlorophenoxy)-	4.42	20	$I = 0.001$	05	Approx.	P60
		Value in mixed solvent is also given						
4693	$C_9H_8O_4S$	Acetic acid, 2-(2-carboxyphenyl)thio-	pK_1 3.01, pK_2 3.99	20	c 0.001, $I = 0.1(KNO_3)$, mixed constants	E3bg	Approx.	P39
			pK_1 2.93, pK_2 4.01	25	$I = 0.1(NaClO_4)$, concentration constants	E3bg,R3f	Approx.	S133
4694	$C_9H_8O_4S$	Acetic acid, 2-(3-carboxyphenyl)thio-	pK_1 3.28, pK_2 4.15	20	$c = 0.001$, $I = 0.1(KNO_3)$, mixed constants	E3bg	Approx.	P39

No.	Molecular formula, name and pK value(s)	$T(^{o}C)$	Remarks	Method	Assessment	Ref.
4695	$C_9H_8O_4S$ Acetic acid, 2-(4-carboxyphenyl)thio- pK$_1$ 3.19, pK$_2$ 4.29	20	c = 0.001, I = 0.1(KNO$_3$), mixed constants	E3bg	Approx.	P39
4696	$C_9H_8O_5N_4$ 6-Pteridinecarboxylic acid, 1,2,3,4,7,8-hexahydro-1,3-dimethyl-2,4,7-trioxo- pK$_1$ 2.1 pK$_2$ 6.30	20		E3bg	Uncert. Approx.	P44
4697	$C_9H_8O_8N_4$ Methane, (2,4-dimethyl-3,5-dinitrophenyl)dinitro- 0.46	20	c ≈ 0.0001 (H$_o$ in HClO$_4$ solutions)	06	Uncert.	T58
4697a	C_9H_9ON Benzenecarbonitrile, 4-hydroxy-3,5-dimethyl- 8.19	25	c = 0.0006	E3bg	Approx.	F16
4697b	C_9H_9ON Methanol, indol-3-yl- 16.50(NH)	25	In aq. KOH, H_ scale	07	Uncert.	Y3
4698	$C_9H_9OF_3$ Ethanol, 2,2,2-trifluoro-1-(p-tolyl)- 12.04	25	I ≈ 0.1, mixed constant	05	Approx.	S116
4699	$C_9H_9O_2N$ 1-Propanone, 2-hydroxyimino-1-phenyl- 9.31 9.10	25	c = 0.0025 I = 0.1(KCl), concentration constant	E3bg,R4	Approx.	M40

No.	Formula	Name	pK	Temp.	Conditions	Method	Assessment	Ref.
4700	$C_9H_9O_2N_5$	Acetamide, N-(4-methoxy-2-pteridinyl)-	1.95	20		O5	Uncert.	P52
4701	$C_9H_9O_2Cl$	Propanoic acid, 3-(4-chlorophenyl)-	4.66	25	c ≈ 0.0008	E3bg	Approx.	T55
4702	$C_9H_9O_2Br$	Propanoic acid, 3-(2-bromophenyl)-	4.58	25	c ≈ 0.01	E3bg	Approx.	B114
4703	$C_9H_9O_2F$	Propanoic acid, 3-(2-fluorophenyl)-	4.60	25	c ≈ 0.01	E3bg	Approx.	B114
4704	$C_9H_9O_2F_3$	Ethanol, 2,2,2-trifluoro-1-(4-methoxyphenyl)-	12.24	25	I ≈ 0.1, mixed constant	O5	Approx.	S116
4705	$C_9H_9O_3N$	Acetic acid, 2-(benzoylamino)- (Glycine, N-benzoyl-)	3.59	40	I = 0.3(KCl)	E3bg	Approx.	S84
4706	$C_9H_9O_3N$	4-Azatricyclo[5.2.1.0(2,6)]dec-8-ene-3,5-dione, 4-hydroxy-	5.9(cis)	30		E3bg	Uncert.	S117
4707	$C_9H_9O_3N$	Benzoic acid, 2-acetylamino-	3.61	20	c ≈ 0.001, 1% ethanol	E3bg	Approx.	P25,P24
4708	$C_9H_9O_3N$	Benzoic acid, 3-acetylamino-	4.03	25	c ≈ 0.001	E3bg	Approx.	B155

No.	Molecular formula, name and pK value(s)	T(°C)	Remarks	Method	Assessment	Ref.
4709	$C_9H_9O_3N$ Propenoic acid, 3-(3-amino-6-hydroxyphenyl)- 3.49(cis)	30	Estimated value of mixed constant		Uncert.	H36
4710	$C_9H_9O_3N_5$ 6-Pteridinecarboxylic acid, 2-ethylamino-1,7-dihydro-7-oxo- pK$_1$ 3.00, pK$_2$ 5.89		Conditions not stated		Uncert.	P58
4711	$C_9H_9O_3Cl$ Acetic acid, 2-(4-chloro-2-methylphenoxy)- 3.107	25	c = 0.0005-0.0026	C2,R1c	Rel.	M58
4712	$C_9H_9O_3I$ 1,2-Benziodoxepin-3(1H)-one, 4,5-dihydro-1-hydroxy- 7.37	23±3	Mixed constant	05	V.Uncert.	W42
4713	$C_9H_9O_4N$ Acetic acid, 2-(2-hydroxybenzoylamino)- 3.41	25	I = 0.1(NaClO$_4$)	E3bg	Approx.	P8
4714	$C_9H_9O_4N$ Benzoic acid, 4-acetylamino-2-hydroxy- pK$_1$ 2.85			05		L56
4715	$C_9H_9O_4N$ Phenol, 2-methoxy-4-(2-nitrovinyl)- 7.03	25	Extrapolated from data obtained in ethanol-water	E3bg	Uncert.	L50
4716	$C_9H_9O_4N$ Propanoic acid, 3-(3-nitrophenyl)- 4.53	25	Values in mixed solvents are also given	E3ag	Approx.	H48

No.	Formula	Name / pK	Temp.	Notes		Reliability	Ref.
4717	$C_9H_9O_4N$	Propanoic acid, 3-(4-nitrophenyl)-					
		4.54	25	c 0.0008	E3bg	Approx.	T55
4718	$C_9H_9O_5N$	Acetophenone, 4-hydroxy-3-methoxy-5-nitro-					
		4.35	25	Extrapolated from data obtained in ethanol-water	05	Uncert.	L50
4719	$C_9H_9O_5N$	Benzoic acid, 2-hydroxy-3,6-dimethyl-5-nitro-					
		pK_1 2.78, pK_2 8.54	25	I = 0.1(KCl), mixed constants	E3bg	Approx.	F35a
4720	$C_9H_9O_5N$	Ethyl 3-hydroxy-4-nitrobenzoate					
		6.11	25	c ≈ 0.00001-0.0001, I ≈ 0.1, mixed constant	05	Approx.	H11
4721	$C_9H_9O_5N$	Ethyl 4-hydroxy-3-nitrobenzoate					
		5.44	25	c = 0.0002, I ≈ 0.5, mixed constant	05	Approx.	R15
4722	$C_9H_9O_6N_3$	Methane, (2,4-dimethyl-5-nitrophenyl)dinitro-					
		0.91	20	c ≈ 0.0001 (H_0 in $HClO_4$ solutions)	06	Uncert.	T58
4723	$C_9H_9O_7N_3$	Phenol, 3-ethyl-5-methyl-2,4,6-trinitro-					
		3.3	not stated	Mixed constant	E3bg	Uncert.	M100
4724	$C_9H_9O_7N_3$	Phenol, 3-isopropyl-2,4,6-trinitro-					
		3.4	not stated	Mixed constant	E3bg	Uncert.	M100

No.	Molecular formula, name and pK value(s)	T(°C)	Remarks	Method	Assessment	Ref.
4725	$C_9H_9N_3S$ 1,2,4-Triazole-3(2H)-thione, 4-methyl-5-phenyl-					
	7.66	25	c ≈ 0.004, mixed constant	E3bg	Approx.	S9
4726	$C_9H_{10}ON_4$ Pteridin-2(1H)-one, 1,6,7-trimethyl-					
	11.74	20	c = 0.005	E3bg	Uncert.	B128
4727	$C_9H_{10}ON_4$ Pteridin-2(1H)-one, 4,6,7-trimethyl-					
	8.5	20	"Anhydrous" species	E3bg	Uncert.	I5
	8.65	20	Equilibrium pK, partial covalent hydration		Approx.	
	11.5	20	Covalently hydrated species		Uncert.	
4728	$C_9H_{10}O_2N_2$ 2-Azaspiro[4,4]nonane-4-carbonitrile, 1,3-dioxo-					
	7.28(NH)	20	c = 0.0001-0.001, 4% ethanol, mixed constant	E3bg	Uncert.	F29
4729	$C_9H_{10}O_2N_2$ Butanamide, 3-oxo-N-(2-pyridyl)-					
	9.72	25		E3bg,R4	Approx.	H19
4730	$C_9H_{10}O_2N_4$ Pteridine-2,4(1H,3H)-dione, 1,6,7-trimethyl-					
	9.06	20	c = 0.0033	E3bg	Approx.	B135
4731	$C_9H_{10}O_2N_4$ Pteridine-2,4(3H,8H)-dione, 6,7,8-trimethyl-					
	9.83	20	c = 0.0025	E3bg	Approx.	B135

No.	Formula	Name / pK	T (°C)	Conditions	Method	Assessment	Ref.
4732	$C_9H_{10}O_2N_4$	Pteridin-2(1H)-one, 4-acetonyl-3,4-dihydro- 12.38	20	$c = 0.0002$ in 0.01M buffer	05	Approx.	A27
4733	$C_9H_{10}O_2N_4$	Pteridin-7(8H)-one, 2-methoxy-6,8-dimethyl- 13	20	$c = 0.00004$	05	Uncert.	J4
4734	$C_9H_{10}O_2S$	Acetic acid, 2-benzylthio- 3.708	25	$c = 0.001$, $I = 0.10(KNO_3)$, concentration constant	E3bg	Approx.	P40
4735	$C_9H_{10}O_2S$	Acetic acid, 2-(2-methylphenyl)thio- 3.38	20	$c = 0.001$, $I = 0.1(KNO_3)$, mixed constant	E3bg	Approx.	P39
4736	$C_9H_{10}O_2S$	Acetic acid, 2-(3-methylphenyl)thio- 3.604	25	$c \approx 0.01$ Value in mixed solvent is also given	E3bg	Approx.	P13
		3.39	20	$c = 0.001$, $I = 0.1(KNO_3)$, mixed constant	E3bg	Approx.	P39
4737	$C_9H_{10}O_2S$	Acetic acid, 2-(4-methylphenyl)thio- 3.689	25	$c \approx 0.01$ Value in mixed solvent is also given	E3bg	Approx.	P13
		3.45	20	$c = 0.001$, $I = 0.1(KNO_3)$, mixed constant	E3bg	Approx.	P39
4738	$C_9H_{10}O_2S_2$	Acetic acid, 2-(2-methylthiophenyl)thio- 3.57	20	$c = 0.001$, $I = 0.1(KNO_3)$, mixed constant	E3bg	Approx.	P39

No.	Molecular formula, name and pK value(s)	T(^{o}C)	Remarks	Method	Assessment	Ref.
4739	$C_9H_{10}O_2S_2$ Acetic acid, 2-(3-methylthiophenyl)thio- 3.468	25	c ≈ 0.01 Value in mixed solvent is also given	E3bg	Approx.	P13
4740	$C_9H_{10}O_2S_2$ Acetic acid, 2-(4-methylthiophenyl)thio- 3.52	20	c ≈ 0.001, I = 0.1(KNO_3), mixed constant	E3bg	Approx.	P39
4741	$C_9H_{10}O_2Se$ Acetic acid, 2-(2-methylphenyl)seleno- 3.76	20	c ≈ 0.001, I = 0.1(KIO_3), mixed constant	E3bg	Approx.	P39
4742	$C_9H_{10}O_2Se$ Acetic acid, 2-(3-methylphenyl)seleno- 3.78	20	c ≈ 0.001, I = 0.1(KIO_3), mixed constant	E3bg	Approx.	P39
4743	$C_9H_{10}O_2Se$ Acetic acid, 2-(4-methylphenyl)seleno- 3.83	20	c ≈ 0.001, I = 0.1(KNO_3), mixed constant	E3bg	Approx.	P39
4744	$C_9H_{10}O_2Se$ 1,3-Butanedione, 2-methyl-1-(2-selenofuryl)- 10.40	room temp	I = 0.1($NaClO_4$)			E7
4745	$C_9H_{10}O_2Se$ 1,3-Pentanedione, 1-(2-selenofuryl)- 8.85	room temp	I = 0.1($NaClO_4$)			E7
4746	$C_9H_{10}O_3N_2$ 6-Indolinone, 5,6-dihydro-3-hydroxy-5-hydroxyimino-1-methyl- 8.6(NOH)	20	c = 0.005, mixed constant	E3d	Uncert.	R20

No.	Formula	Name / pK	Conditions	T (°C)	Method	Reliability	Ref.
4747	$C_9H_{10}O_3N_2$	3-Pyridinecarbonitrile, 2,6-dihydroxy-5-(2-hydroxyethyl)-4-methyl-					
		pK_1 -1.95	c = 0.0001	20	06	Uncert.	S100
		pK_2 1.16, pK_3 13.89	c = 0.0001, I = 0.01	20	05	Approx.	
4748	$C_9H_{10}O_3N_2$	4-Quinazolinecarboxylic acid, 5,6,7,8-tetrahydro-2-hydroxy-					
		pK_1 -0.82	c = 0.00003	20	06	Uncert.	A75
		pK_2 3.36	c = 0.00014, I = 0.01	20	05	Approx.	
		pK_3 10.38		20			
4749	$C_9H_{10}O_3N_4$	Pteridine-2,4,6(1H,3H,5H)-trione, 1,3,7-trimethyl-					
		6.70		20	E3bg	Approx.	P43
4750	$C_9H_{10}O_3N_4$	Pteridine-2,4,7(1H,3H,8H)-trione, 8-ethyl-6-methyl-					
		4.39		20	E3bg	Approx.	P54
		13.4			07	Uncert.	
4751	$C_9H_{10}O_3N_4$	Pteridine-2,4,7(1H,3H,8H)-trione, 1,3,6-trimethyl-					
		3.80		20	E3bg	Approx.	P42
4752	$C_9H_{10}O_3N_4$	Pteridine-2,4,7(1H,3H,8H)-trione, 3,6,8-trimethyl-					
		4.22	c = 0.001	20	E3bg	Approx.	P47
4753	$C_9H_{10}O_3S$	Acetic acid, 2-(2-methoxyphenyl)thio-					
		3.59	c ≈ 0.001, I = 0.1(KNO_3), mixed constant	20	E3bg	Approx.	P39

No.	Molecular formula, name and pK value(s)	T(°C)	Remarks	Method	Assessment	Ref.
4754	$C_9H_{10}O_3S$ Acetic acid, 2-(3-methoxyphenyl)thio-					
	3.39	20	$c \approx 0.001$, $I = 0.1(KNO_3)$, mixed constant	E3bg	Approx.	P39
	3.546	25	$c \approx 0.01$	E3bg	Approx.	P13
			Value in mixed solvent is also given			
4755	$C_9H_{10}O_3S$ Acetic acid, 2-(4-methoxyphenyl)thio-					
	3.54	20	$c \approx 0.001$, $I = 0.1(KNO_3)$, mixed constant	E3bg	Approx.	P39
	3.70	25	$c \approx 0.01$	E3bg	Approx.	P13
			Value in mixed solvent is also given			
4756	$C_9H_{10}O_3S$ Acetic acid, 2-(m-tolyl)sulfinyl-					
	2.756	25	$c \approx 0.01$	E3bg	Approx.	P13
			Value in mixed solvent is also given			
4757	$C_9H_{10}O_3S$ Acetic acid, 2-(p-tolyl)sulfinyl-					
	2.740	25	$c \approx 0.01$	E3bg	Approx.	P13
			Value in mixed solvent is also given			
4758	$C_9H_{10}O_3Se$ Acetic acid, 2-(2-methoxyphenyl)seleno-					
	3.87	20	$c \approx 0.001$, $I = 0.1(KNO_3)$, mixed constant	E3bg	Approx.	P39
4759	$C_9H_{10}O_3Se$ Acetic acid, 2-(3-methoxyphenyl)seleno-					
	3.73	20	$c \approx 0.001$, $I = 0.1(KNO_3)$, mixed constant	E3bg	Approx.	P39

No.	Formula	Name / pK	T(°C)	pK	Remarks	Method	Assessment	Ref.
4760	$C_9H_{10}O_3Se$	Acetic acid, 2-(4-methoxyphenyl)seleno-	20	3.86	$c \approx 0.001$, $I = 0.1(KNO_3)$, mixed constant	E3bg	Approx.	P39
4761	$C_9H_{10}O_4N_4$	4-Pteridinecarboxylic acid, 2-ethoxy-5,6,7,8-tetrahydro-6-oxo-	20	0.81	$c = 0.004$, mixed constants; pK assignment discussed	06	Uncert.	C59
				5.31		E3bg	Approx.	
				12.83		07	Uncert.	
4762	$C_9H_{10}O_4N_4$	Pteridine-2,4,6,7(1$\underline{H}$,3$\underline{H}$,5$\underline{H}$,8$\underline{H}$)-tetrone, 1,3,8-trimethyl-	20	7.64	$c = 0.001$	E3bg	Uncert.	P46
4763	$C_9H_{10}O_4N_4$	Pteridine-2,4,7(1$\underline{H}$,3$\underline{H}$,8$\underline{H}$)-trione, 8-(2-hydroxyethyl)-6-methyl-	20	4.00		E3bg	Approx.	P54
				13.1		06	Uncert.	
4764	$C_9H_{10}O_4S$	Acetic acid, 2-(4-methoxyphenyl)sulfinyl-	25	2.754	$c \approx 0.01$ Value in mixed solvent is also given	E3bg	Approx.	P13
4765	$C_9H_{10}O_4S$	Acetic acid, 2-(m-tolyl)sulfonyl-	25	2.567	$c \approx 0.01$ Value in mixed solvent is also given	E3bg	Approx.	P13
4766	$C_9H_{10}O_4S$	Acetic acid, 2-(p-tolyl)sulfonyl-	25	2.542	$c \approx 0.01$ Value in mixed solvent is also given	E3bg	Approx.	P13

No.	Molecular formula, name and pK value(s)	T(°C)	Remarks	Method	Assessment	Ref.
4767	$C_9H_{10}O_5N_2$ Ethane, 1-benzyloxy-2,2-dinitro- 3.44	20		O5	Approx.	T58a
4768	$C_9H_{10}O_5N_2$ Phenol, 2-isopropyl-4,6-dinitro- 4.536	25	c ≈ 0.0001	O5a	Approx.	R30
4769	$C_9H_{10}O_5N_2$ Propanoic acid, 2-amino-3-(4-hydroxy-3-nitrophenyl)- (3-Nitrotyrosine) pK$_2$ 6.8 pK$_2$ 6.3 7.5	not stated not stated	c = 0.00025, I ≈ 0.9 I ≈ 1.0 value in 8M urea	O5 O5	Uncert. Uncert.	R24 M42
4770	$C_9H_{10}O_5S$ Acetic acid, 2-(4-mesylphenoxy)- 2.85	25 Value in mixed solvent is also given		E3ag	Approx.	H48
4771	$C_9H_{10}O_5S$ Acetic acid, 2-(3-methoxyphenyl)sulfonyl- 2.586	25 Value in mixed solvent is also given	c ≈ 0.01	E3bg	Approx.	P13
4772	$C_9H_{10}O_5S$ Acetic acid, 2-(4-methoxyphenyl)sulfonyl- 2.602	25 Value in mixed solvent is also given	c ≈ 0.01	E3bg	Approx.	P13

No.	Formula	Name	t(°C)	pK	Remarks	Method	Reliability	Ref.
4773	$C_9H_{10}N_4S$	Pteridine-2(8H)-thione, 6,7,8-trimethyl-	20	8.80	c = 0.00004 in 0.01M buffer	05	Approx.	J4
4774	$C_9H_{11}ON$	Phenol, 2-ethyliminomethyl-	25	11.8(OH)		05	Uncert.	G44
4775	$C_9H_{11}O_2N$	Acetamide, N-(2-hydroxyphenyl)-N-methyl-	25±0.5	8.6		E	Uncert.	V13
4776	$C_9H_{11}O_2N$	Acetic acid, 2-(3-aminomethylphenyl)-	25	3.83	Values in mixed solvents are also given	E3ag	Approx.	H49
4777	$C_9H_{11}O_2N$	Acetic acid, 2-(4-aminomethylphenyl)-	25	3.88	Values in mixed solvents are also given	E3ag	Approx.	H49
4778	$C_9H_{11}O_2N$	Benzoic acid, 2-dimethylamino-	25	8.51	I = 0.105, mixed constant	E3bg	Approx.	H20
4779	$C_9H_{11}O_2N$	Bicyclo[2.2.1]heptane-1-carboxylic acid, 4-cyano-	25	4.227	Values in mixed solvents are also given	E3bg	Approx.	W29

No.	Molecular formula, name and pK value(s)	T(°C)	Remarks	Method	Assessment	Ref.
4780	$C_9H_{11}O_2N$ Propanoic acid, 2-amino-3-phenyl- (Phenylalanine)	25	$I = 0.16(KNO_3)$, mixed constant	E3bg	Approx.	M72
	9.02		Thermodynamic quantities are also given			
4781	$C_9H_{11}O_2N_3$ 4-Quinazolinecarboxylic acid, 2-amino-5,6,7,8-tetrahydro-	20	$c = 0.000035$, $I = 0.01$	05	Approx.	A75
	pK_1 0.61, pK_2 4.72					
4782	$C_9H_{11}O_2S$ (cation) Benzoic acid, 3-dimethylsulfonio-	25	Anion = p-toluene sulfonate	E3ag	Approx.	B103
	3.20					
4783	$C_9H_{11}O_2S$ (cation) Benzoic acid, 4-dimethylsulfonio-	25	Anion = p-toluene sulfonate	E3ag	Approx.	B103
	3.30					
4784	$C_9H_{11}O_3N$ Benzoic acid, 4-amino-2-ethoxy-		Conditions not stated	E3bg	Uncert.	V3
	5.32					
4785	$C_9H_{11}O_3N$ Propanohydroxamic acid, 3-hydroxy-2-phenyl-	25	$c = 0.01$, $I = 0.1(KCl)$, mixed constant	E3bg	Approx.	G38
	9.00					
4786	$C_9H_{11}O_3N$ Propanoic acid, 2-amino-3-(p-hydroxyphenyl)- (Tyrosine)	25	$I = 0.1(KCl)$, mixed constant	05	Uncert.	S54
	pK_1 2.25	25		05	Approx.	E20
	pK_2 10.07					
4787	$C_9H_{11}O_3N_3$ 1,3,5-Triazine-2,4,6(1$\underline{H}$,3$\underline{H}$,5$\underline{H}$)-trione, 1,5-diallyl-		Conditions not stated		Uncert.	A73
	7.21					

4788 C$_9$H$_{11}$O$_3$N$_5$ 4-Pteridinecarboxylic acid, 2-dimethylamino-5,6,7,8-tetrahydro-6-oxo-

0.57	20	c = 0.002, mixed constants;	06	Uncert.	C59
7.31		pK assignment discussed	E3bg	Uncert.	C59
>13			07		

4789 C$_9$H$_{11}$O$_3$N$_5$ 6-Pteridinecarboxylic acid, 2-ethylamino-5,6,7,8-tetrahydro-7-oxo-

7.16	Conditions not stated	P58	Uncert.

4790 C$_9$H$_{11}$O$_4$N 1-Oxa-7-azaspiro[4.4]nonane-2,6,8-trione, 9,9-dimethyl-

7.9	Conditions not stated	E3bg	Uncert.	C62

4791 C$_9$H$_{11}$O$_5$N 3-Azapentanedioic acid, 3-furfuryl- (Iminodiacetic acid, N-furfuryl-)

pK$_1$ 2.17, pK$_2$ 8.41	20	I = 0.1(KNO$_3$), concentration constants	E3bg	Approx.	I11

4792 C$_9$H$_{11}$O$_5$P Benzoic acid, 3-dimethoxyphosphinyl-

3.74	25	Conditions not stated	E3bg	Uncert.	T61
		Value in mixed solvent is also given			

4793 C$_9$H$_{11}$O$_5$P Benzoic acid, 4-dimethoxyphosphinyl-

3.61	25	Conditions not stated	E3bg	Uncert.	T61
		Value in mixed solvent is also given			

4794 C$_9$H$_{11}$O$_7$N$_3$ 3-Azapentanedioic acid, 3-(1-methyl-2,4,6-trioxohexahydropyrimidin-5-yl)- (Iminodiacetic acid, N-(1-methyl-2,4,6-trioxohexahydropyrimidin-5-yl)-)

pK$_1$ 1.85, pK$_2$ 2.67, pK$_3$ 9.81	20	I = 0.1((CH$_3$)$_4$N.NO$_3$), titration with (CH$_3$)$_4$NOH	E	F38

(contd)

No.	Molecular formula, name and pK value(s)	T(°C)	Remarks	Method	Assessment	Ref.
4794 (Contd)	$C_9H_{11}O_7N_3$ 3-Azapentanedioic acid, 3-(1-methyl-2,4,6-trioxohexahydropyrimidin-5-yl)- (Iminodiacetic acid, N-(1-methyl-2,4,6-trioxohexahydropyrimidin-5-yl)-)					
	pK$_1$ 1.85	20	I = 0.1$((CH_3)_4N.NO_3)$	E3bg	Uncert. Approx.	I9
	pK$_2$ 2.67, pK$_3$ 9.81					
4795	$C_9H_{11}NS$ 8-Quinolinethiol, 1,2,3,4-tetrahydro-					S123
	pK$_1$ 3.38, pK$_2$ 6.63					
			Microscopic constants also given			
4796	$C_9H_{11}NS_2$ Methanedithioic acid, (N-benzyl-N-methyl)amino- (Dithiocarbamic acid, N-benzyl-N-methyl-)					T65
	4.66		c = 0.1	E		
4797	$C_9H_{12}ON_2$ Propanamidine, 2-hydroxy-2-phenyl- (Atrolactamidine)					G34
	pK$_1$ 10.96, pK$_2$ 12.72	25	I = 0.1(KCl)	E3ag	Uncert.	
			Decomposition occurs; results obtained from extrapolation to zero time			
4798	$C_9H_{12}ON_2$ Quinazolin-4(1H)-one, 5,6,7,8-tetrahydro-2-methyl-					B134
	pK$_1$ 3.73, pK$_2$ 10.50	20	c < 0.001, I = 0.01, mixed constants	O5	Approx.	
4799	$C_9H_{12}O_2N_2$ Benzohydroxamic acid, 2-dimethylamino-					S117
	9.05	30		E3bg	Uncert.	
4800	$C_9H_{12}O_2S$ 1-Propanesulfinic acid, 3-phenyl-					R54
	2.05	20	c = 0.02, mixed constant	E3bg	Approx.	

No.	Formula	Name	pK	Temp.	Conditions	Method	Assessment	Ref.
4801	$C_9H_{12}O_3N_2$	Pyrimidine-2,4(1H,3H)-dione, 5-(2-hydroxycyclopentanyl)-	9.70		Conditions not stated	05	Uncert.	F20
4802	$C_9H_{12}O_3N_2$	Pyrimidine-2,4(1H,3H)-dione, 5-methyl-1-oxolan-2-yl-	9.59		Conditions not stated	05	Uncert.	G17
4803	$C_9H_{12}O_3N_4$	Purine-2,6,8(1H,3H,7H)-trione, 1,3-diethyl-	5.75	not stated	Mixed constant	05	Uncert.	B65
4803a	$C_9H_{12}O_3S$	Phenol, 4-mesyl-3,5-dimethyl-	8.13	25			Uncert.	C76
4804	$C_9H_{12}O_5N_2$	Acetic acid, 2-(1,2,3,4-tetrahydro-5-methoxycarbonyl-1-methyl-2-oxopyrimidin-4-yl)-	4.40		Conditions not stated		Uncert.	S136
4805	$C_9H_{12}O_5N_4$	Acetic acid, 2-(5-amino-4-carboxy-2-ethoxypyrimidin-6-yl)amino-	3.60 7.33	20	c = 0.002, mixed constants; pK assignment discussed	E3bg	Uncert. Approx.	C59
4806	$C_9H_{12}O_6N_2$, pK_3 9.6, pK_4 >13	Pseudouridine A_F (Pyrimidine-2,4(1H,3H)-dione, 5-(α-D-ribopyranosyl)-			Conditions not stated	05	V.Uncert.	C73
4807	$C_9H_{12}O_6N_2$, pK_3 9.6, pK_4 >13	Pseudouridine A_S (Pyrimidine-2,4(1H,3H)-dione, 5-(β-D-ribopyranosyl)-			Conditions not stated	05	V.Uncert.	C73
4808	$C_9H_{12}O_6N_2$	Pseudouridine B (Pyrimidine-2,4(1H,3H)-dione, 5-(α-D-ribofuranosyl)-	9.19	30		05	Approx.	05

No.	Molecular formula, name and pK value(s)	$T(^oC)$	Remarks	Method	Assessment	Ref.
4809	$C_9H_{12}O_6N_2$ Pseudouridine C (Pyrimidine-2,4(1H,3H)-dione, 5-(β-D-ribofuranosyl)-)					
	pK$_1$ 9.10, pK$_2$ >12.5	not stated	I = 0.05	05	Uncert.	D53
	pK$_1$ 8.97	30		05	Approx.	05
	pK$_1$ 9.0		Conditions not stated	E3bg	Uncert.	C27
4810	$C_9H_{12}O_6N_2$ Uridine (Pyrimidine-2,4(1H,3H)-dione, 1-(β-D-ribofuranosyl)-)					
	9.20	20	I = 1.0(NaNO$_3$)	E3bg	Approx.	F19
	9.25	25	titrant (CH$_3$)$_4$NOH	E3bg	Approx.	J16
4811	$C_9H_{12}N_2S$ Quinazoline-4(1H)-thione, 5,6,7,8-tetrahydro-2-methyl-					
	pK$_1$ 2.35	20	c < 0.001, I = 0.01, mixed constant	05	Uncert.	B134
	pK$_2$ 8.91				Approx.	
4812	$C_9H_{13}ON$ Phenol, 2-(3-aminopropyl)-					
	9.51(-OH)	25	I = 0.1(KCl), mixed constant	05	Approx.	E20
4813	$C_9H_{13}ON$ Phenol, 2-(dimethylaminomethyl)-					
	8.62(OH)	25	I = 0.1(KCl), mixed constant	05	Approx.	E19
	8.53	25	I = 0.1(KCl), mixed constant	E3bg	Approx.	E20
4814	$C_9H_{13}ON$ Phenol, 3-(dimethylaminomethyl)-					
	8.88(OH)	25	I = 0.1(KCl), mixed constant	05	Approx.	E20

No.	Formula	Name	pK	T (°C)	Conditions			Ref.
4815	$C_9H_{13}ON$	Phenol, 4-(dimethylaminomethyl)- 8.83(OH)		25	I = 0.1(KCl), mixed constant	Approx.	05	E20
4816	$C_9H_{13}ON_3$	Benzohydrazide, 4-dimethylamino- 13.03(NH)		25	Mixed constant(?)	Uncert.	05	T43, T44
4817	$C_9H_{13}ON_3$	Triazen-3-ol, 3-ethyl-1-(p-tolyl)- 12.67		25	c = 0.00004, I = 0.1(KCl), mixed constant	Uncert.	05	D54
4818	$C_9H_{13}ON_3$	Triazen-3-ol, 1-phenyl-3-propyl- 12.39		25	c = 0.00004, I = 0.1(KCl), mixed constant	Uncert.	05	D54
4819	$C_9H_{13}O_2N$	1,2-Benzenediol, 3-(N,N-dimethylaminomethyl)- pK_1 8.01		25	I = 0.1(KNO_3), mixed constant	Approx.	E3bg	E19
4820	$C_9H_{13}O_2N$	2-Butynoic acid, 4-piperidino- pK_1 1.78, pK_2 8.03		20	mixed constants	Approx.	E3bg	B39
4821	$C_9H_{13}O_2N$	Triazen-3-ol, 3-ethyl-1-(4-methoxyphenyl)- 12.85		25	c = 0.00004, I = 0.1(KCl), mixed constant	Uncert.	C5	D54
4822	$C_9H_{13}O_2N_3$	Pyrimidin-2(1H)-one, 4-amino-5-(2'-hydroxycyclopentyl)- 4.49			Conditions not stated	Uncert.	05	F20
4823	$C_9H_{13}O_2N_3$	4-Quinazolinecarboxylic acid, 2-amino-3,4,5,6,7,8-hexahydro- pK_1 2.71, pK_2 12.22		20	c ≈ 0.0001, I = 0.01	Approx.	05	A75

No.	Molecular formula, name and pK value(s)	$T(^{o}C)$	Method	Assessment	Ref.	Remarks
4824	$C_9H_{13}O_2Br$ Bicyclo[2.2.2]octane-1-carboxylic acid, 4-bromo-		E3bg,D4	Approx.	W29	
	4.619	25				Values in mixed solvents are also given
4825	$C_9H_{13}O_3N$ 1,2,3-Benzenetriol, 5-(N,N-dimethylaminomethyl)-		E3bg	Approx.	E19	
	pK$_1$ 8.26	25				I = 0.1(KIO$_3$), mixed constant
4826	$C_9H_{13}O_3N$ Ethanol, 1-(3,4-dihydroxyphenyl)-2-(methylamino)- (Adrenaline)		E3bg	Approx.	A65	
	pK$_1$ 2.75, pK$_2$ 9.61, pK$_3$ 10.98((-)isomer)	0				I = 0.06(KCl), mixed constants
	pK$_1$ 2.58, pK$_2$ 8.78, pK$_3$ 10.02	25				
	pK$_2$ 9.17, pK$_3$ 10.57((+)isomer)	0				
	pK$_2$ 8.64, pK$_3$ 9.96	25				
4827	$C_9H_{13}O_4N_5$ Acetic acid, 2-(5-amino-4-carboxy-2-dimethylaminopyrimidin-6-yl)amino-		E3bg	Uncert.	C59	
	9.85	20				c = 0.005, mixed constant; pK assignment discussed
4828	$C_9H_{13}O_5N_3$ Cytidine (Pyrimidin-2(1H)-one, 4-amino-1-(β-D-ribofuranosyl)-)		O5,R4	Approx.	W48	
	4.09	20				I = 0.1-1.0

Variation with temperature

10	30	40	50	60	70	80
4.19	3.99	3.89	3.81	3.73	3.65	3.58

Thermodynamic quantities are derived from the results

(contd)

503

No.	Formula / Name	pK values	Temp	Conditions	Method	Assessment	Ref
4828 (Contd)	$C_9H_{13}O_5N_3$ Cytidine (Pyrimidin-2(1H)-one, 4-amino-1-(β-D-ribofuranosyl)-)	4.23	20	I = 1.0(NaNO$_3$), mixed constant	E3bg	Approx.	F19
4829	$C_9H_{13}O_6N$ 2,6-Piperidinedicarboxylic acid, 1-carboxymethyl-	pK$_2$ 2.71, pK$_3$ 9.33	25	I = 0.1(KNO$_3$), concentration constants	E3bg	Approx. Uncert.	T30
		pK$_1$ 1.3					
		pK$_2$ 2.71, pK$_3$ 9.33	25	c = 0.001-0.002, I = 0.1(KNO$_3$), concentration constants	E3bg	Approx. Uncert.	K68
		pK$_1$ 1.3					
4830	$C_9H_{13}O_6N_3$ 4-Imidazolecarboxylic acid, 5-amino-1-(β-D-ribofuranosyl)-	pK$_1$ 3.00, pK$_2$ 6.34	20	c 0.0001, I = 0.10(NaCl)	05	Uncert.	L58
4831	$C_9H_{13}O_8N$ 3-Azaheptanedioic acid, 4-carboxy-3-carboxymethyl-	pK$_1$ 2.1, pK$_2$ 2.7, pK$_3$ 4.8, pK$_4$ 9.66	20		E		E22
4832	$C_9H_{13}NS$ 1-Propanethiol, 2-amino-3-phenyl-	pK$_1$ 7.93, pK$_2$ 10.29	25	c = 0.01, mixed constants	E3bg	Approx.	B147
4833	$C_9H_{13}NS$ 2-Propanethiol, 1-amino-3-phenyl-	pK$_1$ 8.00, pK$_2$ 11.04	25	c = 0.01, mixed constants	E3bg	Approx.	B147
4833a	$C_9H_{14}ON$ (cation) Phenol, 2-trimethylammonio-	7.43	25	I = 0.1(KCl), mixed constant (anion = iodide)	E3bg	Approx.	E20

No.	Molecular formula, name and pK value(s)	T(°C)	Remarks	Method	Assessment	Ref.
4834	$C_9H_{14}ON$ (cation) Phenol, 3-trimethylammonio-					
	8.06	25	(Anion = chloride)	05	Approx.	K40
	8.12	25	I = 0.1(KCl), mixed constant (anion = iodide)	E3bg	Approx.	E20
	8.03	not stated	I = 0.01, (anion = iodide)	E3ag	Uncert.	02
	8.04	25	(Anion = p-toluenesulfonate)	E3ag	Approx.	B103
4835	$C_9H_{14}ON$ (cation) Phenol, 4-trimethylammonio-					
	8.35	25	(Anion = chloride)	05	Approx.	K40
	8.34(H_2O) 8.90(D_2O)	25	(Anion = chloride)	05	Uncert.	W12
	8.30	25	I = 0.1(KCl), mixed constant (anion = iodide)	E3bg	Approx.	E20
	8.21	not stated	I = 0.01, (anion = iodide)	E3ag	Uncert.	02
	8.34	25	(Anion = p-toluenesulfonate)	E3ag	Approx.	B103
4836	$C_9H_{14}O_2N_2$ Pyrimidine-2,4(1H,3H)-dione, 5-tert-butyl-1-methyl-					
	10.8					D48
4837	$C_9H_{14}O_2N_2$ Pyrimidine-2,4(1H,3H)-dione, 5-tert-butyl-3-methyl-					
	10.9					D48

No.	Formula	Name / pK	Conditions	Method	Assessment	Ref.
4838	$C_9H_{14}O_3N_2$	Pyrimidine-2,4,6(1$\underline{H}$,3$\underline{H}$,5$\underline{H}$)-trione, 5,5-diethyl-1-methyl- (Metharbital)				
		8.45	Conditions not stated	05	Uncert.	F32
		8.2	Conditions not stated	05	Uncert.	B189
4839	$C_9H_{14}O_3N_2$	Pyrimidine-2,4,6(1$\underline{H}$,3$\underline{H}$,5$\underline{H}$)-trione, 5-ethyl-5-isopropyl-				
		8.01	25	E3ag	Approx.	K53
4840	$C_9H_{14}O_3N_2$	Pyrimidin-2(1$\underline{H}$)-one, 5-diethoxymethyl-				
		7.8	not stated c = 0.01	E3bg	Uncert.	C58
4841	$C_9H_{15}O_2N$	Butanoic acid, 2-cyano-2-isopropyl-3-methyl-				
		2.556	25	C2	Rel.	I25

Variation with temperature

5	10	15	20	30	35	40	45
2.393	2.432	2.473	2.514	2.598	2.640	2.683	2.726

Thermodynamic quantities are derived from the results

No.	Formula	Name / pK	Conditions	Method	Assessment	Ref.
4842	$C_9H_{15}O_6N$	3,3',3"-Nitrilotripropanoic acid				
		pK_1 3.67, pK_2 4.24, pK_3 9.30	30 I = 0.1(KCl), concentration constants	E3bg	Uncert.	C24
4843	$C_9H_{15}O_6P$	3,3',3"-Phosphinidynetripropanoic acid				
		pK_1 2.89, pK_2 3.81, pK_3 4.70, pK_4 8.42	25 c = 0.01	E3bg,R3	Approx.	L59
4844	$C_9H_{16}O_2N_2$	2-Butanone, 3-hydroxyimino-1-piperidino-				
		pK_1 8.15, pK_2 9.78	25 c = 0.0025-0.005, I = 0.1(KNO$_3$) (probably mixed constants)	E		U9

No.	Molecular formula, name and pK value(s)	T(°C)	Remarks	Method	Assessment	Ref.
4845	$C_9H_{16}O_2N_2$ 2-Butanone, 3-hydroxyimino-4-piperidino-					
	pK$_1$ 7.15	30	c = 0.01, I = 0.1(KCl), mixed constant	E3bg	Approx.	E18
	pK$_1$ 6.96, pK$_2$ 10.65	25	c = 0.0025-0.005, I = 0.1(KNO$_3$)	E		U9
4846	$C_9H_{16}O_2B_{10}$ 1,2-Dicarbadodecaborane(12)-1-carboxylic acid, 2-phenyl- (Barenecarboxylic acid, phenyl-)					
	3.12	25	Mixed constant	E3bg	Uncert.	Z3
4847	$C_9H_{16}O_6N_2$ 3,6-Diazaheptanoic acid, 3-carboxymethyl-7-ethoxy-7-oxo-					
	pK$_1$ 2.2	20	I = 0.1(KCl), concentration constants	E3bg	Uncert.	S40
	pK$_2$ 8.57				Approx.	
4848	$C_9H_{16}O_7N_2$ 3,6-Diazaoctanedioic acid, 3-carboxymethyl-6-hydroxymethyl- (Ethylenedinitrilo-N-hydroxymethyl-N,N',N'-triacetic acid)					
	pK$_1$ 2.39, pK$_2$ 5.37, pK$_3$ 9.93	25	c = 0.01, I = 0.1(KNO$_3$), concentration constants	E3bg	Approx.	M94

Variation with temperature

	15	20	30	35	40
pK$_1$	2.36	2.39	2.38	2.39	2.39
pK$_2$	5.44	5.41	5.33	5.30	5.27
pK$_3$	10.10	10.01	9.83	9.74	9.67

Thermodynamic quantities are derived from the results

No.	Formula	Name	pK	T(°C)	Conditions	Method	Reliability	Ref.
4849	$C_9H_{17}O_2N_3$	2,3-Butanedione dioxime, 1-piperidino- pK$_1$ 8.47, pK$_2$ 10.6		25	c = 0.0025-0.005, I = 0.1(KNO$_3$) (probably mixed constants)	E		U9
4850	$C_9H_{17}O_4N_3$	3,6-Diazanonanoic acid, 8-amino-2,5-dimethyl-4,7-dioxo- (Di-L-alanyl-L-alanine) pK$_1$ 3.36, pK$_2$ 8.08		25	I < 0.16	E3bg	Uncert.	B151
4851	$C_9H_{18}OS_2$	Methanedithioic acid, octyloxy- (Xanthic acid, octyl-) 3.77						H27
4852	$C_9H_{18}O_4N_2$	Nonane, 1,1-dinitro- 5.50		not stated	Extrapolated from values in ethanol/water	E3bg	Uncert.	N36
4853	$C_9H_{19}O_4N_3$	4-Azadecanedioic acid, 2,9-diamino- pK$_1$ <1.5, pK$_2$ 2.2, pK$_3$ 6.5, pK$_4$ 8.8, pK$_5$ 9.9			Conditions not stated	E3bg	Uncert.	B87
4854	$C_9H_{19}NS_2$	Methanedithioic acid, dibutylamino- (Dithiocarbamic acid, N,N-dibutyl-) 5.24			c = 0.1	E		T65
4855	$C_9H_{19}NS_2$	Methanedithioic acid, 2-ethylhexylamino- (Dithiocarbamic acid, N-2-ethylhexyl-) 3.30		21	c ≈ 0.01-0.00001, concentration constant.	O5	Uncert.	T66
		3.40			(Compound unstable in acid solution)	E	Uncert.	

No.	Molecular formula, name and pK value(s)	T(°C)	Remarks	Method	Assessment	Ref.
4856	$C_9H_{20}ON_2$ 2-Butanone oxime, 3-butylamino-3-methyl-					
	9.09	25	I = 0.002	CAL	Uncert.	W8
			Thermodynamic quantities are also given			
4857	$C_9H_{21}O_{17}P_3$ 1'-Glycerylphosphorylinositol 3,4-diphosphoric acid					
	pK$_3$ 5.70, pK$_4$ 8.05	20	I = 0.1((C$_3$H$_7$)$_4$N^I)	E3bg	Uncert.	H31
4858	$C_9H_5ONCl_2$ 8-Quinolinol, 5,7-dichloro-					
	pK$_1$ 2.89, pK$_2$ 7.62	25	c = 0.002, I = 0.5	E3bg	Approx.	T57
			Thermodynamic quantities are also given			
	-3.2	not stated	pK for first excited singlet state	FLU	Uncert.	S29
			In aq. H$_2$SO$_4$, H$_0$ scale			
4859	$C_9H_5ONBr_2$ 8-Quinolinol, 5,7-dibromo-					
	pK$_1$ 5.85, pK$_2$ 9.56			0		A49
	-7.6	not stated	pK for first excited singlet state	FLU	Uncert.	S29
			In aq. H$_2$SO$_4$, H$_0$ scale			
4860	$C_9H_5O_2N_2Cl$ 2-Quinoxalinecarboxylic acid, 3-chloro-					
	pK$_3$ 1.83	25	c ≈ 0.01, I = 0.1(NaClO$_4$), mixed constant	E3bg	Approx.	D61
4861	C_9H_6ONCl 8-Quinolinol, 5-chloro-					
	-9	not stated	pK for first excited singlet state	FLU	Uncert.	S29
			In aq. H$_2$SO$_4$, H$_0$ scale			

No.	Formula	Name	pK	T(°C)	Conditions	Method	Assessment	Ref.
4862	C_9H_6ONBr	8-Quinolinol, 5-bromo-	-7.8	not stated	pK for first excited singlet state. In aq. H_2SO_4, H_O scale	FLU	Uncert.	S29
4863	C_9H_6ONI	8-Quinolinol, 5-iodo-	-6.7	not stated	pK for first excited singlet state. In aq. H_2SO_4, H_O scale	FLU	Uncert.	S29
4864	C_9H_6ONF	8-Quinolinol, 5-fluoro-	-11	not stated	pK for first excited singlet state. In aq. H_2SO_4, H_O scale	FLU	Uncert.	S29
4865	$C_9H_6O_2NBr$	2-Indolecarboxylic acid, 5-bromo-	16.10(NH)	25	In aq. KOH, H_- scale	07	Uncert.	Y2
4866	$C_9H_6O_2NBr$	Isoxazol-5(2$\underline{H}$)-one, 4-bromo-3-phenyl-	2.3	20		05	Uncert.	B110
4867	C_9H_6NFS	8-Quinolinethiol, 5-fluoro-	1.97(NH^+)		I = 0.013	0 & E		B17
			7.44(SH)		I = 0.01	0 & E		
4868	$C_9H_7O_2NS$	Acetic acid, 2-(4-cyanophenyl)thio-	3.12	20	c ≈ 0.001, I = 0.1(KNO_3), mixed constant	E3bg	Approx.	P39

No.	Molecular formula, name and pK value(s)	$T(^{o}C)$	Remarks	Method	Assessment	Ref.
4869	$C_9H_7O_2N_3S$ 1,3-Benzenediol, 4-(2-thiazolylazo)- (TAR)					
	pK_1 0.96, pK_2 6.23, pK_3 9.44	not stated	I = 0.1($NaClO_4$)	O5	Uncert.	H47
	pK_1 1.25, pK_2 6.0, pK_3 9.0			O		M83
	pK_2 6.15, pK_3 9.68	not stated	c = 0.001	O	Uncert.	B183
	pK_2 6.53, pK_3 10.76	25	Mixed constants	E3bg	Uncert.	N30
	pK_3 9.15	not stated	I = 0.1($NaClO_4$), mixed constant	O5	Uncert.	B185
4870	$C_9H_7O_2F_3S$ Acetic acid, 2-(3-trifluoromethylphenyl)thio-					
	3.30	20	c ≈ 0.001, I = 0.1(KNO_3), mixed constant	E3bg	Approx.	P39
4871	$C_9H_7O_3NS_2$ 5-Quinolinesulfonic acid, 8-mercapto-					
	$pK(NH^+)$ 0.92, $pK(SO_3H)$ 1.07, $pK(SH)$ 7.63		Concentration constants	O5		B16
4872	$C_9H_7O_4NS$ 5-Quinolinesulfonic acid, 8-hydroxy-					
	pK_2 3.92, pK_3 8.48	25	I = 0.1(KNO_3)	E3bg	Approx.	M99
	pK_2 3.85, pK_3 8.41	34.9	extrapolated to c = 0, concentration constants	E3bg		
	pK_2 3.84, pK_3 8.23	25	c = 0.002, I = 0.5	E3bg	Approx.	T57
			Thermodynamic quantities are also given			
	pK_2 4.01, pK_3 8.53	25	Thermodynamic quantities are also given	E3bg	Approx.	F40
	pK_2 3.84, pK_3 8.35	25	I = 0.1(KNO_3), concentration constants	E3bg	Approx.	R23
	pK_2 3.98, pK_3 8.47	25	I = 0.1(KNO_3), concentration constants	E3bg	Approx.	L37
	pK_2 3.35, pK_3 7.57	25	I = 2.0(NaCl)	E		C30

No.	Formula	Name	pK	T	Conditions	Code	Reliability	Ref.
4873	$C_9H_7O_4NS$	7-Quinolinesulfonic acid, 8-hydroxy-						
		pK$_2$ 4.27, pK$_3$ 8.76		25		05	Uncert.	B4
		pK$_2$ 3.34, pK$_3$ 7.51		25	$I = 2.0(NaCl)$	E		C30
4874	$C_9H_7O_4N_2F_3$	2,2,2-Trifluoroethyl 4-nitroanilinomethanoate						
		11.9		25	$I = 1.0(KCl)$	05	Uncert.	H26
4875	$C_9H_8O_4NP$	8-Quinolinyl dihydrogen phosphate						
		pK$_1$ 4.17, pK$_2$ 6.42		25	$I = 0.1(KNO_3)$	E3bg	Approx.	M118
		pK$_1$ 4.16, pK$_2$ 6.29		80				M113
4876	$C_9H_9O_2N_3S_2$	N-(2-Thiazolyl)benzenesulfonamide, 4-amino-						
		7.23(-SO$_2$NH-)		20	$c = 0.0025$, $I = 0.1(KCl)$, mixed constant	E3bg	Approx.	W34
4877	$C_9H_{10}ONCl$	Isoquinolin-1(2H)-one, 3-chloro-5,6,7,8-tetrahydro-						
		pK$_1$ -0.28		20	$c = 0.00005$	06	Uncert.	S74
		pK$_2$ 8.81				05	Approx.	
4878	$C_9H_{10}O_2N_4S$	Acetic acid, 2-(6,9-dimethylpurin-2-yl)thio-						
		pK$_1$ $\approx$ -2		20	$c < 0.0001$, $I = 0.01$	06	Uncert.	B138
		pK$_2$ 2.03				05	Uncert.	
4879	$C_9H_{10}O_2SSe$	Acetic acid, 2-(2-methylthiophenyl)seleno-						
		3.80		20	$c = 0.001$, $I = 0.1(KNO_3)$, mixed constant	E3bg	Approx.	P39

No.	Molecular formula, name and pK value(s)	T(°C)	Remarks	Method	Assessment	Ref.
4880	$C_9H_{10}O_2SSe$ Acetic acid, 2-(4-methylthiophenyl)seleno- 3.83	20	c = 0.001, I = 0.1(KNO$_3$), mixed constant	E3bg	Approx.	P39
4881	$C_9H_{10}O_3NF$ Propanoic acid, 2-amino-3-(3-fluoro-4-hydroxyphenyl)- pK$_2$ 8.4	22		O5	Uncert.	A6
4882	$C_9H_{10}O_4Cl_3P$ Ethyl hydrogen (2,4,5-trichlorophenoxymethyl)phosphonate 0.86	20		O5	Approx.	P60
4883	$C_9H_{10}N(Cl)S$ Thieno[3,2-c]pyridinium chloride, 2,4-dimethyl- 6.43	20		E3bg	Uncert.	D51
4884	$C_9H_{10}N(Cl)S$ Thieno[3,2-c]pyridinium chloride, 4,6-dimethyl- 6.75	20		E3bg	Uncert.	D51
4885	$C_9H_{11}ONS$ Acetamide, N-(2-mercaptophenyl)-N-methyl- 5.6	25±0.5	I = 0.01	O5	Uncert.	V13
4886	$C_9H_{11}O_4Cl_2P$ Ethyl hydrogen (2,4-dichlorophenoxymethyl)phosphonate 0.92	20		O5	Approx.	P60
4887	$C_9H_{11}O_5N_2Cl$ Uridine, 5-chloro-2'-deoxy- 7.90	20-22	I = 0.05-0.1	O5	Uncert.	B64

No.	Formula / Name	pK	t (°C)	Conditions	Method	Assessment	Ref.
4888	$C_9H_{11}O_5N_2Br$ Uridine, 5-bromo-2'-deoxy-						
		7.90	20-22	I = 0.05-0.1	05	Uncert.	B64
		8.1		Conditions not stated	05	Uncert.	L8
4889	$C_9H_{11}O_5N_2I$ Uridine, 2'-deoxy-5-iodo-						
		8.20	20-22	I = 0.05-0.1	05	Uncert.	B64
4890	$C_9H_{11}O_5N_2F$ Uridine, 2'-deoxy-5-fluoro-						
		7.80	20-22	I = 0.05-0.1	05	Uncert.	B64
4891	$C_9H_{11}O_6N_2Cl$ Uridine, 5-chloro-						
		8.20	20-22	I = 0.05-0.1	05	Uncert.	B64
4892	$C_9H_{11}O_6N_2Br$ Uridine, 5-bromo-						
		8.20	20-22	I = 0.05-0.1	05	Uncert.	B64
4893	$C_9H_{11}O_6N_2I$ Uridine, 5-iodo-						
		8.50	20-22	I = 0.05-0.1	05	Uncert.	B64
4894	$C_9H_{11}O_6N_2F$ Uridine, 5-fluoro-						
		7.75	20-22	I = 0.05-0.1	05	Uncert.	B64
4895	$C_9H_{12}ONCl$ Phenol, 4-chloro-2-(N,N-dimethylaminomethyl)-						
		pK_1 7.95	25	I = 0.1(KNO_3), mixed constant	E3bg	Approx.	E19

No.	Molecular formula, name and pK value(s)	$T(^oC)$	Remarks	Method	Assessment	Ref.
4896	$C_9H_{12}ON_3Cl$ Triazen-3-ol, 1-(4-chlorophenyl)-3-propyl- 11.74	25	c = 0.00004, I = 0.1(KCl), mixed constant	05	Uncert.	D54
4897	$C_9H_{12}ON_4S$ Purin-2(1H)-one, 9-butyl-7,8-dihydro-8-thioxo- 6.90	20	c = 0.001	E3bg	Approx.	B128
4898	$C_9H_{12}O_4N_2S_2$ Uridine, 2,4-dimercapto- 7.4		Conditions not stated	05	Uncert.	U2
4899	$C_9H_{12}O_4ClP$ Ethyl hydrogen (2-chlorophenoxymethyl)phosphonate 0.96	20		05	Approx.	P60
4900	$C_9H_{12}O_4ClP$ Ethyl hydrogen (4-chlorophenoxymethyl)phosphonate 0.93	20		05	Approx.	P60
4901	$C_9H_{12}O_5N_2S$ Pyrimidin-2(1H)-one, 3,4-dihydro-1-(β-D-ribofuranosyl)-4-thioxo- (4-Thiouridine) 8.2		Conditions not stated	05	Uncert.	L55
4902	$C_9H_{12}O_5N_2S$ Pyrimidin-4(1H)-one, 2,3-dihydro-1-(β-D-ribofuranosyl)-2-thioxo- (2-Thiouridine) 8.8		I = 0.1(tetraethylammonium chloride)	05	Uncert.	L12
4903	$C_9H_{12}O_7N_3P$ Cytidine 2',3'-cyclic phosphate 3.76	25	I = 0.1(KNO_3), concentration constant Thermodynamic quantities are also given	E3bg	Approx.	B3

No.	Formula / Name	pK	T (°C)	Conditions	Method	Reliability	Ref.
4904	$C_9H_{13}ONCl$ (cation) Phenol, 2-chloro-4-trimethylammonio-	6.62	25	$c \approx 0.0001\text{-}0.0002$, $I \approx 0.1$, mixed constant	05	Approx.	S76
4905	$C_9H_{13}ON_2I$ 1-Methylpyridinium iodide, 5-ethyl-2-hydroxyiminomethyl-	8.1	not stated	$c \approx 0.04$	E3bg	Uncert.	W36
4906	$C_9H_{13}O(I)S$ Phenol, 4-dimethylsulfonio-3-methyl-, iodide	7.60	not stated	$I = 0.01$	E3ag	Uncert.	02
4907	$C_9H_{13}O(I)S$ Phenol, 3-dimethylsulfonio-4-methyl-, iodide	8.01		$I = 0.01$	E3ag	Uncert.	02
4908	$C_9H_{13}O_2SP$ S-Propyl hydrogen phenylphosphonothioate	1.60	20	$c = 0.005$, mixed constant	E3bg	Uncert.	N28
4909	$C_9H_{13}O_3NS$ Butyl 2-(4-oxothiazolidin-2-ylidene)acetate	8.20(trans)	not stated	$I = 0.01$	05	Uncert.	T13
4910	$C_9H_{13}O_3N_2(Cl)$ Phenol, 2-nitro-4-trimethylammonio-, chloride	5.03	25	$c = 0.0002$, $I \approx 0.5$, mixed constant	05	Approx.	R15
4911	$C_9H_{13}O_9N_2P$ Pseudouridine-3'-(dihydrogen phosphate) pK_4 9.40, pK_5 >12.5			Conditions not stated	05	Uncert.	D53
4912	$C_9H_{13}O_9N_2P$ Pseudouridine-5'-(dihydrogen phosphate) pK_4 9.60, pK_5 >13			Conditions not stated	05	Uncert.	D53

No.	Molecular formula, name and pK value(s)	T(°C)	Remarks	Method	Assessment	Ref.
4913	$C_9H_{13}O_9N_2P$ Uridine-3'-(dihydrogen phosphate)					
	pK$_3$ 5.74	25	I = 0.2(KNO$_3$)	E3bg	Approx.	A59
4914	$C_9H_{13}O_9N_2P$ Uridine-5'-(dihydrogen phosphate)					
	pK$_3$ 6.63	25		E3bg	Approx.	P61
			Metallic cations replaced by tetra-n-propylammonium ion			
			Thermodynamic quantities are given			
	pK$_4$ 9.71	20	I = 0.015	E3bg	Approx.	A82
	pK$_4$ 9.55	30				
	pK$_4$ 9.38	40	Thermodynamic quantities are derived			
	pK$_4$ 9.25	50	from the results			
4915	$C_9H_{13}O_9N_2P$ Uridine-2'(3')-phosphate					
	pK$_4$ 9.6		Conditions not stated	O5	Uncert.	L23
			Point of attachment of phosphate uncertain			
4916	$C_9H_{14}ON(Cl)$ Phenol, 3-trimethylammonio-, chloride					
	8.06	25		O5	Approx.	K40
			(see also compound 4834)			
4917	$C_9H_{14}ON(Cl)$ Phenol, 4-trimethylammonio-, chloride					
	8.35	25		O5	Approx.	K40

(contd)

No.	Formula / Name	pK	T (°C)	Conditions	Method	Assessment	Ref.
4917 (Contd)	$C_9H_{14}ON(Cl)$ Phenol, 4-trimethylammonio-, chloride						
		8.34(H_2O)	25		05	Uncert.	W12
		8.90(D_2O)	25				
		(See also compound 4835)					
4918	$C_9H_{14}ON(I)$ Phenol, 2-trimethylammonio-, iodide						
		7.43	25	I = 0.1(KCl), mixed constant	E3bg	Approx.	E20
		(See also compound 4833a)					
4919	$C_9H_{14}ON(I)$ Phenol, 3-trimethylammonio-, iodide						
		8.12	25	I = 0.1(KCl), mixed constant	E3bg	Approx.	E20
		8.03	not stated	I = 0.01	E3ag	Uncert.	02
		(See also compound 4834)					
4920	$C_9H_{14}ON(I)$ Phenol, 4-trimethylammonio-, iodide						
		8.30	25	I = 0.1(KCl), mixed constant	E3bg	Approx.	E20
		8.21	not stated	I = 0.01	E3ag	Uncert.	02
		(See also compound 4835)					
4921	$C_9H_{14}ON_2S$ 1,3-Diazaspiro[4.5]decan-4-one, 1-methyl-2-thioxo-						
		9.25	18±3	I = 0.01	05	Uncert.	E4
4922	$C_9H_{14}ON_2S$ 1,3-Diazaspiro[4.5]decan-4-one, 3-methyl-2-thioxo-						
		11.23	18±3	I = 0.01	05	Uncert.	E4

No.	Molecular formula, name and pK value(s)	$T(^oC)$	Remarks	Method	Assessment	Ref.
4923	$C_9H_{14}O_2N_2S$ Pyrimidin-4(1<u>H</u>)-one, 2,3-dihydro-6-hydroxy-5-(1-methylbutyl)-2-thioxo-					S14
	pK_1 2.47, pK_3 11.41	28	2% ethanol	05	Uncert.	
			Assignation of pK values is uncertain			
4924	$C_9H_{14}O_3NP$ Propylphosphonic acid, 3-anilino-					C38
	pK_1 2.10, pK_2 7.15, pK_3 4.25		Conditions not stated	E3bg	Uncert.	
4925	$C_9H_{14}O_3N_3P$ 1-Guanidinophosphonic acid, 3-benzyl-3-methyl-					A48
	pK_1 -0.26	30.5	I = 0.2(NaCl)	KIN	Uncert.	
	pK_2 4.33			E3bg	Approx.	
	pK_3 11.5				Uncert.	
4926	$C_9H_{14}O_8N_3P$ Cytidine-2'-(dihydrogen phosphate)					A59
	pK_2 4.32, pK_3 6.02	25	I = 0.2(KNO_3)	E3bg	Approx.	
4927	$C_9H_{14}O_8N_3P$ Cytidine-3'-(dihydrogen phosphate)					B3
	pK_2 4.21, pK_3 5.79	25	I = 0.1(KNO_3), concentration constants	E3bg	Approx.	
			Thermodynamic quantities are also given			
4928	$C_9H_{14}O_8N_3P$ Cytidine-5'-(dihydrogen phosphate)					W48
	pK_2 4.48	20	I = 0.1-1.0	05,R4	Approx.	

Variation with temperature

10	30	40	50	60	70	80
4.58	4.39	4.31	4.23	4.16	4.09	4.02

Thermodynamic values are derived from the results   (contd)

4928 $C_9H_{14}O_8N_3P$ Cytidine-5'-(dihydrogen phosphate)
(Contd)

pK$_3$ 6.62   25   E3bg   Approx.   P61

Metallic cations replaced by tetrapropylammonium ion

Thermodynamic quantities are also given

4929 $C_9H_{14}O_9N_2P$ 4-Imidazolecarboxylic acid, 5-amino-1-(5'-phosphonoxy-β-D-ribofuranosyl)-

pK$_1$ 3.21, pK$_2$ 6.28   20   c = 0.01, I = 0.12(NaCl), mixed constants   05   Uncert.   L58

4930 $C_9H_{14}O_{12}N_2P_2$ Uridine-5'-(trihydrogen diphosphate)

pK$_4$ 7.16   25   E3bg   Approx.   P61

Metallic cations replaced by tetrapropylammonium ion

Thermodynamic quantities are also given

4931 $C_9H_{15}O_{11}N_3P_2$ Cytidine-5'-(trihydrogen diphosphate)

pK$_4$ 7.18   25   E3bg   Approx.   P61

Metallic cations replaced by tetrapropylammonium ion

Thermodynamic quantities are also given

4932 $C_9H_{15}O_{15}N_2P_3$ Uridine-5'-(tetrahydrogen triphosphate)

pK$_5$ 7.58   25   E3bg   Approx.   P61

Metallic cations replaced by tetrapropylammonium ion

Thermodynamic quantities are also given

pK$_6$ 10.2   25   c = 0.0004   05   Uncert.   S71
      9.6   I = 0.1(NaClO$_4$)

Formation of sodium complexes said to occur in the presence of NaClO$_4$

No.	Molecular formula, name and pK value(s)	T(°C)	Remarks	Method	Assessment	Ref.
4933	$C_9H_{16}O_{14}N_3P_3$ Cytidine-5'-(tetrahydrogen triphosphate) pK_5 7.65	25	Metallic cations replaced by tetrapropylammonium ion Thermodynamic quantities are also given	E3bg	Approx.	P61
4934	$C_9H_{18}O_4N(Cl)$ Butanoic acid, 3-acetoxy-4-trimethylammonio-, chloride (Acetylcarnitine) 3.60	25	c = 0.01	E3bg	Approx.	Y5
4935	$C_9H_{19}O_2N_2(I)$ 2-Butanone, 4-(N,N,N-diethylmethylammonio)-3-hydroxyimino-, iodide (N,N-Diethyl-N-methyl 2-hydroxyimino-3-oxobutylammonium iodide) 6.95	30	c = 0.01, I = 0.1(KCl), mixed constant	E3bg	Approx.	E18
4936	$C_9H_{19}O_4N_2(Br)$ 3-Azapentanedioic acid, 3-(2-trimethylammonioethyl)-, bromide pK_1 2.32, pK_2 5.45	20	I = 0.1(KCl), concentration constants	E3bg	Approx.	S40
4937	$C_9H_{20}O_7N_3P$ 3-Azahexanoic acid, 5-amino-2(4-aminobutyl)-4-oxo-6-phosphonooxy- (O-Phosphono-L-seryl-L-lysine) pK_2 2.99, pK_3 5.34, pK_4 7.58	25	I = 0.15(KCl), concentration constants	E3bg	Approx.	O24
4938	$C_9H_{28}O_{15}N_3P_5$ 2,5,8-Triazanonane-1,9-diphosphonic acid, 2,5,8-tris(phosphonomethyl)- (N-(Phosphonomethyl)-2,2'-iminobis[ethylenenitrilobis(methylphosphonic acid)]) pK_3 2.8, pK_4 4.45, pK_5 5.50, pK_6 6.38, pK_7 7.17, pK_8 8.15, pK_9 10.10, pK_{10} 12.04 pK_1 1.52, pK_2 2.64, pK_3 3.10, pK_4 3.82, pK_5 5.38, pK_6 6.23, pK_7 7.05, pK_8 7.74, pK_9 9.36, pK_{10} 11.12	20	c = 0.002, I = 0.1(KCl) I = 0.1(KCl)	E3bg E3bg,R3g	Uncert. Uncert.	K1 T36

No.	Formula	Name / Conditions	T (°C)	Method	Reliability	Ref.
4939	C_9H_5ONClI	8-Quinolinol, 5-chloro-6-iodo-7- pK for first excited singlet state In aq. H_2SO_4, H_0 scale	not stated	FLU	Uncert.	S29
4940	$C_9H_6O_2N_3BrS$	1,3-Benzenediol, 4-(5-bromo-2-thiazolylazo)- $I = 0.1(NaClO_4)$, mixed constant	not stated	05	Uncert.	B185
		pK$_2$ 10.50				
4941	$C_9H_6O_4NClS$	5-Quinolinesulfonic acid, 7-chloro-8-hydroxy- $I = 2.0(NaCl)$	25	E		C30
		pK$_2$ 2.91, pK$_3$ 6.91				
4942	$C_9H_6O_4NClS$	7-Quinolinesulfonic acid, 5-chloro-8-hydroxy-	25	05	Uncert.	B4
		pK$_2$ 3.87, pK$_3$ 8.46				
		$I = 2.0(NaCl)$	25	E		C30
		pK$_2$ 2.29, pK$_3$ 6.57				
4943	$C_9H_6O_4NBrS$	5-Quinolinesulfonic acid, 7-bromo-8-hydroxy- $I = 2.0(NaCl)$	25	E		C30
		pK$_2$ 3.22, pK$_3$ 6.93				
4944	$C_9H_6O_4NBrS$	7-Quinolinesulfonic acid, 5-bromo-8-hydroxy-	25	05	Uncert.	B4
		pK$_2$ 4.08, pK$_3$ 8.56				
		$I = 2.0(NaCl)$	25	E		C30
		pK$_2$ 3.05, pK$_3$ 6.91				
4945	$C_9H_6O_4NIS$	5-Quinolinesulfonic acid, 8-hydroxy-7-iodo- (Ferron) $c = 0.002$, $I = 0.5$	25	E3bg & 05	Approx.	T57
		pK$_2$ 2.22, pK$_3$ 6.90				
		Thermodynamic quantities are also given				
		$c = 0.005$, $I = 0.1(KCl)$	25	E3bg	Approx.	L5
		pK$_2$ 2.50, pK$_3$ 7.11				

(contd)

No.	Molecular formula, name and pK value(s)	T(°C)	Remarks	Method	Assessment	Ref.
4945 (Contd)	$C_9H_6O_4NIS$ 5-Quinolinesulfonic acid, 8-hydroxy-7-iodo-		(Ferron)			
	pK_2 2.41, pK_3 7.10	25	I = 0.1	0		H56
	pK_2 2.96, pK_3 6.46	25	I = 2.0(NaCl)	E		C30
4946	$C_9H_6O_4NIS$ 7-Quinolinesulfonic acid, 8-hydroxy-5-iodo-					
	pK_2 4.17, pK_3 8.66	25		05	Uncert.	B4
	pK_2 2.99, pK_3 6.49	25	I = 2.0(NaCl)	E		C30
4947	$C_9H_{12}O_9N_2ClP$ Uridine-2'(3')-phosphate, 5-chloro-					
	8.5		Conditions not stated	05	Uncert.	L23
			Point of attachment of phosphate uncertain			
4948	$C_9H_{12}O_9N_2BrP$ Uridine-2'(3')-phosphate, 5-bromo-					
	8.5		Conditions not stated	05	Uncert.	L23
			Point of attachment of phosphate uncertain			
4949	$C_9H_{12}O_9N_2IP$ Uridine-2'(3')-phosphate, 5-iodo-					
	8.8		Conditions not stated	05	Uncert.	L23
			Point of attachment of phosphate uncertain			

No.	Substance / pK	Temp. (°C)	Conditions	Method	Reliability	Ref.
4950	$C_{10}HN_7$ 3-Azapenta-1,4-diene-1,1,2,4,5,5-hexacarbonitrile (Bis(tricyanovinyl)amine)					
	−5.8	25	Thermodynamic quantities are also given	06	Uncert.	B118
4951	$C_{10}H_2N_6$ 1,1,3,3-Propenetetracarbonitrile, 2-dicyanomethyl-					
	pK_1 <−8.5, pK_2 2.5		Conditions not stated	06	Uncert.	B117
4952	$C_{10}H_6O_3$ Cyclobutenedione, 3-hydroxy-4-phenyl-					
	−0.22	not stated	c = 0.00003-0.00005	06	Uncert.	P15
	0.37	not stated		06	Uncert.	S91
4953	$C_{10}H_6O_3$ 1,4-Naphthoquinone, 2-hydroxy- (Lawsone)					
	−5.6(CO)	25±0.2	In aq. H_2SO_4, H_0 scale	06	Uncert.	B41
	4.00(OH)	25±0.2	I = 0.5(NaCl), concentration constant		Approx.	
			Mean of E3bg & 05 determinations			
	pK_2 2.38	not stated	I = 0.1	05	Uncert.	Z16
4954	$C_{10}H_6O_4$ Cyclobutenedione, 3-hydroxy-4-(p-hydroxyphenyl)-					
	pK_1 1.85, pK_2 8.35	20		E3bg	Uncert.	B126
4955	$C_{10}H_6O_8$ 1,2,4,5-Benzenetetracarboxylic acid (Pyromellitic acid)					
	pK_1 1.87, pK_2 2.72, pK_3 4.30, pK_4 5.52	25		E		I2
4956	$C_{10}H_8O$ 1-Naphthol					
	9.34	25	c ≈ 0.0001	05,R4	Approx.	M30
	9.40	25	c ≈ 0.001 (contd)	E3bg	Approx.	B155

No.	Molecular formula, name and pK value(s)	T(°C)	Remarks	Method	Assessment	Ref.
4956 (Contd)	$C_{10}H_8O$ 1-Naphthol					
	9.39	25		05a	Approx.	C79
	9.34	21	I = 0.02	05	Approx.	G13
4957	$C_{10}H_8O$ 2-Naphthol					
	9.57	25	c ≈ 0.001	E3bg	Approx.	B155
	9.58	21	I = 0.02	05	Approx.	G13
	9.51	25	c ≈ 0.0001	05,R4	Approx.	M30
	9.97	20	c = 0.02, mixed constant	E3bg	Approx.	M128
	9.47(H_2O)	25		05	Uncert.	W12
	10.06(D_2O)					
	9.5		Conditions not stated	05	Uncert.	J2
	3.1		First excited singlet state	FLU/05		
	8.1		Lowest triplet state	Flash Photolysis		
	7.7			Phosphorescence		
4958	$C_{10}H_8O_2$ 2,3-Butadienoic acid, 4-phenyl-					
	3.70	25	c = 0.005, I = 0.1(NaCl), mixed constant	E3bg	Approx.	M48
4959	$C_{10}H_8O_2$ 3-Butynoic acid, 4-phenyl-					
	3.44	25	c = 0.01, I = 0.1(NaCl), mixed constant	E3bg	Approx.	M48

No.	Formula / Name	pK values	T(°C)	Medium	Method	Assessment	Ref.
4960	$C_{10}H_8O_2$ 2-Cyclobutenone, 2-hydroxy-3-phenyl-	pK_1 6.25	28		E3d	Uncert.	S73
4961	$C_{10}H_8O_2$ 1,3-Indandione, 2-methyl-	pK_1 6.15	20	$c \approx 0.005$	E3bg	Approx.	S42
		pK_1 6.25	not stated	1% methanol	E3bg	Uncert.	G46
4962	$C_{10}H_8O_2$ 1,3-Naphthalenediol	pK_1 7.22	not stated	$I = 0.5$(KCl), mixed constant	E	Uncert.	H12
4963	$C_{10}H_8O_2$ 1,4-Naphthalenediol	pK_1 9.58, pK_2 11.11	13.9	$I = 0.65$($NaClO_4$), concentration constants	05	Approx.	B38
		pK_1 9.47, pK_2 11.02	20				
		pK_1 9.37, pK_2 10.93	26.5				
	Thermodynamic quantities are derived from the results						
4964	$C_{10}H_8O_2$ 1,8-Naphthalenediol	pK_1 6.71 pK_2 >13	20	$c = 0.0001$, mixed constants	05	Approx.	M128
						Uncert.	
4965	$C_{10}H_8O_2$ 2,3-Naphthalenediol	pK_1 8.68 pK_2 12.5	20	$I = 0.1$(KNO_3), mixed constants	E3bg	Approx.	B27
		pK_1 8.55	25	$I = 0.1$(KCl)	E	Uncert.	A68

No.	Molecular formula, name and pK value(s)	$T(^{\circ}C)$	Remarks	Method	Assessment	Ref.
4966	$C_{10}H_8O_3$ 2H-Chromen-2-one, 5-hydroxy-4-methyl- 8.26	25	c = 0.000087	05	Approx.	M64
4967	$C_{10}H_8O_3$ 2H-Chromen-2-one, 6-hydroxy-4-methyl- 9.14	25	c = 0.000084	05	Approx.	M62
4968	$C_{10}H_8O_3$ 2H-Chromen-2-one, 7-hydroxy-4-methyl- 7.80	25	c = 0.000088	05	Approx.	M62
4969	$C_{10}H_8O_6$ Acetic acid, 2-(2-carboxy-4-hydroxy-3-oxo-1,4,6-cycloheptatrienyl)- pK_1 3.75, pK_2 5.90, pK_3 8.08	25	c = 0.001	E3bg	Approx.	T5
4970	$C_{10}H_{10}O_2$ Benzo[b]furan-4-ol, 2,3-dimethyl- 9.93	20	2% ethanol	05	Approx.	D32
4971	$C_{10}H_{10}O_2$ Benzo[b]furan-4-ol, 2,7-dimethyl- 9.60	20	2% ethanol	05	Approx.	D32
4972	$C_{10}H_{10}O_2$ Benzo[b]furan-5-ol, 2,3-dimethyl- 10.52	20	2% ethanol	05	Approx.	D32
4973	$C_{10}H_{10}O_2$ Benzo[b]furan-6-ol, 2,3-dimethyl- 9.95	20	2% ethanol	05	Approx.	D32

No.	Formula	Name / pK	Temp (°C)	Notes	Method	Assessment	Ref.
4974	$C_{10}H_{10}O_2$	Benzo[b]furan-7-ol, 2,3-dimethyl-					
		9.07	20	2% ethanol	05	Approx.	D32
4975	$C_{10}H_{10}O_2$	Benzoic acid, 3-allyl-					
		4.32	25	Decomposition observed	05	Uncert.	B120
4976	$C_{10}H_{10}O_2$	Benzoic acid, 4-allyl-					
		4.34	25	c = 0.0002	05	Uncert.	B120
4977	$C_{10}H_{10}O_2$	Benzoic acid, 2-cyclopropyl-					
		4.13	20	Mixed constant	E3bg	Uncert.	S61
4978	$C_{10}H_{10}O_2$	Benzoic acid, 4-cyclopropyl-					
		4.44	25	c $\simeq$ 0.0005, mixed constant	E3bg	Approx.	L32
		4.52	20	Mixed constant	E3bg	Uncert.	S61
		4.52		Conditions not stated	05	Uncert.	S62
4979	$C_{10}H_{10}O_2$	Benzoic acid, 4-isopropenyl-					
		4.04	25	c $\simeq$ 0.0005, mixed constant	E3bg	Approx.	L32
4980	$C_{10}H_{10}O_2$	Benzoic acid, 3-(1-propenyl)-					
		4.25	25	Extrapolated from values obtained in 12.2, 25.2, 44.1 and 64.8% w/w ethanol	E3bg	Uncert.	B121

No.	Molecular formula, name and pK value(s)	T(^{o}C)	Remarks	Method	Assessment	Ref.
4981	$C_{10}H_{10}O_2$ Benzoic acid, 4-(1-propenyl)-	25		E3bg	Uncert.	B121
	4.49		Extrapolated from values obtained in 12.2, 25.2, 44.1 and 64.8% w/w ethanol			
4982	$C_{10}H_{10}O_2$ 1,3-Butanedione, 1-phenyl-					
	8.68	20	c = 0.01	E3bg/05	Approx.	R52
	8.90	25	I = 0.1($NaClO_4$), mixed constant	05	Uncert.	Z15
4983	$C_{10}H_{10}O_2$ Cyclopropanecarboxylic acid, 2-phenyl-					
	4.57(trans)	25	c ≈ 0.0008	E3bg	Approx.	T55
4984	$C_{10}H_{10}O_2$ 1,2,3,4-Tetrahydronaphthalen-1-one, 6-hydroxy-					
	7.74	25	c ≈ 0.0002, I = 0.1, mixed constant	05	Approx.	M19
4985	$C_{10}H_{10}O_2$ 1,2,3,4-Tetrahydronaphthalen-1-one, 8-hydroxy-					
	11.31	25	c ≈ 0.0002	05	Approx.	M19
	11.14		I = 0.1, mixed constant			
4986	$C_{10}H_{10}O_3$ Propenoic acid, 3-(2-hydroxy-5-methylphenyl)-					
	3.74	30	I = 1.0(LiCl), mixed constant	05	Approx.	H36
4987	$C_{10}H_{10}O_4$ Acetic acid, 2-(4-acetyl)phenoxy)-					
	2.81	25		E3ag	Approx.	H48
			Value in mixed solvent is also given			

No.	Formula	Name	T (°C)	pK	Conditions	Method	Reliability	Ref.
4938	$C_{10}H_{10}O_4$	Benzoic acid, 3,4-trimethylenedioxy-	25	4.23	c = 0.003, 5% (v/v) ethanol	E3bg	Uncert.	B190
4989	$C_{10}H_{10}O_4$	Bicyclo[2.2.1]hepta-2,5-diene-2-carboxylic acid, 3-methoxycarbonyl-	25	3.02	c = 0.005-0.009	E3bg	Approx.	M11
4990	$C_{10}H_{10}O_4$	Propanedioic acid, benzyl-	25	pK_1 2.91, pK_2 5.87		E3bg,R3d	Approx.	N5
				pK_1 2.56, pK_2 5.22		E		K16
4991	$C_{10}H_{10}O_4$	Propenoic acid, 3-(4-hydroxy-3-methoxyphenyl)- (Ferulic acid)	10	pK_1 4.61, pK_2 9.55		O5,R4	Approx.	K17
			25	pK_1 4.58, pK_2 9.39				
			40	pK_1 4.57, pK_2 9.25				
				Thermodynamic quantities are derived from the results				
			25	pK_1 4.52, pK_2 9.39		O5,R4	Approx.	N34
4992	$C_{10}H_{10}O_5$	Acetic acid, 2-(p-methoxybenzoyl)oxy-	25	3.09		E		C74
4993	$C_{10}H_{10}O_5$	1,2-Benzenedicarboxylic acid, 4-ethoxy-	25	pK_1 2.74, pK_2 5.12	I = 0.1(KNO_3), concentration constants	E3bg	Approx.	Y10
4994	$C_{10}H_{10}O_5$	Benzoic acid, 2-formyl-5,6-dimethoxy- (Opianic acid)	25	3.06	c ≈ 0.0001 (equilibrium value)	O1	Approx.	K48

(contd)

No.	Molecular formula, name and pK value(s)	T(°C)	Remarks	Method	Assessment	Ref.
4994 (Contd)	$C_{10}H_{10}O_5$ Benzoic acid, 2-formyl-5,6-dimethoxy- (Opianic acid)					
	3.07	25	c ≈ 0.005 (equilibrium value)	E3bg	Approx.	K48
			Equilibrium between lactol & open-chain forms			
			"True" pK of open-chain form ca. 2.55			
4995	$C_{10}H_{10}O_5$ 3-Oxapentanedioic acid, 2-phenyl-					
	pK_1 2.87, pK_2 4.39	25		E3ag	Approx.	S92,S93
4996	$C_{10}H_{10}O_6$ 1,2-Phenylenebis(oxyacetic acid)					
	pK_1 2.40, pK_2 3.45	25	I = 0.1($NaClO_4$), concentration constants	E3bg,R3h	Approx.	S130
4997	$C_{10}H_{12}O_2$ Acetophenone, 4-hydroxy-3,5-dimethyl-					
	8.22	25	c = 0.0005	E3bg	Approx.	F16
4998	$C_{10}H_{12}O_2$ Benzoic acid, 4-isopropyl-					
	4.36	25	c ≈ 0.0005, mixed constant	E3bg	Approx.	L32
	4.31	20	Mixed constant	E3bg	Uncert.	S61
	4.30		Conditions not stated	O5	Uncert.	S62
4999	$C_{10}H_{12}O_2$ Benzoic acid, 2,3,4-trimethyl-					
	4.06	25	I = 0.003-0.006	E3bg	Approx.	L76
5000	$C_{10}H_{12}O_2$ Benzoic acid, 2,3,5-trimethyl-					
	4.01	25	I = 0.002-0.005	E3bg	Approx.	L76

No.	Formula	Name	Value	Temp.	Conditions	Method	Assessment	Ref.
5001	$C_{10}H_{12}O_2$	Benzoic acid, 2,3,6-trimethyl-	3.34	25	I = 0.004-0.007	E3bg	Approx.	L76
5002	$C_{10}H_{12}O_2$	Benzoic acid, 2,4,5-trimethyl-	4.38	25	I = 0.007	E3bg	Approx.	L76
5003	$C_{10}H_{12}O_2$	Benzoic acid, 2,4,6-trimethyl-	3.446	25		05	Approx.	W38

Variation with temperature

15	20	30	35	40
3.325	3.394	3.502	3.539	3.579

Thermodynamic quantities are derived from the results

No.	Formula	Name	Value	Temp.	Conditions	Method	Assessment	Ref.
			3.55	25	I = 0.003-0.006	E3bg	Approx.	L76
5004	$C_{10}H_{12}O_2$	Benzoic acid, 3,4,5-trimethyl-	4.52	25	I = 0.003-0.005	E3bg	Approx.	L76
5005	$C_{10}H_{12}O_2$	Butanoic acid, 3-phenyl-	4.4	25		E3bg	Uncert.	C57
5006	$C_{10}H_{12}O_2$	2,4,6-Cycloheptatrien-1-one, 2-hydroxy-4-isopropyl- (Hinokitiol)	6.72	25	I = 2.0	05	Approx.	016
			7.21	30		05,R4		H55

No.	Molecular formula, name and pK value(s)	T(°C)	Remarks	Method	Assessment	Ref.
5007	$C_{10}H_{12}O_2$ 2,4,6-Cycloheptatrien-1-one, 2-hydroxy-5-isopropyl-					
	6.85	25	I = 2.0	05	Approx.	O13
5008	$C_{10}H_{12}O_2$ Phenol, 2-allyl-6-methoxy-					
	10.38	25	c ≈ 0.0003	05	Approx.	B120
5009	$C_{10}H_{12}O_2$ Phenol, 3-allyl-2-methoxy-					
	9.92	25	c = 0.0003	05	Approx.	B120
5010	$C_{10}H_{12}O_2$ Phenol, 3-allyl-6-methoxy- (Chavibetol)					
	10.02	25	c = 0.0003	05	Approx.	B120
5011	$C_{10}H_{12}O_2$ Phenol, 4-allyl-2-methoxy- (Eugenol)					
	10.19	25	c = 0.0002	05	Approx.	B120
	10.15	25		05,R4	Approx.	L49
5012	$C_{10}H_{12}O_2$ Phenol, 2-methoxy-4-(1-propenyl)- (Isoeugenol)					
	9.88(trans)	25	c = 0.00007	05	Uncert.	B120
	9.89(trans)	25		05,R4	Approx.	L49
5013	$C_{10}H_{12}O_2$ Phenol, 2-methoxy-5-(1-propenyl)- (Isochavibetol)					
	9.90	25	c = 0.0003	05	Uncert.	B120
5014	$C_{10}H_{12}O_2$ Phenol, 2-methoxy-6-(1-propenyl)- (o-Isoeugenol)					
	10.20	25	c = 0.0002	05	Uncert.	B120

No.	Formula	Name	pK	T (°C)	Conditions	Method	Reliability	Ref.
5015	$C_{10}H_{12}O_2$	Propanoic acid, 3-(p-tolyl)-	4.74	25	$c \approx 0.0008$	E3bg	Approx.	T55
5016	$C_{10}H_{12}O_3$	Acetic acid, 2-(2,4-dimethylphenoxy)-	3.22	20	$I = 0.0005$	E3bg	Approx.	P31
			3.08		$I = 1.0(KNO_3)$			
5017	$C_{10}H_{12}O_3$	Acetic acid, 2-(3,5-dimethylphenoxy)-	3.20	25	Value in mixed solvent is also given	E3ag	Approx.	H48
5018	$C_{10}H_{12}O_3$	Acetic acid, 2-(2,5-dimethylphenyl)-2-hydroxy-	3.57	25	$c = 0.01$, mixed constant	E3bg	Uncert.	K36
5019	$C_{10}H_{12}O_3$	Acetic acid, 2-(4-ethylphenyl)-2-hydroxy-	3.55	25	$c = 0.01$, mixed constant	E3bg	Uncert.	K36
5020	$C_{10}H_{12}O_3$	Benzoic acid, 2-ethoxy-6-methyl-	3.51	20	$c \approx 0.001$, 1% ethanol	E3bg	Approx.	P23
5021	$C_{10}H_{12}O_3$	Benzoic acid, 3-ethoxy-2-methyl-	3.73	20	$c \approx 0.001$, 1% ethanol	E3bg	Approx.	P22
5022	$C_{10}H_{12}O_3$	Benzoic acid, 5-ethoxy-2-methyl-	3.86	20	$c \approx 0.001$, 1% ethanol	E3bg	Approx.	P22

No.	Molecular formula, name and pK value(s)	T(°C)	Remarks	Method	Assessment	Ref.
5023	$C_{10}H_{12}O_3$ Benzoic acid, 4-ethoxy-2-methyl- 4.54	20	c ≈ 0.001, 1% ethanol	E3bg	Approx.	P23
5024	$C_{10}H_{12}O_3$ Benzoic acid, 2-hydroxy-3-isopropyl- pK_1 2.76	25	I = 0.1(NaClO$_4$)	E3bg	Approx.	P7
5025	$C_{10}H_{12}O_3$ Propanoic acid, 3-(4-methoxyphenyl)- 4.71	25	c ≈ 0.0008	E3bg	Approx.	T55
5026	$C_{10}H_{12}O_3$ 2-Propen-1-ol, 3-(4-hydroxy-3-methoxyphenyl)- (Coniferyl alcohol) 9.75 / 9.54 / 9.32	10 / 25 / 40	Thermodynamic quantities are derived from the results	O5,R4	Approx.	K17
5027	$C_{10}H_{12}O_4$ Bicyclo[2.2.1]hept-2-ene-2-carboxylic acid, 3-methoxycarbonyl- 3.19	25	c = 0.005-0.009	E3bg	Approx.	M11
5028	$C_{10}H_{12}O_4$ Bicyclo[2.2.2]oct-5-ene-1,2-dicarboxylic acid pK_1 4.55, pK_2 6.30(endo cis) pK_1 3.85, pK_2 5.93(trans)	25	c ≈ 0.006	E3bg,R3d	Approx.	C55
5029	$C_{10}H_{12}O_4$ 4H-Pyran-4-one, 2-ethyl-5-ethylcarbonyl-6-hydroxy- 5.38	20	c = 0.001-0.0025	E3bg	Approx.	A37

No.	Formula	Name	pK	T (°C)	Remarks	Method	Assessment	Ref.
5030	$C_{10}H_{12}O_5$	Benzoic acid, 2,4,6-trimethoxy-	3.59	10	Absorbances extrapolated to zero time	05	Approx.	S28
5031	$C_{10}H_{12}O_5$	Benzoic acid, 3,4,5-trimethoxy-	4.12	25		05	Approx.	N35
			4.14	20		C1	Approx.	C36
			4.19	30		E3bg	Uncert.	
			4.30	40	Thermodynamic quantities are derived			
			4.38	50	from the results			
			4.53	60				
			4.65					
			4.24	25		E		C37
5032	$C_{10}H_{12}N_2$	Indole, 3-(2-aminoethyl)- (Tryptamine)	16.60(NH)	25	In aq. KOH, H_- scale	07	V.Uncert.	Y3
5033	$C_{10}H_{13}N_5$	Pyrimidino[2,1-i]purine, 3,7,8,9-tetrahydro-7,7-dimethyl-	7.1	22		E3bg	Uncert.	M53
5034	$C_{10}H_{13}N_5$	Pyrimidino[2,1-i]purine, 3,7,8,9-tetrahydro-9,9-dimethyl-	7.4	22		E3bg	Uncert.	M53
5035	$C_{10}H_{14}O$	Phenol, 2-butyl-	10.58	20	Extrapolated value		Uncert.	D30

No.	Molecular formula, name and pK value(s)	T(°C)	Remarks	Method	Assessment	Ref.
5036	C$_{10}$H$_{14}$O Phenol, 2-tert-butyl-					
	10.623	25	m = 0.00045	05a	Rel.	B93
			Variation with temperature			

5	10	15	20	30	35	40	45	50
10.920	10.838	10.762	10.691	10.559	10.502	10.448	10.400	10.355

Thermodynamic quantities are derived from the results

No.	Molecular formula, name and pK value(s)	T(°C)	Remarks	Method	Assessment	Ref.
	11.33	25	I ≈ 0.05	05	Approx.	R39
	11.16	25	I = 0.1(KCl), mixed constant	E3bg	Approx.	E20
5037	C$_{10}$H$_{14}$O Phenol, 3-tert-butyl-					
	10.119	25	m = 0.0004	05a	Rel.	B93
			Variation with temperature			

10	15	20	30	35	40	45	50
10.348	10.269	10.194	10.052	9.995	9.942	9.892	9.850

Thermodynamic quantities are derived from the results

No.	Molecular formula, name and pK value(s)	T(°C)	Remarks	Method	Assessment	Ref.
	10.08	25		05	Approx.	C71
5038	C$_{10}$H$_{14}$O Phenol, 4-tert-butyl					
	10.390	25	m = 0.00032	05a	Rel.	B93
			Variation with temperature			

10	15	20	30	35	40	45	50
10.604	10.528	10.455	10.326	10.270	10.217	10.168	10.123

Thermodynamic quantities are derived from the results

(contd)

Entry	Formula	Name	pK	T (°C)	Notes	Method	Assessment	Ref
5038 (Contd)	$C_{10}H_{14}O$	Phenol, 4-tert-butyl-	10.23	25		05	Approx.	C70
			10.23	25	$I \approx 0.05$	05	Approx.	R39
			10.25	20	Extrapolated value		Uncert.	D30
5039	$C_{10}H_{14}O$	Phenol, 2-isopropyl-3-methyl-	11.23	20		05	Approx.	D30
5040	$C_{10}H_{14}O$	Phenol, 2-isopropyl-4-methyl-	10.73	20	$c = 0.005$	E3bg	Approx.	D31
5041	$C_{10}H_{14}O$	Phenol, 2-isopropyl-5-methyl-	10.62	20		05	Approx.	D31
5042	$C_{10}H_{14}O$	Phenol, 2-isopropyl-6-methyl-	10.85	20	2% ethanol	05	Approx.	D29
5043	$C_{10}H_{14}O$	Phenol, 4-isopropyl-2-methyl-	10.56	20	$c = 0.005$	E3bg	Approx.	D31
5044	$C_{10}H_{14}O$	Phenol, 4-isopropyl-3-methyl-	10.31	20		05	Approx.	D31
5045	$C_{10}H_{14}O_2$	1,4-Benzenediol, 2,3,5,6-tetramethyl- (Durohydroquinone)	pK_1 11.51	12.1	$I = 0.65(NaClO_4)$, concentration constants	05	Approx.	B38
			pK_1 11.35	18.8				

(contd)

No.	Molecular formula, name and pK value(s)	T(°C)	Remarks	Method	Assessment	Ref.
5045 (Contd)	$C_{10}H_{14}O_2$ 1,4-Benzenediol, 2,3,5,6-tetramethyl- (Durohydroquinone)					
	pK₁ 11.25	25.0				
	pK₂ 13.2	14.9		06	Uncert.	
	pK₂ 12.9	22.5				
	pK₂ 12.7	29.8				
			Thermodynamic quantities are derived from the results			
5046	$C_{10}H_{14}O_4$ Bicyclo[2.2.1]heptane-1-carboxylic acid, 4-methoxycarbonyl-					
	4.494	25		E3bg	Approx.	W29
			Values in mixed solvents are also given			
5047	$C_{10}H_{14}O_4$ Bicyclo[2.2.2]octane-1,4-dicarboxylic acid					
	pK₁ 4.468, pK₂ 5.457	25	Values in mixed solvents are also given	E3bg,R3d	Approx.	W29
5048	$C_{10}H_{14}O_4$ Bicyclo[2.2.2]octane-2,3-dicarboxylic acid					
	pK₁ 4.63, pK₂ 6.90(endo cis)	25	c ≈ 0.006	E3bg,R3d	Approx.	C55
	pK₁ 3.89, pK₂ 6.17(trans)					
5049	$C_{10}H_{14}O_3$ Phenol, 3,5-diethoxy-					
	9.370	25	m = 0.00034	05a	Rel.	B92

Variation with temperature

5	10	15	20	25	30	35	40	45	50	55	60
9.640	9.566	9.495	9.431	9.370	9.316	9.266	9.220	9.178	9.140	9.105	9.075

Thermodynamic quantities are derived from the results

No.	Formula	Name	pK values	T (°C)	c	Assessment	Method	Ref.
5050	$C_{10}H_{16}O_3$	Acetic acid, 2-(3-acetyl-2,2-dimethylcyclobutyl)- (Pinonic acid)	4.82	25-30	$c \approx 0.001$	Uncert.	E3bg	H52
5051	$C_{10}H_{16}O_3$	Bicyclo[3.1.1]heptane-2-carboxylic acid, 2-hydroxy-7,7-dimethyl- (Norpinic acid)	4.61	25-30	$c \approx 0.001$	Uncert.	E3bg	H52
5052	$C_{10}H_{16}O_4$	Cyclobutane-1,3-diacetic acid, 2,2-dimethyl- (Homopinic acid)	pK_1 4.58, pK_2 5.49	25-30	$c \approx 0.0005$	Uncert.	E3bg,R3d	H52
5053	$C_{10}H_{16}O_4$	Cyclooctane-1,2-dicarboxylic acid	pK_1 3.99, pK_2 7.34(cis) pK_1 4.37, pK_2 6.24(trans)	25	$c \approx 0.0005$, mixed constants	Approx.	E3bg E3bg,R3d	S68
5054	$C_{10}H_{16}O_4$	1,3-Cyclopentanedicarboxylic acid, 1,2,2-trimethyl- (cis-d-Camphoric acid)	pK_1 4.71, pK_2 5.83	25-30	$c \approx 0.0005$	Uncert.	E3bg,R3d	H52
5055	$C_{10}H_{16}O_4$	Propanoic acid, 2-(3-carboxycyclopentyl)-2-methyl- (Camphenic acid)	pK_1 4.73, pK_2 5.65	25-30	$c \approx 0.0005$	Uncert.	E3bg,R3d	H52
5056	$C_{10}H_{16}O_5$	Diethyl 2-ethyl-3-oxobutanedioate	6.89				E3d	R51
5057	$C_{10}H_{16}O_6$	Diethyl 2-ethoxy-3-oxobutanedioate	6.72				E3d	R51

No.	Molecular formula, name and pK value(s)	T(°C)	Remarks	Method	Assessment	Ref.
5058	$C_{10}H_{18}O_2$ 3,5-Heptanedione, 2,2,6-trimethyl- 10.72	25	c = 0.00005, 0.5-1% ethanol Apparent pK for keto-enol mixture	O5	Approx.	C5
	pK(enol) 10.02, pK(keto) 10.62	25	Calculated using absorbance data		Uncert.	
5059	$C_{10}H_{18}O_2$ 4,6-Nonanedione, 3-methyl- 9.55	25	c = 0.00005, 0.5-1% ethanol Apparent pK for keto-enol mixture	O5	Approx.	C5
	pK(enol) 8.87, pK(keto) 9.45	25	Calculated from absorbance data		Uncert.	
5060	$C_{10}H_{18}O_2$ 3,5-Octanedione, 2,6-dimethyl- 9.66	25	c = 0.00005, 0.5-1% ethanol Apparent pK for keto-enol mixture	O5	Approx.	C5
	pK(enol) 8.89, pK(keto) 9.58	25	Calculated from absorbance data		Uncert.	
5061	$C_{10}H_{18}O_3$ Acetic acid, 2-(3-(1-hydroxyethyl)-2,2-dimethylcyclobutyl)- (Pinolic acid) 4.83	25-30	c ≈ 0.001	E3bg	Uncert.	H52
5062	$C_{10}H_{18}O_3$ Ethyl 2-acetylhexanoate 13.2	25.7		O5	Uncert.	R53
5063	$C_{10}H_{18}O_4$ Butanoic acid, 2-ethoxycarbonyl-2-isopropyl- 3.87(H_2O) 4.34(D_2O)	25		E3bg	Approx.	G23

(contd)

No.	Compound / pK values	T (°C)	Conditions	Method	Reliability	Ref.
5063 (Contd)	$C_{10}H_{18}O_4$ Butanoic acid, 2-ethoxycarbonyl-2-isopropyl-					
	3.72 Mixed constant	37		E3bg	Uncert.	L33
5064	$C_{10}H_{18}O_4$ Decanedioic acid (Sebacic acid)					
	pK_1 4.40, pK_2 5.22	20	c = 0.002	E3bg,R3d	Approx.	W16
	pK_1 4.58, pK_2 5.54	25	I = 0.1(KNO_3)	E3bg		N32
	pK_1 4.6, pK_2 5.6	20	c = 0.005	E3bg,R3a	Uncert.	D43
	Values in mixed solvents are also given					
5065	$C_{10}H_{18}O_4$ Pentanedioic acid, 3-tert-butyl-3-methyl-					
	pK_1 3.61, pK_2 7.49	25	c = 0.00125, mixed constants	E3bg	Approx.	B146
5066	$C_{10}H_{18}O_4$ Propanedioic acid, 2-ethyl-2-(1-ethylpropyl)-					
	pK_1 2.15, pK_2 7.31	25	I = 0.1(KCl), mixed constants	E3bg	Approx.	M77
5067	$C_{10}H_{18}O_5$ 3-Oxapentanedioic acid, 2,4-diethyl-2,4-dimethyl-					
	pK_1 3.70, pK_2 5.41	20	c = 0.005-0.01	E3bg,R3c	Approx.	K47
5068	$C_{10}H_{18}O_5$ Propanedioic acid, 2-heptyl-2-hydroxy-					
	pK_1 2.77, pK_2 4.70					G37
5069	$C_{10}H_6O_4N_2$ 1,3-Naphthalenediol, 2,4-dinitroso-					
	pK_1 4.62, pK_2 9.50	25	I = 0.1, mixed constants	05	Approx.	M41
	pK_1 4.53, pK_2 9.10					

No.	Molecular formula, name and pK value(s)	$T(^oC)$	Remarks	Method	Assessment	Ref.
5070	$C_{10}H_6O_4N_2$ 2,3-Quinoxalinedicarboxylic acid					
	pK_1 1.80, pK_2 3.31	25	c ≈ 0.01, I = 0.1($NaClO_4$), mixed constants	E3bg	Uncert.	D61
	pK_1 1.62	25	c ≈ 0.005, mixed constants	E3bg	Uncert.	G52
	pK_2 3.64				Approx.	
5071	$C_{10}H_7OCl$ 1-Naphthol, 3-chloro- 8.40	25	c ≈ 0.001	E3bg	Approx.	B155
5072	$C_{10}H_7OCl$ 1-Naphthol, 4-chloro- 8.86	25		O5a	Approx.	C79
5073	$C_{10}H_7OCl$ 2-Naphthol, 4-chloro- 8.82	25	c ≈ 0.001	E3bg	Approx.	B155
5074	$C_{10}H_7OBr$ 1-Naphthol, 3-bromo- 8.34	25	c ≈ 0.001	E3bg	Approx.	B155
5075	$C_{10}H_7OBr$ 1-Naphthol, 4-bromo- 8.72	25		O5a	Approx.	C79
5076	$C_{10}H_7OBr$ 2-Naphthol, 4-bromo- 8.74	25	c ≈ 0.001	E3bg	Approx.	B155
5077	$C_{10}H_7OI$ 1-Naphthol, 3-iodo- 8.36	25	c ≈ 0.001	O5	Approx.	B155

No.	Formula / Name	pK	T (°C)	Conditions	Method	Assessment	Ref.
5078	$C_{10}H_7OI$ 2-Naphthol, 4-iodo-	8.81	25	c ≈ 0.001	E3bg	Approx.	B155
5079	$C_{10}H_7O_2N$ 1-Isoquinolinecarboxylic acid	pK$_2$ 4.97	18±2	I = 0.5(NaCl), concentration constant	05	Uncert.	H50
5080	$C_{10}H_7O_2N$ 1-Naphthol, 2-nitroso-	7.24	25	c = 0.0001, I = 0.1(NaClO$_4$), mixed constant	05	Approx.	D79
		7.38	25				
5031	$C_{10}H_7O_2N$ 1-Naphthol, 4-nitroso-	8.18	25		05a	Approx.	C79
5082	$C_{10}H_7O_2N$ 2-Naphthol, 1-nitroso-	7.77	25		05	Approx.	D79
		7.63	25	c = 0.0001, I = 0.1(NaClO$_4$), mixed constant	05	Approx.	
		7.59	25	c = 0.0001, I = 0.1(NaClO$_4$), mixed constant	E3bg	Approx.	
5083	$C_{10}H_7O_2N$ 2-Quinolinecarboxylic acid (Quinaldic acid)	pK$_1$ 1.45					
		pK$_2$ 4.91	25	c ≈ 0.01, I = 0.1(NaClO$_4$), mixed constants	E3bg	Uncert.	D61
						Approx.	
		pK$_1$ 1.9, pK$_2$ 4.92	18	I = 0.5(NaCl), concentration constants	05	Uncert.	H50
		pK$_2$ 4.96	20	c = 0.03	E3bg	Approx.	W17
		pK$_2$ 4.97	25		E3bg	Approx.	L69
		pK$_2$ 4.49	25	I = 0.1, mixed constant	E3bg	Approx.	S135

No.	Molecular formula, name and pK value(s)	T(°C)	Remarks	Method	Assessment	Ref.
5084	$C_{10}H_7O_2N$ 8-Quinolinecarboxylic acid					
	pK$_1$ 2.01, pK$_2$ 6.84	25	c $\approx$ 0.01, I = 0.1(NaClO$_4$), mixed constants	E3bg	Approx.	D61
	pK$_1$ 2.0, pK$_2$ 6.72	25	c = 0.001, I $\approx$ 0.2, concentration constants	E3bg	Approx.	H50
	pK$_1$ 1.80	25		E3bg	Approx.	L70
5085	$C_{10}H_7O_2F_3$ 1,3-Butanedione, 4,4,4-trifluoro-1-phenyl-					
	6.3	room temp		E3bg	Uncert.	R19
5086	$C_{10}H_7O_3N$ 1-Naphthol, 2-nitro-					
	5.89	25	c $\approx$ 0.0001	O5,R4	Approx.	M30
5087	$C_{10}H_7O_3N$ 1-Naphthol, 3-nitro-					
	7.85	25	c $\approx$ 0.001	E3bg	Approx.	B155
	7.86	25		O5a	Approx.	C79
5088	$C_{10}H_7O_3N$ 1-Naphthol, 4-nitro-					
	5.73	25		O5a	Approx.	C79
5089	$C_{10}H_7O_3N$ 2-Naphthol, 1-nitro-					
	5.93	25	c = 0.000125	O5,R4	Approx.	M30
5090	$C_{10}H_7O_3N$ 2-Naphthol, 4-nitro-					
	8.09	25	c $\approx$ 0.001	E3bg	Approx.	B155

No.	Formula	Name	pK	T (°C)	Conditions	Method	Assessment	Ref.
5091	$C_{10}H_7O_3N$	Propenoic acid, 3-(3-cyano-6-hydroxyphenyl)-	3.46	30	$I = 1.0(LiCl)$, mixed constant	O5	Approx.	H36
5092	$C_{10}H_7O_3Cl$	2H-Benzopyran-2-one, 6-chloro-7-hydroxy-4-methyl-	6.10	25	$c = 0.0001$	O5	Approx.	M63
5093	$C_{10}H_7O_4N_3$	1,2,3-Triazole-4,5-dicarboxylic acid, 1-phenyl-	pK_1 2.13 pK_2 4.93	25		E3bg	Uncert. Approx.	H13
5094	$C_{10}H_6O_6N_3$	Phthalimide, N-(2,2-dinitroethyl)-	3.17	20		O5	Approx.	T58a
5095	$C_{10}H_8O_2N_2$	3-Pyrazolecarboxylic acid, 1-phenyl-	3.60	20	$c = 0.0025$	E3bg	Uncert.	T1
5096	$C_{10}H_8O_2N_2$	4-Pyrazolecarboxylic acid, 1-phenyl-	4.40-4.80	20	Based on value in 50% ethanol	E3bg	V.Uncert.	T1
5097	$C_{10}H_8O_2N_2$	5-Pyrazolecarboxylic acid, 1-phenyl-	2.70	20	$c = 0.0025$	E3bg	Uncert.	T1
5098	$C_{10}H_8O_2N_2$	2-Quinolinecarbaldehyde oxime, 8-hydroxy-	pK_1 1.56, pK_2 9.04	20	$I = 0.2$	DIS*	Uncert.	R50

* (H_2O, $CHCl_3$)

No.	Molecular formula, name and pK value(s)	T(°C)	Method	Assessment	Remarks	Ref.
5099	$C_{10}H_8O_2N_2$ 8-Quinolinol, 2-methyl-5-nitroso- pK_1 3.14, pK_2 7.48	25	E & O			A83
5100	$C_{10}H_8O_3N_2$ Pyrimidine-2,4(1H,3H)-dione, 6-hydroxy-5-phenyl- 3.75		E3bg	Uncert.	Conditions not stated	M68
5101	$C_{10}H_8O_4N_2$ 2,2'-Furil dioxime pK_1 9.86, pK_2 11.53	20	05	Uncert.	$I = 1.0(NaClO_4)$	S65
	pK_1 9.84, pK_2 11.25		0			S15
	Other values in B10 and M75					
5102	$C_{10}H_8O_4S$ 1-Naphthalenesulfonic acid, 2-hydroxy- pK_2 11.38	21	05	Approx.	$I = 0.15$, mixed constant	G13
5103	$C_{10}H_8O_4S$ 1-Naphthalenesulfonic acid, 4-hydroxy- pK_2 8.32	25	E3bg	Approx.	$c = 0.004$	M29
	pK_2 8.47	21	05	Approx.		G13
	pK_2 8.0	20	05	Uncert.	$c = 0.00004$ and 0.00005	S106
5104	$C_{10}H_8O_4S$ 1-Naphthalenesulfonic acid, 5-hydroxy- pK_2 9.11	21	05	Approx.	$I = 0.02$, mixed constant	G13
5105	$C_{10}H_8O_4S$ 1-Naphthalenesulfonic acid, 7-hydroxy- pK_2 9.48	21	05	Uncert.	$I = 0.02$, mixed constant	G13

No.	Formula	Name / pK	t (°C)	Conditions	Code	Reliability	Ref.
5106	$C_{10}H_8O_4S$	1-Naphthalenesulfonic acid, 8-hydroxy-					
		pK_2 13.02	21	$I = 0.1$, mixed constant	05	Approx.	G13
5107	$C_{10}H_8O_4S$	2-Naphthalenesulfonic acid, 1-hydroxy-					
		pK_2 9.47	21	$I = 0.02$, mixed constant	05	Approx.	G13
5108	$C_{10}H_8O_4S$	2-Naphthalenesulfonic acid, 4-hydroxy-					
		pK_2 8.5	20	$c = 0.00004$ & 0.00005, mixed constant	05	Uncert.	S106
5109	$C_{10}H_8O_4S$	2-Naphthalenesulfonic acid, 6-hydroxy-					
		pK_2 9.13	25	$c = 0.004$	E3bg	Approx.	M29
		pK_2 9.14	21	$I = 0.02$, mixed constant	05	Uncert.	G13
		pK_2 8.9	20	$c = 0.00004$ and 0.00005, mixed constant	05	Uncert.	S106
5110	$C_{10}H_8O_4S$	2-Naphthalenesulfonic acid, 7-hydroxy-					
		pK_2 9.35	21	$I = 0.02$, mixed constant	05	Approx.	G13
5111	$C_{10}H_8O_5S$	2-Naphthalenesulfonic acid, 6,7-dihydroxy-					
		pK_2 8.09 pK_3 11.85	25	$I = 0.1(KCl)$	E3bg	Approx.	O11
						Uncert.	
		pK_2 8.19, pK_3 12.16	20-25	$I = 0.1(KNO_3)$	E3bg	Uncert.	B29
5112	$C_{10}H_8O_6N_2$	1,2,3,4-Tetrahydronaphthalen-1-one, 8-hydroxy-5,7-dinitro-					
		5.12	25	$c \approx 0.0002$			
		4.98	25	$I = 0.1$, mixed constant	05	Approx.	M19

No.	Molecular formula, name and pK value(s)	T(°C)	Remarks	Method	Assessment	Ref.
5113	$C_{10}H_8O_7S_2$ 1,3-Naphthalenedisulfonic acid, 7-hydroxy-					
	pK_3 8.7	20	c = 0.00004, I = 0.25, mixed constant	05	Uncert.	S106
	pK_3 8.99	21	I = 0.02, mixed constant	05	Uncert.	G13
5114	$C_{10}H_8O_7S_2$ 1,5-Naphthalenedisulfonic acid, 3-hydroxy-					
	pK_3 8.7	20	c = 0.00004, I = 0.25, mixed constant	05	Uncert.	S106
5115	$C_{10}H_8O_7S_2$ 1,5-Naphthalenedisulfonic acid, 4-hydroxy-					
	pK_3 10.5	20	c = 0.00004, I = 0.25, mixed constant	05	Uncert.	S106
5116	$C_{10}H_8O_7S_2$ 1,6-Naphthalenedisulfonic acid, 4-hydroxy-					
	pK_3 7.6	20	c = 0.00004, I = 0.25, mixed constant	05	Uncert.	S106
5117	$C_{10}H_8O_7S_2$ 2,7-Naphthalenedisulfonic acid, 3-hydroxy-					
	pK_3 9.62	21	I = 0.02, mixed constant	05	Approx.	G13
	pK_3 9.52	25	c = 0.004	E3bg	Approx.	M29
	pK_3 9.23	25	c = 0.02, I = 0.1(NaClO$_4$), mixed constant	E3bg	Approx.	B11
	pK_3 9.3	20	c = 0.00004, I = 0.25, mixed constant	05	Uncert.	S106
5118	$C_{10}H_8O_7S_2$ 2,7-Naphthalenedisulfonic acid, 4-hydroxy-					
	pK_3 8.64	21		05	Approx.	G13
	pK_3 8.1	20	c = 0.00004, I = 0.25, mixed constant	05	Uncert.	S106

5119 $C_{10}H_8O_8S_2$ 2,7-Naphthalenedisulfonic acid, 4,5-dihydroxy-   (Chromotropic acid)

pK	T (°C)	Conditions			Ref.
pK_1 0.61, pK_2 0.73, pK_3 5.45	not stated	I = 0.1, concentration constants	E3bg	Uncert.	B160
pK_4 15.5	20-25		07	V.Uncert.	B29
pK_3 5.36		I = 0.1(KNO_3)	E3bg	Uncert.	B29
pK_4 15.6	25		07	V.Uncert.	B12
pK_3 5.56		I = 0.1($NaClO_4$), mixed constants	E3bg	Approx.	B12
pK_2 12.99	25			V.Uncert.	L37
pK_3 5.38		I = 0.1(KNO_3), concentration constant	E3bg	Approx.	L37
pK_3 5.44	16-23	c = 0.00005, I = 0.1($NaClO_4$), mixed constant	05	Uncert.	S5
pK_3 5.36	20	I = 0.1(KCl), mixed constant	E3bg	Approx.	B31
pK_4 15.6	20		07	V.Uncert.	H28

5120 $C_{10}H_9ON$ Ethanone, 1-(3-indolyl)-   (Indole, 3-acetyl-)

pK	T (°C)	Conditions			Ref.
12.99	25	c ≈ 0.0001	07	Uncert.	T21
12.99(NH)	25	In aq. KOH, H_ scale	07	Uncert.	Y2

5121 $C_{10}H_9ON$ 3-Indolecarbaldehyde, 2-methyl-

pK	T (°C)	Conditions			Ref.
12.47	25	c ≈ 0.0001	05	Approx.	T21

5122 $C_{10}H_9ON$ 1-Naphthol, 3-amino-

pK	T (°C)	Conditions			Ref.
pK_1 4.08, pK_2 9.36	25	c ≈ 0.001	E3bg	Approx.	B155

5123 $C_{10}H_9ON$ 2-Naphthol, 4-amino-

pK	T (°C)	Conditions			Ref.
pK_1 3.45, pK_2 9.84	25	c ≈ 0.001	E3bg	Approx.	B155

No.	Molecular formula, name and pK value(s)	$T(^{o}C)$	Remarks	Method	Assessment	Ref.
5124	$C_{10}H_9ON$ 3-Quinolinol, 2-methyl- pK_1 5.61, pK_2 10.16	25	$c = 0.002$, $I = 0.5$ Thermodynamic quantities are also given	E3bg	Approx.	T57
5125	$C_{10}H_9ON_5$ 6-Purinylamine, N-furfuryl- pK_1 3.8, pK_2 10.0		(N^6-Furfuryladenine) (Kinetin) $c = 0.00006$	05	Uncert.	M78
5126	$C_{10}H_9O_2N$ Acetic acid, 2-(3-indolyl)- pK_1 4.54	20	$I = 0.0005$	E3bg	Approx.	P31
	pK_1 4.36		$I = 1.0(KNO_3)$			
	pK_2(NH) 16.90	25	In aq. KOH, H_- scale	07	V.Uncert.	Y3
5127	$C_{10}H_9O_2N$ Isoxazol-5(2$\underline{H}$)-one, 4-methyl-3-phenyl- 4.73	20		05	Approx.	B110
5128	$C_{10}H_9O_2N$ 2,5-Pyrrolidinedione, 3-phenyl- 8.73	20	$c = 0.0001-0.001$, 4% ethanol, mixed constant	E3bg	Uncert.	F27,F29
5129	$C_{10}H_9O_2N_3$ Isoxazol-5-one, 4,5-dihydro-3-methyl-4-phenylhydrazono- 6.7	20	$c = 0.000002$	05	Uncert.	L16
5130	$C_{10}H_9O_2N_3$ 1,2,4-Triazine-3,5(2$\underline{H}$,4$\underline{H}$)-dione, 6-benzyl- 7.43	not stated	$c = 0.005$	E3bg	Uncert.	C28

No.	Formula	Name	T (°C)	pK	Notes	Method	Assessment	Ref.
5131	$C_{10}H_9O_2N_3$	1,2,3-Triazole-4-carboxylic acid, 5-methyl-1-phenyl-	25	3.73		E3bg	Uncert.	H13
5132	$C_{10}H_9O_2Cl$	Cyclopropanecarboxylic acid, 2-(3-chlorophenyl)-	25	4.51(trans)	$c \approx 0.0008$	E3bg	Approx.	T55
5133	$C_{10}H_9O_2Cl$	Cyclopropanecarboxylic acid, 2-(4-chlorophenyl)-	25	4.53(trans)	$c \approx 0.0008$	E3bg	Approx.	T55
5134	$C_{10}H_9O_3N$	2-Indolecarboxylic acid, 5-methoxy-	25	17.03(NH)	In aq. KOH, $H_=$ scale	07	Uncert.	Y3
5135	$C_{10}H_9O_4N$	Cyclopropanecarboxylic acid, 2-(3-nitrophenyl)-	25	4.41(trans)	$c \approx 0.0008$	E3bg	Approx.	T55
5136	$C_{10}H_9O_4N$	Cyclopropanecarboxylic acid, 2-(4-nitrophenyl)-	25	4.45(trans)	$c \approx 0.0008$	E3bg	Approx.	T55
5137	$C_{10}H_9O_4N$	Propenoic acid, 3-(3-carbamoyl-6-hydroxyphenyl)-	30	3.63(cis)	Estimated value of mixed constant		Uncert.	H36
5138	$C_{10}H_9O_4N$	1,2,3,4-Tetrahydronaphthalen-1-one, 8-hydroxy-5-nitro-	25	8.59 8.41	$c \approx 0.0002$ $I = 0.1$, mixed constant	05	Approx.	M19

No.	Molecular formula, name and pK value(s)	T(°C)	Remarks	Method	Assessment	Ref.
5139	C$_{10}$H$_9$O$_4$N 1,2,3,4-Tetrahydronaphthalen-1-one, 8-hydroxy-7-nitro- 7.95 7.77	25	c ≈ 0.0002 I = 0.1, mixed constant	05	Approx.	M19
5140	C$_{10}$H$_9$O$_4$P 1-Naphthyl dihydrogen phosphate pK$_1$ 0.97 pK$_2$ 5.85	26	c = 0.1, I = 0.1 c = 0.01, I = 0.1 pK$_1$ was determined in 10% dioxane-water Value was corrected to 100% water	E3bg	Uncert. Approx.	C31
5141	C$_{10}$H$_9$O$_4$P 2-Naphthyl dihydrogen phosphate pK$_1$ 1.25 pK$_2$ 5.83	26	c = 0.1, I = 0.1 c = 0.01, I = 0.1 pK$_1$ was determined in 10% dioxane-water Value was corrected to 100% water	E3bg	Uncert. Approx.	C31
5142	C$_{10}$H$_9$O$_6$N Butanedioic acid, 2-(p-nitrophenyl)- pK$_1$ 3.51, pK$_2$ 5.08	20		E3bg	Approx.	F28
5143	C$_{10}$H$_{10}$ON$_2$ 2-Pyrazolin-5-one, 3-methyl-1-phenyl- 7.16	20	c = 0.0025	E3bg	Uncert.	T1
5144	C$_{10}$H$_{10}$ON$_4$ 2-Pyrazolin-5-one, 3-methyl-4-phenylhydrazono- 8.7	20	c = 0.000002	05	Uncert.	L16

No.	Formula	Name	pK	T	Conditions			
5145	$C_{10}H_{10}OS$	Benzo[b]thiophene-4-ol, 2-ethyl-	9.32	20	2% ethanol	05	Approx.	D32
5146	$C_{10}H_{10}OS$	Benzo[b]thiophene-4-ol, 7-ethyl-	9.41	20	2% ethanol	05	Approx.	D32
5147	$C_{10}H_{10}OS$	Benzo[b]thiophene-5-ol, 2,3-dimethyl-	10.03	20	2% ethanol	05	Approx.	D32
5148	$C_{10}H_{10}OS$	Benzo[b]thiophene-5-ol, 3,7-dimethyl-	10.09	20	2% ethanol	05	Approx.	D32
5149	$C_{10}H_{10}OS$	Benzo[b]thiophene-6-ol, 2,3-dimethyl-	9.84	20	2% ethanol	05	Approx.	D32
5150	$C_{10}H_{10}OS$	Benzo[b]thiophene-7-ol, 2,3-dimethyl-	8.80	20	2% ethanol	05	Approx.	D32
5151	$C_{10}H_{10}O_2N_2$	Acetohydroxamic acid, 2-(3-indolyl)-	9.58		I = 0.2(NaCl)	E3bg	Uncert.	C72
5152	$C_{10}H_{10}O_2N_2$	1-Pyrroline-5-carboxylic acid, 2-(2'-pyridyl)- (Pyrimine (anhydrous))	5.2		Conditions not stated	E3bg	Uncert.	S64

The hydrated form of this acid is said to be
2-amino-5-oxo-5(2'-pyridyl)pentanoic acid

No.	Molecular formula, name and pK value(s)	T(°C)	Remarks	Method	Assessment	Ref.
5153	$C_{10}H_{10}O_3N_4$ Ethyl 1,7-dihydro-4-methyl-7-oxopteridine-6-carboxylate 5.74	20	c = 0.0001 in 0.01M buffer	05	Approx.	A39
5154	$C_{10}H_{10}O_4N_4$ Ethyl 2-(3,4,7,8-tetrahydro-4,7-dioxo-6-pteridinyl)acetate pK_1 6.23, pK_2 9.62	20	c = 0.002	E3bg	Approx.	A22
5155	$C_{10}H_{10}O_4S$ 3-Thiapentanedioic acid, 2-phenyl- pK_1 2.98, pK_2 4.65	25		E3ag	Approx.	S92, S93
5156	$C_{10}H_{10}O_5N_2$ 1-Butanone, 4,4-dinitro-1-phenyl- 4.41	25	c ≈ 0.00005 Variation with temperature 5 20 40 60 4.61 4.45 4.29 4.14 Thermodynamic quantities are derived from the results	05	Approx.	S79
5157	$C_{10}H_{10}O_5N_4$ 6-Pteridinecarboxylic acid, 1,2,3,4,7,8-hexahydro-1,3,8-trimethyl-2,4,7-trioxo- 2.82	20		E3bg	Approx.	P44
5158	$C_{10}H_{10}O_5N_4$ 6-Pteridinecarboxylic acid, 1,2,3,4-tetrahydro-7-methoxy-1,3-dimethyl-2,4-dioxo- 2.50	20		E3bg	Approx.	P44
5159	$C_{10}H_{11}O_2N$ Butanamide, 3-oxo-N-phenyl- (Acetoacetanilide) 10.68	20	I = 0.1	05	Approx.	M16

(contd)

No.	Formula	Name	T	Conditions	pK	05/E3bg	Uncert.	Ref.
5159 (Contd)	$C_{10}H_{11}O_2N$	Butanamide, 3-oxo-N-phenyl- (Acetoacetanilide)	22	5% ethanol, mixed constant	10.48		Uncert.	H38
5160	$C_{10}H_{11}O_3N$	Benzoic acid, 2-acetylamino-3-methyl-	20	c = 0.001, 1% ethanol	4.05	E3bg	Approx.	P24,P27
5161	$C_{10}H_{11}O_3N$	Benzoic acid, 2-acetylamino-4-methyl-	20	c ≈ 0.001, 1% ethanol	3.73	E3bg	Approx.	P25
5162	$C_{10}H_{11}O_3N$	Benzoic acid, 2-acetylamino-5-methyl-	20	c = 0.001, 1% ethanol	3.73	E3bg	Approx.	P24,P27
5163	$C_{10}H_{11}O_3N$	Benzoic acid, 2-acetylamino-6-methyl-	20	c ≈ 0.001, 1% ethanol	3.28	E3bg	Approx.	P23
5164	$C_{10}H_{11}O_3N$	Benzoic acid, 3-acetylamino-2-methyl-	20	c ≈ 0.001, 1% ethanol	3.55	E3bg	Approx.	P22
5165	$C_{10}H_{11}O_3N$	Benzoic acid, 3-acetylamino-6-methyl-	20	c ≈ 0.001, 1% ethanol	3.82	E3bg	Approx.	P22
5166	$C_{10}H_{11}O_3N$	Benzoic acid, 4-acetylamino-2-methyl-	20	c ≈ 0.001, 1% ethanol	4.12	E3bg	Approx.	P23
5167	$C_{10}H_{11}O_4N$	Acetic acid, 2-(4-(N-methylamino)benzoyloxy)-			2.39(NMe^+) 3.44(COOH)	E		C74

No.	Molecular formula, name and pK value(s)	T(°C)	Remarks	Method	Assessment	Ref.
5168	$C_{10}H_{11}O_4N$ 3-Azapentanedioic acid, 3-phenyl-		(Iminodiacetic acid, N-phenyl-)			
	pK$_1$ 2.51, pK$_2$ 5.17	30	I = 0.1(KCl), mixed constants	E3bg	Approx.	T32
	pK$_1$ 2.4	20	I = 0.1(KCl), concentration constants	E3bg	Uncert.	S40
	pK$_2$ 4.96				Approx.	
5169	$C_{10}H_{11}O_4N$ Benzoic acid, 4-acetylamino-2-methoxy-					
	4.18			05		L56
5170	$C_{10}H_{11}O_4N$ Bis(methoxycarbonyl)pyridiniomethanide					
	5.12	not stated		05	Uncert.	N31
		Values in mixed solvents are also given				
5171	$C_{10}H_{11}O_4N$ Propanoic acid, 2-amino-3-(3-carboxyphenyl)-					
	pK$_1$ 1.5, pK$_2$ 3.9(L-isomer)		Conditions not stated	E3bg	V.Uncert.	T25
5172	$C_{10}H_{11}O_4N$ Propanoic acid, 2-(2-carboxyphenyl)amino-					
	pK$_1$ 3.12, pK$_2$ 5.01	20	I = 0.1(KCl)	E3bg	Approx.	I18
5173	$C_{10}H_{11}O_4As$ 3-Arsapentanedioic acid, 3-phenyl-		(Arsinodiacetic acid, As-phenyl-)			
	pK$_1$ 3.60, pK$_2$ 5.03	20	I = 0.1(KNO$_3$)	E3bg	Approx.	P38
	pK$_1$ 3.61, pK$_2$ 4.93	25	I = 0.1(K$_2$SO$_4$)	E3bg	Approx.	J5
5174	$C_{10}H_{11}O_5N$ 3-Azapentanedioic acid, 3-(2-hydroxyphenyl)-		(Iminodiacetic acid, N-(2-hydroxyphenyl)-)			
	pK$_1$ 2.98, pK$_2$ 5.43, pK$_3$ 11.08	20	I = 0.1(KCl), concentration constants	E3bg	Approx.	I12
			(contd)			

5174
(Contd) $C_{10}H_{11}O_5N$ 3-Azapentanedioic acid, 3-(2-hydroxyphenyl)-   (Iminodiacetic acid, N-(2-hydroxyphenyl)-)

Values quoted in reference

pK$_1$ 3.20, pK$_2$ 5.22, pK$_3$ 11.06   T16

pK$_1$ 2.98, pK$_2$ 5.43, pK$_3$ 11.08   20   I = 0.1   E   F38

5175 $C_{10}H_{11}O_5N$ Propanoic acid, 2-amino-3-(3-carboxy-4-hydroxyphenyl)-

pK$_1$ 2.0, pK$_2$ 3.4, pK$_3$ 9.3, pK$_4$ >12   E3bg Uncert.   L6

5176 $C_{10}H_{11}O_7N_3$ Phenol, 3-tert-butyl-2,4,6-trinitro-

1.59   not stated   c = 0.003, I = 0.5(NaClO$_4$)   05   Uncert.   P16

5177 $C_{10}H_{12}ON_2$ Ethanol, 1-amino-2-(3-indolyl)-   (L-Tryptophanol)

16.91(NH)   25   In aq. KOH, H$_-$ scale   07   Uncert.   Y2

5178 $C_{10}H_{12}ON_2$ Indol-5-ol, 3-(2-aminoethyl)-   (Serotonine)

18.25(NH)   25   In aq. KOH, H$_-$ scale   07   V.Uncert. Y3

5179 $C_{10}H_{12}O_2N_2$ 2-Azaspiro[4.5]decane-4-carbonitrile, 1,3-dioxo-

7.09(NH)   20   c = 0.0001-0.001, 4% ethanol, mixed   E3bg Uncert.   F29
                         constant

5180 $C_{10}H_{12}O_3N_2$ 3,4'-Bi-3-pyrroline-2,2'-dione, 4-hydroxy-1,1'-dimethyl-

3.46   Conditions not stated   05   Uncert.   M109

5181 $C_{10}H_{12}O_3N_2$ 3,4'-Bi-3-pyrroline-2,2'-dione, 4-hydroxy-5,5'-dimethyl-

3.09   Conditions not stated   05   Uncert.   M109

No.	Molecular formula, name and pK value(s)	T(°C)	Remarks	Method	Assessment	Ref.
5182	$C_{10}H_{12}O_3N_2$ Ethyl (4-hydrazinocarbonyl)benzoate 12.38(NH)			O		T44
5183	$C_{10}H_{12}O_3N_2$ Pentanoic acid, 2-amino-5-oxo-5-(2'-pyridyl)- (Pyrimine hydrate) see compound No. 5152					
5184	$C_{10}H_{12}O_3N_2$ Pyrimidine-2,4,6(1$\underline{H}$,3$\underline{H}$,5$\underline{H}$)-trione, 5,5-diallyl- 7.79	25		E3ag	Approx.	K53
5185	$C_{10}H_{12}O_3N_4$ 6-Indolinone, 3-hydroxy-1-methyl-5-semicarbazono- (DL-Adrenochrome semicarbazone) pK$_2$ 10.8	20	c = 0.005, mixed constant	E3d	Uncert.	R20
5186	$C_{10}H_{12}O_3Se$ Acetic acid, 2-(2-ethoxyphenyl)seleno- 3.90	20	c ≈ 0.001, I = 0.1(KNO$_3$), mixed constant	E3bg	Approx.	P39
5187	$C_{10}H_{12}O_3Se$ Acetic acid, 2-(4-ethoxyphenyl)seleno- 3.86	20	c ≈ 0.001, I = 0.1(KNO$_3$), mixed constant	E3bg	Approx.	P39
5188	$C_{10}H_{12}O_4N_2$ 3-Azapentanedioic acid, 3-(2-pyridylmethyl)- (Iminodiacetic acid, N-(2-pyridylmethyl)-)					
	pK$_1$ 2.6 pK$_2$ 8.21	25	I = 0.1(KNO$_3$), concentration constants	E3bg	Uncert. Approx.	T27
	pK$_1$ 2.85, pK$_2$ 8.25	20	I = 0.1((CH$_3$)$_4$NNO$_3$), concentration constants	E3bg	Approx.	I10
	pK$_1$ 3.06, pK$_2$ 8.22	25	c = 0.01, I = 0.1(KCl), concentration constants	E3bg	Approx.	S97

No.	Formula / Name	pK	t(°C)	Conditions / Notes		Rel.	Ref.
5190	$C_{10}H_{12}O_5N_2$ Phenol, 2-isopropyl-5-methyl-4,6-dinitro-						
		3.958	25	$c \simeq 0.0001$	O5a	Rel.	R30
5191	$C_{10}H_{12}O_5N_4$ Inosine (Purin-6(1H)-one, 9-(β-D-ribofuranosyl)-)						
		pK₁ 8.96	25		E3bg	Approx.	C52
		pK₂ 12.99	10	Thermodynamic quantities are	CAL	Uncert.	
		12.36	25	also given		Approx.	
		11.84	40			Uncert.	
5192	$C_{10}H_{12}O_5N_4$ 1H-Pyrazolo[4,3-d]pyrimidin-7(4H)-one, 3-(β-D-ribofuranosyl)- (Formycin B)						
		pK₁ 8.8		Conditions not stated	E3bg	Uncert.	K52
		pK₂ 10.4		Conditions not stated	E3bg	Uncert.	R27
5193	$C_{10}H_{12}O_6N_4$ Inosine-N(1)-oxide						
		5.40	25	$I = 0.1(NaClO_4)$	O5	Approx.	S70
5194	$C_{10}H_{12}O_6N_4$ Xanthosine (Purine-2,6(1H,3H)-dione, 9-(β-D-ribofuranosyl)-)						
		pK₁ 5.67	25		E3bg	Approx.	C52
		pK₃ 12.85	10	Thermodynamic quantities are	CAL	Approx.	
		12.00	25	also given		Uncert.	
		11.70	40			Uncert.	

No.	Molecular formula, name and pK value(s)	T(°C)	Remarks	Method	Assessment	Ref.
5195	$C_{10}H_{13}OCl$ Phenol, 4-tert-butyl-2-chloro- 8.58	25	c ≈ 0.0001-0.0002, I ≈ 0.1, mixed constant	05	Approx.	S76
5196	$C_{10}H_{13}OCl$ Phenol, 4-chloro-5-methyl-2-isopropyl- 9.98	20		05	Approx.	D31
5197	$C_{10}H_{13}O_2N$ Bicyclo[2.2.2]octane-1-carboxylic acid, 4-cyano- 4.545	25	Values in mixed solvents are also given	E3bg	Approx.	W29
5198	$C_{10}H_{13}O_2N$ Butanohydroxamic acid, N-phenyl- 8.94	25	c ≈ 0.01, I ≈ 0.01	E3bg	Approx.	S67
5199	$C_{10}H_{13}O_2N$ Butanoic acid, 4-amino-2-phenyl- pK_1 3.75, pK_2 10.50	not stated	c = 0.005	E3bg	Uncert.	S87
5200	$C_{10}H_{13}O_2N$ Butanoic acid, 4-amino-3-phenyl- pK_1 4.07, pK_2 9.95	not stated	c = 0.005	E3bg	Uncert.	S87
5201	$C_{10}H_{13}O_2N$ Butanoic acid, 4-amino-4-phenyl- pK_1 4.18, pK_2 9.41	not stated	c = 0.005	E3bg	Uncert.	S87
5202	$C_{10}H_{13}O_2N$ Phenol, 3,5-diethyl-4-nitroso- 8.34	25	c ≈ 0.04, mixed constant	E3bg	Uncert.	V8

No.	Formula / Name	Temp.	pK	Notes	Method	Accuracy	Ref.
5203	$C_{10}H_{13}O_2N$ Propanoic acid, 3-(2-aminomethylphenyl)-	25	4.22	Values in mixed solvents are also given	E3ag	Approx.	H48
5204	$C_{10}H_{13}O_2N$ Propanoic acid, 3-(4-aminomethylphenyl)-	25	4.28	Values in mixed solvents are also given	E3ag	Approx.	H48
5205	$C_{10}H_{13}O_3N$ Phenol, 2,6-diethyl-4-nitro-	25	7.73	$c \approx 0.04$, mixed constant	E3bg	Uncert.	V8
5206	$C_{10}H_{13}O_4N_5$ Adenosine	25	pK$_1$ 3.50	Thermodynamic quantities are also given	E3bg	Approx.	C52
		25	pK$_1$ 3.60	$I = 0.1$	E3bg	Approx.	W3
		25	pK$_1$ 3.5	Conditions not stated	E3bg/05	Uncert.	K34
		25	pK$_2$ 12.35		CAL	Approx.	I33,I34
		25	pK$_2$ 12.34	Thermodynamic quantities are also given	CAL	Approx.	C51
5207	$C_{10}H_{13}O_4N_5$ 1H-Pyrazolo[4,3-d]pyrimidine, 7-amino-3-(β-D-ribofuranosyl)- (Formycin)	25	pK$_1$ 4.4, pK$_2$ 9.7	Conditions not stated	E3bg	Uncert.	K52

No.	Molecular formula, name and pK value(s)	$T(^{\circ}C)$	Remarks	Method	Assessment	Ref.
5208	C$_{10}$H$_{13}$O$_5$N$_5$ Guanosine (Purine-6(1H)-one, 2-amino-9-(β-D-ribofuranosyl)-)					
	pK$_1$ 1.9	25		CAL	Uncert.	C52
	pK$_2$ 9.25	25	Thermodynamic quantities are	E3bg	Approx.	
	pK$_3$ 12.83	10	also given	CAL	Uncert.	
	pK$_3$ 12.33	25			Approx.	
	pK$_3$ 11.60	40			Approx.	
	pK$_1$ 2.20	20	I = 1.0(NaNO$_3$)	E3bg	Uncert.	F19
	pK$_2$ 9.24				Approx.	
	Other values in H29					
5209	C$_{10}$H$_{13}$O$_7$N$_3$ Iminodiacetic acid, N-(1,2,3,4-tetrahydro-6-hydroxy-1,3-dimethyl-2,4-dioxopyrimidin-6-yl)-					
	pK$_1$ 2.05	20	I = 0.1((CH$_3$)$_4$NNO$_3$)	E3bg	Uncert.	I9
	pK$_2$ 2.67, pK$_3$ 10.12				Approx.	
	pK$_1$ 2.05, pK$_2$ 2.67, pK$_3$ 10.12	20	I = 0.1, titration with (CH$_3$)$_4$NOH	E		F38
5210	C$_{10}$H$_{13}$NS$_2$ Methanedithioic acid, (1-methyl-2-phenyl)ethylamino- (Dithiocarbamic acid, N-(1-methyl-2-phenyl)ethyl-)					
	2.82		c = 0.1	E		T65
5211	C$_{10}$H$_{13}$N$_5$(Br) 6-Amino-1-(3-methyl-2-butenyl)purinium bromide					
	7.13	22		E3bg	Approx.	M53

5212 $C_{10}H_{14}ON_2$ Butanamidine, 2-hydroxy-2-phenyl-

pK$_1$ 11.06, pK$_2$ 12.86   25   I = 0.1(KCl)   E3ag   Uncert.   G34

Decomposition occurs; results obtained from extrapolation to zero time

5213 $C_{10}H_{14}O_2S$ Butanesulfinic acid, 4-phenyl-

2.23   20   c = 0.02, mixed constant   E3bg   Approx.   R54

5214 $C_{10}H_{14}O_2Si$ Benzoic acid, 3-trimethylsilyl-

4.089   25   c = 0.0013   05   Rel.   W37

Variation with temperature

15	20	30	35	40
4.140	4.121	4.054	4.032	3.996

Thermodynamic quantities are derived from the results

4.24   25   c = 0.00001   05   Approx.   B60

5215 $C_{10}H_{14}O_2Si$ Benzoic acid, 4-trimethylsilyl-

4.198   25   c = 0.0013   05   Rel.   W37

Variation with temperature

15	20	30	35	40
4.269	4.229	4.150	4.118	4.086

Thermodynamic quantities are derived from the results

4.27   25   c = 0.00001   05   Approx.   B60

No.	Molecular formula, name and pK value(s)	$T(^oC)$	Remarks	Method	Assessment	Ref.
5216	$C_{10}H_{14}O_3N_2$ Pyrimidine-2,4,6(1H,3H,5H)-trione, 5-allyl-5-isopropyl-					
	7.91	25		E3ag	Approx.	K53
5217	$C_{10}H_{14}O_4N_4$ 3-Azapentanoic acid, 5-acetylamino-2-(imidazol-4-yl)methyl-4-oxo- (N-Acetylglycyl-L-histidine)					
	pK$_1$ 2.99, pK$_2$ 7.11	25	I = 0.16(KCl)	E3bg	Approx.	B153
5218	$C_{10}H_{14}O_5N_2$ Uridine, 2'-deoxy-5-methyl- (Thymidine)					
	pK$_2$ 9.65	20	I = 1.0(NaNO$_3$)	E3bg	Approx.	F19
	pK$_2$ 9.8		Conditions not stated	05	Uncert.	L8
5219	$C_{10}H_{14}O_6N_2$ Uridine, 5-methyl- (1-β-D-Ribofuranosylthymine)					
	9.68	not stated	I = 0.05	05	Uncert.	F33
5220	$C_{10}H_{14}O_6N_2$ 1-β-D-Xylofuranosylthymine					
	9.75	not stated	I = 0.05	05	Uncert.	F33
5221	$C_{10}H_{14}O_8S_4$ 3,6-Dithiaoctanedioic acid, 4,5-bis(carboxymethylthio)- (SS'S''S'''-Ethanediylidenetetrakis(thioacetic acid))					
	pK$_1$ 2.89, pK$_2$ 3.54, pK$_3$ 3.96	25	I = 0.1(NaClO$_4$), mixed constants	E3bg	Uncert.	G9
	pK$_4$ 4.572				Approx.	
	pK$_1$ 3.24, pK$_2$ 3.56, pK$_3$ 3.99, pK$_4$ 4.93	20	I = 0.1(NaClO$_4$), mixed constants	E3bg	Uncert.	S3
5222	$C_{10}H_{15}ON$ Phenol, 2-(2-dimethylamino)ethyl-					
	8.75(OH)	25	I = 0.1(KCl), mixed constant	05	Approx.	E20

No.	Formula	Name / pK	T (°C)	Conditions	Method	Reliability	Ref.
5223	$C_{10}H_{15}ON$	Phenol, 3-(2-dimethylamino)ethyl- 9.24(OH)	25	I = 0.1(KCl), mixed constant	05	Approx.	E20
5224	$C_{10}H_{15}ON$	Phenol, 4-(2-dimethylamino)ethyl- 9.46(OH)	25	I = 0.1(KCl), mixed constant	05	Approx.	E20
5225	$C_{10}H_{15}ON$	Phenol, 2-(dimethylamino)methyl-4-methyl- pK$_1$ 8.70	25	I = 0.1(KNO$_3$), mixed constant	E3bg	Approx.	E19
5226	$C_{10}H_{15}ON$	Phenol, 2-(dimethylamino)methyl-6-methyl- pK$_1$ 8.30	25	I = 0.1(KNO$_3$), mixed constant	E3bg	Approx.	E19
5227	$C_{10}H_{15}ON_3$	Triazen-3-ol, 3-propyl-1-(p-tolyl)- 12.79	25	c = 0.00004, I = 0.1(KCl), mixed constant	05	Uncert.	D54
5228	$C_{10}H_{15}ON_5$	2-Butanol, 4-(6-purinyl)amino-2-methyl- 4.96	22		E3bg	Approx.	M53
5229	$C_{10}H_{15}O_2N$	Phenol, 2-(dimethylamino)methyl-4-methoxy- pK$_1$ 8.62	25	I = 0.1(KNO$_3$), mixed constant	E3bg	Approx.	E19
5230	$C_{10}H_{15}O_2N$	Phenol, 2-(dimethylamino)methyl-6-methoxy- pK$_1$ 8.42	25	I = 0.1(KNO$_3$), mixed constant	E3bg	Approx.	E19
5231	$C_{10}H_{15}O_2N_3$	4-Imidazolecarboxylic acid, 5-amino-1-cyclohexyl- pK$_1$ 3.18, pK$_2$ 6.94 pK$_1$ 3.17, pK$_2$ 7.29	20	c ≈ 0.0001, I = 0.10(NaCl) c = 0.01, mixed constants	05 E3bg	Uncert. Approx.	L58

No.	Molecular formula, name and pK value(s)	T(°C)	Remarks	Method	Assessment	Ref.
5232	$C_{10}H_{15}O_4N_5$ 3,6-Diazaoctanoic acid, 8-amino-2-(imidazol-4-yl)methyl-4,7-dioxo- (Glycylglycyl-L-histidine) pK$_1$ 2.84, pK$_2$ 6.87, pK$_3$ 8.22	25	I = 0.16(KCl)	E3bg	Approx.	B153
5233	$C_{10}H_{15}O_5N_3$ Cytidine, N(4)-methyl- (Pyrimidin-2(1H)-one, 4-methylamino-1-(β-D-ribofuranosyl)-) 3.85	22		05	Uncert.	S141
5234	$C_{10}H_{15}O_6N$ 3-Azapentanedioic acid, 3-(1-carboxycyclopentyl)- (Iminodiacetic acid, N-(1-carboxycyclopentyl)-) pK$_1$ 1.61 pK$_2$ 2.56, pK$_3$ 11.69	20	I = 0.1(KCl), concentration constants	E3bg	Uncert. Approx.	I14
5235	$C_{10}H_{15}NS$ Butanethiol, 2-amino-2-phenyl- pK$_1$ 7.65, pK$_2$ 10.15	25	c = 0.01, mixed constants	E3bg	Approx.	B147
5236	$C_{10}H_{15}NS$ 2-Propanethiol, 1-amino-2-methyl-3-phenyl- pK$_1$ 8.00, pK$_2$ 10.86	25	c = 0.01, mixed constants	E3bg	Approx.	B147
5237	$C_{10}H_{16}O_2N_2$ Pyrimidine-2,4(1H,3H)-dione, 5-ethyl-1-methyl-6-propyl- 10.4					D48
5238	$C_{10}H_{16}O_2N_2$ Pyrimidine-2,4(1H,3H)-dione, 5-ethyl-3-methyl-6-propyl- 11.1					D48
5239	$C_{10}H_{16}O_3N_2$ Pyrimidine-2,4,6(1H,3H,5H)-trione, 5-butyl-5-ethyl- 7.92	25		E3ag	Approx.	K53

No.	Formula / Name	pK values	t (°C)	Conditions	Method		Ref.
5240	$C_{10}H_{16}O_8N_2$ 3,6-Diazaoctanedioic acid, 3,6-bis(carboxymethyl)- (Ethylenedinitrilotetraacetic acid) (EDTA)						
		pK_1 0.26, pK_2 0.96	25	c = 0.001, I = 2.0(HCl-NaCl), concentration constants	05	Uncert.	O18
		pK_3 2.70			E3bg		
		pK_3 2.02, pK_4 2.66	20	c = 0.01, I = 0.1(KNO_3), concentration constants	E3bg	Uncert	M93
		pK_5 6.21, pK_6 10.31	20	concentration constants		Approx.	
		pK_5 6.14, pK_6 10.12	30	Thermodynamic quantities are derived			
		pK_5 5.98, pK_6 9.91	40	from the results			
		pK_3 2.0, pK_4 2.67	20	I = 0.1(KNO_3)	E3bg	Uncert.	A54
		pK_5 6.16, pK_6 10.23		Thermodynamic quantities are also given		Approx.	
		pK_3 2.2, pK_4 2.7	20	I = 1.0($(CH_3)_4NCl$), concentration constants	E3bh	Uncert.	A55
		pK_5 6.07, pK_6 10.12		Values in other inert salt solutions are also given		Approx.	
		pK_3 1.99, pK_4 2.67	20	I = 0.1(KCl), concentration constants	E3bg	Uncert.	S34
		pK_5 6.16, pK_6 10.26				Approx.	
		pK_3 2.60, pK_4 2.67	25	I = 0.1(KCl)	E3bg,R3g	Approx.	D67
		pK_5 6.16, pK_6 10.26					
		Other values in I15, S50, T35 and T37					
5241	$C_{10}H_{17}O_2N$ Bicyclo[2.2.1]heptane-1-carboxylic acid, 4-dimethylamino-						
		3.716	25		E3bg	Approx.	W29
		Values in mixed solvents are also given					

No.	Molecular formula, name and pK value(s)	$T(^oC)$	Remarks	Method	Assessment	Ref.
5242	$C_{10}H_{17}O_2N_3$ 1,2,4-Triazine-3,5(2H,4H)-dione, 6-heptyl-					
	7.8	not stated		E3bg	Uncert.	C28
5243	$C_{10}H_{17}O_4N$ 3-Azapentanedioic acid, 3-cyclohexyl- (Iminodiacetic acid, N-cyclohexyl-)					
	pK$_1$ 1.62	25	I = 0.5($NaClO_4$), concentration	E3bg	Uncert.	F4,9
	pK$_2$ 2.22, pK$_3$ 10.59		constants		Approx.	
	pK$_2$ 2.15, pK$_3$ 10.81	20	I = 0.1(KCl), concentration constants	E3bg	Approx.	I12
	pK$_2$ 2.15, pK$_3$ 10.81	20	I = 0.1	E		F38
5244	$C_{10}H_{17}O_4N_5$ Ethyl [5-(ethoxycarbonylaminomethyl)-1-methyl-1,2,3-triazol-4-yl]aminomethanoate					
	12.94	20	c = 0.0001	05	Approx.	A20
5245	$C_{10}H_{17}O_5N$ 3-Azapentanedioic acid, 3-(2-hydroxycyclohexyl)- (Iminodiacetic acid, N-(2-hydroxycyclohexyl)-)					
	pK$_1$ 2.32, pK$_2$ 9.57	20	I = 0.1(KCl), concentration constants	E3bg	Approx.	I12
	pK$_1$ 2.32, pK$_2$ 9.57	20	I = 0.1	E		F38
5246	$C_{10}H_{17}O_5N$ 3-Azapentanedioic acid, 3-(tetrahydropyran-2-yl)methyl- (Iminodiacetic acid, N-(tetrahydropyran-2-yl)methyl-)					
	pK$_1$ 1.88, pK$_2$ 9.04	20	I = 0.1(KNO_3), concentration constants	E3bg	Approx.	I11
5247	$C_{10}H_{17}O_5N$ Diethyl 2-amino-3-oxohexanedioate					
	pK$_1$ 5.82, pK$_2$ 10.4	22		E3bg	Uncert.	L7

5248 $C_{10}H_{18}O_7N_2$ 3,6-Diazaoctanedioic acid, 3-carboxymethyl-6-(2-hydroxyethyl)- (Ethylenedinitrilo-N-(2-hydroxyethyl)-N,N',N'-triacetic acid) (HEDTA)

pK_1 2.51, pK_2 5.31, pK_3 9.86	20	c = 0.01, I = 0.1(KNO_3), concentration constants	E3bg	Approx.	M93
pK_2 5.28, pK_3 9.53	30				
pK_2 5.20, pK_3 9.43	40				

Thermodynamic quantities are derived from the results

pK_1 2.64	29.6	I = 0.1(KCl), concentration constants	E3bg	Uncert.	C26
pK_2 5.33, pK_3 9.73				Approx.	
pK_1 1.83, pK_2 5.24, pK_3 9.13	20	I = 0.5	05	Uncert.	N40

5249 $C_{10}H_{19}O_2N$ 2-Piperidinecarboxylic acid 4-tert-butyl-

pK_1 2.40, pK_2 10.87(cis)	20	c = 0.004	E3bg	Approx.	C3
pK_1 2.59, pK_2 10.74(trans)		c = 0.003			

5250 $C_{10}H_{19}O_4N$ 3-Azapentanedioic acid, 3-(3,3-dimethylbutyl)- (Iminodiacetic acid, N-(3,3-dimethylbutyl)-)

pK_1 2.4	20	I = 0.1(KCl), concentration constants	E3bg	Uncert.	S40
pK_2 10.24				Approx.	

5251 $C_{10}H_{20}O_2N$ (cation) 2-Piperidinecarboxylic acid, N,N-diethyl-

2.08	20	c = 0.004	E3bg	Approx.	C3

5252 $C_{10}H_{20}O_4N_2$ Decane, 1,1-dinitro-

5.55	5	c = 0.005-0.015, mixed constant	05	Uncert.	N38,N39
5.48	20				
5.36	40				

(contd)

No.	Molecular formula, name and pK value(s)	T(°C)	Remarks	Method	Assessment	Ref.
5252 (Contd)	$C_{10}H_{20}O_4N_2$ Decane, 1,1-dinitro- 5.24	60	Thermodynamic quantities are derived from the results		Uncert.	
5253	$C_{10}H_{21}O_2N$ Decanohydroxamic acid 9.66					V12
5254	$C_{10}H_{21}O_3N$ Decanohydroxamic acid, 10-hydroxy- 9.26	30.5?			Uncert.	E17
5255	$C_{10}H_{21}O_3N_3$ Nonanohydroxamic acid, 9-ureido- 9.49	20.5?			Uncert.	E17
5256	$C_{10}H_{21}O_5P$ Acetic acid, 2-dibutoxyphosphinyl- 3.58	25	c = 0.01, mixed constant	E3bg	Uncert.	M52
5257	$C_{10}H_{24}O_4N_2$ 3,6-Diazaoctane-1,8-diol, 3,6-bis(2-hydroxyethyl)- (2,2',2'',2'''-(Ethylenedinitrilo)tetraethanol)					
	pK$_1$ 4.37, pK$_2$ 8.38	25	I = 0.1(NaClO$_4$)	E3bg	Approx.	R42
	pK$_1$ 4.45, pK$_2$ 8.45	25	I = 0.5(KNO$_3$)	E3bg	Approx.	P17
5258	$C_{10}H_{24}O_6P_2$ Decane-1,10-diphosphonic acid pK$_1$ 2.1		Average of values at 25°, 37°, 50°	E3bg	V.Uncert.	I6
	pK$_2$ 3.27, pK$_3$ 7.93, pK$_4$ 8.94		Extrapolated from values at I = 0.1-1.0(N(CH$_3$)$_4$Br)		Uncert.	

No.	Formula	Name	pK	T(°C)	Conditions	Method	Assessment	Ref.
5259	$C_{10}H_6O_8Cl_2S_2$	2,7-Naphthalenedisulfonic acid, 3,6-dichloro-4,5-dihydroxy-	pK_3 3.12	20	c = 0.00005, mixed constant	O5	Uncert.	B34
5260	$C_{10}H_7ONS_2$	Thiazolidin-4-one, 5-benzylidene-2-thioxo-	7.58	20	I = 0.1($NaClO_4$)	DIS	Uncert.	N17
5261	$C_{10}H_7O_5NS$	1-Naphthalenesulfonic acid, 3,4-dihydro-3-hydroxyimino-4-oxo-	pK_2 6.50	25		E3bg		M28
5262	$C_{10}H_7O_5NS$	1-Naphthalenesulfonic acid, 5-hydroxy-6-nitroso-	7.32	25	I = 0.007-1.9	E3bg,R4	Approx.	M35
5263	$C_{10}H_7O_5NS$	1-Naphthalenesulfonic acid, 8-hydroxy-7-nitroso-	pK_2 3.19 pK_2 7.74	25	I = 0.1	E		M36
5264	$C_{10}H_7O_5NS$	2-Naphthalenesulfonic acid, 5-hydroxy-6-nitroso-	pK_2 7.02	25	I = 0.1(KCl)	E		M38
5265	$C_{10}H_7O_5NS$	2-Naphthalenesulfonic acid, 6-hydroxy-5-nitroso-	7.60	25	I = 0.007-0.5	E3bg,R4	Approx.	M35
5266	$C_{10}H_7O_5NS$	2-Naphthalenesulfonic acid, 7-hydroxy-8-nitroso-	pK_2 7.31	25	I = 0.1(KCl)	E		M38

No.	Molecular formula, name and pK value(s)	$T(^oC)$	Remarks	Method	Assessment	Ref.
5267	$C_{10}H_7O_5NS$ 2-Naphthalenesulfonic acid, 8-hydroxy-7-nitroso-					
	pK_2 7.22	25	I = 0.01	E3bg	Approx.	M37
	6.97		I = 0.1, concentration constant			
5268	$C_{10}H_7O_8NS_2$ 1,5-Naphthalenedisulfonic acid, 8-hydroxy-7-nitroso-					
	pK_3 7.32	25		E		M36
	pK_3 6.66		I = 0.1			
5269	$C_{10}H_7O_8NS_2$ 2,7-Naphthalenedisulfonic acid, 3,4-dihydro-4-hydroxyimino-3-oxo-					
	pK_3 7.51	25	c = 0.0045	E3bg	Approx.	M27
	pK_3 7.51	25		05		M28
5270	$C_{10}H_8O_2N_2S$ Pyrimidin-4(1H)-one, 2,3-dihydro-6-hydroxy-5-phenyl-2-thioxo-					
	3.90		Conditions not stated	E3bg	Uncert.	M68
5271	$C_{10}H_8O_2N_2Se$ Pyrimidin-4(1H)-one, 2,3-dihydro-6-hydroxy-5-phenyl-2-selenoxo-					
	4.01		Conditions not stated	E3bg	Uncert.	M68
5272	$C_{10}H_8O_2N_3Cl$ Isoxazol-5-one, 4-(2-chlorophenyl)hydrazono-4,5-dihydro-3-methyl-					
	6.5	20	c = 0.000002	05	Uncert.	L16
5273	$C_{10}H_8O_2N_3Cl$ Isoxazol-5-one, 4-(3-chlorophenyl)hydrazono-4,5-dihydro-3-methyl-					
	6.6	20	c = 0.000002	05	Uncert.	L16

5274 $C_{10}H_8O_2N_3Cl$ Isoxazol-5-one, 4-(4-chlorophenyl)hydrazono-4,5-dihydro-3-methyl-

        6.6   20   c = 0.000002   05   Uncert.   L16

5275 $C_{10}H_9ON_3S$ Phenol, 4-methyl-2-(2-thiazolylazo)-

        8.95   25±0.5   Mixed constant   E3bg   Uncert.   N30

5276 $C_{10}H_9O_2N_3S$ Benzenesulfonamide, N-(2-pyrimidinyl)-

        5.91($-SO_2NH-$)   20   c = 0.0025, I = 0.1(KCl), mixed constant   E3bg   Approx.   W34

5277 $C_{10}H_9O_2N_3S$ Phenol, 5-methoxy-2-(2-thiazolylazo)-   (TAMR)

        pK_1 0.98   not stated   In aq. $HClO_4$ solutions, H_0 scale   06   Uncert.   C56

        pK_2 7.08(OH)   not stated   I = 0.1($NaClO_4$), mixed constant   05   Uncert.   S95

        Value in mixed solvent is also given

5278 $C_{10}H_9O_2N_3S$ Phenol, 4-methoxy-2-(2-thiazolylazo)-

        pK_1 -0.3   not stated   In aq. $HClO_4$ solutions, H_0 scale   06   Uncert.   C56

        pK_2 8.12(OH)   not stated   I = 0.1($NaClO_4$), mixed constant   05   Uncert.   S95

        Value in mixed solvent is also given

5279 $C_{10}H_9O_4NS$ 5-Quinolinesulfonic acid, 8-hydroxy-2-methyl-

        pK_2 4.63, pK_3 8.72   25   c = 0.002, I = 0.5   E3bg   Approx.   T57

        Thermodynamic quantities are also given

5280 $C_{10}H_9O_7NS_2$ 2,7-Naphthalenedisulfonic acid, 3-amino-5-hydroxy-

        pK_3 2.49   25   c = 0.02, I = 0.1($NaClO_4$), mixed constant   E3bg   Approx.   B11

        pK_4 8.54   Uncert.

No.	Molecular formula, name and pK value(s)	T(^{O}C)	Remarks	Method	Assessment	Ref.
5281	$C_{10}H_{10}O_2NCl$ Butanamide, 2-chloro-3-oxo-N-phenyl-					
	8.40	22	5% v/v ethanol, mixed constant	05/E3bg	Uncert.	H38
5282	$C_{10}H_{10}O_2N_4S$ Benzenesulfonamide, 4-amino-N-(2-pyrimidinyl)- (Sulfadiazin)					
	6.35(-SO$_2$ṆH-)	20	c = 0.0025, I = 0.1(KCl), mixed constant	E3bg	Approx.	W34
5283	$C_{10}H_{10}O_7N_2P_2$ P^1,P^2-Di(3-pyridyl) dihydrogen diphosphate (P^1,P^2-Di(3-pyridyl) dihydrogen pyrophosphate)					
	pK$_1$ 3.58, pK$_2$ 4.51	25	I = 0.1(KNO$_3$)	E3bg	Approx.	M118
5284	$C_{10}H_{11}O_4NS$ 3-Azapentanedioic acid, 3-(2-mercaptophenyl)- (Iminodiacetic acid, N-(2-mercaptophenyl)-)					
	pK$_1$ 2.85, pK$_2$ 6.30, pK$_3$ 9.54	20	I = 0.1(KCl)	E3bg	Approx.	I12
	pK$_1$ 2.85, pK$_2$ 6.30, pK$_3$ 9.54	20	I = 0.1	E		F38
5285	$C_{10}H_{11}O_5NS$ 3-Azapentanedioic acid, 3-(2-thenoylmethyl)- (Iminodiacetic acid, N-(2-thenoylmethyl)-)					
	pK$_1$ 1.95	25	I = 0.1(KNO$_3$)	E3bg	Uncert.	A64
	pK$_2$ 7.46				Approx.	
5286	$C_{10}H_{12}ONCl$ Isoquinolin-1(2H)-one, 3-chloro-5,6,7,8-tetrahydro-2-methyl-					
	-0.97(ṆH)	20	c = 0.00005	05	Uncert.	S74
5287	$C_{10}H_{12}O_2N_2S$ Pyrimidin-4(1H)-one, 5-(1-cyclohexenyl)-2,3-dihydro-6-hydroxy-2-thioxo-					
	pK$_1$ 7.11, pK$_2$ >13	28	2% ethanol	05	Uncert.	S14

No.	Formula and name		Temp.	Method	Assessment	Ref.
5288	$C_{10}H_{12}O_2N_2S$ Pyrimidin-4(1H)-one, 5-(2-cyclohexenyl)-2,3-dihydro-6-hydroxy-2-thioxo-					
		2% ethanol	28			
	pK_1 2.35, pK_2 11.31			05	Uncert.	S14
5289	$C_{10}H_{12}N(Cl)S$ Thieno[3,2-c]pyridinium chloride, 2,4,6-trimethyl-		20			
	6.77			E3bg	Uncert.	D51
5290	$C_{10}H_{13}O_3N_2Br$ Pyrimidine-2,4,6(1H,3H,5H)-trione, 5-(2-bromoallyl)-5-isopropyl-		25			
	7.72			E3ag	Approx.	K53
5291	$C_{10}H_{13}O_8N_4P$ Inosine-5'-(dihydrogen phosphate)		25			
	pK_3 6.66			E3bg	Approx.	P61
	Metallic cations replaced by tetrapropylammonium ion					
	Thermodynamic quantities are given					
5292	$C_{10}H_{13}O_9N_4P$ Inosine-5'-(dihydrogen phosphate)-N(1)-oxide					
		$I = 0.1(NaClO_4)$	25			
	pK_1 5.53			05	Approx.	S70
	pK_1 5.43, pK_2 6.31			E3bg	Approx.	
	pK_1 assigned to o-hydroxy-N-oxide group					
	pK_2 assigned to phosphate group					
5293	$C_{10}H_{14}O_3N_2S_2$ Thymidine, 2,4-dithio-					
	Conditions not stated					
	7.70			05	Uncert.	F1
5294	$C_{10}H_{14}O_4N_2S$ Thymidine, 2-thio-					
	Conditions not stated					
	8.70			05	Uncert.	F1

No.	Molecular formula, name and pK value(s)	T(°C)	Remarks	Method	Assessment	Ref.
5295	C$_{10}$H$_{14}$O$_4$N$_2$S$_2$ Uridine, 5-methyl-2,4-dithio-		Conditions not stated	O5	Uncert.	F1
	7.70					
5296	C$_{10}$H$_{14}$O$_6$N$_5$P Adenosine-5'-(dihydrogen phosphate), 2'-deoxy-			E3bg	Approx.	P61
	pK$_3$ 6.65	25	Metallic cations replaced by tetrapropylammonium ion			
			Thermodynamic quantities are given			
5297	C$_{10}$H$_{14}$O$_7$N$_5$P Adenosine-2'-(dihydrogen phosphate)					
	pK$_1$ 4.03, pK$_2$ 6.12	0.4	Extrapolated to c = 0, I = 0.1(KNO$_3$)	E3bg	Approx.	K24
	pK$_1$ 3.88, pK$_2$ 6.07	12	Thermodynamic quantities are derived			
	pK$_1$ 3.71, pK$_2$ 6.01	25	from the results			
	pK$_2$ 3.54, pK$_2$ 5.95	40				
5298	C$_{10}$H$_{14}$O$_7$N$_5$P Adenosine-3'-(dihydrogen phosphate)					
	pK$_1$ 3.95, pK$_2$ 5.93	0.4	Extrapolated to c = 0, I = 0.1(KNO$_3$)	E3bg	Approx.	K24
	pK$_1$ 3.80, pK$_2$ 5.88	12	Thermodynamic quantities are derived			
	pK$_1$ 3.65, pK$_2$ 5.83	25	from the results			
	pK$_1$ 3.49, pK$_2$ 5.78	40				
5299	C$_{10}$H$_{14}$O$_7$N$_5$P Adenosine-5'-(dihydrogen phosphate)					
	pK$_2$ 3.81, pK$_3$ 6.14	20	I = 0.1(KCl)	E3bg	Approx.	S37

(contd)

5299 (Contd) $C_{10}H_{14}O_7N_5P$ Adenosine-5'-(dihydrogen phosphate)

pK values	°C	Conditions / Notes	Method	Reliability	Ref.
pK$_2$ 4.15, pK$_3$ 6.38	0.4	Extrapolated to c = 0, I = 0.1(KNO$_3$)	E3bg	Approx.	K24
pK$_2$ 3.98, pK$_3$ 6.31	12				
pK$_2$ 3.80, pK$_3$ 6.23	25	Thermodynamic quantities are derived			
pK$_2$ 3.62, pK$_3$ 6.16	40	from the results			
pK$_3$ 6.67	25	Metallic cations replaced by	E3bg	Approx.	P61,P62
pK$_3$ 6.70	37	tetrapropylammonium ion			
pK$_3$ 6.47	20	I = 0.1(NaClO$_4$)	E3bg	Approx.	S24
pK$_4$ 13.06	25		CAL	Uncert.	I34

Thermodynamic quantities are also given

5300 $C_{10}H_{14}O_8N_5P$ Adenosine-5'-(dihydrogen phosphate) N(1)-oxide

pK values	°C	Conditions / Notes	Method	Reliability	Ref.
pK$_1$ 2.58, pK$_3$ 12.49	25	I = 0.1(NaClO$_4$)	05	Approx.	S72
pK$_2$ 6.13	25	I = 0.1(NaClO$_4$)	E3bg	Approx.	

5301 $C_{10}H_{14}O_8N_5P$ Guanosine-5'-(dihydrogen phosphate)

pK values	°C	Conditions / Notes	Method	Reliability	Ref.
pK$_3$ 6.66	25	Metallic cations replaced by tetrapropylammonium ion	E3bg	Approx.	P61

Thermodynamic quantities are given

5302 $C_{10}H_{14}O_{11}N_4P_2$ Inosine-5'-(trihydrogen diphosphate)

pK values	°C	Conditions / Notes	Method	Reliability	Ref.
pK$_4$ 7.18	25	Metallic cations replaced by tetrapropylammonium ion	E3bg	Approx.	P61

Thermodynamic quantities are given

No.	Molecular formula, name and pK value(s)	T(°C)	Remarks	Method	Assessment	Ref.
5303	$C_{10}H_{15}O_6N_2P$ 2-[[3-Hydroxy-5-hydroxymethyl-2-methylpyridin-4-yl)methyleneamino]ethyl dihydrogen phosphate (Pyridoxylidene-(2-aminoethyl dihydrogen phosphate))					
	9.96	25	c = 0.0001, I = 1.0(KCl)	05	Approx.	M111
			pK of aldimine			
5304	$C_{10}H_{15}O_{10}N_5P_2$ Adenosine-5'-(trihydrogen diphosphate) (ADP)					
	pK_3 4.20, pK_4 6.51	0.4	I = 0.1(KNO₃), extrapolated to c = 0	E3bg	Approx.	K24
	pK_3 4.09, pK_4 6.48	12				
	pK_3 3.93, pK_4 6.44	25	Thermodynamic quantities are derived			
	pK_3 3.73, pK_4 6.41	40	from the results			
	pK_3 3.99, pK_4 6.35	20	I = 0.1(KCl)	E3bg	Approx.	S37
	pK_3 4.10	20	I = 0.1(NaClO₄)	E3bg	Approx.	S24
	pK_4 7.20	25	Metallic cations replaced by	E3bg	Approx.	P61,P62
	pK_4 7.24	37	tetrapropylammonium ion			
			Thermodynamic quantities are also given			
5305	$C_{10}H_{15}O_{11}N_5P_2$ Guanosine-5'-(trihydrogen diphosphate)					
	pK_4 7.19	25		E3bg	Approx.	P6
			Metallic cations replaced by tetrapropylammonium ion			
			Thermodynamic quantities are given			

No.	Formula / Name				
5306	$C_{10}H_{15}O_{14}N_4P_3$ Inosine-5'-(tetrahydrogen triphosphate) (ITP)				
	pK_5 7.68	25		E3bg Approx.	P61
	Metallic cations replaced by tetrapropylammonium ion				
	Thermodynamic quantities are given				
	pK_6 9.5	25	c = 0.0004	05 Uncert.	S71
	pK_6 9.0		I = 0.1(NaClO$_4$)		
	Formation of sodium complexes is said to occur with NaClO$_4$				
5307	$C_{10}H_{16}ON(I)$ Phenol, 3-methyl-4-trimethylammonio-, iodide				
	8.34	25±1	I = 0.01	E3ag Uncert.	02
5308	$C_{10}H_{16}ON(I)$ Phenol, 4-methyl-3-trimethylammonio-, iodide				
	8.26	25±1	I = 0.01	E3ag Uncert.	02
5309	$C_{10}H_{16}ON(I)$ Phenol, 2-(N,N,N-trimethylammonio)methyl-, iodide				
	8.74	25	I = 0.1(KCl), mixed constant	E3bg Approx.	E20
5310	$C_{10}H_{16}ON(I)$ Phenol, 3-(N,N,N-trimethylammonio)methyl-, iodide				
	8.89	25	I = 0.1(KCl), mixed constant	E3bg Approx.	E20
5311	$C_{10}H_{16}ON(I)$ Phenol, 4-(N,N,N-trimethylammonio)methyl-, iodide				
	8.75	25	I = 0.1(KCl), mixed constant	E3bg Approx.	E20
5312	$C_{10}H_{16}O_2N(I)$ 1,2-Benzenediol, 3-(N,N,N-trimethylammonio)-, iodide				
	pK_1 8.10	25	I = 0.1(KNO$_3$), mixed constant	E3bg Approx.	E19

5313 $C_{10}H_{16}O_2N_2S$ Pyrimidine-4,6(1$\underline{H}$,5$\underline{H}$)-dione, 5-butyl-5-ethyl-2,3-dihydro-2-thioxo-

pK	T	Remarks	Method		Ref.
pK$_1$ 7.37, pK$_2$ 12.07	28	2% ethanol	05	Uncert.	S14

5314 $C_{10}H_{16}O_{13}N_5P_3$ Adenosine-5'-(tetrahydrogen triphosphate) (ATP)

pK	T	Remarks	Method		Ref.
pK$_4$ 3.93, pK$_5$ 6.97	30	I = 0.1((C_2H_5)$_4$NBr)	E3bg	Approx.	O27
pK$_4$ 4.27, pK$_5$ 6.56	0.4	I = 0.1(KNO$_3$), mixed constants	E3bg	Approx.	K23
pK$_4$ 4.14, pK$_5$ 6.54	12				
pK$_4$ 4.06, pK$_5$ 6.53	25	Thermodynamic quantities are derived from			
pK$_4$ 3.87, pK$_5$ 6.52	40	the results			
pK$_4$ 4.05, pK$_5$ 6.50	20	I = 0.1(KCl)	E3bg	Approx.	S37
pK$_5$ 7.68	25	Metallic cations replaced by	E3bg	Approx.	P61,P62
pK$_5$ 7.73	37	tetrapropylammonium ion			
		Thermodynamic quantities are also given			

5315 $C_{10}H_{16}O_{14}N_5P_3$ Guanosine-5'-(tetrahydrogen triphosphate)

pK	T	Remarks	Method		Ref.
pK$_5$ 7.65	25	Metallic cations replaced by	E3bg	Approx.	P61
		tetrapropylammonium ion			
		Thermodynamic quantities are also given			
pK$_6$ 10.1	25	c = 0.0004	05	Uncert.	S71
pK$_6$ 9.5		I = 0.1(NaClO$_4$)			
		Formation of sodium complexes is said to occur with NaClO$_4$			

5316 $C_{10}H_{17}O_{14}N_2P_3$ Thymidine-5'-(tetrahydrogen triphosphate) (TTP)

10.7

10.1

$c = 0.0004$

$I = 0.1(NaClO_4)$

25   05   Uncert.   S71

Formation of sodium complexes is said to occur with $NaClO_4$.

5317 $C_{10}H_{17}O_{16}N_5P_4$ Adenosine-5'-(pentahydrogen tetraphosphate)

pK_5 4.09, pK_6 6.79

$I = 0.1(KCl)$

20   E3bg   Approx.   S37

5318 $C_{10}H_{20}O_5N_2S$ Ethanethioamide, N,N,N',N'-tetrakis(2-hydroxyethyl)-   (Monothiooxamide, N,N'-bis(2-hydroxyethyl)-)

11.38

11.26

$I = 0.1(KCl)$, mixed constant

25   05,R4   Approx.   V13

05

(Values for I = 0.02, 0.05, 0.25, 0.50, 1.00 also given)

5319 $C_{10}H_{21}O_2N_2(I)$ N,N-Diethyl-N-methyl 3-hydroxyimino-4-oxopentylammonium iodide   (2-Pentanone, 5-N,N,N-diethylmethylammonio-3-hydroxyimino-, iodide)

3.60

$c = 0.01$, $I = 0.1(KCl)$, mixed constant

30   E3bg   Approx.   E18

5320 $C_{10}H_{22}O_6N_2S_2$ Piperazine-1,4-bis(propanesulfonic acid)   (1,4-Bis(3-sulfopropyl)piperazine)   (PIPPS)

pK_1 4.05, pK_2 8.1

$c = 0.05$, mixed constants

18   Uncert.   J12

5321 $C_{10}H_{24}O_3NP$ Decylphosphonic acid, 10-amino-

pK_2 8.00, pK_3 11.25

$c = 0.02$

not stated   E3bg   Uncert.   C38

5322 $C_{10}H_{26}O_7N_2P_2$ 6-Oxa-3,9-diazaundecane-2,10-diphosphonic acid, 2,10-dimethyl-

pK(POH) 5.16, pK(POH) 6.45, pK(NH^+) 10.67, pK(NH^+) 11.61

$c = 0.001$, $I = 0.1(KCl)$, concentration constants

25   E3bg   Uncert.   D70

(contd)

No.	Molecular formula, name and pK value(s)	$T(^oC)$	Remarks	Method	Assessment	Ref.
5322 (Contd)	$C_{10}H_{26}O_7N_2P_2$ 6-Oxa-3,9-diazaundecane-2,10-diphosphonic acid, 2,10-dimethyl- pK_1 5.15, pK_2 6.40, pK_3 10.56, pK_4 11.48	not stated		E3bg	Uncert.	K7
5323	$C_{10}H_{27}O_6N_3P_2$ 3,6,9-Triazaundecane-2,10-diphosphonic acid, 2,10-dimethyl- pK_1 5.39, pK_2 6.55, pK_3 10.40, pK_4 11.20	not stated		E3bg	Uncert.	K7
5324	$C_{10}H_{16}O_5NSP$ Diethylphosphoramidate, N-phenylsulfonyl- 2.08	20	c = 0.005, 7% ethanol, mixed constant	E3bg	Uncert.	G19
5325	$C_{10}H_{26}O_6N_2SP$ 6-Thia-3,9-diazaundecane-2,10-diphosphonic acid, 2,10-dimethyl- pK_1 5.51, pK_2 6.34, pK_3 10.38, pK_4 11.48	not stated		E3bg	Uncert.	K7
	$pK(POH)$ 5.21, $pK(POH)$ 6.34, $pK(NH^+)$ 10.42, $pK(NH^+)$ 11.41	25	c = 0.001, I = 0.1(KCl), concentration constants	E3bg	Uncert.	D70
5326	$C_{11}H_6O_4$ 4-Cyclopentene-1,2,3-trione, 2-hydroxy-3-phenyl- 1.64	not stated	c = 0.00003-0.00005	O5	Uncert.	P15
5327	$C_{11}H_8O_2$ 1-Azulenecarboxylic acid 5.5	25	c < 0.00017	E3c & O5	Uncert.	L64

5328	$C_{11}H_8O_2$	1-Naphthocarbaldehyde, 4-hydroxy-			
	6.53	25	05	Approx.	C79

5329	$C_{11}H_8O_2$	1-Naphthoic acid			
	3.597	25	05	Approx.	B125

Variation with temperature

15	20	30	35	40
3.542	3.550	3.628	3.662	3.673

Thermodynamic quantities are derived from the results

	25	05	Uncert.	C79
3.64	25	05	Uncert.	C79
3.73(H_2O)	25	05	Uncert.	W12
4.27(D_2O)				
3.7		05	Uncert.	J2
10-12	FLU/05	First excited singlet state		
3.8	Flash photolysis	Lowest triplet state		
4.6	Phosphorescence			

5330	$C_{11}H_8O_2$	2-Naphthoic acid			
	4.141	25	05	Approx.	B125

Variation with temperature

15	20	30	35	40
4.196	4.166	4.149	4.098	4.056

Thermodynamic quantities are derived from the results

(contd)

No.	Formula	Name / pK	T	Conditions	Assessment	Method	Ref.
5330 (Contd)	$C_{11}H_8O_2$	2-Naphthoic acid					
		4.21(H_2O)	25		Uncert.	05	W12
		4.68(D_2O)					
		4.2			Uncert.	05	J2
		10-12		First excited singlet state		FLU/05	
		4.0		Lowest triplet state		Flash photolysis	
		4.2				Phosphorescence	
5331	$C_{11}H_8O_3$	1,3-Indandione, 2-acetyl-					
		2.9	18±2	1% ethanol, mixed constant	Uncert.	05 or E3d	W25
5332	$C_{11}H_8O_3$	1-Naphthoic acid, 3-hydroxy-					
		pK$_1$ 3.69, pK$_2$ 9.78	25	c ≈ 0.001	Approx.	E3bg	B155
5333	$C_{11}H_8O_3$	2-Naphthoic acid, 3-hydroxy-					
		pK$_1$ 2.79, pK$_2$ 12.84	25	c ≈ 0.0001	Approx.	05	M31
		pK$_1$ 2.71	25	c = 0.0001 to 0.0008	Approx.	L1	B123
				Other values in G58			
5334	$C_{11}H_8O_3$	2-Naphthoic acid, 4-hydroxy-					
		pK$_1$ 4.25, pK$_2$ 9.59	25	c ≈ 0.001	Approx.	E3bg	B155

No.	Formula	Name	T	pK	Conditions	Method		Ref.
5335	$C_{11}H_8O_4$	2H-Benzopyran-2-one, 3-acetyl-4-hydroxy-	25	4.26	I = 0.15, mixed constant			A1a
5336	$C_{11}H_8O_4$	Cyclobutenedione, 3-hydroxy-4-(p-methoxyphenyl)-	20	2.05		E3bg	Uncert.	B126
5337	$C_{11}H_8O_4$	Methyl 1,3-dioxoindanyl-2-carboxylate	not stated	≃2	1% methanol	E3bg		G46
5338	$C_{11}H_{10}O$	1-Naphthol, 4-methyl-	25	9.64		05	Approx.	C79
5339	$C_{11}H_{10}O_3$	Undeca-5,6-diene-8,10-diynoic acid, 4-hydroxy- (Nemotinic acid)		4.80	Conditions not stated	05(?)	Uncert.	B169a
5340	$C_{11}H_{10}O_3$	(-)Undeca-6,8,10-triynoic acid, 4-hydroxy- (Isonemotinic acid)		4.75	Conditions not stated	05(?)	Uncert.	B169b
5341	$C_{11}H_{10}O_4$	1,2-Cyclopropanedicarboxylic acid, 1-phenyl-	20	pK_1 3.48, pK_2 6.08(cis); pK_1 3.72, pK_2 5.14(trans)		E3bg	Approx.	M12
5342	$C_{11}H_{12}O_2$	Benzo[b]furan-4-ol, 2-ethyl-3-methyl-	20	9.89	2% ethanol	05	Approx.	D32

No.	Molecular formula, name and pK value(s)	T(°C)	Remarks	Method	Assessment	Ref.
5343	$C_{11}H_{12}O_2$ Benzo[b]furan-4-ol, 7-ethyl-2-methyl-					
	9.61	20	2% ethanol	05	Approx.	D32
5344	$C_{11}H_{12}O_2$ Benzo[b]furan-7-ol, 2,3,4-trimethyl-					
	9.37	20	2% ethanol	05	Approx.	D32
5345	$C_{11}H_{12}O_2$ Benzo[b]furan-7-ol, 2,3,5-trimethyl-					
	9.11	20	2% ethanol	05	Approx.	D32
5346	$C_{11}H_{12}O_2$ Benzoic acid, 4-cyclobutyl-					
	4.35	20	Mixed constant	05	Uncert.	S62,S61
5347	$C_{11}H_{12}O_2$ 1,3-Butanedione, 2-methyl-1-phenyl-					
	11.13	20	c = 0.01	E3bg/05	Approx.	R52
5348	$C_{11}H_{12}O_2$ Cyclopropanecarboxylic acid, 2-(p-tolyl)-					
	4.60	25	c ≃ 0.0008	E3bg	Approx.	T55
5349	$C_{11}H_{12}O_2$ 2,4-Pentanedione, 3-phenyl-					
	9.32	20	c = 0.01	E3bg/05	Approx.	R52
5350	$C_{11}H_{12}O_3$ Benzoic acid, 2-butyryl-					
	4.55	25		E3bg	Uncert.	H18
	4.61			05	Uncert.	

(contd)

No.	Formula	Name / pK	Temp (°C)	Conditions	Method	Reliability	Ref.
5350 (Contd)	$C_{11}H_{12}O_3$	Benzoic acid, 2-butyryl-					
		4.23		$I = 0.5$	05	Uncert.	
		4.41		$I = 0.5$	KIN	Uncert.	
		Intramolecular enolization occurs; said to be catalyzed by carboxylate anion					
5351	$C_{11}H_{12}O_3$	Cyclopropanecarboxylic acid, 2-(4-methoxyphenyl)-					
		4.62	25	$c \approx 0.0008$	E3bg	Approx.	T55
5352	$C_{11}H_{12}O_3$	Ethyl 3-oxo-3-phenylpropanoate					
		10.53	20		05	Approx.	E26
5353	$C_{11}H_{12}O_3$	Propenoic acid, 3-(2-hydroxy-3,6-dimethylphenyl)-					
		3.84	30	$I = 1.0$(LiCl), mixed constant	05	Approx.	H37
5354	$C_{11}H_{12}O_3$	Propenoic acid, 3-(2-hydroxy-4,6-dimethylphenyl)-					
		3.93	30	$I = 1.0$(LiCl), mixed constant	05	Approx.	H36
5355	$C_{11}H_{12}O_4$	Butanedioic acid, 2-methyl-2-phenyl-					
		pK_1 3.58, pK_2 6.38	20		E3bg	Approx.	F28
5356	$C_{11}H_{12}O_4$	Pentanedioic acid, 3-phenyl-					
		pK_1 4.28, pK_2 5.34	25	$c = 0.00125$, mixed constant	E3bg,R3c	Approx.	B146
5357	$C_{11}H_{12}O_4$	Propanedioic acid, 2-ethyl-2-phenyl-					
		pK_2 7.12	25	$c = 0.005$, $I = 0.1$	E3bg	Approx.	F47
		pK_2 7.16	33				

(contd)

No.	Molecular formula, name and pK value(s)	$T(^{\circ}C)$	Remarks	Method	Assessment	Ref.
5357 (Contd)	$C_{11}H_{12}O_4$ Propanedioic acid, 2-ethyl-2-phenyl- pK_1 1.9 pK_2 7.12	25	I = 0.1, mixed constants	E3bg	Uncert. Approx.	M77
5358	$C_{11}H_{12}O_4$ Propenoic acid, 3-(2,3-dimethoxyphenyl)- 4.37(trans)	25±0.2	c = 0.00001-0.0001, extrapolated to zero concentration	05	Approx.	H54
5359	$C_{11}H_{12}O_4$ Propenoic acid, 3-(2,4-dimethoxyphenyl)- 4.80(trans)	25±0.2	c = 0.00001-0.0001, extrapolated to zero concentration	05	Approx.	H54
5360	$C_{11}H_{12}O_4$ Propenoic acid, 3-(3,4-dimethoxyphenyl)- 4.53(trans)	25±0.2	c = 0.00001-0.0001, extrapolated to zero concentration	05	Approx.	H54
5361	$C_{11}H_{12}O_4$ Propenoic acid, 3-(3,5-dimethoxyphenyl)- 4.36(trans)	25±0.2	c = 0.00001-0.0001, extrapolated to zero concentration. Compound of low solubility	05	Uncert.	H54
5362	$C_{11}H_{12}O_7$ Diethyl 3-hydroxy-4-oxo-4H-pyran-2,6-dicarboxylate 4.85	room temp	Mixed constant	E3bg	Uncert.	E9

5363 $C_{11}H_{14}O_2$ Benzoic acid, 3-tert-butyl-

4.204   25   c = 0.0013   05   Rel.   W37

Variation with temperature

15	20	30	35	40
4.265	4.230	4.165	4.144	4.120

Thermodynamic quantities are derived from the results

5364 $C_{11}H_{14}O_2$ Benzoic acid, 4-tert-butyl-

4.382   25   c = 0.0013   05   Rel.   W37

Variation with temperature

15	20	30	35	40
4.460	4.433	4.356	4.320	4.287

Thermodynamic quantities are derived from the results

5365 $C_{11}H_{14}O_2$ Benzoic acid, 2,3,5,6-tetramethyl-

3.416   25   05   Approx.   W38

Variation with temperature

15	20	30	35	40
3.312	3.365	3.452	3.488	3.503

Thermodynamic quantities are derived from the results

5366 $C_{11}H_{14}O_2$ Pentanoic acid, 4-phenyl-

4.7   25   E3bg   Uncert.   C57

No.	Molecular formula, name and pK value(s)	T(°C)	Remarks	Method	Assessment	Ref.
5367	$C_{11}H_{14}O_2$ Pentanoic acid, 5-phenyl- 4.88	20	c ≈ 0.001, mixed constant	E3bg	Approx.	R55
5368	$C_{11}H_{14}O_3$ Acetic acid, 2-hydroxy-2-(4-isopropylphenyl)- 3.64	25	c = 0.01, mixed constant	E3bg	Uncert.	K36
5369	$C_{11}H_{14}N_2$ Indole, 3-dimethylaminomethyl- (Gramine) 16.00(NH)	25	In aq. KOH, H_ scale	07	V.Uncert.	Y3
5370	$C_{11}H_{16}O$ Phenol, 4-butyl-2-methyl- 10.69	20			Uncert.	D30
5371	$C_{11}H_{16}O$ Phenol, 2-tert-butyl-4-methyl- 11.72	20	Extrapolated value		Uncert.	D30
5372	$C_{11}H_{16}O$ Phenol, 2-tert-butyl-6-methyl- 10.96	20	Extrapolated value	05	Approx.	D30
5373	$C_{11}H_{16}O$ Phenol, 4-tert-butyl-2-methyl- 10.59	20		05	Approx.	D30
5374	$C_{11}H_{16}O$ Phenol, 5-isopropyl-2,4-dimethyl- 10.76	20	2% ethanol	05	Approx.	D29

No.	Formula	Name	pK	Temp.	Conditions			Ref.
5375	$C_{11}H_{16}O$	Phenol, 2-isopropyl-*4,6*-dimethyl-	11.18	20	2% ethanol	05	Approx.	D29
5376	$C_{11}H_{16}O$	Phenol, 4-isopropyl-2,6-dimethyl-	10.90	20	2% ethanol	05	Approx.	D29
5377	$C_{11}H_{16}O$	Phenol, 2-isopropyl-4,5-dimethyl-	10.79	20		05	Approx.	D31
5378	$C_{11}H_{16}O$	Phenol, 2-ethyl 4-isopropyl-	10.77	20	2% ethanol	05	Approx.	D29
5379	$C_{11}H_{16}O$	Phenol, 2-ethyl 5-isopropyl-	10.61	20	2% ethanol	05	Approx.	D29
5380	$C_{11}H_{16}O$	Phenol, 4-ethyl 2-isopropyl-	10.82	20	2% ethanol	05	Approx.	D29
5381	$C_{11}H_{16}O$	Phenol, 2-neopentyl-	10.69	25	$I = 0.1(KCl)$, mixed constant	E3bg	Approx.	E20
5382	$C_{11}H_{16}O_4$	Bicyclo[2.2.2]octane-1-carboxylic acid, 4-methoxycarbonyl-	4.764	25		E3bg	Approx.	W29

Values in mixed solvents are also given

No.	Molecular formula, name and pK value(s)	T(°C)	Remarks	Method	Assessment	Ref.
5383	$C_{11}H_{20}O_2$ 3,5-Heptanedione, 2,2,6,6-tetramethyl-					
	11.77	25	c = 0.0001, I = 0.01$((CH_3)_4NBr)$	O5	Approx.	G64
	11.57	20	c = 0.00002, I = 0.1$(NaClO_4)$	O5	Approx.	K49
5384	$C_{11}H_{20}O_2$ 4,6-Nonanedione, 3,7-dimethyl-					
	9.48	25	c = 0.00005, 0.5-1% ethanol	O5	Approx.	C5
			Apparent pK for keto-enol mixture			
	pK(enol) 8.88, pK(keto) 9.36	25	Calculated using absorbance data		Uncert.	
5385	$C_{11}H_{20}O_4$ Pentanedioic acid, 3,3-diisopropyl-					
	pK_1 3.63, pK_2 7.68	25	c = 0.00125, mixed constants	E3bg	Approx.	B146
5386	$C_{11}H_{20}O_4$ Pentanedioic acid, 3,3-dipropyl-					
	pK_1 3.65, pK_2 7.48	25	c = 0.00125, mixed constants	E3bg	Approx.	B146
5387	$C_{11}H_{20}O_4$ Propanedioic acid, dibutyl-					
	pK_1 1.89	25	I = 0.1(KNO_3)	E3bg	Uncert.	P78
	pK_2 7.19		I = 0.1(KNO_3), concentration constants	E3bg	Approx.	
			pK_2 is extrapolated value at c = 0			
	pK_1 2.01, pK_2 7.21	25	c = 0.001, I = 0.1$(NaClO_4)$, mixed constants	E3bg	Approx.	O23
	pK_2 7.36	25	I = 0.1(KCl), mixed constant	E3bg	Approx.	M77
	pK_1 2.16, pK_2 7.70(H_2O)	25		E3bg	Approx.	G23
	pK_1 2.64, pK_2 8.23(D_2O)					

No.	Formula	Name	T (°C)	pK	Conditions		Qualifier	Method	Ref.
5388	$C_{11}H_7ON$	1-Naphthalenecarbonitrile, 3-hydroxy-	25	8.41	c ≈ 0.001		Approx.	E3bg	B155
5389	$C_{11}H_7ON$	1-Naphthalenecarbonitrile, 4-hydroxy-	25	7.08			Approx.	05	C79
5390	$C_{11}H_7O_2Cl$	1-Naphthoic acid, 4-chloro-	25	3.36			Uncert.	05	C79
5391	$C_{11}H_7O_2Br$	1-Naphthoic acid, 4-bromo-	25	3.37			Uncert.	05	C79
5392	$C_{11}H_7O_2F$	1-Naphthoic acid, 4-fluoro-	25	3.70			Uncert.	05	C79
5393	$C_{11}H_7O_4N$	1-Naphthoic acid, 3-nitro-	25	2.83			Uncert.	05	C79
5394	$C_{11}H_7O_4N$	1-Naphthoic acid, 4-nitro-	25	2.8			Uncert.	05	C79
5395	$C_{11}H_8ON_2$	1H-Naphtho[2,3-d]imidazol-2(3H)-one	20	11.62	c = 0.0000125 in 0.01M buffer		Approx.	05	B127
5396	$C_{11}H_8ON_6$	Pyrimidino[5,4-e]1,2,4-triazin-5(1H)-one, 7-amino-3-phenyl-	20	pK_1 1.00, pK_2 6.77	I = 0.01, mixed constant		Approx.	05	B139

No.	Molecular formula, name and pK value(s)	$T(^oC)$	Remarks	Method	Assessment	Ref.
5397	$C_{11}H_8O_3Se$ 1,3-Propanedione, 1-(2-furyl)-3-(2-selenofuryl)-					E7
	7.88	room temp	$I = 0.1(NaClO_4)$			
5398	$C_{11}H_8O_4N_2$ 4,5-Imidazoledicarboxylic acid, 2-phenyl-					S13
	pK_1 3.00, pK_2 7.68			()		
5399	$C_{11}H_8N_2S$ 1H-Naphtho[2,3-d]imidazole-2(3H)-thione					B127
	9.84	20	c = 0.0000125 in 0.01M buffer	O5	Approx.	
5400	$C_{11}H_9ON_3$ Phenol, 2-(2-pyridylazo)- (o-PAP)					A61
	1.85(NH)	20±2	Mixed constants	O5	Uncert.	
	9.42(OH)			E3bg		
5401	$C_{11}H_9ON_3$ Phenol, 4-(2-pyridylazo)- (p-PAP)					A61
	2.47(NH)	20±2	Mixed constants	O5	Uncert.	
	8.29(OH)			E3bg		
5402	$C_{11}H_9O_2N$ 2-Hydroxyphenyl 1-pyrrolyl ketone					M71
	8.59	25	Imidazole buffers	O5	Uncert.	
5403	$C_{11}H_9O_2N$ 1-Naphthohydroxamic acid					W39
	7.7	20	c = 0.001	E3d	Uncert.	

No.	Formula and Name / pK	T (°C)	Method	Reliability	Remarks	Ref.
5404	$C_{11}H_9O_2N$ 8-Quinolinol, 5-acetyl-					
	pK$_1$ 4.56, pK$_2$ 8.20	25	E			L11
	pK$_1$ 4.21	0				
5405	$C_{11}H_9O_2N_3$ 1,3-Benzenediol, 4-(2-pyridylazo)- (PAR)					
	pK$_1$ 2.69, pK$_2$ 5.50	25	E3bg	Approx.	I = 0.1, concentration constants	G10
	pK$_3$ 12.31		05			
	pK$_1$ 3.1, pK$_2$ 5.6, pK$_3$ 11.9	not stated	05	Uncert.	I = 0.1	H46
	pK$_2$ 5.83	25	E3bg	Approx.		I29
	pK$_3$ 12.5		05	Uncert.		
	Other values in B182 and B185					
5406	$C_{11}H_9O_2N_3$ 1,3-Benzenediol, 4-(3-pyridylazo)- (3-PAR)					
	pK$_3$ 11.40(ortho-OH)	not stated	05	Uncert.	Mixed constant	B182
5407	$C_{11}H_9O_3N$ 2-Furancarbohydroxamic acid, N-phenyl-					
	pK$_1$ 2.22, pK$_2$ 7.69	30	E3bg	Approx.	I = 0.1(NaClO$_4$)	D63
	pK$_2$ 7.69	30	E3bg	Uncert.	c = 0.01, I > 0.1(NaClO$_4$), concentration constant	D62
	pK$_2$ 7.84	25	E3bg	Approx.	c ≈ 0.001, I = 0.1(KCl), mixed constants	F31
	pK$_2$ 7.81				I = 1.0(KCl)	
5408	$C_{11}H_9O_3N$ 2-Naphthohydroxamic acid, 3-hydroxy-					
	9.5	0				M49

No.	Molecular formula, name and pK value(s)	T(^{o}C)	Remarks	Method	Assessment	Ref.
5409	$C_{11}H_9O_4N$ 3-Pyrrolidinecarboxylic acid, 2,5-dioxo-4-phenyl- pK(COOH) 2.90, pK(NH) 8.92	20	c = 0.0001-0.001, 4% ethanol, mixed constants	E3bg	Uncert.	F29, F28
5410	$C_{11}H_9O_4N_3$ Pyrimidin-4(1$\underline{H}$)-one, 2-benzyl-6-hydroxy-5-nitro- 3.49	20	Mixed constant	05	Uncert.	B76
5411	$C_{11}H_9O_6P$ 1-Naphthoic acid, 2-phosphonooxy- pK_1 1.00 pK_2 3.18, pK_3 6.12	26	I = 0.1	E3bg	Uncert. Approx.	C33
5412	$C_{11}H_9O_6P$ 1-Naphthoic acid, 8-phosphonooxy- pK_1 0.97 pK_2 4.08, pK_3 6.57	26	c = 0.1, I = 0.1 c = 0.01, I = 0.1 pK_1 determined in 10% dioxane-water; corrected to 100% water	E3bg	Uncert. Approx.	C31
5413	$C_{11}H_9O_6P$ 2-Naphthoic acid, 1-phosphonooxy- pK_1 1.12 pK_2 3.78, pK_3 7.99	26	I = 0.1	E3bg	Uncert. Approx.	C33
5414	$C_{11}H_9O_6P$ 2-Naphthoic acid, 3-phosphonooxy- pK_1 1.46 pK_2 3.70, pK_3 6.34	26	I = 0.1	E3bg	Uncert. Approx.	C33

No.	Formula	Name	pKa	Temp.	Conditions			Ref.
5415	$C_{11}H_{10}O_2N_2$	Acetic acid, 2-(8-quinolinyl)amino- pK$_1$ 1.9, pK$_2$ 2.98, pK$_3$ 4.26		25	I = 0.1	E3b		T4
5416	$C_{11}H_{10}O_2N_2$	Butanamide, 2-cyano-3-oxo-N-phenyl-	3.40	22	5% (v/v) ethanol, mixed constant	05/E3bg	Uncert.	H38
5417	$C_{11}H_{10}O_2N_2$	Pyrimidin-4(1H)-one, 2-benzyl-6-hydroxy-	5.78	20	Mixed constant	05	Uncert.	B76
5418	$C_{11}H_{10}O_3N_2$	3-Pyrrolidinecarboxamide, 2,5-dioxo-4-phenyl-	7.62	20	c = 0.0001-0.001, 4% ethanol, mixed constant	E3bg	Uncert.	F29,F27
5419	$C_{11}H_{10}O_4N_2$	2,5-Pyrrolidinedione, 3-methyl-3-(p-nitrophenyl)-	8.48	20	4% ethanol	E3bg	Uncert.	F27
5420	$C_{11}H_{11}ON$	Ethanone, 1-(2-methyl-3-indolyl)-	13.24	25	c ≈ 0.0001	07	Uncert.	T21
5421	$C_{11}H_{11}O_2N$	2,5-Pyrrolidinedione, 3-methyl-3-phenyl-	8.80	20	c = 0.0001-0.001, 4% ethanol, mixed constant	E3bg	Uncert.	F29,F27
5422	$C_{11}H_{11}O_3Cl$	Ethyl 3-(4-chlorophenyl)-3-oxopropanoate	10.17	20		05	Approx.	E26

No.	Molecular formula, name and pK value(s)	$T(^oC)$	Remarks	Method	Assessment	Ref.
5423	$C_{11}H_{11}O_3Br$ Ethyl 3-(4-bromophenyl)-3-oxopropanoate 10.14	20		05	Approx.	E26
5424	$C_{11}H_{11}O_4N$ 2-Butenoic acid, 4-(benzyloxy)amino-4-oxo- pK_1 3.8, pK_2 10.0		Conditions not stated	E3bg	Uncert.	A50
5425	$C_{11}H_{11}O_4N$ Propenoic acid, 3-(5-acetylamino-2-hydroxyphenyl)- 3.68(cis)	30	$I = 1.0$(LiCl), mixed constant	05	Approx.	H36
5426	$C_{11}H_{11}O_4Cl$ Propanedioic acid, 2-(3-chlorophenyl)-2-ethyl- pK_2 6.79 pK_2 6.83	25 33	$c = 0.005$, $I = 0.1$	E3bg	Approx.	F47
5427	$C_{11}H_{11}O_4Cl$ Propanedioic acid, 2-(4-chlorophenyl)-2-ethyl- pK_2 6.86 pK_2 6.90	25 33	$c = 0.005$, $I = 0.1$	E3bg	Approx.	F47
5428	$C_{11}H_{11}O_5N$ Ethyl 3-(4-nitrophenyl)-3-oxopropanoate 9.13	20		05	Approx.	E26
5429	$C_{11}H_{11}O_5N_3$ 2,5-Piperazinedione, 3-(4-hydroxy-3-nitrobenzyl)- 6.6(-OH) 7.3	not stated	$I \approx 0.13$ 8M urea	05	Uncert.	M42

599

No.	Formula / Name / pK	Temp.	Conditions	Method	Reliability	Ref.
5430	$C_{11}H_{11}O_6N$ 3-Azapentanedioic acid, 3-(2-carboxyphenyl)- (Iminodiacetic acid, N-(2-carboxyphenyl)-)					
	pK_1 2.20, pK_2 3.01, pK_3 7.75	22	I = 0.1(KCl)	E3bg	Uncert.	U6
	pK_1 2.20, pK_2 3.00, pK_3 7.77	25	I = 0.1(KNO$_3$), concentration constants	E3bg	Approx.	U8
	pK_1 2.33, pK_2 2.98, pK_3 7.75	20	I = 0.1(KCl), concentration constants	E3bg	Approx.	I12
	pK_1 2.33, pK_2 2.98, pK_3 7.75		I = 0.1	E		F38
5431	$C_{11}H_{11}O_6N$ Butanedioic acid, 2-methyl-2-(p-nitrophenyl)-					
	pK_1 3.28, pK_2 5.91	20		E3bg	Approx.	F28
5432	$C_{11}H_{11}O_6N$ Butanoic acid, 4-methoxy-2-(p-nitrophenyl)-4-oxo-					
	3.69	20		E3bg	Approx.	F28
5433	$C_{11}H_{11}O_6N$ Butanoic acid, 4-methoxy-3-(p-nitrophenyl)-4-oxo-					
	4.07	20		E3bg	Approx.	F28
5434	$C_{11}H_{12}ON_2$ 1H-Naphtho[1,2-d]imidazol-2(3H)-one, 6,7,8,9-tetrahydro-					
	13	20		05	Uncert.	B132
5435	$C_{11}H_{12}ON_2$ 1H-Naphtho[2,3-d]imidazol-2(3H)-one, 5,6,7,8-tetrahydro-					
	12.21	20	c = 0.000025 in 0.01M buffer	05	Approx.	B131
5436	$C_{11}H_{12}O_2N_2$ Imidazolidine-2,4-dione, 5-ethyl-5-phenyl- (Nirvanol)					
	8.5		Conditions not stated	05	Uncert.	B189
5437	$C_{11}H_{12}O_2N_2$ Propanoic acid, 2-amino-3-(3-indolyl)- (L-Tryptophan)					
	pK_2 9.28	25	I = 0.16(KNO$_3$), mixed constant	E3bg	Approx.	M72
			Thermodynamic quantities are also given			

(contd)

No.	Molecular formula, name and pK value(s)	T(°C)	Remarks	Method	Assessment	Ref.
5437 (Contd)	$C_{11}H_{12}O_2N_2$ Propanoic acid, 2-amino-3-(3-indolyl)- (L-Tryptophan)					
	pK₃ 16.82(-NH)	25	In aq. KOH, H_ scale	07	Uncert.	Y2
5438	$C_{11}H_{12}O_2N_2$ Propanoic acid, 3-(3-methyl-3H-diazirin-3-yl)-2-phenyl- (Pentanoic acid, 4-(N,N')azo-2-phenyl-)					
	3.9	25		E3bg	Uncert.	C57
5439	$C_{11}H_{12}O_3N_2$ Propanoic acid, 2-amino-3-(5-hydroxyindol-3-yl)-					
	17.95(NH)	25	In aq. KOH, H_ scale	07	V.Uncert.	Y3
5440	$C_{11}H_{12}O_3N_4$ 2,4-Pentanedione, 3-(1,2,3,4-tetrahydro-2-oxopteridin-4-yl)- (Pteridin-2(1H)-one, 4-diacetylmethyl-3,4-dihydro-)					
	8.02	20	c = 0.0002 in 0.01M buffer	05	Approx.	A27
5441	$C_{11}H_{12}O_4S$ 3-Thiapentanedioic acid, 2-methyl-4-phenyl-					
	pK₁ 2.94, pK₂ 4.75	25±0.5	c < 0.001	E3bg	Approx.	S93
5442	$C_{11}H_{12}O_4S_2$ 3,5-Dithiaheptanedioic acid, 4-phenyl- (Benzylidenebis(thioglycolic acid))					
	pK₁ 3.09, pK₂ 3.98	25	I = 0.1(K₂SO₄)	E3bg	Approx.	J5
	pK₁ 3.28		I = 0.3			
	pK₁ 3.31, pK₂ 4.13		I = 0.05			C11
	pK₂ 4.06		I = 0.1			
5443	$C_{11}H_{12}O_5N_4$ Ethyl 1,4,5,6,7,8-hexahydro-6,8-dimethyl-4,5,7-trioxopyrimidino[4,5-c]pyridazine-3-carboxylate					
	7.32	20	c = 0.001	E3bg	Approx.	P50

No.	Formula	Name	pK values	T (°C)	Conditions	Method	Assessment	Ref.
5444	$C_{11}H_{12}O_6N_2$	3-Azapentanedioic acid, 3-(2-nitrobenzyl)- (Iminodiacetic acid, N-(2-nitrobenzyl)-)	pK_1 1.99	25	$c = 0.0008$, $I = 0.1(KNO_3)$, concentration constants	E3bg	Uncert.	A62
			pK_2 8.29				Approx.	
5445	$C_{11}H_{12}O_6N_2$	3-Azapentanedioic acid, 3-(4-nitrobenzyl)- (Iminodiacetic acid, N-(4-nitrobenzyl)-)	pK_1 1.80	25	$c = 0.0008$, $I = 0.1(KNO_3)$, concentration constants	E3bg	Uncert.	A62
			pK_2 7.65				Approx.	
5446	$C_{11}H_{12}O_6N_2$	Propanoic acid, 2-acetylamino-3-(4-hydroxy-3-nitrophenyl)-	pK_2 7.0	not stated	$I \approx 1.0$	05	Uncert.	M42
			pK_2 7.0	not stated	$c = 0.00025$, $I \approx 0.9$	05	Uncert.	R24
5447	$C_{11}H_{12}O_7N_2$	3-Azapentanedioic acid, 3-(2-hydroxy-5-nitrobenzyl)- (Iminodiacetic acid, N-(2-hydroxy-5-nitrobenzyl)-)	pK_1 2.4	20	$I = 0.1$	E3bg	Uncert.	S39
			pK_2 6.18, pK_3 10.22				Approx.	
5448	$C_{11}H_{12}N_2S$	1H-Naphtho[2,3-c]imidazole-2(3H)-thione, 5,6,7,8-tetrahydro-	10.43	20	$c = 0.000006$ in 0.01N buffer	05	Approx.	B131
5449	$C_{11}H_{13}O_2N$	Butanamide, 2-methyl-3-oxo-4-phenyl-	12.40	22	5% v/v ethanol, mixed constant	05/E3bg	Uncert.	H38
5450	$C_{11}H_{13}O_3N$	Benzyl 2-amino-3-oxobutanoate	pK_1 5.20, pK_2 11.0	22		E3bg	Uncert.	L7

No.	Molecular formula, name and pK value(s)	T(°C)	Remarks	Method	Assessment	Ref.
5451	$C_{11}H_{13}O_3N_5$ Ethyl 2-ethylamino-1,7-dihydro-7-oxopteridine-6-carboxylate (6-Pteridinecarboxylic acid ethyl ester, 2-ethylamino-1,7-dihydro-7-oxo-)					P58
	7.00		Conditions not stated		Uncert.	
5452	$C_{11}H_{13}O_3N_5$ 6-Pteridinecarboxylic acid, 2-dimethylamino-8-ethyl-7,8-dihydro-7-oxo-					P58
	3.31		Conditions not stated		Uncert.	
5453	$C_{11}H_{13}O_3N_5$ 6-Pteridinecarboxylic acid, 8-ethyl-2-ethylamino-7,8-dihydro-7-oxo-					P58
	3.35		Conditions not stated		Uncert.	
5454	$C_{11}H_{13}O_4N$ 3-Azapentanedioic acid, 3-benzyl- (Iminodiacetic acid, N-benzyl-)					
	pK_1 1.49, pK_2 2.09	25	$c \approx 0.005$, I = 0.1(KCl)	E3bg	Uncert.	S66
	pK_3 8.96				Approx.	
	pK_2 2.27, pK_3 9.07	10	I = 0.1(KCl)	E3bg	Approx.	E8
	pK_2 2.32, pK_3 8.88	25				
	pK_2 1.90, pK_3 8.76	40				
	pK_2 2.24	25	$c = 0.002$, I = 0.1(KNO_3), concentration constants	E3bg	Uncert.	A62
	pK_3 8.90				Approx.	
	pK_2 2.30, pK_3 8.91	25	$c = 0.01$, I = 0.1(KCl), concentration constants	E3bg	Approx.	S97
5455	$C_{11}H_{13}O_4N$ 3-Azapentanedioic acid, 3-(o-tolyl)- (Iminodiacetic acid, N-(o-tolyl)-)					
	pK_1 2.58, pK_2 5.65	30	I = 0.1(KCl), mixed constants	E3bg	Approx.	T32

No.	Formula / Name	T(°C)	Conditions	Method	Reliability	Ref.
5456	$C_{11}H_{13}O_4N$ Benzoic acid, 3-ethoxycarbonylmethylamino-					
	pK$_1$ 1.15	25	c = 0.005	05	Approx.	S58
	pK$_2$ 4.30	25	c = 0.001	E3bg	Approx.	S58
5457	$C_{11}H_{13}O_4N$ Benzoic acid, 4-ethoxycarbonylmethylamino-					
	pK$_2$ 4.88	25	c = 0.001	E3bg	Approx.	S58
5458	$C_{11}H_{13}O_4N$ Butanedioic acid, benzylamino-					
	pK$_1$ 2.06	25	c ≈ 0.005, I = 0.1(KCl)	E3bg	Uncert.	S66
	pK$_2$ 3.34, pK$_3$ 9.07				Approx.	
5459	$C_{11}H_{13}O_4N$ Butanoic acid, 4-benzyloxyamino-4-oxo- (Succinamic acid, N-benzyloxy-)					
	pK$_1$ 4.6, pK$_2$ 10.2		Conditions not stated	E3bg	Uncert.	A50
5460	$C_{11}H_{13}O_4N$ Propanoic acid, 2-acetylamino-3-(4-hydroxyphenyl)- (N-Acetyltyrosine)					
	pK$_1$ 3.81, pK$_2$ 10.22	25	c = 0.0012	E3bg	Approx.	M69
	pK$_2$ 10.00	25	I = 0.16(KCl)	E3bg	Approx.	K41
	pK$_1$ 2.30	25		05	Uncert.	S54
5461	$C_{11}H_{13}O_5N$ 3-Azapentanedioic acid, 3-(2-hydroxybenzyl)- (Iminodiacetic acid, N-(2-hydroxybenzyl)-)					
	pK$_1$ 2.2(COOH)	20	I = 0.1(KCl)	E3bg	Uncert.	S38
	pK$_2$ 8.17(OH), pK$_3$ 11.79(NH)				Approx.	
5462	$C_{11}H_{13}O_5N$ 3-Azapentanoic acid, 3-(2-carboxyphenyl)-5-hydroxy-					
	pK$_1$ 2.41, pK$_2$ 7.73	22	I = 0.1(KCl)	E3bg	Uncert.	U6

No.	Molecular formula, name and pK value(s)	T(°C)	Remarks	Method	Assessment	Ref.
5463	$C_{11}H_{13}O_5N$ 3-Azapentanedioic acid, 3-(2-methoxyphenyl)- (Iminodiacetic acid, N-(2-methoxyphenyl)-)					
	pK$_1$ 2.69, pK$_2$ 5.58	20	I = 0.1(KCl)	E3bg	Approx.	I12
	pK$_1$ 2.69, pK$_2$ 5.58	20	I = 0.1	E		F38
5464	$C_{11}H_{13}O_5N$ Benzoic acid, 2-(formyl-N-methyloxime)-5,6-dimethoxy- (see Opianic acid, Compound No 4994)					
	2.66	25	c = 0.001	E3bg	Uncert.	K48
	Exists largely as nitrone carboxylic acid					
5465	$C_{11}H_{14}O_3N_2$ Pyrimidine-2,4,6(1H,3H,5H)-trione, 5-(1-cyclohexen-1-yl)-5-methyl- (Nor-hexobarbital)					
	7.9		Conditions not stated	05	Uncert.	B189
5466	$C_{11}H_{14}O_4N_2$ 3-Azapentanedioic acid, 3-(6-methylpyridin-2-yl)methyl- (Iminodiacetic acid, N-(6-methylpyridin-2-yl)methyl-)					
	pK$_1$ 3.46, pK$_2$ 8.30	25	I = 0.1(KNO$_3$), concentration constants	E3bg	Approx.	T27
5467	$C_{11}H_{14}O_4N_2$ Propanoic acid, 2-(3-hydroxy-5-hydroxymethyl-2-methylpyridin-4-yl)methyleneamino- (Pyridoxylidene-DL-alanine)					
	pK$_1$ 6.26, pK$_2$ 9.91	50	Solutions not buffered	05	Uncert.	N1
5468	$C_{11}H_{14}O_8N_2$ Acetic acid, 2-(uridin-5-yl)-					
	pK$_1$ 4.23, pK$_2$ 9.83		Conditions not stated	05	Uncert.	F22
5469	$C_{11}H_{15}ON$ Phenol, 2-butyliminomethyl-					
	12.0	25		05	Uncert.	G40
5470	$C_{11}H_{15}ON$ Phenol, 2-(tert-butyliminomethyl)-					
	13.0(OH)	25		05	Uncert.	G43

| 5471 | $C_{11}H_{15}O_2N$ | Butanohydroxamic acid, N-(p-tolyl)- | | | | |
| | | 9.02 | 25 | Extrapolated from results in dioxane-water mixtures | E3bg | Uncert. | S67 |

5472	$C_{11}H_{15}O_3N$	1,2-Benzenediol, 3-(morpholino)methyl-				
		pK$_1$ 6.75, pK$_2$ 9.66	31.5	c = 0.008, I = 0.2, mixed constants	E3bg	Approx.	K31
		pK$_1$ 6.7, pK$_2$ 9.6			05	Uncert.

| 5473 | $C_{11}H_{15}O_3N_5$ | 6-Pteridinecarboxylic acid, 2-dimethylamino-8-ethyl-5,8-dihydro-7-hydroxy- | | | | |
| | | 7.25 | | Conditions not stated | | Uncert. | P58 |

| 5474 | $C_{11}H_{15}O_3N_5$ | 6-Pteridinecarboxylic acid, 8-ethyl-2-ethylamino-5,8-dihydro-7-hydroxy- | | | | |
| | | 7.14 | | Conditions not stated | | Uncert. | P58 |

| 5475 | $C_{11}H_{15}O_4N_5$ | Adenosine, N(1)-methyl- | | | | |
| | | 8.8 | 22 | | E3bg | Uncert. | M53 |

| 5476 | $C_{11}H_{15}O_4N_5$ | Adenosine, N(6)-methyl- | | | | |
| | | 4.01 | 22 | | E3bg | Approx. | M53 |

| 5477 | $C_{11}H_{15}O_5N$ | Methyl 2-dimethylaminomethyl-3,4,5-trihydroxybenzoate | | | | |
| | | 6.80 | 25 | I = 0.1(KNO$_3$), mixed constant | E3bg | Approx. | E19 |

5478	$C_{11}H_{15}O_5N_5$	Diethyl 2-(4,5,6,7-tetrahydro-5-oxo-1H-1,2,3-triazolo[4,5-d]pyrimidin-7-yl)propanedioate				
		pK$_1$ -1.66	20	c = 0.00002	05	Uncert.	A33
		pK$_2$ 7.73				Approx

No.	Molecular formula, name and pK value(s)	T(°C)	Remarks	Method	Assessment	Ref.
5479	$C_{11}H_{15}O_5N_5$ Guanosine, 7-methyl-					
	6.7	not stated	I = 0.02	05	Uncert.	H29
5480	$C_{11}H_{15}O_5P$ Benzoic acid, 3-diethoxyphosphinyl-			E3bg	Uncert.	T61
	3.73	25	No details			
			Value in mixed solvent is also given			
5481	$C_{11}H_{15}O_5P$ Benzoic acid, 4-diethoxyphosphinyl-			E3bg	Uncert.	T61
	3.63	25	No details			
			Value in mixed solvent is also given			
5482	$C_{11}H_{16}O_2S$ Phenol, 2-tert-butyl-6-methylsulfinyl-			E3bg	Approx.	F25
	9.50	25				
5483	$C_{11}H_{16}O_3N_2$ Pyrimidine-2,4,6(1$\underline{H}$,3$\underline{H}$,5$\underline{H}$)-trione, 5-allyl-5-isobutyl-			E3ag	Approx.	K53
	7.68	25				
5484	$C_{11}H_{16}O_3N_2$ Pyrimidine-2,4,6(1$\underline{H}$,3$\underline{H}$,5$\underline{H}$)-trione, 5-cyclopentyl-5-ethyl-			E3ag	Approx.	K53
	8.09	25				
5485	$C_{11}H_{16}O_3N_2$ Pyrimidine-2,4,6(1$\underline{H}$,3$\underline{H}$,5$\underline{H}$)-trione, 5-ethyl-5-(1-methyl-2-butenyl)-			E3ag	Approx.	K53
	8.00	25				

No.	Formula	Name	Temp (°C)	pK	Conditions	Method	Reliability	Ref.
5486	$C_{11}H_{16}O_3N_2$	Pyrimidine-2,4,6(1$\underline{H}$,3$\underline{H}$,5$\underline{H}$)-trione, 5-(2-methylallyl)-5-propyl-	25	7.82		E3ag	Approx.	K53
5487	$C_{11}H_{16}O_6N_2$	Uridine, 5-ethyl-	20	9.86		05	Uncert.	S138
5488	$C_{11}H_{17}ON$	Phenol, 2-(dimethylaminomethyl)-6-ethyl-	25	pK$_1$ 8.02	I = 0.1(KNO$_3$), mixed constant	E3bg	Approx.	E19
5489	$C_{11}H_{17}O_3N_3$	Pyrimidine-2,4,6(1$\underline{H}$,3$\underline{H}$,5$\underline{H}$)-trione, 5-ethyl-5-piperidyl-	25	7.71		E3ag	Approx.	K53
5490	$C_{11}H_{17}O_5N_3$	Cytidine, N(4),N(4)-dimethyl-	22	pK$_1$ 3.70		05	Uncert.	S141
5491	$C_{11}H_{17}O_5N_3$	Cytidine, N(4),O(2')-dimethyl-		pK$_1$ 3.9	Conditions not stated	05	Uncert.	N29
5492	$C_{11}H_{17}O_6N$	3-Azapentanedioic acid, 3-(1-carboxycyclohexyl)- (Iminodiacetic acid, N-(1-carboxycyclohexyl)-)	20	pK$_1$ 1.63 pK$_2$ 2.59, pK$_3$ 11.24	I = 0.1(KCl), concentration constants	E3bg	Uncert. Approx.	I14
5493	$C_{11}H_{18}O_3N_2$	Pyrimidine-2,4,6(1$\underline{H}$,3$\underline{H}$,5$\underline{H}$)-trione, 5-ethyl-5-(1-ethylpropyl)-	25	6.01		E3ag	Approx.	K53

No.	Molecular formula, name and pK value(s)	T(°C)	Remarks	Method	Assessment	Ref.
5494	C$_{11}$H$_{18}$O$_3$N$_2$ Pyrimidine-2,4,6(1H,3H,5H)-trione, 5-ethyl-5-isopentyl- (Amylobarbitone)					
	7.958	25		05-:	Approx.	B124
			Variation with temperature			
			15 20 30 35 40 45 50			
			8.088 8.027 7.893 7.819 7.746 7.672 7.594			
			Thermodynamic quantities are derived from the results			
	7.94	25		E3ag	Approx.	K53
			Variation with temperature			
			15 20 30 35 40			
			8.19 8.06 7.82 7.70 7.59			
			Thermodynamic quantities are derived from the results			
5495	C$_{11}$H$_{18}$O$_3$N$_2$ Pyrimidine-2,4,6(1H,3H,5H)-trione, 5-ethyl-5-(1-methylbutyl)-					
	8.11	25		E3ag	Approx.	K53
5496	C$_{11}$H$_{18}$O$_3$N$_2$ Pyrimidine-2,4,6(1H,3H,5H)-trione, 5-ethyl-5-pentyl-					
	7.95	25		E3ag	Approx.	K53
5497	C$_{11}$H$_{18}$O$_4$N$_2$ 3,7-Diazabicyclo[3.3.1]nonane-3,7-diacetic acid (Bispidine-N,N'-diacetic acid)					
	pK$_1$ 4.19, pK$_2$ 11.93	not stated	c = 0.005, I = 0.1(KCl), mixed constant	E3bg	Uncert.	S112

5498 $C_{11}H_{18}O_8N_2$ 3,7-Diazanonanedioic acid, 3,7-bis(carboxymethyl)- (Trimethylenedinitrilotetraacetic acid)

T	Conditions	pK values	Method	Ref.	Remark
20	$I = 0.1(KCl)$, concentration constants	pK_1 2.0, pK_2 2.67, pK_3 7.90, pK_4 10.27	E3bg	S50	Uncert. Approx.
20	$I = 0.1(KNO_3)$	pK_1 1.88, pK_2 2.47, pK_3 8.02, pK_4 10.46	E3bg	A54	Uncert. Approx
	Thermodynamic quantities are also given				
20	$I = 1.0((CH_3)_4NCl)$, concentration constants	pK_1 2.3, pK_2 2.55, pK_3 7.80, pK_4 10.23	E3bg	A55	Uncert. Approx.
	Values in other salt solutions are also given				

5499 $C_{11}H_{18}O_8N_2$ 3,6-Diazaoctanedioic acid, 3,6-bis(carboxymethyl)-4-methyl- (Propylenedinitrilotetraacetic acid)

T	Conditions	pK values	Method	Ref.	Remark
30	$I = 0.1(KCl)$, concentration constants	pK_1 2.60, pK_2 3.03, pK_3 6.20, pK_4 10.84	E3bg,R3g	G48	Approx.

5500 $C_{11}H_{18}O_9N_2$ 3,7-Diazanonanedioic acid, 3,7-bis(carboxymethyl)-4-hydroxy- ((2-Hydroxytrimethylenedinitrilo)tetraacetic acid)

T	Conditions	pK values	Method	Ref.	Remark
30	$I = 0.1(KCl)$, concentration constants	pK_1 ≈2, pK_2 3.36, pK_3 6.85, pK_4 9.70	E3bg,R3g	G48	Uncert. Approx.
25	$c ≈ 0.001\text{-}0.002$, $I = 0.1(KNO_3)$, concentration constants	pK_1 ≈1.6, pK_2 2.60, pK_3 6.96, pK_4 9.49	E3bg	T29	Uncert. Approx.
20	$I = 0.1(KCl)$, mixed constants	pK_1 1.85, pK_2 2.90, pK_3 7.10, pK_4 9.70	E3bh	P71	Approx.
20	$I = 0.1(KCl)$, mixed constants	pK_1 2.3, pK_2 2.6, pK_3 7.0, pK_4 9.6	E3bg	D72	Uncert.
20	$I = 0.1(KNO_3)$	pK_1 2.32, pK_2 2.58, pK_3 7.00, pK_4 9.60		D73	

No.	Molecular formula, name and pK value(s)	T(°C)	Remarks	Method	Assessment	Ref.
5501	$C_{11}H_{19}O_2N$ Bicyclo[2.2.2]octane-1-carboxylic acid, 4-dimethylamino- 4.083	25	Values in mixed solvents are also given	E3bg	Approx.	W29
5502	$C_{11}H_{19}O_3N_3$ 1,3,5-Triazine-2,4,6(1H,3H,5H)-trione, 1,3-dibutyl- 7.82		Conditions not stated		Uncert.	A73
5503	$C_{11}H_{19}NS_2$ Methanedithioic acid, N,N-dicyclopentylamino- (Dithiocarbamic acid, N,N-dicyclopentyl-) 9.71		c = 0.1	E		T65
5504	$C_{11}H_{20}O_6N_2$ 4-Azanonane-1,3,9-tricarboxylic acid, 9-amino- (L-Saccharopine) pK₂ 2.6, pK₃ 4.1, pK₄ 9.2, pK₅ 10.3		Conditions not stated	E3bg	V.Uncert.	D10
5505	$C_{11}H_{22}O_5S$ Undecanoic acid, 11-sulfo- pK₂ 5.21	25	c ≈ 0.015-0.02, I ≈ 0.06-0.07, mixed constant	E3bg	Approx.	B58
5506	$C_{11}H_{23}O_4P$ Propanoic acid, 2-(dibutylphosphinyl)-2-hydroxy- pK₁ 2.94, pK₂ 11.6	25	c = 0.01, I = 0.1	E3bg	Uncert.	T53
5507	$C_{11}H_{23}O_5P$ Undecanoic acid, 11-phosphono- pK₁ 5.00(COOH), pK₂ 8.25	not stated	c = 0.04	E3bg	Uncert.	C38

No.	Formula	Name / pK	Temp	Conditions	Method	Reliability	Ref.
5508	$C_{11}H_5O_2N_3Br_4$	1,3-Benzenediol, 2,4-dibromo-6-(3,5-dibromo-2-pyridylazo)-					
		pK_2 12.20	not stated	$I = 0.1(NaClO_4)$, mixed constant	05	Uncert.	B185
5509	$C_{11}H_7O_2N_3Br_2$	1,3-Benzenediol, 2,4-dibromo-6-(2-pyridylazo)-					
		pK_2 12.20	not stated	$I = 0.1(NaClO_4)$, mixed constant	05	Uncert.	B185
5510	$C_{11}H_7O_2N_3Br_2$	1,3-Benzenediol, 4-(3,5-dibromo-2-pyridylazo)-					
		pK_2 12.10	not stated	$I = 0.1(NaClO_4)$, mixed constant	05	Uncert.	B185
5511	$C_{11}H_8ONCl$	Pyridin-2(1H)-one, 6-chloro-4-phenyl-					
		pK_1 -0.10	20	$c = 0.00005$	06	Uncert.	S74
		pK_2 7.32			05	Approx.	
5512	$C_{11}H_8O_2N_3Br$	1,3-Benzenediol, 4-(5-bromo-2-pyridylazo)-					
		pK_2 11.50	not stated	$I = 0.1(NaClO_4)$, mixed constant	05	Uncert.	B185
		pK_2 12.01	not stated	Mixed constant	05	Uncert.	B182
5513	$C_{11}H_9O_2NS$	2-Thiophenecarbohydroxamic acid, N-phenyl-					
		7.54	25	$c \approx 0.001$, $I = 0.1(KCl)$	E3bg	Approx.	F31
		7.50		$I = 1.0(KCl)$			
5514	$C_{11}H_9O_2NS_2$	Thiazolidin-4-one, 5-(4-methoxybenzylidene)-2-thioxo-					
		7.76	20	$I = 0.1(NaClO_4)$	DIS	Uncert.	N17
5515	$C_{11}H_9O_5N_3S$	Benzenesulfonic acid, 4-(5-nitro-2-pyridyl)amino-					
		14.21	20	$c = 0.001$. In aq. NaOH, $H_=$ scale	07	Uncert.	H6

No.	Molecular formula, name and pK value(s)	T(°C)	Remarks	Method	Assessment	Ref.
5516	$C_{11}H_{10}O_2N_2S$ Benzenesulfonamide, N-(2-pyridyl)- 8.20(-SO$_2$NH-)	20	c = 0.0025, I = 0.1(KCl), mixed constant	E3bg	Approx.	W34
5517	$C_{11}H_{11}ON_2(I)$ 1-Methylquinolinium iodide, 4-hydroxyiminomethyl- 8.3	not stated	c ≃ 0.04	E3bg	Uncert.	W36
5518	$C_{11}H_{11}O_2N_3S$ Benzenesulfonamide, 4-amino-N(S)-(2-pyridyl)- 8.48(-SO$_2$NH-)	20	c = 0.0025, I = 0.1(KCl), mixed constant	E3bg	Approx.	W34
5519	$C_{11}H_{11}O_4NI_2$ Propanoic acid, 2-acetylamino-3-(4-hydroxy-3,5-diiodophenyl)- pK$_1$ 4.13, pK$_2$ 7.12	25	c = 0.0012	E3bg	Approx.	M69
5520	$C_{11}H_{12}O_2N_4S$ Benzenesulfonamide, 4-amino-N(S)-(4-methylpyrimidin-2-yl)- 6.84(-SO$_2$NH-)	20	c = 0.0025, I = 0.1(KCl), mixed constant	E3bg	Approx.	W34
5521	$C_{11}H_{12}O_4NI$ Propanoic acid, 2-acetylamino-3-(4-hydroxy-3-iodophenyl)- pK$_1$ 3.19, pK$_2$ 8.83	25	c = 0.0012	E3bg	Approx.	M69
5522	$C_{11}H_{15}O_3N_2Br$ Pyrimidine-2,4,6(1H,3H,5H)-trione, 5-(2-bromoallyl)-5-(1-methylpropyl)- 7.71	25		E3ag	Approx.	K53
5523	$C_{11}H_{15}O_3N_5S$ Adenosine, 5'-methylthio- 3.6		Conditions not stated	E3bg/05	Uncert.	K34

5524 $C_{11}H_{16}O_2N(I)$ N,N,N-Trimethyl (3-carboxymethyl)anilinium iodide (Acetic acid, 2-(3-trimethylammoniophenyl)-, iodide)

3.66 25 Approx. E3ag H49

Values in mixed solvents are also given

5525 $C_{11}H_{16}O_2N(I)$ N,N,N-Trimethyl (4-carboxymethyl)anilinium iodide (Acetic acid, 2-(4-trimethylammoniophenyl)-, iodide)

3.75 25 Approx. E3ag H49

Values in mixed solvents are also given

5526 $C_{11}H_{16}O_2N_2S$ Pyrimidine-4,6(1H,5H)-dione, 5-allyl-5-butyl-2,3-dihydro-2-thioxo-

pK_1 7.27, pK_2 11.94 28 2% ethanol Uncert. 05 S14

5527 $C_{11}H_{16}O_8N_5P$ Guanosine-2'-(dihydrogen phosphate), 7-methyl-

7.0 not stated I = 0.02 Uncert. 05 H29

5528 $C_{11}H_{16}O_8N_5P$ Guanosine-3'-(dihydrogen phosphate), 7-methyl-

6.9 not stated I = 0.02 Uncert. 05 H29

5529 $C_{11}H_{16}O_8N_5P$ Guanosine-5'-(dihydrogen phosphate), 7-methyl-

7.1 not stated I = 0.02 Uncert. 05 H29

5530 $C_{11}H_{17}O_2N_2(Br)$ N,N,N-Dimethyl(hydroxycarbamoylmethyl)benzylammonium bromide (Acetohydroxamic acid, 2-benzyldimethylammonio-, bromide)

6.70 25(?) c = 0.02, I = 0.1(KCl), mixed constant Uncert. E3bg M66

5531 $C_{11}H_{17}O_{11}N_5P_2$ Guanosine-5'-(trihydrogen diphosphate), 7-methyl-

7.2 not stated I = 0.02 Uncert. 05 H29

No.	Molecular formula, name and pK value(s)	T(°C)	Remarks	Method	Assessment	Ref.
5532	$C_{11}H_{18}ON(I)$ N,N,N-Trimethyl 2-(2-hydroxyphenyl)ethylammonium iodide (Phenol, 2-(2-N,N,N-trimethylammonioethyl)-, iodide) 9.57	25	I = 0.1(KCl), mixed constant	E3bg	Approx.	E20
5533	$C_{11}H_{18}ON(I)$ N,N,N-Trimethyl 2-(3-hydroxyphenyl)ethylammonium iodide (Phenol, 3-(2-N,N,N-trimethylammonioethyl)-, iodide) 9.43	25	I = 0.1(KCl), mixed constant	E3bg	Approx.	E20
5534	$C_{11}H_{18}ON(I)$ N,N,N-Trimethyl 2-(4-hydroxyphenyl)ethylammonium iodide (Phenol, 4-(2-N,N,N-trimethylammonioethyl)-, iodide) 9.80	25	I = 0.1(KCl), mixed constant	E3bg	Approx.	E20
5535	$C_{11}H_{18}O_2N_2S$ Pyrimidine-4,6(1H,5H)-dione, 5-ethyl-2,3-dihydro-5-(1-methylbutyl)-2-thioxo- pK_1 7.45, pK_2 12.31	28	2% ethanol	05	Uncert.	S14
5536	$C_{11}H_{18}O_2N_2S$ Pyrimidine-4,6(1H,5H)-dione, 5-ethyl-5-(1-ethylpropyl)-2,3-dihydro-2-thioxo- pK_1 7.46, pK_2 12.36	28	2% ethanol	05	Uncert.	S14
5537	$C_{11}H_{18}O_{13}N_5P_3$ Guanosine-5'-(tetrahydrogen triphosphate), 2'-deoxy-7-methyl- 7.5	not stated	I = 0.02	05	Uncert.	H29
5538	$C_{11}H_{18}O_{14}N_5P_3$ Guanosine-5'-(tetrahydrogen triphosphate), 7-methyl- 7.5	not stated	I = 0.02	05	Uncert.	H29
5539	$C_{11}H_{22}O_4N(Cl)$ N,N,N-Trimethyl 2-butyryloxy-3-carboxypropylammonium chloride (Butanoic acid, 3-butyryloxy-4-trimethylammonio-, chloride) (Butylcarnitine) 3.56	25	c = 0.01	E3bg	Approx.	Y5

5540 $C_{11}H_{26}O_4N_4(Br_2)$ (3,7-Diazonia)nonanedihydroxamic acid, 3,3,7,7-tetramethyl-, dibromide
25(?)   c = 0.02, I = 0.1(KCl), mixed constant   6.20   E3bg   Uncert.   M66

5541 $C_{11}H_{28}O_6N_2P_2$ 3,9-Diazaundecane-2,10-diphosphonic acid, 2,10-dimethyl-
not stated pK_1 5.77, pK_2 6.13, pK_3 >11, pK_4 >11   E3bg   Uncert.   K7
25   c = 0.001, I = 0.1(KCl), concentration constants pK_1 5.34, pK_2 6.43   E3bg   Uncert.   D70

5542 $C_{11}H_{14}O_3NS_2As$ Acetamide, N-[2-hydroxy-5-(4-hydroxymethyl-1,3,2-dithiaarsacyclopent-2-yl)]phenyl-   (Phenol, 2-acetylamino-4-(4-hydroxymethyl-1,3,2-dithiaarsacyclopent-2-yl)-)   (Arsthinol)
Conditions not stated   9.5   E3bg   Uncert.   H44

5543 $C_{12}H_6O_{12}$ Benzenehexacarboxylic acid   (Mellitic acid)
pK_1 ≈0.8   E3bg,R3d,R4   Uncert.   W10
25   I ≈ 0.06-0.35 pK_2 2.26, pK_3 3.52, pK_4 5.15, pK_5 6.52, pK_6 7.71   Approx.

5544 $C_{12}H_8O_4$ 4H-Pyran-2-carboxylic acid, 4-oxo-6-phenyl-
25   1.51   C1   Approx.   S10

5545 $C_{12}H_8O_8$ Cyclooctatetraene-1,2,4,6-tetracarboxylic acid
Conditions not stated pK_1 3.40, pK_2 3.93, pK_3 4.72, pK_4 5.98   E3bg   Uncert.   L22

No.	Molecular formula, name and pK value(s)	T(°C)	Remarks	Method	Assessment	Ref.
5546	$C_{12}H_8O_8$ Cyclooctatetraene-1,3,5,7-tetracarboxylic acid					
	pK_1 4.5		Conditions not stated	E3bg	Uncert.	L22
5547	$C_{12}H_{10}O$ Biphenyl-2-ol					
	10.01	20	c = 0.0001, mixed constant	05	Approx.	M128
	11.27			E		D49
5548	$C_{12}H_{10}O_2$ Acetic acid, 2-(1-naphthyl)-					
	4.23	25	c = 0.002	E3bg	Approx.	B154
	4.22	20	I = 0.0005	E3bg	Approx.	P31
	4.04		I = 1.0(KNO_3)			
5549	$C_{12}H_{10}O_2$ Acetic acid, 2-(2-naphthyl)-					
	4.30	25	c = 0.002	E3bg	Approx.	B154
5550	$C_{12}H_{10}O_2$ Biphenyl-2,2'-diol					
	pK_1 7.56	20	c = 0.02, mixed constants	E3bg	Approx.	M128
	pK_2 11.80				Uncert.	
	pK_1 7.56(H_2O)	25	c ≈ 0.0001	05	Approx.	G31
	pK_1 8.36(D_2O)					
5551	$C_{12}H_{10}O_2$ 1-Naphthoic acid, 2-methyl-					
	3.034	25		05	Approx.	B125

(contd)

5551
(Contd) $C_{12}H_{10}O_2$ 1-Naphthoic acid, 2-methyl-

Variation with temperature

15	20	30	35	40
2.776	2.854	3.227	3.334	3.409

Thermodynamic quantities are derived from the results

5552	$C_{12}H_{10}O_2$	1-Naphthoic acid, 3-methyl-						
		3.72	25		05	Uncert.	C79	
5553	$C_{12}H_{10}O_3$	Methyl 3-hydroxy-1-naphthocarboxylate						
		8.88	25	$c \approx 0.001$	05	Approx.	B155	
5554	$C_{12}H_{10}O_3$	Methyl 4-hydroxy-2-naphthocarboxylate						
		8.75	25	$c \approx 0.001$	05	Approx.	B155	
5555	$C_{12}H_{10}O_3$	1-Naphthoic acid, 4-methoxy-						
		4.31	25		05	Uncert.	C79	
5556	$C_{12}H_{10}O_3$	2-Naphthoic acid, 3-methoxy-						
		pK_1 3.8	25	$c = 0.00004$ to 0.00007	C1	Uncert.	B123	
5557	$C_{12}H_{10}O_4$	2,2',4,4'-Biphenyltetrol						
		pK_1 7.44, pK_2 10.10	20	$c = 0.02$, mixed constants	E3bg	Approx.	M128	
5558	$C_{12}H_{10}O_4$	2H-Chromen-2-one, 8-acetyl-7-hydroxy-4-methyl-						
		7.17	25	$c = 0.0001$	05	Approx.	M63	

No.	Molecular formula, name and pK value(s)	T(°C)	Remarks	Method	Assessment	Ref.
5559	Acetic acid, 2-(7-hydroxy-4-methyl-2-oxo-2H-chromen-6-yl)- pK_2 8.09	25	c = 0.0001	O5	Approx.	M63
5560	2H-Chromen-2-one, 6-ethyl-7-hydroxy-4-methyl- 8.01	25	c = 0.0001	O5	Approx.	M63
5561	$C_{12}H_{12}O_3$ Dodeca-5,6-diene-8,10-diynoic acid, 4-hydroxy- (Odyssic acid) 4.90		Conditions not stated	O5(?)	Uncert.	B169c
5562	$C_{12}H_{12}O_4$ 1,2-Cyclopropanedicarboxylic acid, 1-methyl-2-phenyl- pK_1 3.84, pK_2 6.28(cis) pK_1 3.61, pK_2 5.33(trans)	20		E3bg	Approx.	M12
5563	$C_{12}H_{14}O_2$ Benzoic acid, 4-cyclopentyl- 4.21		Conditions not stated	O5	Uncert.	S62
5564	$C_{12}H_{14}O_3$ 2-Butenoic acid, 3-(2-hydroxy-3,5-dimethylphenyl)- 4.54(cis)	30	Estimated value of mixed constant		Uncert.	H37
5565	Ethyl 3-oxo-2-phenylbutanoate 10.43 10.55	25.7 20.4		O5 E3bg	Uncert.	R53
5566	$C_{12}H_{14}O_3$ Ethyl 3-oxo-3-(p-tolyl)propanoate 10.73	20		O5	Approx.	E26

No.	Formula	Name / pK	T (°C)	Conditions	Method		Ref.
5567	$C_{12}H_{14}O_4$	Butanedioic acid, 2-benzyl-2-methyl- pK_1 4.14, pK_2 6.49	20		E3bg	Approx.	F28
5568	$C_{12}H_{14}O_4$	Butanedioic acid, 2-ethyl-2-phenyl- pK_1 3.47, pK_2 6.52	20		E3bg	Approx.	F28
5569	$C_{12}H_{14}O_4$	Butanoic acid, 3-methoxycarbonyl-3-phenyl- 4.46	20		E3bg	Approx.	F28
5570	$C_{12}H_{14}O_4$	Butanoic acid, 4-methoxy-2-methyl-4-oxo-2-phenyl- 4.15	20		E3bg	Approx.	F28
5571	$C_{12}H_{14}O_4$	Ethyl 3-(p-methoxyphenyl)-3-oxopropanoate 11.12	20		05	Approx.	E26
5572	$C_{12}H_{14}O_4$	Pentanedioic acid, 3-methyl-3-phenyl- pK_1 4.12, pK_2 6.17	25	$c = 0.00125$, mixed constants	E3bg	Approx.	B146
5573	$C_{12}H_{14}O_4$	Propanedioic acid, 2-ethyl-2-(3-methylphenyl)- pK_2 7.21 pK_2 7.24	25 33	$c = 0.005$, $I = 0.1$	E3bg	Approx.	F47
5574	$C_{12}H_{14}O_4$	Propanedioic acid, 2-ethyl-2-(4-methylphenyl)- pK_2 7.26 pK_2 7.29	25 33	$c = 0.005$, $I = 0.1$	E3bg	Approx.	F47

No.	Molecular formula, name and pK value(s)	$T(^{o}C)$	Remarks	Method	Assessment	Ref.
5575	$C_{12}H_{14}O_5$ Propanedioic acid, 2-ethyl-2-(4-methoxyphenyl)-					F47
	pK_2 7.24	25	c = 0.005, I = 0.1	E3bg	Approx.	
	pK_2 7.27	33				
5576	$C_{12}H_{14}O_8$ 3-Cyclooctene-1,2,5,7-tetracarboxylic acid					L22
	pK_1 3.4, pK_2 4.3, pK_3 5.3, pK_4 6.8		Conditions not stated	E3bg	V.Uncert.	
5577	$C_{12}H_{16}O_2$ Hexanoic acid, 5-phenyl-					C57
	4.8	25		E3bg	Uncert.	
5578	$C_{12}H_{16}O_3$ Acetic acid, 2-(4-butylphenyl)-2-hydroxy-					K36
	3.58	25	c = 0.01, mixed constant	E3bg	Uncert.	
5579	$C_{12}H_{16}O_4$ 4<u>H</u>-Pyran-4-one, 3-butyryl-2-hydroxy-6-propyl-					A37
	5.47	20	c = 0.001-0.0025	E3bg	Approx.	
5580	$C_{12}H_{16}O_8$ Cyclooctane-1,2,4,6-tetracarboxylic acid					L22
	pK_1 3.5, pK_2 4.45, pK_3 5.2, pK_4 6.35		Conditions not stated	E3bg	V.Uncert.	
5581	$C_{12}H_{18}O$ Phenol, 4-butyl-2,5-dimethyl-					D30
	10.74	20		05	Approx.	
5582	$C_{12}H_{18}O$ Phenol, 2-tert-butyl-4,6-dimethyl-					D30
	12.04	20		05	Approx.	

No.	Formula	Name / pK	Temp.	Notes	Method		Ref.
5583	$C_{12}H_{18}O$	Phenol, 2,6-diisopropyl- 11.10	20		05	Approx.	D30
5584	$C_{12}H_{18}O$	Phenol, 2,4-dipropyl- 10.91	20		05	Approx.	D30
5585	$C_{12}H_{18}O$	Phenol, 2,6-dipropyl- 11.25	20		05	Approx.	D30
5586	$C_{12}H_{18}O$	Phenol, 2-ethyl-4-isopropyl-3-methyl- 10.85	20	2% ethanol	05	Approx.	D29
5587	$C_{12}H_{18}O$	Phenol, 4-ethyl-2-isopropyl-5-methyl- 10.91	20	2% ethanol	05	Approx.	D29
5588	$C_{12}H_{18}O$	Phenol, 2-hexyl- 10.66	20	Extrapolated value		Uncert.	D30
5589	$C_{12}H_{18}O_5$	Oxybis(1-cyclopentanecarboxylic acid) pK_1 3.30, pK_2 6.00	20	c = 0.005-0.01	E3bg,R3c	Approx.	K47
5590	$C_{12}H_{20}O_4$	1,2-Cyclohexanedicarboxylic acid, 4-tert-butyl- pK_1 4.29, pK_2 6.08(trans)	25	c ≈ 0.0005, mixed constants	E3bg,R3d	Approx.	S68
5591	$C_{12}H_{22}O_4$	Butanedioic acid, 2,3-di-tert-butyl- pK_1 2.20, pK_2 10.25(H_2O) (racemic) pK_1 2.77, pK_2 10.92(D_2O)	25		E3bg	Approx.	G22

No.	Molecular formula, name and pK value(s)	T(^{o}C)	Remarks	Method	Assessment	Ref.
5592	$C_{12}H_{22}O_4$ Butanedioic acid, tetraethyl- pK$_1$ 3.39, pK$_2$ 8.06(H_2O) pK$_1$ 4.29, pK$_2$ 8.50(D_2O)	25		E3bg	Approx.	G22
5593	$C_{12}H_{24}O_2$ Dodecanoic acid (Lauric acid) 5.3	20	Titration of Na-salt below c.m.c.	E3bg	Uncert.	N42
5594	$C_{12}H_{24}N_2$ 2,2'-Bipyrrolidine, 5,5,5',5'-tetramethyl- pK$_1$ 6.95, pK$_2$ 10.3		Conditions not stated	E3bg	Uncert.	B144
5595	$C_{12}H_{24}N_2$ 2,3'-Bipyrrolidine, 5,5,5',5'-tetramethyl- 9.2		Conditions not stated	E3bg	Uncert.	B144
5596	$C_{12}H_8O_3N_4$ Pteridine-2,4,7(1H,3H,8H)-trione, 1-phenyl- pK$_1$ 2.95, pK$_2$ 9.46	20	c = 0.001	E3bg	Approx.	P47
5597	$C_{12}H_8O_3S$ 4H-1-Thiapyran-2-carboxylic acid, 4-oxo-6-phenyl- 2.59	25		Cl	Uncert.	S11
5598	$C_{12}H_8O_3S_2$ Propanoic acid, 2-oxo-3-(5-phenyl-1,2-dithiacyclopent-4-en-1-ylidene)- 2.75	25		Cl	Uncert.	S11
5599	$C_{12}H_8O_5N_2$ Biphenyl-2-ol, 3,5-dinitro- 3.848	25	c ≃ 0.00002	O5a	Approx.	R30

No.	Formula	Name	pK	T (°C)	Conditions			Ref.
5600	$C_{12}H_9OCl$	Biphenyl-3-ol, 4-chloro-	8.07	25	$c \approx 0.0001\text{-}0.0002$, $I \approx 0.1$, 0.5% ethanol, mixed constant	05	Approx.	S76
5601	$C_{12}H_9OBr$	Biphenyl-2-ol, 4'-bromo-	10.96			E		D49
5602	$C_{12}H_9OI$	Biphenyl-2-ol, 4'-iodo-	10.92			E		D49
5603	$C_{12}H_9O_2N$	2,5-Cyclohexadien-1-one, 4-(4-hydroxyphenyl)imino-	8.1	not stated	$c = 0.00004$, 2% dioxane, mixed constant	05	Uncert.	K54
5604	$C_{12}H_9O_3N$	Biphenyl-2-ol, 4'-nitro-	10.42			E		D49
5605	$C_{12}H_9O_3N$	Biphenyl-3-ol, 4-nitro-	6.74	25	$c \approx 0.00001\text{-}0.0001$, $I \approx 0.1$, mixed constant	05	Approx.	H11
5606	$C_{12}H_9O_3N$	Biphenyl-4-ol, 3-nitro-	6.73	25	$c = 0.0002$, $I \approx 0.5$, mixed constant	05	Approx.	R15
5607	$C_{12}H_9O_4N_3$	Benzoic acid, 2-(5-nitro-2-pyridyl)amino-	15.46(-NH)	20	$c = 0.001$, In aq. NaOH, $H_=$ scale	07	Uncert.	H6

No.	Molecular formula, name and pK value(s)	T(°C)	Remarks	Method	Assessment	Ref.
5608	$C_{12}H_9O_4N_3$ Benzoic acid, 3-(5-nitro-2-pyridyl)amino- pK_2 14.59(-NH)	20	c = 0.001 In aq. NaOH, $H_=$ scale	07		H6
5609	$C_{12}H_9O_4N_3$ Benzoic acid, 4-(5-nitro-2-pyridyl)amino- pK_2 14.42(-NH)	20	c = 0.001 In aq. NaOH, $H_=$ scale	07	Uncert.	H6
5610	No entry					
5611	$C_{12}H_{10}O_2N_2$ 1,2-Benzenediol, 4-phenylazo- (Azobenzene-3,4-diol) pK_1 7.50, pK_2 12.46	20±0.5	c = 0.00004, I = 0.1, (May contain 20% ethanol)	05	Uncert.	N27
5612	$C_{12}H_{10}O_2N_2$ Phenol, 2-(2-hydroxyphenylazo)- (Azobenzene-2,2'-diol) pK_1 7.8, pK_2 11.5	not stated	I = 0.1(KCl)	05	Uncert.	D37
5613	$C_{12}H_{10}O_2N_2$ 3-Pyridinecarbohydroxamic acid, N-phenyl- 8.00	not stated	I = 0.2(NaCl)	E3bg	Uncert.	C72
5614	$C_{12}H_{10}O_2N_2$ 2,5-Pyrrolidinedione, 3-cyano-4-methyl-4-phenyl- 6.20	20	c = 0.0001-0.001, 4% ethanol, mixed constant	E3bg	Uncert.	F29,F27

No.	Compound	pK	T(°C)	Conditions	Method	Reliability	Ref.
5615	$C_{12}H_{10}O_2N_4$ Purine-2,6(1H,3H)-dione, 9-benzyl-	5.81	20	c = 0.001	E3bg	Approx.	P53
5616	$C_{12}H_{10}O_3N_2$ 1,3-Benzenediol, 4-(2-hydroxyphenylazo)-	pK_1 6.6, pK_2 8.7, pK_3 12.2	not stated	I = 0.1(KCl)	05	Uncert.	D37
		pK_1 6.60, pK_2 8.80	25±1	c = 0.00002, I = 0.1 (May contain ethanol)	05	Uncert.	N26
5617	$C_{12}H_{11}ON$ Biphenyl-2-ol, 4'-amino-	11.48(OH)			E		D49
5618	$C_{12}H_{11}ON_3$ Triazen-3-ol, 1,3-diphenyl-	11.41	25	c = 0.00004, I = 0.1(KCl), mixed constant	05	Approx.	P91,P89
		11.55	25	c = 0.001 (Solutions unstable)	05	Uncert.	R13
5619	$C_{12}H_{11}O_2N$ Acetamide, N-(3-hydroxy-1-naphthyl)-	9.31	25	c ≈ 0.001	E3bg	Approx.	B155
5620	$C_{12}H_{11}O_2N$ Acetamide, N-(4-hydroxy-2-naphthyl)-	8.87	25	c ≈ 0.001	E3bg	Approx.	B155
5621	$C_{12}H_{11}O_2N_3$ 1,3-Benzenediol, 4-(4-methyl-2-pyridylazo)-	11.38	not stated	Mixed constant for ortho-OH	05	Uncert.	B182
5622	$C_{12}H_{11}O_2N_3$ 1,3-Benzenediol, 4-(5-methyl-2-pyridylazo)-	11.38	not stated	Mixed constant for ortho-OH	05	Uncert.	B182

No.	Molecular formula, name and pK value(s)	$T(^{\circ}C)$	Remarks	Method	Assessment	Ref.
5623	$C_{12}H_{11}O_2P$ Diphenylphosphinic acid					
	1.72	20 ± 2	$c \simeq 0.001$ or 0.0001, $I \approx 0.1$ or $1(NaClO_4)$, concentration constant	DIS	Uncert.	K55
	2.32	not stated	$c = 0.005$, $I \not> 0.025$, mixed constant, 7% ethanol	E3bg	Uncert.	M57
5624	$C_{12}H_{11}O_3N$ 2-Furancarbohydroxamic acid, N-(2-methylphenyl)-					
	7.77	30	$c = 0.01$, $I = 0.1(NaClO_4)$, concentration constant (Position on furan ring not given)	E3bg	Uncert.	D62
5625	$C_{12}H_{11}O_3P$ Phenyl hydrogen phenylphosphonate					
	0.68	20 ± 2	$c \simeq 0.001$ or 0.0001, $I \approx 0.1$ or $1(NaClO_4)$, concentration constant	DIS	Uncert.	K55
5626	$C_{12}H_{11}O_4N$ Ethyl 2,5-dihydro-5-oxo-3-phenylisoxazole-4-carboxylate					
	1.86	25	$I = 1(NaClO_4)$, mixed constant	E3bg	Uncert.	D7
5627	$C_{12}H_{11}O_4N$ 3-Pyrrolidinecarboxylic acid, 4-methyl-2,5-dioxo-4-phenyl-					
	3.02(COOH)	20	$c = 0.0001$-0.001, 4% ethanol, mixed constants	E3bg	Uncert.	F29,F27
	8.96(NH)					

No.	Formula	Name / pK	Temp.	Conditions	Method	Reliability	Ref.
5628	$C_{12}H_{11}O_4N_3$	Pyrimidine-2,4(1H,3H)-dione, 3-benzyl-6-methyl-5-nitro-					
		6.47	23.5		05	Uncert.	B83
5629	$C_{12}H_{11}O_4P$	Diphenyl hydrogen phosphate					
		0.26	20±2	c ≈ 0.001 or 0.0001, I ≈ 0.1 or 1(NaClO₄), concentration constant	DIS	Uncert.	K56
5630	$C_{12}H_{11}O_9N$	3-Azapentanedioic acid, 3-(2,5-dicarboxy-4-hydroxyphenyl)- (Iminodiacetic acid, N-(2,5-dicarboxy-4-hydroxyphenyl)-)					
		pK₁ 1.61	25	I = 0.1(KNO₃), concentration constants	E3bg	Uncert.	U8
		pK₂ 2.20, pK₃ 3.18, pK₄ 8.61				Approx.	
5631	$C_{12}H_{11}S_2P$	Bis(phenyl)phosphinodithioic acid					
		1.75	20	c = 0.005, 7% ethanol, mixed constant	E3bg	Uncert.	K5,K6
5632	$C_{12}H_{12}O_3N_2$	Pyrimidine-2,4,6(1H,3H,5H)-trione, 5-ethyl-5-phenyl- (Phenobarbitone)					
		7.441	25		05	Approx.	B124

Variation with temperature

15	20	30	35	40	45	50
7.594	7.516	7.375	7.314	7.254	7.184	7.129

Thermodynamic quantities are derived from the results

	pK	Temp.	Conditions	Method	Reliability	Ref.
	7.41	25		E3ag	Approx.	K53
	7.29		Conditions not stated	E3bg	Uncert.	M68
	7.30		Conditions not stated	05	Uncert.	B189

No.	Molecular formula, name and pK value(s)	T(°C)	Remarks	Method	Assessment	Ref.
5633	$C_{12}H_{12}O_3N_2$ 3-Pyrrolidinecarboxamide, 4-methyl-2,5-dioxo-4-phenyl- 7.69	20	c = 0.0001-0.001, 4% ethanol, mixed constant	E3bg	Uncert.	F29,F27
5634	$C_{12}H_{12}O_4N_2$ 2,5-Pyrrolidinedione, 3-ethyl-3-(p-nitrophenyl)- 8.49	not stated	4% ethanol	E3bg	Uncert.	F27
5635	$C_{12}H_{12}O_8N_2$ 1,4-Benzenedicarboxylic acid, 2,5-bis(carboxymethylamino)- pK$_1$ 2.8, pK$_2$ 3.45, pK$_3$ 4.40, pK$_4$ 7.00	22	c ≈ 0.003, I = 0.1(KCl), mixed constants	E3bg	Uncert.	U3
5636	$C_{12}H_{13}O_2N$ 2,5-Pyrrolidinedione, 3-benzyl-3-methyl- 9.30	20	c = 0.0001-0.001, 4% ethanol, mixed constant	E3bg	Uncert.	F29,F27
5637	$C_{12}H_{13}O_2N$ 2,5-Pyrrolidinedione, 3-ethyl-3-phenyl- 8.76	20	c = 0.0001-0.001, 4% ethanol, mixed constant	E3bg	Uncert.	F29,F27
5638	$C_{12}H_{13}O_3N$ Butanamide, 2-acetyl-3-oxo-N-phenyl- 6.81	22	5% v/v ethanol, mixed constant	O5/E3bg	Uncert.	H38
5639	$C_{12}H_{13}O_3N$ Cyclopropanecarboxylic acid, 2-(4-acetylaminophenyl)- 4.57(trans)	25	c ≈ 0.0008	E3bg	Approx.	T55

5640	$C_{12}H_{13}O_5N$	3-Azapentanedioic acid, 3-benzoylmethyl- (Iminodiacetic acid, N-benzoylmethyl-)		
	pK$_2$ 1.9			
	pK$_3$ 7.89	25	I = 0.1(KNO$_3$)	E3bg	Uncert.	A62
	pK$_1$ 1.25, pK$_2$ 1.85, pK$_3$ 8.02	25	c ≈ 0.005, I = 0.1(KCl)	E3bg	Approx. Uncert.	S66
5641	$C_{12}H_{13}O_5N$	Pentanedioic acid, 2-benzoylamino-		
	pK$_1$ 3.49, pK$_2$ 4.99	20	c = 0.01	E3bg	Uncert.	A16
5642	$C_{12}H_{13}O_5N$	3-Pyrrolidinecarboxylic acid, 1-hydroxy-4-(2-methoxyphenyl)-2-oxo-		
	pK$_1$ 2.5, pK$_2$ 7.2		Conditions not stated	E3bg	Uncert.	B102
5643	$C_{12}H_{13}O_6N$	3-Azapentanedioic acid, 3-carboxymethyl-2-phenyl-		
	pK$_1$ 1.45	20	c = 0.001, I = 0.1(KCl), concentration constants	E3bg	Uncert.	I13
	pK$_2$ 2.39, pK$_3$ 9.26				Approx.	
5644	$C_{12}H_{13}O_6N$	Butanedioic acid, 2-ethyl-2-(4-nitrophenyl)-		
	pK$_1$ 3.30, pK$_2$ 5.92	20		E3bg	Approx.	F28
5645	$C_{12}H_{13}O_6N$	Butanoic acid, 3-methoxycarbonyl-3-(4-nitrophenyl)-		
	4.27	20		E3bg	Approx.	F28
5646	$C_{12}H_{13}O_6N$	Butanoic acid, 4-methoxy-2-methyl-2-(4-nitrophenyl)-4-oxo-		
	3.70	20		E3bg	Approx.	F28

No.	Molecular formula, name and pK value(s)	T(^{o}C)	Remarks	Method	Assessment	Ref.
5647	$C_{12}H_{13}O_6N$ Butanedioic acid, 2-methyl-2-(4-nitrobenzyl)- pK_1 4.01, pK_2 6.13	20		E3bg	Approx.	F28
5648	$C_{12}H_{14}OS$ Benzo[b]thiophene-4-ol, 2,7-diethyl- 9.56	20	2% ethanol	05	Approx.	D32
5649	$C_{12}H_{14}OS$ Benzo[b]thiophene-7-ol, 4-ethyl-2,3-dimethyl- 8.99	20	2% ethanol	05	Approx.	D32
5650	$C_{12}H_{14}O_2N_2$ Propanoic acid, 2-amino-3-(4-methylindol-3-yl)- 16.90(NH)	25	In aq. KOH, H_- scale	07	Uncert.	Y3
5651	$C_{12}H_{14}O_3N_2$ Propanoic acid, 2-amino-3-(6-methoxyindol-3-yl)- 16.70(NH)	25	In aq. KOH, H_- scale	07	V.Uncert.	Y3
5652	$C_{12}H_{14}N_2S$ 1H-Naphtho[2,3-d]imidazole-2(3H)-thione, 5,6,7,8-tetrahydro-3-methyl- 10.78	20	c = 0.000006 in 0.01M buffer	05	Uncert.	B131
5653	$C_{12}H_{15}OCl$ Phenol, 2-chloro-4-cyclohexyl- 8.66	25	c ≈ 0.0001-0.0002, I ≈ 0.1, mixed constant	05	Approx.	S76
5654	$C_{12}H_{15}O_2N$ Butanamide, 2-ethyl-3-oxo-N-phenyl- 12.93	22	5% v/v ethanol, mixed constant	05/E3bg	Uncert.	H38

5655 $C_{12}H_{15}O_3N_5$ 6-Pteridinecarboxylic acid, 8-ethyl-2-ethylamino-7,8-dihydro-4-methyl-7-oxo-

3.82   Conditions not stated   Uncert.   P58

5656 $C_{12}H_{15}O_4N$ Bis(etnoxycarbonyl)pyridiniomethanide

5.56   not stated   05   Uncert.   N31

Values in mixed solvents are also given

5657 $C_{12}H_{15}O_4N$ Pentanoic acid, 5-benzyloxyamino-5-oxo-

pK$_1$ 4.7, pK$_2$ 10.2   Conditions not stated   E3bg   Uncert.   A50

5658 $C_{12}H_{15}O_5N$ 3-Azapentanedioic acid, 3-(2-hydroxy-5-methylbenzyl)-   (Iminodiacetic acid, N-(2-hydroxy-5-methylbenzyl)-)

pK$_1$ 2.46, pK$_2$ 8.30, pK$_3$ 11.04   c = 0.001   E3bg   Uncert.   T16

5659 $C_{12}H_{15}O_5N$ 3-Azapentanedioic acid, 3-(2-phenoxyethyl)-   (Iminodiacetic acid, N-(2-phenoxyethyl)-)

pK$_3$ 8.49   25   c ≈ 0.005, I = 0.1(KCl)   E3bg   Approx.   S66

5660 $C_{12}H_{15}O_6N_3$ Methyl 5-amino-2-(4-hydroxy-3-nitrobenzyl)-4-oxo-3-azapentanoate   (Glycyltyrosine methyl ester, 3-nitro-)

7.5(-OH)   not stated   8ll urea   05   Uncert.   M42

5661 $C_{12}H_{16}O_2S$ Acetic acid, 2-(4-tert-butylphenyl)thio-

3.758   25   c ≈ 0.01   E3bg   Approx.   P13

Value in mixed solvent is also given

5662 $C_{12}H_{16}O_3N_2$ Pyrimidine-2,4,6(1H,3H,5H)-trione, 5-(1-cyclohexen-1-yl)-1,5-dimethyl-   (Hexobarbital)

8.370   25   05   Approx.   B124

(contd)

No.	Molecular formula, name and pK value(s)	T(°C)	Remarks	Method	Assessment	Ref.
5662 (Contd)	$C_{12}H_{16}O_3N_2$ Pyrimidine-2,4,6(1H,3H,5H)-trione, 5-(1-cyclohexen-1-yl)-1,5-dimethyl- (Hexobarbital)		Variation with temperature			
			15 20 30 35 40 45 50			
			8.494 8.435 8.301 8.229 8.147 8.066 7.972			
			Thermodynamic quantities are derived from the results			
	8.34	25		E3ag	Approx.	K53
	8.3		Conditions not stated	O5	Uncert.	B189
5663	$C_{12}H_{16}O_3N_2$ Pyrimidine-2,4,6(1H,3H,5H)-trione, 5-(1-cyclohexen-1-yl)-5-ethyl-					
	7.50	25		E3ag	Approx.	K53
5664	$C_{12}H_{16}O_3N_6$ 3-Azapentanoic acid, 5-amino-2,6-bis(4-imidazolylmethyl)-4-oxo- (Histidylhistidine)					
	pK$_1$ 2.16	25	I = 0.1(KCl), concentration	E3bg	Uncert.	D47
	pK$_2$ 5.36, pK$_3$ 6.70, pK$_4$ 7.92		constants		Approx.	
	pK$_3$ 6.53, pK$_4$ 8.23	25	I = 0.12(KCl)	E3bg,R3f	Approx.	C39
	pK$_3$ 6.30, pK$_4$ 7.90	35	Thermodynamic quantities are			
	pK$_3$ 6.03, pK$_4$ 7.45	45	derived from the results			
5665	$C_{12}H_{16}O_3S$ Acetic acid, 2-(4-tert-butylphenyl)sulfinyl-					
	2.773	25	c ≈ 0.01	E3bg	Approx.	P13
			Value in mixed solvent is also given			
5666	$C_{12}H_{16}O_4S$ Acetic acid, 2-(4-tert-butylphenyl)sulfonyl-					
	2.598	25	c ≈ 0.01	E3bg	Approx.	P13
			Value in mixed solvent is also given			

No.	Formula / Name	pK	T (°C)	Conditions	Method	Assessment	Ref.
5667	$C_{12}H_{16}O_5N_2$ Butanoic acid, 3-hydroxy-2-(3-hydroxy-5-hydroxymethyl-2-methylpyridin-4-yl)methyleneamino- (Pyridoxylidene-threonine)	9.89	25	c = 0.0001, I = 1.0(KCl)	O5	Approx.	M111
5668	$C_{12}H_{16}O_6N_2$ 1,4-Benzenedicarboxylic acid, 2,5-bis(2-hydroxyethylamino)-	pK_1 3.77, pK_2 6.85	22	c ≈ 0.003, I = 0.1(KCl), mixed constants	E3bg	Approx.	U3
		pK_1 3.71, pK_2 6.79	22	I = 0.1(KCl)	E3bg	Uncert.	U5
5669	$C_{12}H_{16}O_6N_2$ Pseudouridine, O(2'),O(3')-isopropylidene-	pK_1 9.10, pK_2 >12.5	not stated	I = 0.05	O5	Uncert.	D53
5670	$C_{12}H_{16}O_7N_4$ Pteridine-2,4,7(1H,3H,8H)-trione, 6-methyl-8-ribityl-	pK_1 4.04	20		E3bg	Approx.	P54
		pK_2 12.9			O6	Uncert.	U5
5671	$C_{12}H_{17}O_3N_5$ 6-Pteridinecarboxylic acid, 8-ethyl-2-ethylamino-5,8-dihydro-7-hydroxy-4-methyl-	6.96		Conditions not stated		Uncert.	P58
5672	$C_{12}H_{17}O_4N_5$ Adenosine, N(6),N(6)-dimethyl-	4.50	22		E3bg	Approx.	M53
5673	$C_{12}H_{17}O_5N_5$ 3,6,9-Triazaundecanoic acid, 2-(4-imidazolylmethyl)-4,7,10-trioxo- (N-Acetylglycylglycyl-L-histidine)	pK_1 3.08, pK_2 7.18	25	I = 0.16(KCl)	E3bg	Approx.	B153
5674	$C_{12}H_{17}O_5N_5$ 3,6,9-Triazaundecanoic acid, 5-(4-imidazolylmethyl)-4,7,10-trioxo- (Acetylglycyl-L-histidylglycine)	pK_1 3.25, pK_2 6.86	25	I = 0.16(KCl)	E3bg	Approx.	B153

No.	Molecular formula, name and pK value(s)	T(°C)	Remarks	Method	Assessment	Ref.
5675	$C_{12}H_{18}O_3N_2$ Pyrimidine-2,4,6(1H,3H,5H)-trione, 5-allyl-5-(1-methylbutyl)- pK$_1$ 8.08	25		E3ag	Approx.	K53
5676	$C_{12}H_{18}O_3N_2$ Pyrimidine-2,4,6(1H,3H,5H)-trione, 5-isobutyl-5-(2-methylallyl)- pK$_1$ 7.79	25		E3ag	Approx.	K53
5677	$C_{12}H_{18}O_5N_6$ 3,6,9-Triazaundecanoic acid, 11-amino-2-(4-imidazolylmethyl)-4,7,10-trioxo- (Glycylglycylglycyl-L-histidine) pK$_1$ 3.02, pK$_2$ 6.85, pK$_3$ 8.11	25	I = 0.16(KCl)	E3bg	Approx.	B153
5678	$C_{12}H_{18}O_8N_2$ Cyclobutane-1,2-bis(N-iminodiacetic acid) (1,2-Cyclobutylenedinitrilo)tetraacetic acid) pK$_1$ 2.7, pK$_2$ 2.8, pK$_3$ 5.80(trans)					Y8
5679	$C_{12}H_{19}O_4N_3$ Cytidine, 2'-deoxy-N(4)-ethyl-5-methyl- pK$_1$ 4.05, pK$_2$ 13	20		05	Uncert.	K65
5680	$C_{12}H_{19}O_4N_3$ Cytidine, 2'-deoxy-N(4),N(4),5-trimethyl- pK$_1$ 3.92, pK$_2$ 13	20		05	Uncert.	K65
5681	$C_{12}H_{19}O_6N$ 3-Azapentanedioic acid, 3-(1-carboxycycloheptyl)- (Iminodiacetic acid, N-(1-carboxycycloheptyl)-) pK$_1$ 1.60 pK$_2$ 2.61, pK$_3$ 12.06	20	I = 0.1(KCl), concentration constants	E3bg	Uncert. Approx.	I14
5632	$C_{12}H_{20}O_2N_2$ 1,2-Benzenediol, 3,5-bis(N,N-dimethylaminomethyl)- pK$_1$ 7.22 pK$_1$ 7.3, pK$_2$ 9.7	25 31.5	I = 0.1(KNO$_3$), mixed constant I = 0.2(KCl), mixed constants	E3bg E3bg	Approx. Uncert.	E19 S33

5683 $C_{12}H_{20}O_2N_2$ 1,2-Benzenediol, 3,6-bis(N,N-dimethylaminomethyl)-

pK_1 6.35, pK_2 9.65   31.5   I = 0.2(KCl), mixed constants   E3bg   Approx.   S33

pK_1 6.39   25   I = 0.1(KNO$_3$), mixed constant   E3bg   Approx.   E19

5684 $C_{12}H_{20}O_2N_2$ Pyrazin-2(1H)-one, 6-sec-butyl-1-hydroxy-3-isobutyl-   (Aspergillic acid)

5.5   Conditions not stated   E3bg   Uncert.   D59

5685 $C_{12}H_{20}O_3N_2$ Pyrazin-2(1H)-one, 1-hydroxy-6-(1-hydroxy-1-methylpropyl)-3-isobutyl-   (Aspergillic acid, hydroxy-)

4.9   Conditions not stated   E3bg   Uncert.   D59

5686 $C_{12}H_{20}O_3N_2$ Pyrimidine-2,4,6(1H,3H,5H)trione, 5-(1,3-dimethylbutyl)-5-ethyl-

8.14   25   E3ag   Approx.   K53

5687 $C_{12}H_{20}O_3N_2$ Pyrimidine-2,4,6(1H,3H,5H)-trione, 5-ethyl-5-hexyl-

7.79   25   E3ag   Approx.   K53

5688 $C_{12}H_{20}O_8N_2$ 3,6-Diazaoctanedioic acid, 3,6-bis(carboxymethyl)-2,7-dimethyl-   (Ethylenedinitrilo-N,N'-diacetic-N,N'-di-(2-propanoic acid))

pK_1 2.29, pK_2 2.91, pK_3 6.82, pK_4 10.60   30   I = 0.1(KCl), concentration constants   E3bg,R3d   Approx.   G48

pK_1 1.90, pK_2 2.00, pK_3 6.65, pK_4 10.42   20   I = 0.1(KNO$_3$)   E   M25

5689 $C_{12}H_{20}O_8N_2$ 3,6-Diazaoctanedioic acid, 3,6-bis(carboxymethyl)-4,5-dimethyl-

pK_1 1.76, pK_2 2.69, pK_3 6.29, pK_4 11.22(meso)   20   I = 0.1(KCl)   E3bg   Approx.   I16

pK_1 2.38, pK_2 3.44, pK_3 6.05, pK_4 11.61(dl)   concentration constants

pK_1 1.8, pK_2 2.5, pK_3 6.27, pK_4 11.2(meso)   20   E   M23

pK_1 2.4, pK_2 3.5, pK_3 6.12, pK_4 11.7(racemic)

(contd)

No.	Molecular formula, name and pK value(s)	T(°C)	Remarks	Method	Assessment	Ref.
5689 (Contd)	$C_{12}H_{20}O_8N_2$ 3,6-Diazaoctanedioic acid, 3,6-bis(carboxymethyl)-4,5-dimethyl- pK_3 6.3, pK_4 12.3(d) pK_4 11.08(meso)	25	I = 0.5	ROT	Uncert.	C16
5690	$C_{12}H_{20}O_8N_2$ 3,8-Diazadecanedioic acid, 3,8-bis(carboxymethyl)- (Tetramethylenedinitrilotetraacetic acid)					
	pK_1 1.9, pK_2 2.45 pK_3 9.05, pK_4 10.66	20	I = 0.1(KNO_3)	E3bg	Uncert. Approx.	A54
	Thermodynamic quantities are also given					
	pK_1 2.4, pK_2 2.4 pK_3 8.96, pK_4 10.35	20	I = 1.0(KNO_3), concentration constants	E3bg	Uncert. Approx.	A55
	Values in other inert salt solutions are also given					
	pK_1 1.9 pK_2 2.66, pK_3 9.07, pK_4 10.45	20	I = 0.1(KCl), concentration constants	E3bg	Uncert. Approx.	S50
5691	$C_{12}H_{20}O_9N_2$ 6-Oxa-3,9-diazaundecanedioic acid, 3,9-bis(carboxymethyl)- ((2,2'-Oxydiethylene)dinitrilotetraacetic acid)					
	pK_1 1.8 pK_2 2.76, pK_3 8.84, pK_4 9.47	20	I = 0.1(KCl), concentration constants	E3bg	Uncert. Approx.	S50
	pK_1 1.8, pK_2 2.76 pK_3 8.84, pK_4 9.47	20	I = 0.1(KIO_3)	E3bg	Uncert. Approx.	A54
	Thermodynamic quantities are also given					
	pK_1 2.4, pK_2 2.5 pK_3 8.67, pK_4 9.16	20	I = 1.0(KNO_3), concentration constants	E3bg	Uncert. Approx.	A55

5692 $C_{12}H_{20}O_{10}N_2$ 3,8-Diazadecanedioic acid, 3,8-bis(carboxymethyl)-5,6-dihydroxy- (2,3-Dihydroxytetramethylenedinitrilotetraacetic acid)

pK$_1$ 2.20, pK$_2$ 3.18, pK$_3$ 7.52, pK$_4$ 9.26, pK$_5$ 11.24, pK$_6$ 11.79

ca. 20   I = 0.1(KNO$_3$)   E3bg,R3g Uncert. D74

5693 $C_{12}H_{22}O_6N_2$ 3,6-Diazaoctanedioic acid, 3-butyl-6-carboxymethyl- (N-Butyl(ethylenedinitrilo)triacetic acid)

pK$_1$ 1.8

24.2   c = 0.002   E3bg Uncert. B149

pK$_2$ 6.64, pK$_3$ 10.04   Approx.

5694 $C_{12}H_5O_2NBr_4$ 2,5-Cyclohexadien-1-one, 2,6-dibromo-4-(3,5-dibromo-4-hydroxyphenyl)imino-

6.1   not stated   c = 0.00004, 2% dioxane, mixed constant   05 Uncert. K54

5695 $C_{12}H_6O_2NCl_3$ 2,5-Cyclohexadien-1-one, 2,6-dichloro-4-(3-chloro-4-hydroxyphenyl)imino-

5.7   not stated   c = 0.00004, 2% dioxane, mixed constant   05 Uncert. K54

5696 $C_{12}H_6O_2NBr_3$ 2,5-Cyclohexadien-1-one, 2,6-dibromo-4-(3-bromo-4-hydroxyphenyl)imino-

5.4   not stated   c = 0.00004, 2% dioxane, mixed constant   05 Uncert. K54

5697 $C_{12}H_7O_2NCl_2$ 2,5-Cyclohexadien-1-one, 4-(2,6-dichloro-4-hydroxyphenyl)imino-

5.90   26   I = 0.1, mixed constant   05 Approx. A76

5698 $C_{12}H_7O_2NCl_2$ 2,5-Cyclohexadien-1-one, 4-(3,5-dichloro-4-hydroxyphenyl)imino-

5.8   not stated   c = 0.00004, 2% dioxane, mixed constant   05 Uncert. K54

5699 $C_{12}H_7O_2NBr_2$ 2,5-Cyclohexadien-1-one, 4-(3,5-dibromo-4-hydroxyphenyl)imino-

5.8   not stated   c = 0.00004, 2% dioxane, mixed constant   05 Uncert. K54

No.	Molecular formula, name and pK value(s)	T(°C)	Remarks	Method	Assessment	Ref.
5700	$C_{12}H_7O_4N_3Cl_2$ Diphenylamine, 2,4-dichloro-2',6'-dinitro- 12.53(-NH)	20	c = 0.001. In aq. NaOH, $H_=$ scale	O7	Uncert.	H6
5701	$C_{12}H_8O_9N_4S$ Benzenesulfonic acid, 2-(2,4,6-trinitroanilino)- pK_2 11.97(-NH)	20	c = 0.001. In aq. NaOH, $H_=$ scale	O7	Uncert.	H6
5702	$C_{12}H_8O_9N_4S$ Benzenesulfonic acid, 3-(2,4,6-trinitroanilino)- pK_2 10.28(-NH)	20	c = 0.001. In aq. NaOH, $H_=$ scale	O7	Uncert.	H6
5703	$C_{12}H_9ONS_2$ Thiazolidin-4-one, 5-cinnamylidene-2-thioxo- 7.68	20	I = 0.1($NaClO_4$)	DIS	Uncert.	N17
5704	$C_{12}O_2NSe$ 1,3-Propanedione, 1-(2-pyridyl)-3-(2-selenofuryl)- 8.55	room temp	I = 0.1($NaClO_4$)			E7
5705	$C_{12}H_9O_2Cl_2P$ Bis(4-chlorophenyl)phosphinic acid 1.68	not stated	c = 0.005, I ≯ 0.025, mixed constant, 7% ethanol	E3bg	Uncert.	M57
5706	$C_{12}H_9O_4Cl_2P$ Bis(4-chlorophenyl) hydrogen phosphate 0.20	20±2	c ≈ 0.001 or 0.0001, I ≈ 0.1 or 1($NaClO_4$), concentration constant	DIS	Uncert.	K56

No.	Formula	Name	pK	T (°C)	Conditions	Method	Reliability	Ref.
5707	$C_{12}H_9O_7N_3S$	Benzenesulfonic acid, 3-(2,4-dinitroanilino)-	14.05(-NH)	20	c = 0.001. In aq. NaOH, $H_=$ scale	07	Uncert.	H6
5708	$C_{12}H_9O_7N_3S$	Benzenesulfonic acid, 4-(2,4-dinitroanilino)-	pK_2 13.92(-NH)	20	c = 0.001. In aq. NaOH, $H_=$ scale	07	Uncert.	H6
5709	$C_{12}H_9Cl_2S_2P$	Bis(4-chlorophenyl)phosphinodithioic acid	1.79	20	c = 0.005, 7% ethanol, mixed constant	E3bg	Uncert.	K5,K6
5710	$C_{12}H_{10}ON_3Cl$	Triazen-3-ol, 1-(2-chlorophenyl)-3-phenyl-	10.52	25	c = 0.00004, I = 0.1(KCl), mixed constant	05	Approx.	P89,P91
5711	$C_{12}H_{10}ON_3Cl$	Triazen-3-ol, 1-(4-chlorophenyl)-3-phenyl-	10.72	25	c = 0.00004, I = 0.1(KCl), mixed constant	05	Approx.	P89,P91
5712	$C_{12}H_{10}ON_3Cl$	Triazen-3-ol, 3-(4-chlorophenyl)-1-phenyl-	10.65	25	c = 0.00004, I = 0.1(KCl), mixed constant	05	Approx.	P89,P91
5713	$C_{12}H_{10}ON_3Br$	Triazen-3-ol, 1-(4-bromophenyl)-3-phenyl-	10.86	25	c = 0.00004, I = 0.1(KCl), mixed constant	05	Approx.	P89,P91
5714	$C_{12}H_{10}O_4N_2S$	Benzenesulfonamide, 3-nitro-N-phenyl-	7.50	20	c ≈ 0.0001, I = 0.1(KCl), mixed constant	05	Approx.	W33
5715	$C_{12}H_{10}O_4N_2S$	Benzenesulfonamide, 4-nitro-N-phenyl-	7.42	20	c ≈ 0.0001, I = 0.1(KCl), mixed constant	05	Approx.	W33

No.	Molecular formula, name and pK value(s)	T(^{o}C)	Remarks	Method	Assessment	Ref.
5716	$C_{12}H_{10}O_4N_2S$ Benzenesulfonamide, N-(3-nitrophenyl)- 6.94	20	c ≈ 0.0001, I = 0.1(KCl), mixed constant	05	Approx.	W33
5717	$C_{12}H_{10}O_4N_2S$ Benzenesulfonamide, N-(4-nitrophenyl)- 6.20	20	c ≈ 0.0001, I = 0.1(KCl), mixed constant	05	Approx.	W33
5718	$C_{12}H_{10}O_4N_2S$ Benzenesulfonic acid, 4-hydroxy-3-phenylazo- 7.53	not stated	c = 0.00005, mixed constant	05	Uncert.	M20
5719	$C_{12}H_{10}O_5NP$ Phenyl hydrogen p-nitrophenylphosphonate 2.5	not stated	Mixed constant	E3d	V.Uncert.	B175
5720	$C_{12}H_{10}O_5N_2S$ Benzenesulfonic acid, 2-anilino-5-nitro- 15.74(-NH)	20	c = 0.001 In aq. NaOH, H_ scale	07	Uncert.	H6
5721	$C_{12}H_{10}O_5N_2S$ Benzenesulfonic acid, 4-(3,4-dihydroxyphenylazo)- pK$_2$ 7.62, pK$_3$ 12.62	20±0.5	c = 0.00002, I = 0.1	05	Uncert.	N27
5722	$C_{12}H_{10}O_5N_2S$ Benzenesulfonic acid, 2-(4-nitroanilino)- 15.59(-NH)	20	c = 0.001 In aq. NaOH, H_ scale	07	Uncert.	H6
5723	$C_{12}H_{10}O_5N_2S$ Benzenesulfonic acid, 4-(4-nitroanilino)- 15.43(-NH)	20	c = 0.001. In aq. NaOH, H_ scale	07	Uncert.	H6

No.	Formula / Name / pK	T (°C)	Conditions	Method	Reliability	Ref.
5724	$C_{12}H_{10}O_6N_2S$ Benzenesulfonic acid, 4-hydroxy-3-(2,4-dihydroxyphenylazo)-					
	pK_1 0.10	not stated	c = 0.00001, I = 1(KCl), mixed constant	05	Uncert.	F24
	pK_2 6.14, pK_3 7.72		c = 0.00001, mixed constants			
5725	$C_{12}H_{10}O_6N_4S$ Benzenesulfonamide, 4-amino-N-(3,5-dinitrophenyl)-					
	6.19(NH)	27±2	I = 0.2(NaCl), mixed constant	05	Uncert.	Y14
5726	$C_{12}H_{11}ONS$ Acetamide, 2-mercapto-N-(2-naphthyl)- (Thionalide)					
	8.52			SOL		U13
	8.39			SOL		Y4
	8.21					P64
5727	$C_{12}H_{11}OSP$ Diphenylphosphinothioic O-acid					
	1.88	20	c = 0.005, 7% ethanol, 82% thione tautomer	E3bg	V.Uncert.	K6
5728	$C_{12}H_{11}O_2NS$ Benzenesulfonamide, N-phenyl-					
	8.31	20	c ≈ 0.0001, I = 0.1(KCl), mixed constant	05	Approx.	W33
5729	$C_{12}H_{11}O_2S_2P$ O,O-Diphenyl hydrogen phosphorodithioate					
	1.81	20	c = 0.005, 7% ethanol, mixed constant	E3bg	Uncert.	K6,K5
5730	$C_{12}H_{11}O_3SP$ O,O-Diphenyl S-hydrogen phosphorothioate					
	1.50	20	c = 0.005, 7% ethanol, 80% thiol tautomer	E3bg	V.Uncert.	K6

No.	Molecular formula, name and pK value(s)	T(°C)	Remarks	Method	Assessment	Ref.
5731	$C_{12}H_{11}O_4N_2P$ Phenylphosphonic acid, 4-(4-hydroxyphenylazo)- 6.4		Conditions not stated	05	Uncert.	K60
5732	$C_{12}H_{11}O_4N_2As$ Benzenearsonic acid, 4-(4-hydroxyphenylazo)- 7.9		Conditions not stated	05	Uncert.	K60
5733	$C_{12}H_{11}O_4N_3S$ Benzenesulfonamide, 4-amino-N-(3-nitrophenyl)- 7.67(NH)	27±2	I = 0.2(NaCl), mixed constant	05	Uncert.	Y14
5734	$C_{12}H_{11}O_4N_3S$ Benzenesulfonamide, 4-amino-N-(4-nitrophenyl)- 6.97(NH)	27±2	I = 0.2(NaCl), mixed constant	05	Uncert.	Y14
5735	$C_{12}H_{11}O_4N_3S$ Triazen-3-ol, 3-phenyl-1-(4-sulfophenyl)- 9.99(OH)	25	c = 0.00004, I = 0.1(KCl), mixed constant	05	Approx.	P91,P89
5736	$C_{12}H_{12}ON_2S_2$ Thiazolidin-4-one, 5-(4-dimethylaminobenzylidene)-2-thioxo- 8.20	20	I = 0.1(NaClO$_4$)	DIS	Uncert.	N17
5737	$C_{12}H_{12}O_2N_2S$ Benzenesulfonamide, 4-amino-N-phenyl- 8.89(NH)	20	c ≈ 0.0001, I = 0.1(KCl), mixed constant	05	Approx.	W33
	8.97	27	I = 0.2(NaCl), mixed constant	05	Uncert.	Y14
5738	$C_{12}H_{12}O_2N_2S$ Benzenesulfonamide, N-(4-aminophenyl)- 9.05	20	c ≈ 0.0001, I = 0.1(KCl), mixed constant	05	Approx.	W33

No.	Formula	Name	Temp	Conditions	pK	Method	Assessment	Ref.
5739	$C_{12}H_{12}O_2N_2S$	Pyrimidin-4,6(1H,5H)-dione, 5-ethyl-2,3-dihydro-5-phenyl-2-thioxo-		Conditions not stated	6.30	E3bg	Uncert.	M68
5740	$C_{12}H_{12}O_2N_2Se$	Pyrimidine-4,6(1H,5H)-dione, 5-ethyl-2,3-dihydro-5-phenyl-2-selenoxo-		Conditions not stated	6.02	E3bg	Uncert.	M68
5741	$C_{12}H_{12}O_3NP$	Phenyl hydrogen 4-aminophenylphosphonate	not stated	Mixed constant	pK_1 4.0	E3d	V.Uncert.	B175
5742	$C_{12}H_{12}O_6NCl$	3-Azapentanedioic acid, 3-carboxymethyl-2-(4-chlorophenyl)-	20	c = 0.001, I = 0.1(KCl), concentration constants	pK_1 1.55, pK_2 2.31, pK_3 8.88	E3bg	Uncert. Approx.	I13
5743	$C_{12}H_{13}O_4N_3S_2$	Benzenesulfonamide, 4-amino-N(S)-(3-aminosulfonylphenyl)-	27±2	I = 0.2(NaCl), mixed constant	7.81(NH)	05	Uncert.	Y14
5744	$C_{12}H_{13}O_4N_3S_2$	Benzenesulfonamide, 4-amino-N(S)-(4-aminosulfonylphenyl)-	27±2	I = 0.2(NaCl), mixed constant	7.45(NH)	05	Uncert.	Y14
5745	$C_{12}H_{13}O_5N_2Br$	3-Azapentanedioic acid, 3-(4-bromophenylcarbamoyl)methyl-	25	c ≈ 0.005, I = 0.1(KCl)	pK_1 1.60, pK_2 2.20, pK_3 5.90	E3bg	Uncert. Approx.	S66

No.	Molecular formula, name and pK value(s)	$T(^oC)$	Remarks	Method	Assessment	Ref.
5746	$C_{12}H_{14}O_2N_4S$ Benzenesulfonamide, 4-amino-N(S)-(2,6-dimethylpyrimidin-4-yl)- 7.49(-SO$_2$NH-)	20	c = 0.0025, I = 0.1(KCl), mixed constant	E3bg	Approx.	W34
5747	$C_{12}H_{14}O_2N_4S$ Benzenesulfonamide, 4-amino-N(S)-(4,6-dimethylpyrimidin-2-yl)- (Diazil) 7.51(-SO$_2$NH-)	20	c = 0.0025, I = 0.1(KCl), mixed constant	E3bg	Approx.	W34
5748	$C_{12}H_{14}O_4ClAs$ 3,3'-Arsinodipropanoic acid, As-(4-chlorophenyl)- pK$_1$ 4.17, pK$_2$ 5.03	20	I = 0.1(KNO$_3$)	E3bg	Approx.	P38
5749	$C_{12}H_{15}O_6N_2F$ Uridine, 5-fluoro-2',3'-O,O-isopropylidene- 7.50	20-22	I = 0.05-0.1	05	Uncert.	B64
5750	$C_{12}H_{16}O_2N_2S$ Pyrimidine-4,6(1H,5H)-dione, 5-cyclohex-1-enyl-5-ethyl-2,3-dihydro-2-thioxo- pK$_1$ 7.12, pK$_2$ 11.8	28	2% ethanol	05	Uncert.	S14
5751	$C_{12}H_{16}O_5NP$ Cyclohexyl hydrogen 4-nitrophenylphosphonate 2.6	not stated	Mixed constant	E3d	V.Uncert.	B175
5752	$C_{12}H_{16}O_7NCl \equiv C_{12}H_{16}O_3N(ClO_4)$ N,N,N-Trimethyl 3-(2-carboxyvinyl)-4-hydroxyanilinium perchlorate (Propenoic acid, 3-(2-hydroxy-5-trimethylammonio)phenyl-, perchlorate) 3.43(cis)	30	I = 1.0(LiCl), mixed constant	05	Approx.	H36
5753	$C_{12}H_{17}O_3N_2Br$ Pyrimidine-2,4,6(1H,3H,5H)-trione, 5-(2-bromoallyl)-5-(1-ethylpropyl)- 7.70	25		E3ag	Approx.	K53

5754 $C_{12}H_{17}O_8N_2P$ Butanoic acid, 2-(3-hydroxy-5-hydroxymethyl-2-methylpyridin-4-yl)methyleneamino-3-phosphonooxy- (Pyridoxylidene-O-phosphothreonine)

10.48   25   c = 0.0001, I = 1.0(KCl), pK of aldimine   O5   Approx.   M111

5755 $C_{12}H_{18}ON_3(Cl)$ 1-Piperazinium chloride, 1-(p-aminobenzoyl)-4-methyl-

8.9   25   E3bg   Uncert.   F35a

5756 $C_{12}H_{18}O_2N(I)$ N,N,N-Trimethyl 3-(2-carboxyethyl)anilinium iodide   (Propanoic acid, 3-(3-trimethylammoniophenyl)-, iodide)

4.17   25   E3ag   Approx.   H48

Values in mixed solvents are also given

5757 $C_{12}H_{18}O_2N(I)$ N,N,N-Trimethyl 4-(2-carboxyethyl)anilinium iodide   (Propanoic acid, 3-(4-trimethylammoniophenyl)-, iodide)

4.22   25   E3ag   Approx.   H48

Values in mixed solvents are also given

5758 $C_{12}H_{18}O_2N_2S$ Pyrimidine-4,6(1$\underline{H}$,5$\underline{H}$)-dione, 5-allyl-2,3-dihydro-5-(1-methylbutyl)-2-thioxo-

pK$_1$ 7.38, pK$_2$ 12.27   28   2% ethanol   O5   Uncert.   S14

5759 $C_{12}H_{18}O_3NP$ Cyclohexyl hydrogen 4-aminophenylphosphonate

pK$_1$ 3.9   not stated   Mixed constant   E3d   V.Uncert.   B175

5760 $C_{12}H_{19}ON_2Cl$ Phenol, 2,6-bis(dimethylaminomethyl)-4-chloro-

pK$_1$ 5.20   25   I = 0.1(KNO$_3$), mixed constant   E3bg   Approx.   E19

5761 $C_{12}H_{19}ON_2F$ Phenol, 2,6-bis(dimethylaminomethyl)-4-fluoro-

pK$_1$ 5.77   25   I = 0.1(KNO$_3$), mixed constant   E3bg   Approx.   E19

No.	Molecular formula, name and pK value(s)	$T(^oC)$	Method	Assessment	Ref.	Remarks
5762	C$_{12}$H$_{19}$O$_2$N$_2$(Br) N-Benzyl N,N-dimethyl 2-(hydroxycarbamoyl)ethylammonium bromide (Propanohydroxamic acid, 3-(N-benzyl-N,N-dimethylammonio)-, bromide)					
	8.40	25(?)	E3bg	Uncert.	M66	c = 0.02, I = 0.1(KCl), mixed constant
5763	C$_{12}$H$_{20}$ON(I) Phenol, 2-(3-N,N,N-trimethylammoniopropyl)-, iodide					
	9.78	25	E3bg	Approx.	E20	I = 0.1(KCl), mixed constant
5764	C$_{12}$H$_{20}$ON(I) Phenol, 3-(3-N,N,N-trimethylammoniopropyl)-, iodide					
	9.64	25	E3bg	Approx.	E20	I = 0.1(KCl), mixed constant
5765	C$_{12}$H$_{20}$ON(I) Phenol, 4-(3-N,N,N-trimethylammoniopropyl)-, iodide					
	9.80	25	E3bg	Approx.	E20	I = 0.1(KCl), mixed constant
5766	C$_{12}$H$_{20}$O$_8$N$_2$S 6-Thia-3,9-diazaundecanedioic acid, 3,9-bis(carboxymethyl)- ((2,2'-Thiodiethylene)dinitrilotetraacetic acid)					
	pK$_1$ 1.8, pK$_2$ 2.52 pK$_3$ 8.47, pK$_4$ 9.42	20	E3bg	Uncert. Approx.	A54	I = 0.1(KNO$_3$) Thermodynamic quantities are also given
	pK$_1$ 2.1, pK$_2$ 2.5 pK$_3$ 8.29, pK$_4$ 9.29	20	E3bh	Uncert. Approx.	A55	I = 1.0((Me)$_4$NCl), concentration constants Values in other inert salt solutions are also given
	pK$_1$ 1.8 pK$_2$ 2.52, pK$_3$ 8.38, pK$_4$ 9.42	20	E3bg	Uncert. Approx.	S50	I = 0.1(KCl), concentration constants

No.	Formula	Name / pK	T	Conditions	Method	Reliability	Ref.
5767	$C_{12}H_{20}O_8N_2S_2$	6,7-Dithia-3,10-diazadodecanedioic acid, 3,10-bis(carboxymethyl)- (Dithiodiethylenedinitrilotetraacetic acid)					
		pK_1 2.0	20	I = 0.1(KCl), concentration constants	E3bg	Uncert.	S50
		pK_2 2.53, pK_3 8.24, pK_4 9.23				Approx.	
5768	$C_{12}H_{26}O_6N_2S$	Piperazine-1,4-bis(butanesulfonic acid) (1,4-Bis(4-sulfobutyl)piperazine) (PIPBS)					
		pK_1 4.6, pK_2 8.6	18	c = 0.05, mixed constants			J12
5769	$C_{12}H_{29}O_6NP_2$	O,O-Diethyl phosphoramidate, N-dibutoxyphosphinyl-					
		4.73	20	c = 0.005, 7% ethanol, mixed constant	E3bg	Uncert.	G19
5770	$C_{12}H_{29}O_6NP_2$	O,O-Diisopropyl phosphoramidate, N-diisopropoxyphosphinyl-					
		6.32	20	c = 0.005, 7% ethanol, mixed constant	E3bg	Uncert.	G19
5771	$C_{12}H_9OCl_2SP$	Bis(4-chlorophenyl)phosphinothioic acid					
		1.60	20	c = 0.005, 7% ethanol 54% thiol tautomer	E3bg	V.Uncert.	K6
5772	$C_{12}H_9O_5N_2ClS$	Benzenesulfonic acid, 2-(4-chloroanilino)-5-nitro-					
		15.55(-NH)	20	c = 0.001 In aq. NaOH, H_- scale	07		H6
5773	$C_{12}H_9O_5N_2BrS$	Benzenesulfonic acid, 4-(2-bromo-4,5-dihydroxyphenylazo)-					
		pK_2 6.72, pK_3 11.9	20	I = 0.1	05	Uncert.	B32
5774	$C_{12}H_{10}O_2NClS$	Benzenesulfonamide, 4-chloro-N-phenyl-					
		7.98	20	c ≈ 0.0001, I = 0.1(KCl), mixed constant	05	Approx.	W33

No.	Molecular formula, name and pK value(s)	$T(^{o}C)$	Remarks	Method	Assessment	Ref.
5775	$C_{12}H_{10}O_2NClS$ Benzenesulfonamide, N-(4-chlorophenyl)-					
	7.93	20	$c \approx 0.0001$, $I = 0.1(KCl)$, mixed constant	05	Approx.	W33
5776	$C_{12}H_{10}O_2N_2Cl_2S$ Benzenesulfonamide, 4-amino-N(S)-(3,5-dichlorophenyl)-					
	7.54(NH)	27±2	$I = 0.2(NaCl)$, mixed constant	05	Uncert.	Y14
5777	$C_{12}H_{10}O_4N_3ClS$ Benzenesulfonamide, 4-amino-N(S)-(3-chloro-5-nitrophenyl)-					
	6.92(NH)	27±2	$I = 0.2(NaCl)$, mixed constant	05	Uncert.	Y14
5778	$C_{12}H_{10}O_4N_3ClS$ Benzenesulfonamide, 4-amino-N(S)-(4-chloro-3-nitrophenyl)-					
	7.16(NH)	27±2	$I = 0.2(NaCl)$, mixed constant	05	Uncert.	Y14
5779	$C_{12}H_{11}O_2N_2ClS$ Benzenesulfonamide, 4-amino-N(S)-(3-chlorophenyl)-					
	8.28(NH)	27±2	$I = 0.2(NaCl)$, mixed constant	05	Uncert.	Y14
5780	$C_{12}H_{11}O_2N_2ClS$ Benzenesulfonamide, 4-amino-N(S)-(4-chlorophenyl)-					
	8.56(NH)	27±2	$I = 0.2(NaCl)$, mixed constant	05	Uncert.	Y14
5781	$C_{13}H_4N_6$ 1,3,5-Heptatriene-1,1,2,6,7,7-hexacarbonitrile					
	-3.55	25	In aq. $HClO_4$ solutions, H_o scale	06	Uncert.	B116
	-3.90		In aq. H_2SO_4 solutions, H_o scale			

No.	Formula	Name	pK	T (°C)	Notes		Ref.
5782	$C_{13}H_8O_2$	1-Phenalenone, 3-hydroxy- pK_1 -2.1, pK_2 4.30(OH)		0			K13
5783	$C_{13}H_{10}O_2$	Benzophenone, 4-hydroxy- 7.95		25	05	Approx.	C79
5784	$C_{13}H_{10}O_4$	Benzophenone, 2,3',4-trihydroxy- pK_1 7.02, pK_2 9.24, pK_3 12.20		25	I = 0.083	05 Approx.	M14
5785	$C_{13}H_{10}N_2$	Benzimidazole, 2-phenyl- 11.91		25	c = 0.000098	05 Approx.	W1
5786	$C_{13}H_{12}O_2$	1-Naphthoic acid, 2,3-dimethyl- 2.831		25	Variation with temperature	05 Approx.	B125

Variation with temperature

15	20	30	35	40
2.483	2.629	3.021	3.196	3.351

Thermodynamic quantities are derived from the results

No.	Formula	Name	T (°C)	Notes		Ref.
5787	$C_{13}H_{12}O_2$	Biphenyl-2-ol, 2'-methoxy- 10.40	20	c = 0.0001, mixed constant	05 Approx.	M128
5788	$C_{13}H_{12}O_2$	Biphenyl-2-ol, 4'-methoxy- 11.38		E		D49

No.	Molecular formula, name and pK value(s)	$T(^{o}C)$	Remarks	Method	Assessment	Ref.
5789	$C_{13}H_{12}O_2$ Phenol, 2-(2-hydroxybenzyl)- pK_1 7.50, pK_2 11.60	25±1	c = 0.0001, mixed constants	05	Uncert.	F36
5790	$C_{13}H_{12}O_2$ Phenol, 2-(4-hydroxybenzyl)- pK_1 8.00, pK_2 10.90	25±1	c = 0.0001, mixed constants	05	Uncert.	F36
5791	$C_{13}H_{12}O_2$ Phenol, 4-(4-hydroxybenzyl)- pK_1 7.55, pK_2 10.80	25±1	c = 0.0001, mixed constants	05	Uncert.	F36
5792	$C_{13}H_{12}O_3$ Ethyl 3-hydroxy-1-naphthalenecarboxylate 8.89	25	c ≈ 0.001	05	Approx.	B155
5793	$C_{13}H_{12}O_3$ 1,3-Indandione, 2-isobutyryl- 3.2	18±2	1% ethanol, mixed constant	05 or E3d	Uncert.	W25
5794	$C_{13}H_{14}O_2$ Benzoic acid, 4-(bicyclo[3.1.0]hex-1-yl)- 4.02		Conditions not stated	05	Uncert.	S62
5795	$C_{13}H_{16}O_2$ Benzo[b]furan-4-ol, 3,7-diethyl-2-methyl- 9.94	20	2% ethanol	05	Approx.	D32
5796	$C_{13}H_{16}O_2$ Benzoic acid, 4-cyclohexyl- 4.39		Conditions not stated	05	Uncert.	S62

No.	Formula	Name	T(°C)	pKa	Method	Reliability	Ref.
5797	$C_{13}H_{16}O_4$	Butanoic acid, 2-benzyl-4-methoxy-2-methyl-4-oxo-	20	4.65	E3bg	Approx.	F28
5798	$C_{13}H_{16}O_4$	Butanoic acid, 2-methoxycarbonylmethyl-2-phenyl-	20	4.18	E3bg	Approx.	F28
5799	$C_{13}H_{16}O_4$	Pentanedioic acid, 3-ethyl-3-phenyl-	25	pK_1 3.89, pK_2 6.95 $\quad$ c = 0.00125, mixed constants	E3bg	Approx.	B146
5800	$C_{13}H_{16}O_4$	Pentanoic acid, 3-methoxycarbonyl-3-phenyl-	20	4.46	E3bg	Approx.	F28
5801	$C_{13}H_{18}O_2$	Heptanoic acid, 7-phenyl-	20	5.07 $\quad$ c ≈ 0.001. Value obtained by extrapolation from data in ethanol and dioxane solutions	E3bg	Uncert.	R55
5802	$C_{13}H_{24}O_4$	Pentanoic acid, 2-tert-butyl-3-methoxycarbonyl-4,4-dimethyl-	25	5.57(H_2O) (racemic) 6.08(D_2O)	E3bg	Approx.	G22
5803	$C_{13}H_{24}O_4$	Pentanoic acid, 2,2,3-triethyl-3-methoxycarbonyl-	25	5.34(H_2O) 5.83(D_2O)	E3bg	Approx.	G22

No.	Molecular formula, name and pK value(s)	$T(^oC)$	Method	Assessment	Remarks	Ref.
5804	$C_{13}HOF_9$ 4-Biphenylcarboxylic acid, perfluoro- 1.42	25		Uncert.	Low solubility. Estimated from OH stretching frequencies	R56
5805	$C_{13}H_8O_2N_8$ Pteridin-6(1$\underline{H}$)-one, 5,7-dihydro-7-(7',8'-dihydro-7'-oxopteridin-6-ylmethylidene)- pK$_1$ 7.01, pK$_2$ 9.23	20	05	Approx.	c = 0.00001 in 0.01M buffer	A39
5806	$C_{13}H_8O_8N_4$ Benzoic acid, 3,5-dinitro-2-(4-nitroanilino)- pK$_2$ 12.98	20	07	Uncert.	c = 0.001 In aq. NaOH, H$_=$ scale	H6
5807	$C_{13}H_8O_8N_4$ Benzoic acid, 2-(2,4,6-trinitroanilino)- pK$_2$ 12.59(-NH)	20	07	Uncert.	c = 0.001 In aq. NaOH, H$_=$ scale	H6
5808	$C_{13}H_8O_8N_4$ Benzoic acid, 3-(2,4,6-trinitroanilino)- pK$_2$ 10.78(-NH)	20	07	Uncert.	c = 0.001 In aq. NaOH, H$_=$ scale	H6
5809	$C_{13}H_9ON$ 4-Acridinol pK$_1$ 5.31, pK$_2$ 9.84	not stated	05	Uncert.	Mixed constants	I21
5810	$C_{13}H_9O_5N_3$ Benzoic acid, 2-hydroxy-5-(2-nitrophenylazo)- pK$_2$ 11.06		05			M108

No.	Formula	Name	pK	T (°C)	Conditions	Method	Assessment	Ref.
5811	$C_{13}H_9O_5N_3$	Benzoic acid, 2-hydroxy-5-(3-nitrophenylazo)-	pK_2 11.23			05		M108
5812	$C_{13}H_9O_5N_3$	Benzoic acid, 2-hydroxy-5-(4-nitrophenylazo)-	pK_2 10.91			05		M108
5813	$C_{13}H_9O_6N_3$	Benzoic acid, 3-(2,4-dinitroanilino)-	14.57(-NH)	20	c = 0.001 In aq. NaOH, $H_=$ scale	07	Uncert.	H6
5814	$C_{13}H_9O_6N_3$	Benzoic acid, 4-(2,4-dinitroanilino)-	14.28(-NH)	20	c = 0.001 In aq. NaOH, $H_=$ scale	07	Uncert.	H6
5815	$C_{13}H_9O_6N_3$	3-Nitrophenyl 4-nitroanilinomethanoate	11.7	25	I = 1.0(KCl)	05	Uncert.	H26
5816	$C_{13}H_{10}O_2Se$	1,3-Propanedione, 1-phenyl-3-(2-selenofuryl)-	9.05	room temp	I = 0.1(NaClO$_4$)			E7
5817	$C_{13}H_{10}O_3N_2$	Benzoic acid, 2-hydroxy-5-phenylazo-	pK_1 2.30, pK_2 11.64			05		M108
5818	$C_{13}H_{10}O_3N_4$	Pteridine-2,4,7(1H,3H,8H)-trione, 3-methyl-1-phenyl-	3.49	20	c = 0.001	E3bg	Approx.	P47

No.	Molecular formula, name and pK value(s)	T(°C)	Remarks	Method	Assessment	Ref.
5819	$C_{13}H_{10}O_4N_2$ Benzenecarboxamide, N-(4-hydroxyphenyl)-4-nitro- 9.50	25	1.6% CH_3CN	05	Approx.	D46
5820	$C_{13}H_{10}O_4N_2$ Benzohydroxamic acid, 3-nitro-N-phenyl- 7.67	30	c = 0.001, I = 0.1($NaClO_4$), concentration constant	E3bg	Approx.	D62
5821	$C_{13}H_{10}O_4N_2$ Benzoic acid, 2-(3,4-dihydroxyphenylazo)- 3.12	18-20	I = 1		Uncert.	K45
5822	$C_{13}H_{10}O_4N_2$ Benzoic acid, 2-(4-nitroanilino)- 16.23(-NH)	20	c = 0.001 In aq. NaOH, $H_=$ scale	07	Uncert.	H6
5823	$C_{13}H_{10}O_4N_2$ Benzoic acid, 4-(4-nitroanilino)- 15.56(-NH)	20	c = 0.001 In aq. NaOH, $H_=$ scale	07	Uncert.	H6
5824	$C_{13}H_{10}O_4N_2$ Phenyl 4-nitroanilinomethanoate 12.5	25	I = 1.0(KCl)	05	Uncert.	H26
5825	$C_{13}H_{11}ON$ Benzophenone oxime 11.18			05		C26

No.	Formula	Name / pK	Temp (°C)	Notes			Ref
5826	$C_{13}H_{11}ON_3$	Phenol, 2-(benzotriazol-2-yl)-4-methyl-					
		9.97	20±1		DIS	Uncert.	M19
5827	$C_{13}H_{11}O_2N$	Benzenecarboxamide, N-(2-hydroxyphenyl)-					
		7.5	20	1-4% ethanol, mixed constant	05	Uncert.	W24
5828	$C_{13}H_{11}O_2N$	Benzenecarboxamide, N-(4-hydroxyphenyl)-					
		9.54	25	1.6% CH_3CII	05	Approx.	D46
5829	$C_{13}H_{11}O_2N$	Benzohydroxamic acid, N-phenyl-					
		pK_1 2.10, pK_2 8.14	30	$I = 0.1(NaClO_4)$	E3bg	Approx.	D63
		pK_2 8.38	25	$c \approx 0.01$, $I \approx 0.01$	E3bg	Approx.	S67
		pK_2 8.30	25		E3bg	Approx.	D76
		pK_2 8.15		$I = 0.1(NaClO_4)$, concentration constant			
				Other values in C72, D62 and F31			
5830	$C_{13}H_{11}O_2N$	Benzoic acid, 2-anilino- (Anthranilic acid, N-phenyl-)					
		pK_1 -1.28	25	Estimated from solubility in 6-18M H_2SO_4		V.Uncert.	F46
		pK_2 3.86		Estimated from solubility in various buffers		V.Uncert.	
		pK_2 3.99			05	Uncert.	
5831	$C_{13}H_{11}O_2N$	2,5-Cyclohexadien-1-one, 4-(4-hydroxyphenyl)imino-3-methyl-					
		8.8	not stated	$c = 0.00004$, 2% dioxane, mixed constant	05	Uncert.	K54

No.	Molecular formula, name and pK value(s)	T(°C)	Remarks	Method	Assessment	Ref.
5832	$C_{13}H_{11}O_3N$ Benzenecarboxamide, 2,4-dihydroxy-N-phenyl-					
	pK$_1$ 6.8	20	1-4% ethanol, mixed constant	05	Uncert.	W24
5833	$C_{13}H_{11}O_3N$ Benzenecarboxamide, 2,6-dihydroxy-N-phenyl-					
	pK$_1$ 6.3	20	1-4% ethanol, mixed constant	05	Uncert.	W24
5834	$C_{13}H_{11}O_3N$ 1,2-Benzenediol, 3-(2-hydroxyphenyl)iminomethyl-					
	pK$_1$ 7.06, pK$_2$ 9.59	20	I = 0.15(NaClO$_4$), 4% ethanol	E3bg	Uncert.	K8
5835	$C_{13}H_{11}O_3N$ 1,3-Benzenediol, 4-(2-hydroxyphenyl)iminomethyl-					
	pK$_1$ 8.36, pK$_2$ 10.49	20	I = 0.15(NaClO$_4$), 4% ethanol	E3bg	Uncert.	K8
5836	$C_{13}H_{11}O_3N$ 1,4-Benzenediol, 2-(2-hydroxyphenyl)iminomethyl-					
	pK$_1$ 8.41, pK$_2$ 10.42	20	I = 0.15(NaClO$_4$), 4% ethanol	E3bg	Uncert.	K8
5837	$C_{13}H_{11}O_3N$ Benzoic acid, 2-amino-5-phenoxy-					
	pK$_1$ 2.34	25	c = 0.0005, I = 0.1(KCl), mixed constants	05	Uncert.	L15
	pK$_2$ 4.88				Approx.	
5838	$C_{13}H_{11}O_3N_3$ Benzenecarboxamide, N-(4-aminophenyl)-4-nitro-					
	4.48	25	1.6% CH$_3$CN	05	Approx.	D46
5839	$C_{13}H_{11}O_3N_3$ Benzoic acid, 4-(3-hydroxy-3-phenyltriazen-1-yl)-					
	10.97(-OH)	25	c = 0.00004, I = 0.1(KCl), mixed constant	05	Approx.	P91,P89

No.	Formula	Name	pK	T (°C)	Conditions	Method	Assessment	Ref.
5840	$C_{13}H_{11}NS$	Benzenecarbothioamide, N-phenyl-	10.6	not stated	c = 0.00001	05	Uncert.	W4
5841	$C_{13}H_{12}ON_2$	Benzenecarboxamide, N-(4-aminophenyl)-	pK_1 4.55	25	1.6% CH_3CN	05	Approx.	D46
5842	$C_{13}H_{12}ON_4$	Carbazone, 1,5-diphenyl-	8.54	21-23	I = 0.1($NaClO_4$), mixed constant	DIS	Uncert.	B9
5843	$C_{13}H_{12}O_2N_2$	3-Pyrrolidinecarbonitrile, 4-benzyl-4-methyl-2,5-dioxo-	6.69	20	4% ethanol	E3bg	Uncert.	F27,F29
5844	$C_{13}H_{12}O_2N_2$	3-Pyrrolidinecarbonitrile, 4-ethyl-2,5-dioxo-4-phenyl-	5.97	20	c = 0.0001-0.001, 4% ethanol, mixed constant	E3bg	Uncert.	F29,F27
5845	$C_{13}H_{12}O_6S_2$	O,O-Diphenyl methanedisulfonate (Methionol)	7.66	not stated	c = 0.00003, mixed constant	05	Uncert.	W5
5846	$C_{13}H_{13}ON_3$	Triazen-3-ol, 1-phenyl-3-(p-tolyl)-	11.86	25	c = 0.00004, I = 0.1(KCl), mixed constant	05	Approx.	P91,P89
5847	$C_{13}H_{13}ON_3$	Triazen-3-ol, 3-phenyl-1-(o-tolyl)-	11.55	25	c = 0.00004, I = 0.1(KCl), mixed constant	05	Approx.	P89,P91

No.	Molecular formula, name and pK value(s)	T(°C)	Remarks	Method	Assessment	Ref.
5848	$C_{13}H_{13}ON_3$ Triazen-3-ol, 3-phenyl-1-(p-tolyl)- 11.78	25	c = 0.00004, I = 0.1(KCl), mixed constant	05	Approx.	P89,P91
5849	$C_{13}H_{13}O_2N$ Spiro[2,5-pyrrolidinedione-3,1'-1',2',3',4'-tetrahydronaphthalene] (1,2,3,4-Tetrahydronaphthalene-1-spiro-3'- (2',5'-pyrrolidinedione)) 8.85	20	c = 0.0001-0.001, 4% ethanol, mixed constant	E3bg	Uncert.	F29
5850	$C_{13}H_{13}O_2N_3$ Triazen-3-ol, 1-(4-methoxyphenyl)-3-phenyl- 11.95	25	c = 0.00004, I = 0.1(KCl), mixed constant	05	Approx.	P89,P91
5851	$C_{13}H_{13}O_2As$ Benzylphenylarsinic acid 5.65 5.53 5.64	25	I = 0.001 I = 0.1 I = 1.0	E3bg	Approx.	F30
5852	$C_{13}H_{13}O_3P$ Benzylphosphonic acid, α-phenyl- pK$_1$ 2.19, pK$_2$ 7.09	not stated	Probably mixed constants	E3(?)		M51
5853	$C_{13}H_{13}O_4N$ 3-Pyrrolidinecarboxylic acid, 4-benzyl-4-methyl-2,5-dioxo- 2.71(COOH), 9.34(NH)	20	c = 0.0001-0.001, 4% ethanol, mixed constants	E3bg	Uncert.	F29,F27
5854	$C_{13}H_{13}O_4N$ 3-Pyrrolidinecarboxylic acid, 4-ethyl-2,5-dioxo-4-phenyl- 2.90(COOH), 8.95(NH)	20	c = 0.0001-0.001, 4% ethanol, mixed constants	E3bg	Uncert.	F29,F27

No.	Formula	Name / Constants	Temp. (°C)	Conditions	Method	Assessment	Ref.
5855	$C_{13}H_{14}O_3N_2$	Pyrimidine-2,4,6(1H,3H,5H)-trione, 5-ethyl-1-methyl-5-phenyl- (Methylphenobarbitone)			05	Approx.	B124
		8.008					

Variation with temperature

15	20	25	30	35	40	45	50
8.103	8.059	8.008	7.965	7.925	7.877	7.840	7.796

Thermodynamic quantities are derived from the results

No.	Formula	Name / Constants	Temp. (°C)	Conditions	Method	Assessment	Ref.
5856	$C_{13}H_{14}O_3N_2$	3-Pyrrolidinecarboxamide, 4-benzyl-4-methyl-2,5-dioxo-					
		8.11 (NH)	20	c = 0.0001-0.001, 4% ethanol, mixed constant	E3bg	Uncert.	F29,F27
5857	$C_{13}H_{14}O_3N_2$	3-Pyrrolidinecarboxamide, 4-ethyl-2,5-dioxo-4-phenyl-					
		7.71 (NH)	20	c = 0.0001-0.001, 4% ethanol, mixed constant	E3bg	Uncert.	F29,F27
5858	$C_{13}H_{15}O_5N$	Hexanoic acid, 5-amino-6-benzyloxy-4,6-dioxo-					
		pK_1 4.15, pK_2 6.20, pK_3 10.6	22		E3bg	Uncert.	L7
5859	$C_{13}H_{15}O_6N$	3-Azapentanedioic acid, 3-carboxymethyl-2-methyl-2-phenyl- (Iminodiacetic acid, N-(1-carboxy-1-phenyl)ethyl-)					
		pK_1 1.46	20	I = 0.1(KCl), concentration constants	E3bg	Uncert.	I14
		pK_2 2.66, pK_3 11.07				Approx.	
5860	$C_{13}H_{15}O_6N$	3-Azapentanoic acid, 3-carboxymethyl-2-(p-tolyl)-					
		pK_1 1.45	20	c = 0.001, I = 0.1(KCl), concentration constants	E3bg	Uncert.	I13
		pK_2 2.40, pK_3 9.45				Approx.	
5861	$C_{13}H_{15}O_6N$	Butanoic acid, 2-ethyl-4-methoxy-2-(4-nitrobenzyl)-4-oxo-					
		3.70	20		E3bg	Approx.	F28

No.	Molecular formula, name and pK value(s)	$T(^oC)$	Remarks	Method	Assessment	Ref.
5862	$C_{13}H_{15}O_6N$ Butanoic acid, 4-methoxy-2-methyl-2-(4-nitrobenzyl)-4-oxo- 4.40	20		E3bg	Approx.	F28
5863	$C_{13}H_{15}O_6N$ Butanoic acid, 3-methoxycarbonyl-3-(4-nitrobenzyl)- 4.45	20		E3bg	Approx.	F28
5864	$C_{13}H_{15}O_6N$ Pentanoic acid, 3-methoxycarbonyl-3-(4-nitrophenyl)- 4.29	20		E3bg	Approx.	F28
5865	$C_{13}H_{15}O_7N$ 3-Azapentanedioic acid, 3-carboxymethyl-2-(4-methoxyphenyl)- pK_1 1.47 pK_2 2.44, pK_3 9.51	20	$c = 0.001$, $I = 0.1(KCl)$, concentration constants	E3bg	Uncert. Approx.	I13
5866	$C_{13}H_{16}O_4S_2$ 4,6-Dithianonanedioic acid, 5-phenyl- pK_1 3.95 pK_1 3.98, pK_2 4.76 pK_2 4.73		$I = 0.3$ $I = 0.05$ $I = 0.1$			C11
5867	$C_{13}H_{16}O_5N_4$ Diethyl 2-(1,2,3,4-tetrahydro-2-oxopteridin-4-yl)propanedioate (Pteridine, 4-bis(ethoxycarbonyl)methyl-3,4-dihydro-2-hydroxy-) 11.43	20	$c = 0.0001$ in 0.01M buffer (anion hydrolyses readily)	O5	Uncert.	A27

No.	Formula	Name	pK	T (°C)	Conditions	Method	Assessment	Ref.
5868	$C_{13}H_{17}O_4N_3$	6-Azaheptanoic acid, 7-amino-2-benzoylamino-7-oxo- (Pentanoic acid, 2-benzoylamino-5-ureido-)	3.49	25	I = 0.3	O5	Approx.	W35
			3.57		I = 0.03		Uncert.	
			3.57		I = 0.3	E3bg	Uncert.	
			3.60		I = 0.03	O5	Uncert.	
5869	$C_{13}H_{18}O_3N_4$	6-Azaheptanoic acid, 7-amino-2-benzoylamino-7-imino- (α-N-benzoyl-L-arginine)	3.24	25	I = 0.3	O5	Approx.	W23
			3.38	40	I = 0.3(KCl)	E3bg	Approx.	S84
			3.40	25		O5	Uncert.	S54
5870	$C_{13}H_{20}O_3N_2$	Benzohydroxamic acid, 2-(2-diethylamino)ethoxy- pK$_2$ 6.95	8.42	30	c = 0.01, I = 0.1(KCl), mixed constant	E3bg	Approx.	E18
5871	$C_{13}H_{20}O_3N_2$	Pyrimidine-2,4,6(1H,3H,5H)-trione, 5-(2-methylallyl)-5-(2-methylbutyl)-	7.98	25		E3ag	Approx.	K53
5872	$C_{13}H_{20}O_5N_2$	Benzoic acid, 2,6-bis(dimethylaminomethyl)-3,4,5-trihydroxy- pK$_2$ 6.95		25	I = 0.1(KNO$_3$), mixed constant	E3bg	Approx.	E19
5873	$C_{13}H_{20}O_8N_2$	1,2-Cyclopentylenedinitrilotetraacetic acid pK$_1$ 2.4, pK$_2$ 3.3, pK$_3$ 7.56(trans)			Conditions not stated		Uncert.	Y8
5874	$C_{13}H_{21}ON$	Phenol, 4-tert-butyl-2-(dimethylaminomethyl)- pK$_1$ 8.59		25	I = 0.1(KNO$_3$), mixed constant	E3bg	Approx.	E19

No.	Molecular formula, name and pK value(s)	$T(^{o}C)$	Remarks	Method	Assessment	Ref.
5875	$C_{13}H_{22}ON_2$ Phenol, 2,6-bis(dimethylaminomethyl)-4-methyl-					
	pK_1 6.45	25	$I = 0.1(KNO_3)$, mixed constant	E3bg	Approx.	E19
5876	$C_{13}H_{22}O_2N_2$ Phenol, 2,6-bis(dimethylaminomethyl)-4-methoxy-					
	pK_1 6.33	25	$I = 0.1(KNO_3)$, mixed constant	E3bg	Approx.	E19
5877	$C_{13}H_{22}O_8N_2$ 3,9-Diazaundecanedioic acid, 3,9-bis(carboxymethyl)- (Pentamethylenedinitrilotetraacetic acid)					
	pK_1 2.3, pK_2 2.7	20	$I = 0.1(KNO_3)$	E3bg	Uncert.	A54
	pK_3 9.52, pK_4 10.70				Approx.	
			Thermodynamic quantities are also given			
	pK_1 2.2, pK_2 2.6	20	$I = 1.0((Me)_4NCl)$	E3bh	Uncert.	A55
	pK_3 9.42, pK_4 10.49		concentration constants		Approx.	
			Values in other inert salt solutions are also given			
	pK_1 2.3	20	$I = 0.1(KCl)$, concentration constants	E3bg	Uncert.	S50
	pK_2 2.70, pK_3 9.50, pK_4 10.58				Approx.	
5878	$C_{13}H_{23}O_8N_3$ 3,6,9-Triazaundecanedioic acid, 3,9-bis(carboxymethyl)-6-methyl-					
	pK_1 2.8	20	$I = 0.1(KCl)$, concentration constants	E3bg	Uncert.	S50
	pK_2 3.65, pK_3 7.39, pK_4 10.89				Approx.	
5879	$C_{13}H_8O_2NCl_3$ 2,5-Cyclohexadien-1-one, 2-chloro-4-(3,5-dichloro-4-hydroxyphenyl)imino-6-methyl-					
	5.5	not stated	$c = 0.00004$, 2% dioxane, mixed constant	05	Uncert.	K54

No.	Formula	Name / pK values	Conditions		Method	Reference
5880	$C_{13}H_8O_6N_3Cl$	Benzoic acid, 4-(4-chloroanilino)-3,5-dinitro- 13.42(-NH)	$c = 0.001$	20	07	Uncert.	H6
			In aq. NaOH, H_- scale				
5881	$C_{13}H_9ON_3S$	1-Naphthol, 2-(1,3-thiazol-2-ylazo)- pK_1 0.88, pK_2 9.10	$I = 0.05(NaClO_4)$	20	05	Approx.	N16
5882	$C_{13}H_9ON_3S$	2-Naphthol, 1-(1,3-thiazol-2-ylazo)- 8.94	$c = 0.001$		0	Uncert.	B183
5883	$C_{13}H_9O_2NCl_2$	2,5-Cyclohexadien-1-one, 2,6-dichloro-4-(4-hydroxy-2-methylphenyl)imino- 5.9	$c = 0.00004$, 2% dioxane, mixed constant not stated		05	Uncert.	K54
5884	$C_{13}H_9O_2NCl_2$	2,5-Cyclohexadien-1-one, 4-(3,5-dichloro-4-hydroxyphenyl)imino-2-methyl- 5.7	$c = 0.00004$, 2% dioxane, mixed constant not stated		05	Uncert.	K54
5885	$C_{13}H_9O_2NBr_2$	2,5-Cyclohexadien-1-one, 4-(3,5-dibromo-4-hydroxyphenyl)imino-2-methyl- 5.4	$c = 0.00004$, 2% dioxane, mixed constant not stated		05	Uncert.	K54
5886	$C_{13}H_9O_2NBr_2$	2,5-Cyclohexadien-1-one, 4-(3,5-dibromo-4-hydroxyphenyl)imino-3-methyl- 5.6	$c = 0.00004$, 2% dioxane, mixed constant not stated		05	Uncert.	K54
5887	$C_{13}H_9O_2N_3S$	1,3-Benzenediol, 4-(2-benzothiazolylazo)- pK_1 3.4(NH^+), pK_2 5.79, pK_3 10.27			0		B186

No.	Molecular formula, name and pK value(s)	T(^{o}C)	Remarks	Method	Assessment	Ref.
5888	$C_{13}H_9O_3NCl_2$ 2,5-Cyclohexadien-1-one, 2,6-dichloro-4-(4-hydroxy-2-methoxyphenyl)imino- 6.2	not stated	c = 0.00004, 2% dioxane, mixed constant	05	Uncert.	K54
5889	$C_{13}H_9O_3NCl_2$ 2,5-Cyclohexadien-1-one, 4-(3,5-dichloro-4-hydroxyphenyl)imino-2-methoxy- 5.6	not stated	c = 0.00004, 2% dioxane, mixed constant	05	Uncert.	K54
5890	$C_{13}H_9O_3NBr_2$ 2,5-Cyclohexadien-1-one, 4-(3,5-dibromo-4-hydroxyphenyl)imino-2-methoxy- 6.0	not stated	c = 0.00004, 2% dioxane, mixed constant	05	Uncert.	K54
5891	$C_{13}H_9O_3N_2Cl$ Benzoic acid, 5-(4-chlorophenylazo)-2-hydroxy- pK$_1$ 2.50, pK$_2$ 11.56			05		M108
5892	$C_{13}H_9O_3N_2Br$ Benzoic acid, 5-(4-bromophenylazo)-2-hydroxy- pK$_2$ 11.45			05		M108
5893	$C_{13}H_9O_3N_2I$ Benzoic acid, 2-hydroxy-5-(4-iodophenylazo)- pK$_2$ 11.36			05		M108
5894	$C_{13}H_9O_4N_2Cl$ 4-Chlorophenyl 4-nitroanilinomethanoate 12.3	25	I = 1.0(KCl)	05	Uncert.	H26
5895	$C_{13}H_9O_4N_2Br$ 3-Bromophenyl 4-nitroanilinomethanoate 12.0	25	I = 1.0(KCl)	05	Uncert.	H26

No.	Formula	Name / pK	Temp	Conditions	Method	Assessment	Ref.
5896	$C_{13}H_9O_4N_3S$	2-Naphthalenesulphonic acid, 6-hydroxy-5-(1,3-thiazol-2-ylazo)- 8.38	25 ± 0.5	Mixed constant	E3bg	Uncert.	N30
5897	$C_{13}H_9O_7N_3S_3$	2,7-Naphthalenedisulfonic acid, 3-hydroxy-4-(1,3-thiazol-2-ylazo)- pK_3 7.65		Conditions not stated	05	Uncert.	B184
5898	$C_{13}H_{10}O_2NCl$	Benzenecarboxamide, 4-chloro-N-(4-hydroxyphenyl)- 9.50	25	1.6% CH_3CN	05	Approx.	D46
5899	$C_{13}H_{10}O_2NCl$	Benzohydroxamic acid, 2-chloro-N-phenyl- 7.60	30	$c = 0.001$, $I = 0.1(NaClO_4)$, concentration constant	E3bg	Approx.	D62
5900	$C_{13}H_{10}O_6N_2S$	Benzoic acid, 2-hydroxy-5-(4-sulfophenylazo)- pK_1 2.38, pK_2 11.04	25	$I = 0.1(KNO_3)$, mixed constants	E3bg	Approx.	M119
5901	$C_{13}H_{11}ON_2Cl$	Benzenecarboxamide, N-(4-aminophenyl)-4-chloro- 4.54	25	1.6% CH_3CN	05	Approx.	D46
5902	$C_{13}H_{11}O_4NS$	Benzoic acid, 4-phenylsulfonylamino- 7.75($-SO_2NH-$)	20	$c \approx 0.0001$, $I = 0.1(KCl)$, mixed constant	05	Approx.	W33
5903	$C_{13}H_{11}O_5NS$	Benzoic acid, 2-hydroxy-4-phenylsulfonylamino- 7.61($-SO_2NH-$)	20	$c \approx 0.0001$, $I = 0.1(KCl)$, mixed constant	05	Approx.	W33

No.	Molecular formula, name and pK value(s)	T(°C)	Remarks	Method	Assessment	Ref.
5904	$C_{13}H_{12}O_3N_4S$ Formazan-3-sulfonic acid, 1,5-diphenyl-					I17
	pK_1 -0.92(SO_3H)	25	c ≈ 0.00002 in aq. $HClO_4$ solutions, H_0 scale	06	Uncert.	
	pK_2 12.90(NH)	25	c ≈ 0.00002	07		
5905	$C_{13}H_{12}O_4NP$ Phenyl hydrogen N-benzoylphosphoramidate					Z9
	pK_1 1.9	30	I = 0.1(KCl), 10% acetone, mixed constants	E3bg	Uncert.	
	pK_2 13.4			05	Uncert.	
			Solutions of monocyclohexyl ammonium salt used			
5906	$C_{13}H_{12}O_4N_2S$ Benzenesulfonic acid, 4-hydroxy-3-(p-tolylazo)-					M20
	7.72	not stated	c = 0.00005, mixed constant	05	Uncert.	
5907	$C_{13}H_{12}O_4N_2S$ Benzenesulfonic acid, 4-(2-hydroxy-5-methylphenylazo)-					M20
	8.75	not stated	c = 0.00005, mixed constant	05	Uncert.	
5908	$C_{13}H_{13}ON_2(I)$ 1-Methylpyridinium iodide, 2-(α-hydroxyiminobenzyl)-					W36
	8.7	not stated	c ≈ 0.04	E3bg	Uncert.	
5909	$C_{13}H_{13}O_2NS$ Benzenesulfonamide, 4-methyl-N-phenyl-					W33
	8.46	20	c ≈ 0.0001, I = 0.1(KCl), mixed constant	05	Approx.	
5910	$C_{13}H_{13}O_2NS$ Benzenesulfonamide, N-(p-tolyl)-					W33
	8.64	20	c ≈ 0.0001, I = 0.1(KCl), mixed constant	05	Approx.	

No.	Formula	Name	pK	T(°C)	Conditions	Method	Assessment	Ref.
5911	$C_{13}H_{13}O_2N_3S$	Benzothiazole, 2-(3,4-dihydroxyphenylazo)-4,5,6,7-tetrahydro-	pK_1 0.97, pK_2 7.8, pK_3 11.99					K10
5912	$C_{13}H_{13}O_3NS$	Benzenesulfonamide, 4-methoxy-N-phenyl-	8.66	20	$c \approx 0.0001$, I = 0.1(KCl), mixed constant	05	Approx.	W33
5913	$C_{13}H_{13}O_3NS$	Benzenesulfonamide, N-(4-methoxyphenyl)-	8.70	20	$c \approx 0.0001$, I = 0.1(KCl), mixed constant	05	Approx.	W33
5914	$C_{13}H_{14}O_2N_2S$	Acetic acid, 2-[(1H-6,7,8,9-tetrahydronaphtho[2,1-d]imidazol-2-yl)thio]-	pK_1 5.09, pK_2 11.91	20		05	Approx.	B132
5915	$C_{13}H_{14}O_2N_2S$	Benzenesulfonamide, 4-amino-N(S)-(m-tolyl)-	9.05(NH)	27±2	I = 0.2(NaCl), mixed constant	05	Uncert.	Y14
5916	$C_{13}H_{14}O_2N_2S$	Benzenesulfonamide, 4-amino-N(S)-(p-tolyl)-	9.25(NH)	27±2	I = 0.2(NaCl), mixed constant	05	Uncert.	Y14
5917	$C_{13}H_{14}O_3N_2S$	Benzenesulfonamide, 4-amino-N(S)-(3-methoxyphenyl)-	8.72(NH)	27±2	I = 0.2(NaCl), mixed constant	05	Uncert.	Y14
5918	$C_{13}H_{14}O_3N_2S$	Benzenesulfonamide, 4-amino-N(S)-(4-methoxyphenyl)-	9.34(NH)	27±2	I = 0.2(NaCl), mixed constant	05	Uncert.	Y14
5919	$C_{13}H_{14}O_4N_2S$	Thiazolidine-4-carboxylic acid, 3-benzoylaminoacetyl-	3.9(-COOH)	not stated	I = 0.1(KCl)	E3bg	Uncert.	W13

No.	Molecular formula, name and pK value(s)	$T(^oC)$	Remarks	Method	Assessment	Ref.
5920	$C_{13}H_{16}O_2N_2S$ Pyrimidine-4,6(1$\underline{H}$,5$\underline{H}$)-dione, 5-allyl-5-cyclohexen-1-yl-2,3-dihydro-2-thioxo-	28	2% ethanol	05	Uncert.	S14
	pK$_1$ 7.09, pK$_2$ 11.71					
5921	$C_{13}H_{16}O_2N_2S$ Pyrimidine-4,6(1$\underline{H}$,5$\underline{H}$)-dione, 5-allyl-5-cyclohexen-2-yl-2,3-dihydro-2-thioxo-	28	2% ethanol	05	Uncert.	S14
	pK$_1$ 7.25, pK$_2$ 12.16					
5922	$C_{13}H_8O_2N_3BrS$ 1,3-Benzenediol, 4-(6-bromobenzothiazol-2-ylazo)-			0		B189
	pK$_1$ 3.42(NH$^+$), pK$_2$ 5.85, pK$_3$ 10.01					
5923	$C_{13}H_8O_7N_3BrS_3$ 2,7-Naphthalenedisulfonic acid, 4-(5-bromo-1,3-thiazol-2-ylazo)-3-hydroxy-		Conditions not stated	05	Uncert.	B184
	pK$_3$ 7.23					
5924	$C_{14}H_5N_5$ 1,1,2-Ethenetricarbonitrile, 2-(4-dicyanomethylphenyl)- (Williams Blue)			06	Approx.	B118
	0.71	1				
	0.75	10				
	0.75	25	Thermodynamic quantities are derived			
	0.77	40	from these results			
5925	$C_{14}H_8O_4$ 9,10-Anthraquinone, 1,2-dihydroxy-	25±1	c ≈ 0.00001, I = 0.15	05	Uncert.	G18
	pK$_1$ 7.45, pK$_2$ 11.80					

No.	Formula	Name and pK	T (°C)	Conditions	Method	Assessment	Ref.
5926	$C_{14}H_8O_4$	9,10-Anthraquinone, 1,4-dihydroxy- pK_1 9.90, pK_2 11.18	25±1	$c \approx 0.00001$, $I = 0.065$	O5	Uncert.	G18
5927	$C_{14}H_8O_4$	9,10-Anthraquinone, 1,5-dihydroxy- pK_1 9.49, pK_2 11.05	25±1	$c \approx 0.00001$, $I = 0.065$	O5	Uncert.	G18
5928	$C_{14}H_8O_4$	9,10-Anthraquinone, 1,8-dihydroxy- pK_1 8.30, pK_2 12.46	25±1	$c \approx 0.00001$, $I = 0.065$	O5	Uncert.	G18
5929	$C_{14}H_8O_4$	9,10-Anthraquinone, 2,6-dihydroxy- pK_1 6.22, pK_2 8.28	25±1	$c \approx 0.00001$, $I = 0.2$	O5	Uncert.	G18
5930	$C_{14}H_{10}O_3$	Benzoic acid, 2-benzoyl- 3.536	25	$c \approx 0.00015-0.00113$	C2,R1c	Rel.	B122
5931	$C_{14}H_{10}O_3$	1,3-Indandione, 2-furfuryl- 5.71	not stated	$c \approx 0.0001$, 1% methanol	O5	Uncert.	S118
5932	$C_{14}H_{10}O_6$	2,2'-Bi-p-benzoquinone, 3,3'-dihydroxy-5,5'-dimethyl- (Phenicin) pK_1 3.02, pK_2 5.95	20	$c = 0.0001$, mixed constants	O5	Approx.	M128
5933	$C_{14}H_{10}O_6$	2,2'-Bi-p-benzoquinone, 5,5'-dihydroxy-3,3'-dimethyl- pK_1 3.93, pK_2 4.79	20	$c = 0.02$, mixed constants	E3bg	Uncert.	M128
5934	$C_{14}H_{12}O_2$	Biphenyl-2-ol, 3'-acetyl- 10.91			E		D49

No.	Molecular formula, name and pK value(s)	T(°C)	Remarks	Method	Assessment	Ref.
5935	C₁₄H₁₂O₂ Biphenyl-2-ol, 4'-acetyl- 10.78			E		D49
5936	C₁₄H₁₂O₅ 2,5-Cyclohexadiene-1,4-dione, 2-hydroxy-5-(2,4-dihydroxy-6-methylphenyl)-6-methyl- pK₁ 4.37, pK₂ 9.49	20	c = 0.0001, mixed constants	05	Approx.	M128
5937	C₁₄H₁₄O₂ Biphenyl-2-ol, 4'-ethoxy- 11.39			E		D49
5938	C₁₄H₁₄O₃ 1,3-Indandione, 2-isovaleryl- 3.3	18±2	1% ethanol, mixed constant	05 or E3d	Uncert.	I/25
5939	C₁₄H₁₄O₃ 1,3-Indandione, 2-pivaloyl- 4.19			0		T68
5940	C₁₄H₁₄O₄ 2H-Benzopyran-2-one, 8-acetyl-6-ethyl-7-hydroxy-4-methyl- 7.65	25	c = 0.0001	05	Approx.	M63
5941	C₁₄H₁₄O₄ 2,2',4,4'-Biphenyltetrol, 4,4'-dimethyl- pK₁ 8.54, pK₂ 9.30 pK₃ 11.32	20	c = 0.02, mixed constants	E3bg	Uncert. V.Uncert.	M128
5942	C₁₄H₁₄O₄ 2,2',4,4'-Biphenyltetrol, 6,6'-dimethyl- pK₁ 9.34, pK₂ 10.45 pK₃ 11.45	20	c = 0.02, mixed constants	E3bg	Uncert. V.Uncert.	M128

No.	Formula	Name / pK	t(°C)	Conditions	Method	Assessment	Ref.
5943	$C_{14}H_{16}O_2$	Benzoic acid, 4-(bicyclo[4.1.0]hept-1-yl)-					
		4.16		Conditions not stated	05	Uncert.	S62
5944	$C_{14}H_{18}O_4$	Pentanedioic acid, 3-phenyl-3-propyl-					
		pK_1 3.88, pK_2 6.94	25	c = 0.00125, mixed constants	E3bg	Approx.	B146
5945	$C_{14}H_{22}O$	Phenol, 2,4-di-tert-butyl-					
		11.72	20		05	Approx.	D30
		11.56	25	$I \approx 0.05$	05	Approx.	R39
5946	$C_{14}H_{22}O$	Phenol, 2,6-di-tert-butyl-					
		11.70	25	Extrapolated value		Uncert.	C70
5947	$C_{14}H_{22}O$	Phenol, 3,5-di-tert-butyl-					
		10.293	25	m = 0.00042	05a	Rel.	B93

Variation with temperature

10	15	20	30	35	40	45	50
10.513	10.431	10.359	10.233	10.176	10.122	10.073	10.023

Thermodynamic quantities are derived from the results

No.	Formula	Name / pK	t(°C)	Conditions	Method	Assessment	Ref.
5948	$C_{14}H_{24}N_2$	Pyrrolidine, 4,4-dimethyl-2-(4,4,5-trimethyl-1-pyrrolin-2-yl)methylene-					
		12.1		Conditions not stated		Uncert.	B145
5949	$C_{14}H_{28}O_2$	Tetradecanoic acid (Myristic acid)					
		≤6.3	20	Titration of Na-salt below c.m.c.	E3bg	Uncert.	M42

No.	Molecular formula, name and pK value(s)	$T(^{\circ}C)$	Remarks	Method	Assessment	Ref.
5950	$C_{14}H_8O_7S$ 2-Anthracenesulfonic acid, 9,10-dihydro-1,4-dihydroxy-9,10-dioxo-					
	pK_1 7.73, pK_2 9.68	25	c = 0.0001	O5	Uncert.	I23
	pK_2 8.7, $pK_3 \simeq 12$	23	I = 0.1(KIO₃), mixed constant	O5	Uncert.	O29
5951	$C_{14}H_8O_7S$ 2-Anthracenesulfonic acid, 9,10-dihydro-3,4-dihydroxy-9,10-dioxo- (Na-salt = Alizarin S)					
	pK_2 6.07	20	I = 0.1(KNO₃), mixed constants	E3bg	Approx.	B31,B27
	pK_3 11.1				Uncert.	
	pK_2 5.60, pK_3 11.03	25	I = 0.1-1.0	O,R4		B80
	pK_2 5.49, pK_3 10.85	25	c = 0.00002, I = 0.5, mixed constants	O5	Approx.	Z10
			Other values in B28 and O29			
5952	$C_{14}H_{10}ON_4$ 8-Quinolinol, 7-(2-pyridylazo)-					
	7.9	not stated	Mixed constant	O5	Uncert.	B182
5953	$C_{14}H_{10}O_2S$ 1,3-Indandione, 2-(2-thenyl)-					
	5.73		c ≈ 0.0001, 1% methanol	O5	Uncert.	S118
5954	$C_{14}H_{11}O_5N$ Benzoic acid, 2-(2,6-dihydroxybenzoyl)amino-					
	pK_2 7.8	not stated	1% ethanol, mixed constant	O5 or E3d	Uncert.	L9
5955	$C_{14}H_{11}O_5N$ Benzoic acid, 4-(2,6-dihydroxybenzoyl)amino-					
	pK_2 6.6		1% ethanol, mixed constant	O5 or E3d	Uncert.	L9

No.	Formula	Name	pK values	T (°C)	Conditions	Reliability	Assessment	Ref.
5956	$C_{14}H_{11}O_6N$	Benzoic acid, 4-(3,4,5-trihydroxybenzoyl)amino-	pK_1 4.30, pK_2 8.68				E	S85
5957	$C_{14}H_{12}O_2N_2$	1,2-Ethanedione dioxime, 1,2-diphenyl- (Benzil dioxime)	8.50	25		Uncert.	E3bg	B10
5958	$C_{14}H_{12}O_2N_2$	Spiro[naphthalene-1($2\underline{H}$),3'-pyrrolidine]-4'-carbonitrile, 3,4-dihydro-2',5'-dioxo-	6.30	20	c = 0.0001-0.001, 4% ethanol, mixed constant	Uncert.	E3bg	F29
5959	$C_{14}H_{12}O_3N_2$	Benzoic acid, 2-hydroxy-5-(2-methylphenylazo)-	pK_1 2.12, pK_2 11.61				05	M108
5960	$C_{14}H_{12}O_3N_2$	Benzoic acid, 2-hydroxy-5-(3-methylphenylazo)-	pK_2 11.61				05	M108
5961	$C_{14}H_{12}O_3N_2$	Benzoic acid, 2-hydroxy-5-(4-methylphenylazo)-	pK_2 11.63				05	M108
5962	$C_{14}H_{12}O_3N_2$	Benzoic acid, 2-(2-hydroxy-5-methylphenylazo)-	pK_2 11.4	not stated	I = 0.1(KCl)	Uncert.	05	D37
5963	$C_{14}H_{12}O_3N_4$	Pteridine-2,4,7($1\underline{H},3\underline{H},8\underline{H}$)-trione, 8-benzyl-6-methyl-	3.67, 13.1	20		Approx. / Uncert.	E3bg / 06	P54
5964	$C_{14}H_{12}O_4N_2$	Benzoic acid, 2-hydroxy-5-(4-methoxyphenylazo)-	pK_2 11.62				05	M108

No.	Molecular formula, name and pK value(s)	T(°C)	Remarks	Method	Assessment	Ref.
5965	$C_{14}H_{12}O_4N_2$ Ethane, 1,1-dinitro-2,2-diphenyl-					T58a
	4.97	20		05	Approx.	
5966	$C_{14}H_{13}O_2N$ Acetohydroxamic acid, 2,N-diphenyl-					
	8.43	30	c = 0.001, I = 0.1($NaClO_4$), concentration constant	E3bg	Approx.	D62
	7.87		Solubility and partition measurements			Z6
5967	$C_{14}H_{13}O_2N$ Benzenecarboxamide, N-(4-hydroxyphenyl)-4-methyl-					
	9.54	25	1.6% CH_3CN	05	Approx.	D46
5968	$C_{14}H_{13}O_2N$ Benzohydroxamic acid, 2-methyl-N-phenyl-					
	8.08	30	c = 0.001, I = 0.1($NaClO_4$), concentration constant	E3bg	Approx.	D62
	8.41	25	Extrapolated from results in dioxane-water	E3bg	Uncert.	S67
5969	$C_{14}H_{13}O_2N$ Benzohydroxamic acid, 3-methyl-N-phenyl-					
	8.53	25	Extrapolated from results in dioxane-water	E3bg	Uncert.	S67
5970	$C_{14}H_{13}O_2N$ Benzohydroxamic acid, 4-methyl-N-phenyl-					
	pK_1 2.26, pK_2 8.31	30	I = 0.1($NaClO_4$)	E3bg	Approx.	D63
	pK_2 8.31	30	c = 0.001, I = 0.1($NaClO_4$), concentration constant	E3bg	Approx.	D62
	pK_2 8.64	25	Extrapolated from results in dioxane-water	E3bg	Uncert.	S67

No.	Formula / Name	pK	T (°C)	Conditions	Method	Assessment	Ref.
5971	C₁₄H₁₃O₂N Benzohydroxamic acid, N-(o-tolyl)-						
	pK_1 2.20, pK_2 8.18	30	I = 0.1($NaClO_4$)	E3bg	Approx.	D63
	pK_2 8.18	30	c = 0.01, I = 0.1($NaClO_4$), concentration constant	E3bg	Approx.	D62
	pK_2 8.58	25	Extrapolated from results in dioxane-water	E3bg	Uncert.	S67
5972	C₁₄H₁₃O₂N Benzohydroxamic acid, N-(m-tolyl)-						
	pK_2 8.67	25	Extrapolated from results in dioxane-water	E3bg	Uncert.	S67
5973	C₁₄H₁₃O₂N Benzohydroxamic acid, N-(p-tolyl)-						
	pK_1 2.50, pK_2 8.27	30	I = 0.1($NaClO_4$)	E3bg	Approx.	D63
	pK_2 8.27	30	c = 0.01, I = 0.1($NaClO_4$), concentration constant	E3bg	Approx.	D62
	pK_2 8.72	25	Extrapolated from results in dioxane-water	E3bg	Uncert.	S67
5974	C₁₄H₁₃O₂N 2,5-Cyclohexadien-1-one, 4-(4-hydroxy-3,5-dimethylphenyl)imino-						
	8.7	not stated	c = 0.00004, 2% dioxane, mixed constant	05	Uncert.	K54
5975	C₁₄H₁₃O₂N₃ Acetophenone, 4-(3-hydroxy-3-phenyltriazeno)-						
	10.96	25	c = 0.00004, I = 0.1(KCl), mixed constant	05	Approx.	P89,P91
5976	C₁₄H₁₃O₃N Benzenecarboxamide, N-(4-hydroxyphenyl)-4-methoxy-						
	pK_1 9.55	25	1.6% CH_3CN	05	Approx.	D46

No.	Molecular formula, name and pK value(s)	T(°C)	Remarks	Method	Assessment	Ref.
5977	$C_{14}H_{13}O_3N$ Benzohydroxamic acid, 4-methoxy-N-phenyl-					
	pK_1 2.05, pK_2 8.15	30	$I = 0.1(NaClO_4)$	E3bg	Approx.	D63
	pK_2 8.15	30	$c = 0.001$, $I = 0.1(NaClO_4)$, concentration constant	E3bg	Approx.	D62
5978	$C_{14}H_{13}O_3P$ Acetic acid, 2-diphenylphosphinyl-					
	3.62	25	$c = 0.01$, mixed constant	E3bg	Approx.	M52
			Value in mixed solvent is also given			
5979	$C_{14}H_{13}O_4N$ 2,5-Cyclohexadien-1-one, 4-(4-hydroxy-3,5-dimethoxyphenyl)imino-					
	9.0	not stated	$c = 0.00004$, 2% dioxane, mixed constant	05	Uncert.	K54
5980	$C_{14}H_{13}O_4N$ Bis(methoxycarbonyl)quinoliniomethanide					
	4.62 (isomer mp 197)			05	Uncert.	N31
	3.82 (isomer mp 213)					
			Values in mixed solvents are also given			
5981	$C_{14}H_{13}O_4N$ Bis(methoxycarbonyl)isoquinoliniomethanide					
	5.20			.05	Uncert.	N31
			Values in mixed solvents are also given			
5982	$C_{14}H_{14}ON_2$ Acetohydrazide, 2,2-diphenyl-					
	pK_1 2.98	25	Mixed constant(?)	E3bg	Uncert.	T43,T44
	13.00(NH)	25		05	Uncert.	

No.	Formula	Name / pK	T (°C)	Conditions	Method	Assessment	Ref.
5983	$C_{14}H_{14}ON_2$	Benzenecarboxamide, N-(4-aminophenyl)-4-methyl-					
		pK_1 4.52	25	1.6% CH_3CN	05	Approx.	D46
5984	$C_{14}H_{14}O_2N_2$	Benzenecarboxamide, N-(4-aminophenyl)-4-methoxy-					
		pK_1 4.62	25	1.6% CH_3CN	05	Approx.	D46
5985	$C_{14}H_{14}O_2N_4$	Acetamide, N-[4-(3-hydroxy-3-phenyltriazeno)phenyl]-					
		11.66(OH)	25	c = 0.00004, I = 0.1(KCl), mixed constant	05	Approx.	P89,P91
5986	$C_{14}H_{14}O_3N_2$	Pyrimidine-2,4,6(1H,3H,5H)- trione, 5-allyl-5-benzyl-					
		7.21	25	c = 0.005, mixed constant	E3ag	Approx.	K53
5987	$C_{14}H_{14}O_4N_4$	Propanediamide, 2-[(phenyl)(1,2,3,4-tetrahydro-2,4-dioxopyrimidin-5-yl)]methyl-					
		pK_1 6.18	20	c = 0.005, mixed constant	05	Approx.	B142
		pK_2 13.21	20		06	Uncert.	
5988	$C_{14}H_{15}ON_3$	Triazen-3-ol, 1,3-bis(p-tolyl)-					
		12.18	25	c = 0.00004, I = 0.1(KCl), mixed constant	05	Approx.	P89,P91
5989	$C_{14}H_{15}O_2P$	Bis(p-tolyl)phosphinic acid					
		2.47	not stated	c = 0.005, I ≠ 0.025, mixed constant, 7% ethanol	E3bg	Uncert.	M57
5990	$C_{14}H_{15}O_3P$	p-Tolyl hydrogen p-tolylphosphonate					
		0.92	20±2	c ≈ 0.001 or 0.0001, I ≈ 0.1 or 1($NaClO_4$)	DIS	Uncert.	K55

No.	Molecular formula, name and pK value(s)	$T(^{o}C)$	Remarks	Method	Assessment	Ref.
5991	$C_{14}H_{15}O_4P$ Dibenzyl hydrogen phosphate					
	1.0	75.6	c = 0.006-0.014	O2	Uncert.	K66
	0.70	20	c ≈ 0.001 or 0.0001, I ≈ 0.1 or 1($NaClO_4$), concentration constant	DIS	Uncert.	K56
5992	$C_{14}H_{15}O_4P$ Bis(p-tolyl) hydrogen phosphate					
	0.40	20±2	c ≈ 0.001 or 0.0001, I ≈ 0.1 or 1($NaClO_4$), concentration constant	DIS	Uncert.	K56
	1.36	not stated	c = 0.005, I ≠ 0.025, mixed constant, 7% ethanol	E3bg	Uncert.	M57
5993	$C_{14}H_{15}S_2P$ Bis(p-tolyl)phosphinodithioic acid					
	1.81	20	c = 0.005, 7% ethanol	E3bg	V.Uncert.	K6,K5
5994	$C_{14}H_{16}O_2N_4$ 1,3-Cyclohexanedione, 2-(3,4-dihydropteridin-4-yl)-5,5-dimethyl-					
	pK_1 3.55	20	c = 0.00006-0.0007	O5	Approx.	A32
	pK_2 7.69					
5995	$C_{14}H_{16}O_3N_2$ Pyrimidine-2,4,6(1H,3H,5H)-trione, 5-benzyl-5-isopropyl-					
	7.99	25		E3ag	Approx.	K53
5996	$C_{14}H_{16}O_3N_2$ Pyrimidine-2,4,6(1H,3H,5H)-trione, 5-ethyl-5-(2-phenylethyl)-					
	7.90	25		E3ag	Approx.	K53

No.	Formula	Name	T	Conditions	Method	Reliability	Ref.
5997	$C_{14}H_{16}O_8N_2$	1,2-Phenylenedinitrilotetraacetic acid 2.94, 3.83, 5.01, 6.82	30	I = 0.1(KCl), concentration constants	E3bg,R3h	Approx.	G48
5998	$C_{14}H_{16}O_8N_2$	1,3-Phenylenedinitrilotetraacetic acid pK_2 2.92, pK_3 5.07, pK_4 5.82 $pK_1 \simeq 0$	25	I = 0.1(KCl), concentration constants	E3bg,R3g	Approx.	U7
5999	$C_{14}H_{16}O_8N_2$	1,4-Phenylenedinitrilotetraacetic acid pK_1 2.1 pK_2 2.79, pK_3 4.99, pK_4 6.01 pK_1 2.58, pK_2 3.37, pK_3 5.39, pK_4 6.35	25	I = 0.1(KCl), concentration constants	E3bg,R3g	Uncert. Approx.	U7 R25
6000	$C_{14}H_{18}O_7N_2$	Pseudouridine-5'-acetate, O(2'),O(3')-isopropylidene- 9.10, >12.5		Conditions not stated	O5	Uncert.	D53
6001	$C_{14}H_{19}O_2N$	Hexanamide, 2-acetyl-N-phenyl- 12.93	22	5% v/v ethanol, mixed constant	O5/E3bg	Uncert.	H38
6002	$C_{14}H_{20}O_6N_2$	Benzoic acid, 2,5-bis(2-hydroxyethylamino)-4-ethoxycarbonyl- pK_1 2.40, pK_2 5.05	22	c ≈ 0.003, I = 0.1(KCl), mixed constant	E3bg	Approx.	U3
6003	$C_{14}H_{20}O_6N_6$	3,6,9,12-Tetraazatetradecanoic acid, 2-(imidazol-4-yl)methyl-4,7,10,13-tetraoxo- (N-Acetylglycylglycylglycylglycyl-L-histidine) pK_1 3.16, pK_2 7.21	25	I = 0.16(KCl)	E3bg	Approx.	B153

No.	Molecular formula, name and pK value(s)	T(°C)	Remarks	Method	Assessment	Ref.
6004	C$_{14}$H$_{20}$O$_7$N$_2$ 1,4-Benzenedicarboxylic acid, 2-[bis(2-hydroxyethyl)]amino-5-(2-hydroxyethyl)amino- pK$_1$ 3.65, pK$_2$ 8.29	22	I = 0.1(KCl)	E3bg	Uncert.	U5
6005	C$_{14}$H$_{21}$OBr Phenol, 4-bromo-2,6-di-tert-butyl- 10.83	25	Extrapolated value		Uncert.	C70
6006	C$_{14}$H$_{21}$O$_2$N Phenol, 2,6-di-tert-butyl-4-nitroso- 8.18	25		05	Approx.	C70
6007	C$_{14}$H$_{21}$O$_3$N Phenol, 2,6-di-tert-butyl-4-nitro- 6.65	25		05	Approx.	C70
6008	C$_{14}$H$_{22}$O$_4$S Benzenesulfonic acid, 3,5-di-tert-butyl-4-hydroxy- pK$_2$ 10.40	25		05	Approx.	C70
6009	C$_{14}$H$_{22}$O$_8$N$_2$ 1,3-Cyclohexylenedinitrilotetraacetic acid pK$_1$ 2.40, pK$_2$ 3.55, pK$_3$ 6.14, pK$_4$ 11.70	20	I = 0.1(KNO$_3$), concentration constants	E3bg	Approx.	M96

Variation of pK$_3$ and pK$_4$ with temperature

	15	25	30	35	40
pK$_3$	6.15	6.12	6.10	6.08	6.07
pK$_4$	11.83	11.58	11.52	11.43	11.34

	pK$_1$ 2.43, pK$_2$ 3.52 pK$_3$ 6.12, pK$_4$ 11.70	20	I = 0.1(KCl)	E3bg	Uncert. Approx.	S35

(contd)

6009 (Contd) $C_{14}H_{22}O_8N_2$ 1,2-Cyclohexylenedinitrilotetraacetic acid

	T	conditions			
pK_1 1.34, pK_2 3.20, pK_3 5.75, pK_4 9.26(trans)	20	$I = 0.5$	05	Uncert.	N40
pK_3 6.12	20	$I = 0.1(KNO_3)$	E3bg	Approx.	A53
pK_4 12.35			05	Uncert.	

Thermodynamic quantities are also given

pK_3 6.22, pK_4 13.09(ℓ-trans)	25		ROT	Uncert.	C15

Other values in A55 and S26

6010 $C_{14}H_{22}O_8N_2$ 1,3-Cyclohexylenedinitrilotetraacetic acid

pK_1 1.77, pK_2 2.57	20	$I = 0.1(KCl)$	E3bg	Uncert.	S35
pK_3 8.55, pK_4 10.91(trans)				Approx.	

6011 $C_{14}H_{22}O_8N_2$ 1,4-Cyclohexylenedinitrilotetraacetic acid

pK_1 2.07, pK_2 2.52	20	$I = 0.1(KCl)$	E3bg	Uncert.	S35
pK_3 9.04, pK_4 10.86(trans)				Approx.	

6012 $C_{14}H_{22}O_8S_4$ 4,7-Dithiadecanoic acid, 5,6-bis(2-carboxyethylthio)- (Ethanediylidenetetrakis(3-thiopropanoic acid))

pK_1 3.78, pK_2 3.87, pK_3 4.76, pK_4 5.36		$I = 1.10$			C11

6013 $C_{14}H_{22}O_{10}N_4$ 3,6,9,12-Tetraazatetradecanedioic acid, 3,12-bis(carboxymethyl)-5,10-dioxo-

$pK_1 \approx 1.2$	25	$c \approx 0.001$, $I = 0.1(KNO_3)$, concentration	E3bg	V.Uncert.	M104
pK_2 1.93, pK_3 2.66, pK_4 6.18, pK_5 7.30		constants		Uncert.	

No.	Molecular formula, name and pK value(s)	T(^oC)	Remarks	Method	Assessment	Ref.
6014	$C_{14}H_{22}O_{10}N_4$ 3,6,9,12-Tetraazatetradecanedioic acid, 6,9-bis(carboxymethyl)-4,11-dioxo-					M104
	pK$_1$ 2.87, pK$_2$ 3.51, pK$_3$ 4.38, pK$_4$ 7.37	25	c ≈ 0.001, I = 0.1(KNO$_3$), concentration constants	E3bg	Uncert.	
6015	$C_{14}H_{23}O_{10}N_3$ 3,6,9-Triazaundecanedioic acid, 3,6,9-tris(carboxymethyl)- ([[N'-Carboxymethyl-2,2'-iminodiethylene)dinitrilo]tetra-acetic acid) (DTPA)					M97
	pK$_1$ 1.80	20	c = 0.001, I = 0.1(KNO$_3$), concentration constants	E3bg	Uncert. Approx.	
	pK$_2$ 2.55, pK$_3$ 4.33, pK$_4$ 8.60, pK$_5$ 10.58					

Variation of pK$_3$, pK$_4$ and pK$_5$ with temperature

	10	30	40
pK$_3$	4.39	4.30	4.27
pK$_4$	8.72	8.46	8.37
pK$_5$	10.63	10.34	10.23

Thermodynamic quantities are derived from the results

No.	Molecular formula, name and pK value(s)	T(^oC)	Remarks	Method	Assessment	Ref.
	pK$_1$ 1.5, pK$_2$ 2.6	20	I = 0.1(KNO$_3$)	E3bg	Uncert. Approx.	A55
	pK$_3$ 4.27, pK$_4$ 8.60, pK$_5$ 10.58					
	pK$_1$ 2.2, pK$_2$ 2.7		I = 1.0((CH$_3$)$_4$NCl)	E3bh	Uncert. Approx.	
	pK$_3$ 4.14, pK$_4$ 8.41, pK$_5$ 10.46					

Concentration constants

No.	Molecular formula, name and pK value(s)	T(^oC)	Remarks	Method	Assessment	Ref.
	pK$_1$ 2.08, pK$_2$ 2.41	20	c = 0.0006, I = 0.1(KCl), concentration constants	E3bg	Uncert. Approx.	D58
	pK$_3$ 4.26, pK$_4$ 8.60, pK$_5$ 10.55					

(contd)

6015 $C_{14}H_{23}O_{10}N_3$ 3,6,9-Triazaundecanedioic acid, 3,6,9-tris(carboxymethyl)-   ([[N-Carboxymethyl-2,2'-iminodiethylene)dinitrilo]tetra-
acetic acid) (DTPA)
(Contd)

pK₁ 1.79, pK₂ 2.56, pK₃ 4.42, pK₄ 8.76, pK₅ 10.42

25   I = 0.1(KNO₃), mixed constants   E3bg   Uncert.   F44

Other values in M40 and W9

6016 $C_{14}H_{24}O_3N_2$ Pyrimidine-2,4,6(1H,3H,5H)-trione, 5-ethyl-5-(2-ethylhexyl)-

7.89   25   E3ag   Approx.   K53

6017 $C_{14}H_{24}O_6N_2$ 3,6-Diazaoctanedioic acid, 3-carboxymethyl-6-cyclohexyl-   (Ethylenedinitrilo-N-cyclohexyl-N,N',N''-triacetic acid)

pK₁ 1.6   24.2   c = 0.002   E3bg   Uncert.   B149

pK₂ 6.47, pK₃ 10.15   Approx.

6018 $C_{14}H_{24}O_7N_2$ 3,6-Diazaoctanedioic acid, 3-carboxymethyl-6-(2-hydroxycyclohexyl)-   (Ethylenedinitrilo-N-(2-hydroxycyclohexyl)-
N,N',N''-triacetic acid)

pK₁ 2.49, pK₂ 5.69, pK₃ 10.38   20   I = 0.1(KCl)   E3bg   Approx.   S27

6019 $C_{14}H_{24}O_8N_2$ 3,10-Diazadodecanedioic acid, 3,10-bis(carboxymethyl)-   (Hexamethylenedinitrilotetraacetic acid)

pK₁ 2.2, pK₂ 2.7   20   I = 0.1(KNO₃)   E3bg   Uncert.   A54

pK₃ 9.79, pK₄ 10.81   Approx.

Thermodynamic quantities are also given

pK₁ 2.45, pK₂ 2.35   20   I = 1.0(KNO₃), concentration constants   E3bg   Uncert.   A55

pK₃ 9.69, pK₄ 10.56   Approx.

pK₁ 2.2, pK₂ 2.70   20   I = 0.1(KCl), concentration constants   E3bg   Uncert.   S50

pK₃ 9.75, pK₄ 10.65   Approx.

(contd)

No.	Molecular formula, name and pK value(s)	T(°C)	Remarks	Method	Assessment	Ref.
6019 (Contd)	$C_{14}H_{24}O_8N_2$ 3,10-Diazadodecanedioic acid, 3,10-bis(carboxymethyl)- (Hexamethylenedinitrilotetraacetic acid)					
	pK_1 2.41, pK_2 2.91	25	c = 0.005, I = 0.1(KNO_3)	E3bg,R3g	Uncert.	G32
	pK_3 9.65, pK_4 10.72				Approx.	
6020	$C_{14}H_{24}O_9N_2$ 7-oxa-3,11-diazatridecanedioic acid, 3,11-bis(carboxymethyl)- (Oxydi(trimethylene)dinitrilotetraacetic acid)					
	pK_1 2.1, pK_2 2.7	20	I = 0.1(KCl), concentration constants	E3bg	Uncert.	S50
	pK_3 9.67, pK_4 10.17				Approx.	
	pK_1 2.0	20	I = 0.1(KCl), concentration constants	E3bg	Uncert.	I19
	pK_2 2.74, pK_3 9.64, pK_4 10.14				Approx.	
6021	$C_{14}H_{24}O_{10}N_2$ 6,9-Dioxa-3,12-diazatetradecanedioic acid, 3,12-bis(carboxymethyl)- ((Ethylenedioxy)diethylenedinitrilotetraacetic acid) (EGTA)					
	pK_1 2.0, pK_2 2.65	20	I = 0.1(KNO_3), concentration constants	E3bg	Uncert.	A54
	pK_3 8.85, pK_4 9.46				Approx.	
			Thermodynamic quantities are given			
	pK_1 2.4, pK_2 2.5	20	I = 1.0(KNO_3), concentration constants	E3bg	Uncert.	A55
	pK_3 8.67, pK_4 9.22				Approx.	
	pK_1 1.15, pK_2 2.40, pK_3 8.40, pK_4 8.94	20	I = 0.5, mixed constants	O5	Uncert.	N40
	pK_3 8.88, pK_4 9.53	25	I = 0.1(KCl), mixed constants	E3bg,R3d	Approx.	B119

Variation with temperature

	5	15	30
pK_3	9.20	9.01	8.77
pK_4	9.95	9.69	9.38

(Other values in F39 and S50)

6022 $C_{14}H_{25}O_9N_3$ 3,6,9-Triazaundecanedioic acid, 3,9-bis(carboxymethyl)-6-(2-hydroxyethyl)- (N-(2-hydroxyethyl)iminobis(ethylenenitrilo)-tetraacetic acid)

pK₁ 2.58, pK₂ 3.54, pK₃ 8.00, pK₄ 9.3   not stated   c = 0.001, I = 0.1(CaCl₂), mixed constants   E3bg   Uncert.   V7

6023 $C_{14}H_{32}O_4N_2$ 4,7-Diazadecane-2,9-diol, 4,7-bis(2-hydroxypropyl)-

pK₁ 4.24, pK₂ 8.75   25   I = 0.1(NaClO₄)   E3bg   Approx.   R42

6024 $C_{14}H_9O_4N_2F_3$ Benzohydroxamic acid, 5-nitro-N-phenyl-3-trifluoromethyl-

7.85   25   I = 0.01; phosphate buffer/benzene   DIS   Uncert.   L20

6025 $C_{14}H_9O_{12}N_7S_2$ 1,5-Diphenylformazan-3',3''-disulfonic acid, 3-cyano-2',2''-dihydroxy-5',5''-dinitro- (Formazan, 1,5-bis(2-hydroxy-5-nitro-3-sulfophenyl)-3-cyano-)

pK₃ 3.79, pK₄ 5.19   19   c = 0.002, I = 0.1(KCl), mixed constants   E3bg   Approx.   E21
pK₅ 10.01   Uncert.

6026 $C_{14}H_{10}O_2NF_3$ Benzohydroxamic acid, N-phenyl-2-trifluoromethyl-

7.88   25   I = 0.01; phosphate buffer/benzene   DIS   Uncert.   L20

6027 $C_{14}H_{10}O_2NF_3$ Benzohydroxamic acid, N-phenyl-3-trifluoromethyl-

8.15   25   I = 0.01; phosphate buffer/benzene   DIS   Uncert.   L20

6028 $C_{14}H_{10}O_2NF_3$ Benzohydroxamic acid, N-phenyl-4-trifluoromethyl-

7.98   25   I = 0.01; phosphate buffer/chloroform   DIS   Uncert.   L20

6029 $C_{14}H_{11}O_2NCl_2$ 2,5-Cyclohexadien-1-one, 4-(3,5-dichloro-4-hydroxyphenyl)imino-2,5-dimethyl-

6.1   not stated   c = 0.00004, 2% dioxane, mixed constant   O5   Uncert.   K54

No.	Molecular formula, name and pK value(s)	$T(^{\circ}C)$	Remarks	Method	Assessment	Ref.
6030	$C_{14}H_{11}O_2NCl_2$ 2,5-Cyclohexadien-1-one, 4-(3,5-dichloro-4-hydroxyphenyl)imino-2,6-dimethyl- 5.9	not stated	c = 0.00004, 2% dioxane, mixed constant	05	Uncert.	K54
6031	$C_{14}H_{11}O_2NBr_2$ 2,5-Cyclohexadien-1-one, 2,6-dibromo-4-(2-ethyl-4-hydroxyphenyl)imino- 6.4	not stated	c = 0.00004, 2% dioxane, mixed constant	05	Uncert.	K54
6032	$C_{14}H_{11}O_2NBr_2$ 2,5-Cyclohexadien-1-one, 4-(3,5-dibromo-4-hydroxyphenyl)imino-2,6-dimethyl- 5.4	not stated	c = 0.00004, 2% dioxane, mixed constant	05	Uncert.	K54
6033	$C_{14}H_{11}O_2NBr_2$ 2,5-Cyclohexadien-1-one, 4-(3,5-dibromo-4-hydroxyphenyl)imino-2-ethyl- 5.2	not stated	c = 0.00004, 2% dioxane, mixed constant	05	Uncert.	K54
6034	$C_{14}H_{11}O_8N_5S_2$ 1,5-Diphenylformazan-3',3''-disulfonic acid, 3-cyano-2',2''-dihydroxy- (Formazan, 1,5-bis(2-hydroxy-5-sulfophenyl)-3-cyano-) pK_3 6.65, pK_4 8.28 pK_5 11.80	19	c = 0.002, I = 0.1(KCl), mixed constants	E3bg	Approx. Uncert	E21
6035	$C_{14}H_{13}ONS$ Benzenecarbothioamide, N-(4-methoxyphenyl)- 11.0	not stated	c = 0.00001	05	Uncert.	W4
6036	$C_{14}H_{13}O_3NS$ Benzenesulfonamide, N-(p-acetylphenyl)- 6.94	20	c ≈ 0.0001, I = 0.1(KCl), mixed constant	05	Approx.	W33

6037 C₁₄H₁₃O₇NS Iminodiacetic acid, N-(4-sulfonaphth-1-yl)-

$C_{14}H_{13}O_7NS$ Iminodiacetic acid, N-(4-sulfonaphth-1-yl)-

6037 $C_{14}H_{13}O_7NS$ Iminodiacetic acid, N-(4-sulfonaphth-1-yl)-
pK₂ 2.76, pK₃ 3.92   Conditions not stated   05   Uncert.   T17

6038 $C_{14}H_{13}O_8NS$ Iminodiacetic acid, N-(2-hydroxy-4-sulfonaphth-1-yl)-
pK₂ 2.16, pK₃ 3.38, pK₄ 8.85   Conditions not stated   05   Uncert.   T17

6039 $C_{14}H_{13}O_{11}NS_2$ Iminodiacetic acid, N-(8-hydroxy-3,6-disulfonaphth-1-yl)-
pK₃ 2.58, pK₄ 3.87, pK₅ 9.48   Conditions not stated   05   Uncert.   T17

6040 $C_{14}H_{14}O_3N_2S$ Benzenesulfonamide, 4-amino-N(S)-(3-acetylphenyl)-
8.34(NH)   27±2   I = 0.2(NaCl), mixed constant   05   Uncert.   Y14

6041 $C_{14}H_{14}O_3N_2S$ Benzenesulfonamide, 4-amino-N(S)-(4-acetylphenyl)-
7.61(NH)   27±2   I = 0.2(NaCl), mixed constant   05   Uncert.   Y14

6042 $C_{14}H_{16}O_2N_4S$ 1,3-Cyclohexanedione, 2-(1,2,3,4-tetrahydro-2-thioxopteridin-4-yl)-5,5-dimethyl-
12.20   20   c = 0.0001   05   Approx.   A30

6043 $C_{14}H_{17}O_2N_3S$ Benzenesulfonamide, 4-amino-N(S)-(3-dimethylaminophenyl)-
9.01(NH)   27±2   I = 0.2(NaCl), mixed constant   05   Uncert.   Y14

6044 $C_{14}H_{17}O_2N_3S$ Benzenesulfonamide, 4-amino-N(S)-(4-dimethylaminophenyl)-
9.46(NH)   27±2   I = 0.2(NaCl), mixed constant   05   Uncert.   Y14

6045 $C_{14}H_{20}O_5N_6S$ Butanoic acid, 2-amino-4-(5'-thioadenosyl)-   (Homocysteine, S-(5'-adenosyl)-)
pK₁ 1.95(COOH)
pK₂ 3.5(adenine)   not stated   E3bg/05   Uncert.   K34

(contd)

No.	Molecular formula, name and pK value(s)	T(°C)	Remarks	Method	Assessment	Ref.
6045 (Contd)	$C_{14}H_{20}O_5N_6S$ Butanoic acid, 2-amino-4-(5'-thioadenosyl)- (Homocysteine, S-(5'-adenosyl)-)		pK_1 was determined from infra-red measurements in D_2O and corrected to corresponding value in water by subtracting 0.5 pK unit			
6046	$C_{14}H_{23}ON_3(Cl_2)$ Piperazinediium dichloride, 1-(N-acetyl-2,5-dimethylanilino)- pK_1 3.8, pK_2 9.3	25		E3bg	Uncert.	F35a
6047	$C_{14}H_{23}ON_3(Cl_2)$ Piperazinediium dichloride, 1-(N-acetyl-2,6-dimethylanilino)- pK_1 4.2, pK_2 9.4	25		E3bg	Uncert.	F35a
6048	$C_{14}H_{23}O_3N_6S$ (cation) Adenosine, 5'-[S-(3-aminopropyl)-S-methyl]sulfonio- pK_1 3.5(adenine)		Conditions not stated	E3bg/O5	Uncert.	K34
6049	$C_{14}H_{24}O_8N_2S_2$ 6,9-Dithia-3,12-diazatetradecanedioic acid, 3,12-bis(carboxymethyl)- (Ethylenedithio-bis(ethylenenitrilo)tetra- acetic acid) pK_1 1.9 pK_2 2.56, pK_3 8.52, pK_4 9.22	20	$I = 0.1(KCl)$, concentration constants	E3bg	Uncert. Approx.	S50
6050	$C_{14}H_{28}O_5N_2S$ 4,7-Diazadecan-5-one, 4,7-bis(3-hydroxypropyl)-1,10-dihydroxy-6-thioxo- (Ilonothiooxamide, N,N,N',N'-tetrakis(3- hydroxypropyl)-) 11.79 11.67	25	$I = 0.1(KCl)$, mixed constant (Values for $I = 0.02, 0.05, 0.25, 0.50, 1.00$ also given)	O5,R4 O5	Approx. Approx.	C13

6051   C$_{14}$H$_{30}$O$_2$(Cl)P   Tributyl carboxymethylphosphonium chloride   (Acetic acid, 2-(tributylphosphonio)-, chloride)

25	2.34	c = 0.01, mixed constant	E3bg	Uncert.	M52

6052   C$_{14}$H$_9$O$_8$N$_5$Cl$_2$S$_2$ 1,5-Diphenylformazan-3',3''-disulfonic acid, 5',5''-dichloro-3-cyano-2',2''-dihydroxy-

19	pK$_3$ 5.83, pK$_4$ 8.68, pK$_5$ 11.14	c = 0.0002, I = 0.1(KCl), mixed constants	05	Uncert.	E21
19	pK$_3$ 5.98, pK$_4$ 8.74	c = 0.002, I = 0.1(KCl), mixed constants	E3bg	Approx.	
	pK$_5$ 11.13			Uncert.	

6053   C$_{14}$H$_{12}$O$_7$N$_5$SAs   1,5-Diphenylformazan-3'-sulfonic acid, 2''-arsono-6'-hydroxy-

19	pK$_1$ 6.88, pK$_2$ 8.40, pK$_3$ 11.54	c = 0.002, I = 0.1(KCl)	E3bg	Approx.	E21

6054   C$_{15}$H$_{10}$O$_2$ 1,3-Indandione, 2-phenyl-

20	4.13	1% methanol	05	Approx.	L44

Values in mixed solvents are also given

not stated	4.18	1% methanol	E3bg	Uncert.	G46
18±2	4.4	1% ethanol, mixed constant	05 or E3d	Uncert.	W25
	4.18	1% methanol	05		L46
20	4.10		05		S142

6055   C$_{15}$H$_{12}$O$_2$ 1,3-Propanedione, 1,3-diphenyl-

25	8.95	By extrapolation from values in dioxane/water and methanol/water	05	Uncert.	L3

No.	Molecular formula, name and pK value(s)	T(°C)	Remarks	Method	Assessment	Ref.
6056	C$_{15}$H$_{12}$O$_3$ Benzoic acid, 2-(p-toluoyl)-					
	3.644	25	c = 0.000133-0.000746	C2,Rlc	Approx.	B122
6057	C$_{15}$H$_{12}$O$_6$ 2,2'-Bi-p-benzoquinone, 3-hydroxy-3'-methoxy-5,5'-dimethyl-					
	3.04	20	c = 0.0001, mixed constant	O5	Approx.	M128
6058	C$_{15}$H$_{12}$O$_6$ 2,2'-Methylenedioxydibenzoic acid					
	pK$_1$ 3.03, pK$_2$ 4.46,	65	I = 0.1(KCl)	KIN	Uncert.	A60
			Calculated by a nonlinear least-squares procedure which			
			gives the best fit to the experimental data			
6059	C$_{15}$H$_{14}$O$_2$ 1,3-Indandione, 4,5,6,7-tetrahydro-2-phenyl-					
	10.71	not stated	1% methanol	E3bg	Uncert.	G46
	10.2		1% ethanol	O		L45
6060	C$_{15}$H$_{14}$O$_2$ 1,3-Indandione, 5,6,7,7a-tetrahydro-2-phenyl-					
	3.73		1% methanol	O		L45
6061	C$_{15}$H$_{16}$O$_2$ 1,3-Indandione, hexahydro-2-phenyl-					
	4.02		1% methanol	O		L45
6062	C$_{15}$H$_{22}$O$_2$ Benzaldehyde, 3,5-di-tert-butyl-4-hydroxy-					
	8.05	25		O5	Approx.	C70

No.	Formula	Name	Value	Temp.	Notes		Quality	Code
6063	$C_{15}H_{22}O_3$	Benzoic acid, 3,5-di-tert-butyl-4-hydroxy-	10.80	25		05	Approx.	C70
6064	$C_{15}H_{24}O$	Phenol, 2,6-di-tert-butyl-4-methyl-	12.23	25	Extrapolated value		Uncert.	C70
6065	$C_{15}H_{24}O$	Phenol, 2,4,6-tripropyl-	11.59	20		05	Approx.	D30
6066	$C_{15}H_{24}O_2$	Phenol, 2,6-di-tert-butyl-4-methoxy-	12.15	25	Extrapolated value		Uncert.	C70
6067	$C_{15}H_9O_2Cl$	1,3-Indandione, 2-(2-chlorophenyl)-	4.11	20	1% methanol	05	Approx.	L44
		Values in mixed solvents are also given						
6068	$C_{15}H_9O_2Cl$	1,3-Indandione, 2-(3-chlorophenyl)-	3.51	20	1% methanol	05	Approx.	L44
		Values in mixed solvents are also given						
6069	$C_{15}H_9O_2Cl$	1,3-Indandione, 2-(4-chlorophenyl)-	3.72	20	1% methanol	05	Approx.	L44
		Values in mixed solvents are also given						
			4.0	18±2	1% ethanol, mixed constant	05 or E3d	Uncert.	W25
			3.54	20		05		S142

No.	Molecular formula, name and pK value(s)	T(°C)	Remarks	Method	Assessment	Ref.
6070	$C_{15}H_9O_2Br$ 1,3-Indandione, 2-(4-bromophenyl)- 3.54	20		05		S142
6071	$C_{15}H_9O_4N$ 1,3-Indandione, 2-(4-nitrophenyl)- 2.39	20	1% methanol	05	Approx.	L44
	Values in mixed solvents are also given					
	2.39		1% methanol	E3bg	Uncert.	G46
6072	$C_{15}H_{10}O_2N_2$ 2,5-Cyclohexadien-5-one, 4-(8-hydroxyquinolin-5-yl)- 9.5	not stated	c = 0.00004, 2% dioxane, mixed constant	05	Uncert.	K54
6073	$C_{15}H_{10}O_5S$ Benzenesulfonic acid, 4-(1,3-dioxo-2-indanyl)- pK_2 3.70	20	1% methanol	05	Approx.	L44
	Values in mixed solvents are also given					
6074	$C_{15}H_{10}O_5S$ 4-Indansulfonic acid, 1,3-dioxo-2-phenyl- pK_2 3.49		1% methanol	05		L46
6075	$C_{15}H_{11}ON_3$ 1-Naphthol, 4-(2-pyridylazo)- (p-PAN) pK_1 3.0, pK_2 9.1 pK_1 3.1, pK_2 9.5	30-36	I = 0.1($NaClO_4$)	05 DIS	Uncert. V.Uncert.	B68

No.	Formula / Name	pK	Conditions	Method	Reliability	Ref.
6076	$C_{15}H_{11}ON_3$ 2-Naphthol, 1-(2-pyridylazo)- (o-PAN) pK$_1$ 2.9, pK$_2$ 11.2	30-36	I = 0.1(NaClO$_4$)	DIS	V.Uncert.	B67
6077	$C_{15}H_{11}O_3N_3$ 1,3,5-Triazine-2,4,6(1H,3H,5H)-trione, 1,3-diphenyl- 6.51		Conditions not stated		Uncert.	A73
6078	$C_{15}H_{12}O_2N_2$ Imidazolidine-2,4-dione, 5,5-diphenyl- 8.31	not stated	c = 0.0001, 1% ethanol	05	Uncert.	A8
6079	$C_{15}H_{12}O_5S$ Benzenesulfonic acid, 3-(1,3-dioxo-3-phenyl)propyl- 8.30		I = 0.1(NaClO$_4$)	E3b		T56
6080	$C_{15}H_{12}O_5S$ Benzenesulfonic acid, 4-(1,3-dioxo-3-phenyl)propyl- 8.27		I = 0.1(NaClO$_4$)	E3b		T56
6081	$C_{15}H_{12}O_8S_2$ 3,3'-Methylenedicarbonyldibenzenesulfonic acid pK$_3$ 7.95		I = 0.1(NaClO$_4$)	E3b		T56
6082	$C_{15}H_{12}O_8S_2$ 4,4'-Methylenedicarbonyldibenzenesulfonic acid pK$_3$ 7.85		I = 0.1(NaClO$_4$)	E3b		T56
6083	$C_{15}H_{13}O_2N$ Propanamide, 3-oxo-3,N-diphenyl- 9.41(keto)	not stated	Mixed constants	05	Uncert.	B82
	7.22(enol)	not stated		05	Uncert.	

No.	Molecular formula, name and pK value(s)	$T(^\circ C)$	Remarks	Method	Assessment	Ref.
6084	$C_{15}H_{13}O_2N$ Propenohydroxamic acid, 3,N-diphenyl-					
	8.64	30	c = 0.01, I = 0.1($NaClO_4$), concentration constant	E3bg	Approx.	D62
	8.80	20	I = 0.1($NaClO_4$)	No details	Uncert.	Z8
	8.79		I = 0.1(KCl)	0		Z7
6085	$C_{15}H_{13}O_2N$ 2,5-Pyrrolidinedione, 3-methyl-3-(2-naphthyl)-					
	8.73	20	c = 0.0001-0.001, 4% ethanol, mixed constant	E3bg	Uncert.	F29
6086	$C_{15}H_{13}O_4N$ 1,3-Indandione, 4,5,6,7-tetrahydro-2-(p-nitrophenyl)-					
	7.90	not stated	1% methanol	E3bg	Uncert.	G46
	7.90		1% methanol	0		L45
6087	$C_{15}H_{15}O_2N$ Acetohydroxamic acid, 2-phenyl-N-(o-tolyl)-					
	8.46	30	c = 0.001, I = 0.1($NaClO_4$), concentration constant	E3bg	Approx.	D62
6088	$C_{15}H_{15}O_3N$ Benzohydroxamic acid, 4-methoxy-N-(o-tolyl)-					
	3.44	30	c = 0.01, I = 0.1($NaClO_4$), concentration constant	E3bg	Approx.	D62
6089	$C_{15}H_{16}O_6S_2$ Bis(p-tolyl) methanedisulfonate					
	7.84	not stated	Extrapolated from data in acetone/water mixtures	E3bg	Uncert.	W5

6090 $C_{15}H_{17}O_6N$ Hexanoic acid, 5-acetylamino-6-benzyloxy-4,6-dioxo-

pK_1 4.35, pK_2 8.90   22   E3bg   Uncert.   L7

6091 $C_{15}H_{18}O_2N_4$ 1,3-Cyclohexanedione, 2-(3,4-dihydro-2-methylpteridin-4-yl)-5,5-dimethyl-

pK_1 3.59, pK_2 8.51   20   c = 0.000054   O5   Approx.   A32

6092 $C_{15}H_{18}O_3N_2$ Pyrimidine-2,4,6(1H,3H,5H)-trione, 5-benzyl-5-butyl-

7.91   25   E3ag   Approx.   K53

6093 $C_{15}H_{18}O_4N_2$ 2-Piperidinecarboxylic acid, 1-benzoylaminoacetyl-

3.5(-COOH)   not stated   I = 0.1(KCl)   E3bg   Uncert.   W13

6094 $C_{15}H_{20}O_6N_2$ 3,6-Diazaoctanedioic acid, 3-benzyl-6-carboxymethyl-   (Ethylenedinitrilo-N-benzyl-N,N',N'-triacetic acid)

pK_1 1.9   24.2   c = 0.002   E3bg   Uncert.   B149

pK_2 5.10, pK_3 9.84   Approx.

6095 $C_{15}H_{21}ON$ Benzenecarbonitrile, 3,5-di-tert-butyl-4-hydroxy-

8.70   25   O5   Approx.   C70

6096 $C_{15}H_{21}O_4N_5$ Adenosine, N(6)-(3-methyl-2-butenyl)-

3.76   22   E3bg   Approx.   M53

6097 $C_{15}H_{21}O_6N$ Hepta-2,4-diene-3-carboxylic acid, 2-(5-carboxy-4-carboxymethyl-3-pyrrolidinyl)-

pK_1 2.20, pK_2 3.72, pK_3 4.93, pK_4 9.82   Conditions not stated   Uncert.   D3

No.	Molecular formula, name and pK value(s)	$T(^{o}C)$	Remarks	Method	Assessment	Ref.
6098	$C_{15}H_{23}O_2N$ Benzenecarboxamide, 3,5-di-tert-butyl-4-hydroxy- 9.53	25		05	Approx.	C70
6099	$C_{15}H_{23}O_4N$ 2,6-Piperidinedione, 4-[2-(3,5-dimethyl-2-oxocyclohex-1-yl)-2-hydroxy]ethyl- (Actidione) 11.2		Conditions not stated	E3bg	Uncert.	K46
6100	$C_{15}H_{23}O_{12}N_3$ 3,7-Diazanonanedioic acid, 5-[bis(carboxymethyl)]amino-3,7-bis(carboxymethyl)- (1,2,3-Propanetriyltrinitrilohexa-acetic acid) pK_1 2, pK_2 2.43 pK_3 3.52, pK_4 4.30, pK_5 8.26, pK_6 9.88	25	$I = 0.1(KNO_3)$, concentration constants	E3bg	Uncert. Approx.	M92
6101	$C_{15}H_{24}O_6N_2$ Uridine, 5-hexyl- 9.76	20		05	Uncert.	S138
6102	$C_{15}H_{25}O_{10}N_3$ 3,6,9-Triazaundecanedioic acid, 3,9-bis(carboxymethyl)-6-(2-carboxyethyl)- ([[N-2-Carboxyethyliminodiethylene)-dinitrilo]tetraacetic acid) pK_1 2.58, pK_2 2.97, pK_3 4.7, pK_4 8.14, pK_5 9.31	not stated	$c = 0.001$, $I = 0.1(KCl)$, mixed constants	E3bg	Uncert.	V6
6103	$C_{15}H_{27}O_2N_3$ 1,2-Benzenediol, 3,4,6-tris(dimethylaminomethyl)- pK_1 4.95, pK_2 7.10, pK_3 10.35	31.5	$I = 0.2(KCl)$, mixed constants	E3bg	Approx.	S33
	pK_1 4.92	25	$I = 0.1(KNO_3)$, mixed constant	E3bg	Approx.	E19

No.	Formula	Name	pK	T	Conditions		Method	Assessment	Ref.
6104	$C_{15}H_8O_2N_2Cl_2$	2,5-Cyclohexadien-1-one, 2,6-dichloro-4-(8-hydroxyquinolin-5-yl)imino-	5.9	not stated	c = 0.00004, 2% dioxane, mixed constant	05		Uncert.	K54
6105	$C_{15}H_8O_2N_2Br_2$	2,5-Cyclohexadien-1-one, 2,6-dibromo-4-(8-hydroxyquinolin-5-yl)imino-	6.0	not stated	c = 0.00004, 2% dioxane, mixed constant	05		Uncert.	K54
		Determination made on hydrolysed acetate							
6106	$C_{15}H_9O_4NCl_2$	Propenoic acid, 3-[5-(3,5-dichloro-4-oxo-2,5-cyclohexadien-1-ylidene)amino-2-hydroxy]phenyl-	5.6	not stated	c = 0.00004, 2% dioxane, mixed constant	05		Uncert.	K54
6107	$C_{15}H_9O_4NBr_2$	Propenoic acid, 3-[5-(3,5-dibromo-4-oxo-2,5-cyclohexadien-1-ylidene)amino-2-hydroxy]phenyl-	5.7	not stated	c = 0.00004, 2% dioxane, mixed constant	05		Uncert.	K54
6108	$C_{15}H_{11}O_2N_3S$	1,3-Benzenediol, 4-(4-phenyl-1,3-thiazol-2-ylazo)-	pK_2 6.3, pK_3 9.6			0			M83
6109	$C_{15}H_{12}O_3N_4S$	Benzenesulfonamide, 4-(8-hydroxyquinolin-5-ylazo)-	9.5			0			K60a
6110	$C_{15}H_{12}N(Cl)S$	Thieno[2,3-c]pyridinium chloride, 7-styryl-	2.80	20			E3bg	Uncert.	D51
6111	$C_{15}H_{22}O_4N_5(Br)$	Adenosinium bromide, N(1)-(3-methyl-2-butenyl)-	3.47	22			E3bg	Approx.	M53

No.	Molecular formula, name and pK value(s)	T(°C)	Remarks	Method	Assessment	Ref.
6112	$C_{15}H_{23}O_5N_6S$ (cation) Adenosine, 5'-[S-(3-amino-3-carboxypropyl)-S-methyl]sulfonio- (L-Methionine, S-(5'-adenosyl)-)					
	pK_1 1.8(COOH)		Determined from infra-red measurements		Uncert.	K34
			in D_2O and corrected to the corresponding			
			value in water by subtracting 0.5 pK units			
	pK_2 3.4, pK_3 7.8(α-NH$_2$)			E3bg/05		
6113	$C_{15}H_{25}ON_3(Cl)_2$ Piperazinedium dichloride, 1-(N-acetyl-2,6-dimethylanilino)-4-methyl-					
	pK_1 3.8, pK_2 9.0	25		E3bg	Uncert.	F35a
6114	$C_{15}H_{30}O_4N(Cl)$ N,N,N-Trimethyl 3-carboxy-2-octanoyloxypropylammonium chloride (Butanoic acid, 3-octanoyloxy-4-trimethyl-ammonio-, chloride)					
	3.60	25	c = 0.01, c.m.c. ≈ 0.1	E3bg	Approx.	Y5
6115	$C_{15}H_{32}O_2NBr$ N-Decyl-N,N-dimethyl-N-(1-carboxyethyl)ammonium bromide (Propanoic acid, 2-(N-decyl-N,N-dimethyl)ammonio-, bromide)					
	1.75	25	c = 0.002, I = 0.1(NaCl-HCl)	ROT	Approx.	B99
			c.m.c. = 0.01			
6116	$C_{16}H_{10}O_3$ 1,3-Indandione, 2-benzoyl-					
	2.78			0		T68

No.	Formula	Name	pK	Temp (°C)	Conditions	Code	Assessment	Reference
6117	$C_{16}H_{12}O_2$	1,3-Indandione, 2-benzyl-	5.73	not stated	c ≈ 0.0001, 1% methanol	05	Uncert.	S118
6118	$C_{16}H_{12}O_2$	2-Naphthol, 3-(2-hydroxyphenyl)-	pK_1 7.55	20	c = 0.0001, mixed constants	05	Approx.	M128
			pK_2 >13				Uncert.	
6119	$C_{16}H_{12}O_3$	1,3-Indandione, 2-(2-methoxyphenyl)-	5.61	20	1% methanol	05	Approx.	L44
					Values in mixed solvents are also given			
6120	$C_{16}H_{12}O_3$	1,3-Indandione, 2-(3-methoxyphenyl)-	3.98	20	1% methanol	05	Approx.	L44
					Values in mixed solvents are also given			
6121	$C_{16}H_{12}O_3$	1,3-Indandione, 2-(4-methoxyphenyl)-	4.25	20	1% methanol	05	Approx.	L44
					Values in mixed solvents are also given			
			4.21	not stated	1% methanol	E3bg	Uncert.	G46
			4.09	20		05		S142
6122	$C_{16}H_{14}O_5$	Butanedioic acid, 2-hydroxy-3,3-diphenyl-	pK_1 2.22, pK_2 6.35	20	Mixed constants	E3bg	Uncert.	S7

No.	Molecular formula, name and pK value(s)	T(°C)	Remarks	Method	Assessment	Ref.
6123	$C_{16}H_{14}O_5$ 3H-Xanthen-3-one, 2,6,7-trihydroxy-9-propyl- pK$_1$ 3.27, pK$_2$ 6.17, pK$_3$ 10.47, pK$_4$ 11.6	not stated	10% ethanol pK$_1$ (protonated species)	O		N23
6124	$C_{16}H_{18}O_3$ 1,3-Indandione, 2-isoheptanoyl- 3.5	18±2	1% ethanol, mixed constant	05 or E3d	Uncert.	W25
6125	$C_{16}H_{18}O_7$ Undeca-5,6-dien-8,10-diynoic acid, 4-(β-D-xylopyranosyl)- 4.4		Conditions not stated	05(?)	Uncert.	B169d
6126	$C_{16}H_{18}O_9$ Cyclohexanecarboxylic acid, 3-(3',4'-dihydroxycinnamoyl)oxy-1,4,5-trihydroxy- (Chlorogenic acid) pK$_1$ 3.59, pK$_2$ 8.59 pK$_1$ 3.50, pK$_2$ 8.35	25 25	c = 0.0025 I = 0.05(KNO$_3$)	E3bg	Approx.	T39
6127	$C_{16}H_{24}O_2$ Acetophenone, 3,5-di-tert-butyl-4-hydroxy- 8.68			05	Approx.	C70
6128	$C_{16}H_{28}O_7$ Heptanoic acid, 3-(1-chalcosyl)-2,4-dimethyl-6-oxo- 4.5		Conditions not stated	?	Uncert.	W43
6129	$C_{16}H_{11}O_2N$ 2,5-Cyclohexadien-1-one, 4-(4-hydroxynaphth-1-yl)imino- 9.1	not stated	c = 0.00004, 2% dioxane, mixed constant	05	Uncert.	K54
6130	$C_{16}H_{11}O_2N$ Spiro[fluorene-9,3'-pyrrolidine]-2',5'-dione 7.72(NH)	20	c = 0.0001-0.001, 4% ethanol, mixed constant	E3bg	Uncert.	F29

No.	Formula	Name	pK	Temp.	Conditions	Method		Ref.
6131	$C_{16}H_{11}O_2Cl$	1,3-Indandione, 2-(4-chlorobenzyl)-	5.79	not stated	$c \approx 0.0001$, 1% methanol	05	Uncert.	S118
6132	$C_{16}H_{11}O_2Br$	1,3-Indandione, 2-(4-bromobenzyl)-	5.71	not stated	$c \approx 0.0001$, 1% methanol	05	Uncert.	S118
6133	$C_{16}H_{11}O_2I$	1,3-Indandione, 2-(4-iodobenzyl)-	5.90	not stated	$c \approx 0.0001$, 1% methanol	05	Uncert.	S118
6134	$C_{16}H_{11}O_3N$	2,3,5-Pyrrolidinetrione, 4,4-diphenyl-	5.50	20	Mixed constant	E3bg	Uncert.	H17
6135	$C_{16}H_{11}O_4N$	1,3-Irdandione, 2-(4-nitrobenzyl)-	5.06	not stated	$c \approx 0.0001$, 1% methanol	05	Uncert.	S118
6136	$C_{16}H_{12}O_2N_2$	2-Naphthol, 1-(2-hydroxyphenylazo)-	pK_1 7.7, pK_2 12.4	not stated	$I = 0.1(KCl)$	05	Uncert.	D37
6137	$C_{16}H_{12}O_2N_2$	3-Pyrrolidinecarbonitrile, 4-methyl-4-(1-naphthyl)-2,5-dioxo-	6.20	20	$c = 0.0001-0.001$, 4% ethanol	E3bg	Uncert.	F27,F29
6138	$C_{16}H_{12}O_2N_2$	3-Pyrrolidinecarbonitrile, 4-methyl-4-(2-naphthyl)-2,5-dioxo-	6.25	20	$c = 0.0001-0.001$, 4% ethanol, mixed constant	E3bg	Uncert.	F29
6139	$C_{16}H_{13}O_2N$	2,5-Pyrrolidinedione, 3,3-diphenyl-	8.25	20	$c = 0.0001-0.001$, 4% ethanol, mixed constant	E3bg	Uncert.	F29,F27

No.	Molecular formula, name and pK value(s)	T(°C)	Remarks	Method	Assessment	Ref.
6140	$C_{16}H_{13}O_3N$ 2,5-Pyrrolidinedione, 3-hydroxy-4,4-diphenyl- (Malic acid imide, α,α-diphenyl) 7.65	20	Mixed constant	E3bg	Uncert.	S7
6141	$C_{16}H_{13}O_4N$ 3-Pyrrolidinecarboxylic acid, 4-methyl-4-(1-naphthyl)-2,5-dioxo- 2.94(COOH) 8.93(NH)	20	c = 0.0001-0.001, 4% ethanol, mixed constants	E3bg	Uncert.	F29,F27
6142	$C_{16}H_{14}ON_4$ 2-Pyrazolin-5-one, 3-methyl-1-phenyl-4-phenylhydrazono- 8.9	20	c = 0.000002	05	Uncert.	L16
6143	$C_{16}H_{14}O_3N_2$ 3-Pyrrolidinecarboxamide, 4-methyl-4-(1-naphthyl)-2,5-dioxo- 7.60	20	c = 0.0001-0.001, 4% ethanol, mixed constant	E3bg	Uncert.	F29,F27
6144	$C_{16}H_{14}O_3N_2$ 3-Pyrrolidinecarboxamide, 4-methyl-4-(2-naphthyl)-2,5-dioxo- 7.70	20	c = 0.0001-0.001, 4% ethanol, mixed constant	E3bg	Uncert.	F29
6145	$C_{16}H_{14}O_4S$ 3-Thiapentanedioic acid, 2,4-diphenyl- pK_1 2.61, pK_2 4.22	25±0.5	c < 0.001	E3bg	Approx.	S93
6146	$C_{16}H_{14}O_5N_2$ Benzoic acid, 5-(4-ethoxycarbonylphenyl)azo)-2-hydroxy- pK_2 11.19			05		M108

| 6147 | $C_{16}H_{15}O_2N$ | Butanamide, 3-oxo-2,N-diphenyl- | | | | | | |
| | | | 22 | 10.00 | 5% v/v ethanol; mixed constant | 05/E3bg | Uncert. | H38 |

| 6148 | $C_{16}H_{15}O_4N$ | Ethyl 4-(N-hydroxy-N-phenyl)carbamoylbenzoate (Benzohydroxamic acid, 4-ethoxycarbonyl-N-phenyl) | | | | | | |
| | | | 20 | 8.00 | $c \approx 0.001$, $I = 0.1(KCl)$ | E3bg | Approx. | F28 |

6149	$C_{16}H_{15}O_7N$	Iminodiacetic acid, N-(3-carboxy-2-hydroxynaphth-1-yl)-						
			not stated	pK_1 0.7, pK_2 2.7, pK_3 8.5		05	Uncert.	B168
				pK_4 9.7, pK_5 13.8				

6150	$C_{16}H_{16}O_2N_2$	Ethylenedinitrilobis(2-hydroxybenzylidene) (N,N'-Bis(2-hydroxybenzylidene)-1,2-diaminoethane)						
			20	pK_1 4.4, pK_2 4.8, pK_3 8.1, pK_4 11.4	$I = 0.3$	05	Uncert.	S122
					Theorell-Steinhagen buffer			

| 6151 | $C_{16}H_{16}O_4N_2$ | Ethane, 1,1-bis(o-tolyl)-2,2-dinitro- | | | | | | |
| | | | 20 | 5.80 | | 05 | Approx. | T58a |

6152	$C_{16}H_{17}O_4N$	Bis(ethoxycarbonyl)isoquinoliniomethanide						
			not stated	5.50		05	Uncert.	N31
				Values in mixed solvents are also given				

6153	$C_{16}H_{17}O_4N$	Bis(ethoxycarbonyl)quinoliniomethanide						
			not stated	5.03		05	Uncert.	N31
				Values in mixed solvents are also given				

No.	Molecular formula, name and pK value(s)	$T(^oC)$	Remarks	Method	Assessment	Ref.
6154	$C_{16}H_{20}O_2N_4$ 1,3-Cyclohexanedione, 2-(3,4-dihydro-6,7-dimethylpteridin-4-yl)-5,5-dimethyl- pK_1 3.62, pK_2 8.32	20	c = 0.00003-0.000067	05	Approx.	A32
6155	$C_{16}H_{20}O_8N_2$ 3,6-Diazaoctanedioic acid, 3,6-bis(carboxymethyl)-4-phenyl- (Ethylenedinitrilotetraacetic acid, 1-phenyl-) pK_1 1.87 pK_2 3.21, pK_3 5.42, pK_4 9.60	25	c = 0.001, I = 0.1(KCl/KNO$_3$), concentration constants	E3bg	Uncert. Approx.	(PEDTA) O17
6156	$C_{16}H_{20}O_{10}N_2$ 1,4-Benzenedicarboxylic acid, 2,5-bis[N-(carboxymethyl)-N-(2-hydroxyethyl)amino]- pK_1 1.95, pK_2 2.68, pK_3 5.25, pK_4 9.06	22	I = 0.1(KCl)	E3bg	Uncert.	U6
6157	$C_{16}H_{22}O_2N_2$ 2,3-Naphthalenediol, 1,4-bis(dimethylaminomethyl)- pK_1 5.07 pK_2 9.7	31.5	I = 0.2(KCl), mixed constants	E3bg	Approx. Uncert.	S33
6158	$C_{16}H_{24}O_4N_2$ 1,2-Benzenediol, 3,6-bis(morpholinomethyl)- pK_1 5.65, pK_2 7.32, pK_3 9.86 pK_1 5.6, pK_2 7.3, pK_3 9.8	31.5 ≈27	c = 0.008, I = 0.2	E3bg 05	Approx. Uncert.	K31
6159	$C_{16}H_{24}O_5N_2$ 1,2,3-Benzenetriol, 4,6-bis(morpholinomethyl)- pK_1 6.25, pK_2 7.66, pK_3 9.65 pK_1 6.0, pK_2 7.6, pK_3 9.6	31.5 ≈27	c = 0.008, I = 0.2	E3bg 05	Approx. Uncert.	K31

6160 $C_{16}H_{24}O_8N_2$ 1,4-Benzenedicarboxylic acid, 2,5-bis[bis(2-hydroxyethyl)amino]-

pK(NH$^+$) 2.01, pK$_2$ 5.07, pK$_3$ 8.01   22   c 0.003, I = 0.1(KCl), mixed constants   E3bg   Approx.   U3

pK$_2$ 5.04, pK$_3$ 7.98   22   I = 0.1(KCl)   E3bg   Uncert.   U5

6161 $C_{16}H_{28}ON_2$ Phenol, 2,6-bis(dimethylaminomethyl)-4-tert-butyl-

pK$_1$ 6.30   25   I = 0.1(KHO$_3$), mixed constant   E3bg   Approx.   E19

6162 $C_{16}H_{28}O_8N_2$ 3,12-Diazatetradecanedioic acid, 3,12-bis(carboxymethyl)-   ((Octamethylenedinitrilo)tetraacetic acid)

pK$_1$ 2.0, pK$_2$ 2.75   20   I = 0.1(KNO$_3$)   E3bg   Uncert.   A54
pK$_3$ 9.91, pK$_4$ 10.77   Approx.

Thermodynamic quantities are also given

pK$_1$ 2.0, pK$_2$ 2.75   20   I = 0.1(KCl), concentration constants   E3bg   Uncert.   S50
pK$_3$ 9.94, pK$_4$ 10.75   Approx.

6163 $C_{16}H_{28}O_9N_2$ 8-Oxa-3,13-diazapentadecanedioic acid, 3,13-bis(carboxymethyl)-   (Oxy-bis(tetramethylenenitrilo)tetraacetic acid)

pK$_1$ 2.1, pK$_2$ 2.7   20   I = 0.1(KCl), concentration constants   E3bg   Uncert.   S50
pK$_3$ 9.52, pK$_4$ 10.29   Approx.

6164 $C_{16}H_{30}O_6N_2$ 3,6-Diazaoctanedioic acid, 3-(carboxymethyl)-6-octyl-   (N-Octyl(ethylenedinitrilo)triacetic acid)

pK$_1$ 1.9   24.2   c = 0.002   E3bg   Uncert.   B149
pK$_2$ 6.53, pK$_3$ 9.76   Approx.

6165 $C_{16}H_{30}O_8N_4$ 3,6,9,12-Tetraazatetradecanedioic acid, 3,12-bis(carboxymethyl)-6,9-dimethyl-

pK$_1$ 3.05, pK$_2$ 5.15   20   I = 0.1(KCl), concentration constants   E3bg   Approx.   S50
pK$_3$ 8.99, pK$_4$ 10.54

No.	Molecular formula, name and pK value(s)	T(°C)	Remarks	Method	Assessment	Ref.
6166	$C_{16}H_{32}O_5S$ Hexadecanoic acid, 2-sulfo- (Palmitic acid, α-sulfo-) pK_1 2.43, pK_2 5.15					I22
6167	$C_{16}H_{35}O_4P$ Bis(2-ethylhexyl) hydrogen phosphate 2.85	25				S140
6168	$C_{16}H_9O_2NCl_2$ 2,5-Cyclohexadien-1-one, 2,6-dichloro-4-(4-hydroxynaphth-1-yl)imino- 6.8	not stated	c = 0.00004, 2% dioxane, mixed constant	05	Uncert.	K54
6169	$C_{16}H_9O_2NBr_2$ 2,5-Cyclohexadien-1-one, 2,6-dibromo-4-(4-hydroxynaphth-1-yl)imino- 6.9	not stated	c = 0.00004, 2% dioxane, mixed constant	05	Uncert.	K54
6170	$C_{16}H_{10}O_2N_2Cl_2$ 2,5-Cyclohexadien-1-one, 2,6-dichloro-4-(8-hydroxy-2-methylquinolin-5-yl)imino- 5.9	not stated	c = 0.00004, 2% dioxane, mixed constant	05	Uncert.	K54
6171	$C_{16}H_{10}O_2N_2Br_2$ 2,5-Cyclohexadien-1-one, 2,6-dibromo-4-(8-hydroxy-2-methylquinolin-5-yl)imino- 6.0	not stated	c = 0.00004, 2% dioxane, mixed constant	05	Uncert.	K54
6172	$C_{16}H_{10}O_{12}N_4S_2$ 2,7-Naphthalenedisulfonic acid, 3-hydroxy-4-(2-hydroxy-3,5-dinitrophenylazo)- pK(OH-benzene) 2.02, pK(OH-naphthalene)12.20	not stated	I ≈ 0.2, mixed constants	05	Uncert.	S6
6173	$C_{16}H_{11}ON_2Cl$ 2-Naphthol, 1-(4-chlorophenylazo)- pK_2 9.3	20	c = 0.00004, mixed constant	05	Uncert.	S106

6174 $C_{16}H_{11}O_6N_3S$ 2-Naphthalenesulfonic acid, 5-hydroxy-8-(3-nitrophenylazo)-

7.47   20   c = 0.000004, I = 0.25, mixed constant   05   Uncert.   R43

6175 $C_{16}H_{11}O_6N_3S$ 2-Naphthalenesulfonic acid, 5-hydroxy-8-(4-nitrophenylazo)-

8.08   20   c = 0.000004, I = 0.25, mixed constant   05   Uncert.   R43

6176 $C_{16}H_{11}O_7N_3S$ Benzenesulfonic acid, 4-hydroxy-3-(3-hydroxynaphth-1-ylazo)-5-nitro-   (Solochrome Fast Grey R.A.)

pK$_1$ 3.83(SO$_3$H), pK$_2$ 10.53(4-OH), pK$_3$ 10.7(3'-OH)

not stated   c = 0.0002, mixed constants   E3bg   Uncert.   K18

6177 $C_{16}H_{11}O_{10}N_3S_2$ 2,7-Naphthalenedisulfonic acid, 4,5-dihydroxy-3-(2-nitrophenylazo)-

pK$_3$ 8.96   25   c = 0.000025, I = 0.1(NaClO$_4$), mixed constant   05   Approx.   T46

6178 $C_{16}H_{11}O_{10}N_3S_2$ 2,7-Naphthalenedisulfonic acid, 4,5-dihydroxy-3-(3-nitrophenylazo)-

pK$_3$ 8.60   25   c = 0.000025, I = 0.1(NaClO$_4$), mixed constant   05   Approx.   T46

6179 $C_{16}H_{11}O_{10}N_3S_2$ 2,7-Naphthalenedisulfonic acid, 4,5-dihydroxy-3-(4-nitrophenylazo)-   Disodium salt = (Chromotrope 2B)

pK$_1$ 3.86(SO$_3$H), pK$_2$ 6.8(SO$_3$H), pK$_3$ 9.27(OH), pK$_4$ 9.86(OH)

not stated   c = 0.0002, mixed constants   E3bg   Uncert.   K21

pK$_3$ 8.77   25   c = 0.000025, I = 0.1(NaClO$_4$), mixed constant   05   Approx.   T46

No.	Molecular formula, name and pK value(s)	T(°C)	Remarks	Method	Assessment	Ref.
6180	$C_{16}H_{11}O_{10}N_3S_2$ 2,7-Naphthalenedisulfonic acid, 3-hydroxy-4-(2-hydroxy-4-nitrophenylazo)- pK(OH-benzene) 6.59, pK(OH-naphthalene) 12.95	not stated	I ≈ 0.2, mixed constants	05	Uncert.	S6
6181	$C_{16}H_{11}O_{13}N_3S_3$ 2,7-Naphthalenedisulfonic acid, 3-hydroxy-4-(2-hydroxy-3-nitro-5-sulfophenylazo)- pK(OH-benzene) 4.03, pK(OH-naphthalene) 12.60	not stated	I ≈ 0.1, mixed constants	05	Uncert.	S6
6182	$C_{16}H_{11}O_{13}N_3S_3$ 2,7-Naphthalenedisulfonic acid, 3-hydroxy-4-(2-hydroxy-5-nitro-3-sulfophenylazo)- pK(OH-benzene) 4.29, pK(OH-naphthalene) 11.91	not stated	I ≈ 0.1, mixed constants	05	Uncert.	S6
6183	$C_{16}H_{12}O_4N_2S$ Benzenesulfonic acid, 4-(2-hydroxynaphth-1-ylazo)- 11.4					S106
6184	$C_{16}H_{12}O_4N_2S$ Benzenesulfonic acid, 4-(4-hydroxynaphth-1-ylazo)- 8.2					S106
6185	$C_{16}H_{12}O_4N_2S$ 2-Naphthalenesulfonic acid, 5-hydroxy-8-phenylazo- 7.30	20	c = 0.000004, I = 0.25, mixed constant	05	Uncert.	R43

6186 $C_{16}H_{12}O_5N_2S$ Benzenesulfonic acid, 4-hydroxy-3-(2-hydroxynaphth-1-ylazo)- (Solochrome Violet R) (CI Mordant Violet S)
pK_2 7.03, pK_3 13.04   20   c = 0.000013, I = 0.1   05   Approx.   C67

Variation with temperature

	30	40	50	60
pK_2	6.94	6.88	6.83	6.79
pK_3	12.78	12.56	12.36	12.19

Thermodynamic quantities are derived from the results

pK_2 7.22(-OH)   25   c ≈ 0.00003, extrapolation of values   05   Uncert.   C68
obtained in dioxane/water mixtures
pK_3 13.39(-OH)   25   c ≈ 0.0004, in $(CH_3)_4$NOH   07

6187 $C_{16}H_{12}O_5N_2S$ Benzenesulfonic acid, 4-hydroxy-3-(3-hydroxynaphth-1-ylazo)-   Sodium salt is (Solochrome Violet R.S.)
pK_2 7.4(4-OH), pK_3 9.45(3-OH)   not stated   c = 0.0002, mixed constants   E3bg   Uncert.   K19

6188 $C_{16}H_{12}O_5N_2S$ Benzenesulfonic acid, 4-hydroxy-3-(4-hydroxynaphth-1-ylazo)-   (Solochrome Black P.V.)
pK_2(naphthalene OH) 7.0, pK_3(benzene OH) 11.3   not stated   c = 0.0002, mixed constants   E3bg   Uncert.   K20

6189 $C_{16}H_{12}O_6N_2S$ 1-Naphthalenesulfonic acid, 3-(2,4-dihydroxyphenylazo)-4-hydroxy-
pK_2 7.00, pK_3 9.20   25±1   c = 0.00002, I = 0.1   05   Uncert.   N26

6190 $C_{16}H_{12}O_8N_2S_2$ 2,7-Naphthalenedisulfonic acid, 4,5-dihydroxy-3-phenylazo-
pK_3 9.15   25   c = 0.002, I = 0.1(KNO_3), concentration constant   E3bg   Approx.   N10

(contd)

No.	Molecular formula, name and pK value(s)	T(°C)	Remarks	Method	Assessment	Ref.
6190 (Contd)	$C_{16}H_{12}O_8N_2S_2$ 2,7-Naphthalenedisulfonic acid, 4,5-dihydroxy-3-phenylazo- pK$_3$ 9.30(-OH)	25	c = 0.00002, I = 0.1(NaClO$_4$), mixed constant	05	Uncert.	M87
6191	$C_{16}H_{12}O_8N_2S_2$ 2,7-Naphthalenedisulfonic acid, 3-hydroxy-4-(2-hydroxyphenylazo)- pK(OH-benzene) 7.30, pK(OH-naphthalene) 12.62	not stated	I ≈ 0.2, mixed constants	05	Uncert.	S6
6192	$C_{16}H_{12}O_9N_2S_2$ 2,7-Naphthalenedisulfonic acid, 4,5-dihydroxy-3-(2-hydroxyphenylazo)- pK$_3$ 7.60, pK$_4$ 10.60	25	I = 0.1(NaClO$_4$), mixed constants	05	Approx.	T48
	pK$_3$ 7.60, pK$_4$ 10.60	25	I = 0.1(KNO$_3$), mixed constants	E3bg	Approx.	M9,M10
	pK$_3$ 7.55, pK$_4$ 9.28, pK$_5$ 12.40					T51
6193	$C_{16}H_{12}O_{10}N_2S_3$ 1,5-Naphthalenedisulfonic acid, 3-hydroxy-4-(4-sulfophenylazo)- pK$_3$ 10.5	20	c = 0.00004, I = 0.25, mixed constant	05	Uncert.	S106
6194	$C_{16}H_{12}O_{10}N_2S_3$ 1,6-Naphthalenedisulfonic acid, 4-hydroxy-3-(4-sulfophenylazo)- pK$_3$ 7.2	20	c = 0.00004, I = 0.25, mixed constant	05	Uncert.	S106
6195	$C_{16}H_{12}O_{10}N_2S_3$ 2,5-Naphthalenedisulfonic acid, 4-hydroxy-3-(4-sulfophenylazo)- pK$_3$ 9.4	20	c = 0.00004, I = 0.25, mixed constant	05	Uncert.	S106
6196	$C_{16}H_{12}O_{10}N_2S_3$ 2,7-Naphthalenedisulfonic acid, 3-hydroxy-4-(4-sulfophenylazo)- pK$_3$ 10.6	20	c = 0.00004, I = 0.25, mixed constant	05	Uncert.	S106

No.	Formula / Name / pK	T(°C)	Conditions	Method	Reliability	Ref.
6197	$C_{16}H_{12}O_{10}N_2S_3$ 2,7-Naphthalenedisulfonic acid, 4-hydroxy-3-(4-sulfophenylazo)- pK_3 10.4	20	c = 0.00004, I = 0.25, mixed constant	05	Uncert.	S106
6198	$C_{16}H_{12}O_{11}N_2S_3$ 2,7-Naphthalenedisulfonic acid, 4,5-dihydroxy-3-(2-sulfophenylazo)- pK_4 9.35	25	c = 0.002, I = 0.1(KNO$_3$), concentration constant	E3bg	Approx.	N10
	pK_4 9.76	25	c = 0.00003, I = 0.1(NaClO$_4$), mixed constant	05	Uncert.	M88
6199	$C_{16}H_{12}O_{11}N_2S_3$ 2,7-Naphthalenedisulfonic acid, 4,5-dihydroxy-3-(3-sulfophenylazo)- pK_4 8.85(OH)	25	c = 0.002, I = 0.1(KNO$_3$), concentration constant	E3bg	Approx.	N10
6200	$C_{16}H_{12}O_{11}N_2S_3$ 2,7-Naphthalenedisulfonic acid, 4,5-dihydroxy-3-(4-sulfophenylazo)- pK_4 8.90(OH)	25	c = 0.002, I = 0.1(KNO$_3$), concentration constant	E3bg	Approx.	N10
6201	$C_{16}H_{12}O_{12}N_2S_3$ 2,7-Naphthalenedisulfonic acid, 4,5-dihydroxy-3-(2-hydroxy-4-sulfophenylazo)- pK_4 6.70, pK_5 10.20, pK_6 14.6	20-22	I = 0.2(NaNO$_3$)	05	Uncert.	B71
6202	$C_{16}H_{13}O_6N_3S$ 2,7-Naphthalenedisulfonic acid, 5-amino-4-hydroxy-3-phenylazo- 12.72(OH)	25	I = 0.1(KNO$_3$), mixed constant	05	Approx.	E29
6203	$C_{16}H_{13}O_8N_3S_2$ 1,3-Naphthalenedisulfonic acid, 4-amino-5-hydroxy-6-(2-hydroxyphenylazo)- pK_3 7.4, pK_4 11.6	not stated	I = 0.1(KCl)	05	Uncert.	D37

No.	Molecular formula, name and pK value(s)	T(°C)	Remarks	Method	Assessment	Ref.
6204	$C_{16}H_{13}O_8N_3S_2$ 2,7-Naphthalenedisulfonic acid, 4-amino-5-hydroxy-6-(2-hydroxyphenylazo)-					
	pK_3 8.4, pK_4 11.9	not stated	I = 0.1(KCl)	05	Uncert.	D37
6205	$C_{16}H_{13}O_8N_3S_2$ 2,7-Naphthalenedisulfonic acid, 4-(4-amino-2-hydroxyphenylazo)-3-hydroxy-					
	pK(OH-benzene) 7.28, pK(OH-naphthalene) 12.51					
		not stated	I ≈ 0.2, mixed constants	05	Uncert.	S6
6206	$C_{16}H_{14}O_3N_6S$ Benzenesulfonamide, N-amidino-4-(8-hydroxyquinolin-5-ylazo)-					
	8.4			0		K60a
6207	$C_{16}H_{14}O_7N_4S_2$ 2,7-Naphthalenedisulfonic acid, 4-amino-6-(4-aminophenylazo)-5-hydroxy-					
	12.34(OH)	25	I = 0.1(KIO$_3$), mixed constant	05	Approx.	E29
6208	$C_{16}H_{16}O_6N_2S_4$ 2,5-Diazahexane-1,6-diyl-bis(3-benzenesulfonic acid), 3,4-dithioxo- (Dithiooxamide, N,N'-bis(3-sulfobenzyl)-)					
	pK_1 11.29	25	I = 0.01-1.0(KCl)	05,R4	Approx.	G28
	pK_2 13.63	25	I = 1.0, concentration constant	05	Uncert.	
6209	$C_{16}H_{16}O_6N_2S_4$ 2,5-Diazahexane-1,6-diyl-bis(4-benzenesulfonic acid), 3,4-dithioxo- (Dithiooxamide, N,N'-bis(4-sulfobenzyl)-)					
	pK_1 11.27	25	I = 0.01-1.0(KCl)	05,R4	Approx.	G28
	pK_2 13.51	25	I = 1.0(KCl), concentration constant	05	Uncert.	
6210	$C_{16}H_{16}O_7N_2S_3$ 2,5-Diazahexane-1,6-diyl-bis(4-benzenesulfonic acid), 3-oxo-4-thioxo- (Monothiooxamide, N,N'-bis(4-sulfobenzyl)-)					
	11.67	25	c = 0.0001, I = 0.01-0.1(KCl)	05,R4	Approx.	G29

6211 $C_{16}H_{18}O_4N_2S$ 4,1-Thiaazabicyclo[3.2.0]heptane-2-carboxylic acid, 6-benzylcarbonylamino-3,3-dimethyl-7-oxo-   (Benzylpenicillinic

acid)   (Penicillin G)

  2.8(-COOH)   not stated   I = 0.1(KCl)   E3bg   Uncert.   W13

6212 $C_{16}H_{19}O_9N_2Cl$ 1,3-Phenylenebis(methylnitrilodiacetic acid), 5-chloro-2-hydroxy-

  $pK_1 \simeq 2$, pK_2 2.2   20   I = 0.1(KCl)   E3bg   Uncert.   S38

  pK_3 5.95, pK_4 9.60   Approx.

  pK_5 11.6   Uncert.

6213 $C_{16}H_{20}O_{12}N_2S$ 1,3-Phenylenebis(methylnitrilodiacetic acid), 2-hydroxy-5-sulfo-

  $pK_2 \simeq 2$, pK_3 2.6   20   I = 0.1(KCl)   E3bg   Uncert.   S38

  pK_3 5.49, pK_4 10.06   Approx.

  pK_5 11.9   Uncert.

6214 $C_{16}H_{22}O_8N_2P_2$ 2,5-Diazahexane-1,6-diphosphonic acid, 1,6-bis(2-hydroxyphenyl)-

  pK_1 4.16(POH), pK_2 6.57(POH), pK_3 7.38(NH^+), pK_4 9.61(NH^+)

     25   c = 0.008, I = 0.5($NaClO_4$), concentration   E3bg   Approx.   F26

     constants

  pK_5 11.99(OH), pK_6 12.30(OH)   25   O5   Uncert.

6215 $C_{16}H_{10}O_{10}N_3ClS_2$ 2,7-Naphthalenedisulfonic acid, 4-(3-chloro-2-hydroxy-5-nitrophenylazo)-3-hydroxy-

  pK(OH-benzene) 3.62, pK(OH-naphthalene) 11.77

     not stated   I ≈ 0.2, mixed constants   O5   Uncert.   S6

No.	Molecular formula, name and pK value(s)	$T(^{o}C)$	Remarks	Method	Assessment	Ref.
6216	$C_{16}H_{11}O_4N_2ClS$ 1-Naphthalenesulfonic acid, 3-(4-chlorophenylazo)-4-hydroxy- pK_2 7.8	20	c = 0.00004, mixed constant	05	Uncert.	S106
6217	$C_{16}H_{11}O_4N_2ClS$ 1-Naphthalenesulfonic acid, 8-(4-chlorophenylazo)-7-hydroxy- pK_2 12.1	20	c = 0.00004, mixed constant	05	Uncert.	S106
6218	$C_{16}H_{11}O_4N_2ClS$ 2-Naphthalenesulfonic acid, 3-(4-chlorophenylazo)-4-hydroxy- pK_2 11.3	20	c = 0.00004, mixed constant	05	Uncert.	S106
6219	$C_{16}H_{11}O_4N_2ClS$ 2-Naphthalenesulfonic acid, 5-(4-chlorophenylazo)-6-hydroxy- pK_2 10.3	20	c = 0.00004, mixed constant	05	Uncert.	S106
6220	$C_{16}H_{11}O_4N_2ClS$ 2-Naphthalenesulfonic acid, 8-(3-chlorophenylazo)-5-hydroxy- 7.49	20	c = 0.000004, I = 0.25, mixed constant	05	Uncert.	R43
6221	$C_{16}H_{11}O_4N_2ClS$ 2-Naphthalenesulfonic acid, 8-(4-chlorophenylazo)-5-hydroxy- 7.40	20	c = 0.000004, I = 0.25, mixed constant	05	Uncert.	R43

6222 $C_{16}H_{11}O_7N_2ClS_2$ 1,3-Naphthalenedisulfonic acid, 8-(4-chlorophenylazo)-7-hydroxy-

pK$_3$ 11.0   20   c = 0.00004, I = 0.25,   05   Uncert.   S106

mixed constant

6223 $C_{16}H_{11}O_8N_2ClS_2$ 2,7-Naphthalenedisulfonic acid, 4-(2-chloro-6-hydroxyphenylazo)-3-hydroxy-

pK(OH-benzene) 7.46, pK(OH-naphthalene) 12.50

not stated   I ≈ 0.2, mixed constants   05   Uncert.   S6

6224 $C_{16}H_{11}O_9N_2ClS_2$ 2,7-Naphthalenedisulfonic acid, 3-(3-chloro-6-hydroxyphenylazo)-4,5-dihydroxy-

(Erio SE) is the di-sodium salt

(Plasmocorinth) is the di-sodium salt

pK$_3$ 8.0, pK$_4$ 10.5, pK$_5$ 11.9   not stated   Mixed constants   E3bg   B150

pK$_3$ 7.56, pK$_4$ 10.35   25   I = 0.1(KNO$_3$), mixed constants   E3bg   Approx.   N9

pK$_3$ 7.15, pK$_4$ 10.48   25   c = 0.00003, I = 0.1(NaClO$_4$), mixed   05   Uncert.   M89

constants

6225 $C_{16}H_{11}O_{11}N_2ClS_3$ 2,7-Naphthalenedisulfonic acid, 4-(3-chloro-6-hydroxy-5-sulfophenylazo)-3-hydroxy-

pK(OH-benzene) 6.62, pK(OH-naphthalene) 12.45

not stated   I ≈ 0.2,   05   Uncert.   S6

mixed constants

No.	Molecular formula, name and pK value(s)	T(°C)	Remarks	Method	Assessment	Ref.
6226	$C_{16}H_{13}O_{10}N_2S_2As$ 2,7-Naphthalenedisulfonic acid, 4-(2-arsonophenylazo)-3-hydroxy- (Thoron)					A14
	pK_1 2.37, pK_2 4.35, pK_3 8.26, pK_4 11.18		c ≈ 0.001	05		
6227	$C_{16}H_{13}O_{11}N_2S_2As$ 2,7-Naphthalenedisulfonic acid, 3-(2-arsonophenylazo)-4,5-dihydroxy- (Arsenazo I)					B160
	pK_1 0.6, pK_2 0.8	not stated	Concentration constants	E3bg	Uncert.	
	pK_3 3.52		I = 0.1, concentration constant	E3bg	Uncert.	
	pK_4 8.20, pK_5 11.62		I = 0.1, concentration constants	05	Uncert.	
	pK_6 15.0		Concentration constant	07	Uncert.	
	pK_3 2.97(AsO_3H_2), pK_4 7.57(AsO_3H^-), pK_5 9.98(-OH)	25	c = 0.002, I = 0.1(KNO_3), concentration constants	E3bg	Approx.	N10
6228	$C_{16}H_{14}O_6N_3SAs$ 1-Naphthalenesulfonic acid, 4-amino-3-(2-arsonophenylazo)-					B159
	pK_2 3.90, pK_3 6.35, pK_4 9.30	not stated	Mixed constants	05	Uncert.	

6229 $C_{16}H_{12}O_{11}N_2ClS_2P$ 2,7-Naphthalenedisulfonic acid, 3-(4-chloro-2-phosphonophenylazo)-4,5-dihydroxy-

				B160
pK_1 0.6, pK_2 0.8	not stated	Concentration constants	E3bg	Uncert.
pK_3 1.85		I = 0.1, concentration constant	E3bg	Uncert
pK_4 7.28, pK_5 11.75		I = 0.1, concentration constants	05	Uncert.
pK_6 15.2		Concentration constant	07	Uncert.

No.	Molecular formula, name and pK value(s)	T(°C)	Remarks	Method	Assessment	Ref.
6230	$C_{17}H_{10}O_9$ 4,4'-Carbonylbis(1,2-benzenedicarboxylic acid)					
	pK_1 2.12, pK_2 2.42, pK_3 4.02, pK_4 4.34	20	c = 0.01-0.02, I = 1.0(NaClO$_4$), concentration constants	E3bg	Uncert.	K52
	pK (protonated form) -7.37	20	c = 0.001 in aq.H$_2$SO$_4$ solutions, H$_0$ scale	06	Uncert.	
6231	$C_{17}H_{12}O_2$ 1-Naphthol, 4-benzoyl-					
	7.33	25		05	Approx.	C79
6232	$C_{17}H_{12}O_4$ 2H-Benzopyran-2-one, 8-benzoyl-7-hydroxy-4-methyl-					
	6.44	25	c = 0.0001	05	Approx.	M63
6233	$C_{17}H_{12}O_4$ 1,3-Indandione, 2-(1,3-benzodioxol-5-yl)methyl-					
	5.89	not stated	c ≈ 0.0001, 1% methanol	05	Uncert.	S118
6234	$C_{17}H_{12}O_4$ Spiro[cyclopropane-1,9'-fluorene]-2,3-dicarboxylic acid					
	pK_1 2.91, pK_2 6.66 (cis)	10.5	I = 0.097	E3bg	Approx.	H20
	pK_1 2.90, pK_2 6.54	24.9	I = 0.095			
	pK_1 2.98, pK_2 6.78 (H$_2$O)	25		E3bg	Approx.	H21
	pK_1 3.44, pK_2 7.14 (D$_2$O)					
6235	$C_{17}H_{14}O_3$ 1,3-Indandione, 2-(4-methoxybenzyl)-					
	5.94	not stated	c ≈ 0.0001 in 1% methanol	05	Uncert.	S118
	4.21		1% methanol	C5		L46

No.	Compound	T	Conditions	Method	Reliability	Ref.
6236	$C_{17}H_{14}O_4$ 1,2-Cyclopropanedicarboxylic acid, 3,3-diphenyl-					
	pK$_1$ 2.25, pK$_2$ 9.04 (cis)	10.7	I = 0.115, mixed constants	E3bg	Approx.	H20
	pK$_1$ 2.22, pK$_2$ 8.96	25				
	pK$_1$ 2.30, pK$_2$ 9.20 (H$_2$O)	25		E3bg	Approx.	H21
	pK$_1$ 2.58, pK$_2$ 9.57 (D$_2$O)					
6237	$C_{17}H_{16}O_4$ Pentanedioic acid, 3,3-diphenyl-					
	pK$_1$ 4.02, pK$_2$ 6.81	25	c = 0.00125, mixed constant	E3bg	Approx.	B146
6238	$C_{17}H_{18}O_3$ Tetracyclo[7.5.1(1,12).0(1,9).0(3,8)]pentadeca-3,5,7-triene-2-carboxylic acid, 12-methyl-13-oxo- (Gibberic acid)					
	4.24	25	c = 0.00032	E3bg	Approx.	T33
6239	$C_{17}H_{26}O_3$ Phenol, 2,6-di-tert-butyl-4-ethoxycarbonyl-					
	9.50	25		05	Approx.	C70
6240	$C_{17}H_{32}O_4$ Propanedioic acid, diheptyl-					
	pK$_2$ 7.45	25	I = 0.1(KCl), mixed constant	E3bg	Approx.	M77
6241	$C_{17}H_{10}O_2N_2$ Spiro[fluorene-9,3'-pyrrolidine]-4'-carbonitrile, 2',5'-dioxo-					
	pK(NH) 5.60	20	c = 0.0001-0.001, 4% ethanol, mixed constant	E3bg	Uncert.	F29
6242	$C_{17}H_{11}O_5N_3$ 2-Naphthoic acid, 1-hydroxy-4-(2-nitrophenylazo)-					
	pK$_2$ 11.69			05		M108

No.	Molecular formula, name and pK value(s)	T(°C)	Remarks	Method	Assessment	Ref.
6243	$C_{17}H_{11}O_5N_3$ 2-Naphthoic acid, 1-hydroxy-4-(3-nitrophenylazo)- pK₂ 10.84			05		M108
6244	$C_{17}H_{12}O_2N_2$ 3-Pyrrolidinecarbonitrile, 2,5-dioxo-4,4-diphenyl- 5.18	20	c = 0.0001-0.001, 4% ethanol, mixed constant	E3bg	Uncert.	F29,F27
6245	$C_{17}H_{12}O_3N_2$ Benzoic acid, 2(2-hydroxy-1-naphthylazo)- pK₂ 12.0	not stated	I = 0.1(KCl)	05	Uncert.	D37
6246	$C_{17}H_{12}O_3N_2$ 2-Naphthoic acid, 1-hydroxy-4-phenylazo- pK₂ 11.40			05		M108
6247	$C_{17}H_{13}O_3N$ Acetamide, N-[4-(1,3-dioxo-indan-2-yl)phenyl]- 3.82	20	1% methanol Values in mixed solvents are also given	05	Approx.	L44
6248	$C_{17}H_{13}O_4N$ 3-Pyrrolidinecarboxylic acid, 2,5-dioxo-4,4-diphenyl- pK₁ 2.85, pK₂ 8.76	20	c = 0.0001-0.001, 4% ethanol, mixed constants	E3bg	Uncert.	F29,F27
6249	$C_{17}H_{14}O_2N_2$ 2-Pyrazolin-5-one, 4-benzoyl-3-methyl-1-phenyl- pK₁ 0.28	25	I = 1.0	05	Uncert.	Z13
	pK₂ 4.11				Approx.	
	pK₂ 4.11	22	I = 1.0(NaClO₄)	0		Z12
	pK₂ 4.04	20	c = 0.000005-0.00005, I = 0.01, mixed constant	05	Approx.	S78

No.	Formula	Name / pK	Temp (°C)	Conditions	Code	Reliability	Ref.
6250	$C_{17}H_{14}O_3N_2$	3-Pyrrolidinecarboxamide, 2,5-dioxo-4,4-diphenyl-					
		7.44	20	c = 0.0001-0.001, 4% ethanol, mixed constant	E3bg	Uncert.	F29,F27
6251	$C_{17}H_{15}O_2N$	2,5-Pyrrolidinedione, 3-benzyl-3-phenyl-					
		8.40	20	c = 0.0001-0.001, 4% ethanol, mixed constant	E3bg	Uncert.	F29,F27
6252	$C_{17}H_{17}O_2N$	Butanamide, 2-benzyl-3-oxo-N-phenyl-					
		11.89	22	5% v/v ethanol; mixed constant	05/E3bg	Uncert.	H38
6253	$C_{17}H_{20}O_6N_4$	Benzo[g]pteridine-2,4(3H,10H)-dione, 7,8-dimethyl-10-ribityl- (Riboflavine)					
		9.93	20	c = 0.001	E3bg	Approx.	A17
6254	$C_{17}H_{22}O_9N_2$	1,3-Phenylenebis(methylnitrilodiacetic acid), 2-hydroxy-5-methyl-					
		pK_1 (COOH) ca 2.0, pK_2 (COOH) 2.9	20	I = 0.1(KCl)	E3bg	Uncert.	
		pK_3 (OH) 6.65, pK_4 (NH) 9.74				Approx.	
		pK_5 (NH) 11.4				Uncert.	S38
6255	$C_{17}H_{25}O_3N$	2-(Dimethylamino)ethyl [α(1-hydroxycyclopentyl)-α-phenyl]acetate					
		7.93 (NH^+)		Conditions not stated	E3bg	Uncert.	W7
6256	$C_{17}H_{30}ON$(cation)	Phenol, 2,6-di-tert-butyl-4-trimethylammonio-					
		9.84	25		05	Approx.	C70
6257	$C_{17}H_{11}O_3N_2Cl$	2-Naphthoic acid, 4-(4-chlorophenylazo)-1-hydroxy-					
		pK_2 11.25			05		M108

No.	Molecular formula, name and pK value(s)	T($^{\circ}C$)	Method	Assessment	Ref.
6258	$C_{17}H_{11}O_3N_2I$ 2-Naphthoic acid, 1-hydroxy-4-(4-iodophenylazo)- pK_2 11.28		05		M108
6259	$C_{17}H_{11}O_4N_3S$ 2-Naphthalenesulfonic acid, 8-(4-cyanophenylazo)-5-hydroxy- 7.81 c = 0.000004, I = 0.25, mixed constant	20	05	Uncert.	R43
6260	$C_{17}H_{11}O_4N_3S_2$ 2-Naphthalenesulfonic acid, 5-(2-benzothiazolylazo)-6-hydroxy- 8.41 Mixed constant	25±0.5	E3bg	Uncert.	N30
6261	$C_{17}H_{12}O_{10}N_2S_2$ Benzoic acid, 2-(1,8-dihydroxy-3,6-disulfonaphth-2-ylazo)- pK_3 (COOH) 3.71, pK_4 (OH) 9.94 I = 0.1(KNO_3), c= 0.002, concentration constants	25	E3bg	Approx.	N10
	pK_3 3.20, pK_4 10.17, (pK_5 > 14) I = 0.1($NaClO_4$), mixed constants	25	05	Approx.	T45
6262	$C_{17}H_{13}O_5N_3S$ Benzoic acid, 2-(1-amino-4-sulfonaphth-2-ylazo)- 3.95 Mixed constant	not stated	05	Uncert.	B159
6263	$C_{17}H_{14}O_4N_2S$ 2-Naphthalenesulfonic acid, 5-hydroxy-8-(4-methylphenylazo)- 7.30 c = 0.000004, I = 0.25, mixed constant	20	05	Uncert.	R43
6264	$C_{17}H_{14}O_4N_4S$ Benzenesulfonamide, N-acetyl-4-(8-hydroxyquinolin-5-ylazo)- 8.7		0		K60a

No.	Formula	Name		Temp.	Conditions			Ref.
6265	$C_{17}H_{14}O_5N_2S$	1-Naphthalenesulfonic acid, 3-hydroxy-4-(2-hydroxy-5-methylphenylazo)-	pK$_3$ 8.14, pK$_4$ 12.35	not stated	I = 0.1(KCl)	05	Uncert.	L51
6266	$C_{17}H_{14}O_5N_2S$	2-Naphthalenesulfonic acid, 5-hydroxy-8-(3-methoxyphenylazo)-	7.45	20	c = 0.000004, I = 0.25, mixed constant	05	Uncert.	R43
6267	$C_{17}H_{14}O_5N_2S$	2-Naphthalenesulfonic acid, 5-hydroxy-8-(4-methoxyphenylazo)-	7.30	20	c = 0.000004, I = 0.25, mixed constant	05	Uncert.	R43
6268	$C_{17}H_{14}O_9N_2S_2$	2,7-Naphthalenedisulfonic acid, 4,5-dihydroxy-3-(2-methoxyphenylazo)-	pK$_3$ 10.00	25	I = 0.10(KNO$_3$), mixed constant	E3bg	Approx.	N9
6269	$C_{17}H_{15}ON_2(Cl)$	1-Methylpyridinium chloride, 2-[2-(8-hydroxy-2-quinolyl)vinyl]-	pK(NH$^+$) 2.7, pK(OH) 9.1	not stated	Mixed constants	05	Uncert.	F4
6270	$C_{17}H_{15}ON_2(Cl)$	1-Methylpyridinium chloride, 3-[2-(8-hydroxy-2-quinolyl)vinyl]-	pK(NH$^+$) 3.5, pK(OH) 9.7	not stated	Mixed constant	05	Uncert.	F4
6271	$C_{17}H_{15}ON_2(Cl)$	1-Methylpyridinium chloride, 4-[2-(8-hydroxy-2-quinolyl)vinyl]-	pK(NH$^+$) 2.7, pK(OH) 9.2	not stated	Mixed constant	05	Uncert.	F4
6272	$C_{17}H_{15}O_7N_3S_2$	2,7-Naphthalenedisulfonic acid, 4-amino-5-hydroxy-6-(4-methylphenylazo)-	12.64(OH)	25	I = 0.1(KNO$_3$), mixed constant	05	Uncert.	E29
6273	$C_{17}H_{26}O_3N_3(Br)$	Piperidinium bromide, 3-diethylcarbamoyl-1-(m-nitrobenzyl)-	6.90	25	c = 0.0015	E3bg	Approx.	Q2

No.	Formula	Name					
6274	$C_{17}H_{26}O_3N_3(Br)$	Piperidinium bromide, 3-diethylcarbamoyl-1-(p-nitrobenzyl)-					
	6.90		25	c = 0.0015	E3bg	Approx.	Q2
6275	$C_{17}H_{27}ON_2(Br)$	Piperidinium bromide, 1-benzyl-3-diethylcarbamoyl-					
	7.81		25	c = 0.0015	E3bg	Approx.	Q2
6276	$C_{17}H_{34}O_4N(Cl)$	Trimethyl 3-carboxy-2-decanoyloxypropylammonium chloride	(Butanoic acid, 3-decanoyloxy-4-trimethylammonio-, chloride)				
	3.65		25	c = 0.01	E3bg	Approx.	Y5
				c.m.c. ≈ 0.011-0.015 (varies with pH)			
6277	$C_{17}H_{26}ON_2Cl(Br)$	Piperidinium bromide, 1-(m-chlorobenzyl)-3-diethylcarbamoyl-					
	7.38		25	c = 0.0015	E3bg	Approx.	Q2
6278	$C_{17}H_{26}ON_2Cl(Br)$	Piperidinium bromide, 1-(p-chlorobenzyl)-3-diethylcarbamoyl-					
	7.55		25	c = 0.0015	E3bg	Approx.	Q2
6279	$C_{18}H_{12}O_3$	4-Cyclopentene-1,3-dione, 2-(9-xantheny1)-					
	9.46			1% methanol	0		L45
6280	$C_{18}H_{12}O_6$	1,4-Benzoquinone, 2,5-dihydroxy-3,6-diphenoxy-					
	pK_1 2.00, pK_2 3.00		25	I = 0.5(HCl-NaCl), concentration constants	05	Uncert.	B42

No.	Molecular formula, name and pK value(s)	$T(^{\circ}C)$	Remarks	Method	Assessment	Ref.
6281	$C_{18}H_{14}O_3$ 1,3-Indandione, 2-(2,4-dimethylbenzoyl)- 3.8	18 ± 2	1% ethanol, mixed constant	O5 or E3d	Uncert.	W25
6282	$C_{18}H_{14}O_4$ Spiro[cyclopropane-1,9'-fluorene]-1-carboxylic acid, 2-methoxycarbonyl- 3.78(H_2O) (cis) 4.62(D_2O)	25		E3bg	Approx.	H21
6283	$C_{18}H_{16}O_4$ Cyclopropanecarboxylic acid, 2-methoxycarbonyl-3,3-diphenyl- 4.43(H_2O) (cis) 5.17(D_2O)	25		E3bg	Approx.	H21
6284	$C_{18}H_{16}O_4$ 1,3-Indandione, 2-(3,4-dimethoxybenzyl)- 5.91	not stated	$c \approx 0.0001$, 1% methanol	O5	Uncert.	S118
6285	$C_{18}H_{22}O_2$ Estra-1,3,5(10),6-tetraene-3,17β-diol (6-Dehydroestradiol) 9.74	25	$c \approx 0.00004$, I = 0.03	O5	Approx.	K33
6286	$C_{18}H_{22}O_2$ Estra-1,3,5(10)-trien-17-one, 3-hydroxy- (Estrone) 10.91	25	$c \approx 0.00001$, I = 0.03	O5	Approx.	K33
6287	$C_{18}H_{22}O_3$ Estra-1,3,5(10)-trien-6-one, 3,17-dihydroxy- (6-Ketoestradiol) 9.08	25	$c \approx 0.0001$, I = 0.03	O5	Approx.	K33
6288	$C_{18}H_{24}O_2$ Estra-1,3,5(10)-triene-3,17α-diol (Estradiol-17α) 10.10	25	$c \approx 0.0001$, I = 0.03	O5	Uncert.	K33

No.	Formula / Name / pK	Temp.	Remarks	Method	Reliability	Ref.
6289	$C_{18}H_{24}O_2$ Estra-1,3,5(10)-triene-3,17β-diol (Estradiol-17β) 10.08	25	$c \approx 0.00005$, I = 0.03	05	Approx.	K33
6290	$C_{18}H_{26}O_4$ Tetracyclo[7.5.1(1,12).0(1,9).0(3,8)]pentadecane-2-carboxylic acid, 12-hydroxy-4,13-dimethyl-5-oxo- 4.73	25	$c = 0.00025$	E3bg	Approx.	T33
6291	$C_{18}H_{30}O$ Phenol, 2,4,6-tri-tert-butyl- 12.19	25	Extrapolated value		Uncert.	C70
6292	$C_{18}H_{32}O_2$ 9,12-Octadecadienoic acid (Linoleic acid) 5.16 4.77	0.7 25	Solubility and conductivity determinations			M1
6293	$C_{18}H_{34}O_6$ Octadecanedioic acid, 9,10-dihydroxy- (Phloionic acid) pK_2 5.74 pK_1 not determinable due to precipitation of the mono-K salt		Titration of di-K salt.	E		S57
6294	$C_{18}H_{36}O_5$ Octadecanoic acid, 9,10,18-trihydroxy- (Phloionolic acid) 4.95			E		S57
6295	$C_{18}H_{14}O_2N_2$ 2,5-Pyrrolidinedione, 3-benzyl-4-cyano-3-phenyl- 5.38	20	$c = 0.0001-0.001$, 4% ethanol, mixed constant	E3bg	Uncert.	F29,F27
6296	$C_{18}H_{14}O_3N_2$ 1,3-Benzenediol, 4-[(2-hydroxy-5-phenyl)phenylazo]- pK_1 6.7, pK_2 8.1, pK_3 11.4	not stated	I = 0.1(KCl)	05	Uncert.	D37

No.	Molecular formula, name and pK value(s)	T(°C)	Method	Assessment	Ref.
6297	$C_{18}H_{14}O_3N_2$ 2-Naphthoic acid, 1-hydroxy-4-(2-methylphenylazo)- pK$_2$ 11.55		05		M108
6298	$C_{18}H_{14}O_3N_2$ 2-Naphthoic acid, 1-hydroxy-4-(3-methylphenylazo)- pK$_2$ 11.43		05		M108
6299	$C_{18}H_{14}O_3N_2$ 2-Naphthoic acid, 1-hydroxy-4-(4-methylphenylazo)- pK$_2$ 11.49		05		M108
6300	$C_{18}H_{14}O_4N_2$ 2-Naphthoic acid, 1-hydroxy-4-(4-methoxyphenylazo)- pK$_2$ 11.61		05		M108
6301	$C_{18}H_{15}O_4N$ 3-Pyrrolidinecarboxylic acid, 4-benzyl-2,5-dioxo-4-phenyl- pK(COOH) 2.97, pK(NH) 9.10 c = 0.0001-0.001, 4% ethanol, mixed constants	20	E3bg	Uncert.	F29,F27
6302	$C_{18}H_{16}O_3N_2$ 3-Pyrrolidinecarboxamide, 4-benzyl-2,5-dioxo-4-phenyl- 7.58 c = 0.0001-0.001, 4% ethanol, mixed constant	20	E3bg	Uncert.	F29,F27
6303	$C_{18}H_{17}O_2N$ 1,3-Indandione, 2-(p-dimethylaminobenzyl)- ≈ 5.5 c ≈ 0.0001, 1% methanol	not stated	05	Uncert.	S118
6304	$C_{18}H_{17}O_2N$ 2,5-Pyrrolidinedione, 3,3-dibenzyl- 8.90 c = 0.0001-0.001, 4% ethanol, mixed constant	20	E3bg	Uncert.	F29,F27

No.	Compound / pK	Temp	Conditions	Method	Reliability	Ref.
6305	$C_{18}H_{18}O_2N$ (cation) 1,3-Indandione, 2-(p-trimethylammoniophenyl)-	20	1% methanol	05	Approx.	L44
	2.66		Values in mixed solvents are also given			
6306	$C_{18}H_{18}O_2N_2$ 3,5-Pyrazolidinedione, 4-isopropyl-1,2-diphenyl-		No details	05	Uncert.	B181
	5.5					
6307	$C_{18}H_{18}O_3N_2$ Propenoic acid, 2-benzoylamino-3-(4-dimethylaminophenyl)-	not stated	c = 0.0008	E3bg	Uncert.	K59
	4.75					
6308	$C_{18}H_{19}O_2N$ Butanamide, 2-acetyl-4,N-diphenyl-	22	5% v/v ethanol, mixed constant	05/E3bg	Uncert.	H38
	12.25					
6309	$C_{18}H_{20}ON_2$ NN-Dimethyl (5-benzyloxyindol-3-yl)methylamine (Indole, 5-benzyloxy-3-dimethylaminomethyl-)	25	In aq.KOH, H_- scale	07	V.uncert.	Y3
	pK(NH) 16.90					
6310	$C_{18}H_{20}O_4N_2$ Ethylenedinitrilobis(2-hydroxy-3-methoxybenzylidene) (N,N'-Bis(2-hydroxy-3-methoxybenzylidene)-1,2-diaminoethane)	20	I = 0.3, Theorell-Steinhagen buffer	05	Uncert.	S122
	pK_1 3.3, pK_2 4.4, pK_3 7.8, pK_4 10.9					
6311	$C_{18}H_{20}O_6N_2$ 3,6-Diazaoctanedioic acid, 2,7-bis(2-hydroxyphenyl)- (Ethylenediimino-N,N'-bis[2-(o-hydroxyphenyl)acetic acid])	25	I = 0.1(KNO_3)	E3bg	Approx.	F45
	pK_1 6.32, pK_2 8.64					
	pK_3 10.24, pK_4 11.68		concentration constants		Uncert.	
	pK_1 6.39, pK_2 8.78	20	I = 0.1(KNO_3), concentration	E3bg	Approx.	A56
	pK_3 10.56, pK_4 11.85		constants	05		

No.	Molecular formula, name and pK value(s)	T(°C)	Remarks	Method	Assessment	Ref.
6312	$C_{18}H_{22}O_4N_4$ 3,6-Diazaoctanedioic acid, 3,6-bis(2-pyridylmethyl)- (Ethylenedinitrilo-NN'-bis(2-pyridylmethyl)-NN'-diacetic acid)					L1
	pK_1 2.34, pK_2 3.02	25	$I = 0.1(KNO_3)$,	E3bg	Uncert.	
	pK_3 5.63, pK_4 8.84		concentration constants		Approx.	
6313	$C_{18}H_{22}O_8N_2$ 1,2,3,4-Tetrahydronaphthalene-2,3-diyldinitrilotetraacetic acid					Y6
	pK_1 2.2 (trans)	25	$c = 0.001$, $I = 0.1(KNO_3)$	E3bg,R3g	V.uncert.	
	pK_1 1.9			R3f	V.uncert.	
	pK_2 3.48, pK_3 5.96, pK_4 10.26			R3g	Approx.	
	pK_2 3.54, pK_3 5.99, pK_4 10.30			R3f	Approx.	
6314	$C_{18}H_{24}ON_2$ Phenol, 2,6-bis(dimethylaminomethyl)-4-phenyl-					E19
	pK_1 5.65	25	$I = 0.1(KNO_3)$, mixed constant	E3bg	Approx.	
6315	$C_{18}H_{27}O_{12}N_3$ 1,3,5-Cyclohexanetriyltrinitrilohexaacetic acid					G48
	pK_4 6.3, pK_5 8.5, pK_6 10.6	30	$I = 0.1(KCl)$, concentration constants	E3bg,R3d	Approx.	
	$-\log(K_1K_2K_3) = 6.6$					
6316	$C_{18}H_{28}O_2N_2$ 1,2-Benzenediol, 3,6-bis(piperidinomethyl)-					K32
	pK_1 6.39, pK_2 9.87		Conditions not stated	E3bg	Uncert.	
6317	$C_{18}H_{30}O_{12}N_4$ Nitrilotris(ethylenenitrilodiacetic acid) (TTAHA)					H53
	pK_4 2.38, pK_5 2.48, pK_6 3.22, pK_7 8.29, pK_8 8.67, pK_9 10.09	25	$I = 0.1(KCl)$	E3bg	Uncert.	

No.	Formula	Name / pK values	T (°C)	Conditions	Method	Remarks	Ref.
6318	$C_{18}H_{30}O_{12}N_4$	3,6,9,12-Tetraazatetradecanedioic acid, 3,6,9,12-tetrakis(carboxymethyl)- (TTHA.) [Ethylenedinitrilo-N,N'-diacetic-N,N'-bis(ethylenenitrilodiacetic acid)]					
			25	c = 0.002, I = 0.1(KNO$_3$), concentration constants	E3bg,R3g	Uncert.	B88,B89
						Approx.	B89
		pK_1 2.42, pK_2 2.95					
		pK_3 4.16, pK_4 6.16, pK_5 9.40, pK_6 10.19					
			30	I = 0.1(KCl), concentration constants	E3bg,R3d	Uncert.	G48
						Approx.	
		pK_1 2.46, pK_2 2.52					
		pK_3 4.00, pK_4 5.98, pK_5 9.35, pK_6 10.33					
			25	I = 0.1(KNO$_3$), mixed constants	E3bg	Uncert.	F44
		pK_2 2.64, pK_3 4.08, pK_4 6.26, pK_5 9.67, pK_6 10.82					
6319	$C_{18}H_{11}ONCl_2$	2,5-Cyclohexadien-1-one, 4-(3,5-dichloro-4-hydroxyphenyl)imino-2-phenyl-					
		5.8	not stated	c = 0.00004, 2% dioxane, mixed constant	05	Uncert.	K54
6320	$C_{18}H_{11}O_2NBr_2$	2,5-Cyclohexadien-1-one, 2,6-dibromo-4-(6-hydroxybiphenyl-3-yl)imino-					
		6.1	not stated	c = 0.00002, 2% dioxane, mixed constant	05	Uncert.	K54
6321	$C_{18}H_{13}O_3N_5S_2$	Benzenesulfonamide, 4-(8-hydroxyquinolin-5-ylazo)-N-(1,3-thiazol-2-yl)-					
		7.15			0		K60a
6322	$C_{18}H_{14}O_5N_2S$	2-Naphthalenesulfonic acid, 8-(3-acetylphenylazo)-5-hydroxy-					
		7.48	20	c = 0.000004, I = 0.25, mixed constant	05	Uncert.	R43
6323	$C_{18}H_{14}O_5N_2S$	2-Naphthalenesulfonic acid, 8-(4-acetyl]phenylazo)-5-hydroxy-					
		7.94	20	c = 0.000004, I = 0.25, mixed constant	05	Uncert.	R43

No.	Molecular formula, name and pK value(s)	T(°C)	Method	Assessment	Remarks	Ref.
6324	$C_{18}H_{14}O_{11}N_2S_2$ Acetic acid, 2-[2-(1,8-dihydroxy-3,6-disulfonaphth-2-ylazo)phenoxy]-					M90,T49
	pK(COOH) 2.99, pK(OH) 9.76	25	05	Approx.	I = 0.1(NaClO$_4$), mixed constants	
6325	$C_{18}H_{14}O_{11}N_2S_2$ Acetic acid, 2-[3-(1,8-dihydroxy-3,6-disulfonaphth-2-ylazo)phenoxy]-					T50
	pK(OH) 9.48	25	05	Uncert.	c = 0.00005, I = 0.1(NaClO$_4$), mixed constant	
6326	$C_{18}H_{14}O_{11}N_2S_2$ Acetic acid, 2-[4-(1,8-dihydroxy-3,6-disulfonaphth-2-ylazo)phenoxy]-					T50
	pK(COOH) 3.04, pK(OH) 9.03	25	05	Uncert.	c = 0.00005, I = 0.1(NaClO$_4$), mixed constants	
6327	$C_{18}H_{14}O_{11}N_2S_2$ Acetic acid, 2-[2-(1,8-dihydroxy-3,6-disulfonaphth-2-ylazo)phenyl]-2-hydroxy-					T47
	pK$_3$(COOH) 3.1, pK$_4$(OH) 9.7	7-9	05	Uncert.	Mixed constants	
	pK$_3$ 3.10, pK$_4$ 9.55	25	05	Approx.	I = 0.1(NaClO$_4$), mixed constants	M89
6328	$C_{18}H_{14}O_{11}N_2S_2$ Acetic acid, 2-[3-(1,8-dihydroxy-3,6-disulfonaphth-2-ylazo)phenyl]-2-hydroxy-					T47
	pK$_3$(COOH) 3.4, pK$_4$(OH) 9.7	7-9	05	Uncert.	Mixed constants	
6329	$C_{18}H_{14}O_{11}N_2S_2$ Acetic acid, 2-[4-(1,8-dihydroxy-3,6-disulfonaphth-2-ylazo)phenyl]-2-hydroxy-					T47
	pK$_3$(COOH) 3.9, pK$_4$(OH) 9.2	7-9	05	Uncert.	Mixed constants	
6330	$C_{18}H_{15}O_3SP$ Benzenesulfonic acid, 3-diphenylphosphino-					W47
	pK$_2$ 0.13	25	05	Uncert.	c = 0.000123, I → 0(HClO$_4$)	
					Protonation on phosphorus atom; compound	
					undergoes self-association when c > 0.01.	
					pK varies linearly with concentration of HClO$_4$; pK' = 0.13+0.5c$_{HClO_4}$	

No.	Formula / Name	pK	T	Conditions	Method	Reliability	Ref.
6331	$C_{18}H_{15}O_8N_3S_2$ 2,7-Naphthalenedisulfonic acid, 4-acetylamino-5-hydroxy-3-phenylazo-	10.65(OH)	25	$I = 0.1(KNO_3)$, mixed constant	05	Uncert.	E29
6332	$C_{18}H_{16}O_8N_4S_2$ 2,7-Naphthalenedisulfonic acid, 6-(4-acetylaminophenylazo)-4-amino-5-hydroxy-	12.49(OH)	25	$I = 0.1(KNO_3)$, mixed constant	05	Uncert.	E29
6333	$C_{18}H_{20}O_9N_2S$ 3,6-Diazaoctanedioic acid, 3-carboxymethyl-6-(4-sulfonaphth-1-yl)- (Ethylenedinitrilo-N-(4-sulfonaphth-1-yl)-NN'N'-triacetic acid).	pK_2 2.91, pK_3 4.37, pK_4 9.01		Conditions not stated	05	Uncert.	T17
6334	$C_{18}H_{23}O_{14}N_4P$ 3'-Uridinyl-5'-uridinyl hydrogen phosphate	pK_2 9.33	20	$I = 0.1(NaCl)$	05	Approx.	S75
6335	$C_{18}H_{29}ON_2(Br)$ Piperidinium bromide, 3-diethylcarbamoyl-1-(m-methylbenzyl)-	7.86	25	$c = 0.0015$	E3bg	Approx.	Q2
6336	$C_{18}H_{29}ON_2(Br)$ Piperidinium bromide, 3-diethylcarbamoyl-1-(p-methylbenzyl)-	7.91	25	$c = 0.0015$	E3bg	Approx.	Q2
6337	$C_{18}H_{29}O_2N_2(Br)$ Piperidinium bromide, 3-diethylcarbamoyl-1-(m-methoxybenzyl)-	7.71	25	$c = 0.0015$	E3bg	Approx.	Q2
6338	$C_{18}H_{29}O_2N_2(Br)$ Piperidinium bromide, 3-diethylcarbamoyl-1-(p-methoxybenzyl)-	7.98	25	$c = 0.0015$	E3bg	Approx.	Q2

No.	Molecular formula, name and pK value(s)	T(°C)	Remarks	Method	Assessment	Ref.
6339	$C_{18}H_{36}O_5N_2S$ 5,8-Diazadodecan-6-one, 5,8-bis(4-hydroxybutyl)-1,12-dihydroxy-7-thioxo- [Monothiooxamide, NNN'N'-tetrakis(4-hydroxybutyl)-]					
	12.02	25		O5,R4	Approx.	C13
	11.91	25	I = 0.1(KCl), mixed constant (Values for I = 0.02, 0.05, 0.25, 0.50, 1.00 are also given)	O5	Approx.	
6340	$C_{18}H_{15}O_8N_6SAs$ Benzenesulfonic acid, 4-[[4-[3-(2-arsono-4-nitrophenyl)-2-triazeno]phenyl]azo]- (Azobenzene-4-sulfonic acid, 4'-[3-(2-arsono-4-nitrophenyl)-2-triazeno]-) (Sulfarsazen)					
	pK_1 5.1, pK_2 8.5, pK_3 11.7	20	I = 0.08(KCl)	E3bg	Uncert.	P10
6341	$C_{19}H_{12}O_2$ 1,3-Indandione, 2-(1-naphthyl)-					
	3.74	20		O5		S142
6342	$C_{19}H_{12}O_2$ 1,3-Indandione, 2-(2-naphthyl)-					
	3.74	20		O5		S142
6343	$C_{19}H_{12}O_5$ 3H-Xanthen-3-one, 2,6,7-trihydroxy-9-phenyl-					
	pK_1(cation) 2.11, pK_2 6.28, pK_3 10.21, pK_4 11.7		10% ethanol	O		N23

No.	Formula	Name	pK / value	t(°C)	Conditions	Method		Ref.
6344	$C_{19}H_{12}O_6$	3H-Xanthen-3-one, 2,6,7-trihydroxy-9-(2-hydroxyphenyl)-	pK$_1$(cation) 3.51, pK$_2$ 6.25	25	c = 0.00002, extrapolated to I = 0 from values in LiCl solutions	05	Approx.	N22
6345	$C_{19}H_{14}O_3$	2,5-Cyclohexadien-1-one, 4-[bis(p-hydroxyphenyl)]methylene- (Rosolic acid) (Aurin)	pK$_1$ 3.11, pK$_2$ 8.62	25	0	0	Approx.	L43
6346	$C_{19}H_{22}O_5$	16-Oxapentacyclo[9.3.2.1(4,7).0(2,11).0(4,10)]heptadec-12-ene-3-carboxylic acid, 14-hydroxy-1-methyl-6-methylene-15-oxo- (Gibberellin A$_7$)	4.11	25	c = 0.00056	E3bg	Approx.	T33
6347	$C_{19}H_{22}O_5$	16-Oxapentacyclo[9.3.2.1(4,7).0(2,11).0(4,10)]heptadec-13-ene-3-carboxylic acid, 7-hydroxy-1-methyl-6-methylene-15-oxo- (Gibberellin A$_5$)	4.09	25	c = 0.00016	E3bg	Approx.	T33
6348	$C_{19}H_{22}O_6$	16-Oxapentacyclo[9.3.2.1(4,7).0(2,11).0(4,10)]heptadecane-3-carboxylic acid, 7-hydroxy-1-methyl-6-methylene-15-oxo-13,14-epoxy- (Gibberellin A$_6$)	4.03	25	c = 0.00014	E3bg	Approx.	T33
6349	$C_{19}H_{22}O_6$	16-Oxapentacyclo[9.3.2.1(4,7).0(2,11).0(4,10)]heptadec-12-ene-3-carboxylic acid, 7,14-dihydroxy-1-methyl-6-methylene-15-oxo- (Gibberellin A$_3$)	3.97	25	c = 0.0005	E3bg	Approx.	T33
6350	$C_{19}H_{24}O_4$	16-Oxapentacyclo[9.3.2.1(4,7).0(2,11).0(4,10)]heptadecane-3-carboxylic acid, 1-methyl-6-methylene-15-oxo- (Gibberellin A$_9$)	4.26	25	c = 0.00012	E3bg	Approx.	T33

No.	Molecular formula, name and pK value(s)	$T(^{o}C)$	Method	Assessment	Ref.	Remarks
6351	$C_{19}H_{24}O_5$ 16-Oxapentacyclo[9.3.2.1(4,7).0(2,11).0(4,10)]heptadecane-3-carboxylic acid, 14-hydroxy-1-methyl-6-methylene-15-oxo- (Gibberellin A_4) pK$_1$ 4.21	25	E3bg	Approx.	T33	c = 0.00044
6352	$C_{19}H_{24}O_6$ 16-Oxapentacyclo[9.3.2.1(4,7).0(2,11).0(4,10)]heptadecane-3-carboxylic acid, 7,14-dihydroxy-1-methyl-6-methylene-15-oxo- (Gibberellin A_1) 4.09	25	E3bg	Approx.	T33	c = 0.0003
6353	$C_{19}H_{24}O_6$ 16-Oxapentacyclo[9.3.2.1(4,7).0(2,11).0(4,10)]heptadecane-3-carboxylic acid, 14-hydroxy-1,7-dimethyl-6,15-dioxo- 4.05	25	E3bg	Approx.	T33	c = 0.00024
6354	$C_{19}H_{24}O_7$ 16-Oxapentacyclo[9.3.2.1(4,7).0(2,11).0(4,10)]heptadecane-3-carboxylic acid, 7,13,14-trihydroxy-1-methyl-6-methylene-15-oxo- (Gibberellin A_8) 4.04	25	E3bg	Approx.	T33	c = 0.0002
6355	$C_{19}H_{26}O_6$ 16-Oxapentacyclo[9.3.2.1(4,7).0(2,11).0(4,10)]heptadecane-3-carboxylic acid, 6,14-dihydroxy-1,6-dimethyl-15-oxo- (Gibberellin A_2) 4.21	25	E3bg	Approx.	T33	c = 0.00049
6356	$C_{19}H_{11}O_7N$ 3H-Xanthen-3-one, 2,6,7-trihydroxy-9-(4-nitrophenyl)- pK$_1$ 2.26, pK$_2$ 6.06, pK$_3$ 9.95, pK$_4$ 11.6		O		N23	10% ethanol, pK$_1$ (protonated species)
6357	$C_{19}H_{12}O_8S$ 3H-Xanthen-3-one, 4,5,6-trihydroxy-9-(2-sulfophenyl)- (Pyrogallol red) pK$_1$ 2.56, pK$_2$ 6.28, pK$_3$ 9.75, pK$_4$ 11.94	not stated	05	Uncert.	S125	c = 0.00002, I = 0.2(KCl), mixed constants

6358 $C_{19}H_{14}O_5S$ 4,4'-(3$\underline{H}$-2,1-Benzoxathiol-3-ylidene)diphenol, S,S-dioxide   (Phenol red)

pK values	Temp	Conditions	Method	Reliability	Ref
pK_1 1.33, pK_2 7.92		Conditions not stated	0		P72
pK_1 1.03			05	Uncert	G61
pK_2 8.08	25	c = 0.00002	05	Approx.	S56
pK_2 7.66	25	I = 0.15(NaCl), mixed constant			
pK_3 7.57	37				

6359 $C_{19}H_{14}O_7S$ 4,4'-(3$\underline{H}$-2,1-Benzoxathiol-3-ylidene)bis(benzene-1,2-diol), S,S-dioxide   (Pyrocatechol violet)

pK values	Temp	Conditions	Method	Reliability	Ref
pK_2 7.90, pK_3 9.94, pK_4 11.82	25	c = 0.00001, I = 0.1(NaClO$_4$), mixed constants	05	Approx.	B81
pK_2 7.45, pK_3 9.56, pK_4 11.42		I = 0.5(NaClO$_4$)			
pK_2 6.92, pK_3 9.10, pK_4 10.80		I = 1.0(NaClO$_4$)			
pK_2 7.82, pK_3 9.76, pK_4 11.73	room temp.	I = 0.2, mixed constants	E3bg	Uncert.	R57
pK_2 7.81, pK_3 9.80			05		
pK_4 12.50	not stated	c = 0.00002, mixed constant	05	Uncert.	M44

6360 $C_{19}H_{15}O_8N$ Iminodiacetic acid, N-(9,10-dihydro-3,4-dihydroxy-9,10-dioxoanthracen-2-yl)methyl-   (Alizarin fluorine blue)

pK values	Temp	Conditions	Method	Reliability	Ref
pK_1 4.89, pK_2 7.55, pK_3 10.43, pK_4 11.19	25	I = 0.1(KNO$_3$), mixed constants	E3bg	Uncert.	L1a

6361 $C_{19}H_{16}O_2N_2$ 3-Pyrrolidinecarbonitrile, 4,4-dibenzyl-2,5-dioxo-

pK values	Temp	Conditions	Method	Reliability	Ref
5.80	20	c = 0.0001-0.001, 4% ethanol, mixed constant	E3bg	Uncert.	F29,F30

6362 $C_{19}H_{17}O_4N$ 3-Pyrrolidinecarboxylic acid, 4,4-dibenzyl-2,5-dioxo-

pK values	Temp	Conditions	Method	Reliability	Ref
pK_1 3.09	20	In 4% ethanol	E3bg	Uncert.	F30

No.	Molecular formula, name and pK value(s)	$T(^{\circ}C)$	Remarks	Method	Assessment	Ref.
6363	$C_{19}H_{18}O_3N_2$ 3-Pyrrolidinecarboxamide, 4,4-dibenzyl-2,5-dioxo-					
	pK(NH) 7.51	20	c = 0.0001-0.001, 4% ethanol, mixed constant	E3bg	Uncert.	F29,F30
6364	$C_{19}H_{19}O_4N_3$ 3,5-Pyrazolidinedione, 4-butyl-1-(4-nitrophenyl)-2-phenyl-					
	3.2		Conditions not stated	05	Uncert.	B181
6365	$C_{19}H_{19}O_6N_7$ Pentanedioic acid, 2-[4-[[(2-amino-3,4-dihydro-4-oxopteridin-6-yl)methylamino]benzoylamino]- (Folic acid)					
	pK$_3$ 8.26	20	c = 0.002	E3bg	Uncert.	A17
	pK$_1$ 4.65, pK$_2$ 6.75, pK$_3$ 9.00	30		E3bg,R4	Uncert.	N20
	pK$_1$ 4.30, pK$_2$ 6.55, pK$_3$ 8.80		I = 0.01(KNO$_3$)	E3bg		
	pK$_1$ 2.95, pK$_2$ 5.70, pK$_3$ 8.45		I = 0.05(KNO$_3$)			
	pK$_1$ 2.35, pK$_2$ 5.35, pK$_3$ 8.30		I = 0.10(KNO$_3$)			
6366	$C_{19}H_{20}O_2N_2$ 3,5-Pyrazolidinedione, 4-butyl-1,2-diphenyl- (Phenylbutazone)					
	4.5		Conditions not stated	05	Uncert.	B181
6367	$C_{19}H_{20}O_3N_2$ 3,5-Pyrazolidinedione, 4-butyl-1-(4-hydroxyphenyl)-2-phenyl-					
	4.7		Conditions not stated	05	Uncert.	B181
6368	$C_{19}H_{20}O_3N_2$ 3,5-Pyrazolidinedione, 4-(3-hydroxybutyl)-1,2-diphenyl-					
	4.0		Conditions not stated	05	Uncert.	B181
6369	$C_{19}H_{23}O_5N_5$ 3,6,9-Triazaundecanoic acid, 5-(1-benzylimidazol-4-yl)methyl-4,7,10-trioxo-					
	pK$_1$ 3.30, pK$_2$ 6.37	25	I = 0.16(KCl)	E3bg	Approx.	B153

| 6370 | $C_{19}H_{10}O_5Br_4S$ | 4,4'-(3H-2,1-Benzoxathiol-3-ylidene)bis(2,6-dibromophenol), S,S-dioxide (Bromophenol blue) | | | | |
| | pK_2 3.62 | | | 0 | | P72 |

6371	$C_{19}H_{10}O_8Br_2S$	Benzenesulfonic acid, 2-(2,7-dibromo-4,5,6-trihydroxy-3-oxo-9-xanthenyl)- (Bromopyrogallol red)				
	pK_1 0.16, pK_2 4.39, pK_3 9.13, pK_4 11.27	not stated	c = 0.00002, I = 0.2(KCl), mixed constants	05	Uncert.	S125
	pK_1 2.93, pK_2 5.21, pK_3 9.15, pK_4 11.48		c ≈ 0.001	05		A14

| 6372 | $C_{19}H_{12}O_5Br_2S$ | 4,4'-(3H-2,1-Benzoxathiol-3-ylidene)bis(2-bromophenol), S,S-dioxide (Bromophenol red) | | | | |
| | pK_1 1.51, pK_2 6.89 | | | 0 | | P72 |

| 6373 | $C_{19}H_{13}O_7N_3S_2$ | 2,7-Naphthalenedisulfonic acid, 3-hydroxy-4-(quinolin-8-ylazo)- | | | | |
| | pK_3 1.48, pK_4(OH) 11.49 | 20 | c = 0.0001, I = 0.1, mixed constants | 05 | V.uncert. | B33 |

| 6374 | $C_{19}H_{16}O_3N_6S_2$ | Benzenesulfonamide, N-(5-ethyl-1,2,4-thiadiazol-3-yl)-4-(8-hydroxyquinolin-5-ylazo)- | | | | |
| | 7.70 | | | 0 | | K60a |

| 6375 | $C_{19}H_{17}O_2N_3S_2$ | Benzothiazol-7-one, 4,5,6-7-tetrahydro-2-(2-hydroxy-1-naphthylazo)-5,5-dimethyl- | | | | |
| | pK_1 -1.10(NH), pK_2 8.46(OH) | | | | | I1 |

| 6376 | $C_{19}H_{17}O_8N_3S_2$ | 2,7-Naphthalenedisulfonic acid, 4-acetylamino-5-hydroxy-6-(4-methylphenylazo)- | | | | |
| | 10.43(OH) | 25 | I = 0.1(KNO_3), mixed constant | 05 | Approx. | E29 |

| 6377 | $C_{19}H_{19}ON_2(I)$ | 1-Methylpyridinium iodide, 3-ethyl-6-[2-(8-hydroxy-5-quinolyl)vinyl]- | | | | |
| | pK(NH^+) 3.8, pK(OH) 7.9 | not stated | Mixed constants | 05 | Uncert. | F4 |

No.	Molecular formula, name and pK value(s)	$T(^oC)$	Remarks	Method	Assessment	Ref.
6378	$C_{19}H_{24}O_{12}N_7P$ 3'-Adenosyl 5'-uridinyl hydrogen phosphate					
	pK_1 3.85	20	I = 0.1(NaCl)	05	Uncert.	S75
	pK_2 9.35				Approx.	
6379	$C_{19}H_{24}O_{12}N_7P$ 5'-Adenosyl 3'-uridinyl hydrogen phosphate					
	pK_1 3.70	20	I = 0.1(NaCl)	05	Uncert.	S75
	pK_2 9.48				Approx.	
6380	$C_{20}H_{14}O_2$ 1,3-Indandione, 2-(1-naphthyl)methyl-					
	5.32	not stated	c ≈ 0.0001, 1% methanol	05	Uncert.	S118
6381	$C_{20}H_{28}O_2$ Dehydroabietic acid					
	5.7	20	Titration of Na-salt below c.m.c.	E3bg	Uncert.	N43
6382	$C_{20}H_{30}O_2$ Abietic acid					
	6.4	20	Titration of Na-salt below c.m.c.	E3bg	Uncert.	N43
	5.27					S119

No.	Formula	Name / pK values	Temp.	Conditions	Method	Reliability	Ref.
6383	$C_{20}H_{12}O_{11}S_2$	2-Anthracenesulfonic acid, 9,10-dihydro-1,4-dihydroxy-9,10-dioxo-3-(4-sulfophenoxy)- pK_3 7.3, $pK_4 \approx 12.4$	23	$I = 0.1(KNO_3)$, mixed constants	05	Uncert.	029
6384	$C_{20}H_{15}O_4P$	Di(2-naphthyl) hydrogen phosphate 0.74	20±2	$c \approx 0.001$ or 0.0001, $I \approx 0.1$ or 1($NaClO_4$), concentration constant	DIS	Uncert.	K56
6385	$C_{20}H_{16}O_2N_2$	2,2'-[1,2-Phenylenebis(iminomethyl)]diphenol pK_1 5.4, pK_2 6.2, pK_3 8.1, pK_4 12	20	$I = 0.3$, Theorell-Steinhagen buffer Compound decomposed by acid and alkali	05	Uncert. V.uncert.	S122
6386	$C_{20}H_{16}O_5N_2$	2-Naphthoic acid, 4-(4-ethoxycarbonylphenylazo)-1-hydroxy- pK_2 11.1			05	Uncert.	M108
6387	$C_{20}H_{19}O_8N$	1,4,4a,5,5a,6,11,12a-Octahydronaphthacene-2-carboxamide, 3,6,10,12,12a-pentahydroxy-6-methyl-1,11-dioxo- (Desmethylaminotetracycline) pK_1 5.97, pK_2 8.56		Conditions not stated	E3bg	Uncert.	L14
6388	$C_{20}H_{21}O_2N$	1,3-Indandione, 2-(4-diethylaminobenzyl)- 5.75	not stated	$c \approx 0.0001$, 1% methanol	05	Uncert.	S118
6389	$C_{20}H_{24}O_6N_2$	Ethylenedinitrilo-NN'-bis(2-hydroxybenzyl)-NN'-diacetic acid pK_3 4.64, pK_4 8.32, pK_5 11.00, pK_6 12.46	25	$I = 0.1(KNO_3)$	E3bg 05	Approx. Approx.	L19

No.	Molecular formula, name and pK value(s)	T(°C)	Remarks	Method	Assessment	Ref.
6390	C$_{20}$H$_{36}$O$_8$N$_2$ (Dodecamethylenedinitrilo)tetraacetic acid					A55
	pK$_1$ 2.6, pK$_2$ 2.6	20	I = 0.1(KNO$_3$)	E3bg	Uncert.	
	pK$_3$ 9.96, pK$_4$ 10.55				Approx.	
	pK$_1$ 2.5, pK$_2$ 2.5		I = 1.0(KNO$_3$)	E3bg	Uncert.	
	pK$_3$ 9.95, pK$_4$ 10.62				Approx.	
			Concentration constants			
6391	C$_{20}$H$_{38}$O$_6$N$_2$ Ethylenedinitrilo-N-dodecyl-NN'N'-triacetic acid					B149
	pK$_1$ 2.0	24.2	c = 0.002	E3bg	Uncert.	
	pK$_2$ 6.56, pK$_3$ 9.77				Approx.	
6392	C$_{20}$H$_{13}$O$_7$N$_3$S 1-Naphthalenesulfonic acid, 3-hydroxy-4-(1-hydroxy-2-naphthylazo)-7-nitro- (Eriochrome Black T) (C I Mordant Black 11)					C68
	pK$_2$ 5.81	25	c ≈ 0.000002. Extrapolated from values in dioxan/water mixtures	05	Uncert.	
	pK$_2$ 6.3, pK$_3$ 11.55	not stated	I = 0.008 (Reference S41 gives structure of Eriochrome Black T as the 8-nitro compound)	05	Uncert.	S41
6393	No entry					

6394	$C_{20}H_{13}O_7N_3S$	1-Naphthalenesulfonic acid, 3-hydroxy-4-(2-hydroxy-1-naphthylazo)-8-nitro-		(Eriochrome Black A)	
	pK_2 6.2, pK_3 13.0		not stated	$I = 0.008$		05	Uncert.	S41
	pK_2 6.22		25	$c = 0.000002$, by extrapolation from		05	Uncert.	C68
				values in dioxan/water mixtures		
	pK_3 13.10			$c = 0.0000025$ in $(CH_3)_4NOH$		07	

6395	$C_{20}H_{14}O_5N_2S$	1-Naphthalenesulfonic acid, 3-hydroxy-4-(1-hydroxy-2-naphthylazo)-		(Eriochrome Black B)		(CI Mordant Black 3)	
	pK_2 6.2, pK_3 12.5		not stated	$I = 0.008$		05	Uncert.	S41
	pK_2 6.50		25	$c = 0.00004$, by extrapolation by values		05	Uncert.	C68
				in dioxan/water mixtures		
	pK_3 12.81			$c \approx 0.0003$ in $(CH_3)_4NOH$		07	

6396	$C_{20}H_{14}O_5N_2S$	1-Naphthalenesulfonic acid, 3-hydroxy-4-(2-hydroxy-1-naphthylazo)-		(Eriochrome Black R)		(Calcon)	(CI Mordant Black 17)	
	pK_2 7.36, pK_3 13.5		18-22	$I = 0.1(KCl)$		05	Uncert.	H40
	pK_2 7.0, pK_3 13.5		not stated			05	Uncert.	S41
	pK_2 7.31		25	$c = 0.00003$, by extrapolation of		05	Uncert.	C68
				values in dioxan/water mixtures		
	pK_3 13.80			$c \approx 0.00003$ in $(CH_3)_4NOH$		07	

6397	$C_{20}H_{14}O_{15}N_2S_4$	2,7-Naphthalenedisulfonic acid, 4,6-dihydroxy-3-(8-hydroxy-3,6-disulfonaphth-1-ylazo)-			
	pK_3 2.39, pK_4 4.43, pK_5 5.71, pK_6 7.01, pK_7 10.88		20	$I = 0.2(KCl)$		05	Uncert.	A3

6398	$C_{20}H_{18}O_2(Cl)P$	Triphenyl(carboxymethyl)phosphonium chloride		(Acetic acid, triphenylphosphonio-, chloride)	
	1.77		25	$c = 0.01$, mixed constant		E3bg	Uncert.	M52

No.	Molecular formula, name and pK values(s)	T(^{o}C)	Remarks	Method	Assessment	Ref.
6399	$C_{20}H_{18}O_9N_4S_2$ 2,7-Naphthalenedisulfonic acid, 4-acetylamino-6-(4-acetylaminophenylazo)-5-hydroxy-					
	10.41(OH)	25	I = 0.1(KNO_3), mixed constant	05	Uncert.	E29
6400	$C_{20}H_{25}O_{10}N_{10}P$ 3'-Adenosyl 5'-adenosyl hydrogen phosphate					
	3.50	20	I = 0.1(NaCl)	05	Approx.	S75
6401	$C_{21}H_{16}O_6$ 2,2'-(Phenylmethylenedioxy)dibenzoic acid					
	pK_1 3.32, pK_2 4.48	25	I = 0.1(KCl)	KIN	Uncert.	A60
			Calculated from a nonlinear least-squares			
			procedure which gives the best fit to the			
			experimental data			
6402	$C_{21}H_{26}O_9$ 15-Oxatetracyclo[8.3.2.0(2,10).0(4,9)]pentadecane-3-carboxylic acid, 13-acetoxy-4-methoxycarbonylmethyl-1-methyl-6,14-dioxo-					
	4.29	25	c = 0.00021	E3bg	Approx.	T33
6403	$C_{21}H_{18}O_5S$ 4,4'-(3$\underline{H}$-2,1-Benzoxathiol-3-ylidene)bis(2-methylphenol), S,S-dioxide (cresol red)					
	pK_1 1.56, pK_2 8.18			0		P72
	pK_1 1.05		Conditions not stated	05	Uncert.	G61

No.	Formula / Name / pK values	Temp	Conditions	Method	Assessment	Ref.
6404	$C_{21}H_{22}O_2N_4$ 2,7-Naphthalenediol, 1-[5-(1-methylpiperidin-2-yl)-2-pyridylazo]- pK$_1$ 1.69, pK$_2$ 7.88, pK$_3$ 8.88	18	Mean of three methods			S86
6405	$C_{21}H_{24}O_2N_2$ 3,5-Pyrazolidinedione, 1,2-bis(p-tolyl)-4-butyl- 4.9		Conditions not stated	05	Uncert.	B181
6406	$C_{21}H_{33}O_5N_3$ 1,2-Benzenediol, 3,4,6-tris(morpholinomethyl)- pK$_1$ 3.75, pK$_2$ 5.80, pK$_3$ 7.90 pK$_1$ 3.7, pK$_2$ 5.7, pK$_3$ 7.8	31.5 ca 27	c = 0.008, I = 0.2	E3bg 05	Approx. Uncert.	K31
6407	$C_{21}H_{14}O_5Br_4S$ 4,4'-(3H-2,1-Benzoxathiol-3-ylidene)bis(2,6-dibromo-3-methylphenol), S,S-dioxide (Bromocresol green) pK$_2$ 4.51			0		P72
6408	$C_{21}H_{16}O_5Br_2S$ 4,4'-(3H-2,1-Benzoxathiol-3-ylidene)bis(2-bromo-6-methylphenol), S,S-dioxide (Bromocresol purple) pK$_1$ -0.78 pK$_1$ -2.15	not stated	In aq.HCl, H$_0$ scale In aq.H$_2$SO$_4$, H$_0$ scale	05	Uncert.	G61
6409	$C_{21}H_{18}O_3N_6S$ Benzenesulfonamide, N-(4,6-dimethylpyrimidin-2-yl)-4-(8-hydroxyquinolin-5-ylazo)- 7.15			0		K60a
6410	$C_{21}H_{18}O_6N_4S$ Benzoic acid, 2-[1-(2-methoxy-5-sulfophenyl)-3-phenylformazan-5-yl]- pK(COOH) 3.6, pK(NH) 14.4			0		U1
6411	$C_{21}H_{19}O_8N_3S$ Iminodiacetic acid, N-[2-hydroxy-4-(4'-sulfophenylazo)naphth-2-yl]methyl- (Methyl naphthol orange) pK$_2$ 2.40, pK$_3$ 4.51, pK$_4$ 10.49	not stated	c = 0.0002, I = 0.2(NaNO$_3$)	05	Uncert.	B157

No.	Molecular formula, name and pK value(s)	T(°C)	Remarks	Method	Assessment	Ref.
6412	$C_{21}H_{24}O_6N_2S_2$ 3,5-Pyrazolidinedione, 1,2-bis(4-mesylphenyl)-4-butyl- 2.7		Conditions not stated	05	Uncert.	B181
6413	$C_{21}H_{27}O_{14}N_7P_2$ 5'-Adenosyl (3-carbamoylpyridinio-β-D-ribofuranos-5'-yl) hydrogen diphosphate (Diphosphopyridine nucleotide) (DPN) (Nadide) 3.67	20	I = 0.1	E3bg	Approx.	W3
6414	$C_{22}H_{14}O_9$ 1,4-Cyclohexadiene-1-carboxylic acid, 3-[bis(3-carboxy-4-hydroxyphenyl)methylene]-6-oxo- (Aluminon) pK_4 8.65, pK_5 8.85	25	I = 0.1(NaClO$_4$)	E3bg,R3g	Uncert.	B13
	pK_4 8.93, pK_5 9.76	25	I = 0.02(NaCl)	E3bg,R3h	Uncert.	A81
6415	$C_{22}H_{16}O_2N_2$ 2-Naphthol, 1-(4-hydroxy-3-biphenylylazo)- pK_1 8.0, pK_2 11.8	not stated	I = 0.1(KCl)	05	Uncert.	D37
6416	$C_{22}H_{24}O_8N_2$ Ethylenedinitrilotetraacetic acid, 1,2-diphenyl- (dl-DPEDTA) pK_1 2.18 pK_2 3.73, pK_3 5.42, pK_4 9.91	25	c = 0.001, I = 0.1(KCl/KNO$_3$), concentration constants	E3bg	Uncert. Approx.	O17
6417	$C_{22}H_{24}O_8N_2$ 1,4,4a,5,5a,6,11,12a-Octahydronaphthacene-2-carboxamide, 4-dimethylamino-3,6,10,12,12a-pentahydroxy-6-methyl-1,11-dioxo- (Tetracycline) pK_1 3.33, pK_2 7.75, pK_3 9.61		Conditions not stated	E3bg	Uncert.	L14

No.	Formula	Name	pK data	Conditions	E3bg	Uncert.	Ref.
6418	$C_{22}H_{27}O_3N_3$	Benzyl 4-(N-acetyl-2,6-dimethylanilino)piperazine-1-carboxylate	4.5	25		Uncert.	F35a
6419	$C_{22}H_{37}O_{14}N_5$	3,6,9,12,15-Pentaazaheptadecanedioic acid, 3,6,9,12,15-pentakis(carboxymethyl)- (TPHA)	pK_3 2.79, pK_4 3.82, pK_5 5.56, pK_6 8.85, pK_7 9.95		E		D66
6420	$C_{22}H_{12}O_{18}N_8S_2$	2,7-Naphthalenedisulfonic acid, 3,6-bis(2-hydroxy-3,5-dinitrophenylazo)-4,5-dihydroxy- (Picramine S)	pK_1 2.55, pK_2 10.2	not stated Mixed constant values for naphthalene hydroxyl groups	05	Uncert.	M106
6421	$C_{22}H_{14}O_{12}N_6S_2$	2,7-Naphthalenedisulfonic acid, 3,6-bis(3-nitrophenylazo)-4,5-dihydroxy-	8.44(OH)	not stated	05	Uncert.	P37
			13.72(OH)		07	Uncert.	
6422	$C_{22}H_{14}O_{12}N_6S_2$	2,7-Naphthalenedisulfonic acid, 3,6-bis(4-nitrophenylazo)-4,5-dihydroxy-	pK_3 8.1	not stated	05	Uncert.	P37
			pK_4 13.47		07	Uncert.	
			pK_3 8.00	not stated $c \simeq 0.00002$, mixed constant	05	Uncert.	S16
6423	$C_{22}H_{14}O_{16}N_6S_3$	2,7-Naphthalenedisulfonic acid, 4,5-dihydroxy-3-(2-hydroxy-3,5-dinitrophenylazo)-6-(3-sulfophenylazo)- (Picramine M)	pK_1 2.19, pK_2 8.9	not stated Mixed constant values for naphthalene hydroxyl groups	05	Uncert.	M106
6424	$C_{22}H_{15}O_{10}N_5S_2$	2,7-Naphthalenedisulfonic acid, 4,5-dihydroxy-3-(4-nitrophenylazo)-6-phenylazo-	8.89(OH)	not stated	05	Uncert.	P37
			13.67(OH)		07	Uncert.	

No.	Molecular formula, name and pK value(s)	T(°C)	Remarks	Method	Assessment	Ref.
6425	$C_{22}H_{15}O_{13}N_5S_3$ 2,7-Naphthalenedisulfonic acid, 4,5-dihydroxy-3-(4-nitrophenylazo)-6-(4-sulfophenylazo)-					
	pK_3 8.5(OH)	not stated		05	Uncert.	P37
	pK_3 8.50	not stated	$c \approx 0.00002$, mixed constant	05	Uncert.	S16
6426	$C_{22}H_{16}O_8N_4S_2$ 2,7-Naphthalenedisulfonic acid, 3,6-bis(phenylazo)-4,5-dihydroxy-					
	pK_1 -1.1, pK_2 -1.1	not stated	H_0 scale	06	Uncert.	B160
	pK_3 0.58, pK_4 0.81		concentration constants	E3bg	Uncert.	
	pK_5 8.76		$I = 0.1$, concentration constant	E3bg	Uncert.	
	pK_6 14.5		concentration constant	07	Uncert.	
	pK_5 8.94	not stated		05	Uncert.	P37
	pK_6 14.34			07	Uncert.	
	pK_5 9.10	not stated	$c \approx 0.00002$, mixed constant	05	Uncert.	S16
6427	$C_{22}H_{16}O_9N_4S_2$ 2,7-Naphthalenedisulfonic acid, 4,5-dihydroxy-3-(4-hydroxyphenylazo)-6-phenylazo-					
	10.32(OH)	not stated		05	Uncert.	P37
6428	$C_{22}H_{16}O_{10}N_4S_2$ 2,7-Naphthalenedisulfonic acid, 3,6-bis(4-hydroxyphenylazo)-4,5-dihydroxy-					
	pK(OH-naphthalene) 10.35	not stated	$c \approx 0.00002$, mixed constant	05	Uncert.	S16
	pK 10.42		Conditions not stated	05	Uncert.	P37
6429	$C_{22}H_{16}O_{11}N_4S_3$ 2,7-Naphthalenedisulfonic acid, 4,5-dihydroxy-3-phenylazo-6-(2-sulfophenylazo)-					
	9.94(OH)	not stated		05	Uncert.	P37
	14.72(OH)			07	Uncert.	

6430 $C_{22}H_{16}O_{11}N_4S_3$ 2,7-Naphthalenedisulfonic acid, 4,5-dihydroxy-3-phenylazo-6-(4-sulfophenylazo)-

pK 9.6(OH)

14.17(OH)

not stated

05	Uncert.	
07	Uncert.	P37

6431 $C_{22}H_{23}O_8N_2Cl$ 1,4,4a,5,5a,6,11,12a-Octahydronaphthacene-2-carboxamide, 7-chloro-4-dimethylamino-3,6,10,12,12a-pentahydroxy-6-methyl-1,11-dioxo- (Chlortetracycline)

pK_1 3.27, pK_2 7.36, pK_3 9.22

Conditions not stated

E3bg	Uncert.	L14

6432 $C_{22}H_{16}O_{14}N_4S_4$ 2,7-Naphthalenedisulfonic acid, 3,6-bis(2-sulfophenylazo)-4,5-dihydroxy- (Sulfonaza III)

pK_1 -2.0, pK_2 0.3, pK_3 0.9

pK_4 1.9, pK_5 2.3, pK_6 2.9, pK_7 11.7

pK_8 14.5

not stated H_0 scale

I = 0.2(NaNO$_3$)

07		
06	Uncert.	B167
05	Uncert.	

pK_1 -0.3, pK_2 -0.3

pK_3 0.6, pK_4 0.8

pK_5 2.4, pK_6 2.8

pK_7 11.61

pK_8 14.4

not stated H_0 scale

concentration constants

I = 0.1, concentration constants

concentration constants

07		
E3bg	Uncert.	B160
05	Uncert.	

6433 $C_{22}H_{16}O_{14}N_4S_4$ 2,7-Naphthalenedisulfonic acid, 3,6-bis(4-sulfophenylazo)-4,5-dihydroxy-

$pK(OH)$ 9.50

$pK(OH)$ 13.95

$pK(OH)$ 9.50

not stated

c ≈ 0.00002, mixed constant

05	Uncert.	
07	Uncert.	P34
05	Uncert.	S16

749

No.	Molecular formula, name and pK value(s)	T(^{o}C)	Remarks	Method	Assessment	Ref.
6434	$C_{22}H_{44}O_5N_2S$ 6,9-Diazatetradecan-7-one, 6,9-bis(5-hydroxypentyl)-1,14-dihydroxy-8-thioxo- [Monothiooxamide, NNN'N'-tetrakis(5-hydroxypentyl)-]					
	12.19	25		05,R4	Approx.	C13
	12.04		I = 0.1(KCl), mixed constant			
			Values for I = 0.02,0.05,0.25,0.50,1.0 are also given			
6435	$C_{22}H_{14}O_{16}N_4Cl_2S_4$ 2,7-Naphthalenedisulfonic acid, 3,6-bis(3-chloro-6-hydroxy-5-sulfophenylazo)-4,5-dihydroxy- (Chlorosulphophenol S)					
	pK_1 -3.6, pK_2 -2.5, pK_3 -0.5, pK_4 0.5, pK_5 1.3, pK_6 2.5, pK_7 7.1, pK_8 9.7, pK_9 11.9, pK_{10} 14.5	25	I = 0.2(KNO_3)	0	Uncert.	B166
6436	$C_{22}H_{15}O_{16}N_6S_2As$ 2,7-Naphthalenedisulfonic acid, 3-(2-arsonophenylazo)-4,5-dihydroxy-6-(2-hydroxy-3,5-dinitrophenylazo)- (Picramine arsenazo)					
	pK_1 2.02, pK_2 12	not stated	Mixed constant values for naphthalene hydroxyl groups	05	Uncert.	M106
6437	$C_{22}H_{17}O_{11}N_4S_2As$ 2,7-Naphthalenedisulfonic acid, 3-(2-arsonophenylazo)-4,5-dihydroxy-6-phenylazo- (Monoarsenazo III)					
	pK_3 6.64, pK_4 8.39, pK_5 12.46		Conditions not stated	05	Uncert.	B158
6438	$C_{22}H_{18}O_{14}N_4S_2P_2$ 2,7-Naphthalenedisulfonic acid, 3,6-bis(2-phosphonophenylazo)-4,5-dihydroxy- (Phosphonazo III)					
	pK_1 -2.0, pK_2 -0.4, pK_3 0.3, pK_4 0.6, pK_5 1.7	25	I = 0.2(KNO_3)	05,06	Uncert.	B165
	pK_6 4.5, pK_7 7.2, pK_8 9.6, pK_9 11.3, pK_{10} 14.6		H_o values used for pK_1 and pK_2	07		

6439 $C_{22}H_{18}O_{14}N_4S_2As_2$ 2,7-Naphthalenedisulfonic acid, 3,6-bis(2-arsonophenylazo)-4,5-dihydroxy- (Arsenoazo III)

not stated H_o scale

pK_1 -2.7, pK_2 -2.7

pK_3 0.6, pK_4 0.8, pK_5 1.6, pK_6 3.4

pK_7 6.27, pK_8 9.05, pK_9 11.98

pK_{10} 15.1

06	Uncert.	B160
E3bg	Uncert.	
05	Uncert.	
07	Uncert.	

Concentration constants.

Other values in B156 and S101

6440 $C_{22}H_{16}O_{14}N_4Cl_2S_2P_2$ 2,7-Naphthalenedisulfonic acid, 3,6-bis(4-chloro-2-phosphonophenylazo)-4,5-dihydroxy-

not stated H_o scale

pK_1 -1.1, pK_2 -1.1

pK_3 0.6, pK_4 0.8, pK_5 1.5, pK_6 2.5

pK_7 5.47, pK_8 7.20, pK_9 12.5

pK_{10} 15.3

06	Uncert.	B160
E3bg	Uncert.	
05	Uncert.	
07	Uncert.	

Concentration constants

Other values in B165

6441 $C_{22}H_{17}O_{14}N_4ClS_2P_2$ 2,7-Naphthalenedisulfonic acid, 3-(4-chloro-2-phosphonophenylazo)-4,5-dihydroxy-6-(2-phosphonophenylazo)-

25 I = 0.2(KNO_3)

pK_1 -2, pK_2 -0.4, pK_3 0.3, pK_4 0.6

pK_5 1.6, pK_6 4.3, pK_7 8.1, pK_8 9.5, pK_9 11.2

pK_{10} 14.6

06	Uncert.	B165
05	Uncert	
07	Uncert	

No.	Molecular formula, name and pK value(s)	T(°C)	Remarks	Method	Assessment	Ref.
6442	$C_{23}H_{16}O_2$ 4-Cyclopentene-1,3-dione, 2,4,5-triphenyl- 8.40		1% methanol	0		L45
6443	$C_{23}H_{15}O_3Cl$ 1,3-Indandione, 2-[α-(4-chlorophenyl)-α-phenyl]acetyl- 2.3			0		T68
6444	$C_{23}H_{15}O_3F$ 1,3-Indandione, 2-[α-(4-fluorophenyl)-α-phenyl]acetyl- 2.0			0		T68
6445	$C_{23}H_{15}O_6Cl_3$ 1,4-Cyclohexadiene-1-carboxylic acid, 3-[(3-carboxy-4-hydroxy-5-methyl)phenyl](2,3,6-trichlorophenyl)methylene]-5- methyl-6-oxo- (Radiochrome blue B) pK$_2$ 3.1, pK$_3$ 5.0, pK$_4$ 12.4	not stated	I = 0.1	05	Uncert.	M129
6446	$C_{23}H_{16}O_6Cl_2$ 1,4-Cyclohexadiene-1-carboxylic acid, 3-[(3-carboxy-4-hydroxy-5-methyl)phenyl](2,6-dichlorophenyl)methylene]-5- methyl-6-oxo- (Chromoxane pure pale blue B) pK$_2$ 3.2, pK$_3$ 4.7, pK$_4$ 12.4	not stated	I = 0.1	05	Uncert.	M129
6447	$C_{23}H_{18}O_4N_2$ 3,5-Pyrazolidinedione, 1-(4-hydroxyphenyl)-2-phenyl-4-phenylacetyl- 2.0		Conditions not stated	05	Uncert.	B181
6448	$C_{23}H_{18}O_9S$ 3,3'-(3H-2,1-Benzoxathiol-3-ylidene)bis(6-hydroxy-5-methylbenzoic acid), S,S-dioxide pK$_2$ 1.83, pK$_3$ 5.74, pK$_4$ 11.83	not stated	c = 0.00004, I = 0.2(KNO$_3$), mixed constant	05	Uncert.	S126
	pK$_1$ -4.9		H$_o$ scale	06	Uncert.	S94
	pK$_2$ 2.23, pK$_3$ 5.47, pK$_4$ 11.85		I = 0.1(NaClO$_4$)	05	Uncert.	S94
	pK$_4$ 11.90			0		P72

6449   No entry

6450 $C_{23}H_{21}O_4P$ Bis(methoxycarbonyl)(triphenylphosphinio)methanide

1.96	not stated	2% ethanol	05	Uncert.	N31

Value in mixed solvents is also given

6451 $C_{23}H_{24}O_5N_2$ 3,7-Diazabicyclo[3.3.1]nonane-3,7-diacetic acid, 9-oxo-1,5-diphenyl-   (Bispidine N,N'-diacetic acid, 9-oxo-1,5-diphenyl-)

pK_1 4.01, pK_2 10.38   not stated $c = 0.001$, $I = 0.1$(KCl), mixed constants   E3bg   Uncert.   S112

6452 $C_{23}H_{14}O_{15}N_6S_2$ Benzoic acid, 2-[1,8-dihydroxy-7-(2-hydroxy-3,5-dinitrophenylazo)-3,6-disulfonaphth-2-ylazo]-   (Picramine K)

pK_1 2.79, pK_2 8.8   not stated   Mixed constant values for   05   Uncert.   M106

naphthalene hydroxyl groups

6453 $C_{23}H_{15}O_{12}N_5S_2$ Benzoic acid, 2-[1,8-dihydroxy-7-(2-nitrophenylazo)-3,6-disulfonaphth-2-ylazo]-

11.03(OH)   Conditions not stated   05   Uncert.   P37

14.54(OH)   07   Uncert.

6454 $C_{23}H_{15}O_{12}N_5S_2$ Benzoic acid, 2-[1,8-dihydroxy-7-(2-nitrophenylazo)-3,6-disulfonaphth-2-ylazo]-

9.8(OH)   Conditions not stated   05   Uncert.   P37

14.38(OH)   07   Uncert.

6455 $C_{23}H_{15}O_{12}N_5S_2$ Benzoic acid, 2-[1,8-dihydroxy-7-(4-nitrophenylazo)-3,6-disulfonaphth-2-ylazo]-

9.91(OH)   Conditions not stated   05   Uncert.   P37

14.25(OH)   07   Uncert.

10.45(OH)   not stated $c \simeq 0.00002$, mixed constant   05   Uncert.   S16

No.	Molecular formula, name and pK value(s)	T(°C)	Remarks	Method	Assessment	Ref.
6456	$C_{23}H_{16}O_9Cl_2S$ 1,4-Cyclohexadiene-1-carboxylic acid, 3-[(3-carboxy-4-hydroxy-5-methylphenyl)(2,6-dichloro-3-sulfophenyl)methylene]-5-methyl-6-oxo- (Chromazurol S)					
	pK_2 2.55, pK_3 4.71, pK_4 11.81 (pK$_1$ < 0)	20	I = 0.1(KCl)	05	Approx.	L4
	pK_1 -4.8	not stated	H_o scale	06	V.uncert.	S94
	pK_2 2.37, pK_3 4.88, pK_4 11.79			05	Uncert.	
	pK_2 2.25, pK_3 4.88, pK_4 11.75	25	I = 0.1(NaClO$_4$)	05	Uncert.	B5
	pK_2 2.28, pK_3 4.92, pK_4 12.21	20	I ≈ 0.1, mixed constants	05	Uncert.	A5
			Other values in M43 and M129			
6457	$C_{23}H_{16}O_{10}N_4S_2$ Benzoic acid, 2-(1,8-dihydroxy-7-phenylazo-3,6-disulfonaphth-2-ylazo)-					
	pK(OH) 10.30	not stated	c ≈ 0.00002, mixed constant	05	Uncert.	S16
	pK(OH) 10.19		Conditions not stated	05	Uncert.	P37
	pK(OH) 15.15			07		
6458	$C_{23}H_{16}O_{13}N_4S_3$ Benzoic acid, 2-[1,8-dihydroxy-7-(3-sulfophenylazo)-3,6-disulfonaphth-2-ylazo]-					
	10.43(OH)		Conditions not stated	05	Uncert.	P37
	14.33(OH)			07	Uncert.	
	10.45(OH)	not stated	c ≈ 0.00002, mixed constant	05	Uncert.	S16

No.	Formula	Name	pK values	Conditions	T	Method	Ref.
6459	$C_{23}H_{16}O_{13}N_4S_3$	Benzoic acid, 2-[1,8-dihydroxy-7-(4-sulfophenylazo)-3,6-disulfonaphth-2-ylazo]-	10.55(OH) 14.36(OH)	Conditions not stated	05 07	Uncert.	P37
6460	$C_{23}H_{20}O_2N_2S$	3,5-Pyrazolidinedione, 1,2-diphenyl-4-[2-(phenylthio)ethyl]-	3.9	Conditions not stated	05	Uncert.	B181
6461	$C_{23}H_{20}O_3N_2S$	3,5-Pyrazolidinedione, 1,2-diphenyl-4-[2-(phenylsulfinyl)ethyl]-	2.8	Conditions not stated	05	Uncert.	B181
6462	$C_{23}H_{27}O_8N_2(I)$	1,4,4a,5,5a,6,11,12a-Octahydronaphthacene-2-carboxamide, 4-dimethylammonio-3,6,10,12,12a-pentahydroxy-6-methyl-1,11-dioxo-, iodide (Tetracycline methiodide)	pK_1 3.56, pK_2 7.80	Conditions not stated	E3bg	Uncert.	L14
6463	$C_{23}H_{17}O_5N_4ClS$	Benzenesulfonic acid, 2-chloro-5-[3-hydroxy-6-methyl-4-(2-hydroxy-1-naphthylazo)]phenylazo- (Solochrome Green V) is the sodium salt	pK_1(SO_3H) 3.05, pK_2(3-OH) 9.30, pK_3 (2-OH) 10.12	not stated c = 0.001	E3bg	Uncert.	A51
6464	$C_{24}H_{14}O_4$	2,2'-(1,4-Phenylene)bis(1,3-indandione)	4.58	20 ± 0.5	05		S142
6465	$C_{24}H_{18}O_3$	1,3-Indandione, 2-[α-phenyl-α-(p-tolyl)]acetyl-	2.58		0		T68

No.	Molecular formula, name and pK value(s)	T(°C)	Remarks	Method	Assessment	Ref.
6466	$C_{24}H_{40}O_4$ Desoxycholic acid 5.15	20	c = 0.001 to 0.005 (Micelle formation above c = 0.005)	E3bg	Uncert.	E12
6467	$C_{24}H_{40}O_5$ Cholic acid 4.98	20	c = 0.003 to 0.014 (Micelle formation above c = 0.014)	E3bg	Uncert.	E12
6468	$C_{24}H_{26}O_{10}N_2$ Stilbene-3,5-diylbis(methylenenitrilo)tetraacetic acid, 4,4'-dihydroxy- pK_1 2.62, pK_2 3.83(COOH) pK_3 7.96, pK_4 9.18(Zwitterions) pK_5 10.52, pK_6 11.0(OH)	not stated	c = 0.001	E3bg	Uncert.	T19
6469	$C_{24}H_{39}O_8N_5$ 2-Amino-6-[4-(4-amino-4-carboxybutyl)-3,5-bis(3-amino-3-carboxypropyl)-1-pyridinio]hexanoate (Desmosine) pK_1 1.7, pK_2 2.4, pK_3 8.8, pK_4 9.9, pK_5 > 11.5		Conditions not stated	E3bg	V.uncert.	T24
6470	$C_{24}H_{16}O_{12}N_4S_2$ 2,2'-(1,8-Dihydroxy-3,6-disulfonaphth-2,7-diylbisazo)dibenzoic acid pK(OH) 10.6		Conditions not stated	05	Uncert.	P37
	pK(OH) 15.13			07		
	pK(OH) 10.60	not stated	c ≈ 0.00002, mixed constant	05	Uncert.	S16
6471	$C_{24}H_{16}O_{12}N_4S_2$ 2,3'-(1,8-Dihydroxy-3,6-disulfonaphth-2,7-diylbisazo)dibenzoic acid 9.83(OH)		Conditions not stated	05	Uncert.	P37
	14.61(OH)			07	Uncert.	

No.	Formula	Name / pK values	Conditions		Uncert.	Ref.
6472	$C_{24}H_{16}O_{12}N_4S_2$	2,4'-(1,8-Dihydroxy-3,6-disulfonaphth-2,7-diylbisazo)dibenzoic acid				
		9.9(OH)	Conditions not stated	05	Uncert.	P37
		14.54(OH)		07		
6473	$C_{24}H_{16}O_{12}N_4S_2$	3,3'-(1,8-Dihydroxy-3,6-disulfonaphth-2,7-diylbisazo)dibenzoic acid				
		pK(OH) 9.70	Conditions not stated	05	Uncert.	P37
		pK(OH) 14.28		07	Uncert.	
		pK(OH) 9.55	not stated c ≈ 0.00002, mixed constant	05	Uncert.	S16
6474	$C_{24}H_{16}O_{12}N_4S_2$	4,4'-(1,8-Dihydroxy-3,6-disulfonaphth-2,7-diylbisazo)dibenzoic acid				
		9.25(OH)	Conditions not stated	05	Uncert.	P37
		14.16(OH)		07	Uncert.	
6475	$C_{24}H_{18}O_{10}N_4S_2$	Benzoic acid, 2-[1,8-dihydroxy-7-(2-methylphenylazo)-3,6-disulfonaphth-2-ylazo]-				
		11.17(OH)	Conditions not stated	05	Uncert.	P37
		14.91(OH)		07	Uncert.	
6476	$C_{24}H_{18}O_{10}N_4S_2$	Benzoic acid, 2-[1,8-dihydroxy-7-(3-methylphenylazo)-3,6-disulfonaphth-2-ylazo]-				
		10.33(OH)	Conditions not stated	05	Uncert.	P37
		14.82(OH)		07		
6477	$C_{24}H_{18}O_{10}N_4S_2$	Benzoic acid, 2-[1,8-dihydroxy-7-(4-methylphenylazo)-3,6-disulfonaphth-2-ylazo]-				
		10.34(OH)	Conditions not stated	05	Uncert.	P37
		14.67(OH)		07	Uncert.	

No.	Molecular formula, name and pK value(s)	T(°C)	Remarks	Method	Assessment	Ref.
6478	$C_{24}H_{18}O_{11}N_4S_2$ Benzoic acid, 2-[1,8-dihydroxy-7-(4-methoxyphenylazo)-3,6-disulfonaphth-2-ylazo]-					
	10.47(OH)		Conditions not stated	05	Uncert.	P37
	14.93(OH)			07	Uncert.	
6479	$C_{24}H_{20}O_8N_4S_2$ 2,7-Naphthalenedisulfonic acid, 3,6-bis(p-tolylazo)-4,5-dihydroxy-					
	9.15(OH)		Conditions not stated	05	Uncert.	P37
	14.34(OH)			07	Uncert.	
	9.15(OH)	not stated	c ≈ 0.00002, mixed constant	05	Uncert.	S16
6480	$C_{24}H_{20}O_{10}N_4S_2$ 2,7-Naphthalenedisulfonic acid, 3,6-bis(4-methoxyphenylazo)-4,5-dihydroxy-					
	9.44(OH)		Conditions not stated	05	Uncert.	P37
	14.55(OH)			07		
6481	$C_{24}H_{20}O_{14}N_4S_4$ 2,7-Naphthalenedisulfonic acid, 3,6-bis(4-methyl-2-sulfophenylazo)-4,5-dihydroxy-					
	pK_1 -0.5, pK_2 -0.5	not stated	H_0 scale	06	Uncert.	B160
	pK_3 0.6, pK_4 0.8, pK_5 2.5, pK_6 2.9			E3bg	Uncert.	
	pK_7 11.77			05	Uncert.	
	pK_8 14.4			07	Uncert.	
			Concentrations constants			
6482	$C_{24}H_{18}O_{18}N_4S_2As_2$ 3,3'-(1,8-Dihydroxy-3,6-disulfonaphth-2,7-diylbisazo)dibenzoic acid, 6,6'-diarsono-					
	pK_1 -5.7, pK_2 -2.5, pK_3 -1.7, pK_4 -0.9, pK_5 2.8, pK_6 3.3, pK_7 5.0, pK_8 6.9, pK_9 6.9, pK_{10} 9.0, pK_{11} 11.5, pK_{12} 14.7					
		25	I = 0.2	0	Uncert.	B163

6483 $C_{25}H_{20}O_{14}N_4Cl_2S_2P_2$ 2,7-Naphthalenedisulfonic acid, 3,6-bis(4-chloro-3-methyl-5-phosphonophenylazo)-4,5-dihydroxy-
(Methylchlorophosphonazo III)

                         $I = 0.2(KNO_3)$ 25   05,06,07 Uncert.   B165

pK_1 -1.9, pK_2 -0.4, pK_3 0.3, pK_4 0.6, pK_5 1.6,
pK_6 4.3, pK_7 7.2, pK_8 9.6, pK_9 11.2, pK_{10} 14.6

6484 $C_{25}H_{20}O_9$ 1,4-Cyclohexadiene-1-carboxylic acid, 3-[(3-carboxy-2-hydroxy-5-methylphenyl)(3-carboxy-4-hydroxy-5-methylphenyl)methylene]-5-methyl-6-oxo-
(Chromoxane violet B)

        not stated $I = 0.1$ 05   Uncert.   M129

pK_2 3.0, pK_3 5.2, pK_4 12.7

6485 $C_{25}H_{20}O_9$ 1,4-Cyclohexadiene-1-carboxylic acid, 3-[(3-carboxy-5-hydroxy-4-methylphenyl)(3-carboxy-4-hydroxy-5-methylphenyl)methylene]-5-methyl-6-oxo-
(Chromoxane violet R)

        not stated $I = 0.1$ 05   Uncert.   M129

pK_2 2.9 pK_3 4.2 pK_4 11.3, pK_5 13.3

6486 $C_{25}H_{48}O_8N_6$ 3,9,14,20,25-Pentaazatriacontane-2,10,13,21,24-pentone, 30-amino-3,14,25-trihydroxy- (Deferoxamine)

                     20    $I = 0.1(NaNO_3)$ E3bg   Approx.   S49

pK_1 8.39, pK_2 9.03, pK_3 9.70, $pK_4 > 11$

6487 $C_{25}H_{30}O_{11}N_2SP_2$ 3H-2,1-Benzoxathiol-3-ylidenebis[(6-hydroxy-5-methyl-m-phenylene)methyleneimino]bis(methylphosphonic acid), S,S-dioxide

    not stated   c= 0.0002, $I = 0.2(NaNO_3)$, mixed constants   05   Uncert.   B162

pK_1 1.8, pK_2 3.4, pK_3 4.9, pK_4 7.3, pK_5 10.4, pK_6 12.8

No.	Molecular formula, name and pK value(s)	T(°C)	Remarks	Method	Assessment	Ref.
6488	$C_{26}H_{22}O_9$ 1,4-Cyclohexadiene-1-carboxylic acid, 3-[(3-carboxy-4-hydroxy-5-methylphenyl)(3-carboxy-4-hydroxy-2,6-dimethylphenyl)methylene]-5-methyl-6-oxo- (Chromoxane violet 5)					
	pK_2 2.7, pK_3 5.2, pK_4 12.7	not stated	I = 0.1	05	Uncert.	M129
6489	$C_{26}H_{18}O_9N_4S_2$ 2-Naphthalenesulfonic acid, 6-hydroxy-5-[2-hydroxy-3-(2-hydroxynaphth-1-ylazo)-5-sulfophenylazo]- (Alizarin acid black)					
	pK_3 5.79, pK_4 12.80, pK_5 > 14		I = 0.1(NaNO$_3$)	05	Approx.	R44
6489a	$C_{26}H_{20}O_{10}N_4S_2$ 2,2'-Stilbenedisulfonic acid, 4,4'-bis(3,4-dihydroxyphenylazo)- (Stilbazo) is diammonium salt					
	pK_3 6.43, pK_4 7.34, pK_5 9.44, pK_6 10.56		c ≈ 0.001	05		A14
6490	$C_{27}H_{20}O_{13}S_2$ 1,4-Cyclohexadiene-1-carboxylic acid, 3-[(3-carboxy-4-hydroxy-5-methylphenyl)(2-hydroxy-3,6-disulfonaphth-1-yl)methylene]-5-methyl-6-oxo- (Chromoxane blue R)					
	pK_4 2.6, pK_5 4.8, pK_6 12.6	not stated	I = 0.1	05	Uncert.	M129
6491	$C_{27}H_{45}O_{12}N_9$ Desferri-ferrichrome					
	pK_1 8.11, pK_2 9.00, pK_3 9.83	20	I = 0.1(NaNO$_3$)	E3bg	Approx.	A57
6492	$C_{27}H_{48}O_9N_6$ Desferri-ferrioxamin E (Nocardamin)					
	pK_1 8.65, pK_2 9.42, pK_3 9.89	20	I = 0.1(NaNO$_3$)	E3bg	Approx.	A57

6493 $C_{27}H_{50}O_9N_6$ Desferriferrioxamin B, N-acetyl-

pK$_1$ 8.50, pK$_2$ 9.24, pK$_3$ 9.69   20   I = 0.1(NaNO$_3$)   E3bg   Approx.   A58

6494 $C_{27}H_{28}O_5Br_2S$ 4,4'-(3H-2,1-Benzoxathiol-3-ylidene)bis(2-bromo-6-isopropyl-3-methylphenol), S,S-dioxide   (Bromothymol blue)

pK$_1$ -0.66

pK$_2$ 6.99   not stated   c = 0.000016   E3d   V.uncert.   G60

0   P72

6495 $C_{27}H_{28}O_9N_2S$ 3H-2,1-Benzoxathiol-3-ylidenebis[(6-hydroxy-5-methyl-m-phenylene)methyleneimino]diacetic acid, S,S-dioxide

(Glycine cresol green)

pK$_1$ -1.8, pK$_2$ 0.0, pK$_3$ 2.46, pK$_4$ 4.92, pK$_5$ 7.12, pK$_6$ 10.77, pK$_7$ 12.43

not stated   I = 0.2(NaNO$_3$)   05,06   Uncert.   B161

6496 $C_{27}H_{36}O_{17}N_2SP_4$ 3H-2,1-Benzoxathiol-3-ylidenebis[[(6-hydroxy-5-methyl-m-phenylene)methylenenitrilo]tetrakis(methylphosphonic acid), S,S-dioxide

pK$_1$ 1.7, pK$_2$ 3.3, pK$_3$ 4.6, pK$_4$ 5.6, pK$_5$ 7.4, pK$_6$ 10.4, pK$_7$ 12.8

not stated   c = 0.0002, I = 0.2(NaNO$_3$), mixed constants   05   Uncert.   B162

6497 $C_{28}H_{16}O_4$ 2,2'-(1,4-Naphthalenediyl)bis(1,3-indandione)

3.68   20±0.5   05   S142

No.	Molecular formula, name and pK value(s)	T(^{o}C)	Method	Assessment	Ref.
6498	$C_{29}H_{18}O_6$ 1,4-Dihydronaphthalene-2-carboxylic acid, 4-[(3-carboxy-4-hydroxynaphth-1-yl)(phenyl)]methylene-1-oxo- (Naphthochrome Green G)				A4
	pK$_2$ 6.3	20	05	V.uncert.	
	pK$_3$ 11.7			Uncert.	
	c = 0.00015, I = 0.1, mixed constant				
6499	$C_{30}H_{18}O_4$ 2,2'-(4,4'-Biphenyldiyl)bis(1,3-indandione)				S142
	pK$_1$ 4.11, pK$_2$ 4.54	20±0.5	05		
	pK$_7$ 7.10, pK$_8$ 11.5				
6500	$C_{30}H_{18}O_{21}N_6S_6$ Cyclo[7,8'(naphth-1"(7),7"(8')-diylbisazo)-1,2'-azonaphthalene]-3,3',3",6,6',6"-hexasulfonic acid, 1',8,8"-trihydroxy- [Cyclo-tris-7-(1-azo-8-hydroxynaphthalene-3,6-disulfonic acid)] (Calcichrome)				B69
		18-22	05	Uncert.	
	I = 0.2(NaNO$_3$)				
6501	$C_{31}H_{32}O_{13}N_2S$ 3H-2,1-Benzoxathiol-3-ylidenebis[(6-hydroxy-5-methyl-m-phenylene)methylenenitrilo]tetraacetic acid, S,S-dioxide (Xylenol Orange)				R17
	pK$_1$ -1.74, pK$_2$ -1.09 (H$_2$SO$_4$-KNO$_3$)		06	Uncert.	
	pK$_3$ 2.58, pK$_4$ 3.23, pK$_5$ 6.46		E3bg	Uncert.	
	pK$_5$ 6.37, pK$_6$ 10.46, pK$_7$ 12.28 I = 0.2(KNO$_3$)		05	Uncert.	

6502 $C_{32}H_{28}O_{10}N_8S_2$ 2,7-Naphthalenedisulfonic acid, 3,6-bis(2,3-dimethyl-5-oxo-1-phenyl-3-pyrazolin-4-ylazo)-4,5-dihydroxy-

pK_1 -1.0, pK_2 0.5, pK_3 1.1, pK_4 2.3, pK_5 4.7, pK_6 7.1, pK_7 11.0, pK_8 14.6

not stated $I = 0.2(KNO_3)$ 0   Uncert.   B72

6503 $C_{32}H_{30}O_{16}N_6S_2$ 1,8-Dihydroxy-3,6-disulfonaphth-3,7-diylbisazobis(o-phenylenemethylenenitrilo)tetraacetic acid   (Aminomethylazo III)

pK_1 -3.3, pK_2 -1.0, pK_3 1.3, pK_4 2.9, pK_5 3.8, pK_6 4.9

pK_7 6.3, pK_8 9.3, pK_9 10.3, pK_{10} 12.2, pK_{11} 13.3, pK_{12} 14.1

not stated $c = 0.0001$, $I = 0.2(KNO_3)$ 05,06,07   Uncert.   B164

6504 $C_{35}H_{42}O_{13}N_4S$ 3H-2,1-Benzoxathiol-3-ylidenebis[(6-hydroxy-5-methyl-m-phenylene)methylenenitrilo]di-NN'-acetic-bis-NN'-2-ethylaminoacetic acid, S,S-dioxide

pK_1 2.6, pK_2 3.8, pK_3 5.2, pK_4 7.3, pK_5 10.4, pK_6 12.8

not stated $c = 0.0002$, $I = 0.2(NaNO_3)$, mixed constants   05   Uncert.   B162

6505 $C_{37}H_{44}O_{13}N_2S$ 3H-2,1-Benzoxathiol-3-ylidenebis[(6-hydroxy-5-isopropyl-2-methyl-m-phenylene)methylenenitrilo]tetraacetic acid, S,S-dioxide   (Methylthymol blue)

pK_1 3.0, pK_2 3.3, pK_3 3.8, pK_4 7.4

not stated $I = 0.2(NaNO_3)$ E3bg   Uncert.   T20

pK_4 7.2, pK_5 11.15, pK_6 13.4

not stated $I = 0.2$ 05   Uncert.   K44

6506 $C_{38}H_{44}O_{12}N_2$ 7-Oxo-2,7-dihydrobenzo[c]furan-2-ylidenebis[(6-hydroxy-5-isopropyl-2-methyl-m-phenylene)methylenenitrilo]tetraacetic acid   (Thymolphthalein complexone)

pK_1 7.35, pK_2 12.25   18-22 $I = 0.2(NaNO_3)$ 05   Uncert.   B70

No.	Molecular formula, name and pK value(s)	$T(^{\circ}C)$	Method	Assessment	Remarks	Ref.
6507	Macromolecule Albumin, bovine					
	10.3(-OH)	25	05	Uncert.		T8
6508	Macromolecule Albumin, human serum					
	Nitrotyrosyl residues					
	pK(-OH) 5.5 to 5.8		05	Uncert.	Conditions not stated	M42
	7.3				8M urea	
	Tyrosyl residues					
	pK(-OH) 10.6				8M urea	
6509	Macromolecule Aquocobalamine complex					
	pK$_3$(-NH) 10.44	20.0	05	Approx.	I = 0.042	H9
	10.27	26.5			I = 0 (calc)	
	10.25	25.0			Thermodynamic quantities are derived from these results	
6510	Macromolecule Carboxypeptidase, nitro-					
	6.3	not stated	05	Uncert.	I ≈ 0.9. Carboxypeptidase nitrated with	R24
	7.0				4-fold excess of tetranitromethane (TNM).	
					Conditions as above. Spectra measured in	
					0.1M β-phenylpropionate	
	6.9				I ≈ 0.9. Nitration with 32-fold excess of TNM. Spectra measured in 0.1M	
					β-phenylpropionate. (Values refer to nitrotyrosyl residues)	

No.	Macromolecule		t	Conditions / Notes	Method	Reliability	Ref.
6511	Macromolecule	Ferrimyoglobin complex					
	pK$_3$(-NH) 10.67		15.3	I = 0.2	05	Uncert.	G14
		10.56	20.0	Thermodynamic quantities are derived from			
		10.40	25.0	these results			
		10.11	35.0				
6512	Macromolecule	7-Methyl-dGdC polymer					
	> 9.5			I = 0.02	05	Uncert.	H29
6513	Macromolecule	Papain					
	10.3			Titration of phenolic groups of 11 to 12 tyrosine residues	E	V.uncert.	G25
6514	Macromolecule	Pepsinogen					
	pK 10.97 (-OH)		25	Mixed constant (Tyrosine residues) Thermodynamic quantities are given	05	Uncert.	P29
6515	Macromolecule	Phosphatidylserine					
	4.42, 9.93		20	I = 0.1 [(C$_3$H$_7$)$_4$NI]	E3bg	Uncert.	H31
6516	Macromolecule	Poly-7-methylguanylic acid					
	> 9.8			I = 0.02	05	Uncert.	H29

765

No.	Molecular formula, name and pK value(s)	$T(^{\circ}C)$	Remarks	Method	Assessment	Ref.
6517	Macromolecule Ribonuclease					
	10.27	6	$I = 0.15$	05	Uncert.	T7
	9.92	25	Molecular weight taken as 13,895			
			Values refer to ionization of phenolic groups			
			Thermodynamic quantities are given			
6518	Macromolecule Triphosphoinositide					
	pK_1 6.38, pK_2 8.45	20	$I = 0.1$ $[(C_3H_7)_4NI]$	E3bg	Uncert.	H31
6519	Macromolecule Uridylic polynucleotide, 5-bromo-					
	8.8			05	Uncert.	L23
6520	Macromolecule Uridylic polynucleotide, 5-chloro-					
	8.7			05	Uncert.	L23
6521	Macromolecule Uridylic polynucleotide, 5-iodo-					
	9.0			05	Uncert.	L23

I.C.O.A.A.—AA*

IV. REFERENCES

A

A1   G.Ackermann, D.Hesse and P.Volland, Z.Anorg.Allg.Chem. 377, 92 (1970).

A1a   T.Adachi and A.Ejima, Eisei Shikenjo Hokoku No.81, 20(1963); CA 62, 8439d.

A2   L.P.Adamovich and M.S.Kravchenko, J.Anal.Chem.USSR (Engl.Transl.) 25, 1434 (1970).

A3   L.P.Adamovich and A.P.Mirnaya, J.Anal.Chem.USSR (Engl.Transl.) 18, 259 (1963).

A4   L.P.Adamovich, A.P.Mirnaya and A.V.Starchenko, J.Anal.Chem. USSR (Engl.Transl.)
        18, 369 (1963).

A5   L.P.Adamovich, O.V.Morgyl-Meshkova and B.V.Yutsis, J.Anal.Chem.USSR (Engl.Transl.)
        17, 673 (1962).

A6   G.E.Adams, J.L.Redpath, R.H.Bisby and R.B.Cundall, J.Chem.Soc., Faraday Trans.1
        69, 1608 (1973).

A7   H.G.Adolph and M.J.Kamlet, J.Am.Chem.Soc. 88, 4761 (1966).

A8   S.P.Agarwal and M.I.Blake, J.Pharm.Sci. 57, 1434 (1968).

A9   J.Aggett and Eng Sing Ow, Aust.J.Chem. 24, 1839 (1971).

A10   Y.K.Agrawal and J.P.Shukla, Aust.J.Chem. 26, 913 (1973).

A11   A.Agren, Acta Chem.Scand. 9, 49 (1955).

A12   A.Agren and G.Schwarzenbach, Helv.Chim.Acta 38, 1920 (1955).

A13   No entry.

A14   M.K.Akhmedli and F.B.Imamverdieva, Azerb.Khim.Zh. 1966, 122; CA 66, 6127q.

A15   G.Aksnes and S.I.Snaprud, Acta Chem.Scand. 15, 457 (1961).

A16   A Albert, Biochem.J. 50, 690 (1952).

A17   A.Albert, Biochem.J. 54, 646 (1953).

A18   A.Albert, J.Chem.Soc. 1960, 1020.

A19   A.Albert, J.Chem.Soc.(B) 1966, 438.

A20   A.Albert, J.Chem.Soc.,Perkin Trans.1 1973, 1634.

A21   A.Albert, J.Chem.Soc., Perkin Trans.1 1973, 2659.

A22   A.Albert and D.J.Brown, J.Chem.Soc. 1953, 74.

A23   A.Albert and D.J.Brown, J.Chem.Soc. 1954, 2060.

A24   A.Albert, D.J.Brown and G.Cheeseman, J.Chem.Soc. 1951, 474.

A25   A.Albert, D.J.Brown and G.Cheeseman, J.Chem.Soc. 1952, 4219.

A26   A.Albert, D.J.Brown and H.C.S.Wood, J.Chem.Soc. 1954, 3832.

A27   A.Albert and C.F.Howell, J.Chem.Soc. 1962, 1591.

A28   A.Albert, Y.Inoue and D.D.Perrin, J.Chem.Soc. 1963, 5151.

A29   A.Albert, J.H.Lister and C.Pedersen, J.Chem.Soc. 1956, 4612.

A30   A.Albert and J.J.McCormack, J.Chem.Soc.(C) 1968, 63.

A31   A.Albert and S.Matsuuara, J.Chem.Soc. 1962, 2162.

A32   A.Albert and H.Mizuno, J.Chem.Soc., Perkin Trans.1 1973, 1974.

A33   A.Albert and W.Pendergast, J.Chem.Soc., Perkin Trans.1 1972, 457.

A34   A.Albert and W.Pendergast, J.Chem.Soc., Perkin Trans.1 1973, 1620.

A35   A.Albert and W.Pendergast, J.Chem.Soc., Perkin Trans.1 1973, 1625.

A36   A.Albert and J.N.Phillips, J.Chem.Soc. 1956, 1294.

A37   A.Albert, C.W.Rees and A.J.H.Tomlinson, Br.J.Exp.Pathol. 37, 500 (1956).

A38   A.Albert and F.Reich, J.Chem.Soc. 1961, 127.

A39   A.Albert and E.P.Serjeant, J.Chem.Soc. 1964, 3357.

A40   A.Albert and H.Taguchi, J.Chem.Soc., Perkin Trans 1 1972, 449.

A41   A.Albert and H.Taguchi, J.Chem.Soc., Perkin Trans.1 1973, 1629.

A42   A.Albert and H.Yamamoto, J.Chem.Soc.(C) 1968, 1181.

A43   T.Alfrey and H.Morawetz, J.Am.Chem.Soc. 74, 436 (1952).

A44   I.P.Alimarin, N.P.Borzenkova and R.I.Shmatko, J.Anal.Chem. USSR (Engl.Transl.)
      18, 302 (1963).

A45   I.P.Alimarin and S.Hanshi, Russ.J.Inorg.Chem.(Engl.Transl.) 6, 1054 (1961).

A46   I.P.Alimarin and H-H.Shen, Vestn.Mosk.Univ., Ser.II, Khim.15, 38 (1960);
      CA 56, 995 g.

A47   G.F.Allen, R.A.Robinson and V.E.Bower, J.Phys.Chem. 66, 171 (1962).

A48   G.W.Allen and P.Haake, J.Am.Chem.Soc. 95, 8080 (1973).

A49   I.Ambro, Nehezvegyip.Kut.Intez.Kozlem. 3, 105 (1966); CA 69, 39198v.

A50   D.E.Ames and T.F.Grey, J.Chem.Soc. 1955, 631.

A51   A.M.Amin, H.Khalifa and A.S.Moustafa, Fresenius' Z.Anal.Chem. 173, 138 (1960).

A52   G.Anderegg, Helv.Chim.Acta 46, 1011 (1963).

A53   G.Anderegg, Helv.Chim.Acta 46, 1833 (1963).

A54   G.Anderegg, Helv.Chim.Acta 47, 1801 (1964).

A55   G.Anderegg, Helv.Chim.Acta 50, 2333 (1967).

A56   G.Anderegg and F.L'Eplattenier, Helv.Chim.Acta 47, 1067 (1964).

A57   G.Anderegg, F.L'Eplattenier and G.Schwarzenbach, Helv.Chim.Acta 46, 1400 (1963).

A58   G.Anderegg, F.L'Eplattenier and G.Schwarzenbach, Helv.Chim.Acta 46, 1409 (1963).

A59   D.G.Anderson, G.G.Hammes and F.G.Walz, Biochemistry 7, 1637 (1968).

A60   E.Anderson and T.H.Fife, J.Am.Chem.Soc. 95, 6437 (1973).

A61   R.G.Anderson and G.Nickless, Anal.Chim.Acta 39, 469 (1967).

A62   T.Ando, Bull.Chem.Soc.Jpn. 35, 1395 (1962).

A63   T.Ando, Bull.Chem.Soc.Jpn. 36, 1593 (1963).

A64   T.Ando and K.Ueno, Inorg.Chem. 4, 375 (1965).

A65   A.C.Andrews, T.D.Lyons and T.D.O'Brien, J.Chem.Soc. 1962, 1776.

A66   K.P.Ang, J.Phys.Chem. 62, 1109 (1958).

A67   K.P.Ang, J.Chem.Soc. 1959, 3822.

A68   P.J.Antikainen, Suom.Kemistil.B 43, 31 (1970); CA 72,78255t.

A69   P.J.Antikainen and H.Oksanen, Acta Chem.Scand. 22, 2867 (1968).

A70   P.J.Antikainen and V.M.K.Rossi, Suom.Kemistil.B 36, 132 (1963).

A71   P.J.Antikainen and K.Tevanen, Suom.Kemistil.B 35, 224 (1962).

A72   P.J.Antikainen, M.Viro and L.R.Sahlstrom, Suom.Kemistil.B 42, 178 (1969); CA 70, 118667w.

A73   P.A.Argabright and B.L.Phillips, J.Heterocycl.Chem. 7, 999 (1970).

A74   J.Armand, Bull.Soc.Chim.Fr. 1966, 882.

A75   W.L.F.Armarego and B.A.Milloy, J.Chem.Soc., Perkin Trans.1 1973, 2814.

A76   J.McD.Armstrong, Biochim.Biophys.Acta 86, 194 (1964).

A77   R.Arnaud, Bull.Soc.Chim.Fr. 1967, 4541.

A78   R.Arnold, S.Ticktin, G.Monseair and A.M.Stephen, J.Chem.Soc. 1963, 5810.

A79   K.I.Aspila, S.J.Joris and C.L.Chakrabarti, J.Phys.Chem. 74, 3625 (1970).

A80   V.T.Athavale, L.H.Prabhu and D.G.Vartak, J.Inorg.Nucl.Chem. 28, 1237 (1966).

A81   B.K.Avinashi, C.D.Dwivedi and S.K.Banerji, J.Inorg.Nucl.Chem., 32, 2641 (1970).

A82   N.N.Aylward, J.Chem.Soc.B 1967, 401.

A83   J.Aznarez Alduan, Rev.Acad.Cienc.Exact.Fis.-Quim.Nat.Zaragoza 15, 41 (1960); CA 59, 14549h.

                                        B

B1   S.P.Bag, Q.Fernando and H.Freiser, Inorg.Chem. 1, 887 (1962).

B2   S.P.Bag, Q.Fernando and H.Freiser, Anal.Chem. 35, 719(1963).

B3   J.T.Bahr, R.E.Cathou and G.G.Hammes, J.Biol.Chem. 240, 3372 (1965).

B4   K.Balachandran and S.K.Banerji, J.Inorg.Nucl.Chem. 32, 3333 (1970).

B5   W.G.Baldwin and D.R.Stranks, Aust.J.Chem. 21, 603 (1968).

B6   P.Ballinger and F.A.Long, J.Am.Chem.Soc. 81, 1050 (1959).

B7   P.Ballinger and F.A.Long, J.Am.Chem.Soc. 81, 2347 (1959).

B8   P.Ballinger and F.A.Long, J.Am.Chem.Soc. 82, 795 (1960).

B9   S.Balt and E.Van Dalen, Anal.Chim.Acta 27, 188 (1962).

B10   M.A.Bambenek and R.T.Pflaum, Inorg.Chem. 2, 289 (1963).

B11   A.Banerjee and A.K.Dey, Anal.Chim.Acta 42, 473 (1968).

B12   A.Banerjee and A.K.Dey, J.Inorg.Nucl.Chem. 30, 995 (1968).

B13   A.Banerjee and A.K.Dey, J.Inorg.Nucl.Chem. 30, 3134 (1968).

B14   A.Banerjee, S.Mandal, T.Singh and A.K.Dey, Indian J.Chem. 7, 733 (1969).

B15   L.Banford and W.J.Geary, J.Chem.Soc. 1964, 378.

B16   J.Bankovskis, J.Asaks and A.Ievins, Latv.PSR Zinat.Akad.Vestis, Kim.Ser. 1966,
      533; CA 67, 68919v.

B17   J.Bankovskis, M.Buka, G.Mezaraups, A.Ievins and M.Abolina, Latv.PSR Zinat.Akad.Vestis,
      Kim.Ser. 1967, 243; CA 68, 100261h.

B18   C.V.Banks and A.B.Carlson, Anal.Chim.Acta 7, 291 (1952).

B19   C.V.Banks, J.P.LaPlante and J.J.Richard, J.Org.Chem. 23, 1210 (1958).

B20   C V.Banks and R.S.Singh, J.Am.Chem.Soc. 81, 6159 (1959).

B21   C.V.Banks and J.Zimmerman, J.Org.Chem. 21, 1439 (1956).

B22   F.Barbulescu, G.Mincu-Diaconescu and D.A.Isaescu, Rev.Roumaine Chim. 10, 599 (1965);
      Studii Cercetari Chim. 13, 633 (1965); CA 64, 9538e.

B23   T.J.Bardos and T.I.Kalman, J.Pharm.Sci. 55, 606 (1966).

B24   F.Baroncelli and G.Grossi, J.Inorg.Nucl.Chem. 27, 1085 (1965).

B25   W.Bartok, R.H.Hartman and P.J.Lucchesi, Photochem. Photobiol. 4, 499 (1965).

B26   M.Bartusek, Collect.Czech.Chem.Commun. 32, 116 (1967).

B27   M.Bartusek, Collect.Czech.Chem.Commun. 32, 757 (1967).

B28   M.Bartusek and L.Havelkova, Collect.Czech.Chem.Commun. 33, 385 (1968).

B29   M.Bartusek and L.Sommer, J.Inorg.Nucl.Chem. 27, 2397 (1965).

B30   M.Bartusek and O.Stankova, Collect.Czech.Chem.Commun. 30, 3415 (1965).

B31   M.Bartusek and J.Zelinka, Collect.Czech.Chem.Commun. 32, 992 (1967).

B32   N.N.Basargin, M.K.Akhmedli and A.A.Kafarova, J.Anal.Chem. USSR (Engl.Transl.)
      25, 1292 (1970).

B33   N.N.Basargin, A.V.Kadomtseva and V.I.Petrashen, J.Anal.Chem.USSR (Engl.Transl.)
      25, 25 (1970).

B34   N.N.Basargin and T.V.Petrova, J.Anal.Chem.USSR (Engl.Transl.) 19, 775 (1964).

B35   I.M.Batyaev, S.V.Larionov and V.M.Shulman, Russ.J.Inorg.Chem.(Engl.Transl.)
      6, 75 (1961).

B36   R.F.Bauer and W.M.Smith, Can.J.Chem. 43, 2755 (1965).

B37   K.Baum, J.Org.Chem. 35, 1203 (1970).

B38   J.H.Baxendale and H.R.Hardy, Trans.Faraday Soc. 49, 1140 (1953).

B39   P.M.Beart and G.A.R.Johnston, Aust.J.Chem. 25, 1359 (1972).

B40   A.Beauchamp and R.L.Benoit, Can.J.Chem. 42, 2161 (1964).

B41   A.Beauchamp and R.L.Benoit, Can.J.Chem. 44, 1607 (1966).

B42   A.Beauchamp and R.Benoit, Bull.Soc.Chim.Fr. 1967, 672.

B43   M.T.Beck and M.Halmos, Nature (London) 186, 388 (1960).

B44   E.J.Behrman, M.J.Biallas, H.J.Brass, J.O.Edwards and M.Isaks, J.Org.Chem. 35,
        3063 (1970).

B45   J.S.Belew and L.G.Hepler, J.Am.Chem.Soc. 78, 4005 (1956).

B46   E.A.Bell, Biochim.Biophys.Acta 47, 602 (1961); Nature (London) 194, 91 (1962).

B47   R.P.Bell, Trans.Faraday Soc. 39, 253 (1943).

B48   R.P.Bell and B.G.Cox, J.Chem.Soc.(B) 1971, 652.

B49   R.P.Bell and G.G.Davis, J.Chem.Soc. 1965, 353.

B50   R.P.Bell and M.B.Jensen, Proc.Chem.Soc.,London 1960, 307.

B51   R.P.Bell and A.T.Kuhn, Trans.Faraday Soc. 59, 1789 (1963).

B52   R.P.Bell and P.T.McTigue, J.Chem.Soc. 1960, 2983.

B53   R.P.Bell and W.B.T.Miller, Trans.Faraday Soc. 59, 1147 (1963).

B54   R.P.Bell and D.P.Onwood, Trans.Faraday Soc. 58, 1557 (1962).

B55   R.P.Bell and P.W.Smith, J.Chem.Soc.(B) 1966, 241.

B56   R.P.Bell and B.A.Timimi, J.Chem.Soc., Perkin Trans.2 1973, 1518.

B57   R.P.Bell and D.C.Vogelsong, J.Chem.Soc. 1958, 243.

B58   R.P.Bell and G.A.Wright, Trans.Faraday Soc. 57, 1377 (1961).

B59   R.E.Benesch and R.Benesch, J.Am.Chem.Soc. 77, 5877 (1955).

B60   R.A.Benkeser and H.R.Krysiak, J.Am.Chem.Soc. 75, 2421 (1953).

B61   F.W.Bennett, H.J.Emeleus and R.N.Haszeldine, J.Chem.Soc. 1954, 3598.

B62   W.E.Bennett and D.O.Skovlin, J.Inorg.Nucl.Chem. 28, 591 (1966).

B63   M.Beran and S.Havelka, Collect.Czech.Chem.Commun. 32, 2944 (1967).

B64   K.Berens and D.Shugar, Acta Biochim.Pol. 10, 25 (1963).

B65   F.Bergmann and S.Dikstein, J.Am.Chem.Soc. 77, 691 (1955).

B66   F.Bergmann, M.Kleiner, Z.Neiman and M.Rashi, Isr.J.Chem. 2, 185 (1964).

B67   D.Betteridge, Q.Fernando and H.Freiser, Anal.Chem. 35, 294 (1963).

B68   D.Betteridge, P.K.Todd, Q.Fernando and H.Freiser, Anal.Chem. 35, 729 (1963).

B69   A.Bezdekova and B.Budesinsky, Collect.Czech.Chem.Commun. 30, 811 (1965).

B70   A.Bezdekova and B.Budesinsky, Collect.Czech.Chem.Commun. 30, 818 (1965).

B71   A.Bezdekova and B.Budesinsky, Collect.Czech.Chem.Commun. 31, 199 (1966).

B72   A.Bezdekova and B.Budesinsky, Collect.Czech.Chem.Commun. 33, 4178 (1968).

B73   A.S.Bhaduri and N.N.Gosh, Z.Anorg.Allg.Chem. 297, 73 (1958).

B74   I.M.Bhatt, K.P.Soni and A.M.Trivedi, J.Indian Chem.Soc. 45, 354 (1968).

B75   U.C.Bhattacharyya, S.C.Lahiri and S.Aditya, J.Indian Chem.Soc. 46, 247 (1969).

B76   M.E.C.Biffin, D.J.Brown and T.C.Lee, Aust.J.Chem. 20, 1041 (1967).

B77   A.I.Biggs and R.A.Robinson, J.Chem.Soc. 1961, 388.

B78   E.H.Binns, Trans.Faraday Soc. 55, 1900 (1959).

B79   J.M.Birchall and R.N.Haszeldine, J.Chem.Soc. 1959, 3653.

B80   E.A.Biryuk and R.V.Ravitskaya, Zh.Vses.Khim.Obshchest. 14, 461 (1969); CA 71 105890j.

B81   E.A.Biryuk and R.V.Ravitskaya, J.Anal.Chem.USSR (Engl.Transl.) 25, 494 (1970).

B82   C.A.Bishop and L.K.J.Tong, J.Phys.Chem. 66, 1034 (1962).

B83   H.U.Blank and J.J.Fox, J.Heterocycl.Chem. 7, 735 (1970).

B84   G.G.Blinova and E.G.Sochilin, J.Gen.Chem.USSR (Engl.Transl.) 40, 2745 (1970).

B85   J.J.Bloomfield and R.Fuchs, J.Chem.Soc.(B) 1970, 363.

B86   H.Bode, W.Eggeling and V.Steinbrecht, Fresenius' Z.Anal.Chem. 216, 30 (1966).

B87   Z.Bohak, J.Biol.Chem. 239, 2878 (1964).

B88   T.A.Bohigian and A.E.Martell, Inorg.Chem. 4, 1264 (1965).

B89   T.A.Bohigian and A.E.Martell, J.Am.Chem.Soc. 89, 832 (1967).

B90   P.D.Bolton, J.Ellis and F.M.Hall, J.Chem.Soc.(B) 1970, 1252.

B91   P.D.Bolton, K.A.Fleming and F.M.Hall, J.Am.Chem.Soc., 94,1033 (1972).

B92   P.D.Bolton, F.M.Hall and J.Kudrynski, Aust.J.Chem. 21, 1541 (1968).

B93   P.D.Bolton, F.M.Hall and J.Kudrynski, Aust.J.Chem. 25, 75 (1972).

B94   P.D.Bolton, F.M.Hall and I.H.Reece, J.Chem.Soc.(B) 1966, 717.

B95   P.D.Bolton, F.M.Hall and I.H.Reece, Spectrochim.Acta 22, 1149 (1966).

B96   P.D.Bolton, F.M.Hall and I.H.Reece, Spectrochim.Acta 22, 1825 (1966).

B97   P.D.Bolton, F.M.Hall and I.H.Reece, J.Chem.Soc.(B) 1967, 709.

B98   S.Bolton and R.I.Ellin, J.Pharm.Sci. 51, 533 (1962).

B99   S.Bonkoski and J.H.Perrin, J.Pharm.Sci. 57, 1784 (1968).

B100 O.D.Bonner, H.B.Flora and H.W.Aitken, J.Phys.Chem. 75, 2492 (1971).

B101 R.Bonnett, R.F.C.Brown, V.M.Clark, I.O.Sutherland and A.Todd, J.Chem.Soc. 1959, 2094.

B102 R.Bonnett, V.M.Clark and A.Todd, J.Chem.Soc. 1959, 2102.

B103 F.G.Bordwell and P.J.Boutan, J.Am.Chem.Soc. 78, 87 (1956).

B104 F.G.Bordwell and P.J.Boutan, J.Am.Chem.Soc. 78, 854 (1956).

B105 F.G.Bordwell and P.J.Boutan, J.Am.Chem.Soc. 79, 717 (1957).

B106 M.S.Borisov, A.A.Elesin, I.A.Lebedev, E.M.Piskunov, V.T.Filimonov and
      G.N.Yakovlev, Radiokhimiya 9, 166 (1967); CA 68, 72865n.

B107 E.Bottari, Monatsh.Chem. 99, 176 (1968).

B108 E.Bottari and A.Rufolo, Monatsh.Chem. 99, 2383 (1968).

B109 R.S.Bottei and W.A.Joern, J.Chem.Eng.Data 13, 522 (1968).

B110 A.J.Boulton and A.R.Katritzky, Tetrahedron 12, 41 (1961).

B111 W.J.Bover and P.Zuman, J.Chem.Soc.,Perkin Trans.2 1973, 786.

B112 K.Bowden, Can.J.Chem. 43, 3354 (1965).

B113 K.Bowden, M.Hardy and D.C.Parkin, Can.J.Chem. 46, 2929 (1968).

B114 K.Bowden and D.C.Parkin, Can.J.Chem. 46, 3909 (1968).

B115 V.E.Bower and R.A.Robinson, J.Phys.Chem. 64, 1078(1960).

B116 R.H.Boyd, J.Am.Chem.Soc. 83, 4288 (1961).

B117 R.H.Boyd, J.Phys.Chem. 67, 737 (1963).

B118 R.H.Boyd and C-H.Wang, J.Am.Chem.Soc. 87, 430 (1965).

B119 S.Boyd, A.Bryson, G.H.Nancollas and K.Torrance, J.Chem.Soc. 1965, 7353.

B120 G.M.Brauer, H.Argentar and G.Durany, J.Res.Nat.Bur.Stand., Sect.A 68A, 619 (1964).

B121 G.M.Brauer, G.Durany and H.Argentar, J.Res.Nat.Bur.Stand., 71A, 379 (1967).

B122 L.G.Bray, J.F.J.Dippy and S.R.C.Hughes, J.Chem.Soc. 1957, 265.

B123 L.G.Bray, J.F.J.Dippy, S.R.C.Hughes and L.W.Laxton, J.Chem.Soc. 1957, 2405.

B124 A.G.Briggs, J.E.Sawbridge, P.Tickle and J.M.Wilson, J.Chem.Soc.(B) 1969, 802.

B125 A.G.Briggs, P.Tickle and J.M.Wilson, Spectrochim.Acta, Part A 26A, 1399 (1970).

B126 W.Broser and M.Seekamp, Tetrahedron Lett. 1966, 6337.

B127 D.J.Brown, J.Chem.Soc. 1958, 1974.

B128 D.J.Brown, J.Appl.Chem. 9, 203 (1959).

B129 D.J.Brown, J.Chem.Soc. 1959, 3647.

B130 D.J.Brown and J.S.Harper, J.Chem.Soc. 1961, 1298.

B131 D.J.Brown and R.J.Harrisson, J.Chem.Soc. 1959, 3332.

B132 D.J.Brown and R.J.Harrisson, J.Chem.Soc. 1960, 1837.

B133 D.J.Brown, E.Hoerger and S.F.Mason, J.Chem.Soc. 1955, 211.

B134 D.J.Brown and J.A.Hoskins, Aust.J.Chem. 25, 2641 (1972).

B135 D.J.Brown and N.W.Jacobsen, J.Chem.Soc. 1961, 4413.

B136 D.J.Brown and N.W.Jacobsen, J.Chem.Soc. 1965, 1175.

B137 D.J.Brown and R.L.Jones, Aust.J.Chem. 25, 2711 (1972).

B138 D.J.Brown, R.L.Jones, A.M.Angyal and G.W.Grigg, J.Chem.Soc.,Perkin Trans.1 1972, 1819.

B139 D.J.Brown and J.R.Kershaw, J.Chem.Soc.,Perkin Trans.1 1972, 2316.

B140 D.J.Brown and S.F.Mason, J.Chem.Soc. 1956, 3443.

B141 D.J.Brown and T.Sugimoto, Aust.J.Chem. 24, 633 (1971).

B142 D.J.Brown and T.Teitei, Aust.J.Chem. 17, 567 (1964).

B143 R.D.Brown, A.S.Buchanan and A.A.Humffray, Aust.J.Chem. 18, 1527 (1965).

B144 R.F.C.Brown, V.M.Clark, M.Lamchen and A.Todd, J.Chem.Soc. 1959, 2116.

B145 R.F.C.Brown, V.M.Clark, I.O.Sutherland and A.Todd, J.Chem.Soc. 1959, 2109.

B146 T.C.Bruice and W.C.Bradbury, J.Am.Chem.Soc. 87, 4851 (1965).

B147 Yu.A.Bruk, A.A.Derzhavets, L.V.Pavlova, F.Yu.Rachinskii and N.M.Slavachevskaya,
        J.Gen.Chem.USSR (Engl.Transl.) 40, 2300 (1970).

B148 A.P.Brunetti, M.C.Lim and G.H.Nancollas, J.Am.Chem.Soc. 90, 5120 (1968).

B149 A.J.Bruno, S.Chaberek and A.E.Martell, J.Am.Chem.Soc. 78, 2723 (1956).

B150 J.S.Brush, Anal.Chem. 33, 798 (1961).

B151 G.F.Bryce and F.R.N.Gurd, J.Biol.Chem. 241, 1439 (1966).

B152 G.F.Bryce, J.M.H.Pinkerton, L.K.Steinrauf and F.R.N.Gurd, J.Biol.Chem. 240,
        3829 (1965).

B153 G.F.Bryce, R.W.Roeske and F.R.N.Gurd, J.Biol.Chem. 240, 3837 (1965).

B154 A.Bryson, J.Am.Chem.Soc. 82, 4862 (1960).

B155 A.Bryson and R.W.Matthews, Aust.J.Chem. 16, 401 (1963).

B156 B.Budesinsky, Collect.Czech.Chem.Commun. 28, 2902 (1963).

B157 B.Budesinsky, Fresenius' Z.Anal.Chem. 195, 324(1963).

B158 B.Budesinsky, Fresenius' Z.Anal.Chem. 207, 105 (1965).

B159 B.Budesinsky, Fresenius' Z.Anal.Chem. 207, 241 (1965).

B160 B.Budesinsky, Talanta 16, 1277 (1969).

B161 B.Budesinsky and J.Gurovic, Collect.Czech.Chem.Commun. 28, 1154 (1963).

B162 B.Budesinsky and K.Haas, Collect.Czech.Chem.Commun. 29, 1006 (1964).

B163 B.Budesinsky and K.Haas, Fresenius' Z.Anal.Chem. 210, 263 (1965).

B164 B.Budesinsky and K.Haas, Fresenius' Z.Anal.Chem. 214, 325 (1965).

B165 B.Budesinsky, K.Haas and A.Bezdekova, Collect.Czech.Chem.Commun. 32, 1528 (1967).

B166 B.Budesinsky and S.B.Savvin, Fresenius' Z.Anal.Chem. 214, 189 (1965).

B167 B.Budesinsky and D.Vrzalova, Fresenius' Z.Anal.Chem. 210, 161 (1965).

B168 B.Budesinsky and T.S.West, Anal.Chim.Acta 42, 455 (1968).

B169 O.Budevski and E.Ruseva, Izv.Inst.Obshta Neorg.Khim., Bulg.Akad.Nauk. 4, 5 (1966); CA 66, 69417w.

B169a J.D.Bu'Lock, E.R.H.Jones and P.R.Leeming, J.Chem.Soc. 1955, 4270.

B169b J.D.Bu'Lock, E.R.H.Jones, P.R.Leeming and J.M.Thompson, J.Chem.Soc. 1956, 3767.

B169c J.D.Bu'Lock, E.R.H.Jones and P.R.Leeming, J.Chem.Soc. 1957, 1097.

B169d J.D.Bu'Lock and H.Gregory, J.Chem.Soc. 1960, 2280.

B170 J.W.Bunting and D.D.Perrin, Aust.J.Chem. 19, 337 (1966).

B171 C.A.Bunton and H.Chaimovich, J.Am.Chem.Soc. 88, 4082 (1966).

B172 C.A.Bunton, E.J.Fendler and J.H.Fendler, J.Am.Chem.Soc. 89, 1221 (1967).

B173 C.A.Bunton, E.J.Fendler, E.Humeres and Kui-Un Yang, J.Org.Chem. 32, 2806 (1967).

B174 C.A.Bunton, D.R.Llewellyn, K.G.Oldham and C.A.Vernon, J.Chem.Soc. 1958, 3574.

B175 A.Burger and J.J.Anderson, J.Am.Chem.Soc. 79, 3575 (1957).

B176 K.Burger, Talanta 8, 77 (1961).

B177 K.Burger, I.Egyed and I.Ruff, J.Inorg.Nucl.Chem. 28, 139 (1966).

B178 K.Burger and I.Ruff, Acta Chim.Acad.Sci.Hung. 49, 1 (1966); CA 65, 14502c.

B179 J.F.Burke and M.W.Whitehouse, Biochem.Pharmacol. 14, 1039 (1965).

B180 R.K.Burkhard, D.E.Sellers, F.DeCou and J.L.Lambert, J.Org.Chem. 24, 767 (1959).

B181 J.J.Burns, T.F.Yu, P.Dayton, L.Berger, A.B.Gutman and B.B.Brodie, Nature (London) 182, 1162 (1958).

B182 A.I.Busev and V.M. Ivanov, J.Anal.Chem.USSR (Engl.Transl.) 22, 332 (1967).

B183 A.I.Busev, V.M.Ivanov and L.S.Krysina, Vestn.Mosk.Univ.,Khim. 23, 80 (1968); CA 69, 106599s.

B184 A.I.Busev, L.S.Krysina, T.N.Zholondkovskaya, G.A.Pribylova and E.P.Krysin, J.Anal.Chem.USSR (Engl.Transl.) 25, 1359 (1970).

B185 A.I.Busev, Z.I.Nemtseva and V.M.Ivanov, J.Anal.Chem.USSR (Engl.Transl.) 24, 1111 (1969).

B186 A.I.Busev, Z.I.Nemtseva and V.M.Ivanov, Vestn.Mosk.Univ.,Khim. 24, 35 (1969); CA 70, 119801x.

B187 D.I.Bustin, J.E.Earley and A.A.Vlcek, Inorg.Chem. 8, 2065 (1969).

B188 A.R.Butler, J.Chem.Soc.(B) 1970, 867.

B189 T.C.Butler, J.Am.Pharm.Assoc.,Sci.Ed. 44, 367 (1955).

B190 M.M.Byrne and N.H.P.Smith, J.Chem.Soc.(B) 1968, 809.

C

C1   D.K.Cabbiness and E.S.Amis, Bull.Chem.Soc.Jpn. 40, 435 (1967).

C2   I.Cadariu, E.Gavrila and G.Niac, Z.Phys.Chem.(Leipzig) 242, 391 (1969).

C3   D.E.Caddy and J.H.P.Utley, J.Chem.Soc.,Perkin Trans.2 1973, 1258.

C4   J.Calmon, Y.Cazaux-Maraval and P.Maroni, Bull.Soc.Chim.Fr. 1968, 3779.

C5   J.Calmon and P.Maroni, Bull.Soc.Chim.Fr. 1965, 2525.

C6   C.Calzolari, Univ.Studi Trieste,Fac.Sci.Ist.Chim. 1954, 4; CA 49, 7338.

C7   D.L.Campbell and T.Moeller, J.Inorg.Nucl.Chem. 31, 1077 (1969).

C8   E.Campi, G.Ostacoli, N.Cibrario and G.Saini, Gazz.Chim.Ital. 91, 361 (1961).

C9   E.Campi, G.Ostacoli, M.Meirone and G.Saini, J.Inorg.Nucl.Chem. 26, 553 (1964).

C10   E.Campi, G.Ostacoli and A.Vanni, Gazz.Chim.Ital. 95, 796 (1965).

C11   J.Canonne, J.Nicole and G.Tridot, Chim.Anal.(Paris) 51, 317 (1969); CA 71 74808g.

C12   G.L.van de Cappelle and M.A.Herman, Anal.Chim.Acta 43, 89 (1968).

C13   G.L.van de Cappelle, M.A.Herman and Z.Eeckhaut, Bull.Soc.Chim.Belg. 79, 421 (1970).

C14   B.Carlqvist and D.Dyrssen, Acta Chem.Scand. 16, 94 (1962).

C15   J.D.Carr and D.G.Swartzfager, Anal.Chem. 43, 1520 (1971).

C16   J.D.Carr and D.G.Swartzfager, J.Am.Chem.Soc. 95, 3569 (1973).

C17   R.L.Carroll and R.R.Irani, Inorg.Chem. 6, 1994 (1967).

C18   R.P.Carter, R.L.Carroll and R.R.Irani, Inorg.Chem. 6, 939 (1967).

C19   R.P.Carter, M.M.Crutchfield and R.R.Irani, Inorg.Chem. 6, 943 (1967).

C20   M.L.Casey, D.S.Kemp, K.G.Paul and D.D.Cox, J.Org.Chem. 38, 2294 (1973).

C21   J.Cason and M.J.Kalm, J.Org.Chem. 19, 1947 (1954).

C22   D.Cavallino, C.DeMarco, B.Mondovi and F.Stirpe, Atti accad.nazl.Lincei Rend.
      Classe sci.fis.mat.e nat. 18, 552 (1958); CA 52, 17101h.

C23   M.Cefola, A.S.Tompa, A.V.Celiano and P.S.Gentile, Inorg.Chem. 1, 290 (1962).

C24   S.Chaberek and A.E.Martell, J.Am.Chem.Soc. 75, 2888 (1953).

C25   S.Chaberek and A.E.Martell, J.Am.Chem.Soc. 76, 215 (1954).

C26   S.Chaberek and A.E.Martell, J.Am.Chem.Soc. 77, 1477 (1955).

C27   R.W.Chambers, Prog.Nucleic Acid Res.Mol.Biol. 5, 349 (1966).

C28   P.K.Chang, J.Org.Chem. 23, 1951 (1958).

C29   P.K.Chang, J.Org.Chem. 26, 1118 (1961).

C30   T-H.Chang, J-T.Lin, T-I.Chou and S-K.Yang, J.Chinese Chem.Soc.(Taiwan) 11, 125 (1964);
      CA 62, 9859e.

C31   J.D.Chanley and E.Feageson, J.Am.Chem.Soc. 77, 4002 (1955).

C32   J.D.Chanley, E.M.Gindler and H.Sobotka, J.Am.Chem.Soc. 74, 4347 (1952).

C33   J.D.Chanley and E.M.Gindler, J.Am.Chem.Soc. 75, 4035 (1953).

C34   M.C.Chattopadhyaya and R.S.Singh, Indian J.Chem. 8, 1126 (1970).

C35   R.K.Chaturvedi, Dinkar and B.Biswas, Proc.Nat.Acad.Sci., India Sect.A 34A, 22 (1964).

C36   R.K.Chaturvedi and S.S.Katiyar, Bull.Chem.Soc.Jpn. 35, 1416 (1962).

C37   R.K.Chaturvedi and S.S.Katiyar, J.Sci.Ind.Res. (India) 21B, 47 (1962); CA 57, 9726h.

C38   V.Chavane, Ann.Chim.(Paris) [12], 4, 383 (1949).

C39   I.D.Chawla and A.C.Andrews, J.Inorg.Nucl.Chem. 32 91 (1970).

C40   D.T.Y.Chen and K.J.Laidler, Trans.Faraday.Soc. 58, 480 (1962).

C41   G.E.Cheney, Q.Fernando and H.Freiser, J.Phys.Chem. 63, 2055 (1959).

C42   B.D.Chernokalskii, V.S.Gamayurova and G.K.Kamai. J.Gen.Chem.USSR (Engl.Transl.)
      36, 1674 (1966).

C43   L.I.Chernomorchenko and A.G.Akhmetshin, J. Anal.Chem.USSR (Engl.Transl.) 25, 192(1970).

C44   M.V.Chidambaram and P.K.Bhattacharya, J.Inorg.Nucl.Chem. 32, 3271 (1970).

C45   N.M.Chopra, W.Cocker, B.E.Cross, J.T.Edward, D.H.Hayes and H.P.Hutchison,
      J.Chem.Soc. 1955, 588.

C46   G.Choux and R.L.Benoit, Bull.Soc.Chim.Fr. 1967, 2920.

C47   G.Choux and R.L.Benoit, J.Org.Chem. 32, 3974 (1967).

C48   H.N.Christensen, J.Biol.Chem. 160, 425 (1945).

C49   J.J.Christensen, R.M.Izatt and L.D.Hansen, J.Am.Chem.Soc. 89, 213 (1967).

C50   J.J.Christensen, J.L.Oscarson and R.M.Izatt, J.Am.Chem.Soc. 90, 5949 (1968).

C51   J.J.Christensen, J.H.Rytting and R.M.Izatt, J.Am.Chem.Soc. 88, 5105 (1966).

C52   J.J.Christensen, J.H.Rytting and R.M.Izatt, Biochemistry 9, 4907 (1970).

C53   J.J.Christensen, J.H.Rytting and R.M.Izatt, J.Chem.Soc.(B) 1970, 1646.

C54   J.J.Christensen, M.D.Slade, D.E.Smith, R.M.Izatt and J.Tsang, J.Am.Chem.Soc. 92,
      4164 (1970).

C55   H.Christol and M.Gaignon, J.Chim.Phys.Phys.-Chim.Biol. 57, 707 (1960).

C56   V.Chromy and L.Sommer, Talanta 14, 393 (1967).

C57   R.F.R.Church and M.J.Weiss, J.Org.Chem. 35, 2465 (1970).

C58   M.Claesen and H.Vanderhaeghe, Bull.Soc.Chim.Belg. 66, 292 (1957).

C59   J.Clark, W.Kernick and A.J.Layton, J.Chem.Soc. 1964, 3215.

C60   M.E.Clark and J.L.Bear, J.Inorg.Nucl.Chem. 32, 3569 (1970).

C61   R.J.H.Clark and A.J.Ellis, J.Chem.Soc. 1960, 247.

C62   V.M.Clark, A.W.Johnson, I.O.Sutherland and A.Todd, J.Chem.Soc. 1958, 3283.

C63   E.R.Clarke and A.E.Martell, J.Inorg.Nucl.Chem. 32, 911 (1970).

C64   J.H.R.Clarke and L.A.Woodward, Trans.Faraday Soc. 62, 2226 (1966).

C65   G.L.Closs and L.E.Closs, J.Am.Chem.Soc. 83, 1003 (1961).

C66   G.L.Closs and E.L.Closs, J.Am.Chem.Soc. 85, 2022 (1963).

C67   E.Coates and B.Rigg, Trans.Faraday Soc. 57, 1088 (1961).

C68   E.Coates, B.Rigg and D.L.Smith, Trans.Faraday Soc. 64, 3255 (1968).

C69   W.Cocker, W.J.Davis, T.B.H.McMurry and P.A.Start, Tetrahedron 7, 299 (1959).

C70   L.A.Cohen and W.M.Jones, J.Am.Chem.Soc. 85, 3397 (1963).

C71   L.A.Cohen and W.M.Jones, J.Am.Chem.Soc. 85, 3402 (1963).

C72   W.Cohen and B.F.Erlanger, J.Am.Chem.Soc. 82, 3928 (1960).

C73   W.E.Cohn, J.Biol.Chem. 235, 1488 (1960).

C74   C.Concilio and A.Bongini, Ann.Chim.(Rome) 56, 417 (1966); CA 65, 3082c.

C75   J.W.Cook, A.R.Gibb, R.A.Raphael and A.R.Somerville, J.Chem.Soc. 1951, 503.

C76   E.J.Corey, H.Konig and T.H.Lowry, Tetrahedron Lett. 1962, 515.

C77   A.K.Covington and T.H.Lilley, Trans.Faraday Soc. 63, 1749 (1967).

C78   A.K.Covington, M.Paabo, R.A.Robinson and R.G.Bates, Anal.Chem. 40, 700 (1968).

C79   L.K.Creamer, A.Fischer, B.R.Mann, J.Packer, R.B.Richards and J.Vaughan,
       J.Org.Chem. 26, 3148 (1961).

C80   F.T.Crimmins, C.Dymek, M.Flood and W.F.O'Hara, J.Phys.Chem. 70, 931 (1966).

C81   D.G.Crosby and R.V.Berthold, J.Org.Chem.23, 1377 (1958).

C82   W.D.Crow and I.Gosney, Aust.J.Chem. 20, 2729 (1967).

C83   C.A.Crutchfield, W.M.McNabb and J.F.Hazel, J.Inorg.Nucl.Chem. 24, 291 (1962).

C84   B.Csiszar, M.Halmos, M.T.Beck and P.Szarvas, Magy.Kem.Folyoirat. 70, 214 (1964);
       CA 61, 5012d.

D

D1   D.B.Dahlberg and F.A.Long, J.Am.Chem.Soc. 95, 3825 (1973).

D2   G.Dahlgren and F.A.Long, J.Am.Chem.Soc. 82, 1303 (1960).

D3   K.Daigo, J.Pharm.Soc.Jpn. 79, 353 (1959).

D4   G.Dalman and G.Gorin, J.Org.Chem. 26, 4682 (1961).

D5   D.Dal Monte and E.Sandri, Ann.Chim.(Rome) 54, 486 (1964).

D6   D.Dal Monte and E.Sandri, Boll.Sci.Fac.Chim.Ind.Bologna 22, 33 (1964); CA 62, 1542g.

D7   G.D'Amore, F.Corigliano and S.Di Pasquale, Ann.Chim.(Rome) 54, 317 (1964).

D8   J.P.Danehy and C.J.Noel, J.Am.Chem.Soc. 82, 2511 (1960).

D9   J.P.Danehy and K.N.Parameswaran, J.Chem.Eng.Data 13, 386 (1968).

D10   S.Darling and P.O.Larsen, Acta Chem.Scand. 15, 743 (1961).

D11   S.N.Das and D.J.G.Ives, Proc.Chem.Soc., London 1961, 373.

D12   S.P.Datta and A.K.Grzybowski, Biochem.J. 69, 218 (1958).

D13   N.K.Davidenko, J.Gen.Chem.USSR (Engl.Transl.) 28, 832 (1957).

D14   N.K.Davidenko, Russ.J.Inorg.Chem.(Engl.Transl.) 9, 965 (1964).

D15   N.K.Davidenko and V.F.Deribon, Russ.J.Inorg.Chem.(Engl.Transl.) 11, 53 (1966).

D16   M.M.Davis and H.B.Hetzer, J.Phys.Chem. 61, 123 (1957).

D17   M.M.Davis and H.B.Hetzer, J.Phys.Chem. 61, 125 (1957).

D18   M.M.Davis, M.Paabo and R.A.Robinson, J.Res.Nat.Bur.Stand., Sect.A 64A, 531 (1960).

D19   M.M.Davis and M.Paabo, J.Res.Nat.Bur.Stand.,Sect.A 64A, 533 (1960).

D20   M.M.Davis and M.Paabo, J.Res.Nat.Bur.Stand.,Sect.A 67A, 241 (1963).

D21   H-S.Dawn, I.H.Pitman, T.Higuchi and S.Young, J.Pharm.Sci. 59, 955 (1970).

D22   H.Deelstra and F.Verbeek, Bull.Soc.Chim.Belg. 72, 612 (1963).

D23   D.DeFilippo and F.Momicchioli, Tetrahedron 25, 5733 (1969).

D24   D.DeFilippo and C.Preti, Gazz.Chim.Ital. 95, 707 (1965); CA 63, 12398d.

D25   D.DeFilippo and C.Preti, Gazz.Chim.Ital. 96, 60 (1966); CA 64, 18500g.

D26   C.Degani and M.Halmann, J.Am.Chem.Soc. 88, 4075 (1966).

D27   G.Degischer and G.H.Nancollas, Inorg.Chem. 9, 1259 (1970).

D28   P.DeMaria, A.Fini and F.M.Hall, J.Chem.Soc.,Perkin Trans.2 1973, 1969.

D29   P.Demerseman, J.P.Lechartier, A.Cheutin, R.Reynaud, R.Royer and P.Rumpf,
      Bull.Soc.Chim.Fr. 1962, 1700.

D30   P.Demerseman, J.P.Lechartier, R.Reynaud, A.Cheutin, R.Royer and P.Rumpf,
      Bull.Soc.Chim.Fr. 1963, 2559.

D31   P.Demerseman, R.Reynaud, A.Cheutin, R.Royer and P.Rumpf, Bull.Soc.Chim.Fr.1960, 1508.

D32   P.Demerseman, R.Reynaud, A.Cheutin, J.P.Lechartier, C.Pène, A.M.Laval-Jeantet,
      R.Royer and P.Rumpf, Bull.Soc.Chim.Fr. 1965, 1464.

D33   M.Deneux, R.Meilleur and R.L.Benoit, Can.J.Chem. 46, 1383 (1968).

D34   A.G.Desai and R.M.Milburn, J.Am.Chem.Soc. 91, 1958 (1969).

D35   M.J.Deschamps-Dutrieux and M.Chaillet, Tetrahedron Lett. 1968, 1171.

D36   S.Deswarte,C.R.Acad.Sci.,Ser.C 263C, 628 (1966).

D37   H.Diehl and J.Ellinboe, Anal.Chem. 32, 1120 (1960).

D38   S.Dikstein and F.Bergmann, J.Am.Chem.Soc. 77, 4671 (1955).

D39   R.H.Dinius and G.R.Choppin, J.Phys.Chem. 66, 268 (1962).

D40   J.F.J.Dippy, S.R.C.Hughes and L.G.Bray, J.Chem.Soc. 1959, 1717.

D41   J.F.J.Dippy, S.R.C.Hughes and A.Rozanski, J.Chem.Soc. 1959, 2492.

D42   D.G.Dittmer, O.B.Ramsay and R.E.Spalding, J.Org.Chem. 28, 1273 (1963).

D43   M.L.Dondon, J.Chim.Phys.Phys.-Chim.Biol. 54, 290 (1957).

D44   M.L.Dondon, J.Chim.Phys.Phys.-Chim.Biol. 54, 304 (1957).

D45   M.L.Dondon and M.L.Paris, J.Chim.Phys.Phys.-Chim.Biol. 58, 222 (1961).

D46   J.A.Donohue, R.M.Scott and F.M.Menger, J.Org.Chem. 35, 2034 (1970).

D47   M.A.Doran, S.Chaberek and A.E.Martell, J.Am.Chem.Soc. 86, 2129(1964).

D48   M.Draminski and B.Fiszer, Khim.Geterotsikl.Soedin. 6, 859 (1970); CA 74 52854d.

D49   T.Drapala, Rocz.Chem. 42, 1287 (1968); CA 70, 10945w.

D50   D.A.Drapkina, V.G.Brudz, Y.S.Ryabokobylko, A.V.Chekunov and V.A.Inshakova,
       J.Org.Chem.USSR (Engl.Transl.) 3, 1566 (1967).

D51   M.L.Dressler and M.M.Joullie, J.Heterocycl.Chem. 7, 1257 (1970).

D52   J.S.Driscoll, W.Pfleiderer and E.C.Taylor, J.Org.Chem. 26, 5230 (1961).

D53   A.Dugaiczyk, Biochemistry 9, 1557 (1970).

D54   S.M.Dugar and N.C.Sogani, Z.Naturforsch.Teil B 21, 657 (1966).

D55   G.Dumas and P.Rumpf, C.R.Acad.Sci.,Ser.C 242, 2574 (1956).

D56   G.E.Dunn and F.L.Kung, Can.J.Chem. 44, 1261 (1966).

D57   G.E.Dunn, P.Leggate and I.E.Scheffler, Can.J.Chem. 43, 3080 (1965).

D58   E.J.Durham and D.P.Ryskiewich, J.Am.Chem.Soc. 80, 4812 (1958).

D59   J.D.Dutcher, J.Biol.Chem. 232, 785 (1958).

D60   N.K.Dutt and P.Bose, Z.Anorg.Allg.Chem. 295, 131 (1958).

D61   N.K.Dutt, G.S.Sanyal and K.Nag, J.Indian Chem.Soc. 45, 334 (1968).

D62   N.K.Dutt and T.Seshadri, Bull.Chem.Soc.Jpn. 40, 2280 (1967).

D63   N.K.Dutt and T.Seshadri, J.Inorg.Nucl.Chem. 31, 2153 (1969).

D64   R.L.Dutta and S.Ghosh, J.Indian Chem.Soc. 44, 820 (1967).

D65   B.L.Dyatkin, E.P.Mochalina and I.L.Knunyants, Tetrahedron 21,2991 (1965).

D66   N.M.Dyatlova, Y.F.Belugin and V.Y.Temkina, Tr.Soveshch.po Fiz.Metodam Issled.
       Organ.Soedin. i Khim.Protsessov, Akad.Nauk Kirg.SSR, Inst.Org.Khim., Frunze
       1962, 55; CA 62, 2288c.

D67   N.M.Dyatlova, M.I.Kabachnik, T.Y.Medved, M.V.Rudomino and Y.F.Belugin,
       Dokl.Chem.(Engl.Transl.) 161, 307 (1965).

D68   N.M.Dyatlova, V.V.Medyntsev, T.Y.Medved and M.I.Kabachnik, J.Gen.Chem.USSR (Engl.
       Transl.) 38, 1025 (1968).

D69   N.M.Dyatlova, V.V.Medyntsev, T.Y.Medved and M.I.Kabachnik, J.Gen.Chem.USSR (Engl.
       Transl.) 38, 1030 (1968).

D70   N.M.Dyatlova, V.V.Medyntsev, T.Y.Medved and M.I.Kabachnik, J.Gen.Chem.USSR (Engl. Transl.) 38, 1035 (1968).

D71   N.M.Dyatlova, V.V.Medyntsev, T.M.Balashova, T.Y.Medved and M.I.Kabachnik, J.Gen.Chem.USSR (Engl.Transl.) 39, 309 (1969).

D72   N.M.Dyatlova, I.A.Seliverstova, V.G.Yashunskii and O.I.Samoilova, J.Gen.Chem.USSR (Engl.Transl.) 34, 4061 (1964).

D73   N.M.Dyatlova, I.A.Seliverstova and N.A.Dobynina, Tr., Vses.Nauch.-Issled. Inst.Khim.Reaktivov Osobo Chist.Khim.Veshchestv 28, 270 (1966); CA 67, 15448j.

D74   N.M.Dyatlova, I.A.Seliverstova, O.I.Samoilova and V.G.Yashunskii, Dokl.Chem.(Engl.Transl.) 172, 4 (1967).

D75   J.R.Dyer, H.B.Hayes, E.G.Miller and R.F.Nassar, J.Am.Chem.Soc. 86, 5363 (1964).

D76   D.Dyrssen, Acta Chem.Scand. 10, 353 (1956).

D77   D.Dyrssen, Acta Chem.Scand. 11, 1771 (1957).

D78   D.Dyrssen, S.Ekberg and D.H.Liem, Acta Chem.Scand. 18, 135 (1964).

D79   D.Dyrssen and E.Johansson, Acta Chem.Scand. 9, 763 (1955).

D80   A.F.Dzhelassi and A.M.Egorov, Russ.J.Phys.Chem. (Engl.Transl.) 43, 128 (1969).

E

E1   L.Eberson, Acta.Chem.Scand. 18, 1276 (1964).

E2   M.Eden and R.G.Bates, J.Res.Nat.Bur.Stand. 62, 161 (1959).

E3   J.T.Edward and O.J.Chin, Can.J.Chem. 41, 1650 (1963).

E4   J.T.Edward and S.Nielsen, J.Chem.Soc. 1957, 5075.

E5   J.T.Edward and I.C.Wang, Can.J.Chem. 40, 399 (1962).

E6   J.O.Edwards and R.J.Sederstrom, J.Phys.Chem. 65, 862 (1961).

E7   I.P.Efimov, O.D.Lagunova, N.N.Magdesieva, V.V.Titov, Y.K.Yur'ev and V.M.Peshkova, Vestn.Mosk.Univ.,Ser.II,Khim. 18, 49 (1963); CA 60, 1564f.

E8   C.Eger, J.A.Marinsky and W.M.Anspach, J.Inorg.Nucl.Chem. 30, 1911 (1968).

E9   F.Eiden and J.Pluckhan, Arch.Pharm.(Weinheim.Ger.) 302, 622 (1969).

E10   M.O.Eidhin and S.O.Cinneide, J.Inorg.Nucl.Chem. 30, 3209 (1968).

E11   M.L.Eidinoff, J.Am.Chem.Soc. 67, 2072 (1945).

E12   P.Ekwall, T.Rosendahl and N.Lofman, Acta Chem.Scand. 11, 590 (1957).

E13   A.J.Ellis, J.Chem.Soc. 1963, 2299.

E14   A.J.Ellis and D.W.Anderson, J.Chem.Soc. 1961, 1765.

E15   H.J.Emeleus, R.N.Haszeldine and R.C.Paul, J.Chem.Soc. 1954, 881.

E16   H.J.Emeleus, R.N.Haszeldine and R.C.Paul, J.Chem.Soc. 1955, 563.

E17   G.Endres, quoted in R.Swidler, R.E.Plapinger and G.M.Steinberg, J.Am.Chem.Soc.
       81, 3271 (1959).

E18   J.Epstein, P.L.Cannon, H.O.Michel, B.E.Hackley and W.A.Mosher, J.Am.Chem.Soc.
       89, 2937 (1967).

E19   J.Epstein, H.O.Michel, D.H.Rosenblatt, R.E.Plapinger, R.A.Stephani and E.Cook,
       J.Am.Chem.Soc. 86, 4959 (1964).

E20   J.Epstein, R.E.Plapinger, H.O.Michel, J.R.Cable, R.A.Stephani, R.J.Hester,
       C.Billington and G.R.List, J.Am.Chem.Soc. 86, 3075 (1964).

E21   M.I.Ermakova and N.I.Latosh, J.Gen.Chem.USSR(Engl.Transl.) 38, 2314 (1968).

E22   L.Erdey, J. Gyimesi and T.Meisel, Acta Chim.Acad.Sci.Hung. 21, 327 (1959);
       CA 54, 14137g.

E23   Z.L.Ernst and F.G.Herring, Trans.Faraday Soc. 60, 1053 (1964).

E24   Z.L.Ernst and J.Menashi, Trans.Faraday Soc. 59, 230 (1963).

E25   Z.L.Ernst and J.Menashi, Trans.Faraday Soc. 59, 1803 (1963).

E26   B.A.Ershov, A.V.Popova and E.E.Emelina, J.Org.Chem.USSR (Engl.Transl.) 6, 1148 (1970).

E27   A.J.Everett and G.J.Minkoff, Trans.Faraday Soc. 49, 410 (1953).

E28   E.M.Evleth, L.D.Freeman and R.I.Wagner, J.Org.Chem. 27, 2192 (1962).

E29   E.Eyring, L.D.Rich, L.L.McCoy, R.C.Graham and N.Taylor, Adv.Chem.Phys. 21, 237 (1971).

                                          F

F1   P.Faerber and K.H.Scheit, Chem.Ber. 103, 1307 (1970).

F2   G.P.Faerman and V.S.Kozeya, Uspekhi Nauch.Fot.,Akad.Nauk.SSSR, Otdel.Khim.Nauk
       5, 107(1957); CA 52, 937d.

F3   G.P.Faerman and A.B.Simkina, Uspekhi Nauch.Fot., Akad.Nauk.SSSR, Otdel.Khim.Nauk
       5, 81 (1957); CA 52, 843b.

F4   J.W.Faller, A.Mueller and J.P.Phillips, J.Org.Chem. 29, 3450 (1964).

F5   M.I.Fauth, A.C.Richardson and G.W.Nauflett, Anal.Chem. 38, 1947 (1966).

F6   F.S.Feates and D.J.G.Ives, J.Chem.Soc. 1956, 2798.

F7   I.Feldman and L.Koval, Inorg.Chem. 2, 145 (1963).

F8   L.P.Fernandez and L.G.Hepler, J.Am.Chem.Soc. 81, 1783 (1959).

F9   Q.Fernando and H.Freiser, J.Am.Chem.Soc. 80, 4928 (1958).

F10   F.Ferranti, Ann.Chim.(Rome) 56, 1595 (1966); CA 66, 108898a.

F11   A.F.Ferris and I.G.Marks, J.Org.Chem. 19, 1975 (1954).

F12   M.M.Fickling, A.Fischer, B.R.Mann, J.Packer and J.Vaughan, J.Am.Chem.Soc.
       81, 4226 (1959).

F13   T.H.Fife and S.Milstien, J.Org.Chem. 34, 4007 (1969).

F14   R.Filler and R.M.Schure, J.Org.Chem. 32, 1217 (1967).

F15   H.L.Finkbeiner and M.Stiles, J.Am.Chem.Soc. 85, 616 (1963).

F16   A.Fischer, G.J.Leary, R.D.Topsom and J.Vaughan, J.Chem.Soc.(B) 1966, 782.

F17   A.Fischer, G.J.Leary, R.D.Topsom and J.Vaughan, J.Chem.Soc.(B) 1967, 686.

F18   A.Fischer, B.R.Mann and J.Vaughan, J.Chem.Soc. 1961, 1093.

F19   A.M.Fiskin and M.Beer, Biochemistry 4, 1289 (1965).

F20   J.D.Fissekis and B.A.Markert, J.Org.Chem. 31, 2945 (1966).

F21   J.D.Fissekis, A.Myles and G.B.Brown, J.Org.Chem. 29, 2670 (1964).

F22   J.D.Fissekis and F.Sweet, Biochemistry 9, 3136 (1970).

F23   M.Flavin, C.Delavier-Klutchko and C.Slaughter, Science 143, 50 (1964).

F24   M.H.Fletcher, Anal.Chem. 32, 1822 (1960).

F25   U.Folli, D.Iarossi and F.Taddei, J.Chem.Soc.,Perkin Trans.2 1973, 848.

F26   D.Forest and G.Thomas, Bull.Soc.Chim.Fr. 1968, 3441.

F27   A.Foucaud and M.Duclos, C.R.Acad.Sci.,Ser.C 256, 4033 (1963).

F28   A.Foucaud and P.LeGuellec, C.R.Acad.Sci.,Ser.C 252, 3063 (1961).

F29   A.Foucaud, H.Person and M.Duclos, Bull.Soc.Chim.Fr. 1965, 2552.

F30   K.F.Fouché and J.G.V.Lessing, J.Inorg.Nucl.Chem. 32, 2357 (1970).

F31   K.F.Fouché, H.J.leRoux and F.Phillips, J.Inorg.Nucl.Chem. 32, 1949 (1970).

F32   J.J.Fox and D.Shugar, Bull.Soc.Chim.Belg. 61, 44 (1952).

F33   J.J.Fox, N.Yung, J.Davoll and G.B.Brown, J.Am.Chem.Soc. 78, 2117 (1956).

F34   J.J.Fox, N.Yung and I.Wempen, Biochim.Biophys.Acta 23, 295 (1957).

F35   W.O.Foye, M.D.Blum and D.A.Williams, J.Pharm.Sci. 56, 332 (1967).

F35a   W.O.Foye, H.B.Levine and W.L.McKenzie, J.Med.Chem. 9, 61 (1966).

F36   D.J.Francis and L.M.Yedanapalli, Indian J.Chem. 2, 166 (1964).

F37   No entry.

F38   J.J.R.Fruasto da Silva, Rev.Port.Quim. 7, 88 (1965); CA 64, 12526h.

F39   J.J.R.Frausto da Silva and M.deL.S.Simoes, Rev.Port.Quim. 7, 137 (1965); CA 65, 90f.

F40   B.F.Freasier, A.G.Oberg and W.W.Wendlandt, J.Phys.Chem. 62, 700 (1958).

F41   V.Frei, Collect.Czech.Chem.Commun. 30, 1402 (1965).

F42   M.Friedman, J.F.Cavins and J.S.Wall, J.Am.Chem.Soc. 87, 3672 (1965).

F43   F.Fringuelli, G.Marino and G.Savelli, Tetrahedron 25, 5815 (1969).

F44   A.E.Frost, Nature (London) 178, 322 (1956).

F45   A.E.Frost, H.H.Freedman, S.J.Westerback and A.E.Martell, J.Am.Chem.Soc. 80, 530 (1958).

F46   N.S.Frumina, K.G.Petrikova, E.G.Tregub and S.V.Pletnev, J.Anal.Chem.USSR (Engl.Transl.)
       25, 373 (1970).

F47   F.T.Fueno, O.Kajimoto, Y.Nishigaki and T.Yoshioka, J.Chem.Soc.,Perkin Trans.2
       1973, 738.

F48   R.W.Furlanetto and E.T.Kaiser, J.Am.Chem.Soc. 95, 6786 (1973).

F49   A.Furlani, M.Maltese, E.Mantovani and C.Maremmani, Gazz.Chim.Ital. 97, 1423 (1967).

G

G1   D.M.Gardner, R.E.Oesterling and F.L.Scott, J.Org.Chem. 28, 2650 (1963).

G2   E.R.Garrett and D.J.Weber, J.Pharm.Sci. 59, 1383 (1970).

G3   R.Gary, R.G.Bates and R.A.Robinson, J.Phys.Chem. 69, 2750 (1965).

G4   G.Gattow and V.Hahnkamm, Z.Anorg.Allg.Chem. 365, 70 (1969).

G5   A.P.Gavrish, Ukr.Khim.Zh. 29, 900 (1963); CA 60, 6264b.

G6   O.Gawron and F.Draus, J.Am.Chem.Soc. 80, 5392 (1958).

G7   O.Gawron, J.Fernando, J.Keil and T.J.Weismann, J.Org.Chem. 27, 3117 (1962).

G8   O.Gawron and A.J.Glaid, J.Am.Chem.Soc. 77, 6638 (1955).

G9   W.J.Geary and D.E.Malcolm, J.Chem.Soc.(A) 1970, 797.

G10   W.J.Geary, G.Nickless and F.H.Pollard, Anal.Chim.Acta 27, 71 (1962).

G11   H.Gehlen and J.Rinck, Z.Phys.Chem.(Leipzig) 237, 388 (1968).

G12   E.Gelles and R.W.Hay, J.Chem.Soc. 1958, 3673.

G13   J.T.van Gemert, Aust.J.Chem. 22, 1883 (1969).

G14   P.George, G.I.H.Hanania, D.H.Irvine and I.Abu-Issa, J.Chem.Soc. 1964, 5689.

G15   A.Gergely, P.Szarvas and I.Korondan, Acta Chim.Acad.Sci.Hung. 26, 313 (1961);
       CA 55, 20752f.

G16   R.D.Gillard, H.M.Irving, R.M.Parkins, N.C.Payne and L.D.Pettit, J.Chem.Soc.(A)
       1966, 1159.

G17   S.A.Giller, R.A.Zhuk and M.Y.Lidak, Dokl.Chem.(Engl.Transl.) 176, 798 (1967).

G18   H.Gillet and J.C.Pariaud, Bull.Soc.Chim.Fr. 1966, 2624.

G19   V.A.Gilyarov, E.N.Tsvetkov and M.I.Kabachnik, J.Gen.Chem.USSR (Engl.Transl.)
       36, 285 (1966).

G20   A.Giner-Sorolla and A.Bendich, J.Am.Chem.Soc. 80, 5744 (1958).

G21   P.K.Glasoe, J.Phys.Chem. 69, 4416 (1965).

G22   P.K.Glasoe and L.Eberson, J.Phys.Chem. 68, 1560 (1964).

G23   P.K.Glasoe and J.R.Hutchison, J.Phys.Chem. 68, 1562 (1964).

G24   P.K.Glasoe and F.A.Long, J.Phys.Chem. 64, 188 (1960).

G25   A.N.Glazer and E.L.Smith, J.Biol.Chem. 236, 2948 (1961).

G26   D.R.Goddard and S.I. Nwankwo, J.Chem.Soc.(A) 1967, 1371.

G27   D.P.Goel, Y.Dutt and R.P.Singh, J.Inorg.Nucl.Chem. 32, 3119 (1970).

G28   A.M.Goeminne and M.A.Herman, Anal.Chim.Acta 41, 400 (1968).

G29   A.M.Goeminne and M.A.Herman, Bull.Soc.Chim.Belg. 77, 227 (1968).

G30   J.Goerdeler, H.Groschopp and U.Sommerlad, Chem.Ber. 90, 182 (1957).

G31   J.E.Gordon and S.L.Johnson, J.Phys.Chem. 66, 534 (1962).

G32   I.P.Gorelov and M.K.Kolosova, Russ.J.Inorg.Chem.(Engl.Transl.) 14, 1416 (1969).

G33   H.Gottschling and C.Heidelberger, J.Mol.Biol. 7, 541 (1963).

G34   R.O.Gould and R.F.Jameson, J.Chem.Soc. 1962, 296.

G35   M.A.Gouveia and R.G.DeCarvahlo, J.Inorg.Nucl.Chem. 30, 2219 (1968).

G36   R.J.Grabenstetter, O.T.Quimby and T.J.Flautt, J.Phys.Chem. 71, 4194 (1967).

G37   J.Grandjean, Bull.Soc.Roy.Sci.Liege 38, 288 (1969); CA 72, 42684t.

G38   A.L.Green, G.L.Sainsbury, B.Saville and M.Stanfield, J.Chem.Soc. 1958, 1583.

G39   A.L.Green and B.Saville, J.Chem.Soc. 1956, 3887.

G40   R.W.Green and P.W.Alexander, Aust.J.Chem. 18, 329 (1965).

G41   R.W.Green and I.R.Freer, J.Phys.Chem. 65, 2211 (1961).

G42   R.W.Green and G.K.S.Ooi, Aust.J.Chem. 15, 786 (1962).

G43   R.W.Green and M.J.Rogerson, Aust.J.Chem. 21, 2427 (1968).

G44   R.W.Green and R.J.Sleet, Aust.J.Chem. 19, 2101 (1966).

G45   S.B.Greenbaum, J.Am.Chem.Soc. 77, 3221 (1955).

G46   E.Y.Gren and G.Y.Vanag, J.Org.Chem.USSR (Engl.Transl.) 1, 1(1965).

G47   V.V.Grigor'eva, Ukrain.Khim.Zhur. 23, 306 (1957); CA 52, 1735b.

G48   J.H.Grimes, A.J.Huggard and S.P.Wilford, J.Inorg.Nucl.Chem. 25, 1225 (1963).

G49   A.A.Grinberg and K.K.Khakimov, Russ.J.Inorg.Chem.(Engl.Transl.) 6, 71 (1961).

G50   E.Grunwald and J.F.Haley, J.Phys.Chem. 72, 1944 (1968).

G51   J.Guilleme and B.Wojtkowiak, Bull.Soc.Chim.Fr. 1969, 3007.

G52   W.F.Gum and M.M.Jouille, J.Org.Chem. 30, 3982 (1965).

G53   B.P.Gupta, Y.Dutt and R.P.Singh, J.Indian Chem.Soc. 43, 610 (1966).

G54   B.P.Gupta, Y.Dutt and R.P.Singh, Indian J.Chem. 5, 322 (1967).

G55   S.L.Gupta and M.C.Gupta, J.Indian Chem.Soc. 40, 321 (1963).

G56   S.L.Gupta and R.N.Soni, J.Indian Chem.Soc. 42, 377 (1965).

G57   S.L.Gupta and R.N.Soni, J.Inst.Chem.,Calcutta 40, 154 (1968); CA 69, 90440v.

G58   S.L.Gupta and R.N.Soni, J.Inst.Chem.,Calcutta 41, 129 (1969); CA 71, 117109z.

G59   S.L.Gupta and R.N.Soni, J.Inst.Chem.,Calcutta 41, 156 (1969); CA 71, 117110t.

G60   V.D.Gupta and D.E.Cadwallader, J.Pharm.Sci. 57, 2140 (1968).

G61   V.D.Gupta and J.B.Reed, J.Pharm.Sci. 59, 1683 (1970).

G62   J.Gut, M.Prystas, J.Jonas and F.Sorm, Collect.Czech.Chem.Commun. 26, 974 (1961).

G63   J.Gut, M.Prystas and J.Jonas, Collect.Czech.Chem.Commun. 26, 986 (1961).

G64   G.A.Guter and G.S.Hammond, J.Am.Chem.Soc. 78, 5166 (1956).

H

H1   C.L.Habraken, E.C.Westra, G.H.Bomhoff and R.A.Heytink, Rec.Trav.Chim.Pays-Bas
       85, 1194 (1966).

H2   B.E.Hackley, R.Plapinger, M.Stolbert and T.Wagner-Jauregg, J.Am.Chem.Soc. 77,
       3651 (1955).

H3   J.A.Haines, C.B.Reese and A.Todd, J.Chem.Soc. 1962, 5281.

H4   R.A.Haines, D.E.Ryan and G.E.Cheney, Can.J.Chem. 40, 1149 (1962).

H5   T.N.Hall, J.Org.Chem. 29, 3587 (1964).

H6   J.C.Halle, F.Terrier and R.Schaal, Bull.Soc.Chim.Fr. 1969, 4569.

H7   S.D.Hamann and W.Strauss, Trans.Faraday Soc. 51, 1684 (1955).

H8   G.I.H.Hanania and D.H.Irvine, J.Chem.Soc. 1962, 2745.

H9   G.I.H.Hanania and D.H.Irvine, J.Chem.Soc. 1964, 5694.

H10   G.I.H.Hanania, D.H.Irvine and F.Shurayh, J.Chem.Soc. 1965, 1149.

H11   C.K.Hancock and A.D.H.Clague, J.Am.Chem.Soc. 86, 4942 (1964).

H12   E.S.Hand and R.M.Horowitz, J.Org.Chem. 29, 3088 (1964).

H13   L.D.Hansen, B.D.West, E.J.Baca and C.L.Blank, J.Am.Chem.Soc. 90, 6588 (1968).

H14   R.L.Hansen, J.Phys.Chem. 66, 369 (1962).

H15   H.Harada, Nippon Kagaku Zasshi 90, 267 (1969); CA 71, 12391g.

H16   F.E.Hardy and J.P.Johnston, J.Chem.Soc.,Perkin Trans.2 1973, 742.

H17   M.K.Hargreaves, E.A.Stevinson and J.Evans, J.Chem.Soc. 1965, 4582.

H18   E.T.Harper and M.L.Bender, J.Am.Chem.Soc. 87, 5625 (1965).

H19   H.J.Harries, J.Inorg.Nucl.Chem. 29, 2484 (1967).

H20   J.L.Haslam, E.M.Eyring, W.W.Epstein, G.A.Christiansen and M.H.Miles, J.Am.Chem.Soc.
        87, 1 (1965).

H21   J.L.Haslam, E.M.Eyring, W.E.Epstein, R.P.Jensen and C.W.Jaget, J.Am.Chem.Soc.
        87, 4247 (1965).

H22   C.J.Hawkins and D.D.Perrin, Inorg.Chem. 2, 839 (1963).

H23   C.J.Hawkins and D.D.Perrin, Inorg.Chem. 2, 843 (1963).

H24   R.W.Hay and S.J.Harvie, Aust.J.Chem. 18, 1197 (1965).

H25   R.W.Hay, P.J.Morris & D.D.Perrin, Aust.J.Chem. 21, 1073 (1968).

H26   A.F.Hegarty and L.N.Frost, J.Chem.Soc.,Perkin Trans.2 1973, 1719.

H27   V.Hejl and F.Pechar, Chem.Zvesti 21, 261 (1967); CA 67, 85514b.

H28   J.Heller and G.Schwarzenbach, Helv.Chim.Acta 34, 1876 (1951).

H29   S.Hendler, E.Furer and P.R.Srinivasan, Biochemistry 9, 4141 (1970).

H30   H.S.Hendrickson, Anal.Chem.39, 998 (1967).

H31   H.S.Hendrickson and J.G.Fullington, Biochemistry 4, 1599 (1965).

H32   E.F.G.Herington and W.Kynaston, Trans.Faraday Soc. 53, 138 (1957).

H33   J.Hermans, S.J.Leach and H.A.Scheraga, J.Am.Chem.Soc. 85, 1390 (1963).

H34   R.Herscovitch, J.J.Charette and E.de Hoffmann, J.Am.Chem.Soc. 95, 5135 (1973).

H35   R.Hershfield and G.L.Schmir, J.Am.Chem.Soc. 95, 3994 (1973).

H36   R.Hershfield and G.L.Schmir, J.Am.Chem.Soc. 95, 7359 (1973).

H37   R.Hershfield and G.L.Schmir, J.Am.Chem.Soc. 95, 8032 (1973).

H38   L.Hevesi, P.Van Brandt and A.Bruylants, Bull.Soc.Chim.Fr. 1970, 3971.

H39   F.Hibbert, J.Chem.Soc.,Perkin Trans.2 1973, 1289.

H40   G.P.Hildebrand and C.N.Reilly, Anal.Chem. 29, 258 (1957).

H41   J.Hine, J.G.Houston and J.H.Jensen, J.Org.Chem. 30, 1184 (1965).

H42   J.Hine, J.C.Philips and J.I.Maxwell, J.Org.Chem. 35, 3943 (1970).

H43   R.C.Hirt and R.G.Schmitt, Spectrochim.Acta 12, 127 (1958).

H44   C.F.Hiskey and F.F.Cantwell, J.Pharm.Sci.57, 2105 (1968).

H45   D.I.Hitchcock, J.Phys.Chem. 62, 1337 (1958).

H46   M.Hnilickova and L.Sommer, Collect.Czech.Chem.Commun. 26, 2189 (1961).

H47   M.Hnilickova and L.Sommer, Talanta 13, 667 (1966).

H48   A.J.Hoefnagel, J.C.Monshouwer, E.C.G.Snorn and B.M.Wepster, J.Am.Chem.Soc.
        95, 5350 (1973).

H49   A.J.Hoefnagel and B.M.Wepster, J.Am.Chem.Soc., 95, 5357 (1973).

H50   F.Holmes and W.R.C.Crimmin, J.Chem.Soc. 1955, 1175.

H51   C.van Hooidonk and L.Ginjaar, Recl.Trav.Chim.Pays-Bas 86, 449 (1967).

H52   H.Howell and G.S.Fisher, J.Am.Chem.Soc. 80, 6316 (1958).

H53   E.Hoyer, D.Wagler and J.Anton, Z.Phys.Chem.(Leipzig) 241, 65 (1969).

H54   I.Hsu and L.F.Koons, J.Indian Chem.Soc. 44, 540 (1967).

H55   T.M.Hsu, J.Chinese Chem.Soc.(Taiwan) 6, Ser II. 38 (1959); CA 57, 384d.

H56   T.M.Hsu and L.S.Chen, J.Chin.Chem.Soc.(Taipei) 13, 150 (1966); CA 67, 68157v.

H57   V.L.Hughes and A.E.Martell, J.Am.Chem.Soc. 78, 1319 (1956).

H58   A.A.Humffray, J.J.Ryan, J.P.Warren and Y.H.Yung, Chem.Commun. 1965, 610.

H59   J.E.C.Hutchins and T.H.Fife, J.Am.Chem.Soc. 95, 2282 (1973).

I

I1   A.Iesalniecs, Latv.PSR Zinat.Akad.Vestis, Kim.Ser. 3, 287 (1970); CA 73, 114703u.

I2   H.Imai, Y.Yaehashi and Y.Mizuno, Nippon Kagaka Zasshi 90, 1048 (1969); CA 72, 21178r.

I3   E.Imoto and R.Motoyama, Bull.Naniwa Univ. 2A, 127 (1954); CA 49, 9614.

I4   Y.Inoue, K.Kurosawa, K.Nakanishi and H.Obara, J.Chem.Soc. 1965, 3339.

I5   Y.Inoue and D.D.Perrin, J.Chem.Soc. 1962, 2600.

I6   R.R.Irani and K.Moedritzer, J.Phys.Chem. 66, 1349 (1962).

I7   D.T.Ireland and H.F.Walton, J.Phys.Chem. 71, 751 (1967).

I8   H.M.Irving and J.J.R.F. da Silva, J.Chem.Soc. 1963, 448.

I9   H.M.Irving and J.J.R.F. da Silva, J.Chem.Soc. 1963, 458.

I10   H.M.Irving and J.J.R.F.da Silva, J.Chem.Soc. 1963, 945.

I11   H.M.Irving and J.J.R.F.da Silva, J.Chem.Soc. 1963, 1144.

I12   H.M.Irving and J.J.R.F.da Silva, J.Chem.Soc. 1963, 3308.

I13   H.M.Irving and M.G.Miles, J.Chem.Soc.(A) 1966, 727.

I14   H.M.Irving and M.G.Miles, J.Chem.Soc.(A) 1966, 1268.

I15   H.M.Irving, M.G.Miles and L.D.Pettit, Anal.Chim Acta 38, 475 (1967).

I16   H.M.Irving and R.Parkins, J.Inorg.Nucl.Chem. 28, 1629 (1966).

I17   H.M.Irving, D.C.Rupainwar and S.S.Sahota, Anal.Chim.Acta 45, 249 (1969).

I18   H.M.Irving, R.Shelton and R.Evans, J.Chem.Soc. 1958, 3540.

I19   H.M.Irving and M.H.Stacey, J.Chem.Soc. 1961, 2019.

I20   R.J.Irving, L.Nelander and I.Wadso, Acta Chem.Scand. 18, 769 (1964).

I21   M.Ishibashi, Y.Yamamoto and H.Yamada, Bull.Chem.Soc.Jpn. 32, 1064 (1959).

I22   T.Ishiguro and T.Asahara, Yukaguku 8, 27 (1959); CA 54, 15964b.

I23 I.M.Issa, R.M.Issa, M.S.El-Ezabey and Y.Z.Ahmed, Z.Phys.Chem.(Leipzig) 242, 169(1969).

I24 A.I.Ivanov, V.I.Slovetskii, S.A.Shevelev, V.I.Erashko, A.A.Fainzilberg
     and S.S.Novikov, Russ.J.Phys.Chem.(Engl.Transl.) 40, 1234 (1966).

I25 D.J.G.Ives and P.D.Marsden, J.Chem.Soc. 1965, 649.

I26 D.J.G.Ives and P.G.N.Moseley, J.Chem.Soc.(B) 1970, 1655.

I27 D.J.G.Ives and D.Prasad, J.Chem.Soc. (B) 1970, 1649.

I28 D.J.G.Ives and D.Prasad, J.Chem.Soc.(B) 1970, 1652.

I29 T.Iwamoto, Bull.Chem.Soc.Jpn. 34, 605 (1961).

I30 I.Iwasaki and S.R.B.Cooke, J.Phys.Chem. 68, 2031 (1964).

I31 R.M.Izatt, W.C.Fernelius and B.P.Block, J.Phys.Chem. 59, 235 (1955).

I32 R.M.Izatt, C.G.Haas, B.P.Block and W.C.Fernelius, J.Phys.Chem. 58, 1133 (1954).

I33 R.M.Izatt, L.D.Hansen, J.H.Rytting and J.J.Christensen, J.Am.Chem.Soc. 87, 2760(1965).

I34 R.M.Izatt, J.H.Rytting, L.D.Hansen and J.J.Christensen, J.Am.Chem.Soc. 88, 2641(1966).

J

J1   K.E.Jabalpurivala and R.M.Milburn, J.Am.Chem.Soc. 88, 3224 (1966).

J2   G.Jackson and G.Porter, Proc. R.Soc.London, Ser.A A260, 13 (1961).

J3   W.A.Jacob and M.A.Herman, Anal.Chim.Acta 33, 229 (1965).

J4   N.W.Jacobsen, J.Chem.Soc.(C) 1966, 1065.

J5   B.R.James and R.J.P.Williams, J.Chem.Soc. 1961, 2007.

J6   C.Janion and D.Shugar, Acta Biochim.Pol. 7, 309 (1960).

J7   C.Janion and D.Shugar, Acta Biochim.Pol. 12, 337 (1965).

J8   C.Janion and D.Shugar, Acta.Biochim.Pol. 15, 261 (1968).

J9   M.J.Janssen, Recl.Trav.Chim.Pays-Bas 82, 931 (1963).

J10 E.S.Jayadevappa and P.B.Hukkeri, J.Inorg.Nucl.Chem. 30, 157 (1968).

J11 W.P.Jencks and J.Carriuolo, J.Am.Chem.Soc. 82, 1778 (1960).

J12 M.A.Jermyn, Aust.J.Chem. 20, 183 (1967).

J13 E.A.Johnson, Biochem.J. 51, 133 (1952).

J14 S.L.Johnson and K.A.Rumon, Biochemistry 9, 847 (1970).

J15 V.Jokl, J.Majer, H.Scharf and H.Kroll, Mikrochim.Acta 1966, 63; CA 65, 3309 h.

J16 J.Jonas and J.Gut, Collect.Czech.Chem. Commun. 27, 716 (1962).

J17 J.Jonas and J.Gut, Collect.Czech.Chem.Commun. 27, 1886 (1962).

J18 R.A.Jones and A.R.Katritzky, J.Chem.Soc. 1960, 2937.

J19 R.H.Jones and D.I.Stock, J.Chem.Soc. 1960, 102.

K

K1 M.I.Kabachnik, I.M.Dyatlova, T.A.Medved, Y.F.Belucin and V.V.Sidorenko, Dokl.Chem.(Engl.Transl.) 175, 621 (1967).

K2 M.I.Kabachnik, V.A.Gilyarov, C.Cheng-tieh and E.I.Matrosov, Bull.Acad.Sci.USSR, Div.Chem.Sci. 1962, 1504.

K3 M.I.Kabachnik, S.T.Ioffe and T.A.Mastryukova, Zh.Obshch.Khim. 25, 684 (1955); CA 50, 3851a.

K4 M.I.Kabachnik, R.P.Lastovskii, T.Y.Medved, V.V.Medyntsev, I.D.Kolpakova and N.M.Dyatlova, Dokl.Chem.(Engl.Transl.) 177, 1060 (1967).

K5 M.I.Kabachnik, T.A.Mastryukova, G.A.Balneva, E.E.Kugucheva, A.E.Shipov and T.A.Melent'eva, J.Gen.Chem.USSR (Engl.Transl.) 31, 132 (1961).

K6 M.I.Kabachnik, T.A.Mastryukova, A.E.Shipov and T.A.Melentyeva, Tetrahedron 9, 10 (1960).

K7 M.I.Kabachnik, T.Y.Medved, N.M.Dyatlova, M.N.Rusina and M.V.Rudomino, Bull.Acad.Sci.USSR, Div.Chem.Sci.(Engl.Transl.) 1967, 1450.

K8 L.Kabrt and Z.Holzbecher, Collect.Czech.Chem.Commun. 33, 3734 (1968).

K9 K.Kalfus, Collect.Czech.Chem.Commun. 33, 2962 (1968).

K10 J.Kalina, J.Putnins, E.Gudriniece and D.Kreicberga, Latv.PSR Zinat.Akad.Vestis, Kim.Ser. 4, 412 (1970); CA 73, 135482m.

K11 R.C.Kapoor, O.P.Agrawal and T.D.Seth, Can.J.Chem. 39, 2236 (1961).

K12 R.Karlicek, J.Majer and J.Polakovicova, Chem.Zvesti 24, 161 (1970); CA 74, 57852m.

K13 G.Karlsone, E.Gudriniece and J.Linabergs, Latv.PSR Zinat.Akad.Vestis, Kim.Ser. 1965, 537; CA 64, 8104a.

K14 R.Kavcic and B.Plesnicar, Vestn.Slov.Kem.Drus. 13, 39 (1966); CA 69, 86566y.

K15 R.Kavcic and B.Plesnicar, Bull.Sci.,Cons.Acad.Sci.Arts RSF Youoslavie, Sect.A 13, 71 (1968); CA 71, 38539m.

K16 T.C.Kawassiades, T.A.Kouimtzis and J.A.Tossidis, Chim.Cron.,A 33, 1 (1968); CA 69, 5686n.

K17 J.Kenttamaa, S.Raisanen, L.Auterinen and J.J.Lindberg, Suom.Kemistil.B 43B, 333 (1970).

K18 H.Khalifa, A.M.Amin and F.A.Osman, Fresenius' Z.Anal.Chem. 172, 167 (1960).

K19 H.Khalifa and S.W.Bishara, Fresenius' Z.Anal.Chem. 178, 184 (1960).

K20 H.Khalifa and S.W.Bishara, Fresenius' Z.Anal.Chem. 183, 108 (1961).

K21 H.Khalifa, M.A.Khater and A.A.El-Sirafy, Fresenius' Z.Anal.Chem. 237, 111 (1968).

K22 M.K.A.Khan and K.J.Morgan, Tetrahedron 21, 2197 (1965).

K23   M.M.T.Khan and A.E.Martell, J.Am.Chem.Soc. 88, 668 (1966).

K24   M.M.T.Khan and A.E.Martell, J.Am.Chem.Soc. 89, 5585 (1967).

K25   M.M.T.Khan and A.E.Martell, J.Am.Chem.Soc. 89, 7104 (1967).

K26   M.M.T.Khan and A.E.Martell, J.Am.Chem.Soc. 91, 4668 (1969).

K27   M.K.Kim and A.E.Martell, J.Am.Chem.Soc. 88, 914 (1966).

K28   C.V.King and A.P.Marion, J.Am.Chem.Soc. 66, 977 (1944).

K29   E.J.King, J.Am.Chem.Soc. 82, 3575 (1960).

K30   E.J.King and J.E.Prue, J.Chem.Soc. 1961, 275.

K31   R.D.Kinget and M.A.Schwartz, J.Pharm.Sci. 57, 1916 (1968).

K32   R.D.Kinget and M.A.Schwartz, J.Pharm.Sci. 58, 1102 (1969).

K33   R.Kirdani and M.J.Burgett, Arch.Biochem. Biophys. 118, 33 (1967).

K34   W.A.Klee and S.H.Mudd, Biochemistry 6, 988 (1967).

K35   D.S.Klett and B.F.Freasier, J.Chem.Soc. 1962, 4741.

K36   J.J.Klingenberg, J.P.Thole and R.D.Lingg, J.Chem.Eng.Data 11, 94 (1966).

K37   I.L.Knunyants and B.L.Dyatkin, Bull.Acad.Sci.USSR. Div.Chem.Sci.(Engl.Transl.)
       1964, 863.

K38   I.L.Knunyants, B.L.Dyatkin, E.P.Mochalina and L.T.Lantseva, Bull.Acad.Sci.USSR,
       Div.Chem.Sci.(Engl.Transl.) 1966, 164.

K39   I.L.Knunyants, L.S.German and I.N.Rozhkov, Bull.Acad.Sci.USSR, Div.Chem.Sci.
       (Engl.Transl.) 1964, 1013.

K40   H.C.Ko, W.F.O'Hara, T.Hu and L.G.Hepler, J.Am.Chem.Soc. 86, 1003 (1964).

K41   W.L.Koltun, L.Ng and F.R.N.Gurd, J.Biol.Chem. 238, 1367 (1963).

K42   N.P.Komar and A.K.Khukhryanskii, Russ.J.Inorg.Chem.(Engl.Trans.) 11, 614 (1966).

K43   N.Konopik and W.Luf, Monatsh.Chem. 101, 1591 (1970).

K44   J.Korbl and B.Kakac, Collect.Czech.Chem. Commun. 23, 889 (1958).

K45   I.M.Korenman, F.R.Sheyanova and Z.M.Gur'eva, Russ.J.Inorg.Chem.(Engl.Transl.)
       11, 1485 (1966).

K46   E.C.Kornfeld, R.G.Jones and T.V.Parke, J.Am.Chem.Soc. 71, 150 (1949).

K47   I.K.Korobitsyna, L.S.Gurevich and I.V.Zerova, J.Org.Chem.USSR(Engl.Transl.) 4,
       1949 (1968).

K48   B.H.Korsch and N.V.Riggs, Aust.J.Chem. 16, 709 (1963).

K49   H.Koshimura and T.Okubo, Anal.Chim.Acta 49, 67 (1970).

K50   G.M.Kosolapoff and R.F.Struck, J.Chem.Soc. 1957, 3739.

K51   A.N.Kost, P.B.Terentev, L.A.Golovleva and A.A.Stolyarchuk, Khim.-Farm.Zh.
       1, 3 (1967); CA 68, 29556a.

K52   G.Koyama and H.Umezawa, J.Antibiot.,Ser.A 18, 175 (1965).

K53   M.E.Krahl, J.Phys.Chem. 44, 449 (1940).

K54   D.N.Kramer, R.M.Gamson and F.M.Miller, J.Org.Chem. 24, 1742 (1959).

K55   F.Krasovec, Croat.Chem.Acta 37, 107 (1965).

K56   F.Krasovec and J.Jan, Croat.Chem.Acta 35, 183 (1963).

K57   M.M.Kreevoy, B.E.Eichinger, F.E.Stary, E.A.Katz and J.H.Sellstedt, J.Org.Chem.
        29, 1641 (1964).

K58   M.M.Kreevoy, E.T.Harper, R.E.Duvall, H.S.Wilgus and L.T.Ditsch, J.Am.Chem.Soc.
        82, 4899 (1960).

K59   S.U.Kreingold and E.A.Bozhevolnov, J.Anal.Chem.USSR (Engl.Transl.) 18, 818 (1963).

K60   V.P.Kreiter and D.Pressman, Biochemistry 2, 97 (1963).

K60a N.I.Krikova and G.M.Pisichenko, Khim.Geterosikl.Soedin 1967, 317; CA 68, 14051h.

K61   B.Krishna and E.V.Sundaram, Indian J.Chem. 6, 591 (1968).

K62   G.G.Kryukova, Y.I.Turyan and A.V.Bondarenko, J.Gen.Chem.USSR (Engl.Transl.)
        38, 2108 (1968).

K63   W.Kuchen and H.Meyer, Z.Anorg.Allg.Chem. 333, 71 (1964).

K64   L.Kugel and M.Halmann, J.Org.Chem. 32, 642 (1967).

K65   T.Kulikowski, B.Zmudzka and D.Shugar, Acta Biochim.Pol. 16, 201 (1969).

K66   J.Kumamoto and F.H.Westheimer, J.Am.Chem.Soc. 77, 2515 (1955).

K67   W.D.Kumler and J.J.Eiler, J.Am.Chem.Soc. 65, 2355 (1943).

K68   S.K.Kundra and L.C.Thompson, J.Inorg.Nucl.Chem. 30, 1847 (1968).

K69   J.L.Kurz and J.M.Farrar, J.Am.Chem.Soc. 91, 6057 (1969).

K70   J.L.Kurz and J.C.Harris, J.Org.Chem. 35, 3086 (1970).

K71   P.M.Kuznesof and W.L.Jolly, Inorg.Chem. 7, 2574 (1968).

                                      L

L1   R.G.Lacoste, C.V.Christoffers and A.E.Martell, J.Am.Chem.Soc. 87, 2385 (1965).

L1a   C.K.Laird and M.A.Leonard, Talanta 17, 173 (1970).

L2   E.Lakanen, Suom.Kemistil.B 43, 226 (1970).

L3   M.Laloi-Diard and M.Rubinstein, Bull.Soc.Chim.Fr. 1965, 310.

L3a   M.Laloi-Diard and P.Rumpf, Bull.Soc.Chim.Fr. 1961, 1645.

L4   F.J.Langmyhr and K.S.Klausen, Anal.Chim.Acta 29, 149 (1963).

L5   F.J.Langmyhr and A.R.Storm, Acta Chem.Scand. 15, 1461 (1961).

L6   P.O.Larsen and A.Kjaer, Acta Chem.Scand. 16, 142 (1962).

L7   W.G.Laver, A.Neuberger and J.J.Scott, J.Chem.Soc. 1959, 1483.

L8   P.D.Lawley and P.Brookes, J.Mol.Biol. 4, 216 (1962).

L9   J.E.Leader and M.W.Whitehouse, Biochem.Pharmacol. 15, 1379 (1966).

L10   J.Le Coarer, M.Wone and A.Broche, Ann.Fac.Sci.,Univ.Dakar 6, 25 (1961); CA 58, 6244f.

L11   D.H.Lee, Daehan Hwahak Hwoejee 9, 33 (1965); CA 64, 5820g.

L12   H.J.Lee and P.W.Wigler, Biochemistry 7, 1427 (1968).

L13   W.F.Lee, N.K.Shastri and E.S.Amis, Talanta 11, 685 (1964).

L14   L.J.Leeson, J.E.Krueger and R.A.Nash, Tetrahedron Lett. 1963, 1155.

L15   P.Leggate and G.E.Dunn, Can.J.Chem. 43, 1158 (1965).

L16   K.Lehtonen and L.A.Summers, Aust.J.Chem. 23, 1699 (1970).

L17   W.J.Le Noble, J.Am.Chem.Soc. 83, 3897 (1961).

L18   G.R.Lenz and A.E.Martell, Inorg.Chem. 4, 378 (1965).

L19   F.L'Eplattenier, I.Murase and A.E.Martell, J.Am.Chem.Soc. 89, 837 (1967).

L20   H.J.Le Roux and K.F.Fouché, J.Inorg.Nucl.Chem. 32, 3059 (1970).

L21   P.Lesfauries and P.Rumpf, Bull.Chim.Soc.Fr. 1950, 542.

L22   J.R.Leto and M.F.Leto, J.Am.Chem.Soc. 83, 2944 (1961).

L23   R.Letters and A.M.Michelson, J.Chem.Soc. 1962, 71.

L24   D.L.Leussing, J.Am.Chem.Soc. 80, 4180 (1958).

L25   D.L.Leussing and G.S.Alberts, J.Am.Chem.Soc. 82, 4458 (1960).

L26   D.L.Leussing and K.S.Bai, Anal.Chem. 40, 575 (1968).

L27   D.L.Leussing and E.M.Hanna, J.Am.Chem.Soc. 88, 693 (1966).

L28   D.L.Leussing and E.M.Hanna, J.Am.Chem.Soc. 88, 696 (1966).

L29   D.L.Leussing and J.Jayne, J.Phys.Chem. 66, 426 (1962).

L30   D.L.Leussing, R.E.Laramy and G.S.Alberts, J.Am.Chem.Soc. 82, 4826 (1960).

L31   D.L.Leussing and D.C.Schultz, J.Am.Chem.Soc. 86, 4846 (1964).

L32   R.Y.Levina, P.A.Gembitskii, L.P.Guseva and P.K.Agasyan, J.Gen.Chem.USSR (Engl.Transl.)
        34, 144 (1964).

L33   M.Levy and P.J.Magoulas, J.Am.Chem.Soc. 84, 1345 (1962).

L34   S.Lewin and M.A.Barnes, J.Chem.Soc.(B) 1966, 478.

L35   S.Lewin and N.W.Tann, J.Chem.Soc. 1962, 1466.

L36   E.S.Lewis and M.D.Johnson, J.Am.Chem.Soc. 81, 2070 (1959).

L37   G.A.L'Heureux and A.E.Martell, J.Inorg.Nucl.Chem. 28, 481 (1966).

L38   N.C.Li, P.Tang and R.Mathur, J.Phys.Chem. 65, 1074 (1961).

L39   D.Lichtenberg, F.Bergmann and Z.Neiman, J.Chem.Soc.(C) 1971, 1676.

L40   D.Lichtenberg, F.Bergmann and Z.Neiman, J.Chem.Soc.,Perkin Trans.2 1972, 1676.

L41   D.Lichtenberg, F.Bergmann and Z.Neiman, J.Chem.Soc.,Perkin Trans 1 1973, 2445.

L42   D.Lichtenberg, F.Bergmann, M.Rahat and Z.Neiman, J.Chem.Soc.,Perkin Trans 1 1972,2950.

L43   T.H.Lin, J.Chinese Chem.Soc.(Taiwan) 9, 220 (1962); CA 60, 7509b.

L44   Y.Linaberg, O.Nieland, A.Veis and G.Vanag, Dokl.Chem.(Engl.Trans.) 154, 184 (1964).

L45   J.Linabergs, E.Grens, G.Vanags and A.Veiss, Latv.PSR Zinat.Akad.Vestis,
       Kim.Ser. 1963, 325; CA 60, 9128e.

L46   J.Linabergs and A.Veis, Latv.PSR Zinat.Akad.Vestis, Kim.Ser. 1962, 213; CA 68, 8443f.

L47   B.J.Lindberg, Acta Chem.Scand. 24, 2852 (1970).

L48   J.J.Lindberg and C.G.Nordstrom, Acta Chem.Scand. 17, 2346 (1963).

L49   J.J.Lindberg, C.G.Nordstrom and R.Lauren, Suom.Kemistil.B 35, 182 (1962).

L50   J.J.Lindberg, F.Odeyemi and K.Penttinen, Suom.Kemistil.B 41, 237 (1968).

L51   F.Lindstrom and H.Diehl, Anal.Chem. 32, 1123 (1960).

L52   C.L.Liotta, K.H.Leavell and D.F.Smith, J.Phys.Chem. 71, 3091 (1967).

L53   C.L.Liotta and D.F.Smith, Chem.Commun. 1968, 416.

L54   F.Lipmann and L.C.Tuttle, Arch.Biochem. 31, 373 (1947).

L55   M.N.Lipsett, J.Biol.Chem. 240, 3975 (1965).

L56   A.M.Liquori and A.Ripamonti, Ricerca Sci. 25, 1132 (1955); CA 50, 7086h.

L57   E.P.Lira, J.Heterocycl.Chem. 6, 955 (1969).

L58   G.J.Litchfield and G.Shaw, J.Chem.Soc.(C) 1971, 817.

L59   D.Litchinsky, N.Purdie, M.B.Tomson and W.D.White, Anal.Chem. 41, 1726 (1969).

L60   C.Liteanu, Acad.rep.populare Romane (Cluj), Studii cercetari stiint. 3, 76 (1952);
       CA 50, 11775h.

L61   A.K.Livshits, A.V.Glembotskii and S.M.Gurvich, Tsvet.Metal. 41, 19 (1968);
       CA 69, 80926u.

L62   L.N.Lomakina and I.P.Alimarin, Vestn.Mosk.Univ.,Ser.II, Khim. 20, 58 (1965);
       CA 64, 7426e.

L63   F.A.Long and P.Ballinger, Electrolytes, Proc.Intern.Symp.,Trieste,Yugoslavia 1959,
       152 (Publ.1962); CA 61, 10098.

L64   J.L.Longridge and F.A.Long, J.Am.Chem.Soc. 90, 3088 (1968).

L65   N.G.Lordi and E.M.Cohen, Anal.Chim.Acta 25, 281 (1961).

L66   K.D.Louise, Trans.Faraday Soc. 56, 1633 (1960).

L67   D.A.Lown, H.R.Thirsk and W.F.K.Wynne-Jones, Trans.Faraday Soc. 66, 51 (1970).

L68   S.Lukkari, K.Paakkonen and E.Huttunen, Farm.Aikak. 79, 28 (1970); CA 74 16270b.

L69   P.O.Lumme, Suom.Kemistil.B 29, 220 (1956).

L70   P.O.Lumme, Suom.Kemistil.B 29, 223 (1956).

L71   P.O.Lumme, Suom.Kemistil.B 30, 176 (1957).

L72   P.O.Lumme, Suom.Kemistil.B 30, 194 (1957).

L73   P.Lumme, Suom.Kemistil.B 33, 87 (1960).

L74   P.Lumme and N.Seppalainen, Suom.Kemistil.B 35, 123 (1962).

L75   W.Lund and E.Jacobsen, Acta Chem.Scand. 19, 1783 (1965).

L76   B.Luning, Acta Chem.Scand. 14, 321 (1960).

L77   S.J.Lyle and S.J.Naqvi, J.Inorg.Nucl.Chem. 28, 2993 (1966).

L78   B.M.Lynch, R.K.Robins and C.C.Cheng, J.Chem.Soc. 1958, 2973.

M

M1   A.F.Mabrouk and L.R.Dugan, J.Am.Oil Chemists' Soc. 38, 9 (1961); CA 55, 6121h.

M2   A.McAuley and G.H.Nancollas, Trans.Faraday Soc. 56, 1165 (1960).

M3   A.McAuley and G.H.Nancollas, J.Chem.Soc. 1961, 2215.

M4   E.T.McBee, W.F.Marzluff and O.R.Pierce, J.Am.Chem.Soc. 74, 444 (1952).

M5   W.A.E.McBryde, Can.J.Chem.46, 2385 (1968).

M6   W.A.E.McBryde, Analyst (London) 94, 337 (1969).

M7   W.A.E.McBryde and G.F.Atkinson, Can.J.Chem. 39, 510 (1961).

M8   W.A.E.McBryde, J.L.Rohr, J.S.Penciner and J.A.Page, Can.J.Chem. 48, 2574 (1970).

M9   K.S.McCallum and W.D.Emmons, J.Org.Chem. 21, 367 (1956).

M10   E.A.McCoy and L.L.McCoy, J.Org.Chem. 33, 2354 (1968).

M11   L.L.McCoy, J.Am.Chem.Soc. 89, 1673 (1967).

M12   L.L.McCoy and G.Nachtigall, J.Am.Chem.Soc. 85, 1321 (1963).

M13   D.J.MacDonald, J.Org.Chem. 33, 4559 (1968).

M14   P.D.McDonald and G.A.Hamilton, J.Am.Chem.Soc. 95, 7752 (1973).

M15   R.W.McGilvery, Biochemistry 4, 1924 (1965).

M16   V.Machacek, J.Panchartek, V.Sterba and M.Vecera, Collect.Czech.Chem.Commun. 35,
         844 (1970).

M17   J.A.Maclaren, Aust.J.Chem. 21, 1891 (1968).

M18   L.B.Magnusson, C.A.Craig and C.Postmus, J.Am.Chem.Soc. 86, 3958 (1964).

M19   L.B.Magnusson, C.Postmus and C.A.Craig, J.Am.Chem.Soc. 85, 1711 (1963).

M20   H.L.Mahalaha, M.K.Dave and D.D.Sharma, J.Indian Chem.Soc. 45, 413 (1968).

M21   L.A.Mai, J.Gen.Chem.USSR (Engl.Transl.) 30, 1109 (1960).

M22   S.G.Mairanovskii, N.Y.Grigoreva, N.V.Barashkova and V.F.Kucherov,

      Bull.Acad.Sci.USSR, Div.Chem.Sci.(Engl.Transl.) 1963, 219.

M23   J.Majer and E.Dvorakova, Chem.Zvesti 17, 402 (1963); CA 64, 7652e.

M24   J.Majer, R.Karlicek and B.Kopecka, Collect.Czech.Chem.Commun. 35, 1066 (1970).

M25   J.Majer, M.Kotoucek and E.Dvorakova, Chem.Zvesti. 20, 242 (1966); CA 65, 3304h.

M26   L.Majs, Zhur.Obshch.Khim. 26, 3206 (1956); CA 51, 7111c.

M27   O.Makitie, Suom.Kemistil.B 33, 207 (1960).

M28   O.Makitie, Maatalouden Tutkimuskeskus Maantutkimuslaitos 1961, (79); CA 56, 2035b.

M29   O.Makitie, Suom.Kemistil.B 35, 1 (1962).

M30   O.Makitie, Suom.Kemistil.B 39, 23 (1966); CA 64, 18508f.

M31   O.Makitie, Suom.Kemistil.B 39, 26 (1966).

M32   O.Makitie, Suom.Kemistil.B 39, 282 (1966).

M33   O.Makitie, Acta Chem.Scand. 22, 2703 (1968).

M34   O.Makitie and V.Konttinen, Acta Chem.Scand. 23, 1459 (1969).

M35   O.Makitie and H.Saarinen, Anal.Chim.Acta 46, 314 (1969).

M36   O.Makitie and H.Saarinen, Suom.Kemistil.B 42, 394 (1969); CA 72, 25710f.

M37   O.Makitie and H.Saarinen, J.Inorg.Nucl.Chem. 32, 2800 (1970).

M38   O.Makitie, H.Saarinen and H.Mattinen, Suom.Kemistil.B 43, 340 (1970); CA 74, 7145n.

M39   O.Makitie, K.Soininen and H.Saarinen, Suom.Kemistil.B 41, 246 (1968); CA 69, 54844w.

M40   O.Makitie, K.Soininen and H.Saarinen, Suom.Kemistil.B 41, 261 (1968).

M41   O.Makitie and E.Toivanen, Acta Chem.Scand. 24, 2247 (1970).

M42   P.G.Malan and H.Edelhoch, Biochemistry 9, 3205 (1970).

M43   M.Malat, Anal.Chim.Acta 25, 289 (1961).

M44   M.Malat, Collect.Czech.Chem.Commun. 26, 1877 (1961).

M45   S.Mandel and A.K.Dey, J.Inorg.Nucl.Chem. 30, 1221 (1968).

M46   F.Manok, C.Varhelyi and I.Mikulas, Stud.Univ.Babes-Bolyai, Ser.Chim. 15, 139 (1970);

      CA 75, 26064x.

M47   K.R.Manolov, Nauch.Tr.,Vissh Inst.Khranit.Vkusova Prom., Plovdiv. 14, 247 (1967);

      CA 72, 54572t.

M48   G.H.Mansfield and M.C.Whiting, J.Chem.Soc., 1956, 4761.

M49   A.L.Markman, G.B.Ostrobrod and E.K.Abdusalamova, Izv.Vyssh.Ucheb.Zaved.,Khim.Khim.

      Tekhnol 12, 97, (1969); CA 71, 7058z.

M50   P.Maroni and J.P.Calmon, Bull.Soc.Chim.Fr. 1964, 519.

M51   D.J.Martin and C.E.Griffin, J.Organomet.Chem. 1, 292 (1964).

M52   D.J.Martin and C.E.Griffin, J.Org.Chem. 30, 4034 (1965).

M53   D.M.G.Martin and C.B.Reese, J.Chem.Soc.(C) 1968, 1731.

M54   R.P.Martin and R.A.Paris, Bull.Soc.Chim.Fr. 1963, 1600.

M55   R.P.Martin and R.A.Paris, Bull.Soc.Chim.Fr. 1964, 80.

M56   S.F.Mason, J.Chem.Soc., 1959, 1247.

M57   T.A.Mastryukova, T.A.Melenteva, A.E.Shipov and M.I.Kabachnik, J.Gen.Chem.USSR.
      (Engl.Transl.) 29, 2145 (1959).

M58   M.Matell and S.Lindenfors, Acta Chem.Scand. 11, 324 (1957).

M59   K.S.Math, K.A.Venkatachalam and M.B.Kabadi, J.Indian Chem.Soc. 36, 65 (1959).

M60   J.G.Mather and J.Shorter, J.Chem.Soc. 1961, 4744.

M61   M.Matsukawa, M.Ohta, S.Takata and R.Tsuchiya, Bull.Chem.Soc.Jpn. 38, 1235 (1965).

M62   B.N.Mattoo, Trans.Faraday Soc. 53, 760 (1957).

M63   B.N.Mattoo, Z.Phys.Chem.(Frankfurt am Main) 12, 232 (1957).

M64   B.N.Mattoo, Trans.Faraday Soc. 54, 19 (1958).

M65   B.N.Mattoo, Z.Phys.Chem.(Frankfurt am Main) 22, 187 (1959).

M66   B.V.Matveev and G.G.Tsybaeva, J.Gen. Chem.USSR (Engl.Transl.) 34, 2512 (1964).

M67   H.G.Mautner, J.Am.Chem.Soc. 78, 5292 (1956).

M68   H.G.Mautner and E.M.Clayton, J.Am.Chem.Soc. 81, 6270 (1959).

M69   W.E.Mayberry, J.E.Rall, M.Berman and D.Bertoli, Biochemistry 4, 1965 (1965).

M70   R.Meilleur and R.L.Benoit, Can.J.Chem. 47, 2569 (1969).

M71   F.M.Menger and J.A.Donohue, J.Am.Chem.Soc. 95, 432 (1973).

M72   J.L.Meyer and J.E.Bauman, J.Chem.Eng.Data, 15, 404 (1970).

M73   W.J.Middleton and R.V.Lindsey, J.Am.Chem.Soc. 86, 4948 (1964).

M74   P.K.Migal and A.V.Ivanov, J.Gen.Chem.USSR (Engl.Transl.) 37, 355 (1967).

M75   P.B.Mikhelson and V.I.Kozachek, J.Anal.Chem.USSR (Engl.Transl.) 21, 1112 (1966).

M74   R.M.Milburn and L.M.Venanzi, Inorg.Chim.Acta 2, 97 (1968).

M77   M.H.Miles, E.M.Eyring, W.W.Epstein and R.E.Ostlund, J.Phys.Chem. 69, 467 (1965).

M78   C.O.Miller, F.Skoog, F.S.Okumura, M.H.Von Saltza and F.M.Strong, J.Am.Chem.Soc.
      78, 1375 (1956).

M79   D.M.Miller and R.A.Latimer, Can.J.Chem. 40, 246 (1962).

M80   F.J.Millero, J.C.Ahluwalia and L.G.Hepler, J.Chem.Eng.Data, 10, 199 (1965).

M81   S.Milstien and T.H.Fife, J.Am.Chem.Soc. 89, 5820 (1967).

M82   I.Minamida, Y.Ikeda, K.Uneyama, W.Tagaki and S.Oae, Tetrahedron 24, 5293 (1968).

M83   J.Minczewski, E.Grzegrzolka and K.Kasiura, Chem.Anal.(Warsaw) 13, 601 (1968);
      CA 70, 25293f.

M84   J.Minczewski and Z.Skorko-Trybula, Talanta 10, 1063 (1963).

M85   S.Misumi and M.Aihara, Bull.Chem.Soc.Jpn. 39, 2677 (1966).

M86   S.Miyamoto and E.Brochmann-Hanssen, J.Pharm.Sci. 51, 552 (1962).

M87   H.Miyata, Bull.Chem.Soc.Jpn. 36, 382 (1963).

M88   H.Miyata, Bull.Chem.Soc.Jpn. 36, 386 (1963).

M89   H.Miyata, Bull.Chem.Soc.Jpn. 37, 426 (1964).

M90   H.Miyata, Bull.Chem.Soc.Jpn. 40, 1875 (1967).

M91   H.Miyata, Bull.Chem.Soc.Jpn. 40, 2815 (1967).

M92   M.Miyazaki, Y.Moriguchi and K.Ueno, Bull.Chem.Soc.Jpn. 41, 838 (1968).

M93   T.Moeller and S.K.Chu, J.Inorg.Nucl.Chem. 28, 153 (1966).

M94   T.Moeller and R.Ferrus, J.Inorg.Nucl.Chem. 20, 261 (1961).

M95   T.Moeller and R.Ferrus, Inorg.Chem. 1, 49, (1962).

M96   T.Moeller and T.M.Hseu, J.Inorg.Nucl.Chem. 24, 1635 (1962).

M97   T.Moeller and L.C.Thompson, J.Inorg.Nucl.Chem. 24, 499 (1962).

M98   D.Monnier and C.Jegge, Helv.Chim.Acta 40, 513 (1957).

M99   G.E.Mont and A.E.Martell, J.Am.Chem.Soc. 88, 1387 (1966).

M100 C.E.Moore and R.Peck, J.Org.Chem. 20, 673 (1955).

M101 E.G.Moorhead and N.Sutin, Inorg.Chem. 5, 1866 (1966).

M102 S.Morazzani-Pelletier and S.Meriaux, J.Chim.Phys.Phys.-Chim.Biol. 63, 278 (1966).

M103 W.A.Mosher, R.N.Hively and F.H.Dean, J.Org.Chem. 35, 3689 (1970).

M104 R.J.Motekaitis and A.E.Martell, J.Am.Chem.Soc. 92, 4223 (1970).

M105 L.Moyne and G.Thomas, Anal.Chim.Acta 31, 583 (1964).

M106 A.A.Muk and M.B.Pravica, Anal.Chim.Acta 45, 534 (1969).

M107 L.M.Mukherjee and E.Grunwald, J.Phys.Chem. 62, 1311 (1958).

M108 N.A.Mukhina, K.P.Tetenchuk and M.M.Kaganskii, Zh.Vses.Khim.Obshchest. 15, 351(1970);
      CA 73, 55465t.

M109 T.P.C.Mulholland, R.Foster and D.B.Haycock, J.Chem.Soc.,Perkin Trans.1 1972, 2121.

M110 Y.Murakami, Bull.Soc.Chem.Jpn. 35, 52 (1962).

M111 Y.Murakami, H.Kondo and A.E.Martell, J.Am.Chem.Soc. 95, 7138 (1973).

M112 Y.Murakami, K.Nakamura and M.Tokunaga, Bull.Soc.Chem.Jpn. 36, 669 (1963).

M113 Y.Murakami and J.Sunamoto, Bull.Chem.Soc.Jpn. 44, 1939 (1971).

M114 Y.Murakami, J.Sunamoto and H.Ishizu, Bull.Chem.Soc.Jpn. 45, 590 (1972).

M115 Y.Murakami, J.Sunamoto and N.Kanamoto, Bull.Chem.Soc.Jpn. 46, 871 (1973).

M116 Y.Murakami, J.Sunamoto and N.Kanamoto, Bull.Chem.Soc.Jpn. 46, 1730 (1973).

M117 Y.Murakami, J.Sunamoto, S.Kinuwaki and H.Honda, Bull.Chem.Soc.Jpn. 46, 2187 (1973).

M118 Y.Murakami, J.Sunamoto, H.Sadamori, H.Kondo and M.Takagi, Bull.Chem.Soc.Jpn.
      43, 2518 (1970).

M119 Y.Murakami and M.Takagi, Bull.Soc.Chem.Jpn. 37, 268 (1964).

M120 Y.Murakami and M.Takagi, J.Phys.Chem. 72, 116 (1968).

M121 Y.Murakami and M.Takagi, J.Am.Chem.Soc. 91, 5130 (1969).

M122 Y.Murakami, M.Takagi and H.Nishi, Bull.Chem.Soc.Jpn. 39, 1197 (1966).

M123 Y.Murakami and M.Tokunaga, Bull.Soc.Chem.Jpn. 37, 1562 (1964).

M124 A.Murata, T.Suzuki and T.Ito, Bunseki Kaguku 14, 630 (1965); CA 63, 12296b.

M125 C.B.Murphy and A.E.Martell, J.Biol.Chem. 226, 37 (1957).

M126 J.Murto, Acta Chem.Scand. 18, 1043 (1964).

M127 O.S.Musailov, B.A.Dunai and N.P.Komar, J.Anal.Chem.USSR (Engl.Transl.) 23, 123 (1968)

M128 H.Musso and H.G.Matthies, Chem.Ber. 94, 356 (1961).

M129 I.S.Mustafin, A.N.Ivanova and N.F.Lisenko, J.Anal.Chem. USSR (Engl.Transl.) 20,
      13 (1965).

                                        N

N1   K.Nagano and D.E.Metzler, J.Am.Chem.Soc. 89, 2891 (1967).

N2   K.Nagata, A.Umayahara and R.Tsuchiya, Bull.Chem.Soc.Jpn. 38, 1059 (1965).

N3   V.S.K.Nair, Trans.Faraday Soc. 57, 1988 (1961).

N4   V.S.K.Nair and G.H.Nancollas, J.Chem.Soc. 1961, 4367.

N5   V.S.K.Nair and S.Parthasarathy, J.Inorg.Nucl.Chem. 32, 3289 (1970).

N6   V.S.K.Nair and S.Parthasarathy, J.Inorg.Nucl.Chem. 32, 3293 (1970).

N7   V.S.K.Nair and S.Parthasarathy, J.Inorg.Nucl.Chem. 32, 3297 (1970).

N8   K.Nakanishi, N.Suzuki and F.Yamazaki, Bull.Chem.Soc.Jpn. 34, 53 (1961).

N9   S.Nakashima, H.Miyata and K.Toei, Bull.Chem.Soc.Jpn. 40, 870 (1967).

N10   S.Nakashima, H.Miyata and K.Toei, Bull.Soc.Chem.Jpn. 41, 2632 (1968).

N11   G.H.Nancollas and A.C.Park, Inorg.Chem. 7, 58 (1968).

N12   G.H.Nancollas and D.J.Poulton, Inorg.Chem. 8, 680 (1969).

N13   A.Napoli, Talanta 15, 189 (1968).

N14   A.Napoli, J.Inorg.Nucl.Chem. 32, 1907 (1970).

N15   R.Nasanen, Suom.Kemistil.B 30, 61 (1957).

N16   O.Navratil, Collect.Czech.Chem.Commun. 29, 2490 (1964).

N17   O.Navratil and J.Kotas, Collect.Czech.Chem.Commun. 30, 2736 (1965).

N18   O.Navratil and J.Liska, Collect.Czech.Chem.Commun. 33, 987 (1968).

N19   O.Navratil and J.Liska, Collect.Czech.Chem.Commun. 33, 991 (1968).

N20   R.Nayan and A.K.Dey, Z.Naturforsch.,Teil B 25, 1453 (1970).

N21   V.A.Nazarenko and G.V.Flyantikova, J.Gen.Chem.USSR (Engl.Transl.) 35, 213 (1965).

N22   V.A.Nazarenko, G.V.Flyantikova and G.G.Shitareva, Zh.Vses.Khim.Obshchest. 13,
       233 (1968); CA 69, 70645n.

N23   V.A.Nazarenko, N.V.Lebedeva and E.A.Biryuk, Zh.Vses.Khim.Obshchestva im
       D.I.Mendeleeva 9, 589 (1964); CA 62, 2695f.

N24   V.A.Nazarenko, N.V.Lebedeva and L.I.Vinarova, Russ.J.Inorg.Chem.(Engl.Transl.)
       15, 330 (1970).

N25   V.A.Nazarenko, N.V.Lebedeva and L.I.Vinarova, Russ.J.Inorg.Chem.(Engl.Transl.)
       15, 1557 (1970).

N26   V.A.Nazarenko, L.Ngoc-Thu and R.M.Dranitskaya, J.Anal.Chem.USSR (Engl.Transl.)
       22, 302 (1967).

N27   V.A.Nazarenko and S.Y.Vinkovetskaya, J.Anal.Chem.USSR (Engl.Transl.) 22, 153 (1967).

N28   A.A.Neimysheva, V.I.Savchuk and I.L.Knunyants, J.Gen.Chem.USSR(Engl.Transl.)
       36, 520 (1966).

N29   J.L.Nichols and B.G.Lane, Can.J.Biochem. 44, 1633 (1966).

N30   G.Nickless, F.H.Pollard and T.J.Samuelson, Anal.Chim.Acta 39, 37 (1967).

N31   O.Y.Nieland and S.V.Kalnin, J.Org.Chem.USSR (Engl.Transl.) 4, 132 (1968).

N32   A.Ninomiya and K.Toei, Nippon Kagaku Zasshi 90, 655 (1969).

N33   F.F.Noe and L.Fowden, Biochem.J. 77, 543 (1960).

N34   C.G.Nordstrom and J.J.Lindberg, Suom.Kemistil.B 38, 291 (1965).

N35   C.G.Nordstrom, J.J.Lindberg and L.J.Karumaa, Suom.Kemistil.B 36, 105 (1963).

N36   S.S.Novikov, V.M.Belikov, A.A.Fainzilberg, L.V.Ershova, V.I.Slovetskii and
       S.A.Shevelev, Bull.Acad.Sci.USSR, Div.Chem.Sci.(Engl.Transl.) 1959, 1773.

N37   S.S.Novikov, I.S.Ivanov, G.V.Bogdanova, T.A.Alekseeva and Y.V.Konnova,
       Bull.Acad.Sci.USSR, Div.Chem.Sci.(Engl.Transl.) 1966, 719.

N38   S.S.Novikov, V.N.Slovetskii, V.M.Belikov, I.M.Zavilovich and L.V.Epishina,
       Bull.Acad.Sci.USSR, Div.Chem.Sci.(Engl.Transl.) 1962, 480.

N39   S.S.Novikov, V.I.Slovetskii, S.A.Shevelev and A.A.Fainzilberg,
       Bull.Acad.Sci.USSR, Div.Chem.Sci.(Engl.Transl.) 1962, 552.

N40   T.Nozaki and K.Koshiba, Nippon Kagaku Zasshi 88, 1287 (1967).

N41   M.H.T.Nyberg, M.Cefola and D.Sabine, Arch.Biochem.Biophys. 85, 82 (1959).

N42   V.Nyren and E.Back, Acta Chem.Scand. 12, 1305 (1958).

N43   V.Nyren and E.Back, Acta Chem.Scand. 12, 1516 (1958).

O

O1   S.Oae, N.Furukawa, T.Watanabe, Y.Otsuji and M.Hamada, Bull.Chem.Soc.Jpn.
      38, 1247 (1965).

O2   S.Oae and C.C.Price, J.Am.Chem.Soc. 80, 3425 (1958).

O3   N.E Ockerbloom and A.E.Martell, J.Am.Chem.Soc. 78, 267 (1956).

O4   N.Ockerbloom and A.E.Martell, J.Am.Chem.Soc. 80, 2351 (1958).

O5   J.Ofengand and H.Schaefer, Biochemistry 4, 2832 (1965).

O6   W.F.O'Hara and L.G.Hepler, J.Phys.Chem. 65, 2107 (1961).

O7   W.F.O'Hara, T.Hu and L.G.Hepler, J.Phys.Chem. 67, 1933 (1963).

O8   G.Ojelund and I.Wadso, Acta Chem.Scand. 21, 1408 (1967).

O9   Y.Oka and H.Harada, Nippon Kagaku Zasshi 86, 1158 (1965).

O10   Y.Oka and M.Hirai, Nippon Kagaku Zasshi 89, 589 (1968).

O11   Y.Oka, N.Nakazawa and H.Hirada, Nippon Kagaku Zasshi 86, 1162 (1965).

O12   Y.Oka, M.Umehara and T.Nozoe, Nippon Kagaku Zasshi 83, 703 (1962).

O13   Y.Oka, M.Umehara and T.Nozoe, Nippon Kagaku Zasshi 83, 1197 (1962).

O14   Y.Oka and K.Yamamoto, Nippon Kagaku Zasshi 85, 779 (1964).

O15   Y.Oka and M.Yanai, Nippon Kagaku Zasshi 86, 929 (1965).

O16   Y.Oka, M.Yanai and C.Suzuki, Nippon Kagaku Zasshi 85, 873 (1964).

O17   N.Okaku, K.Toyoda, Y.Moriguchi and K.Ueno, Bull.Chem.Soc.Jpn. 40, 2326 (1967).

O18   D.C.Olson and D.W.Magerum, J.Am.Chem.Soc. 82, 5602 (1960).

O19   G.K.S.Ooi and R.J.Magee, J.Inorg.Nucl.Chem. 32, 3115 (1970).

O20   V.V.Orda, L.M.Yagupolskii, V.F.Bystrov and A.U.Stepanyants, J.Gen.Chem.USSR
      (Engl.Transl.) 35, 1631 (1965).

O21   G.Ostacoli, E.Campi, N.Cabrario and G.Saini, Gazz.Chim.Ital.91, 349 (1961).

O22   G.Ostacoli, E.Campi and M.C.Gennaro, Gazz.Chim.Ital. 96, 741 (1966).

O23   G.Ostacoli, A.Vanni and E.Roletto, Gazz.Chim.Ital. 100, 350 (1970).

O24   R.Osterberg, Acta Chem.Scand. 16, 2434 (1962).

O25   R.Osterberg, Acta Chem.Scand. 19, 1445 (1965).

O26   R.Osterberg and B.Sjoberg, J.Biol.Chem. 243, 3038 (1968).

027 W.J.O'Sullivan and D.D.Perrin, Biochem.Biophys.Acta 52, 612 (1961).

028 Y.Otsuji, T.Kimura, Y.Sugimoto, E.Imoto, Y.Omori and T.Okawara, Nippon Kagaku Zasshi 80, 1021 (1959).

029 E.G.Owens and J.H.Yoe, Talanta 8, 505 (1961).

P

P1   M.Paabo, R.G.Bates and R.A.Robinson, J.Res.Nat.Bur.Stand., Sect.A 67A, 573 (1963).

P2   M.Paabo, R.G.Bates and R.A.Robinson, J.Phys.Chem. 70, 540 (1966).

P3   M.Paabo, R.G.Bates and R.A.Robinson, J.Phys.Chem. 70, 2073 (1966).

P4   V.V.Pal'chevskii, M.S.Zakharevskii and E.A.Malinina, Vestnik Leningrad Univ., 15. No.16, Ser Fiz i Khim. No.3 95 (1960); CA 55, 1170b.

P5   R.Palmaeus and P.Kierkegaard, Acta Chem.Scand. 18, 2226 (1964).

P6   M.R.Paris and C.Gregoire, Anal.Chim.Acta 42, 439 (1968).

P7   M.V.Park, Nature (London) 197, 283 (1963).

P8   M.V.Park, J.Chem.Soc.(A) 1966, 816.

P9   V.H.Parker, Biochem.J. 97, 658 (1965).

P10 M.Z.Partashnikova and I.G.Shafran, J.Anal.Chem.USSR (Engl.Transl.) 20, 258 (1965).

P11 I.Pascal and D.S.Tarbell, J.Am.Chem.Soc. 79, 6015 (1957).

P12 D.J.Pasto and R.Kent, J.Org.Chem. 30, 2684 (1965).

P13 D.J.Pasto, D.McMillan and T.Murphy, J.Org.Chem. 30, 2688 (1965).

P14 E.Patton and R.West, J.Phys.Chem. 74, 2512 (1970).

P15 E.Patton and R.West, J.Am.Chem.Soc. 95, 8703 (1973).

P16 P.J.Pearce and R.J.J.Simkins, Can.J.Chem. 46, 241 (1968).

P17 K.H.Pearson and K.H.Gayer, Inorg.Chem. 3, 476 (1964).

P18 R.G.Pearson and R.L.Dillon, J.Am.Chem.Soc. 75, 2439 (1953).

P19 K.J.Pedersen, Acta Chem.Scand. 9, 1634 (1955).

P20 K.J.Pedersen, Acta Chem.Scand. 12, 919 (1958).

P21 E.Pelizzetti and C.Verdi, J.Chem.Soc., Perkin Trans.2 1973, 808.

P22 D.Peltier, C.R.Acad.Sci.,Ser.C 241, 57 (1955).

P23 D.Peltier, C.R.Acad.Sci., Ser.C 241, 1467 (1955).

P24 D.Peltier, Bull.Soc.Chim.Fr. 1958, 994. Values also quoted from D.Peltier, Thesis, Rennes, (1956); C.R.Acad.Sci.,Ser.C 243, 2086 (1956); 244, 2811 (1957); 245, 436 (1957).

P25 D.Peltier and M.Conti, C.R.Acad.Sci.Ser.C 244, 2811 (1957).

P26   D.Peltier and M.Kerdavid, C.R.Acad.Sci.,Ser.C 243, 2086 (1956).

P27   D.Peltier and A.Pichevin, C.R.Acad.Sci.,Ser.C 245, 436 (1957).

P28   D.J.Perkins, Biochem.J. 51, 487 (1952).

P29   G.E.Perlmann, J.Biol.Chem. 239, 3762 (1964).

P30   D.D.Perrin, Nature (London) 182, 741 (1958).

P31   D.D.Perrin, Nature (London) 191, 253 (1961).

P32   D.D.Perrin and I.G.Sayce, J.Chem.Soc.(A) 1967, 82.

P33   D.D.Perrin and I.G.Sayce, J.Chem.Soc.(A) 1968, 53.

P34   D.D.Perrin, I.G.Sayce and V.S.Sharma, J.Chem.Soc.(A) 1967, 1755.

P35   D.D.Perrin and V.S.Sharma, J.Chem.Soc.(A) 1967, 724.

P36   V.M.Peshkova and P.Ang, Russ.J.Inorg.Chem.(Engl.Transl.) 7, 765 (1962).

P37   T.V.Petrov, N.Khakimkhodzhaev and S.B.Savvin, J.Anal.Chem.USSR (Engl.Transl.)
        25, 188 (1970).

P38   L.D.Pettit and H.M.Irving, J.Chem.Soc. 1964, 5336.

P39   L.D.Pettit, A.Royston, C.Sherrington and R.J.Whewell, J.Chem.Soc.(B) 1968, 588.

P40   L.D.Pettit and C.Sherrington, J.Chem.Soc.(A) 1968, 3078.

P41   W.Pfleiderer, Chem.Ber. 90, 2582 (1957).

P42   W.Pfleiderer, Chem.Ber. 90, 2588 (1957).

P43   W.Pfleiderer, Chem.Ber. 90, 2604 (1957).

P44   W.Pfleiderer, Chem.Ber. 90, 2617 (1957).

P45   W.Pfleiderer, Chem.Ber. 90, 2624 (1957).

P46   W.Pfleiderer, Chem.Ber. 90, 2631 (1957).

P47   W.Pfleiderer, Chem.Ber. 91, 1671 (1958).

P48   W.Pfleiderer, Chem.Ber. 92, 3190 (1959).

P49   W.Pfleiderer, Justus Liebigs Ann.Chem. 647, 161 (1961).

P50   W.Pfleiderer and H.Ferch, Justus Liebigs Ann.Chem. 615, 48 (1958).

P51   W.Pfleiderer and H.Ferch, Justus Liebigs Ann.Chem. 615, 52 (1958).

P52   W.Pfleiderer, E.Liedek, R.Lohrmann and M.Rukwied, Chem.Ber. 93, 2015 (1960).

P53   W.Pfleiderer and G.Nubel, Justus Liebigs Ann.Chem.631, 168 (1960).

P54   W.Pfleiderer and G.Nubel, Chem.Ber. 93, 1406 (1960).

P55   W.Pfleiderer and G.Nubel, Justus Liebigs Ann.Chem. 647, 155 (1961).

P56   W.Pfleiderer and M.Rukwied, Chem.Ber. 94, 118 (1961).

P57   W.Pfleiderer and M Rukwied, Chem.Ber. 94, 1 (1961).

P58   W.Pfleiderer and E.C.Taylor, J.Am.Chem.Soc. 82, 3765 (1960).

P59   W.Pfleiderer and K.H.Schundehutte, Justus Liebigs Ann.Chem. 615, 42 (1958).

P60   J.N.Phillips, J.Chem.Soc. 1958, 4271.

P61   R.C.Phillips, P.Eisenberg, P.George and R.J.Rutman, J.Biol.Chem. 240, 4393 (1965).

P62   R.C.Phillips, P.George and R.J.Rutman, Biochemistry 2, 501 (1963).

P63   P.Pichet and R.L.Benoit, Inorg.Chem. 6, 1505 (1967).

P64   A.T.Pilipenko, A.P.Kostyshina and L.N.Kudritskaya, Ukr.Khim.Zh.28, 109 (1962);
      CA 57, 4095h.

P65   A.T.Pilipenko and N.N.Maslei, Ukr.Khim.Zh. 33, 831 (1967); CA 68, 6920a.

P66   A.T.Pilipenko and O.P.Ryabushko, Ukr.Khim.Zh. 28, 955 (1962); CA 59, 2141b

P67   A.T.Pilipenko and O.P.Ryabushko, Ukr.Khim.Zh. 32, 622 (1966); CA 65, 11412g.

P68   P.Pino and R.Piacenti, Rend.ist.lombardo sci., Pt 1, Classe sci.mat.e nat. 87,
      183 (1954); CA 50, 5527h.

P69   A.Piskala and J.Gut, Collect.Czech.Chem.Commun. 28, 2376 (1963).

P70   J.Pitha, P.Fiedler and J.Gut, Collect.Czech.Chem.Commun. 31, 1864 (1966).

P71   I.V.Podgornaya, A.A.Ivakin and K.N.Klyachina, J.Gen.Chem.USSR (Engl.Transl.)
      36, 2044 (1966).

P72   G.Popa, C.Luca and E.Iosif, Omagiu Raluca Ripan 1966, 457; CA 67, 94493w.

P73   L.J.Porter and D.D.Perrin, Aust.J.Chem. 22, 267 (1969).

P74   C.Postmus, I.A.Kaye, C.A.Craig and R.S.Matthews, J.Org.Chem. 29, 2693 (1964).

P75   L.Van Poucke and M.Herman, Anal.Chim.Acta 30, 569 (1964).

P76   J.E.Powell, A.R.Chughtai and J.W.Ingemanson, Inorg.Chem. 8, 2216 (1969).

P77   J.E.Powell and J.L.Farrell, J.Inorg.Nucl.Chem. 30, 2135 (1968).

P78   J.E.Powell, J.L.Farrell, W.F.S.Neillie and R.Russell, J.Inorg.Nucl.Chem.
      30, 2223 (1968).

P79   J.E.Powell, R.S.Kolat and G.S.Paul, Inorg.Chem. 3, 518 (1964).

P80   J.E.Powell and W.F.S.Neillie, J.Inorg.Nucl.Chem. 29, 2371 (1967).

P81   J.E.Powell and D.L.G.Rowlands, Inorg.Chem. 5, 819 (1966).

P82   J.E.Powell and Y.Suzuki, Inorg.Chem. 3, 690 (1964).

P83   I.G.Pozharliev and K.Zakharieva, Izv.Otd.Khim.Nauki,Bulg.Akad.Nauk. 2, 341 (1969);
      CA 72, 78121w.

P84   P.W.Preisler, L.Berger and E.S.Hill, J.Am.Chem.Soc. 69, 326 (1947).

P85   O.E.Presnyakova and R.S.Prishchepo, J.Gen.Chem.USSR (Engl.Transl.) 39, 2340 (1969).

P86   C.Preti and D.De Filippo, Gazz.Chim.Ital. 97, 819 (1967); CA 67, 76729y.

P87   J.E.Prue and A.J.Read, Trans.Faraday Soc. 62, 1271 (1966).

P88   N.Purdie and M.B.Tomson, J.Am.Chem.Soc. 95, 48 (1973).

P89   D.N.Purohit and N.C.Sogani, J.Chem.Soc. 1964, 2820.

P90   D.N.Purohit and N.C.Sogani, J.Indian Chem.Soc. 41, 160 (1964).

P91   D.N.Purohit and N.C.Sogani, Bull.Acad.Pol.Sci.,Ser.Sci.Chim., 15, 423 (1967);
      CA 69, 46467.

                                      Q

Q1   L.D.Quin and M.R.Dysart, J.Org.Chem. 27, 1012 (1962).

Q2   R.P.Quintana and W.R.Smithfield, J.Med.Chem. 10, 1178 (1967).

                                      R

R1   J.C.Rabinowitz and W.E.Pricer, J.Biol.Chem. 218, 189 (1956).

R2   K.S.Rajan and A.E.Martell, J.Inorg.Nucl.Chem. 26, 789 (1964).

R3   K.S.Rajan and A.E.Martell, J.Inorg.Nucl.Chem. 26, 1927 (1964).

R4   K.S.Rajan and A.E.Martell, Inorg.Chem. 4, 462 (1965).

R5   K.S.Rajan and A.E.Martell, J.Inorg.Nucl.Chem. 29, 523 (1967).

R6   K.S.Rajan, I.Murase and A.E.Martell, J.Am.Chem.Soc. 91, 4408 (1969).

R7   E.V.Raju and H.B.Mathur, J.Inorg.Nucl.Chem. 30, 2181 (1968).

R8   E.V.Raju and H.B.Mathur, J.Inorg.Nucl.Chem. 31, 425 (1969).

R9   N.A.Ramaiah and R.K.Chaturvedi, Proc.Indian Acad.Sci. 51A, 177 (1960).

R10   N.A.Ramaiah and S.L.Gupta, J.Indian Chem.Soc. 33, 535 (1956).

R11   S.Ramamoorthy and M.Santappa, Bull.Chem.Soc.Jpn. 41, 1330 (1968).

R12   M.M.Ramel and M.R.Paris, Bull.Soc.Chim.Fr. 1967, 1359.

R13   R.W.Ramette and T.R.Blackburn, J.Phys.Chem. 61, 378 (1957).

R14   J.M.Rao and U.V.Seshaiah, Bull.Chem.Soc.Jpn. 39, 2668 (1966).

R15   M.Rapoport, C.K.Hancock and E.A.Meyers, J.Am.Chem.Soc. 83, 3489 (1961).

R16   H.Rapoport and C.D.Willson, J.Org.Chem. 26, 1102 (1961).

R17   B.Rehak and J.Korbl, Collect.Czeck.Chem.Commun. 25, 797 (1960).

R18   U.Reichman, F.Bergmann and D.Lichtenberg, J.Chem.Soc.,PerkinTrans.1 1973, 2647.

R19   J.C.Reid and M.Calvin, J.Am.Chem.Soc. 72, 2948 (1950).

R20   A.L.Remizov, J.Gen.Chem.USSR (Engl.Transl.) 28, 3364 (1958).

R21   C.Ressler and H.Ratzkin, J.Org.Chem. 26, 3356 (1961).

R22   R.Riccardi and P.Franzosini, Ann.Chim.(Rome) 47, 977 (1957); CA 52, 842d

R23   C.F.Richard, R.L.Gustafson and A.E.Martell, J.Am.Chem.Soc. 81, 1033 (1959).

R24   J.F.Riordan, M.Sokolovsky and B.L.Vallee, Biochemistry 6, 358 (1967).

R25   B.R.Rios, A.P.Uruena and A.M.Perez, Anales Real Soc.Espan.Fis.Quim.(Madrid) Ser.B
        61, 717 (1965); CA 64, 8991f.

R26   C.W.Roberts, E.T.McBee and C.E.Hathaway, J.Org.Chem. 21, 1369 (1956).

R27   R.K.Robins, L.B.Townsend, F.Cassidy, J.F.Gerster, A.F.Lewis and R.L.Miller,
        J.Heterocycl.Chem. 3, 110 (1966).

R28   R.A.Robinson, J.Res.Nat.Bur.Stand.,Sect.A 68A, 159 (1964).

R29   R.A.Robinson, J.Res.Nat.Bur.Stand., Sect.A 71A, 213 (1967).

R30   R.A.Robinson, J.Res.Nat.Bur.Stand.,Sect.A 71A, 385 (1967).

R31   R.A.Robinson, J.Chem.Eng.Data 14, 247 (1969).

R32   R.A.Robinson and K.P.Ang, J.Chem.Soc. 1959, 2314.

R33   R.A.Robinson and A.I.Biggs, Aust.J.Chem. 10, 128 (1957).

R34   R.A.Robinson, M.M.Davis, M.Paabo and V.E.Bower, J.Res.Nat.Bur.Stand.,Sect.A
        64A, 347 (1960).

R35   R.A.Robinson and A.K.Kiang, Trans.Faraday Soc. 52, 327 (1956).

R36   R.A.Robinson, M.Paabo and R.G.Bates, J.Res.Nat.Bur.Stand.,Sect.A 73A, 299 (1969).

R37   R.A.Robinson and A.Peiperl, J.Phys.Chem. 67, 1723 (1963).

R38   R.A.Robinson and A.Peiperl, J.Phys.Chem., 67, 2860 (1963).

R39   C.H.Rochester, J.Chem.Soc. 1965, 4603.

R40   S.Rogers and J.B.Neilands, Biochemistry 2, 6 (1963).

R41   G.A.Ropp, J.Am.Chem.Soc. 82, 4252 (1960).

R42   D.B.Rorabacher, T.S.Turan, J.A.Defever and W.G.Nickels, Inorg.Chem. 8, 1498 (1969).

R43   A.N.Roseira, O.A.Stamm, A.Zenhausern and H.Zollinger, Chimia 13, 366 (1959).

R44   G.Ross, D.A.Aikens and C.N.Reilley, Anal.Chem. 34, 1766 (1962).

R45   A.Roulet, J.Feux and T.Vu Duc, Helv.Chim.Acta 52, 2154 (1969).

R46   R.Roulet, J.Feuz and T.Vu Duc, Helv.Chim.Acta 53, 1876 (1970).

R47   R.Roulet and T.Vu Duc, Helv.Chim.Acta 53, 1873 (1970).

R48   R.Rowland and C.E.Meloan, Anal.Chem. 36, 1997 (1964).

R49   R.N.Roy, R.A.Robinson and R.G.Bates, J.Am.Chem.Soc. 95, 8231 (1973).

R50   N.P.Rudenko, I.N.Kremenskaya and V.N.Avilina, Russ.J.Inorg.Chem.(Engl.Transl.)
        10, 627 (1965).

R51   P.Rumpf, Colloques intern.centre natl.recherche sci. l'hydroxycarbonylation,
        Paris 56, 14 (1954)(publ.1955); CA 50, 11776i.

R52   P.Rumpf and R.La Riviere, C.R.Acad.Sci.,Ser.C 244, 902 (1957).

R53   P.Rumpf and R.Reynaud, C.R.Acad.Sci.,Ser.C 250, 1501 (1960).

R54 P.Rumpf and J.Sadet, Bull.Soc.Chim.Fr. 1958, 447.

R55 P.Rumpf and J.Sadet, Bull.Soc.Chim.Fr. 1958, 450.

R56 M.T.Ryan and K.J.Berner, Spectrochim.Acta, Part A 25A, 1155 (1969).

R57 O.Ryba, J.Cifka, M.Malat and V.Suk, Collect.Czech.Chem.Commun. 21, 349 (1956).

R58 P.Rys and H.Zollinger, Helv.Chim.Acta 49, 1406 (1966).

S

S1   H.Saarinen, M.Melnik and O.Makitie, Acta Chem.Scand. 23, 2542 (1969).

S2   G.Saini, G.Ostacolli, E.Campi and N.Cibrario, Gazz.Chim.Ital. 91, 242 (1961).

S3   G.Saini, G.Ostacolli, E.Campi and N.Cibrario, Gazz.Chim.Ital. 91, 904 (1961).

S4   T.St.Pierre and W.P.Jencks, J.Am.Chem.Soc. 90, 3817 (1968).

S5   M.Sakaguchi, A.Mizote, H.Miyata and K.Toei, Bull.Chem.Soc.Jpn. 36, 885 (1963).

S6   V.D.Salikhov, Y.M.Dedkov and M.Z.Yampolskii, J.Anal.Chem.USSR (Engl.Transl.)
     23, 449 (1968).

S7   F.Salmon-Legagneur and Y.Olivier, Bull.Soc.Chem.Fr. 1965, 1392.

S8   A.Sandell, Acta Chem.Scand. 15, 190 (1961).

S9   J.Sandstrom and I.Wennerbeck, Acta Chem.Scand. 20, 57 (1966).

S10 M.Sanesi, Ann.Chim.(Rome), 47, 203 (1957); CA 51, 11013.

S11 M.Sanesi, Ann.Chim.(Rome), 50, 997 (1960); CA 55, 12398.

S12 R.P.Saper, Glasnik Hem.Drustva, Beograd 25-26, 277 (1960-1961); CA 59, 1196b.

S13 R.P.Saper, Glasnik Hem.Drustva, Beograd. 25-26, 287 (1960-1961); CA 59, 1196c.

S14 Y.Sato, Nippon Kagaku Zusshi 78, 921 (1957); CA 54, 4598.

S15 V.M.Savostina, E.K.Astakhova and V.M.Peshkova, Vestn.Mosk.Univ.,Ser.II : Khim. 18,
     43 (1963); CA 62, 15397h.

S16 S.B.Savvin, T.V.Petrova and E.L.Kuzin, Bull.Acad.Sci.USSR, Div.Chem.Sci.(Engl.Transl.)
     1969, 247.

S17 R.S.Saxena and P.Singh, J.Indian Chem.Soc. 47, 1076 (1970).

S18 J.M.Sayer and W.P.Jencks, J.Am.Chem.Soc. 91, 6353 (1969).

S19 J.M.Sayer and W.P.Jencks, J.Am.Chem.Soc. 95, 5637 (1973).

S20 W.P.Schaefer, Inorg.Chem. 4, 642 (1965).

S21 H.Schmid and H.Sofer, Monatsh.Chem. 98, 469 (1967).

S22 H.Schmid, H.Sofer and H.Pleschberger, Monatsh.Chem. 98, 353 (1967).

S23 H.Schmid and H.Sofer, Monatsh.Chem. 97, 1742 (1966).

S24   P.W.Schneider, H.Brintzinger and H.Erlenmeyer, Helv.Chim.Acta 47, 992 (1964).

S25   G.R.Schonbaum and M.L.Bender, J.Am.Chem.Soc. 82, 1900 (1960).

S26   K.Schroder, Talanta 15, 1035 (1968).

S27   J.Schubert, G.Anderegg and G.Schwarzenbach, Helv.Chim.Acta 43, 410 (1960).

S28   W.M.Schubert, R.E.Zahler and J.Robins, J.Am.Chem.Soc. 77, 2293 (1955).

S29   S.Schulman and Q.Fernando, J.Phys.Chem. 71, 2668 (1967).

S30   H.Schurmans, H.Thun and F.Verbeek, J.Inorg.Nucl.Chem. 29, 1759 (1967).

S31   H.Schurmans, H.Thun and F.Verbeek, J.Electroanal.Chem. Interfacial Electrochem.,
        26, 299 (1970).

S32   L.M.Schwartz and L.O.Howard, J.Phys.Chem. 74, 4374 (1970).

S33   M.A.Schwartz, J.Pharm.Sci. 54, 1308 (1965).

S34   G.Schwarzenbach and H.Ackermann, Helv.Chim.Acta 30, 1798 (1947).

S35   G.Schwarzenbach and H.Ackermann, Helv.Chim.Acta 32, 1682 (1949).

S36   G.Schwarzenbach, H.Ackermann and P.Ruckstuhl, Helv.Chim.Acta 32, 1175 (1949).

S37   G.Schwarzenbach and G.Anderegg, Helv.Chim.Acta 40, 1229 (1957).

S38   G.Schwarzenbach, G.Anderegg and R.Sallmann, Helv.Chim.Acta 35, 1785 (1952).

S39   G.Schwarzenbach, G.Anderegg and R.Sallmann, Helv.Chim.Acta 35, 1794 (1952).

S40   G.Schwarzenbach, G.Anderegg, W.Schneider and H.Senn, Helv.Chim.Acta 38, 1147 (1955).

S41   G.Schwarzenbach and W.Biedermann, Helv.Chim.Acta 31, 678 (1948).

S42   G.Schwarzenbach and E.Felder, Helv.Chim.Acta 27, 1701 (1944).

S43   G.Schwarzenbach and E.Freitag, Helv.Chim.Acta 34, 1492 (1951).

S44   G.Schwarzenbach and H.Gysling, Helv.Chim.Acta 32, 1314 (1949).

S45   G.Schwarzenbach, E.Kampitsch and R.Steiner, Helv.Chim.Acta 28, 1133 (1945).

S46   G.Schwarzenbach, E.Kampitsch and R.Steiner, Helv.Chim.Acta 29, 364 (1946).

S47   G.Schwarzenbach and K.Lutz, Helv.Chim.Acta 23, 1162 (1940).

S48   G.Schwarzenbach, P.Ruckstuhl and J.Zurc, Helv.Chim.Acta 34, 455 (1951).

S49   G.Schwarzenbach and K.Schwarzenbach, Helv.Chim.Acta 46, 1390 (1963).

S50   G.Schwarzenbach, H.Senn and G.Anderegg, Helv.Chim.Acta 40, 1886 (1957).

S51   G.Schwarzenbach and A.Willi, Helv.Chim.Acta 34, 528 (1951).

S52   G.Schwarzenbach and C.Wittwer, Helv.Chim.Acta 30, 663 (1947).

S53   G.Schwarzenbach and J.Zuro, Monatsh.Chem. 81, 202 (1950).

S54   G.W.Schwert and Y.Takenaka, Biochim.Biophys.Acta 16, 570 (1955).

S55   M.Scrocco and R.Nicolaus, Atti accad.naz.Lincei, Rend.,Classe sci.fis.,mat.e nat.
        22, 311 (1957); CA 52, 50i.

S56   J.Sendroy and F.L.Rodkey, Clin.Chem. 7, 646 (1961).

S57   E.Seoane and J.Roque Buscato, Quim.Ind.Madrid 16, 3 (1970); CA 74, 116680d.

S58   E.P.Serjeant, Aust.J.Chem. 22, 1189 (1969).

S59   T.Seshadri, Indian J.Chem. 8, 282 (1970).

S60   B.Sethuram and E.V.Sundaram, Indian J.Chem. 7,415 (1969).

S61   Y.S.Shabarov, V.K.Potapov and R.Y.Levina, J.Gen.Chem.USSR (Engl.Transl.) 34,
        2865 (1964).

S62   Y.S.Shaborov, T.P.Surikova and R.Y.Levina, J.Org.Chem.USSR (Engl.Transl.) 4,
        1131 (1968).

S63   E.Shaw, J.Am.Chem.Soc. 71, 67 (1949).

S64   R.Shiman and J.B.Neilands, Biochemistry 4, 2233 (1965).

S65   V.I.Shlenskaya, T.I.Tikhvinskaya, A.A.Biryukov and I.P.Alimarin,
        Bull.Acad.Sci.USSR, Div.Chem.Sci.(Engl.Transl.) 1967, 2063.

S66   G.Shtacher, J.Inorg.Nucl.Chem. 28, 845 (1966).

S67   J.P.Shukla and S.G.Tandon, Aust.J.Chem. 24, 2701 (1971).

S68   J.Sicher, F.Sipos and J.Jonas, Collect.Czech.Chem.Commun. 26, 262 (1961).

S69   S.Siegel and J.M.Komarmy, J.Am.Chem.Soc. 82, 2547 (1960).

S70   H.Sigel, Helv.Chim.Acta 48, 1513 (1965).

S71   H.Sigel, Eur.J.Biochem. 3, 530 (1968).

S72   H.Sigel and H.Brintzinger, Helv.Chim.Acta 47, 1701 (1964).

S73   E.F.Silversmith and J.D.Roberts, J.Am.Chem.Soc. 80, 4083 (1958).

S74   G.Simchen, Chem.Ber. 103, 398 (1970).

S75   H.Simpkins and E.G.Richards, Biochemistry 6, 2513 (1967).

S76   H.N.Simpson, C.K.Hancock and E.A.Meyers, J.Org.Chem. 30, 2678 (1965).

S77   M.E.Sitzmann, H.G.Adolph and M.J.Kamlet, J.Am.Chem.Soc. 90, 2815 (1968).

S78   N.T.Sizonenko and Y.A.Zolotov, J.Anal.Chem.USSR (Engl.Transl.) 24, 1053 (1969).

S79   V.I.Slovetskii and I.S.Ivanov, Bull.Acad.Sci.USSR, Div.Chem.Sci.(Engl.Transl.)
        1968, 1641.

S80   V.I.Slovetskii, L.V.Okhlobystina, A.A.Fainzil'berg, A.I.Ivanov, L.I.Biryukova
        and S.S.Novikov, Bull.Acad.Sci.USSR, Div.Chem.Sci.(Engl.Transl.) 1965, 2032.

S81   V.I.Slovetskii, S.A.Shevelev, V.I.Erashko, L.I.Biryukova, A.A.Fainzilberg and
        S.S.Novikov, Bull.Acad.Sci.USSR, Div.Chem.Sci.(Engl.Transl.) 1966, 621.

S82   V.I.Slovetskii, S.A.Shevelev, A.A.Fainzil'berg and S.S.Novikov, Zh.Vses.Khim.Obshch.
        im D.I.Mendeleeva 6, 599 (1961); CA 56, 6712.

S83   V.I.Slovetskii, S.A.Shevelev, A.A.Fainzil'berg and S.S.Novikov, Zh.Vses.Khim.
        Obshch. im.D.I.Mendeleeva 6, 707 (1961); CA 56, 14986e.

S84　L.A.A.E.Sluyterman, Biochim.Biophys.Acta 85, 305 (1964).

S85　A.T.Slyusarev and A.L.Gershuns, Ukr.Khim.Zh. 28, 309 (1962); CA 57, 9282h.

S86　N.G.Smaglyuk, R.K.Dzhiyanbaeva and S.T.Talipov, Uzb.Khim.Zh. 14, 24 (1970);
　　　CA 73, 39313m.

S87　A.A.Smirnova, V.V.Perekalin and V.A.Shcherbakov, J.Org.Chem.USSR (Engl.Transl.)
　　　4, 2166 (1968).

S88　B.S.Smolyakov, Izv.Sib.Otd.Akad.Nauk.SSSR, Ser.Khim.Nauk 1966, 22; CA 66, 32397p.

S89　B.S.Smolyakov and M.P.Primanchuk, Russ.J.Phys.Chem.(Engl.Transl.) 40, 463 (1966).

S90　B.S.Smolyakov and M.P.Primanchuk, Russ.J.Phys.Chem.(Engl.Transl.) 40, 989 (1966).

S91　E.J.Smutny, M.C.Caserio and J.D.Roberts, J.Am.Chem.Soc. 82, 1793 (1960).

S92　A.Solladie-Cavallo, C.R.Acad.Sci.,Ser.C 263, 93 (1966).

S93　A.Solladie-Cavallo and P.Vieles, J.Chim.Phys.Phys.-Chim.Biol. 64, 1593 (1967).

S94　L.Sommer and V.Kuban, Collect.Czech.Chem.Commun. 32, 4355 (1967).

S95　L.Sommer, T.Sepel and V.M.Ivanov, Talanta 15, 949 (1968).

S96　K.P.Soni and A.M.Trivedi, J.Indian Chem.Soc. 37, 349 (1960).

S97　P.Souchay, N.Israily and P.Gouzerh, Bull.Soc.Chim.Fr. 1966, 3917.

S98　I.D.Spenser and A.D.Notation, Can.J.Chem. 40, 1374 (1962).

S99　D.Spinelli and G.Guanti, Ric.Sci.38, 1048 (1968); CA 71 25195w.

S100　E.Spinner and G.B.Yeoh, Aust.J.Chem. 24, 2557 (1971).

S101　P.K.Spitsyn and V.S.Shvarev, J.Anal.Chem.USSR (Engl.Transl.) 25, 1297 (1970).

S102　V.Springer, R.Karlicek and J.Majer, Collect.Czech. Chem.Commun. 32, 774 (1967).

S103　K.C.Srivastava, Bull.Chem.Soc.Jpn. 39, 1591 (1966).

S104　K.C.Srivastava, J.Prakt.Chem. 35, 118 (1967).

S105　W.R.Stagg and J.E.Powell, Inorg.Chem., 3, 242 (1964).

S106　O.A.Stamm, A.Zenhausern and H.Zollinger, Chimia 19, 224 (1965).

S107　V.I.Stanko and A.I.Klimova, J.Gen.Chem.USSR (Engl.Transl.) 36, 165 (1966).

S108　G.R.Stark, Biochemistry 4, 588 (1965).

S109　G.R.Stark, Biochemistry, 4, 1030 (1965).

S110　G.R.Stark, Biochemistry 4, 2363 (1965).

S111　H.Staude and M.Teupel, Z.Elektrochem. 61, 181 (1957).

S112　H.Stetter and K.Dieminger, Chem.Ber. 92, 2658 (1959).

S113　D.B.Stevancevic, Glasnik Hem.Drustva, Beograd. 27, 367 (1962); CA 61, 43d.

S114　R.Stewart and M.M.Mocek, Can.J.Chem. 41, 1160 (1963).

S115　R.Stewart and L.G.Walker, Can.J.Chem. 35, 1561 (1957).

S116 R.Stewart and R.Van der Linden, Can.J.Chem. 38, 399 (1960).

S117 M.A.Stolberg and W.A.Mosher, J.Am.Chem.Soc. 79, 2618 (1957).

S118 Y.Stradyn, E.Ermane, T.Dumpis, Y.Linaberg and G.Vanag, J.Org.Chem.USSR (Engl.Transl.) 1, 380 (1965).

S119 E.Strazdins and E.H.Sheers, Tappi 41, 658 (1958); CA 53, 7589d.

S120 A.Streitwieser and H.S.Klein, J.Am.Chem.Soc. 85, 2759 (1963).

S121 R.D.Strickland and M.M.Anderson, Anal.Chem. 38, 980 (1966).

S122 I.Stronski, A.Zielinski, A.Samotus, Z.Stasick and B.Budesinky, Fresenius' Z.Anal.Chem. 222, 14 (1966).

S123 A.Sturis and J.Bankovskis, Latv.PSR Zinat.Akad.Vestis.Kim.Ser. 1968, 250; CA 69, 90436y.

S124 L.Sucha, Z.Urner and M.Suchanek, Collect.Czech.Chem.Commun. 35, 3651 (1970).

S125 V.Suk, Collect.Czech.Chem.Commun. 31, 3127 (1966).

S126 V.Suk and V.Miketukova, Collect.Czech.Chem.Commun. 24, 3629 (1959).

S127 P.J.Sun, Q.Fernando and H.Freiser, Anal.Chem. 36, 2485 (1964).

S128 J.Sunkel and H.Staude, Ber.Bunsenges.Phys.Chem. 72, 567 (1968).

S129 R.Sureau, Bull.Soc.Chim.Fr. 1956, 622.

S130 K.Suzuki, T.Hattori and K.Yamasaki, J.Inorg.Nucl.Chem. 30, 161 (1968).

S131 K.Suzuki, C.Karaki, S.Mori and K.Yamasaki, J.Inorg.Nucl.Chem. 30, 167 (1968).

S132 K.Suzuki, I.Nakano and K.Yamasaki, J.Inorg.Nucl.Chem. 30, 545 (1968).

S133 K.Suzuki and K.Yamasaki, J.Inorg.Nucl.Chem. 24, 1093 (1962).

S134 K.Suzuki and K.Yamasaki, J.Inorg.Nucl.Chem. 28, 473 (1966).

S135 K.Suzuki, M.Yasuda and K.Yamasaki, J.Phys.Chem. 61, 229 (1957).

S136 F.Sweet and J.D.Fissekis, J.Am.Chem.Soc. 95, 8741 (1973).

S137 R.Swidler, R.E.Plapinger and G.M.Steinberg, J.Am.Chem.Soc. 81, 3271 (1959).

S138 M.Swierkowski and D.Shugar, Acta Biochim.Pol. 16, 263 (1969).

S139 A.Y.Sychev and N.S.Mitsul, Russ.J.Inorg.Chem.(Engl.Transl.) 12, 1120 (1967).

S140 E.Szabo and J.Szabon, Acta Chim.Acad.Sci.Hung. 48, 299 (1966); CA 65, 11412a.

S141 W.Szer and D.Shugar, Acta Biochim.Pol. 13, 177 (1966).

S142 W.Sztark, Diss.Pharm.Pharmacol. 19, 429 (1967); CA 68, 95569h.

T

T1   S.Tabak, I.I.Grandberg and A.N.Kost, Tetrahedron 22, 2703 (1966).

T1a   T.Takemoto and T.Nakajima, J.Pharm.Soc.Jpn. 84, 1230 (1964).

T2   T.Takemoto, T.Nakajima and T.Yokobe, J.Pharm.Soc.Jpn. 84, 1232 (1964).

T3   D.E.Tallman and D.L.Leussing, J.Am.Chem.Soc. 91, 6253 (1969).

T4   T.Tanabe, K.Kimura and S.Takamoto, Nippon Zaguka Zasshi 90, 598 (1969).

T5   N.Tanaka, I.T.Oiwa, T.Kurosawa and T.Nozoe, Bull.Chem.Soc.Jpn. 32, 92 (1959).

T6   S.G.Tandon, J.Phys.Chem. 65, 1644 (1961).

T7   C.Tanford, J.D.Hauestein and D.G.Rands, J.Am.Chem.Soc. 77, 6409 (1955).

T8   C.Tanford and G.L.Roberts, J.Am.Chem.Soc. 74, 2509 (1952).

T9   A.G.Tarasenko and V.M.Fedoseev, Vestn.Mosk.Univ.,Ser.II : Khim. 21, 75 (1966);
      CA 66, 32396n.

T10   V.M.Tarayan and A.N.Pogosyan, Arm.Khim.Zh. 22, 569 (1969); CA 71 117057f.

T11   S.S.Tate, A.K.Grzybowski and S.P.Datta, J.Chem.Soc. 1964, 1372.

T12   S.S.Tate, A.K.Grzybowski and S.P.Datta, J.Chem.Soc. 1965, 3905.

T13   M.E.Taylor and R.J.Robinson, Talanta 8, 518 (1961).

T14   P.J.Taylor, Spectrochim.Acta,Part A A26, 165 (1970).

T15   P.H.Tedesco and H.F.Walton, Inorg.Chem. 8, 932 (1969).

T16   V.Y.Temkina, N.M.Dyatlova, M.N.Rusina, N.V.Tsirulnikova, B.V.Zhadanov and
      R.P.Lastovskii, Dokl.Chem.(Engl.Transl.) 180, 521 (1968).

T17   V.Y.Temkina, N.M.Dyatlova, G.F.Yaroshenko, O.Y.Lavrova and R.P.Lastovskii,
      J.Anal.Chem.USSR (Engl.Transl.) 24, 135 (1969).

T18   V.Y.Temkina, N.M.Dyatlova, G.F.Yaroshenko, M.N.Rusina, L.M.Timakova and
      R.P.Lastovskii, Dokl.Akad.Nauk.SSSR. 194, 602 (1970) [Chem]; CA 74 68385n.

T19   V.Y.Temkina, N.M.Dyatlova, G.F.Yaroshenko, M.N.Rusina, L.M.Timakova and
      R.P.Lastovskii, Dokl.Chem.(Engl.Transl.) 194, 697 (1970).

T20   G.S.Tereshin, A.R.Rubinshtein and I.V.Tananev, J.Anal.Chem.USSR (Engl.Transl.)
      20, 1138 (1965).

T21   F.Terrier, F.Millot and R.Schaal, Bull.Soc.Chim.Fr. 1969, 3003.

T22   N.V.Thakur, S.M.Jogdeo and C.R.Kanekar, J.Inorg.Nucl.Chem. 28, 2297 (1966).

T23   G.F.Thiers, L.C.Van Poucke and M.A.Herman, J.Inorg.Nucl.Chem. 30, 1543 (1968).

T24   J.Thomas, D.F.Elsden and S.M.Partridge, Nature (London) 200, 651 (1963).

T25   J.F.Thompson, C.J.Morris, S.Asen and F.Irreverre, J.Biol.Chem. 236, 1183 (1961).

T26   L.C.Thompson, J.Inorg.Nucl.Chem. 24, 1083 (1962).

T27 L.C.Thompson, Inorg.Chem. 3, 1015 (1964).

T28 L.C.Thompson, Inorg.Chem. 3, 1319 (1964).

T29 L.C.Thompson and S.K.Kundra, J.Inorg.Nucl.Chem. 28, 2945 (1966).

T30 L.C.Thompson and S.K.Kundra, Inorg.Chem. 7, 338 (1968).

T31 L.C.Thompson and J.A.Loraas, Inorg.Chem. 2, 594 (1963).

T32 R.M.Tichane and W.E.Bennett, J.Am.Chem.Soc. 79, 1293 (1957).

T33 B.K.Tidd, J.Chem.Soc. 1964, 1521.

T34 L.I.Tikhonova, Russ.J.Inorg.Chem. (Engl.Transl.) 10, 70 (1965).

T35 L.I.Tikhonova, Zh.Prikl.Khim. 40, 1887 (1967); CA 68, 6918f.

T36 L.I.Tikhonova, Russ.J.Inorg.Chem.(Engl.Transl.) 13, 1384, (1968).

T37 M.J.L.Tillotson and L.A.K.Staveley, J.Chem.Soc. 1958, 3613.

T38 C.F.Timberlake, J.Chem.Soc. 1957, 4987.

T39 C.F.Timberlake, J.Chem.Soc. 1959, 2795.

T40 C.F.Timberlake, J.Chem.Soc. 1964, 1229.

T41 C.F.Timberlake, J.Chem.Soc. 1964, 5078.

T42 C.Tissier and M.Tissier, Bull.Soc.Chim.Fr. 1970, 3752.

T43 E.V.Titov, L.M.Kapkan, V.I.Rybachenko and N.G.Korzhenevskaya, Org.React.(USSR)
     5, 273 (1968).

T44 E.V.Titov, N.G.Korzhenevskaya and V.I.Rybachenko, Ukr.Khim.Zh. 34, 1253 (1968);
     CA 70, 109624c.

T45 K.Toei, H.Miyata and T.Harada, Bull.Chem.Soc.Jpn. 40, 1141 (1967).

T46 K.Toei, H.Miyata and H.Kimura, Bull.Chem.Soc.Jpn. 40, 2085 (1967).

T47 K.Toei, H.Miyata and T.Mitsumata, Bull.Chem.Soc.Jpn. 38, 1050 (1965).

T48 K.Toei, H.Miyata, S.Nakashima and S.Kiguchi, Bull.Chem.Soc.Jpn. 40, 1145 (1967).

T49 K.Toei, H.Miyata, T.Shibata and S.Miyamura, Bull.Chem.Soc.Jpn. 38, 334 (1965).

T50 K.Toei, H.Miyata and T.Shibata, Bull.Chem.Soc.Jpn. 40, 1273 (1967).

T51 V.N.Tolmachev and G.G.Lomakina, Russ.J.Phys.Chem.(Engl.Transl.) 31, 1027 (1957).

T52 P.E.Toren and I.M.Kolthoff, J.Am.Chem.Soc. 77, 2061 (1955).

T53 V.F.Toropova, A.Kh.Miftakhova, I.V.Gur'yanova, M.G.Zimin and A.N.Pudovik,
     J.Gen.Chem.USSR (Engl.Transl.) 40, 2159 (1970).

T54 V.F.Totopova, M.K.Saikina and R.S.Aleshov, Russ.J.Inorg.Chem.(Engl.Transl.)
     11, 605 (1966).

T55 E.N.Trachtenberg and G.Odian, J.Am.Chem.Soc. 80, 4018 (1958).

T56 C.Troeltzsch, Z.Chem. 6, 387 (1966); CA 66, 25606c.

T57 J.Tsau, S.Matsouo, P.Clerc and R.Benoit, Bull.Soc.Chim.Fr. 1967, 1039.

T58   I.V.Tselenskii, G.I.Kolesetskaya and A.S.Kesmynina, Org.React.(USSR) 6, 100 (1969).

T58a I.V.Tselenskii, A.S.Kosmynina, V.N.Dronov and I.N.Shokhor, Org.React.(USSR)

      7, 20 (1970).

T59   Y.Tsuchitani, T.Ando and K.Ueno, Bull.Chem.Soc.Jpn. 36, 1534 (1963).

T60   K.Tsuji, J.Sci.Research Inst.(Tokyo) 48, 126 (1954); CA 49, 4379.

T61   E.N.Tsvetkov, D.I.Lobanov, L.A.Izosenkova and M.I.Kabachnik, J.Gen.Chem.USSR (Engl.

      Transl.) 39, 2126 (1969).

T62   E.N.Tsvetkov, M.M.Makhamatkhanov, D.D.Lobanov and M.I.Kabachnik,

      J.Gen.Chem.USSR (Engl.Transl.) 40, 465 (1970).

T63   E.R.Tucci, E.Doody and N.C.Li, J.Phys.Chem. 65, 1570 (1961).

T64   E.R.Tucci, F.Takahashi, V.A.Tucci and N.C.Li, J.Inorg.Nucl.Chem. 26, 1263 (1964).

T65   F.M.Tulyupa, V.S.Barbalov and Y.I.Usatenko, Khim.Tekhnol. 12, 61 (1969);

      CA 72 21184q.

T66   F.M.Tulyupa, Yu.I.Usatenko and V.S.Barkalov, J.Appl.Spectrosc.(Engl.Transl.)

      9, 664 (1968).

T67   D.Turnbull and S.H.Maron, J.Am.Chem.Soc. 65, 212 (1943).

T68   I.V.Turovskii, V.P.Kadysh, V.Zelmeme and J.Stradins, Latv.PSR.Zinat.Akad.Vestis,

      Kim.Ser. 6, 749 (1969); CA 72, 120882x.

T69   C.A.Tyson and A.E.Martell, J.Am.Chem.Soc. 90, 3379 (1968).

                    U

U1   A.Uchiumi and H.Iida, Nippon Kagaku Zasshi 89, 776 (1968).

U2   T.Ueda, Y.Iida, K.Ikeda and Y.Mizuno, Chem.Pharm.Bull. 16, 1788 (1968).

U3   E.Uhlig, Chem.Ber. 93, 679 (1960).

U4   E.Uhlig, Z.Anorg.Allg.Chem. 311, 249 (1961).

U5   E.Uhlig, Z.Anorg.Allg.Chem. 312, 332 (1961).

U6   E.Uhlig, Z.Anorg.Allg.Chem. 320, 283 (1963).

U7   E.Uhlig and D.Herrmann, Z.Anorg.Allg.Chem. 359, 135 (1968).

U8   E.Uhlig and R.Krannich, J.Inorg.Nucl.Chem. 29, 1164 (1967).

U9   E.Uhlig and D.Linke, Z.Chem. 5, 232 (1965); CA 63, 6826d.

U10   H.E.Ungnade, G.Fritz and L.W.Kissinger, Tetrahedron 19, Suppl.1, 235 (1963).

U11   H.E.Ungnade, L.W.Kissinger, A.Narath and D.C.Barham, J.Org.Chem. 28, 134 (1963).

U12   Y.I.Usatenko and A.S.Sukhoruchkina, J.Anal.Chem.USSR (Engl.Transl.) 18, 1123 (1963).

U13   Y.I.Usatenko and T.V.Zamorskaya, Ukr.Khim.Zh. 29, 925 (1963); CA 60, 7513h.

V

V1   J.Vaissermann and M.Quintin, J.Chim.Phys.Phys.-Chim.Biol. 63, 731 (1966).

V2   J.M.Vandenbelt, C.Henrich and S.G.Vanden Berg, Anal.Chem. 26, 726 (1954).

V3   H.Vanderhaeghe and G.Derudder, Bull.Soc.Chim.Belg. 61, 310 (1952).

V4   V.P.Vasilev and L.A.Kochergina, Russ.J.Phys.Chem.(Engl.Transl.) 41, 1496 (1967).

V5   V.P.Vasilev, L.S.Mukhina and A.F.Belyakova, Izv.Vyssh.Ucheb.Zaved.,Khim.Khim.Tekhnol.
       9, 879 (1966); CA 66, 119454d.

V6   V.F.Vasileva, O.Y.Lavrova, N.M.Dyatlova and V.G.Yashunskii, J.Gen.Chem.USSR (Engl.
       Transl.) 36, 688 (1966).

V7   V.F.Vasileva, O.Y.Lavrova, N.M.Dyatlova and V.G.Yashunskii, J.Gen.Chem.USSR.
       (Engl.Transl.) 38, 468 (1968).

V8   W.R.Vaughan and G.K.Finch, J.Org.Chem. 21, 1201 (1956).

V9   R.C.Vickery, J.Chem.Soc. 1954, 385.

V10   P.Vieles and N.Israily, J.Chim.Phys.Phys.-Chim.Biol. 59, 671 (1962).

V11   O.N.Vlasov, B.N.Rybakov and L.M.Kogan, Zh.Prikl.Khim. 41, 373 (1968);
       CA 69, 54722e.

V12   V.G.Voden, M.E.Obukhova and M.F.Pushlenkov, Radiokhimiya 11, 644 (1969);
       CA 72, 125627c.

V13   H.Vorsanger, Bull.Soc.Chim.Fr. 1967, 556.

W

W1   H.Walba and R.W.Isensee, J.Am.Chem.Soc. 77, 5488 (1955).

W2   K.Wallenfels and F.Freidrich, Chem.Ber. 93, 3070 (1960).

W3   K.Wallenfels and H.Sund, Biochem.Z. 329, 41 (1957).

W4   W.Walter and R.F.Becker, Justus Liebigs Ann.Chim. 727, 71 (1969).

W5   W.Walter and H.L.Weidemann, Justus Liebigs Ann.Chem. 685, 29 (1965).

W6   H.F.Walton and A.A.Schilt, J.Am.Chem.Soc. 74, 4995 (1952).

W7   E.S.N.Wang and E.R.Hammarlund, J.Pharm.Sci. 59, 1559 (1970).

W8   J.C.Wang, J.E.Baumann and R.K.Murmann, J.Phys.Chem. 68, 2296 (1964).

W9   E.Wanninen, Suom.Kemistil.B 28, 146 (1955).

W10   A.G.Warner and E.P.Serjeant, unpublished results.

W10a O.W.Webster, J.Am.Chem.Soc. 88, 3046 (1966).

W11   E.L.Wehry and L.B.Rogers, J.Am.Chem.Soc. 87, 4234 (1965).

W12   E.L.Wehry and L.B.Rogers, J.Am.Chem.Soc. 88, 351 (1966).

W13   A.Weiss, S.Fallab and H.Erlenmeyer, Helv.Chim.Acta 40, 611 (1957).

W14   No entry.

W15   P.R.Wells and W.Adcock, Aust.J.Chem. 18, 1365 (1965).

W16   P.E.Wenger and I.Kapetanidis, Recl.Trav.Chim.Pays-Bas 79, 567 (1960).

W17   P.E.Wenger, D.Monnier and L.Epars, Helv.Chim.Acta 35, 396 (1952).

W18   P.E.Wenger, O.Monnier and I.Kapetanidis, Helv.Chim.Acta 40, 1456 (1957).

W19   R.L.Wershaw, M.C.Goldberg and D.J.Pinckney, Water Resour.Res. 3, 511 (1967);
         CA 68, 77218p.

W20   S.J.Westerback and A.E.Martell, Nature (London) 178, 321 (1956).

W21   S.Westerback, K.S.Rajan and A.E.Martell, J.Am.Chem.Soc. 87, 2567 (1965).

W22   F.W.Wheland and J.Farr, J.Am.Chem.Soc. 65, 1433 (1943).

W23   J.R.Whitaker and M.L.Bender, J.Am.Chem.Soc. 87, 2728 (1965).

W24   M.W.Whitehouse and P.D.G.Dean, Biochem.Pharmacol. 14 557 (1965).

W25   M.W.Whitehouse and J.E.Leader, Biochem.Pharmacol. 16, 537 (1967).

W26   J.M.Whiteley and F.M.Huennekens, Biochemistry 6, 2620 (1967).

W27   K.B.Wiberg and B.R.Lowry, J.Am.Chem.Soc. 85, 3188 (1963).

W28   K.B.Wiberg and V.Z.Williams, J.Org.Chem. 35, 369 (1970).

W29   C.F.Wilcox and C.Leung, J.Am.Chem.Soc. 90, 336 (1968).

W30   A.V.Willi and P.Mori, Helv.Chim.Acta 47, 155 (1964).

W31   A.Willi and G.Schwarzenbach, Helv.Chim.Acta 34, 528 (1951).

W32   A.V.Willi and J.F.Stocker, Helv.Chim.Acta 38, 1279 (1955).

W33   A.V.Willi, Helv.Chim.Acta 39, 46 (1956).

W34   A.V.Willi and W.Meier, Helv.Chim.Acta 39, 54 (1956).

W35   D.C.Williams and J.R.Whitaker, Biochemistry 6, 3711 (1967).

W36   I.B.Wilson, S.Ginsburg and C.Quan, Arch.Biochem.Biophys. 77, 286 (1958).

W37   J.M.Wilson, A.G.Briggs, J.E.Sawbridge, P.Tickle and J.J.Zuckerman, J.Chem.Soc.(A)
         1970, 1024.

W38   J.M.Wilson, N.E.Gore, J.E.Sawbridge and F.Cardenas-Cruz, J.Chem.Soc.(B) 1967, 852.

W39   W.M.Wise and W.W.Brandt, J.Am.Chem.Soc. 77, 1058 (1955).

W40   H.F.Van Woerden, Recl.Trav.Chim.Pays-Bas. 83, 920 (1963).

W41   F.Wold and C.E.Ballou, J.Biol.Chem. 227, 301 (1957).

W42   W.Wolf, J.C.J.Chen and L.L.J.Hsu, J.Pharm.Sci. 55, 68 (1966).

W43   P.W.K.Woo, H.W.Dion and Q.R.Bartz, J.Am.Chem.Soc. 84, 1512 (1962).

W44   R.B.Woodward and G.Small, J.Am.Chem.Soc. 72, 1297 (1950).

W45   E.M.Wooley, R.W.Wilton and L.G.Hepler, Can.J.Chem. 48, 3249 (1970).

W46   D.P.Wrathall, R.M.Izatt and J.J.Christensen, J.Am.Chem.Soc. 86, 4779 (1964).

W47   G.Wright and J.Bjerrum, Acta Chem.Scand. 16, 1262 (1962).

W48   A.Wrobel, A.Rabczenko and D.Shugar, Acta Biochim.Pol. 17, 339 (1970).

Y

Y1   R.P.Yaffe and A.F.Voigt, J.Am.Chem.Soc. 74, 2941 (1952).

Y2   G.Yagil, J.Phys.Chem. 71, 1034 (1967).

Y3   G.Yagil, Tetrahedron 23, 2855 (1967).

Y4   P.Y.Yakovlev and R.D.Malinina, Zavod.Lab. 33, 267 (1967); CA 67, 36783e.

Y5   S.H.Yalkowsky and G.Zografi, J.Pharm.Sci. 59, 798 (1970).

Y6   T.Yano, H.Kobayashi and K.Ueno, Bull.Chem.Soc.Jpn. 43, 3167 (1970).

Y7   V.G.Yashunskii, O.I.Samoilova, N.M.Dyatlova and O.Y.Lavrova, J.Gen.Chem.USSR (Engl.
       Transl.) 32, 3309 (1962).

Y8   V.G.Yashunskii and M.N.Shchukina, Khim.Nauka i Prom. 2, 662 (1957); CA 52, 6209b.

Y9   M.Yasuda, Z.Phys.Chem.(Frankfurt am Main) 27, 333 (1961).

Y10   M.Yasuda, K.Suzuki and K.Yamasaki, J.Phys.Chem. 60, 1649 (1956).

Y11   M.Yasuda and K.Yamasaki, Naturwissenschaften 45, 84 (1958).

Y12   M.Yasuda, K.Yamasaki and H.Ohtaki, Bull.Chem.Soc.Jpn. 33, 1067 (1960).

Y13   R.W.Yip and W.D.Riddell, Can.J.Chem. 48, 987 (1970).

Y14   M.Yoshioka, H.Hamamato and T.Kubota, Bull.Chem.Soc.Jpn. 35, 1723 (1962).

Y15   N.Yui, Sci.Repts.Tohoku Univ.,First Ser. 40, 102 (1956); CA 51, 17354d.

Y16   N.Yui, Sci.Repts.Tohoku Univ.,First Ser. 40, 114 (1956); CA 51, 17354e.

Z

Z1   W.L.Zahler and W.W.Cleland, J.Biol.Chem. 243, 716 (1968).

Z2   R.Zahradnik, Z.Phys.-Chem.(Leipzig) 213, 318 (1960).

Z3   L.I.Zakharkin, Y.A.Chapooskii and V.I.Stanko, Bull.Acad.Sci.USSR, Div.Chem.Sci.(Engl.
       Transl.) 1964, 2105.

Z4   A.A.Zavitsas, J.Chem.Eng.Data 12, 94 (1967).

Z5   B.V.Zhadanov, V.Y.Temkina, N.M.Dyatlova and V.V.Medyntsev, Russ.J.Phys.Chem.
       (Engl.Transl.) 43, 130 (1969).

Z6   F.G.Zharovskii and M.S.Ostrovskaya, Ukr.Khim.Zh. 32, 893 (1966); CA 66, 8011e.

Z7   F.G.Zharovskii and R.I.Sukhomlin, Ukr.Khim.Zh.33, 509 (1967); CA 67, 76728x.

Z8   F.G.Zharovskii, R.I.Sukhomlin and M.S.Ostrovskaya, Russ.J.Inorg.Chem.(Engl.Transl.)
        12, 1306 (1967).

Z9   C.Zioudrou, Tetrahedron 18, 197 (1962).

Z10   H.E.Zittel and T.M.Florence, Anal.Chem. 39, 320 (1967).

Z11   H.Zollinger and C.Wittwer, Helv.Chim.Acta 39, 347 (1956).

Z12   Y.A.Zolotov and V.G.Lambrev, Khim.Osnovy Ekstraktsion.Metoda Razdeleniya Elementov,
        Akad.Nauk.SSSR, Inst.Geokhim.i Analit.Khim. 1966, 65; CA 65, 9808a.

Z13   Y.A.Zolotov, V.G.Lambrev, M.K.Chmutova and N.T.Sizonenko, Dokl.Chem.(Engl.Transl.)
        165, 1075 (1965).

Z14   S.Zommer and J.Szuszkiewicz, Chem.Anal.(Warsaw) 14, 1075 (1969); CA 72 89616n.

Z15   A.P.Zozulya, N.N.Mezentseva, V.M.Peshkova and Y.K.Yurev, J.Anal.Chem.USSR (Engl.
        Transl.) 14, 15 (1959).

Z16   A.P.Zozulya and V.M.Peshkova, Russ.J.Inorg.Chem.(Engl.Transl.) 4, 168 (1959).

Z17   O.E.Zvyagintsev and V.P.Tikhonov, Russ.J.Inorg.Chem.(Engl.Transl.) 9, 865 (1964).

V. INDEX OF COMPOUNDS

The number appearing after each entry is the compound number. Numbers less than 2,000 refer to the original compilation, Kortum et al., Dissociation Constants of Organic Acids in Aqueous Solution, Butterworths (1961).

Acetic acid, 2-[2-(1,8-dihydroxy-3,6-disulfonaphth-2-ylazo)phenyl]-2-hydroxy-, 6327

— , 2-[3-(1,8-dihydroxy-3,6-disulfonaphth-2-ylazo)phenyl]-2-hydroxy-, 6328

— , 2-[4-(1,8-dihydroxy-3,6-disulfonaphth-2-ylazo)phenyl]-2-hydroxy-, 6329

— , 2-(3,4-dihydroxyphenyl)-, 4011

— , 2-(3,4-dimethoxyphenyl)-, 591

— , 2-(2,4-dimethylphenoxy)-, 5016

— , 2-(2,6-dimethylphenoxy)-, 219

— , 2-(3,5-dimethylphenoxy)-, 5017

— , 2-(2,5-dimethylphenyl)-2-hydroxy-, 5018

— , dimethylphenylsilyl-, 403

— , 2-(6,9-dimethylpurin-2-yl)thio-, 4878

— , 2-(3,5-dimethyl-1-pyrazolyl)-, 3771

— , dimethylsulfonio-, 2454

— , 2-(2,4-dinitrophenyl)-, 582

— , diphenyl-, 526

— , 2-diphenylphosphinyl-, 5978

— , (dithiocarboxy)amino-, 2287

— , ethoxy-, 2325, 207

— , 2-ethoxycarbonyl-, 2563

— , ethoxycarbonylamino-, 255

— , 2-(2-ethoxyphenyl)seleno-, 5186

— , 2-(4-ethoxyphenyl)seleno-, 5187

— , ethylamino-, 246

— , 2-(4-ethylphenyl)-, 523

— , 2-(4-ethylphenyl)-2-hydroxy-, 5019

— , ethylthio-, 2434

— , fluoro-, 2067, 159

— , 2-(2-fluorophenoxy)-, 319

— , 2-(3-fluorophenoxy)-, 320

— , 2-(4-fluorophenoxy)-, 321

— , 2-(3-fluorophenyl)-, 4178

— , 2-(4-fluorophenyl)-, 4179, 556

— , 2-(2-fluorophenyl)-2-hydroxy-, 4194

— , 2-(3-fluorophenyl)-2-hydroxy-, 600

— , 2-(4-fluorophenyl)-2-hydroxy-, 4195

— , 2-(3-fluorophenyl)sulfinyl-, 4433

Adenosine-5'-(dihydrogen phosphate), 2'-deoxy-, 5296

Adenosine-5'-(dihydrogen phosphate) N(1)-oxide, 5300

Adenosine, N(6),N(6)-dimethyl-, 5672

— , N(1)-methyl-, 5475

— , N(6)-methyl-, 5476

— , N(6)-(3-methyl-2-butenyl)-, 6096

— , 5'-methylthio-, 5523

Adenosine-5'-(pentahydrogen tetraphosphate), 5317

Adenosine-5'-(tetrahydrogen triphosphate), 5314

Adenosine-5'-(trihydrogen diphosphate), 5304

Adenosinium bromide, N(1)-(3-methyl-2-butenyl)-, 6111

3'-Adenosyl 5'-adenosyl hydrogen phosphate, 6400

5'-Adenosyl (3-carbamoylpyridinio-β-D-ribofuranos-5'-yl) hydrogen diphosphate, 6413

3'-Adenosyl 5'-uridinyl hydrogen phosphate, 6378

5'-Adenosyl 3'-uridinyl hydrogen phosphate, 6379

Adipamic acid, 354

Adipic acid, 2880, 25

ADP, 5304

Adrenaline, 4826

DL-Adrenochrome semicarbazone, 5185

α-Alanine, 2233, 257

β-Alanine, 2234, 263

Di-L-alanyl-L-alanine, 4850

L-Alanylglycine, 2715

L-Alanylglycylglycine, 3812

DL-α-Alanyl-DL-serine, 3214

Albumin, bovine, 6507

Albumin, human serum, 6508

Alizarin acid black, 6489

Alizarin fluorine blue, 6360

Alizarin S, 5951

Allomaltol, 2836

Allyl alcohol, 2141

Allylsulfonic acid, 2220

Aluminon, 6414

Benzoic acid, 4-cyclopentyl-, 5563

- , 2-cyclopropyl-, 4977

- , 4-cyclopropyl-, 4978

- , 3,5-dibromo-2-hydroxy-, 3509

- , 3,5-di-tert-butyl-4-hydroxy-, 6063

- , 2,3-dichloro-, 3495

- , 2,4-dichloro-, 3496

- , 2,5-dichloro-, 3497

- , 2,6-dichloro-, 3498

- , 3,4-dichloro-, 3499

- , 3,5-dichloro-, 3500

- , 3,5-dichloro-2-hydroxy-, 3507

- , 3,5-dichloro-4-hydroxy-, 3508

- , 3-diethoxyphosphinyl-, 5480, 516

- , 4-diethoxyphosphinyl-, 5481, 517

- , 2,3-dihydroxy-, 3404

- , 2,4-dihydroxy-, 3405, 461

- , 2,5-dihydroxy-, 3406

- , 2,6-dihydroxy-, 3407, 462

- , 3,4-dihydroxy-, 3408

- , 3,5-dihydroxy-, 3409

- , 2-(2,6-dihydroxybenzoyl)amino-, 5954

- , 4-(2,6-dihydroxybenzoyl)amino-, 5955

- , 2-(1,8-dihydroxy-3,6-disulfonaphth-2-ylazo)-, 6261

- , 2-[1,8-dihydroxy-7-(2-hydroxy-3,5-dinitrophenylazo)-3,6-disulfonaphth-2-ylazo]-, 6452

- , 2-[1,8-dihydroxy-7-(4-methoxyphenylazo)-3,6-disulfonaphth-2-ylazo]-, 6478

- , 2-[1,8-dihydroxy-7-(2-methylphenylazo)-3,6-disulfonaphth-2-ylazo]-, 6475

- , 2-[1,8-dihydroxy-7-(3-methylphenylazo)-3,6-disulfonaphth-2-ylazo]-, 6476

- , 2-[1,8-dihydroxy-7-(4-methylphenylazo)-3,6-disulfonaphth-2-ylazo]-, 6477

- , 2-[1,8-dihydroxy-7-(2-nitrophenylazo)-3,6-disulfonaphth-2-ylazo]-, 6453

- , 2-[1,8-dihydroxy-7-(3-nitrophenylazo)-3,6-disulfonaphth-2-ylazo]-, 6454

- , 2-[1,8-dihydroxy-7-(4-nitrophenylazo)-3,6-disulfonaphth-2-ylazo]-, 6455

- , 2-(3,4-dihydroxyphenylazo)-, 5821

- , 2-(1,8-dihydroxy-7-phenylazo-3,6-disulfonaphth-2-yl)azo-, 6457

- , 2-dihydroxyphosphino-, 3706

- , 2-[1,8-dihydroxy-7-(3-sulfophenylazo)-3,6-disulfonaphth-2-ylazo]-, 6458

Benzoic acid, 3-ethoxy-2-methyl-, 5021

— , 4-ethoxy-2-methyl-, 5023

— , 5-ethoxy-2-methyl-, 5022

— , 2-ethyl-, 415

— , 4-ethyl-, 416

— , 3,4-ethylenedioxy-, 4542

— , 2-fluoro-, 444

— , 3-fluoro-, 3551, 445

— , 4-fluoro-, 3552, 446

— , 3-formyl-, 3962

— , 4-formyl-, 3963

— , 2-formyl-5,6-dimethoxy-, 4994

— , 2-(formyl-N-methyloxime)-5,6-dimethoxy-, 5464

— , 2-hydroxy-, 3397, 458

— , 3-hydroxy-, 3398, 459

— , 4-hydroxy-, 3399, 460

— , 2-hydroxy-3,5-diiodo-, 3510

— , 4-hydroxy-3,5-dimethoxy-, 4605

— , 2-hydroxy-3,6-dimethyl-, 4573

— , 2-hydroxy-3,6-dimethyl-5-nitro-, 4719

— , 2-hydroxy-3,5-dinitro-, 3518

— , 2-hydroxy-3-iodo-, 3565

— , 2-hydroxy-4-iodo-, 3566

— , 2-hydroxy-5-iodo-, 3567

— , 2-hydroxy-5-(4-iodophenylazo)-, 5893

— , 2-hydroxy-3-isopropyl-, 5024

— , 2-hydroxy-4-methoxy-, 4013

— , 2-hydroxy-5-methoxy-, 4014

— , 3-hydroxy-4-methoxy-, 4015

— , 4-hydroxy-3-methoxy-, 4016

— , 2-hydroxy-3-methoxy-5-nitro-, 4221

— , 4-hydroxy-3-methoxy-5-nitro-, 4222

— , 2-hydroxy-5-(4-methoxyphenylazo)-, 5964

— , 2-hydroxy-3-methyl-, 3995

— , 2-hydroxy-4-methyl-, 3996

— , 2-hydroxy-5-methyl-, 3997

O,O-Dimethyl O-hydrogen phosphorothioate, 2127

N,N,N-Dimethyl(hydroxycarbamoylmethyl) benzylammonium bromide, 5530

O,O-Dimethylphosphoramidate, N-methylsulfonyl-, 2298

Dimethylpropiothetin, 2729

Di(2-naphthyl) hydrogen phosphate, 6384

3,3-Dinitropropyl acetate, 2681

cis-2,9-Dioxabicyclo[4.3.0]nonane-t-4,t-6-dicarboxylic acid, 4624

6,9-Dioxa-3,12-diazatetradecanedioic acid, 3,12-bis(carboxymethyl)-, 6021

3,6-Dioxaoctanedioic acid, 2889

Diphenylamine, 2,4-dichloro-2',6'-dinitro-, 5700

1,5-Diphenylformazan-3',3"-disulfonic acid, 3-cyano-2',2"-dihydroxy-, 6034

— , 3-cyano-2',2"-dihydroxy-5',5"-dinitro-, 6025

— , 5',5"-dichloro-3-cyano-2',2"-dihydroxy-, 6052

1,5-Diphenylformazan-3'-sulfonic acid, 2"-arsono-6'-hydroxy-, 6053

Diphenyl hydrogen phosphate, 5629

O,O-Diphenyl hydrogen phosphorodithioate, 5729

O,O-Diphenyl S-hydrogen phosphorothioate, 5730

O,O-Diphenyl methanedisulfonate, 5845

Diphenylphosphinic acid, 5623

Diphenylphosphinothioic O-acid, 5727

Diphosphopyridine nucleotide, 6413

Dipicolinic acid, 3581

Dipropyl hydrogen phosphate, 3248

O,O-Dipropyl hydrogen phosphorodithioate, 3365

O,O-Dipropyl hydrogen phosphorothioate, 3367

Dipropylphosphinic acid, 3246

Dipropylphosphinodithioic acid, 3250

Dipropylphosphinothioic O-acid, 3359

P^1,P^2-Di(3-pyridyl) dihydrogen diphosphate, 5283

P^1,P^2-Di(3-pyridyl) dihydrogen, pyrophosphate, 5283

1,3-Dithiacycloheptane-1,1,3,3-tetraoxide, 2720

1,3-Dithiacyclohexane-1,1,3,3-tetraoxide, 2443

1,3-Dithiacyclooctane 1,1,3,3-tetraoxide, 3219

1,3-Dithiacyclopentane 1,1,3,3-tetraoxide, 2222

3,8-Dithiadecanedioic acid, 361

4,7-Dithiadecanoic acid, 5,6-bis(2-carboxyethylthio)-, 6012

1-Methylpyridinium iodide, 4-formyl-, 3908

— , 2-(N-hydroxycarbamoyl)-, 3938

— , 3-(N-hydroxycarbamoyl)-, 3939

— , 4-(N-hydroxycarbamoyl)-, 3940

— , 3-hydroxy-2-hydroxyiminomethyl-, 3937

— , 3-hydroxyiminoacetyl-, 4463

— , 4-hydroxyiminoacetyl-, 4464

— , 2-(α-hydroxyiminobenzyl)-, 5908

— , 2-(1-hydroxyimino)ethyl-, 4482

— , 2-hydroxyiminomethyl-, 3933

— , 3-hydroxyiminomethyl-, 3934

— , 4-hydroxyiminomethyl-, 3935

— , 2-hydroxyiminomethyl-3-methoxy-, 4487

— , 2-hydroxyiminomethyl-3-methyl-, 4483

— , 2-hydroxyiminomethyl-6-methyl-, 4484

(6-Methylpyridin-2-yl)methyl dihydrogen phosphate, 3947

1-Methylquinolinium iodide, 4-hydroxyiminomethyl-, 5517

Methyl Red, 910

Methyl salicylate, 4010

Methyl (1,2,3,6-tetrahydro-2,6-dioxopyrimidin-4-yl)carboxylate, 3077

Methyl tetrahydrogen triphosphate, 2022

Methyl 4,4,6,6-tetranitrohexanoate, 3780

Methylthioglycolate, 2214

Methylthymol blue, 6505

Methyl 3,4,5-trihydroxybenzoate, 4029

DL-Mimosine, 4326

Monoarsenazo III, 6437

Monothiooxamide, 2123

— , N,N'-bis(2-hydroxyethyl)-, 5318

— , N,N'-bis(4-sulfobenzyl)-, 6210

— , N,N'-bis(2-sulfoethyl)-, 3354

— , N,N,N',N'-tetrakis(4-hydroxybutyl)-, 6339

— , N,N,N',N'-tetrakis(5-hydroxypentyl)-, 6434

— , N,N,N',N'-tetrakis(3-hydroxypropyl)-, 6050

Morpholine-4-carbodithioic acid, 2786

Mucic acid, 2890

Mucochloric acid, 2339

Murexide, 4111

Myristic acid, 5949

Nadide, 6413

1-Naphthalenecarbonitrile, 3-hydroxy-, 5388

— , 4-hydroxy-, 5389

2,3-Naphthalenedicarboxylic acid, 1,2,3,4-tetrahydro-, 134, 135

1,3-Naphthalenediol, 4962

— , 2,4-dinitroso-, 5069

1,4-Naphthalenediol, 4963, 670

1,8-Naphthalenediol, 4964

2,3-Naphthalenediol, 4965

2,3-Naphthalenediol, 1,4-bis(dimethylaminomethyl)-, 6157

2,7-Naphthalenediol, 1-[5-(1-methylpiperidin-2-yl)-2-pyridylazo]-, 6404

1,3-Naphthalenedisulfonic acid, 4-amino-5-hydroxy-6-(2-hydroxyphenylazo)-, 6203

— , 8-(4-chlorophenylazo)-7-hydroxy-, 6222

— , 7-hydroxy-, 5113

1,5-Naphthalenedisulfonic acid, 3-hydroxy-, 5114

— , 4-hydroxy-, 5115

— , 8-hydroxy-7-nitroso-, 5268

— , 3-hydroxy-4-(4-sulfophenylazo)-, 6193

1,6-Naphthalenedisulfonic acid, 4-hydroxy-, 5116

— , 4-hydroxy-3-(4-sulfophenylazo)-, 6194

2,5-Naphthalenedisulfonic acid, 4-hydroxy-3-(4-sulfophenylazo)-, 6195

2,7-Naphthalenedisulfonic acid, 4-acetylamino-6-(4-acetylaminophenylazo)-5-hydroxy-, 6399

— , 4-acetylamino-5-hydroxy-6-(4-methylphenylazo)-, 6376

— , 4-acetylamino-5-hydroxy-3-phenylazo-, 6331

— , 6-(4-acetylaminophenylazo)-4-amino-5-hydroxy-, 6332

— , 4-amino-6-(4-aminophenylazo)-5-hydroxy-, 6207

— , 3-amino-5-hydroxy-, 5280

— , 4-amino-5-hydroxy-6-(2-hydroxyphenylazo)-, 6204

— , 4-amino-5-hydroxy-6-(4-methylphenylazo)-, 6272

— , 5-amino-4-hydroxy-3-phenylazo-, 6202

2,7-Naphthalenedisulfonic acid, 4-(4-amino-2-hydroxyphenylazo)-3-hydroxy-, 6205

— , 3-(2-arsonophenylazo)-4,5-dihydroxy-, 6227

— , 3-(2-arsonophenylazo)-4,5-dihydroxy-6-(2-hydroxy-3,5-dinitrophenylazo)-, 6436

— , 3-(2-arsonophenylazo)-4,5-dihydroxy-6-phenylazo-, 6437

— , 4-(2-arsonophenylazo)-3-hydroxy-, 6226

— , 3,6-bis(2-arsonophenylazo)-4,5-dihydroxy-, 6439

— , 3,6-bis(3-chloro-6-hydroxy-5-sulfophenylazo)-4,5-dihydroxy-, 6435

— , 3,6-bis(4-chloro-3-methyl-5-phosphonophenylazo)-4,5-dihydroxy-, 6483

— , 3,6-bis(4-chloro-2-phosphonophenylazo)-4,5-dihydroxy-, 6440

— , 3,6-bis(2,3-dimethyl-5-oxo-1-phenyl-3-pyrazolin-4-ylazo)-4,5-dihydroxy-, 6502

— , 4-(5-bromo-1,3-thiazol-2-ylazo)-3-hydroxy-, 5923

— , 4-(3-chloro-2-hydroxy-5-nitrophenylazo)-3-hydroxy-, 6215

— , 3-(3-chloro-6-hydroxyphenylazo)-4,5-dihydroxy-, 6224

— , 4-(2-chloro-6-hydroxyphenylazo)-3-hydroxy-, 6223

— , 4-(3-chloro-6-hydroxy-5-sulfophenylazo)-3-hydroxy-, 6225

— , 3-(4-chloro-2-phosphonophenylazo)-4,5-dihydroxy-, 6229

— , 3-(4-chloro-2-phosphonophenylazo)-4,5-dihydroxy-6-(2-phosphonophenylazo)-, 6441

— , 3,6-dichloro-4,5-dihydroxy-, 5259

— , 3,4-dihydro-4-hydroxyimino-3-oxo-, 5269

— , 4,5-dihydroxy-, 5119

— , 3,6-bis(2-hydroxy-3,5-dinitrophenylazo)-4,5-dihydroxy-, 6420

— , 3,6-bis(4-hydroxyphenylazo)-4,5-dihydroxy-, 6428

— , 3,6-bis(4-methoxyphenylazo)-4,5-dihydroxy-, 6480

— , 3,6-bis(4-methyl-2-sulfophenylazo)-4,5-dihydroxy-, 6481

— , 3,6-bis(3-nitrophenylazo)-4,5-dihydroxy-, 6421

— , 3,6-bis(4-nitrophenylazo)-4,5-dihydroxy-, 6422

— , 3,6-bis(phenylazo)-4,5-dihydroxy-, 6426

— , 3,6-bis(2-phosphonophenylazo)-4,5-dihydroxy-, 6438

— , 3,6-bis(2-sulfophenylazo)-4,5-dihydroxy-, 6432

— , 3,6-bis(4-sulfophenylazo)-4,5-dihydroxy-, 6433

— , 3,6-bis(p-tolylazo)-4,5-dihydroxy-, 6479

— , 4,5-dihydroxy-3-(2-hydroxy-3,5-dinitrophenylazo)-6-(3-sulfophenylazo)-, 6423

— , 4,6-dihydroxy-3-(8-hydroxy-3,6-disulfonaphth-1-ylazo)-, 6397

— , 4,5-dihydroxy-3-(2-hydroxyphenylazo)-, 6192

— , 4,5-dihydroxy-3-(4-hydroxyphenylazo)-6-phenylazo-, 6427

— , 4,5-dihydroxy-3-(2-hydroxy-4-sulfophenylazo)-, 6201

1-Naphthalenesulfonic acid, 3-hydroxy-4-(1-hydroxy-2-naphthylazo)-, 6395

— , 3-hydroxy-4-(2-hydroxy-1-naphthylazo)-, 6396

— , 3-hydroxy-4-(1-hydroxy-2-naphthylazo)-7-nitro-, 6392

— , 3-hydroxy-4-(2-hydroxy-1-naphthylazo)-8-nitro-, 6394

— , 5-hydroxy-6-nitroso-, 5262

— , 8-hydroxy-7-nitroso-, 5263

— , 3-hydroxy-6-(p-tolylsulfonyloxy)-, 765

— , 6-hydroxy-3-(p-tolylsulfonyloxy)-, 766

2-Naphthalenesulfonic acid, 8-(3-acetylphenylazo)-5-hydroxy-, 6322

— , 8-(4-acetylphenylazo)-5-hydroxy-, 6323

— , 5-(2-benzothiazolylazo)-6-hydroxy-, 6260

— , 3-(4-chlorophenylazo)-4-hydroxy-, 6218

— , 5-(4-chlorophenylazo)-6-hydroxy-, 6219

— , 8-(3-chlorophenylazo)-5-hydroxy-, 6220

— , 8-(4-chlorophenylazo)-5-hydroxy-, 6221

— , 8-(4-cyanophenylazo)-5-hydroxy-, 6259

— , 6,7-dihydroxy-, 5111

— , 1-hydroxy-, 5107

— , 4-hydroxy-, 5108

— , 6-hydroxy-, 5109

— , 7-hydroxy-, 5110

— , 6-hydroxy-5-[2-hydroxy-3-(2-hydroxynaphth-1-ylazo)-5-sulfophenylazo]-, 6489

— , 5-hydroxy-8-(3-methoxyphenylazo)-, 6266

— , 5-hydroxy-8-(4-methoxyphenylazo)-, 6267

— , 5-hydroxy-8-(4-methylphenylazo)-, 6263

— , 5-hydroxy-8-(3-nitrophenylazo)-, 6174

— , 5-hydroxy-8-(4-nitrophenylazo)-, 6175

— , 5-hydroxy-6-nitroso-, 5264

— , 6-hydroxy-5-nitroso-, 5265

— , 7-hydroxy-8-nitroso-, 5266

— , 8-hydroxy-7-nitroso-, 5267

— , 5-hydroxy-8-phenylazo-, 6185

— , 6-hydroxy-5-(1,3-thiazol-2-ylazo)-, 5896

1-Naphthocarbaldehyde, 4-hydroxy-, 5328

Naphthochrome Green G, 6498

Oxydiacetic acid, 2319

(2,2'-Oxydiethylene)dinitrilotetraacetic acid, 5691

2,2'-Oxydipropionic acid, 2887

Oxydi(trimethylene)dinitrilotetraacetic acid, 6020

Oxytetracycline, 740

Palmitic acid, α-sulfo-, 6166

PAM-2, 3933

PAM-3, 3934

PAM-4, 3935

o-PAN, 6076

p-PAN, 6075

o-PAP, 5400

p-PAP, 5401

Papain, 6513

PAR, 5405

3-PAR, 5406

Parabanic acid, 992

PAS, 3681

PEDTA, 6155

Penicillamine, 2794

Penicillin G, 6211

3,6,9,12,15-Pentaazaheptadecanedioic acid, 3,6,9,12,15-pentakis(carboxymethyl)-, 6419

3,9,14,20,25-Pentaazatriacontane-2,10,13,21,24-pentone, 30-amino-3,14,25-trihydroxy-, 6486

Pentaerythritol, 2582

Pentamethylenedinitrilotetraacetic acid, 5877

Pentamethylenephosphinic acid, 2730

Pentanamide, 5,5-dinitro-, 2702

Pentane, 2,2-bis(difluoramino)-5,5-dinitro-, 2785

— , 1,1-dinitro-, 2719

— , 1,1,3,3-tetranitro-, 2686

— , 1,1,4,4-tetranitro-, 2687

— , 1,1,1-trichloro-5,5-dinitro-, 2782

— , 1,1,3-trinitro-, 2706

Propanal oxime, 3-nitro-, 2219

Propanamide, 3,3-dinitro-, 2206

— , 3-oxo-3,N-diphenyl-, 6083

Propanamidine, 2-hydroxy-2-phenyl-, 4797

Propane, 1,1-dinitro-, 2221

— , 1,1-dinitro-3-nitrooxy-, 2209

— , 2-methyl-1,1-dinitro-, 2441

— , 3-methyl-1,1-dinitro-, 2442

— , 1-nitro-, 2230

— , 2-nitro-, 2231

— , 1,1,3,3-tetranitro-, 2187

— , 1,1,3-trinitro-, 2207

Propanedial, 2,2-dihydroxy-, 924

Propanedial dioxime, 2-nitro-, 2203

Propanediamide, 2-[(phenyl)(1,2,3,4-tetrahydro-2,4-dioxopyrimidin-5-yl)]methyl-, 5987

— , 2-(4,5,6,7-tetrahydro-5-oxo-1H-1,2,3-triazolo[4,5-d]pyrimidin-7-yl)-, 3763

Propanedinitrile, 2131

Propanedioic acid, 2137, 22

— , amino-, 2202

— , benzyl-, 4990

— , tert-butyl-, 3478

— , 2-butyl-2-ethyl-, 4644

— , 2-butyl-2-hydroxy-, 3483

— , 2-(3-chlorophenyl)-2-ethyl-, 5426

— , 2-(4-chlorophenyl)-2-ethyl-, 5427

— , 2-cyclohexyl-2-hydroxy-, 4635

— , dibutyl-, 5387

— , diethyl-, 3479, 35

— , diheptyl-, 6240

— , diisopropyl-, 4645

— , dimethyl-, 2566, 33

— , dipropyl-, 4646, 37

— , ethyl-, 2567, 30

— , 2-ethyl-2-(1-ethylpropyl)-, 5066

— , 2-ethyl-2-isopropyl-, 4088

— , 2-ethyl-2-(4-methoxyphenyl)-, 5575

Thymolphthalein complexone, 6506

Tiglic acid, 67

Tiron, 3085

p-Toluenesulfonamide, N-chloro-, 984

o-Toluic acid, 405

m-Toluic acid, 406

p-Toluic acid, 407

o-Tolylarsonic acid, 870

m-Tolylarsonic acid, 871

p-Tolylarsonic acid, 872

p-Tolyl dihydrogen phosphate, 3770

p-Tolyl hydrogen p-tolylphosphonate, 5990

o-Tolylphosphonic acid, 796

m-Tolylphosphonic acid, 797

p-Tolylphosphonic acid, 798

m-Tolylselenonic acid, 894

p-Tolylselenonic acid, 895

TPHA, 6419

2,5,8-Triazanonane-1,9-diphosphonic acid, 2,5,8-tris(phosphonomethyl)-, 4938

3,6,9-Triazaundecanedioic acid, 3,9-bis(carboxymethyl)-6-(2-carboxyethyl)-, 6102

— , 3,9-bis(carboxymethyl)-6-(2-hydroxyethyl)-, 6022

— , 3,9-bis(carboxymethyl)-6-methyl-, 5878

— , 3,6,9-tris(carboxymethyl)-, 6015

3,6,9-Triazaundecane-2,10-diphosphonic acid, 2,10-dimethyl-, 5323

3,6,9-Triazaundecanoic acid, 11-amino-2-(4-imidazolylmethyl)-4,7,10-trioxo-, 5677

— , 11-amino-4,7,10-trioxo-, 4382

— , 5-(1-benzylimidazol-4-yl)methyl-4,7,10-trioxo-, 6369

— , 2-(4-imidazolylmethyl)-4,7,10-trioxo-, 5673

— , 5-(4-imidazolylmethyl)-4,7,10-trioxo-, 5674

Triazen-3-ol, 1,3-bis(p-tolyl)-, 5988

— , 1-(4-bromophenyl)-3-phenyl-, 5713

— , 1-(2-chlorophenyl)-3-ethyl-, 4469

— , 1-(4-chlorophenyl)-3-ethyl-, 4470

— , 1-(2-chlorophenyl)-3-methyl-, 3909

— , 1-(4-chlorophenyl)-3-methyl-, 3910

— , 1-(2-chlorophenyl-3-phenyl-, 5710

Uridine-3'-(dihydrogen phosphate), 4913

Uridine-5'-(dihydrogen phosphate), 4914

Uridine, 2,4-dimercapto-, 4898

— , 5-ethyl-, 5487

— , 5-fluoro-, 4894

— , 5-fluoro-2',3'-O,O-iscpropylidene-, 5749

— , 5-hexyl-, 6101

— , 5-iodo-, 4893

— , 5-methyl-, 5219

— , 5-methyl-2,4-dithio-, 5295

Uridine-2'(3')-phosphate, 4915

— , 5-bromo-, 4948

— , 5-chloro-, 4947

— , 5-iodo-, 4949

Uridine-5'-(tetrahydrogen triphosphate), 4932

Uridine-5'-(trihydrogen diphosphate), 4930

3'-Uridinyl-5'-uridinyl hydrogen phosphate, 6334

Uridylic polynucleotide, 5-bromo-, 6519

— , 5-chloro-, 6520

— , 5-iodo-, 6521

Valeric acid, 5-amino-, 2728, 275

Valine, 2724, 276

— , 4,4,4-trifluoro-, 390

Vanillic acid, 4016

Vanillin, 3994, 722

o-Vanillin, 723

Vinylacetic acid, 2311, 60

Violuric acid, 2175

Viomycidine, 3184

Williams Blue, 5924